MECHANICAL DESIGN OF ELECTRONIC SYSTEMS

JAMES W. DALLY
University of Maryland at College Park

PRADEEP LALL
Auburn University

JEFFREY C. SUHLING
Auburn University

College House Enterprises, LLC
Knoxville, Tennessee

The manuscript was prepared using Word 2002 with 11 point Times New Roman font. Publishing and Printing Inc., Knoxville, TN printed this book from pdf files.

College House Enterprises, LLC.
5713 Glen Cove Drive
Knoxville, TN 37919, U. S. A.
Phone and FAX (865) 558 6111
Email jwd@collegehousebooks.com
http://www.collegehousebooks.com

10 Digit ISBN 0-9762413-3-1

13 Digit ISBN-978-0-9762413-3-1

ABOUT THE AUTHORS

James W. Dally obtained a Bachelor of Science and a Master of Science degree, both in Mechanical Engineering from the Carnegie Institute of Technology. He obtained a Doctoral degree in mechanics from the Illinois Institute of Technology. He has taught at Cornell University, Illinois Institute of Technology, the U. S. Air Force Academy and served as Dean of Engineering at the University of Rhode Island. He is currently a Glenn L. Martin Institute Professor of Engineering (Emeritus) at the University of Maryland, College Park.

Dr. Dally has also held positions at the Mesta Machine Co., IIT Research Institute and IBM. He is a fellow of the American Society for Mechanical Engineers, Society for Experimental Mechanics, and the American Academy of Mechanics. He was appointed as an honorary member of the Society for Experimental Mechanics in 1983 and elected to the National Academy of Engineering in 1984. Professor Dally was selected by his peers to receive the Senior Faculty Outstanding Teaching Award in the College of Engineering and the Distinguish Scholar Teacher Award from the University of Maryland. He was also a member of the University of Maryland team receiving the 1996 Outstanding Educator Award sponsored by the Boeing Co.

Professor Dally has co-authored several other books: *Experimental Stress Analysis, Photoelastic Coatings, Instrumentation for Engineering Measurements, Packaging of Electronic Systems, Production Engineering and Manufacturing, and Introduction to Engineering Design, Books 1, 2, 3, 4, 5, 6, 7, 8 and 9.* He has authored or coauthored about 200 scientific papers and holds five patents.

Pradeep Lall is the Thomas Walter Professor with the Department of Mechanical Engineering and Associate Director of NSF Center for Advanced Vehicle Electronics at Auburn University. He received the MS and Ph.D. degrees from the University of Maryland and the M.B.A. from Kellogg School of Management at Northwestern University.

Dr. Lall has ten years of industry experience. He was previously with Motorola's Wireless Technology Center. Professor Lall has published extensively in the area of electronic packaging with emphasis on modeling and predictive techniques. He is the senior author of the book titled "Influence of Temperature on Microelectronic and System Reliability".

Dr. Lall is recipient of the Samuel Ginn College of Engineering Senior Faculty Research Award in 2007. Dr. Lall is a Six-Sigma Black-Belt in Statistics. He is the recipient of Three-Motorola Outstanding Innovation Awards and he holds several U.S.-patents. He is an Associate Editor for the ASME Journal of Electronic Packaging, IEEE Transactions on Components and Packaging Technologies, and IEEE Transactions on Electronics Packaging Manufacturing.

Jeffrey C. Suhling is the Quina Distinguished Professor with the Department of Mechanical Engineering and Director of the NSF Center for Advanced Vehicle Electronics at Auburn University. He received the B.S. degree in Applied Mathematics, Engineering, and Physics (AMEP); and the M.S. and Ph.D. degrees in Engineering Mechanics, all from the University of Wisconsin, Madison, WI.

At Auburn, he was selected Outstanding Mechanical Engineering Faculty Member by the undergraduate students; and has received both the College of Engineering Birdsong Superior Teaching Award and the College of Engineering Senior Research Award. Dr. Suhling has authored or co-authored over 225 technical publications, including four papers selected as Best of Conference, and one award winning journal paper. He has advised over 50 graduate students.

Professor Suhling has served as Chair of the Electrical and Electronic Packaging Division of ASME, and as a member of the Executive Committee. He was Program Chair of the ASME InterPACK '07 Conference; and currently serves as Vice-Chair if the Professional Development Course Committee of the IEEE Electronic Components and Technology Conference (ECTC), as well as the General Chair of the InterPACK '09 Conference.

PREFACE

This book has been written for engineers to serve as a first text on the packaging of electronic systems. The material has been written for an engineering student or for a practicing professional working as a mechanical or electrical engineer with a company producing electronic products or systems. The engineering student should have completed fundamental courses in the engineering sciences, thermal sciences and materials as prerequisites. The practicing professional will probably be at the early stages of his or her career and be more concerned with the technical details of the design rather than the business strategy of a product line.

This book is an introduction to the design of electronic systems from a mechanical engineering perspective, although attention is given to circuit analysis that may be of more interest to the electrical engineer. As such it covers a very broad range of topics from the physics of semiconductors to the design of advanced high performance heat exchangers. To accommodate this breadth, we have divided the text into four independent parts. The first, Part 1, which includes three chapters, deals with foundations for design. Chapter 1 provides an overview of the entire field, covering in some detail the mechanical design issues that arise in developing an electronic system. Also covered are business aspects of this industry, particularly as they are related to a rapidly changing technology, which leads to early obsolescence of an existing product line, and exacting requirements for investing in new product development. Chapter 2 describes electronic components with emphasis on semiconductor devices. Importance is placed on silicon technology and the new developments in logic and memory circuits with higher levels of integration. We try to show in this chapter the dynamics of the technology and the emerging design problems associated with the ever increasing scale of integration. The most important problems are handling high I/O counts and dissipating very large heat flux with exceedingly small temperature differences. Chapter 3 covers circuit analysis. The conventional methods of analysis of ordinary ac and dc circuits are briefly reviewed. However, the emphasis is placed on transmission lines where inductance, resistance and capacitance are distributed along the length of the line. This development is new to most mechanical engineers and it is critical to their understanding of propagation delay, line charging and pulse reflections that must be taken into account in the design electronic systems with even moderate performance specifications.

Part 2 deals with the different levels of packaging electronic components and the methods commonly used in manufacturing and assembly. The chip carrier, the first level package is treated in detail in Chapter 4. Through-hole packages, surface-mounted chip carriers, chip scale packages, flip-chip and ball-grid-arrays are described in some detail. The printed circuit board, the second level package, is described in Chapter 5. The treatment is often descriptive and provides the opportunity to introduce the terminology used in the industry to depict circuit board features. Some of the very difficult problems associated with component placement and trace routing are introduced in this chapter with simple examples. The manufacturing processes associated with production of circuit boards are described in Chapter 6. These processes must be well understood if one is to design circuit boards suitable for a quality product. The production methods employed in assembling electronic products are described in Chapter 7. These methods include the technologies associate with surface mounting components. It also includes descriptions of solder processes and materials. The chapter concludes with a discussion of quality assurance programs and reworking processes. Chapter 8, the final chapter in Part 2, deals with third level packaging. Third level packaging is a loose term that describes the many components encountered when enclosing a number of printed circuit boards in instrument cases or cabinets. We treat connectors, back panels, cabling, power supplies, cooling hardware and both commercial and military enclosures in some detail.

Analysis methods commonly employed to predict performance of systems are covered in the six chapters included in Part 3. While much of the analysis in industry is performed using specialized software programs, we have emphasized a more theoretical approach that leads to closed form solutions for simple problems. Chapter 9 deals with heat transfer by conduction. Basic conduction equations are reviewed and applications to problems arising in conducting the heat generated on a chip to the heat sink are given. Heat transfer by radiation and convection are treated in Chapter 10. We have only limited coverage of heat transfer by boiling, because its use in the industry to date has been limited. Chapter 11 presents the basic methods for calculating stresses and predicting failure by yielding or fatigue. This chapter also presents methods for the analysis of fasteners. Methods for determining stresses imposed by temperature changes are described in Chapter 13. These methods include closed form solutions for solder joints and die stresses in molded packages. Bi-material stresses and PCB warpage are also covered. The chapter includes descriptions of finite element solutions that illustrate the approach often used in industry to study stresses in solder joints and in molded chip carriers. Results from a recent study that enable a prediction of the thermo-reliability of BGAs has been included. The final chapter in Part 3 covers vibration of components and circuit boards. Here the emphasis is on simple analytical procedures, which give insight on methods to reduce transmissibility coefficients and, if possible, to avoid resonance conditions. We recognize that failure by fatigue is a major problem in systems exposed to vibration, and methods for treating of failure by fatigue are covered. Advanced methods for dealing with shock and vibration of fine pitch BGAs are presented. Examples of finite element analysis are presented together with results from an extensive test program involving drop testing.

Part 4 of this text deals with reliability, with the theory of reliability supporting this important topic presented in Chapter 15. In addition to the basic coverage of the various measures of reliability, the chapter includes a description of reliability models. Also included is a discussion of statistical methods for predicting failure including both the normal distribution function and the Weibull distribution function. Chapter 16, the final chapter discusses design methods to improve reliability. Failure mechanisms are described to provide a better understanding of why failures occur after some period of time in operation. Reliability improvement is achieved by preferred part selection, derating and stress management, screening and accelerated testing.

The book does differ to some degree from the conventional design textbook commonly found in engineering colleges. We have attempted to introduce some business aspects of design. Of particular importance here is the fact that the market price of a product drives the design. We have also tried to cover some of the thought process associated with design by describing both design and manufacturing aspects (good and bad) in the narrative parts of the text. The exercises that are given at the end of each chapter are markedly different from those found in the usual engineering textbook. They often require the student to do much more than plug and chug. The students are required to sketch, to prepare graphs describing solution space, to conceive new designs better than existing designs, to write engineering briefs, to interpret solutions and draw conclusions related to design merits and to make judgments based on business aspects. These exercises will stretch the experience base of most students and, indeed, they may stretch the experience base of some instructors. Care should be taken in assigning the exercises because they range in difficulty from trivial to impossible. Hopefully, the more difficult problems can be used to stimulate classroom discussion and the easier ones will not lull the student into a false sense of security.

ACKNOWLEDGEMENTS

We want to thank IBM for their permission to use the cover photograph, which beautifully illustrates the complexity of advanced packaging methods. This photograph was used on the cover of the IBM Journal of Research and Development, Volume 46, No. 6, November 2002. This issue also describes many of the advances made to develop a high performance server. We have employed other drawings and photographs from our notes, a previous textbook by one of the authors, and industrial literature found on the Internet. We thank all those who have supplied this material.

We appreciate the feedback from students taking our courses as the content in this book was developed. The feedback they provided was essential in developing the sequence and scope of the material selected for inclusion in the text.

Finally, we appreciate our families who tolerate our absences as we teach, perform research and spend countless hours at the keyboard writing, editing and drawing.

CONTENTS

PART 2 PACKAGING

CHAPTER 4 FIRST LEVEL PACKAGING — THE CHIP CARRIER

CHAPTER 5 SECOND LEVEL PACKAGING: SUBSTRATES AND PRINTED CIRCUIT BOARDS

CHAPTER 6 PRODUCTION OF PRINTED CIRCUIT BOARDS

CHAPTER 7 ELECTRONICS MANUFACTURING: CHIP CARRIER TO SUBSTRATE

CHAPTER 8 THIRD LEVEL PACKAGING: CONNECTORS, CABLES, MODULES, CARD CAGES AND CABINETS

PART 3 ANALYSIS METHODS

CHAPTER 9 THERMAL ANALYSIS METHODS: CONDUCTION

CHAPTER 10 THERMAL ANALYSIS METHODS: RADIATION AND CONVECTION

CHAPTER 11 STRESS AND FAILURE ANALYSIS OF MECHANICAL COMPONENTS

CHAPTER 12 THERMO-MECHANICAL ANALYSIS

CHAPTER 13 ANALYSIS OF VIBRATION OF ELECTRONIC EQUIPMENT

PART 4 RELIABILITY

CHAPTER 14 THEORY OF RELIABILITY

CHAPTER 15 DESIGN TO IMPROVE RELIABILITY

APPENDIX A

LIST OF SYMBOLS

Upper Case

A = Acceleration factor
= Area
= Constant
A_c = Area, contact
A_v = Area, voids
= Area, vias
$\boldsymbol{A}$ = Acceleration factor
= Amplitude transmission coefficient
B = Bandwidth
= Constant
C = Capability
= Capacitance
= Coefficient, damping
= Degrees, Celsius
C_c = Critical damping coeff.
C_v = Coefficient of variation
C' = Capacitance/unit length
$[C]$ = Damping matrix
D = Diameter
= Diameter, hydraulic
= Ductility
= Flexural rigidity
D_s = Substrate density
$\{D\}$ = Displacement vector
DNP = Distance neutral point
E = Electric field
= Emission energy
= Energy, radiation
= Modulus of elasticity
E_a = Energy, activation
$E_\lambda{}^b$ = Energy radiated
$\boldsymbol{E}_k$ = Energy, kinetic
$\boldsymbol{E}_p$ = Energy, potential
$\boldsymbol{E}_s$ = Energy, stored
F = Correlation factor
= Force
= Unreliability
F_e = Etch Factor
= Force, extraction
F_i = Force, insertion
= Force, preload
F_w = Force, wiping
F_L = Loss factor
F^s = Unreliability, system
FIT = Failure in time
$\boldsymbol{F}$ = Force transmission coeff.
= Shape factor
G = acceleration, dimensionless
= Gain
= Material constant
= Shear modulus
= Velocity, mass
= Volume flow rate, GPM
$G' = 1/R'$
GR = Grashof number
H = Hardness
= Height
= Irradiation
$\boldsymbol{H}$ = Hazard funct, cumulative
HR = Hazard rate
I = Current
= Moment of inertia
$\boldsymbol{I}$ = Amplitude ratio, isolated
J = Current density
= Polar moment of inertia
= Radiosity
K = Conduction, thermal
= Constant
= stress concentration factor
$[K]$ = Stiffness matrix
L = Length
= Inductance
L' = Inductance/unit length
$\boldsymbol{L}$ = Length
M = Material constant
= Moment
$[M]$ = Mass matrix
MR = Median rank
MTTF = Mean time to failure
N = Number
N_f = Cyclic life
= Number of failures
N_s = Number of successes
Nu = Nusselt's number
P = Force
= Power
= Probability
P_{cr} = Load, critical
Pr = Prandtl's number
PRN = Priority ranking number
Q = Shear force
= Volume flow rate
R = Radius
= Resistance, electrical
= Reliability
R_T = Resistance, thermal
R^s = Reliability, system
R' = Resistance/unit length
$\boldsymbol{R}$ = Range
RA = Reduction in area
Ra = Rayleigh's number
Re = Reynold's number
$\{R\}$ = Force vector
S = Shape factor
S_x = Standard deviation of x
$S_{\overline{x}}$ = Standard deviation of $\overline{x}$
S_e = Strength, endurance
S_f = Strength, fatigue
S_a = Strength, fatigue modified
S_p = Strength, proof
S_u = Strength, tensile
S_y = Strength, yield
S_{ys} = Strength, yield in shear
$\boldsymbol{S}$ = Safety factor
SF = Safety factor
T = Temperature
T_a = Temperature, absolute
= Temperature, ambient
T_j = Temperature, junction
T_0 = Temperature, reference
T = Torque
U = Energy, distortion
= Velocity, free stream
V = Force, shear
= Voltage
V_{CB} = Voltage, collector-base
$\boldsymbol{V}$ = Volume
W = Channel width
= Strain energy density
= Weight
WD = Wiring demand
Z = Impedance
Z_0 = Characteristic Impedance

Lower case

a = Acceleration
= Dimension, plate
= Radius
b = Dimension, plate
= Pad width, beam width
= Radius
c = ½ wire diameter
= Specific heat
= Wave velocity
d = Diameter
= Differential expansion
= Distance
d_x = Deviation in x
e = Electron charge
= Exponential number (2.718)
f = Failure density function
= Function, relative frequency
f_n = Frequency, natural
f = Friction factor
g = Gravitational constant
h = Coefficient, convection
= Head, hydraulic
= Height, beam
= Thickness
h_c = Coefficient, contact
h_r = Coefficient, radiation
j = Imaginary number (−1)
k = Boltzmann's constant
= Coeff. thermal conductivity
= Exponent
= Number
= Spring rate
m = Mass
= Slope
$\dot{m}$ = Mass flow rate
n = Index number
p = Perimeter
= Pressure
= Pressure, contact
q = Charge
= Exponent
= Load per unit length
= Rate, heat transfer
q_g = Rate, heat generated
r = Radial coordinate
= Radius
s = Manhattan distance
= Spacing
t = Thickness
= Time
t_f = Time to failure
t_0 = Location parameter
t_w = Time, warm-up
u = Dimension — under size
= Velocity, convection
v = Velocity
w = Width
= Displacement, out-of-plane
$\ddot{w}$ = Acceleration, out-of-plane
x, y, z Rectangular coordinates
x = Arbitrary quantity
$\overline{x}$ = Mean of some quantity
y = Displacement
= Distance, vertical
$\dot{y}$ = Velocity
z = Height
= Material constant

Greek Symbols

α = Absorbed, radiation ratio
= Coefficient
= Diffusivity, thermal
= Exponent
= Scale parameter
= Temp. coeff. expansion
β = Coeff. thermal expansion
= Coefficient
= Shape factor
γ = Constant
= Coefficient
= Material constant
= Shear strain
δ = Boundary layer thickness
= Deflection, elongation
= Loss angle
ε = Emissivity
= Permittivity
= Strain
ε_p = Strain, plastic
ε_r = Dielectric constant
ζ = Dummy variable
η = Carrier mobility
ξ = Damping ratio
θ = Angle
θj-a = Resistance, thermal
κ = Curvature
= Parameter
λ = failure rate
= Real or complex number
= Wave length
μ = Friction coefficient
= Permeability
= Viscosity
ν = Frequency
= Poisson's ratio
= Viscosity, kinematic
π = Number Pi (3.1416)
ρ = Density, mass
= Density, mass/unit volume
= Radius coordinate
= Radius of curvature
= Reflected, radiation ratio
= Resistivity
σ = RMS surface roughness
= Stefan-Boltzmann's const.
= Stress
$\sigma_1, \sigma_2, \sigma_3$ = Stress, principal
σ_a = Stress, alternating
σ_m = Stress, mean
τ = Stress, shear
= Time constant
= Transmitted, radiation ratio
τ_w = Shear stress, wall
ϕ = Angle, phase
= Angle, twist
ψ = Fin factor
ω = Angular frequency
= Density
= Velocity, angular
ω_n = Natural frequency

Δ = Boundary layer thickness
∇^2 = Laplace's operator

PART 1

MECHANICAL DESIGN

OF

ELECTRONIC SYSTEMS

PART 1

MECHANICAL DESIGN OF ELECTRONIC SYSTEMS

CHAPTER 1 INTRODUCTION

1.1 OVERVIEW AND OBJECTIVES

This book has been prepared to serve as a text for students and entry level persons beginning to design electronic systems. The coverage begins at the interface between electrical engineering and mechanical engineering and pertains to the mechanical, materials, manufacturing and assembly issues which arise in developing a new electronic or computer system. The treatment is quite broad starting with integrated circuits (the chip) and proceeding through the many levels of packaging involved in developing a complete electronic system. The material is often highly descriptive, particularly when compared to the mathematical treatments presented in more mature subjects such as mechanics or the design of mechanical components. However, the descriptive material is important in introducing the essential vocabulary, which is full of acronyms, and to present the wide array of electronic components that mechanical and electrical engineers must deal with in the design process.

This book is divided into four parts to organize the material and to facilitate understanding by the reader. Part 1 covers background material including semiconductor physics, analog circuit theory, digital circuit theory and transmission line theory. The objective of this background information is to give the engineers involved in mechanical design of electronic systems, often called packaging, the basic understanding as to why and how electronic circuits operate. Experience has shown that this information greatly enhances communication between the mechanical and electrical engineering functions in a development project, and enhances the opportunity for successful concurrent product development instead of sequential development.

Part 2 involves packaging beginning at the first level where the chip is housed in its carrier, and then extending through the higher levels to the design of the cabinets and instrument panels. The word packaging is poorly understood by the engineering community where it is often confused with the design of a container to prevent damage during shipment of a product. Packaging of electronic systems refers to the placement and connection of many electronic and electro-mechanical components (sometimes thousands) in an enclosure which protects the system from the environment and provides easy access for routine maintenance. An example which describes some of the important features of packaging is illustrated in Fig. 1.1. The packaging process starts with a chip which has been diced from a wafer of silicon that was fabricated using photolithographic semiconductor fabrication processes. The chip contains electronic devices (e.g. transistors, resistors, etc.) that are interconnected in a planned manner to form integrated circuits (IC's) that perform a desired electrical function. After testing, the chip is housed in a chip carrier and small wires or solder balls are used to electrically connect the chip to the carrier. The chip carrier or component is often referred to as the first level of packaging, while the electrical connections to the chip carrier are called the first level of interconnects. Next, several chip carriers are placed on a circuit board or substrate (second level of packaging), and connected together with wiring traces (second level of interconnects) which have been formed by photoetching the circuit board. Edge connectors on the circuit boards are then inserted into contacts on a back panel (third level of packaging) which carries the higher level connections that permit communication from one circuit board to the next. Cables are shown which connect the power supply to the back panel and which bring the input and output (I/O) signals to and from the unit. Finally, the entire array of circuit boards, back panels, power supplies and cables are housed in a cabinet (fourth level of packaging). The packaging aspect of the design of an electronic system is extremely important as often more than one-half of the cost of a system is involved in packaging.

Part 2 also describes materials and manufacturing methods used to fabricate printed circuit boards including those made from glass reinforced plastics and multilayered ceramics. Mechanical, electrical and thermal properties of metals, glasses, polymers and ceramics used in electronic systems are discussed. The coverage includes all aspects of the manufacturing process including material selection, tooling, drilling and cleaning. Photolithography used to pattern the footprint of each layer is described in detail. Finally, solder materials and processes used to connect the terminals of the electronic components to the bonding pads on the printed circuit boards are also covered.

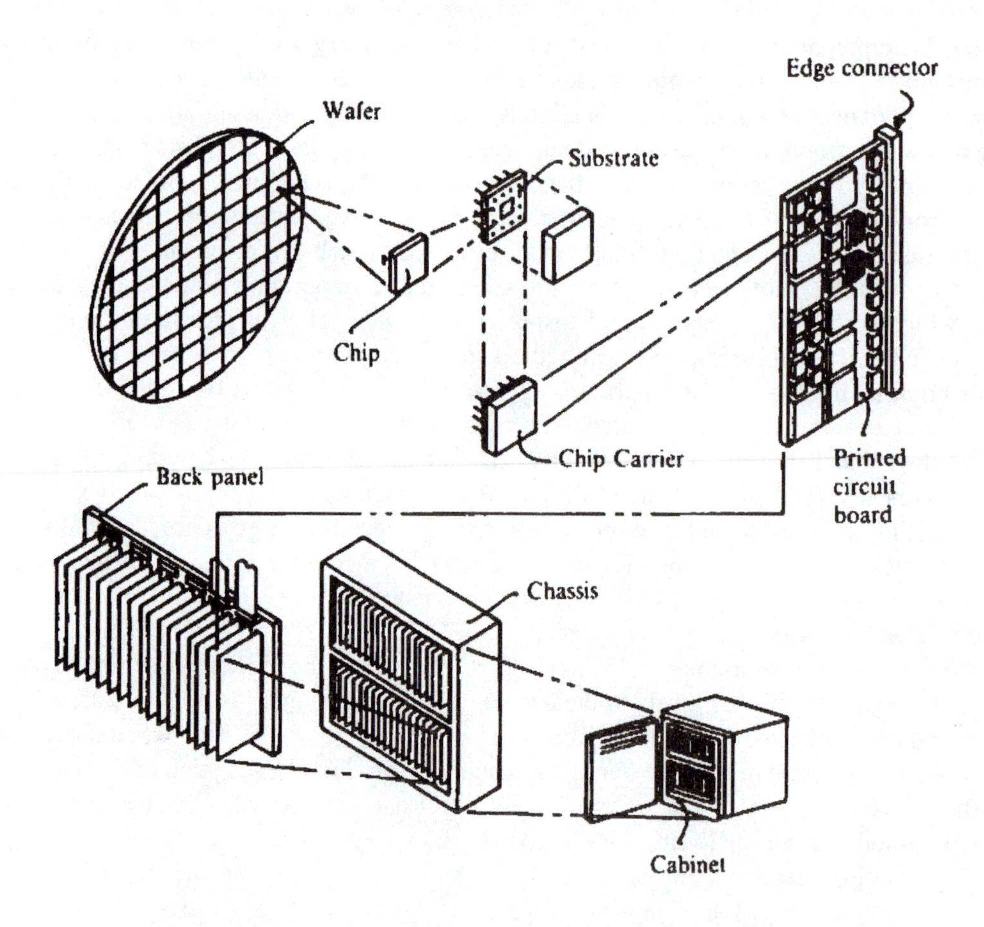

Fig. 1.1 An illustration of several levels of packaging involved in an electronic system.

Part 3 deals with analytical, experimental, and numerical methods used to insure the reliability of electronic systems exposed to harsh environments. In addition to carefully packing potentially thousands of electronic components into a stylish and functional cabinet, packaging involves the protection of these components from the environment. Thermal management is one of the most important of the environmental considerations, as the operating temperatures of the chips markedly affect the reliability of the circuits and the availability of a system. Thermo-mechanical loadings and stresses in the assembly will result due to temperature changes caused by powering the system on and off, or by changes in the local ambient environment. For electronic systems destined for use in the field by either the military or industry, shock, vibration and humidity represent harsh environments which must be accommodated in the design of the hardware. Finally, noise is an important consideration particularly in air cooled equipment where one or more fans are used to move significant volumes of air.

Part 4 deals with designing an electronic system for reliability. This treatment has two main components—a discussion of failure mechanisms, and analysis emphasizing the use of reliability theory and statistics. Techniques used to enhance reliability or system availability are also discussed.

In this presentation, simple examples will be selected which emphasize basic theoretical approaches and which illustrate sound design concepts. In actual practice, the real problems will be much more complex but the applicable theory and the design concepts are the same. In many organizations performing the design of electronic systems, special codes or software exist for handling some of the complexity associated with very large electronic systems. We will recognize some of these approaches in this book, but we will not describe them in any detail because they change rapidly with time.

1.2 FUNCTIONS INVOLVED IN MECHANICAL DESIGN OF ELECTRONIC SYSTEMS

Electronic packaging serves four major functions in the performance of electronic systems, which include:

1. Interconnection of electrical signals at several levels of packaging.
2. Mechanical and environmental protection of the components, circuits, and devices
3. Distribution of power to the electronic circuits and devices.
4. Dissipation of the heat generated by these devices.

These four functions are the basic drivers advancing the state of the art in packaging technology.

It is important to understand the differences in the advancement of semiconductor and packaging technologies. Semiconductor technology has been generally following Moore's law, which predicts the rate of scaling of the minimum feature size on semiconductor devices. With reduced feature sizes, more devices and more functions can be fabricated on a specified chip area for each succeeding generation of semiconductor devices. The ability of the semiconductor industry to follow Moore's law is discussed later in this chapter (see Fig. 1.16). Although some packaging elements have been scaled to match the reduced dimensions of the semiconductor input and output (I/O) pads, the dimensions of most of the other packaging elements cannot be scaled down at the same rate.

There are several reasons for the inability of the packaging industry to markedly reduce the features size in the different levels of packaging. First, at the external inputs/outputs, the connections must be sufficiently large to be able to be handled by human scale peripheral devices. Second, excessive reduction of the cross-sectional area of the circuit lines on the second and third level circuit boards results in increasing line resistance that degrades the signal as it propagates and increases the heat that must be dissipated. Third, increased I/O at the chip level has generated the need for more expensive area array interconnections, such as flip-chip solder balls, to replace less expensive wire bonds to perimeter pads; thus, causing a general increase in the cost of packaging a system. Fourth, increased power densities from high-performance chips have placed severe demands on the power distribution system to deliver high currents and then to dissipate high power levels from individual chips. The first level chip carriers must house chips that are becoming larger in area with each new generation. Increase density of transistors on each chip results in increased power levels that require more aggressive cooling technologies. Finally, the mechanical and environmental protection mechanisms do not scale.

As the microelectronic industry continues to develop chips with more dense circuits that require increasing amounts of power, efficiently packaging the array of new products will be a significant challenge. Packaging costs, which often exceed the cost of the associated microelectronic chips, will become even more important in the design of all but the highest performing systems.

1.3 MECHANICAL DEVELOPMENT OF ELECTRONIC SYSTEMS

A mechanical development department is usually responsible for packaging electronic systems in most industrial firms that produce an extensive line of electronic products. Because large electronic systems are usually quite complex involving several different engineering disciplines, multi-department organizations are usually established to handle the logical flow of paper, CAD files and other information generated in a typical development project. An example of one organization is shown in Fig. 1.2. The development process is initiated (in large firms at least) by corporate planners who identify customer needs and predict market trends and technological advances. With this information as a basis, they prepare the specifications for a new product indicating performance requirements, market estimates on volume, costs, price, and other higher level design objectives. Development budgets and schedules are also part of the output from the corporate planners.

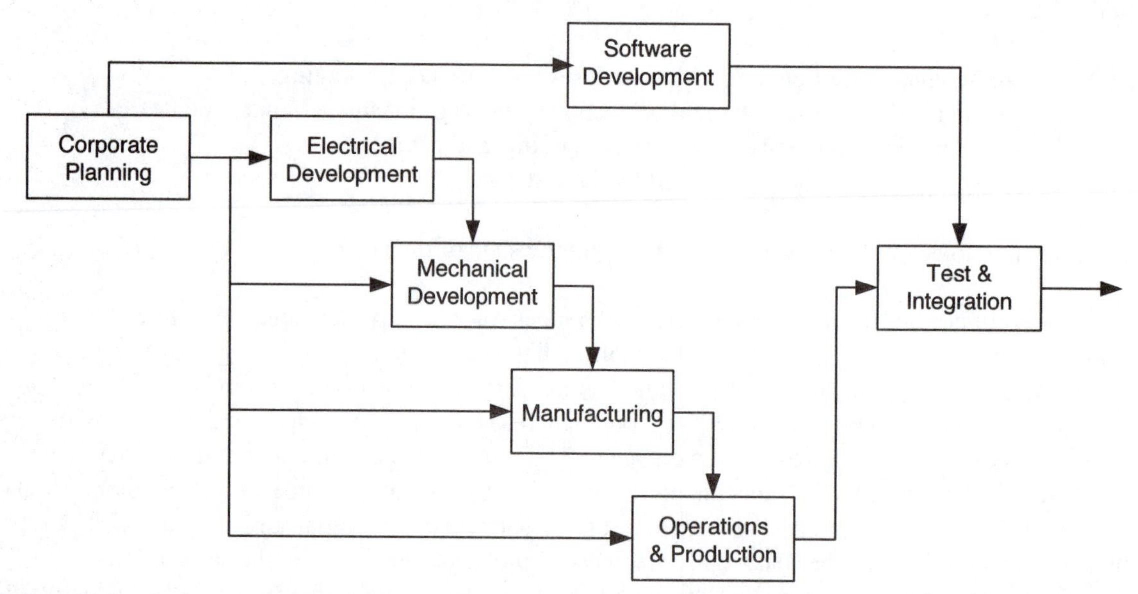

Fig. 1.2 Typical organization used in developing an electronic product showing information flow and discipline interfaces.

The project moves from the planners to the electrical or computer engineering department where the system architecture is established and then the detailed analog and digital circuits are designed. The output from electrical or computer engineering department is passed on to mechanical engineering department in the form of circuit diagrams, logic diagrams, component lists, wiring lists and ground rules for component layout and shielding requirements for the cables. The output from the electrical or computer engineering department also goes to the computer science department where the code necessary to operate the system is written.

The mechanical development department packages the system by sizing and designing circuit boards, placing components on each board and connecting the appropriate pins on each component. Back panels and power distribution busses are designed in conjunction with gates, chassis drawers and cabinets. Connectors and cables are designed to ensure connectivity as well as access for routine maintenance. A cooling system is designed to carry away the heat that is dissipated by the many different electronic devices employed in constructing high performance systems. Operating panels that are appreciated by the operator and the customer are developed to be attractive, functional and ergonomic. The system is designed to accommodate the operating and storage environments as described by the controlling specifications. The output from the mechanical development department is in the form of drawings, CAD files, and reports describing analyses predicting response of the system to the environment and its reliability. This output is transmitted to the manufacturing and operations departments.

Manufacturing engineers design and build the tooling necessary to manufacture the components, sub-assemblies, circuit boards, cables, etc. The circuit boards, back panels, sub-assemblies, cabinets, etc., are fabricated and then assembled into the final product. This manufacturing process is quite challenging particularly during the assembly and test of the first few units. Operation engineers insure that all materials and components are available on time for the assembly of the system. They also oversee the production process and control both the input and output inventory. Design errors are discovered during the early production period and close cooperation among mechanical development, manufacturing and operations is essential to develop design modifications that correct the errors without incurring high costs. Design for enhanced manufacturing is a critical objective in developing a product known for its quality as well as for its performance.

The prototype moves from operations to unit test where the performance of the circuits and the software are evaluated. This phase, often called debugging, requires close cooperation between the electrical engineers and the software personnel. Mechanical engineers are involved in recording the electrical changes (so called soft or white wires) so that the changes in connections can be made on revised circuit board layouts and cable drawings. Unit evaluation may involve shock, vibration, temperature cycling or other accelerated life type testing for product qualification and reliability evaluation. If failure occurs with the hardware, the mechanical engineer is responsible for the diagnostics and for the design modifications required to improve the package to adequately protect the system from the environment imposed during product qualification.

The design of smaller electronic systems is usually accomplished by a team of engineers who are all located in close proximity. The engineering disciplines described in the previous paragraphs are all represented on this focused development team. Because the system under consideration is relatively small, only a few engineers from each discipline are required to complete the development in a timely manner. The advantage of the team structure over the functional organization, described previously, is that the design process is less formal and rapid communication is enhanced. The disadvantage is that the depth of knowledge and expertise of the team members, in one or more disciplines, may not be equivalent to that found in the more heavily populated departments in a functional organization.

1.4 MECHANICAL DESIGN ASPECTS OF PACKAGING

There are ten main issues involved in the design of an electronic system as indicated below:

1. Connections
2. Thermal management
3. Cost
4. Performance
5. Manufacturing
6. Maintenance
7. Thermo-mechanical deformations
8. Shock and vibration
9. Reliability
10. Ergonomics

The priority order of the issues is dependent on the product under consideration. In many high end products, performance is more important than cost. In other applications, reliability (or system availability) is more important than performance. In high volume products, cost and performance are usually balanced. In products used on military platforms, the ability to survive in intense shock and vibration environments is critical. In most products, each of these issues markedly affects the design, and close attention is necessary in every area to insure that the design specifications are met or exceeded.

1.4.1 Connections

In advanced electronic systems, it is easy to define at least six levels of connections. We start with the bonding pads on the chip and connect these pads to the I/O leads on the chip carrier. This normally accomplished using small wires or solder balls. For example, the photograph in Fig. 1.3 shows small diameter gold wires that have been thermosonically bonded to the I/O pads on the chip and its associated chip carrier. Second, the leads from the chip carrier are connected to the printed circuit board (PCB). These connections normally involve the design of solder joints and the trace routing on the PCB; both are very important tasks. Third, connections are made between the different PCB's which exist in the system. There are several different approaches to connecting PCB's depending primarily on the complexity and size of the system. For small simple systems containing only a few PCB's, the connections are made with edge card connectors, which are cabled together. However, for large systems, which may contain many PCB's, the connections are made with edge card connectors, which are inserted into a back panel or motherboard. Many layers of circuit traces that are photo etched on each layer of the back panel serve as a very compact form of cabling. Fourth, the back panels, which usually are housed in a cage, drawer or gate, must be connected together to permit one sub-system to communicate with another. The number of connections, at this fourth level, is reduced to primarily sub-system I/O and power; hence, the signal connections are usually made with cables and the power connections are made with heavy wires or bus bars capable of handling high currents. The fifth level of connection involves wiring the back panels to the I/O connectors within the cabinet or enclosure. Cable harnesses for signal I/O and bus bars for power distribution are commonly employed. The sixth and final level of connection occurs in very large systems where cabinets and or work stations are connected together. The methods employed for cabinet to cabinet connections depend largely on the distance between cabinets and the speed required in transmitting signals. For short distances, these connections are usually made with twisted pairs of wire or coaxial cable. For longer distances, only coaxial or fiber optic cables are sufficient. For very long distances a dedicated set of cables becomes very expensive and transmission is accomplished with common carriers such as phone lines, microwave transmission or fiber optic transmission lines.

Fig. 1.3 Enlargement of a corner of a chip showing small diameter gold wires connecting the chip I/O to the lead wires on the chip carrier.

Very large digital systems can contain billions of semiconductor devices and several hundred chips, all of which must be connected together. For example the IBM eServer z990 [1, 2], which was introduced in June of 2003, is based on a modular design that enables the customer to expand the system by plugging "blades" into a common back panel. Each blade contains several processor cores on a multi-chip module (MCM), up to 64 GB of memory on two memory cards and up to 12 input cables. Each of the MCM's contains 16 chips, which dissipate up to 800 W of power (heat). A total of 55,000 C4 connections[1] are necessary to electrically connect the silicon chips within each MCM used in a blade. The length of the electrical pathways used for the connections in the MCM is 378 m. In addition to the C4 connections to the chips, a complete system consisting of four blades employs a variety of connectors with a total pin count of 44,032. A photograph illustrating the advanced technology incorporated in a multi-chip-module used in a high performance server is illustrated in Fig. 1.4.

Fig. 1.4 Several levels of inter connections are illustrated in this cut away view of a multi-chip module.

Clearly the task of connecting a lead or a wire from point A to point B becomes extremely difficult when we must consider up to 100,000 wiring points and several thousand pins on either connectors or components. In addition, and even more important, is the reliability of the connections. If even one of them fails over the life of the product, the entire board will fail and the system may malfunction. With the number of solder joints exceeding several thousand in even very simple systems, meticulous attention must be given to every aspect of the design and manufacture of the connection system.

1.4.2 Thermal Management

Heat is generated at several locations in electronic systems. The power supplies, where the AC line voltage is converted to the various DC supply voltages required, are significant sources of the heat generation because of the relatively low efficiency of the AC/DC conversion process. Also, the Ohmic (I^2R) losses which occur at each chip and along the wiring result in additional heat generation. The heat load to be dissipated depends to a significant degree on the type of product. For high performance computers and signal processors, the heat load is large and elaborate cooling methods are employed. In

---

[1] C4 is a term used to designate controlled-collapse chip connection, a flip chip solder joint technology which will be described in more detail in Chapter 4.

small relatively simple systems the heat generated is small and natural or free convection is often sufficient to transfer the heat from the system to the environment. In moderate sized systems, fans are employed to enhance air-cooling. Regardless of the complexity, the heat management system employed is extremely important because this system controls the temperatures of the microcircuits on the chips. These chip temperatures in turn markedly affect the reliability of the electronic system.

An equation that often accurately predicts the failure rate for an individual component is the Arrhenius model given by:

$$\lambda(T_j) = A + Be^{-C[1/T_j - 1/T_0]} \tag{1.1}$$

where A, B and C are constants that depend upon the chip and chip carrier technologies, and T_j and T_0 are the junction and reference temperatures, respectively in degrees Kelvin. In this terminology, the junction is the device location on the silicon chip surface. In studies of failure rate, it is common practice to introduce an acceleration factor $\boldsymbol{A}_\lambda$ that is defined by:

$$\boldsymbol{A}_\lambda = \frac{\lambda(T_j)}{\lambda(T_0)} \tag{1.2}$$

For most electronic components, the acceleration factor $\boldsymbol{A}_\lambda$ for the failure rate increases as an exponential function of the junction temperature as shown in Fig. 1.5. It is evident that it is essential to minimize the silicon die temperature to reduce the failure rate and to increase the mean time between failures for an electronic system. Examination of Fig. 1.5 shows that increasing the junction temperature from 25 to 100 °C increases the acceleration factor from 1.0 to 4.5.

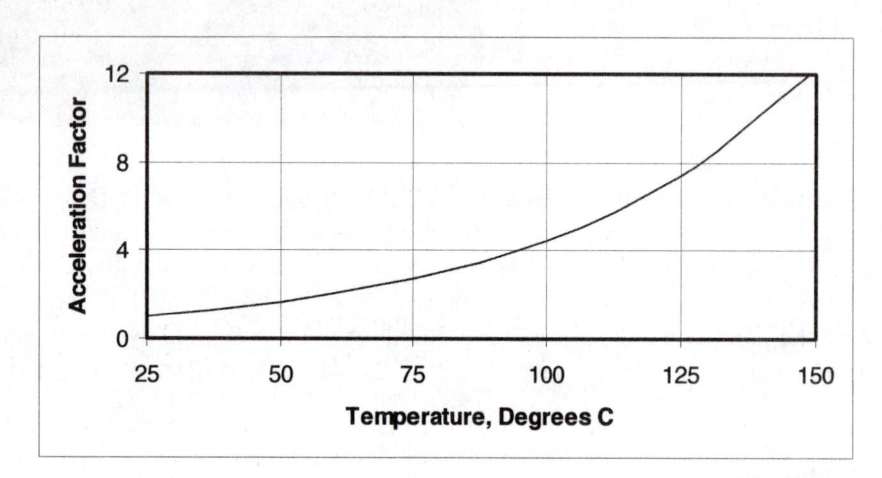

Fig. 1.5 Acceleration factor $\boldsymbol{A}_\lambda$ as a function of junction temperature showing the marked effect of temperature on failure rate.

With the very large number of components used in a typical electronic system, it is essential that the junction temperatures of the most critical of the devices be minimized to enhance reliability of the system and to improve the availability. Availability is the percent of the total time that the system remains up and operational.

The methods of cooling used in design depend upon the product price and performance. For simple low cost systems with small heat loads (for example a typical facsimile machine), the power is dissipated with natural convection, and no additional costs are involved. For slightly more complex systems with higher packing densities and higher heat loads, small fans are employed to give the improved heat transfer coefficients associated with forced convection. As we move up the price and performance scale, conduction cooling is employed where chilled water or a refrigerant is used in cold plates to transfer the heat to the environment. The high performance IBM eServer z990 introduced earlier utilizes a hybrid system with air cooling (fans) used to cool the memory chips and refrigerant to cool the high powered MCM modules [3].

Applications such as communication and surveillance satellites require deployment of electronic systems for extended periods of time in space. In the vacuum of space, heat transfer to the environment by convection is not possible and radiation methods must be employed for the final step in dissipating the heat.

All of these heat management systems must be designed in conjunction with an electrical system, which often requires that electrical insulation be placed in the path of the heat flow. The conflict in the electrical requirements for insulation, which are poor conductors of heat, and the mechanical requirements for good conductors make the design of an efficient heat transfer system quite challenging.

1.4.3 Cost

Costs are almost always an important parameter in the successful marketing of any product. The exceptions are in those cases where competition does not exist because of patent protection, trade secrets or sole source bidding. The product specification and the design of a system markedly affect its price. Corporate planners usually provide the product specifications as well as cost guidelines for both the product and the development. Product specifications requiring higher levels of performance will drive up costs. For example, electronic systems with increased performance will require faster more densely packed components that have been pre-screened for reliability. These will naturally cost more than low frequency more bulky components that have not been screened to eliminate the infant-mortality type of failures.

Production costs also depend upon the design of the PCBs, selection of connectors and cabling, and the type of enclosure used to house the electronics. Clearly many design trade-offs are made as the product is developed that affect the cost. It is important that costs of the PCBs, which depend largely on the number of layers required for signal and power wiring, and the costs of the connectors be minimized. Finally, assembly costs, which include insertion and placement of component as well as soldering operations, must be minimized.

The volume of product to be produced is a very important parameter. With high volume products, such as the systems produced for control of automobile engines, special assembly lines can be developed that enable very high production rates to drive down the assembly costs while improving system reliability. In this case, the assembly and warranty costs are traded-off against significant tooling costs. With relatively low volume products, general purpose machines are used for insertion or placement and subsequent assembly operations. With limited production, efficiencies are lost. Also, components can be damaged in the production processes resulting in increased costs and decreased yield and reliability.

1.4.4 Performance

There are many measures of performance, and a typical electronic system performance usually is specified with several parameters. Customers of computers are usually interested in processing speed that is measured in million instructions per second (MIPS) or billion floating-point operations per second (GFLOPS). Memory chips are ranked by storage capacity measured in mega-bytes (Mb) or Giga-bytes (Gb) as well as access speed. High performance chips and their first level packages are also ranked based on their switching speeds (i.e. their rise and fall times). Circuit density is also an important factor because more transistors can be placed on a specific chip thereby increasing speed while increasing the ability of a single chip to perform more functions. Circuit density also reduces the number of chips required in the design of a product, which in turn reduces cost of higher level packages.

A measure of circuit density is the I/O count, with higher counts for chips with dense circuits and larger areas. Improvements in performance coincide with I/O count. There has been a marked increase in I/O count over the last 40 years as indicated in Fig. 1.6. Single chips with I/O counts of several thousand are used in high performance systems. Multi-chip modules with I/O counts exceeding tens of thousands are also employed.

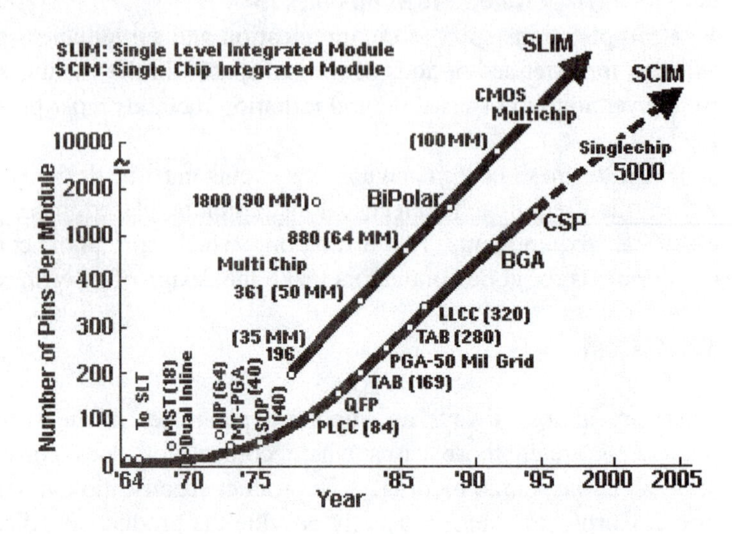

Fig. 1.6 I/O count increasing with respect to time resulting in enhanced system performance.

Thermal performance involves the ability of the thermal management system to dissipate the heat generated by the power supplies and the circuitry while maintaining the junction temperatures on the chip at specified values. Several strategies briefly described in Section 1.4.2 are employed to enhance thermal performance. Thermal performance is critically important because it markedly affects reliability and system availability.

Mechanical performance involves the ability of the product to withstand various loads during both shipment and operation. In systems with changing temperature, thermo-mechanical strains will result in the various materials of the assembly due to mismatches in the thermal expansion coefficients. This effect is of most concern in solder joints, where fatigue damage can lead to interconnect failures. For office and home entertainment equipment, vibration and shock are primarily encountered in shipping. However, for military equipment, shock and vibration conditions can be severe during normal use. Special design of higher level packaging is necessary to insure that the equipment operates without interruption during the vibratory disturbances. Other performance requirements for electronic systems designed for military and factory applications include the need for enclosures that protect the equipment from dust, humidity, large temperature variations and electric fields. Clearly, performance, whether electrical, thermal or mechanical, presents a significant challenge in designing cutting edge products that can compete in a global market.

1.4.5 Manufacturing

We noted in Fig 1.2 that mechanical development and operations which includes manufacturing are usually two separate organizations. However, the design of a product crosses all of the organizational interfaces. It is particularly important that the design of each sub-assembly in a product be accomplished so that it is compatible with each step in the manufacturing process. This is much more difficult to achieve than might be imagined, because many designers have little experience in manufacturing and many manufacturing engineers have never designed a product.

An obvious example of design for manufacturing is the use of common first level packages on a circuit board. First level packages (chip carriers) are available with several different types of lead structures such as pin in hole, gull wing surface mount, J lead surface mount, leadless, ball grid arrays, etc. An example of a chip carrier showing the chip, its wire bonds, leads and encapsulation is presented in Fig. 1.7. If different types of chip carriers are used on the same circuit board, the manufacturing process becomes much more difficult. Each type of component can require different assembly equipment or additional accessories (part feeders) for a single piece of equipment. Also, different soldering equipment and/or procedures can be required so that the circuit board might have to pass through several different assembly machines or even different assembly lines to complete the manufacturing process. These added processing steps are certain to increase cost, decrease yield and degrade the quality of the board.

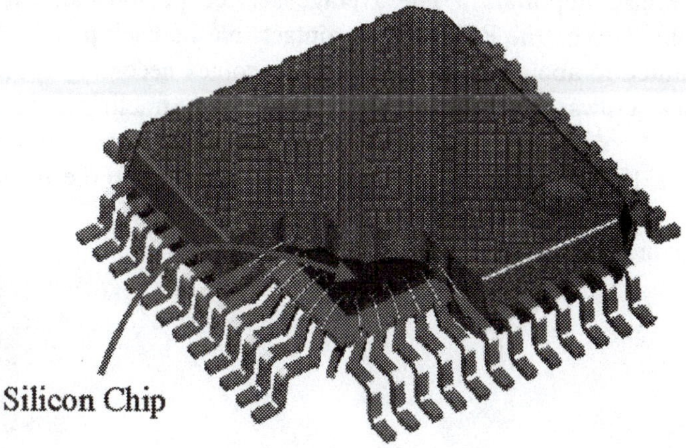

Silicon Chip

Fig. 1.7 Cut away view of a gull-wing chip carrier.

Layout details represent another example of design for manufacturing. Are the first level packages placed on the PCB with the same orientation and with sufficient space between the packages? The orientation is important as some placement tools cannot rotate a component prior to insertion or placement. If the orientation is not consistent, it could be necessary to rotate the board and this requires a second pass through the placement machine. When sufficient space is not provided between the chip carriers, the placement tools on the assembly machine can interfere with adjacent components so that assembly is impossible. In this case it is necessary to redesign the circuit board.

Dimensioning is another area that affects manufacturing in a significant way. Drawings should be dimensioned to accommodate the user. For example, the drawing of a frame which supports a circuit card during assembly, should be prepared and dimensioned for a tool maker who is responsible for the dies used to fabricate the frame. A second drawing should be prepared and dimensioned for the inspector who will certify the dimensional accuracy of the parts after they are produced. The dimensioning on these two drawings will be different so as to facilitate the tasks of the tool maker and the inspector.

Examples of design for manufacturing are numerous and they all illustrate that close cooperation between manufacturing, operations and mechanical development is essential if a high quality and cost effective electronic system is to be produced on schedule.

1.4.6 Maintenance

Mechanical design markedly affects the ability to properly service a product in the field. This is particularly important for a surprisingly large number of different types of electronic systems. Have you ever been in line in a large retail store when its computer went down? How long did you wait for it to become operational? Failures will occur at random when the reliability of the system is not adequate. To minimize delay and inconvenience, prompt service is essential. Prompt service is no accident and minimizing down time requires careful design, extensive training of service personnel, and an adequate inventory of spare parts.

The general design approach to quality maintenance is to design the system with a number of different field replaceable units (FRU's). These may be PCB's or power supplies or other sub-assemblies. One does not attempt to repair a single component that failed because it takes too long to find the flawed component, and the number of spare parts required for field service is too large to be practical. Time to access the FRU is important and that time is often controlled by the mechanical design. In the chassis of a complex computing system, drawers which open to expose banks of PCB's are often employed to reduce access time to a minimum. Extraction and insertion forces for the PCB are also important. The average service person can exert about 35 lb in engaging the edge card connector on the PCB into its contacts on the back panel. At about 6 oz/pin for edge connectors one is limited to about 100 pins before it becomes necessary to incorporate some form of assist (lever and/or jack screws) to aid the service personnel in extracting and inserting the cards.

An example of a field replaceable unit is the automotive engine control module presented in Fig. 1.8. If this unit fails it is replaced and returned to the manufacturer for repair. Replacement is simple because only three bolts are removed and then the module is unplugged.

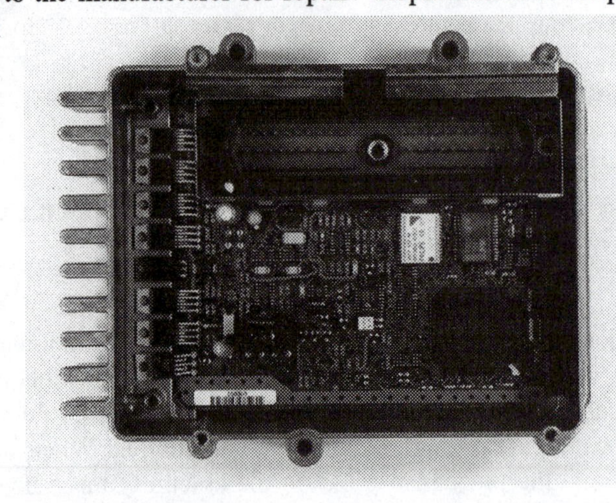

Fig. 1.8 Electronic module from an under the hood automotive application. Note that the cover has been removed to display the internal components and the circuit card.

Cable wiring is another area where long and unacceptable delays are frequently encountered in servicing. Cable harnesses often contain hundreds of wires which lead from one difficult to access location to another (see Fig. 1.9). After finding the failed cable, the service person does not want to take the time to cut the faulty cable from the harness, replace it, and then retie the harness. Instead, the design should incorporate several spare wires which can be used as substitutes to facilitate repair. In this case, repair consists of disconnecting the ends of the faulty wire and reconnecting the ends of one of the available spare wires.

Fig. 1.9 Cable harness for both power and signal wiring in a full height electronic enclosure.

Thermal warning systems are another area of design for maintenance. When forced convection or conduction methods are used to cool an electronic system, failure of a fan, pump or refrigeration unit is possible. This failure can result in over heating of major portions of the system within a short period of time. To prevent the resulting damage, a thermal warning system, which alerts the operator to initiate a controlled shutdown, is incorporated into the system. For example, all modern personal computers employ temperature sensors on the main central processing unit (CPU) chip. Another automatic system that can be incorporated initiates an uncontrolled shutdown in the event of operator error. These safety systems should be installed on high-performance highly-priced products.

Other examples could be cited to illustrate the importance of designing the enclosures, drawers, cages and racks in the system to permit rapid and complete servicing. Close cooperation between mechanical development and the field service organization is essential during the design process to insure a product capable of quality servicing.

1.4.7 Thermo-Mechanical Deformations

Perhaps the most important mechanical problem in the design of electronic systems is the generation of strains in the solder joints due to temperature changes that occur in shipping, storage, and operation. The chip carriers are soldered to printed circuit boards to mount them mechanically and to connect them electrically. Unfortunately there is often a large mismatch in the effective coefficient of thermal expansion between the chip carrier and the printed circuit board. The coefficient of thermal expansion of the PCB is higher than that of the chip carrier ($\alpha_{PCB} > \alpha_{cc}$). As a consequence, temperature changes ΔT produce a differential expansion d (or contraction) as illustrated in the simplified/idealized chip carrier geometry shown in Fig. 1.10. This differential expansion (or contraction) produces a shearing strain γ in the solder joints that can be estimated by the expression:

$$\gamma = \frac{L(\alpha_{PCB} - \alpha_{cc})\Delta T}{2h} \qquad (1.3)$$

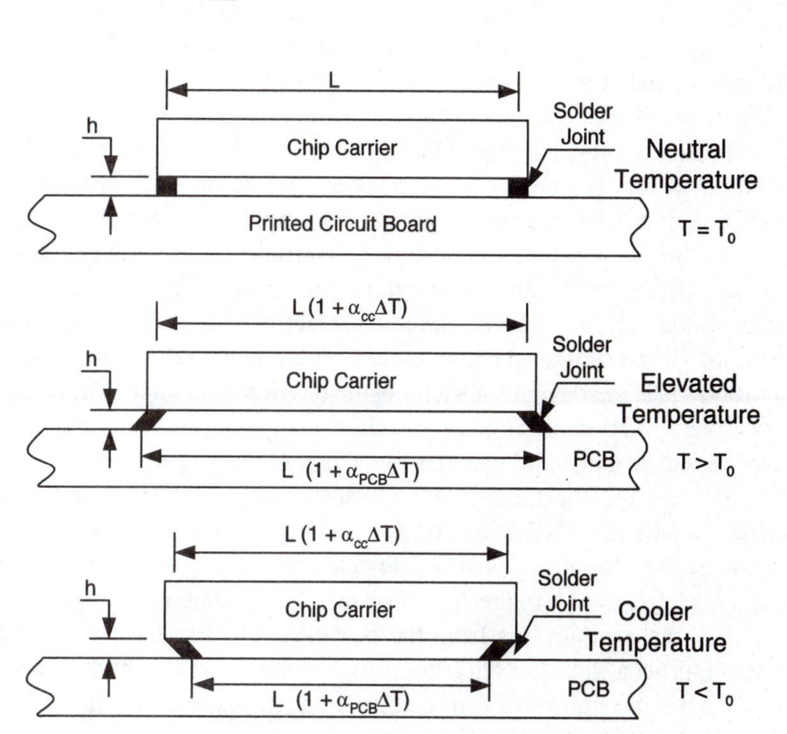

Fig. 1.10 Two-dimensional schematic showing expansion and contraction of a chip carrier and a PCB due to temperature changes and a mismatch in coefficient of thermal expansion.

Electronic systems can undergo hundreds or thousands of start-up and shut-down cycles over their lifetime. The temperature of the components increases after start-up and decreases after shut-down. For this reason, the solder joints are subjected to cyclic thermal shear strains that can result in thermal fatigue failure of the solder joints. An example of a failed solder ball joint found on a flip chip type chip carrier is presented in Fig. 1.11. The fatigue crack typically initiates after hundreds of thermal cycles, and it then grows slowly until the crack crosses the joint and failure occurs.

Fig. 1.11 Fatigue crack developed in a solder ball type solder joint of a flip chip package.

1.4.8 Shock and Vibration

Shock and vibration are important at three different times in the life of an electronic system. The first occurs during the manufacturing and assembly process, when the product or its sub assemblies are being moved from one manufacturing station to another. Abusive handling or dropping the product can produce very high accelerations (shock loadings), which will deform or fracture devices and cause failure of the system to function. The second occurs in transporting the system through the distribution chain until it is finally installed at the customer's location. If the distribution chain is long, the product will be handled at several locations. In these situations, the normal approach is to enclose the product in a shock proof box. Such a container deforms under impact when the product is dropped, and limits the accelerations (g-levels) and transient forces (g-loads) transmitted through the box to the product. Finally, it is recognized that both shock and vibration environments may be encountered in service. Of course, in typical office surroundings a computer system is not subjected to adverse environments. However, modern handheld consumer electronics (e.g. cellular phones and portable music players) as well as military hardware can be exposed to severe vibrations and or shocks on repeated occasions and for extended periods of time. Similarly, electronic systems for automotive applications are subjected to very hostile environments that include high temperature, dust, humidity in addition to shock and vibration. It is the design for such hostile environments that will be covered in this textbook.

In specifying a product for military applications, the procuring agency will define the vibratory environment through the use of a standard military specification. For example, ship board electronic systems intended for submarine applications are often designed to satisfy the vibration specification defined in MIL-STANDARD 167-1. This mil-standard describes the g-levels and frequencies to which the electronic equipment must be subjected to and survive during qualification tests, as indicated in Fig. 1.12. The frequency range for shipboard equipment is relatively low 4-34 Hz, because the primary source of the excitation is from the propeller which rotates at a relatively low angular velocity. The design approach in this application is to construct stiff structures with relatively high natural frequencies to avoid the possibility of a resonance frequency occurring in the cabinet, any circuit board or major sub-assembly within the frequency range specified.

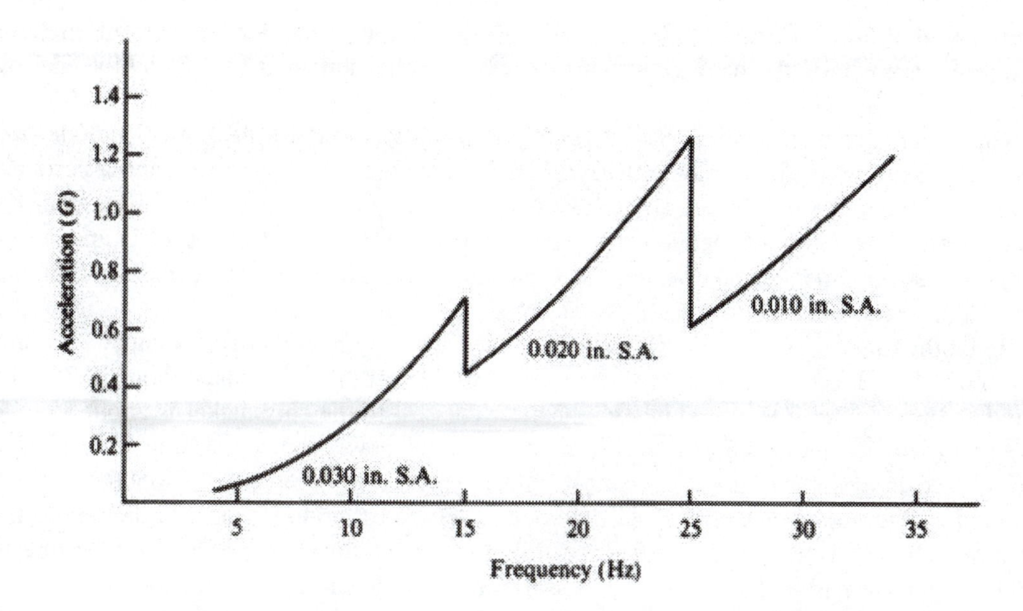

Fig. 1.12 Vibration specification from MIL-STD-167-1 for equipment installed on naval vessels.

The specifications on the vibratory environment for aircraft electronics are entirely different. The range of frequencies is much wider—from 15-2000 Hz. Moreover, with the weight limitations imposed on aircraft electronics, it is impossible to design the enclosures and PCBs with sufficient rigidity to avoid resonances. For this reason, the design approach is to accept resonances and limit the force and displacement transmission factors. With limited forces it is then possible to design the circuit boards and sub-assemblies with sufficient fatigue endurance to avoid failure during extended periods of exposure.

Upon completion of the prototype of a system, it is subjected to a closely controlled qualification testing where the prototype is exposed to a specified shock or vibration environment. Damage that often occurs includes fatigue of structural components which support the electronic sub assemblies; wiring, cable and solder joint failures due to fatigue; pin breakage in connectors; cable chafing; and shear of bolts and pins due to shock imposed loads. Clearly the design of military, automotive and factory hardware with demanding shock and/or vibration specifications is a challenging problem which requires an extensive knowledge of this subject. We will show the basics of the theory covering shock, vibration and fatigue in treating this area of design.

1.4.9 Reliability

System reliability is measured by the ability of a system to meet its specification for an extended period of time without failure. The reliability of a system is affected by both hardware and software because either can fail. If your computer crashes, the software has failed and it is necessary to reboot your computer or to reload an application program. Usually this is not a serious issue if the delay is short and only one person is affected. However, if the accounting system of a major bank crashes, hundreds of people standing in line have to wait for their accounts to be serviced. While system failures (crashes) due to software are more frequent, failures due to hardware problems are more serious. It takes much more time to locate and replace a faulty circuit board or failing connector than to reboot a system. In this textbook, we will focus on hardware issues that affect reliability.

Reliability controls system availability, which is critically important for many applications including airline reservation systems, bank account management systems, communication systems, health care systems, air traffic management, etc. In many applications, redundant systems are employed so that if one system fails the back-up system is automatically activated to seamlessly take over the tasks of the primary system.

The development of a reliable electronic system entails many aspects that include designing against failure, insuring manufacturing process are under statistical quality control, and ascertaining that maintenance procedures are adequate and are being performed on schedule. To design against failure requires the correct selection of materials and components, careful layout of the PCB and selection of connectors, the development of an effective heat management system and the selection of an enclosure that provides protection for environmental parameters.

Reliability also affects cost, as illustrated by the following hypothetical example. An integrated circuit (IC) that fails a test on the wafer and is rejected might cost the chip maker about $0.25. If that IC is detected as flawed after it is placed in a chip carrier the cost to the chip maker is about $5 to $10. If the flawed IC is found on the PCB, the cost of replacement increases to about $20 to $50. If the flawed IC is discovered in system testing, the cost of replacement increases to $200 to $500. If the flaw is discovered after the product is shipped and field personnel are called in to perform the repair, the costs are in the neighborhood of several hundreds of dollars. Clearly, it is cost effective to insure that the components and materials are flaw free before assembly and shipment of a product.

Problem areas affecting reliability such as solder joint failures are studied by using analytical, computational and experimental methods. An example of such a study is presented in Fig. 1.13, which shows a solder joint failure that occurred in a thermal cycling test of a ceramic chip resistor mounted on a metal-backed glass-epoxy composite printed circuit board. A finite element model of the solder joint, the ceramic resistor and the PCB was also developed. The prediction of the finite element model for the distribution of the plastic shearing strains in the solder joint is shown in Fig. 1.14. The results from the test program and the analysis employing the finite element model were in close correspondence.

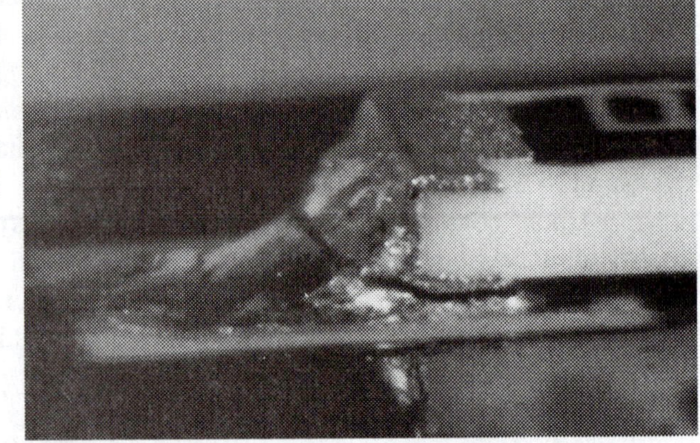

Fig. 1.13 Solder joint failure of a ceramic resistor due to thermal cycling.

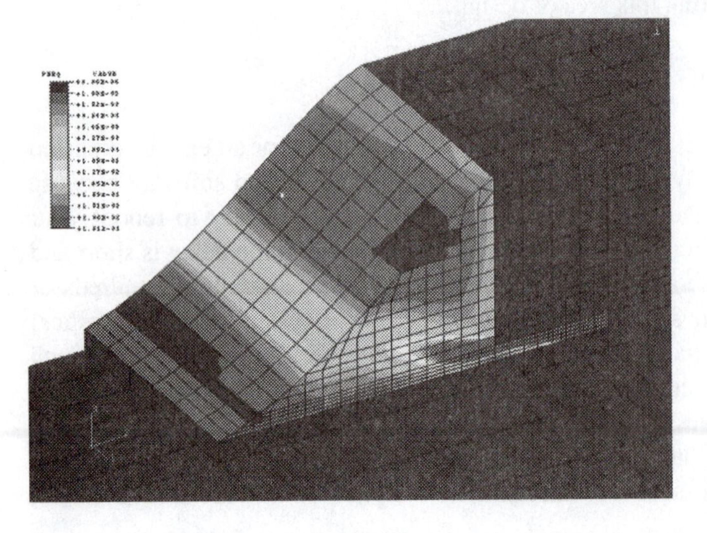

Fig. 1.14 Results from a finite element analysis showing the shearing strain distribution due to a mismatch in the temperature induced thermal expansion of the ceramic resistor and the PCB.

1.4.10 Ergonomics

Ergonomics involves the relationship between people and the machine or in this case the electronic system. The basic idea is to design the system to reduce the physical effort required by the operator when using the system. Examples of ergonomics in design of electronic systems are numerous. The design of a keyboard for a computer terminal should clearly take into account the need for concave key shape with tactile feedback. Spacing between the keys, the angle of the keyboard, the ability to change this angle and the height of the keyboard all affect the productivity of the operator and the quality of the output.

There are three main areas of ergonomics which markedly affect mechanical design—the operator panel layout, the operator workstation and the environment. The environment pertains to the office, vehicle or factory and includes noise, lighting and temperature. Layout involves placing switches, meters, lights, and control knobs in logical positions where the operator's task is made as easy as possible. For example, the meter which displays the state of a variable should be located adjacent to the control knob for that variable. This placement permits the operator to monitor the meter as the control knob adjustment is made. Controls should follow common convention--clockwise to increase voltage or current. Switches toggle upward in the on position in the US, but remember that they toggle downward for the on position if the product is designed for a customer in Europe.

Workstation design involves operator comfort and is much more important than generally considered, particularly if the operator is expected to be on station for extended periods of time. If possible, the operator should be comfortably seated as it increases his or her attention span and alertness. The chair should be designed with sufficient adjustments to adapt to the user and not vice-versa. While mechanical designers may not have the responsibility for designing chairs, they should be sufficiently knowledgeable in anthropometrics to select an appropriate chair and then to design the operator-machine interface to accommodate the range in the size and weight of the operators which will occupy that chair. An example of a workstation design for a computer operator is illustrated in Fig. 1.15. Note the dimensions indicating the position of the monitor relative to the eyes of the operator, the height of the table and chair and the position and tilt of the keyboard.

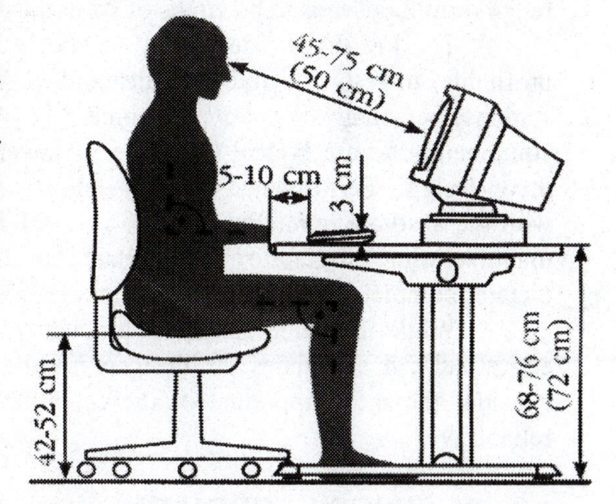

Fig. 1.15 Important dimensions in the design of a workstation for a computer operator.

The office, factory or vehicle environment involves noise, lighting and temperature. Noise can be very distracting and prevent complete concentration on the task at hand. If the noise is sufficiently loud and the operator is exposed for extended periods of time, hearing losses and other physical disorders result. Lighting involves the proper intensity and distribution of light at the work station. Of particular importance is glare from instrument panels and displays such as CRT's. Glare can be avoided by the use of anti-reflection coatings or by the positioning of the light sources to avoid the reflections producing glare. When glare is permitted, displays cannot be monitored with out eye strain and/or error. Temperatures either too high or too low result in discomfort and dissatisfaction of the operator. Clearly sweating or shivering operators are less effective and efficient.

1.5 RANGE OF PRODUCTS

The products produced and marketed as electronic systems cover an extremely wide range. Some products are advanced, complex and high in price. Others are very simple, produced in millions of units and very competitive in cost. These differences in the price and volume of the products require a change in the design approach and in some cases major changes in the methods employed.

At the risk of over simplification, we will divide the entire market of electronic products into three general classes, namely the high-end, intermediate- and low-end products. As the name implies, the high-end products are high-performance, high-cost, long-life systems usually produced in relatively low volume. The primary design objective with the high-end product is performance, and while cost is also important, it is secondary to performance. Large mainframe computers, supercomputers, massive servers and advanced signal processors for military applications are products in this category. Extremely high reliability, often achieved with redundant sub-systems, is a characteristic of these products. Prices are often above one million dollars per unit with a relatively low volume of unit sales. Select laboratory instrumentation usually associated with health care facilities or major national laboratories also fall into this category.

The intermediate product line is much broader with lower cost products included in this category. Mid size computers, less critical military systems, most laboratory instruments and most special purpose data processing systems are the typical products. Both performance and cost are quite important. Any gain in performance must be carefully balanced by the extra expenditure required to achieve the incremental gain in speed or reliability. Product volume is larger, and hence the design is coordinated even more closely with manufacturing. The product life is in the intermediate range so the design tends toward a flexibility that permits product upgrades by making periodic changes in the model. Reliability remains important but usually not at the cost of major redundant sub-systems. Prices usually range from thousands to hundreds of thousands of dollars per unit in the intermediate range product.

The low-end product is by far the largest segment of the market, the most competitive, least profitable, lowest cost and most demanding from the design point of view. Products such as work stations, personal computers, consumer electronics, automobile electronics, home appliances, office equipment, etc. are typical examples. Annual product volume is very high and manufacturing costs drive design. Performance is always important, but in these products performance must be achieved with the absolute minimum increase in cost. Quality is achieved by the close coupling of design and manufacturing. Reliability is important, but often this design goal is supplemented with a policy that dictates complete replacement of the unit for a day or more during servicing.

While this wide variety of products complicates the task of writing a textbook to address the entire field, it stimulates the designer to address development in a market-oriented manner and to consider the most important of the essential design characteristics—quality, cost, performance and reliability.

1.6 BUSINESS ASPECTS

The electronics system business is and has been the fastest growing segment of the manufacturing and service industries in the US during the past forty years. Annual sales, depending upon the definitions imposed, are in the neighborhood of about a trillion dollars per year and growing at an annual rate of nearly 10%. It is a very competitive business because it requires relatively little capital[2] to introduce a new product and to start a new company. Also, funds may be provided by venture capitalists seeking to share the equity in the firm. New ideas for new products are essential. The successful ideas and products developed by existing and/or new companies drive the growth in the industry.

---

[2] Many small business can be started with an initial investment of $10 to 20 million.

Paramount to the growth of the business has been the technological advances which seem to occur on a schedule similar to the development of new and advanced integrated circuits (IC's). This progress, illustrated in Fig. 1.16, shows the steady increase in the number of transistors which can be placed on the small area of a silicon chip. As more and more devices and circuits are placed on a single chip, several advantages result which permit the development of entirely new products and the marked improvement of existing products. For example, each new generation of memory chips includes more memory capacity in the same or smaller physical area, with the same or lower cost. Also, as chips become increasingly dense, the speed of the switching is improved and the processing rate of the product increases.

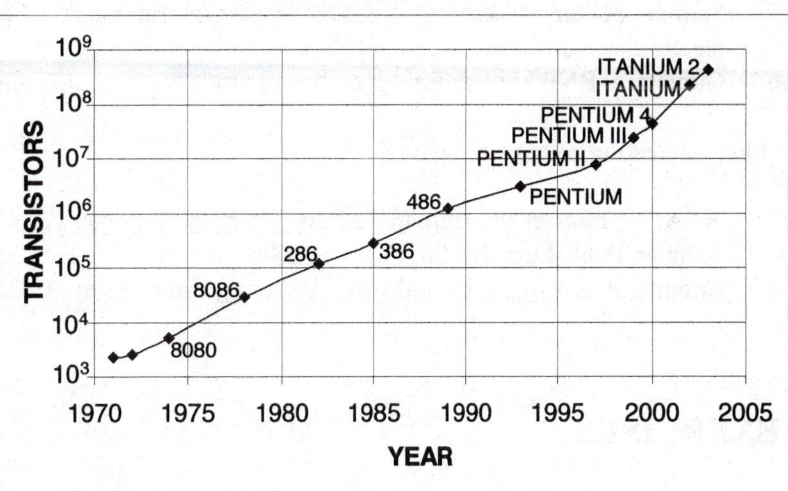

Fig. 1.16 Continuing progress made in increasing the number of transistors on a single chip. Data from Intel.

We are experiencing a rapidly changing business world that is driven by new developments in IC technology, by new ideas for product development, by new requirements for information processing on the Internet, and by Global facilities for research, development and manufacturing. Consider the rapid changes occurring in products introduced in the last decade such as DVD players and writers, cellular phones, digital cameras, hard drive recorders (TiVo), and portable music players (iPod). These developments will lead to improvements in the way we process and store information, and will provide the basis for continuous enhancement in the standard of living for society as a whole.

REFERENCES

[1] Winkle, T.-M., et al, "First and Second Level Packaging of the z990 Processor Cage," IBM Journal of Research and Development, Vol. 48, No. 3/4, pp. 379-394, 2004.

[2] Parrilla, J. C., et al, "Packaging the IBM eServer z990 Central Electronic Complex," IBM Journal of Research and Development, Vol. 48, No. 3/4, pp. 395-408, 2004.

[3] Goth, G. F., et al, "Hybrid Cooling with Cycle Steering in the IBM eServer z990," IBM Journal of Research and Development, Vol. 48, No. 3/4, pp. 409-423, 2004.

[4] Pecht, M. ed., Handbook of Electronic Package Design, Marcel Dekker, Inc. New York, NY, 1991.

[5] Dally, W. J. and J. W. Poulton, Digital systems Engineering, Cambridge University Press, Cambridge, UK, 1998.

The following periodicals provide current articles which will aid you in understanding the acronyms and in identifying the issues involved in mechanical design of electronic systems:

1. IBM Journal of Research and Development, International Business Machine Corporation, Armonk, NY.
2. Electronic Design, Penton Media, Inc. find it online at http://www.elecdesign.com.
3. Surface Mount Technology (SMT), Penn Well Corp. Find it on line at http://www.pennnet.com.
4. Printed Circuit Design and Manufacture, UP Media Group, Find it on line at http://www.pcdandm.com.
5. Semiconductor Packaging, Reed Business International, Find it on line at http://www.read-electronics.com.
6. Circuitree, BNP Media. Find it on line at http://www.circuiTree.com.

Other interesting references include:

7. N. A. Stanton et al, Human Factors Methods: A Practical Guide for Engineering and Design, Ashgate Publishing, Burlington, VT, 2005.
8. Cunniff, P. F. Environmental Noise Pollution, John Wiley, 1979.

EXERCISES

1.1 Inspect the computer system which you use most of the time. Describe the key board features which help you type more accurately and those features which help you from becoming tired as you type. What do you think of the keyboard layout? Is it optimized in any sense? Clearly it is not, but do you know the story behind why the keyboard layout is so poor.

1.2 Why are some of the lead wires to your PC placed on the rear panel instead of on the front panel where they would be more accessible? Why are these leads placed in a location which is not easily accessible?

1.3 Why are the vents for the cooling air placed on a side or back panel of your PC? The top of the PC provides a larger area and would permit including more vents for enhanced air flow with a lower head loss. Please explain this apparent conflict in the placement of the vents.

1.4 The sound pressure level issued by a product is measured in decibels. What is the typical sound pressure level produced by a rock and roll band? This is clearly too high for any product, but what are acceptable sound pressure levels? What common product usually exceeds the acceptable limits in an office environment?

1.5 List the different types of connectors with which you are familiar.

1.6 Take the cover off of your PC and identify as many of the sub-assemblies as possible.

1.7 Without removing a circuit card identify as many of the components on that card as possible. If you don't follow these instructions and you do remove the card for a better view make sure you wear a grounding strap on your wrist. The strap will prevent electrostatic damage to one or more of the components.

1.8 Estimate the number of solder joints and pins for connections on this PC circuit card.

1.9 Describe the thermal management system in five different electronic products which you encounter on a day to day basis.

1.10 Prepare a graph of the failure rate, as a function of junction temperature for T_j ranging from 20 to 150 °C. Let A = B = C = 1 for this initial determination. Discuss the effect of the higher temperatures on the failure rate.

1.11 Give an example of good design for manufacturing of an electronic product. If you can't think of one for an electronic product, give an example for an automotive product.

1.12 Derive an equation for the shearing strain in a solder joint due to temperature changes. Reference Fig. 1.9 for drawings of the chip carrier and definition of the symbols.

1.13 Give an example of poor design for manufacturing for electronic products.

1.14 If you drop a box on the floor from say a height of 1.2 m, determine the deceleration during impact. What assumptions did you have to make in your analysis?

1.15 What is your opinion of the ergonomics design of the chair in your office or your class room? Why do you think management selected the chair you used?

1.16 Give an example of a high-end product and estimate its cost.

1.17 Give an example of an intermediate product and estimate its cost.

1.18 Give an example of a low-end product and estimate its cost.

1.19 What is the least expensive electronic product that you can identify? Is it produced in a large volume? How many electronic components does it contain? Is it designed for ease of manufacturing? Is it designed for ease of maintenance?

1.20 What is a venture capitalist? Do they serve an important function in the development of small business in the US?

1.21 Samsung Electronics, a world leader in advanced memory technology, announced in December of 2004 that it has developed a 512Mb DRAM device. It will begin to market this high density memory chip in 2005. Samsung introduced the 256 Mb DRAM chip in November of 2003. . When do you estimate that they will announce the development of the 1 Gb memory chip?

1.22 If your PC memory card contains 1 Gb how many memory chips are installed on the board if the chips are:

(a) 64 Mb
(b) 128 Mb
(c) 256 Mb
(d) 512 Mb

CHAPTER 2

ELECTRONIC COMPONENTS AND SEMICONDUCTOR DEVICES

2.1 INTRODUCTION

In packaging electronic systems, it is often necessary to design cabinets, drawers, cages or gates to house the components and to permit easy and rapid access for maintenance. Printed circuit boards (PCB's) are designed to support hundreds of small components with thousands of interconnections. A cooling system is incorporated into the cabinets to dissipate the heat generated by the electronic devices. Interconnections and heat removal with small thermal penalties are the two most important features in the design of highly reliable products. There is a wide variety of hardware both mechanical and electrical which must be included in the design of even a relatively simple electronic system. A partial listing of some of the most common components used in design is given below:

Discrete transistors	Power supplies	Logic type IC's
Transformers	Memory type IC's	Plasma panels
Diodes	Liquid crystal displays	Resistors
Cathode ray tubes	Capacitors	Vacuum tubes
Inductors	Disk drives	Potentiometers
Tape drives	Relays	Lamps
Switches	Fans	Circuit breakers
Cold plates	Connectors	Cable harnesses

The emphasis in this textbook is placed on the packaging of microelectronic circuits which are used extensively in the design of high performance digital systems, laboratory instrumentation and automated manufacturing systems. To package microelectronic circuits, it is necessary to understand in at least a qualitative manner the functional behavior of microelectronic devices. For this reason, the basic principles of semiconductor theory are reviewed and the operation of semiconductor diodes and transistors is described. The coverage is then extended to introduce logic gates and the Metal-Oxide-Semiconductor-Field-Effect-Transistor (MOSFET) that are used in most electronic systems today. The last part of the chapter treats the scale of integration with projections into the future which will markedly affect packaging strategies in the later years of this decade. Of particular importance are the significant changes in packaging design that will occur because of the introduction of new electronic devices. These changes in packaging design will be driven by higher I/O count and higher heat dissipation that will be typical of the newer high performance components. These changes will require designers to create PCBs with higher density pads and circuit lines and to develop more effective heat dissipation methods in order to fully utilize the logic devices produced with new chip fabricating technologies.

2.2 CONDUCTORS, INSULATORS AND SEMICONDUCTORS

Conductivity in materials depends on the structure of the atoms of elements which are combined to give an alloy and the resulting atomic bonding between these atoms. Consider the metal aluminum with the atomic structure illustrated in Fig. 2.1. The aluminum atom has a full inner shell (the K shell) and a full L shell with 2 and 8 electrons filling these shells. However, the outer M shell contains only 3 electrons and they are loosely bound to the nucleus. These nearly free M shell electrons act as negative charge carriers in conducting current in a wire made of aluminum. The resistance R of a conductor in the shape of a wire or a rectangular conductor (line) formed on the surface of a PCB is given by:

$$R = \frac{\rho L}{A} \tag{2.1}$$

where L is the conductor length (cm), A is its cross section area (cm) and ρ is its resistivity (ohm-cm).

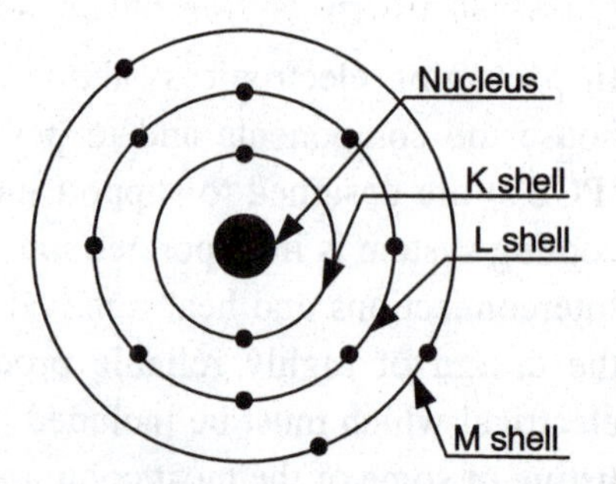

Fig. 2.1 Atomic structure of aluminum.

The resistivity ρ depends on the atomic structure of the element. If there are a large number of loosely bound electrons in the outer shell, the resistivity is low as indicated in Table 2.1. It is evident from the results shown in this table that the resistivity of different types of materials can vary over 18 orders of magnitude. For metal conductors ρ is of order 10^{-6} Ω-cm but for insulators ρ is of order 10^{12} Ω-cm or higher.

Table 2.1
Resistivity of select materials

Material	Classification	Resistivity (Ω-cm)
Silver	Conductor	1.63×10^{-6}
Copper	Conductor	1.72×10^{-6}
Aluminum	Conductor	2.83×10^{-6}
Nickel	Conductor	6.9×10^{-6}
Platinum	Conductor	9.8×10^{-6}
Silicon	Semiconductor	1.56×10^{5}
Aluminum oxide	Insulator	1×10^{15}
Silicon dioxide	Insulator	1×10^{14}
Epoxy	Insulator	1×10^{15}
Polyethylene	Insulator	1×10^{18}

The atomic structure of insulating materials clearly shows the reason for the large values of resistivity ρ. Consider the atomic structure of silicon dioxide SiO_2 shown in Fig. 2.2. Silicon has a full K shell, a full L shell and four electrons in the outer M shell. Oxygen has a full K shell and six of eight electrons necessary to fill the L shell. The silicon atom combines with two oxygen atoms to provide the correct number of electrons to fill the outer shells of the two oxygen atoms when the four silicon electrons are shared. The shared electrons are covalently bonded, and the free silicon electrons are not available to act as charge carriers. Materials like SiO_2, Al_2O_3, MgO, BN, Si_3N_4 and BeO, which are classified as ceramics, are all covalently bonded with complete outer shells. They act as insulators with very high resistivity.

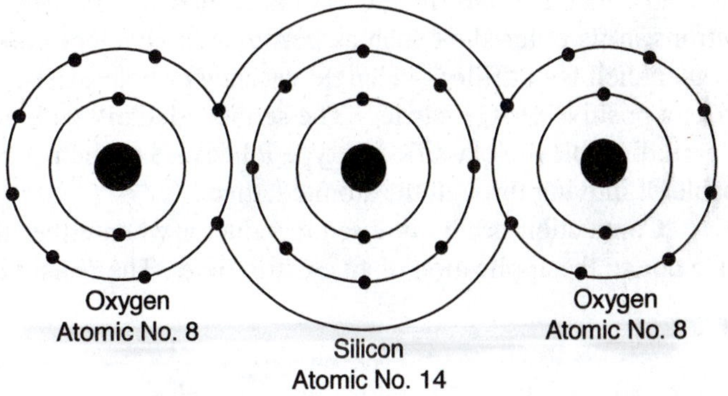

Fig. 2.2 Atomic structure of SiO_2.

Oxygen
Atomic No. 8
Silicon
Atomic No. 14
Oxygen
Atomic No. 8

Elements such as carbon, silicon and germanium each have four electrons in their outer shells. The atoms in these elements form covalent bonds with four neighboring atoms to form stable structures similar to that shown for silicon in Fig. 2.3. The electrons are tightly bound due to the covalent structure and free electrons are not available to act as charge carriers. However, the atomic structure is not perfect and some of the electrons have sufficient energy to jump from the valence state to the conduction state. For this reason, materials such as silicon, germanium and gallium arsenide are classified as semiconductors and exhibit resistivity much higher than the metal conductors but significantly lower than the ceramic insulators. The resistivity of silicon is 1.56×10^5 Ω-cm and is due to the presence of one electron out of 2×10^{13} which has sufficient energy (1.1eV) to jump from the valence band to the conduction band. The ability of these electrons in the conduction band to carry charge is termed intrinsic conduction.

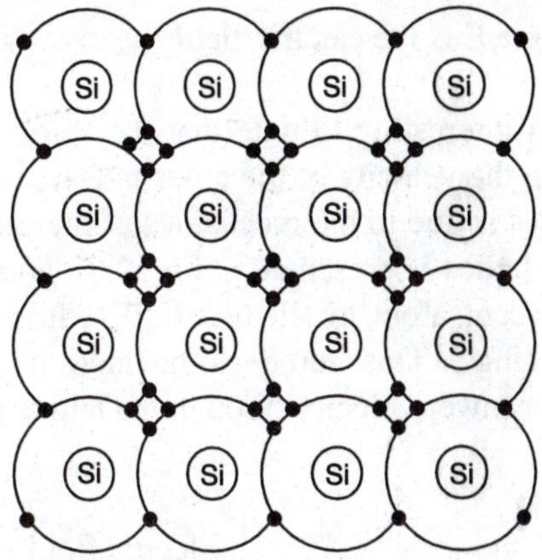

Fig. 2.3 Covalent bonding of electrons in silicon.

Si Si Si Si
Si Si Si Si
Si Si Si Si
Si Si Si Si

2.3 EXTRINSIC SEMICONDUCTORS

The intrinsic semiconducting capability of silicon was described in the previous section. The semiconducting properties of silicon, geranium or gallium-arsenide can be modified by changing the atomic structure of single crystals of these materials. To show the modification of the crystal lattice, consider the structure of silicon with its covalent bonding as shown in Fig. 2.3. If one or more of the silicon atoms in the lattice is replaced with an impurity atom, the conductivity of the modified structure is changed. Impurity atoms called dopants are of two basic types:

1. Elements from the V column of the periodic table with five valence electrons in their outer shell such as phosphorus, arsenic and antimony
2. Elements from the III column of the periodic table with three valence electrons including boron, aluminum and gallium in their outer shells.

For example, if phosphorus with its five valence electrons is introduced into the silicon lattice, the fifth electron is not used to complete covalent bonding. This extra electron, an extrinsic charge, is relatively free to carry current. Hence, the resistivity of the doped silicon is lower than that of pure silicon. Silicon with dopant elements from the V column of the periodic table are classified as type N semiconductors, because they contain extra electrons which are negative charge carriers.

If silicon atoms in the crystal lattice are replaced with a dopant atom with three valence electrons in its outer shell such as boron, then covalent bonding occurs between silicon and boron but the outer shell is not filled—a single vacancy or hole exists. This hole is an accepter of electrons, and it acts as a positive charge carrier. The semiconductors with dopant elements selected from column III in the periodic table are classified as type P because of the positive charge, which is carried by the holes, is capable of moving through the atomic lattice.

Conduction occurs in a semiconductor when either the electrons or the holes move through the lattice due to the application of an electric field. The resistivity of the semiconductor is given by:

$$\rho = \frac{1}{eN\eta} \tag{2.2}$$

where N is the number of charge carriers, e is the charge on the carrier and η is the carrier's mobility.

The velocity v of the charge carrier is dependent upon:

$$v = \eta E \tag{2.3}$$

where E is the electric field.

It is interesting to note that the velocity of the electrons in N type silicon is about three times greater than the velocity of the holes in P type silicon under the same applied field. The reduced mobility of the holes is due to the mechanics of the motion of the holes. The manner in which the holes move through the lattice is presented in Fig. 2.4. The movement of a hole is due to the jumping of an electron from an adjacent atom to fill that hole, while at the same time creating a new hole displaced by one lattice spacing. This cumbersome motion of the holes accounts for their lower velocity through the lattice structure.

Electric field + ⇒ –
Electron direction ⇐
Hole direction ⇒
Current direction ⇒

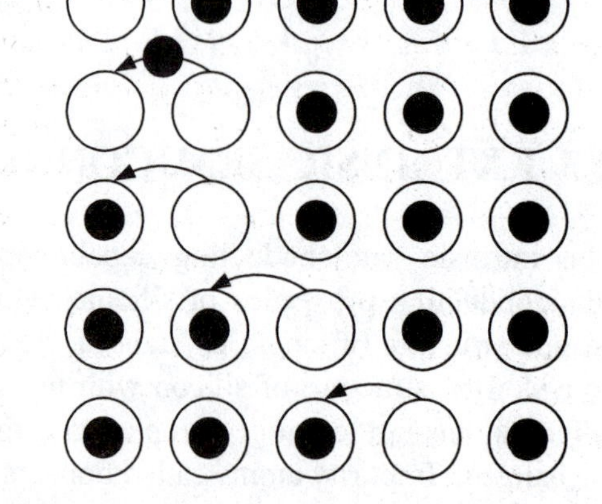

Fig. 2.4 Movement of a hole to the right by a sequence of electron movements to the left.

The resistivity of both N and P type silicon can be controlled over five orders of magnitude 10^2 to 10^{-3} (Ω-cm) by adjusting the amount of dopant (its concentration) added to the silicon as indicated in Fig. 2.5. The ability to control the resistivity is of critical importance in fabricating diodes, transistors or resistors in silicon. Methods have been developed for producing single crystal silicon with the dopant impurities of different concentrations introduced at local sites on the surface of a wafer cut from a single crystal.

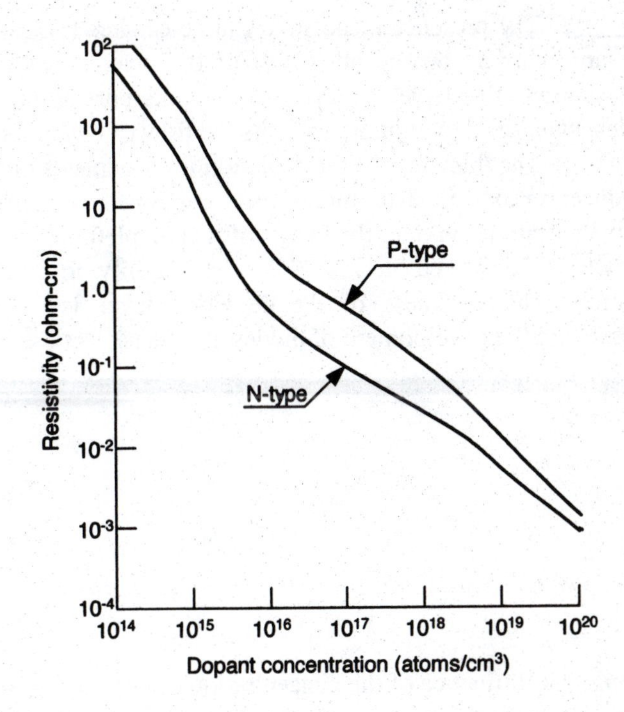

Fig. 2.5 Resistivity of P- and N-type silicon as a function of dopant concentration.

2.4 THE P-N JUNCTION

An interface between P and N type silicon can be formed by taking N type material and implanting boron into one of its surfaces to create a local region of P type material. This procedure produces a P-N junction, which is illustrated schematically in Fig. 2.6a. The P type material contains holes which act as accepters for electrons and the N type material contains electrons which serve as donors to fill these holes.

At the interface of the P and N type semiconductors, the holes and electrons combine and eliminate each other forming a thin region free of charge carriers. The layer where the holes have accepted the donor electrons is termed the depletion region indicating that the available charge carriers have been depleted. The remaining holes in the P material create a negative charge on the electrode opposite the junction and the remaining electrons in the N material create a positive charge on its electrode. In this state the junction is electrically neutral.

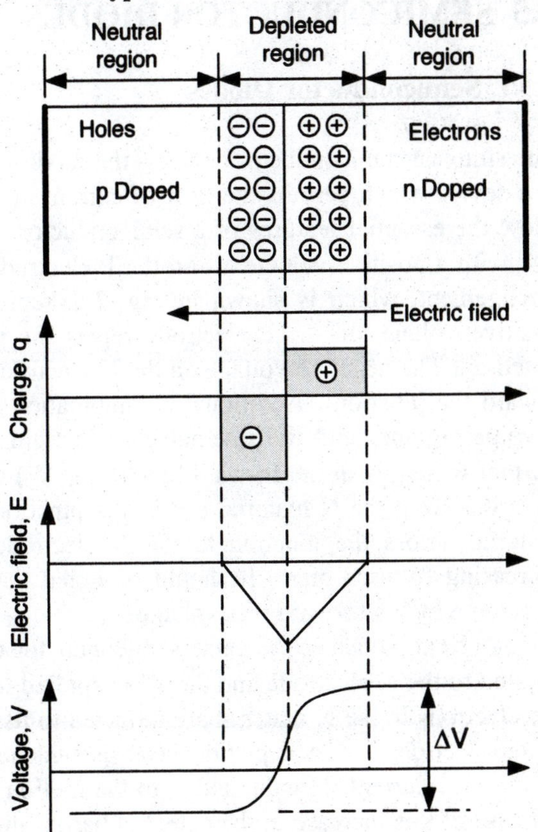

Fig. 2.6 Electrical characteristics of a P-N junction.

The presence of positive charge carriers (holes) on one side of the depletion layer and negative charge carriers on the other side of the layer develops an internal electric field across the junction. Essentially the depletion layer behaves like pure silicon and is capable of only intrinsic conduction. In this state, the depletion layer limits the flow of electrons across the junction.

The thickness of the depletion layer depends on the concentration of the impurity dopants. High concentrations lead to thin layers and lower concentrations yield thick layers. This influence of concentration on the thickness of the depletion layer is illustrated in Fig. 2.7. The ability to control the thickness of the depletion layer is useful in the development of diodes and transistors.

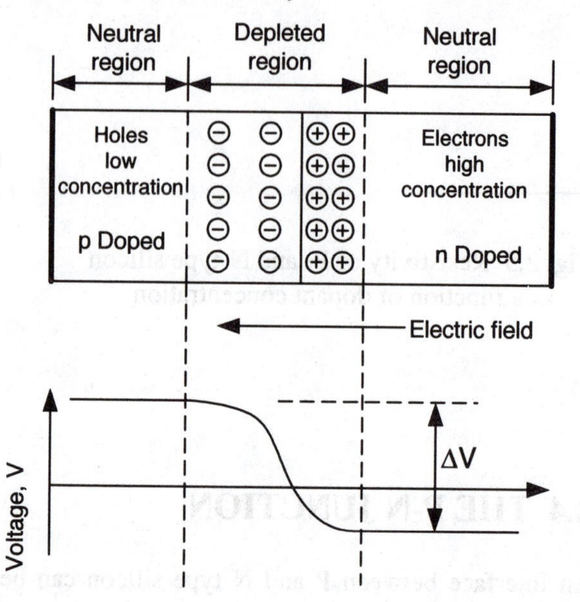

Fig. 2.7 Influence of the concentration of dopant on the thickness of the P-N junction.

2.5 SEMICONDUCTOR DIODES AND RESISTORS

2.5.1 Semiconductor Diodes

The simplest semiconductor device—the diode—is based on the properties of a P-N junction. A diode is a device which permits the flow of current in one direction and blocks current flow in the other. To show the essential features of a semiconductor diode, consider the P-N junction, described in Section 2.4, with a positive voltage V^+ on the P electrode and a negative voltage V^- on the N electrode. This arrangement, which is shown in Fig. 2.8, represents the diode under a forward bias voltage. The positive voltage on the P electrode repels the positively charged holes and drives them toward the junction. The negative voltage on the N electrode repels the negatively charged electrons driving them toward the junction. Additional recombination of holes and electrons take place, the depletion layer becomes thinner and the internal electric field which serves as a barrier to conduction is reduced. Further increases in the forward bias ($V^+ - V^-$) overcomes the potential barrier permitting the flow of electrons from the N material across the junction to the P electrode and the flow of holes from the P material across the junction to the N electrode. The forward current increases non-linearly with increasing forward bias. It should be noted that current does not flow until the potential barrier is overcome by the forward bias voltage.

Next, consider the same semiconductor diode but with the polarity reversed so that the V^+ is applied to the N electrode and the V^- is applied to the P electrode as indicated in Fig. 2.9. In this case, the electrons in the N material are attracted to its electrode and the holes in the P material are attracted to its electrode. The depletion layer is widened and the potential barrier due to the internal field increases. Current flow due either to the motion of electrons or holes across the junction cannot occur because of this increase in the potential barrier due to the reverse bias. Some leakage current flows but this is due to intrinsic conduction resulting from lattice imperfections. Typical leakage currents are measured in nanoamperes; whereas, the forward currents are measured in milliamperes. Forward and reversed currents are shown as a function of voltage bias for a typical semiconductor diode in Fig. 2.10.

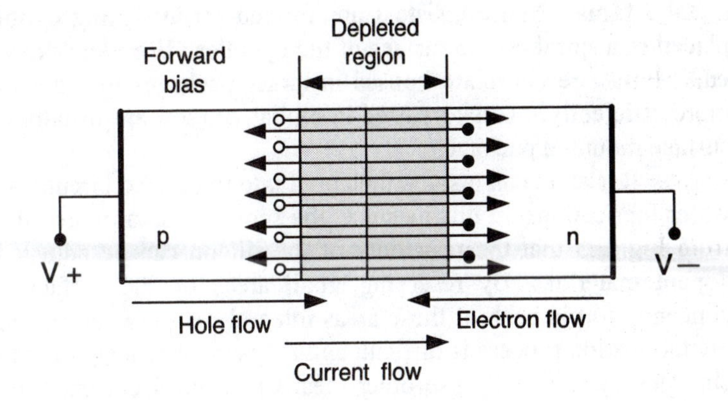

Fig. 2.8 Characteristics of a semiconductor diode under forward bias voltage.

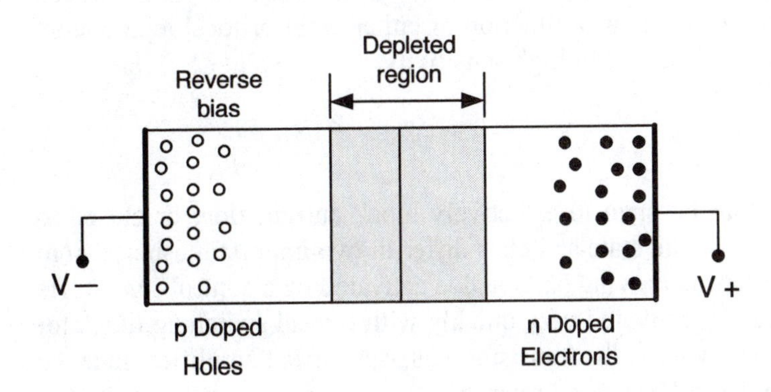

Fig. 2.9 Characteristics of a semiconductor diode under reverse bias.

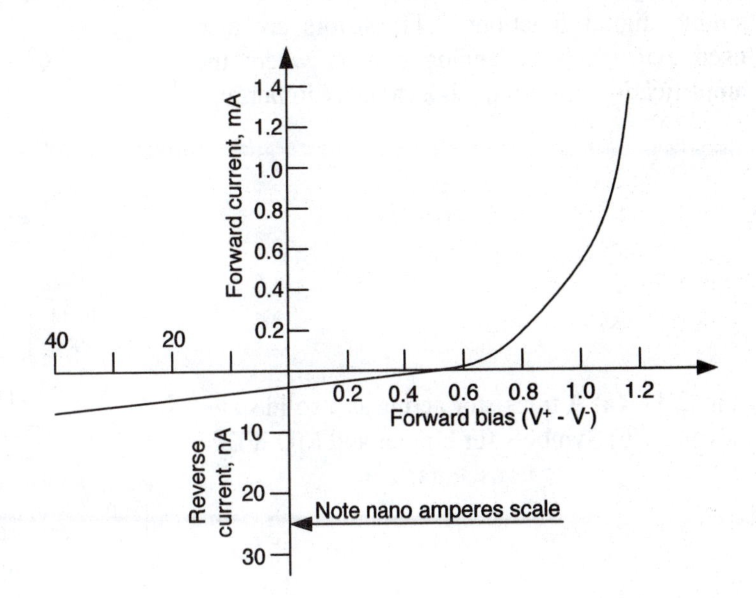

Fig. 2.10 Forward and leakage current as a function of voltage bias on a P-N junction diode.

2.5.2 Semiconductor Resistors

Resistors are usually discrete devices added to the circuit to control the characterisitc impedance of signal linmes or to limit currents. These discrete resistors are fabricarted from wire, carbon, or thin metalic films. Many resistors are formed on insulating cylindrical cores with the resistive material placed in a spiral on the surface of the cylinder. Wire leads extend from the axis of the cylinder at each end. Power resistors are housed in larger packages in order to better dissipate the heat they generate more efficiently. Resistors used in digital systems are usually much smaller and are housed in leadless surface mounted packages.

Resistors can also be inplanted into integrated circuits as part of the chip making process during wafer fabrication. In this instance, the semiconductor material employed to form the a resistor. Recall from Fig. 2.5 that the resistivity of the silicon can be adjusted by changing the concentration of the dopant material. By reserving small areas on the surface of the chip, and adjusting the dopant concentration in each of these areas integral resistors can be fabricated directly on the chip. However, the fabrication process is difficult and they consume a signifiant amount of chip area. For these reasons, chip designers usually use other means to control current flow on the chip and discrete resistors are mounted on the printed circuit board to control impedance and off-chip current flows.

In some applications semiconductor resistors serve as sensors because their resistance is markedly affected by both strain and temperature. In some research projects these sensors are implanted on a chip to study the strains developed in the silicon as a function of either globl or local temperature changes. The semiconductor sensors have an advantage of high sensitivity.

2.6 TRANSISTORS

Transistors are solid state switches which can be open to effectively block current flow or closed to permit current flow as illustrated in Fig. 2.11. Solid state switches differ in two important aspects from mechanical switches. Firstly, the transistor has no moving parts and is activated by a control signal. As a consequence of this feature, the switch can be activated very quickly with typical switching times for modern chip designs of the order of 0.1 ns. Secondly, the transistor acts as a current amplifier since the current passing through the switch $I_c + I_b$ is 10 to 100 times larger than the control current I_b required to activate the transistor. These two characteristics of the transistor permit it to be used extensively in digital logic circuits where logic design requires an extremely large number of switches to execute even simple digital functions. Transistors are also used extensively in analog circuits where the amplification of currents is extremely important.

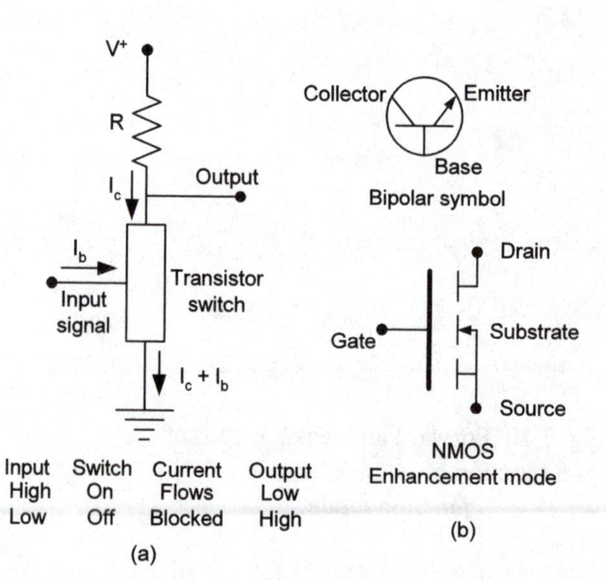

Fig. 2.11 (a) A transistor acting as a solid state switch. (b) Symbols for bipolar and MOSFET transistors.

There are two basic types of transistors that are in use today—the bipolar type and the metal oxide semiconductor field effect transistor (MOSFET). Key features of both of these basic transistors will be described in the next section.

2.6.1 Bipolar Transistors

Consider the N-P-N bipolar transistor, defined in Fig. 2.12a, which has three elements with solid state interfaces. Metal electrodes are placed on each element to provide for electrical connections to the emitter, base and collector. Note, that the P type silicon in the base is relatively thin compared to the N material used in the collector and emitter. This arrangement corresponds to two back-to-back diodes which share a common base.

The base is lightly doped and the depletion regions at the P-N junctions on both sides of the base are wide and effectively cover the entire thickness of the base when V^+ is applied to the collector and V^- is applied to the emitter. The depleted P material in the base acts as a potential barrier to conduction between the emitter and the collector as shown in Fig. 2.12b. In this state, the transistor acts as an open switch, because the base-collector diode is operating in reverse bias. Also the forward bias on the base-emitter diode is less than the threshold voltage required to conduct the forward current.

If a positive voltage is applied to the base electrode, the forward bias on the emitter-base junction increases and the potential barrier at this interface is overcome. Electrons from the emitter are injected into the base. These electrons find the base depleted, diffuse through it and then are attracted to the positively charged collector electrode. Thus, it is clear that a small positive voltage on the base turns on the transistor switch, electrons flow from the emitter to the collector and current flows from the collector to the emitter. The simple circuit, presented in Fig. 2.13, shows the action of a N-P-N transistor as a switch. The potentiometer at the left permits the base to emitter voltage V_{BE} to be varied. If V_{BE} is less than the threshold voltage of the transistor, the switch will be open the forward current blocked and V_{CE} equals V_s. However, when V_{BE} is greater than the threshold voltage, the switch will close, the forward current flows and V_{CE} approaches zero.

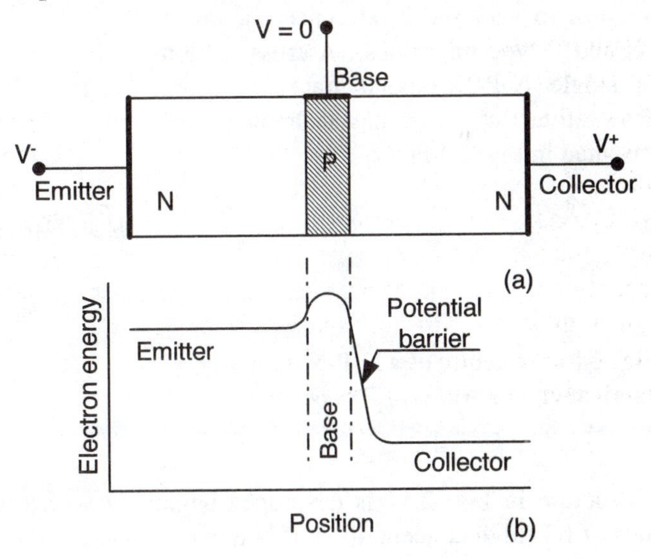

Fig. 2.12 A bipolar transistor acting as an open switch.

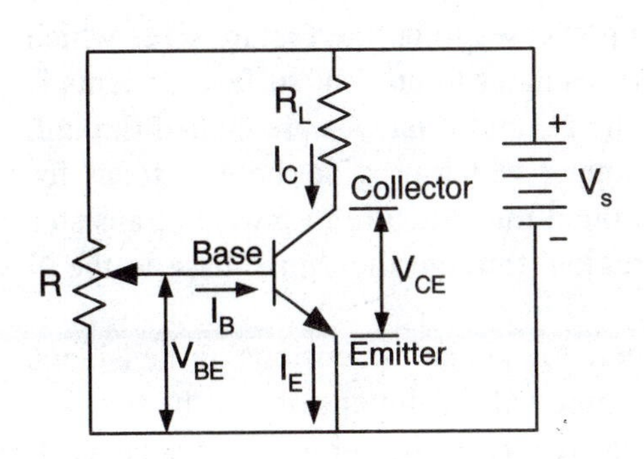

Fig. 2.13 Circuit showing the adjustment of the base voltage until the transistor acts as a closed switch.

An analysis of the circuit shown in Fig. 2.13 using Kirchhoff's law gives the current I_E as:

$$I_E = I_B + I_C \quad (2.4)$$

The current gain G for the transistor is defined as:

$$G = I_E / I_B \quad (2.5)$$

The current gain is dependent primarily on the thickness and area of the base and the concentration of the doping on both sides of the P-N junctions. Current gains of 10 to 100 are commonly achieved in bipolar transistors.

The illustration of the N-P-N bipolar transistor, presented in Fig. 2.12, is a schematic used to show the concept of back to back diodes with forward and reverse bias. The actual construction details of a transistor differ significantly from this simple diagram. Production techniques for integrated circuits are based on placing 10^4 to 10^8 transistors as well as other components such as resistors, capacitors and diodes on a wafer cut from a very large single crystal of silicon. The wafers used in production facilities today are usually from 4 to 12 in. (100 to 300 mm) in diameter and 21 to 31 mils (525 to 775 μm) thick. The transistors are arranged in a planar array on one surface of the wafer. The N-P-N structure, described previously, is formed through the surface of the wafer using ion implantation techniques to vary the local concentration of the N and P type impurities. A cross section of a single N-P-N bipolar transistor more representative of current technology is represented in Fig. 2.14a.

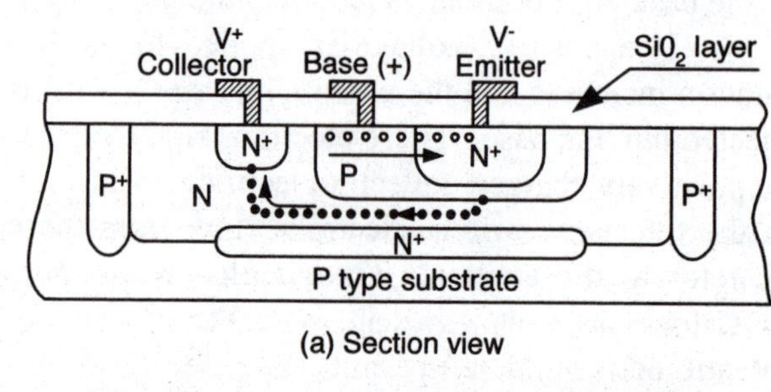

(a) Section view

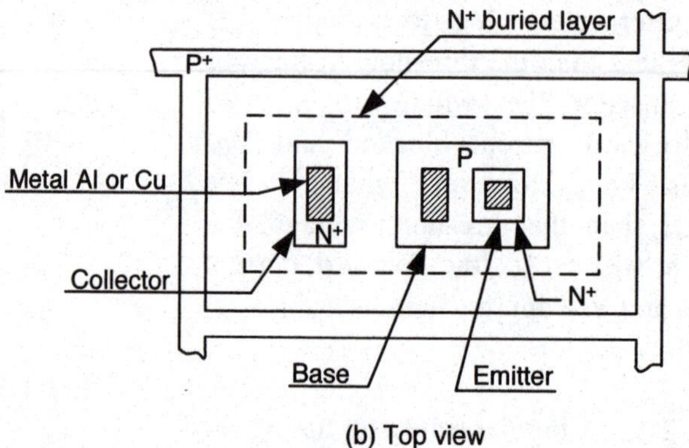

(b) Top view

Fig. 2.14 Structure of a N-P-N transistor fabricated on a wafer of P doped silicon.

The structure in Fig. 2.14 is developed beginning with a wafer cut from a P doped silicon crystal. Islands of N^+ (high concentration of N impurities) and more localized islands of N and P are formed by a series of production steps which utilize advanced lithographic processes to define feature sizes which are measured in nanometers. A layer of SiO_2 which serves as an insulator to prevent surface currents is formed by oxidizing the top surface of the wafer. Channels to the N^+ and P islands are etched through the SiO_2 layer and filled with a vapor deposited aluminum or copper which provide the electrodes for the collector, base and emitter. Deep islands of P^+ are placed around the structure to provide transistor to transistor isolation. The electron flow from the N^+ emitter region, through the thin P base to the N island and N^+ collector, is shown in Fig. 2.14a.

A top view of the structure is illustrated in Fig. 2.14b where the planar dimensions of the islands of doped silicon are indicated. As manufacturing processes are improved the dimensions of the features are reduced and the overall planar area required for a transistor, capacitor, resistor or a gate is decreased.

With feature sizes on gate lengths of 40 nm and the pitch on metal lines of about 170 nm, the area of silicon necessary for a bipolar transistor with its isolating boundary is of the order of 25 x 10^{-14} m^2. The very small feature sizes[1] permit the placement of about 400 million transistors on a 10 × 10 mm size chip.

Feature size is a common measure of the technology used in developing and manufacturing a chip. Feature size is defined in technical dictionaries as half the distance between cells in a dynamic RAM memory chip. For example, in 2006 the technology used to produce a dynamic RAM yielded a half pitch of 65 nm. This fact implies that the smallest feature size on a chip manufactured with this technology in 2006 is smaller than the dynamic RAM half pitch of 65 nm. Hence, the gate lengths on this generation of chips will be smaller than 65 nm.

2.6.2 Metal Oxide Semiconductor Field Effect Transistor

The MOSFET (metal-oxide-semiconductor-field-effect-transistors) devices were developed after the bipolar transistor. As the name implies, these transistors utilize field effects to control the flow of electrons or holes to perform the switching action. A typical structure of a NMOS transistor, which is fabricated by implanting N type islands in a P type substrate, is presented in Fig. 2.15a. After the two islands are implanted in the substrate, the entire surface is covered with an oxide layer. The P substrate between the two islands forms the channel region. A metal electrode is placed over the SiO_2 layer to form the gate, which is essentially a small capacitor coupled to the channel region. The electrode to one N island (the source) is connected to the P substrate and then it is grounded. The electrode to the other N island (the drain) is connected to a positive supply voltage. The voltage on the gate controls the switching action of the transistor. When the gate voltage is zero, the positive voltage on the drain attracts the electrons in the N material at the source, but these electrons cannot flow through the P channel because it contains many holes that combine with the electrons to produce a depletion layer, which blocks the further flow of electrons. With zero or negative gate voltage, the NMOS transistor acts as an open switch. However, if a positive voltage is applied to the gate, the holes in the P channel are repelled, the electrons from the source travel across the P channel to the drain and current flows in the opposite sense. With a positive gate voltage the transistor acts as a closed switch. This type of NMOS transistor action is called the enhancement mode, because positive gate voltage enhances the flow of electrons from the source to the drain.

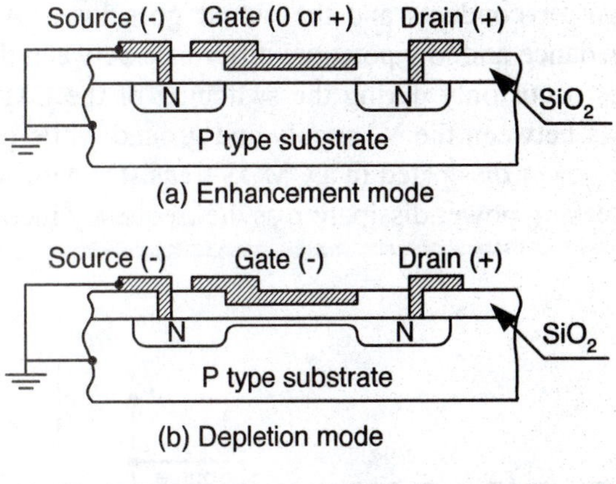

Fig. 2.15 Section view of a NMOS transistor showing key features.

NMOS transistors are fabricated differently if they operate in the depletion mode. The depletion mode NMOS transistor is illustrated in Fig. 2.15b. The primary difference between the fabrication details is the presence of a thin layer of N material under the gate which connects the source and the drain. In this configuration, a zero gate voltage permits the normal passage of electrons through the N channel from the source to the drain. However, the application of a negative voltage to the gate

[1] The data on feature sizes was taken from the International Technology Roadmap for Semiconductors and reflect their predictions for Application Specific Integrated Circuit (ASIC) chips produced in 2006.

repels the electrons from the thin N channel converting it to an insulating channel blocking the flow of electrons from the source to the drain and effectively opening the transistor switch.

PMOS transistors are also designed to operate in a manner similar to the NMOS transistors as illustrated in Fig. 2.16. In this case, two islands of P material are placed in an N substrate. With PMOS, the polarities of the source, gate and the drain are reversed and the charge carriers are holes rather than electrons. The PMOS transistors are not commonly employed, because their switching time is 2 to 3 times longer than the NMOS transistors. The longer switching time is due to the lower mobility (velocity) of the holes in switching PMOS in comparison to the velocity of the electrons that are utilized in NMOS switching.

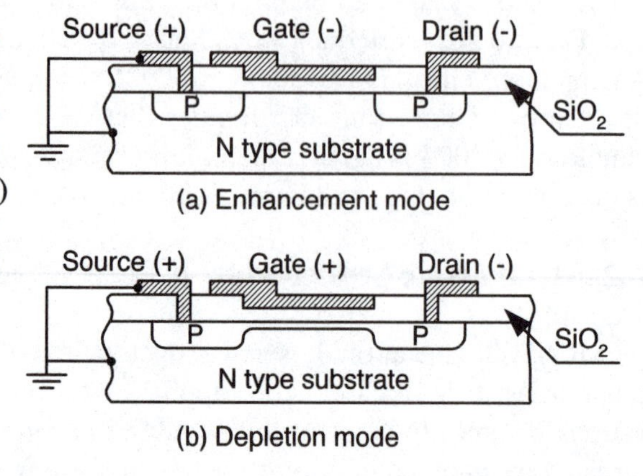

Fig. 2.16 Section view of a PMOS transistor showing key features.

A third type of MOS transistor is the complementary MOS device known as CMOS. This device incorporates both a NMOS and a PMOS transistor connected in series as shown in Fig. 2.17. The two gates are connected together and are activated by a single control voltage. The two drains are also connected together to provide an output signal, which can be either high or low. The two sources are wired separately with the source on the NMOS transistor grounded and the source on the PMOS transistor connected to the V^+ supply. This series connection of the two transistors results in very small power dissipation in the steady state. The two transistors require control signals of opposite polarity to conduct; consequently, one or the other of the transistor switches is always open. Because of this fact, the V^+ supply and ground are not connected together except for the period during switching from one state to the other. When the control signal is negative, the PMOS transistor conducts and the output goes high; however, the current flow is very small because of the of the high impedance and the low voltage difference between the source and the output. When the control signal is positive, the NMOS transistor conducts and the output goes low. Again the current flow is minimized due to the high impedance and low potential difference between the gate and the source. Large current flow and power loses occur only during the switching of the CMOS transistor[2]. During switching, a conducting path exists between the V^+ supply and ground while one transistor is turned on and the other is turned off. The power dissipated in a CMOS transistor depends on the frequency of operation of the transistor with increasing power dissipation as the frequency increases.

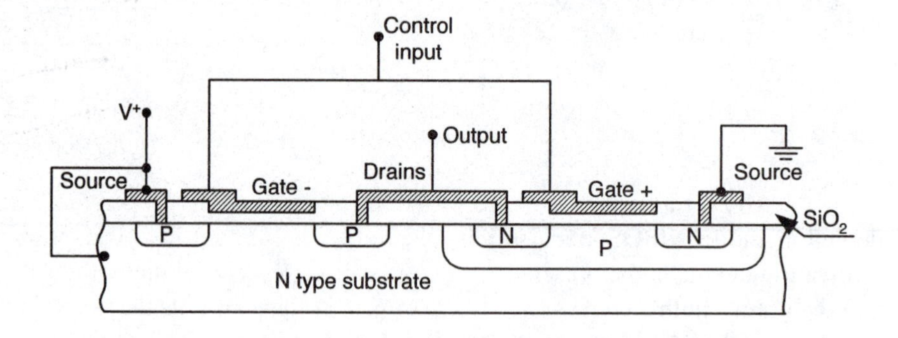

Fig. 2.17 A CMOS transistor employs a series connected pair of NMOS and PMOS transistors.

[2] Power loses that occur due to leakage current have been ignored in this discussion.

2.7 COMPARISON OF TRANSISTOR TYPES

Bipolar transistors are low impedance devices which require relatively high currents to operate and dissipate large amounts of heat. Bipolar transistors switch rapidly, and for equal feature size and power density, they operate at a higher frequency than the MOS devices. Bipolar transistors must be isolated to prevent interaction with adjacent devices on the chip. This isolation is achieved by using isolation barriers as demonstrated in Fig. 2.14; however, placement of these barriers utilizes a sizable area on the chip. This placement reduces the number of bipolar transistors that can be placed on a chip. Consequently, the density (number of components/area) is lower with bipolar than with MOSFET technology. Because of the very high heat dissipation and the large chip area required for a bipolar transistor, MOSFET technology has become dominant. In particular, the CMOS transistor with its low power dissipation is the most widely employed technology in a wide range of products.

The MOSFET devices exhibit high impedance and consequently they operate with lower currents. The isolation of the MOSFET transistors is inherent and no isolation barriers are required. This feature saves chip area and component density is higher than that which can be achieved with bipolar technology. Processing of NMOS or PMOS is generally easier than with bipolar and the design scales readily to smaller feature sizes. The channels are shallow and parallel to the surface and their length is dependent upon feature size. This lateral deployment adds capacitance and limits the switching speed. The very shallow diffusion layers yield small cross-sectional areas for current flow near the surface at relatively high current density.

PMOS transistors are rarely used because their switching speed is low relative to NMOS transistors. The mobility of holes as charge carriers is much less than the mobility of electrons and because PMOS depends largely on the movement of holes it is inherently slow. NMOS operates with electrons as charge carriers and is much faster than PMOS. The primary advantages of NMOS over bipolar is density, ease in scaling to smaller feature sizes and less complexity in manufacturing.

CMOS transistors consist of two series connected transistors, one PMOS and the other NMOS. In this arrangement, the forward current from the V^+ supply to ground occurs only during the switching operation and not during steady state in either the high or the low mode. This feature greatly reduces their power dissipation. Also relatively high switching speed can be achieved at low power levels. Feature size is critical in determining MOSFET switching speeds, and as feature size has progressively decreased in the past four decades, CMOS has become the most commonly used technology in the design and manufacturing of new chips for new products.

2.8 LOGIC GATES

In most digital instrumentation or computers, logic gates are employed in large numbers to perform complex operations at extremely high frequencies. The circuits involved are large often containing 10^5 to 10^8 logic gates. However, while these circuits are large, they are simple because they contain only three different types of basic gates. The basic gates include the AND, OR and NOT. Other more complex logic elements are often used, but they consist of combinations of these three gates.

The AND gate may be represented by the circuit shown in Fig. 2.18 where two switches A and B are placed in the line from the source to the load. The voltage V_s is applied to the load Z_L only if switch A and switch B are both closed. The possibilities for the AND gate are listed in a truth table shown below in Table 2.2. Note that (0) is used to represent a false statement and (1) to represent a true statement. With regard to the voltage applied to the load, the number (1) indicates true—V_s is applied to the load.

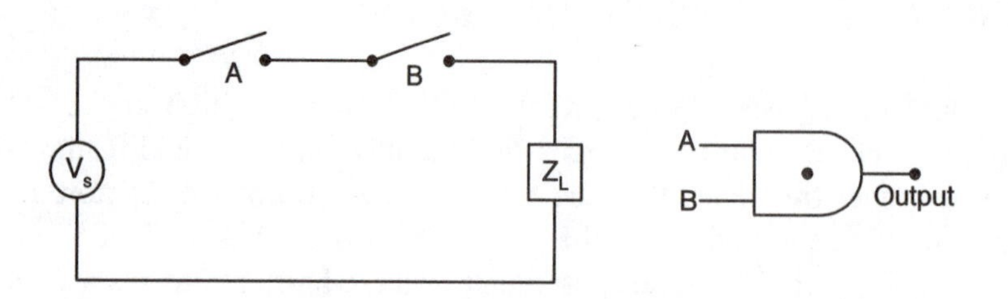

Fig. 2.18 Circuit representing the AND gate and its symbol.

Table 2.2
Truth table for the AND gate A – B = T

Switches or inputs		Output
A	B	T
0	0	0
0	1	0
1	0	0
1	1	1

The OR gate is represented by the circuit given in Fig. 2.19, where two switches A and B are placed in parallel in the line between the voltage source and the load. When A is closed **or** when B is closed, the voltage is applied to the load Z_L and T = 1.
The truth table for an OR gate with two switches is presented in Table 2.3.

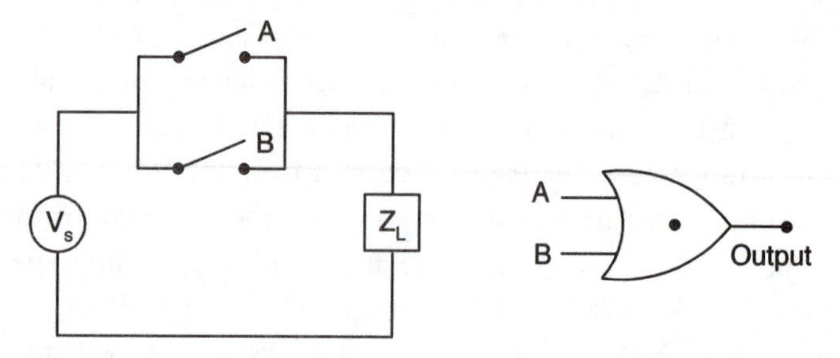

Fig. 2.19 Circuit representing the OR gate and its symbol.

Table 2.3
Truth table for the OR gate A + B = T

Switches or inputs		Output
A	B	T
0	0	0
0	1	1
1	0	1
1	1	1

The NOT gate which is illustrated in Fig. 2.20 is an inverter. In this case, the mechanical switch has been replaced by a transistor which is turned on (closed) by a positive input voltage. If the input signal to the transistor is (0), the transistor act as an open switch, no current flows and the output voltage is V_s or (1). When the input signal goes to (1), the transistor conducts acting like a closed switch and the output is grounded giving the low state or (0). It is clear from this circuit that the input is high (A) the output is low ($\overline{A}$) and changing the input to low ($\overline{A}$) results in an output which is high (A).

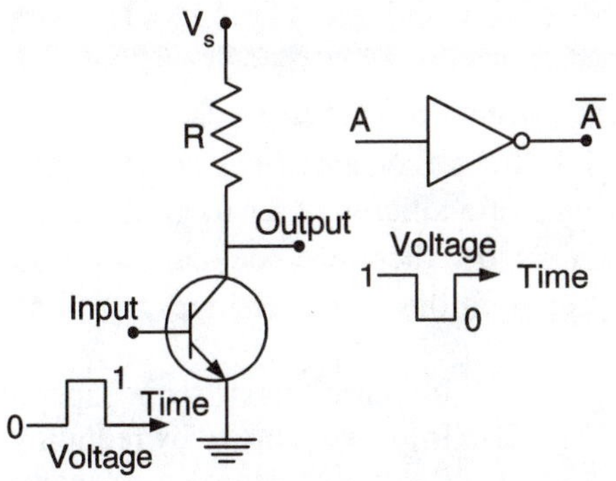

Fig. 2.20 Circuit representing the NOT gate and its symbol.

These basic gates are arranged in circuits to perform digital functions. For example, the arrangement of AND, OR and NOT gates shown in Fig. 2.21 represents a binary adder. A digital system is composed of many of these digital functions and may contain a million or more of the simple basic gates. The number of chips use to build the logic circuits depends on the scale of integration used to fabricate the chip. With ULSI (ultra large scale integration) and larger chip sizes, it is possible to place of the order of 10^8 gates on a single chip thus permitting the development of extremely large digital systems with relatively few chips.

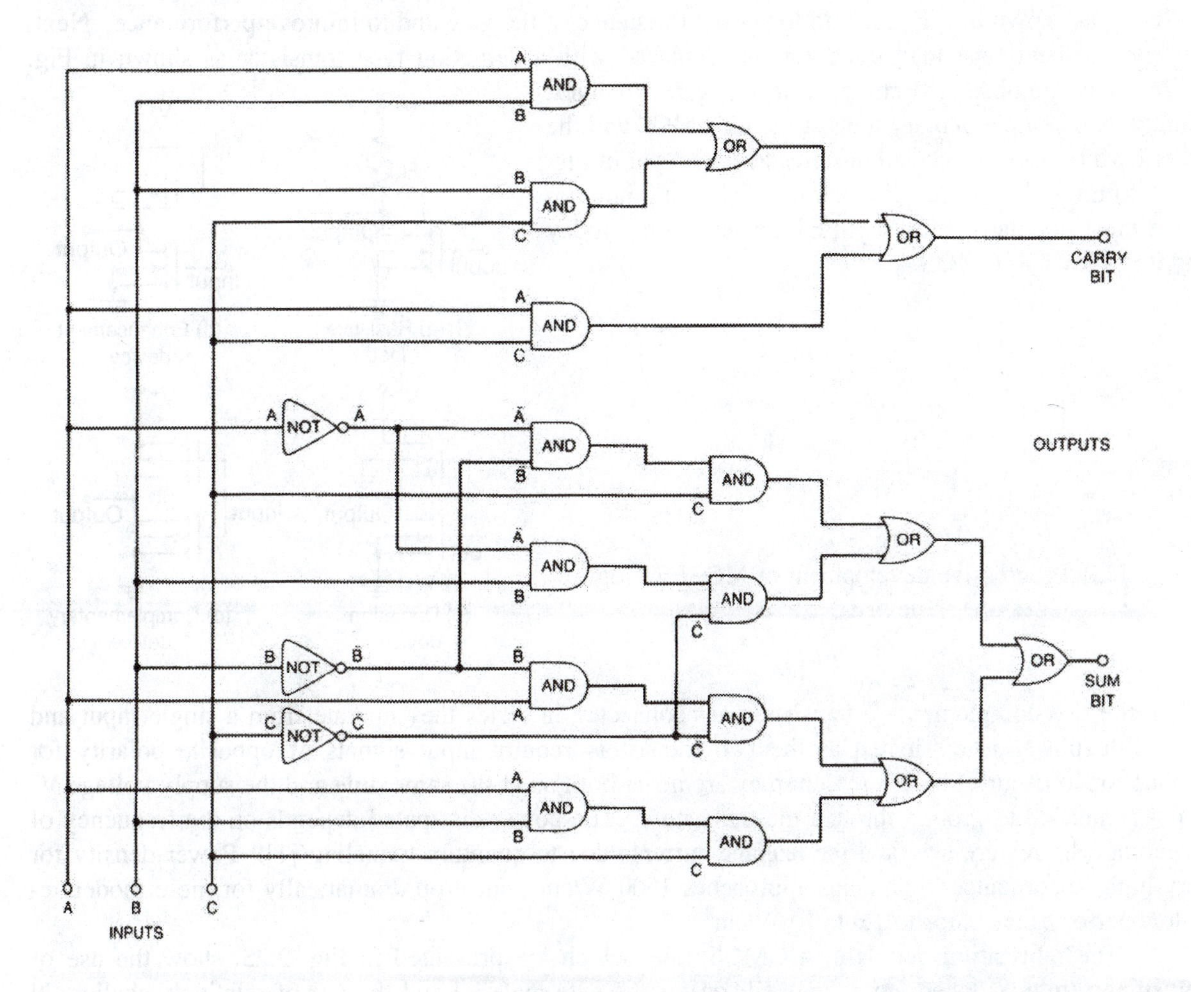

Fig. 2.21 The binary adder, a functional element of a digital system, is constructed from many simple AND, OR and NOT gates.

2.9 GATE TECHNOLOGIES

Transistors are used to replace mechanical switches shown in the logic gates represented in Fig. 2.19 and 2.20. Other electrical components such as resistors, capacitors and diodes are implanted on the surface of a silicon wafer to produce an integrated circuit (IC). Since the advent of the transistor in the early 1950's, there has been an evolution in the design of logic gates to improve their performance. The objective of the design changes have been to:

1. Increase the switching speed of the gate.
2. Improve density by reducing the number of components required for the gate
3. Reduce the chip area needed for the gate components.
4. Reduce the noise generated in switching.
5. Improve the drive capability.
6. Reduce power requirements.

2.9.1 MOSFET Gates

MOSFET circuits were introduced for switches in gates in the late sixties, with a FET gate incorporating a resistive load as shown in Fig. 2.22a. The resistor of this circuit was replaced with a second MOSFET transistor as shown in Fig. 2.22b to lower the resistance of the gate and to improve performance. Next, the enhancement type load transistor was replaced with a depletion type transistor as shown in Fig. 2.22c to accomplish switching at lower gate voltages. Finally, two complementary transistors one PMOS and the other NMOS were connected in series as illustrated in Fig. 2.22d to minimize power dissipation. This configuration is the most popular of the MOSFET devices in use today and it is known as CMOS.

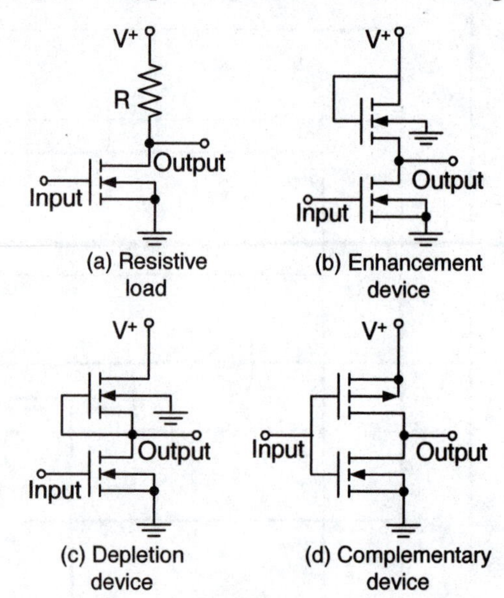

Fig. 2.22 Progressive development of MOSFET logic devices.

When the two complementary transistors are connected in series they operate from a single input and very little power is dissipated as the two transistors require input signals of opposite polarity for conduction to occur. For this reason, they are never both on at the same time and the supply voltage V^+ is not connected to ground through the transistors. The power dissipated depends on the frequency of switching, the feature size and the leakage currents due to quantum tunneling [1]. Power density for very-high-performance logic chips approaches 1000 W/cm^2, but drop dramatically for more moderate-to-low performance chips to 1.0 to 10 W/cm^2.

The fabrication details of a CMOS gate, which are presented in Fig. 2.23, show the use of polysilicon (highly doped amorphous silicon) as the gate material and the use of relatively shallow N and P channels. Because isolation channels between transistors are not required, the density (gates/unit area) that can be achieved in MOSFET technology is higher than that which can be obtained with bipolar technology.

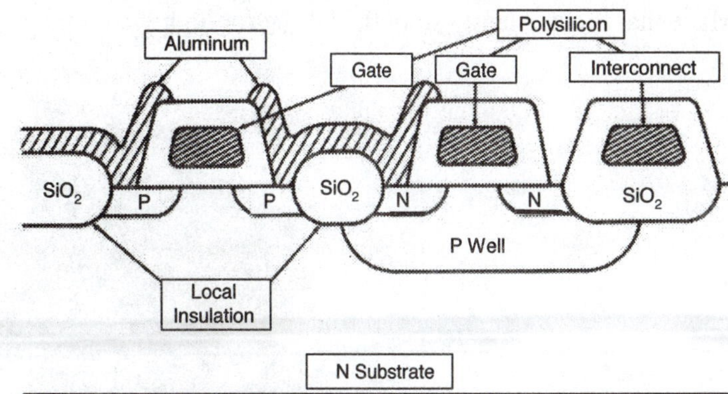

Fig. 2.23 Fabrication details revealed in a section view of a CMOS transistor.

2.10 CHIP AND WAFER FABRICATION

Most chips are fabricated from doped silicon to form products like a central processing unit (CPU) or an Application Specific Integrated Circuit (ASIC). An example of the Intel P-6 CPU with 5.5×10^6 transistors is presented in Fig. 2.24. The chips are small rectangular pieces of silicon that are cut from a wafer of silicon. The wafers are double-side polished and have a notch to identify the orientation of the grain structure. The thickness of the wafer varies with its diameter with the thickness increasing with the diameter. The 300 mm thick wafers are supplied with thickness ranging from 0.75 to 0.80 mm.

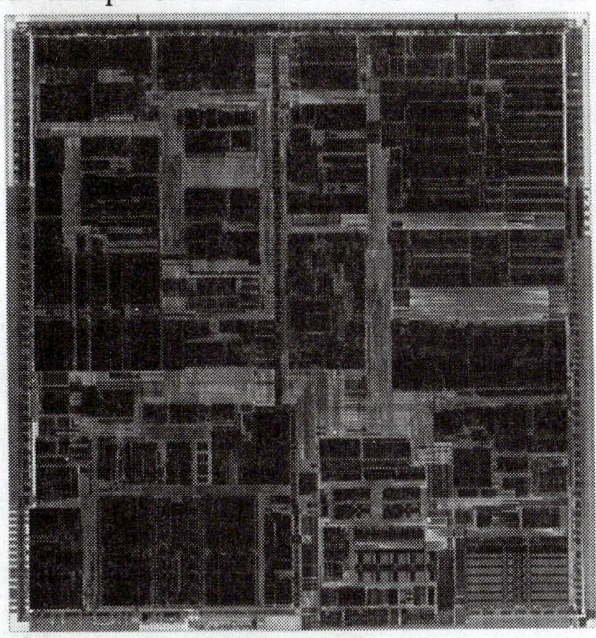

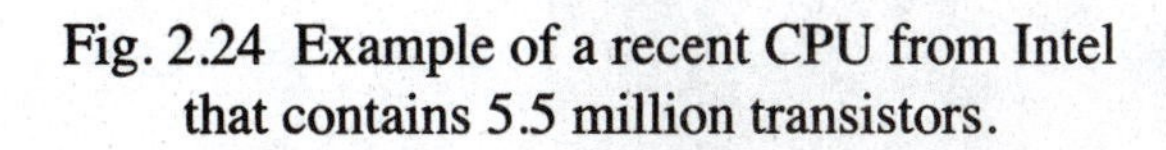

Fig. 2.24 Example of a recent CPU from Intel that contains 5.5 million transistors.

2.10.1 Wafer Fabrication

Single crystal ingots of silicon are usually produced by the Czochralski (CZ) method. In the CZ process, polycrystalline silicon, purified to less than 1 part per billion of trace impurities, is molten together with dopants in a quartz crucible. The dopants, such as boron and phosphorus, are used for adjusting the resistivity of the silicon. A small single crystal silicon rod (seed), about 5 mm in diameter by 200 mm long, is placed in the molten silicon with an inert gas atmosphere at a temperature of about 1421° C. The seed is slowly rotated and pulled up from the melt forming a single crystalline ingot with the same orientation as the seed. An illustration of growing a crystal by the CZ process is presented in Fig. 2.25.

Ingot diameters vary considerably with 200 to 300 mm commonly employed by most semi-conductor manufactures[3]. The ingots are ground to the finished diameter and then sliced into thin

[3] It is anticipated that wafer diameters will increase to 450 mm by 2012.

wafers with a diamond saw. The slices are sorted and those with the same thickness are then lapped and etched to remove the surface damage caused by the saw. Finally, the wafer is polished to produce a surface that is sufficiently smooth and flat for high yield optical photolithography.

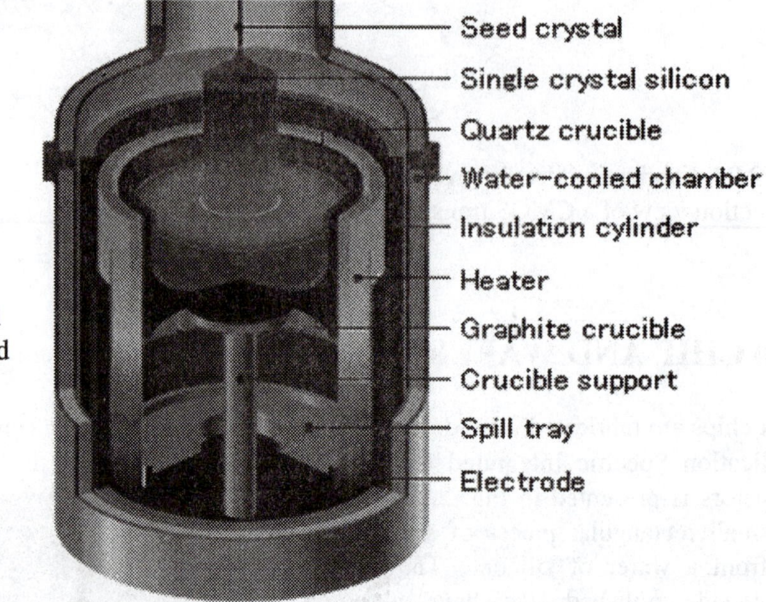

Fig. 2.25 Growing single crystal silicon ingots in a CZ crucible and furnace.

A photograph of a several wafers cut from a finished ingot of silicon is presented in Fig. 2.26.

Fig. 2.26 Photograph of wafers cut from an ingot of single crystal silicon.

Mechanical properties of interest for silicon are presented in Table 2.4.

Table 2.4
Mechanical and thermal properties of Silicon.

Property	Value
Density	2.33 g/cm^3
Tensile Strength	7,000 MPa
Modulus of Elasticity	190 GPa
Poisson's Ratio	0.17
Knoop Hardness	850 kg/mm^2
Coefficient of Thermal Expansion	2.59×10^{-6}/°C
Thermal Conductivity at 300 °K	1.56 W/(cm-°C)
Thermal Diffusivity	0.9 cm^2 /s
Specific Heat	0.70 J/(g-°C)

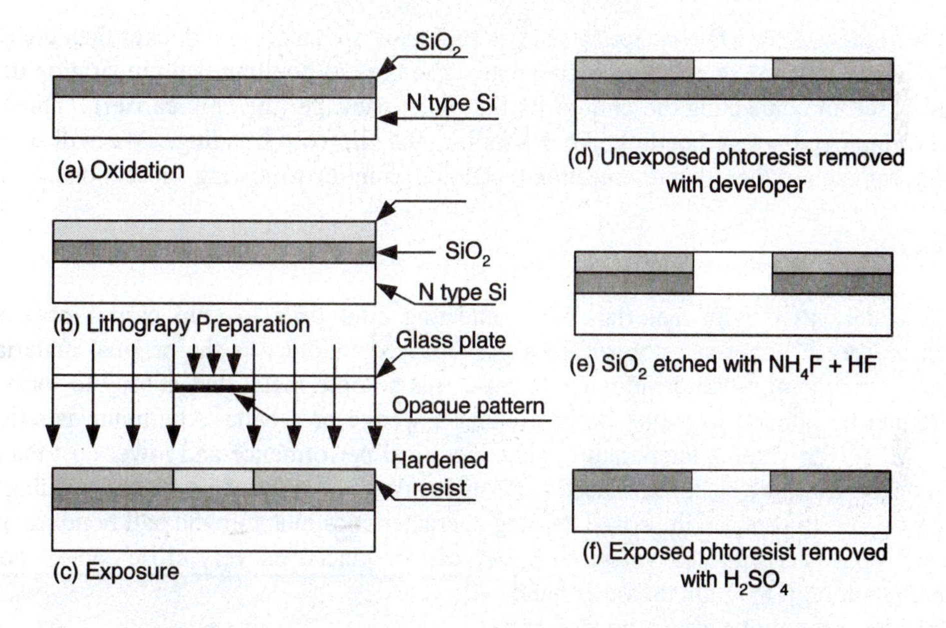

Fig. 2.27 Illustration of several steps in the lithographic process for producing the pattern of the features on a chip.

2.10.2 Masks and Lithography

Chips are produced on wafers by using a lithographic process that was originally developed to produce printing plates. This process, illustrated in Fig. 2.27, has been refined and is used today to produce chips as well as printed circuit boards. The steps in the process include:

1. Form a thin layer of oxide over the entire surface of the silicon wafer.
2. Apply a layer of photoresist over this surface.
3. Position a mask with the chip's surface features over this photosensitive layer.
4. Project UV light through the mask to image the surface features onto the photoresist.
5. Develop the photoresist opening small regions where chip features are to be applied.
6. Etch away the layer of oxide with hydrofluoric acid.
7. Remove the photoresist with sulfuric acid.

After the feature regions of the chip are defined, the P and N regions on the chip are formed with diffusion doping or with ion implantation methods. At this stage in the production process, it is necessary to apply the wiring to connect the various P and N islands to form gates and to wire the gates together to produce the logic or memory elements. Photolithography is also used in making the required interconnections. The surface of the wafer is covered with photoresist and a mask is used that has the image of the wiring lines and the bonding pads. The photoresist is exposed with UV light to form an image and of the wiring lines and bonding pads. The wafer is then placed in a vacuum chamber and aluminum is sputtered over the entire surface of the wafer. When the photoresist is removed a single layer of wiring connecting many of the features remains. To complete the wiring a layer of polyimide is applied to insulate the first wiring layer from the second. Second, third and additional wiring layers are added as required to complete the interconnections of the gates and on-chip capacitors and to connect the I/O with the bonding pads.

2.10.3 Bonding Pad Preparation

Bonding pads are use for I/O connections and as such they are larger and thicker than the normal wiring used for interconnections of on-chip components. The type of bonding pad preparation depends on the technology used in connecting the chip to its first level package (the chip carrier). Three methods are commonly employed—wire bonding, TAB bonding and flip-chip bonding. We will describe the pad preparation for each of these interconnection technologies in the following subsections.

Wire Bonding

The most widely used wire materials for connecting chip pads to chip carrier pads are gold and aluminum; however, silver and copper are also employed. Gold was the original material used when wire bonding technology was developing because it is a noble metal that is easy to form and to weld. Gold wire can be bonded in the shape of a either a wedge or a ball. Aluminum is still widely used because of its lower welding temperature, good electrical performance and lower cost than gold. More recently copper wire has been introduced for ball bonding, in spite of its higher welding temperature, because of its lower costs and improved looping characteristics in automatic ball bonding machines.

Ball bonds have the advantage that they can be placed on very close centers consistent with newer chips with high I/O. On the other hand, aluminum wire can only be wedge bonded and the use of aluminum is limited when the bonding pads are on close centers. An example of ball bonds formed from gold wire on two rows of chip bonding pads is presented in Fig. 2.28.

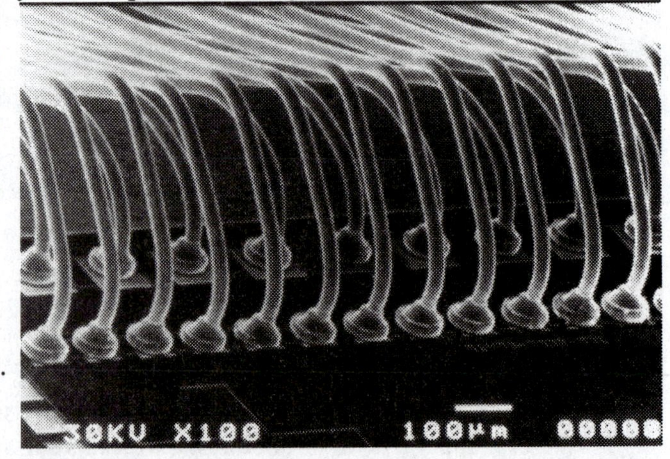

Fig. 2.28 Gold wires welded to chip bonding pads to make connections with the chip carrier.

The pad preparation depends on the wire material and the bonding process. The thermo-sonic process is used with gold wires and currently comprises about 90% of all wire bonding. It is performed at temperatures ranging from 100°C to 240°C depending on the wire and pad materials. The weld is formed when the ultrasonic energy combines with the capillary action of thermo-compression bonding.

An illustration showing the capillary tube and the clamp used with this process is presented in Fig. 2.29. In this instance, the bonding pads are square and they have been built up to increase their thickness by sputtering additional aluminum onto the pad surfaces. In some cases the aluminum pads are plated with gold to improve the thermo-compression bonding process.

Fig. 2.29 A wire bonding machine.

Tape Automated Bonding (TAB)

Tape-automated bonding (TAB) is another method for connecting the bonding pads to the leadframe of a chip carrier. The interconnecting lines are patterned on a multilayer polyimide tape, as indicated in Fig. 2.30. The tape is positioned above the chip with ends of the copper lines in line with the bonding pads on the chip.

Fig. 2.30 Tape automated bonding (TAB) used to make connections between the chip bonding pads and the lead frame of a chip carrier.

The connections are made in a compression welding process, where a "thermode" is pressed down on the inner-lead-ends forcing them into contact with the bumped pads around the periphery of the chip, as indicated in Fig. 2.31. This process is rapid as all of the connections are made simultaneously; however, the bonding pads on the chip must be bumped to provide the clearance required by the presence of the polyimide carrier.

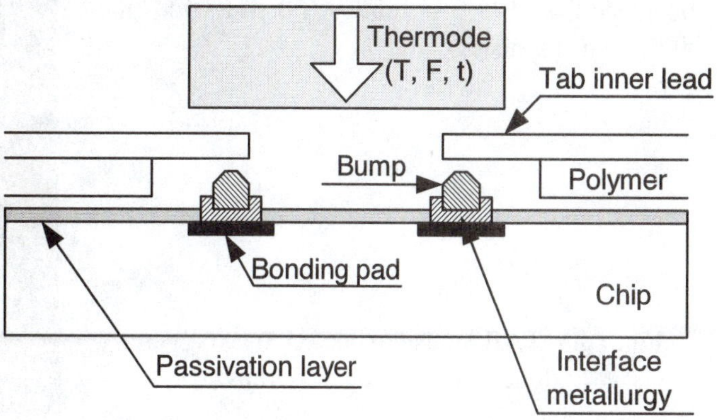

Fig. 2.31 Thermo-compression welding of copper leads to the bumped bonding pads about the periphery of a chip.

The process for bumping the chip bonding pads is illustrated in Fig. 2.32. After the chip pads have been coated with a layer of aluminum, the surface of the chip is coated with a passivation layer that is usually polyimide. The next step is to apply two or three layers of interface metallurgy (sometimes called Under Bump Metallization (UBM). The interface metallurgy often consists of three layers of metal, which include:

1. An adhesion layer to provide a strong, low-stress mechanical and electrical connection to the bond pad metal and the adjacent passivation.
2. A diffusion barrier layer to constrain the diffusion of solder into the adhesion layer or the aluminum.
3. A solder wettable layer offers a wettable surface that is compatible with either gold or solder used to form the bumps.

As indicated in Fig. 2.32, photoresist is used to delineate the bonding pads after the application of the interface metallurgy. Either gold or silver bumps are applied by plating; however, high-lead solder bumps may be applied by evaporation. For TAB connections, the top of the bumps are finished with a relatively flat top.

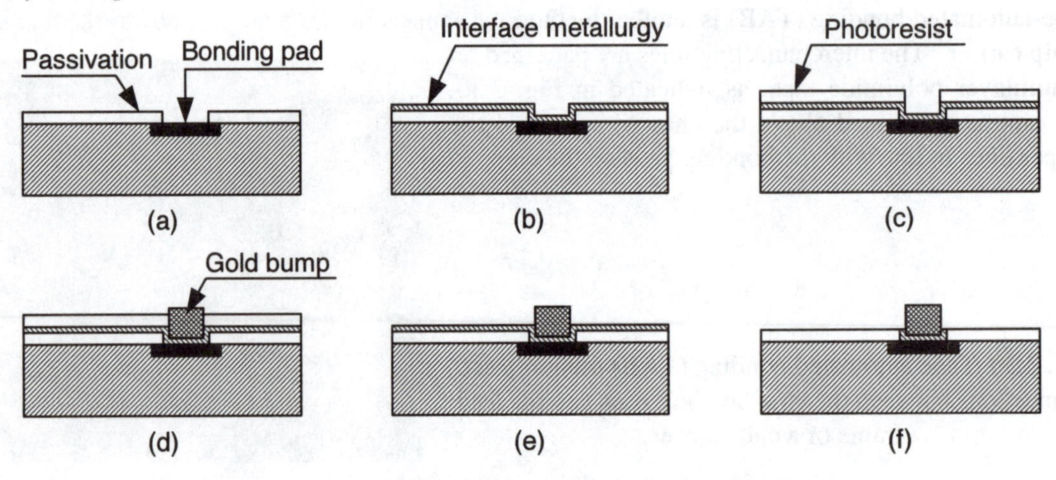

Fig. 2.32 Six steps in a process used for chip bumping.

An example of TAB connections to gold bumps is illustrated in Fig. 2.33. Note the staggered double row of solder bumps. This design permits additional I/O to be deployed about the perimeter of the chip. The TAB leads with the staggered I/O design can be placed on 100 μm centers (50 μm wide lines with 50 μm spaces).

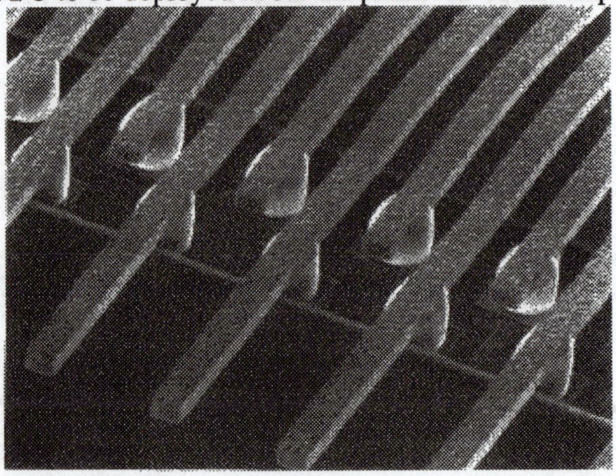

Fig. 2.33 TAB leads connected to a staggered double row of solder bumps.

Flip-Chip Bumping

There are several steps in fabricating flip-chip packages, which include bumping the wafer, cutting the dies from the wafer, attaching the bumped chip to a substrate and filling the remaining space under the die with an adhesive. The conductive bump, the attachment materials, and the processes employed differ with the types of flip-chip packages.

The bump serves four different functions in a flip-chip package. First, it provides the electrical and thermal conductive path from chip to substrate. Second, the bump serves to mechanically attach the chip to the substrate. Third, it acts as a spacer preventing electrical contact between the chip and substrate conductors. Finally, the bumps serve as very short leads that relieve the thermally induced shearing strains, which develop between board and substrate due to temperature changes.

The first step in solder bumping is preparing the bond pads located on the chip. This step involves the removal of the insulating aluminum oxide layer from the pads to prepare them for several layers of metals that are applied to enable good mechanical and electrical connection to the solder bump. Next a ball-limiting-metal (BLM) is placed on the chip bond pads by either sputtering or plating. The BLM, which usually consists of several layers, defines and limits the solder-wetted area. The adhesion layer adheres well to the bond pad metal and the surrounding passivation, providing strong, low-stress mechanical and electrical connections. The diffusion barrier layer limits the diffusion of the bump material into the underlying material. The solder wettable layer offers an easily wettable surface to the molten solder during assembly to enhance bonding of the solder to the underlying metal. A protective layer may be used to prevent oxidation of this underlying layer.

There are several processes employed in applying metal to form the flip-chip bumps. Wet chemical processes to etch away the aluminum oxide and plate conductive nickel-gold bumps onto the chip bond pads. After plating the required thickness of nickel, a gold layer is added for corrosion protection. The gold stud bump process employs a slight modification of the wire bonding process. A gold ball is produced by melting the end of a gold wire forming a sphere. The gold ball is attached to the chip bond pad as the first part of a wire bond. The modified wire bonder breaks off the wire after attaching the ball to the chip bond pad. This gold ball provides a permanent connection through the aluminum oxide layer to the underlying metal.

High lead solder (97% Pb and 3% Sn) is one of the most popular materials for chip bumping. The IBM C4 process was based on evaporation of high lead solder to form the bumps. Conventional tin-lead eutectic solder cannot be evaporated because of the difference in vapor pressures of the two metals. Bump size and position is determined by the solder mask openings and spacing; however, the solder is subsequently reflowed forming spherical solder bumps.

Electroplating solder to form the bumps is less costly and this manufacturing technique is more flexible than evaporation. The BLM consists of an adhesion layer of titanium tungsten, a chrome/copper layer, a copper wetting layer and a gold protective layer[4]. After a photoresist layer is patterned to form the bump sites, solder[5] is electroplated over these sites, as shown in Fig. 2.34. The photoresist is then removed and the layers of BLM are etched away. The solder bumps are reflowed to form spheres that are attached to the chip pads, as illustrated in Fig. 2.35 and Fig. 2.36.

---

[4] A number of different metallurgies are employed in sputtering the BLM layers including—TiW/CrCu/Cu/Ni/Au, Cr/CrCu/Cu/Au, Ti/Cu/Ni/Au, Ti/Cu and Cr/Cu/Cu/Ni/Au.

[5] Several different solder compositions may be employed including—Pb/Sn (97/3), Pb/Sn (37/63) and the Sn/Ag/Cu family of lead free solders.

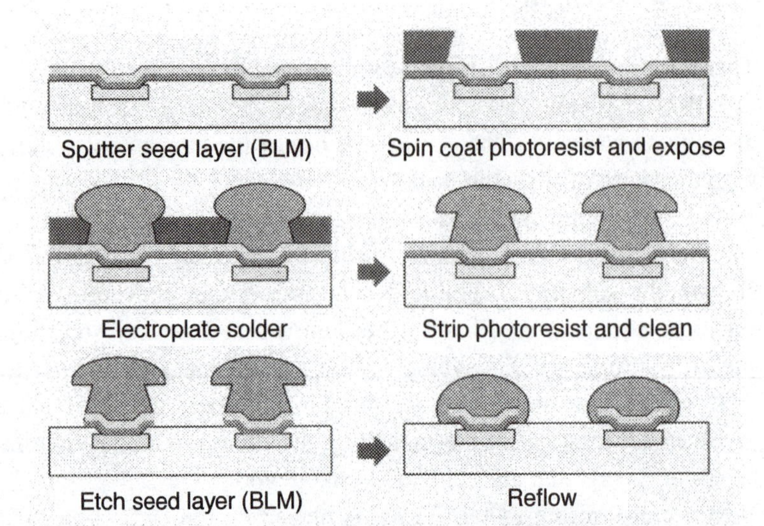

Fig. 2.34 Fabrication steps in producing solder ball bumps for flip-chip bonding.

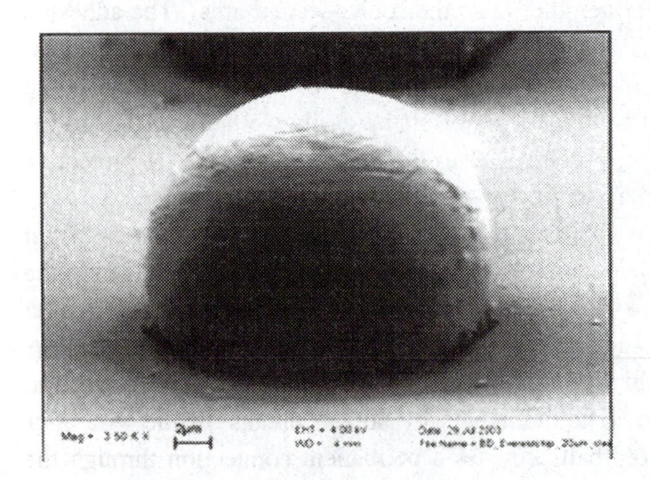

Fig. 2.35 Photomicrograph of a micro-solder bump 25 μm in diameter on a 50 μm pitch.

Fig. 2.36 Staggered array of solder balls on the back side of a flip-chip.

2.10.4 Advanced Chip Structures

As chips are designed with millions of transistors that operate at increasing frequencies the number of on-chip interconnections and the number of I/O are increasing dramatically. The cost and the development time for a line of chips are also increasing markedly. These facts are drivers that are changing both design and manufacturing methods used in the production of chips. One important advance has been the development of a process for using copper to replace the aluminum used to interconnect the transistors, gates and capacitors on integrated circuits. In the late 1990s, IBM introduced a technology that allows the use of copper lines, in place of the aluminum interconnects in chips. IBM has developed electroplating process for copper that is known as damascene. The process begins by coating the surface of the wafer with a seed layer, or plating base, whose function is to conduct the current from the anode located at the wafer edge to all points on the wafer where a deposit of copper is specified. One approach to accommodate the seed layer with complex inner connection structures is illustrated in Fig. 2.37. Damascene plating involves depositing the seed layer over a patterned material, which is the insulator that must remain in place because it is a functional part of the circuit. After plating, copper covers the entire surface of the wafer. Complete coverage causes major problems because all of the wiring lines are shorted out. It is necessary to remove the excess copper by chemical-mechanical polishing (CMP) which planarizes the surface of the wafer.

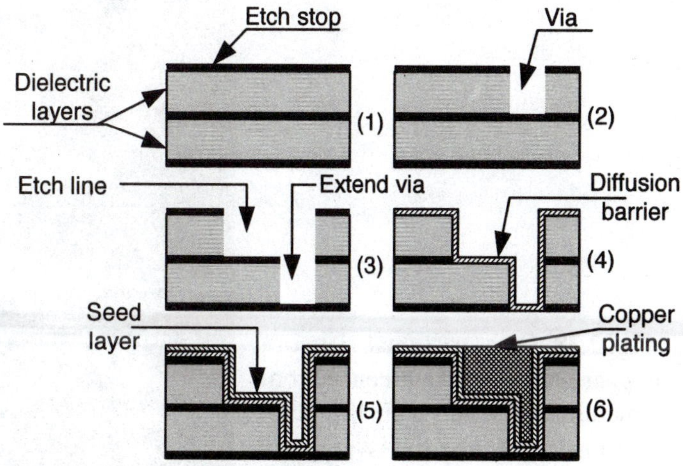

Fig. 2.37 Schematic illustration of several of the steps involved in the damascene plating process adapted to making multi-layer inner connections on chips.

The production steps illustrated in Fig. 2.37 are described below:

1. Deposit insulation with a low dielectric constant and separate layers with an etch stop (silicon nitride).
2. Define vias and etch first layer of insulation.
3. Define lines and via extensions into second layer and etch.
4. Deposit barrier layer to prevent diffusion of copper into the thin dielectric layer.
5. Deposit seed layer of copper over the entire surface of the wafer.
6. Plate copper filling vias and lines.
7. Not shown in Fig. 2.37 is the chemical-mechanical polishing (CMP) of the wafer surface necessary to remove the copper from the surface. The only copper remain is in via holes and trenches for the wiring lines.

Damascene electroplating is ideally suited for the fabrication of multi-layer interconnection structures, because it is possible to apply copper to fill via holes and adjacent trenches to form wiring lines without voids or seams. This plating process is compatible with the requirement for a barrier layer between the seed layer and the insulating (dielectric) layer. The purpose of the barrier layer is to prevent interaction between the copper and the insulator.

Examples of copper multi-layer interconnection structures are presented in Fig. 2.38 and Fig. 2.39. In Fig. 2.38, it is clear that six layers of copper wiring lines have been applied over the initial layer of tungsten interconnections. The layers are applied in sequence with chemical-mechanical polishing (CMP) of the wafer surface after each layer of copper has been deposited by the damascene plating process. The photomicrograph shown in Fig. 2.39 provides a view of the copper layers after the dielectric insulation has been removed. The photomicrograph indicates the complexity of the wiring necessary to inner connect the dense high-end chips developed with feature sizes less than 180 nm. The primary reason for using copper metallization is the reduced cross sectional area of the wiring lines increases the parasitic resistance to an unacceptable level if aluminum lines are employed. Companies such as IBM and Intel are already using this process with six or seven layers of copper wiring for inner connections. It is believed that nearly all semiconductor manufacturers will employ copper interconnections in the near future.

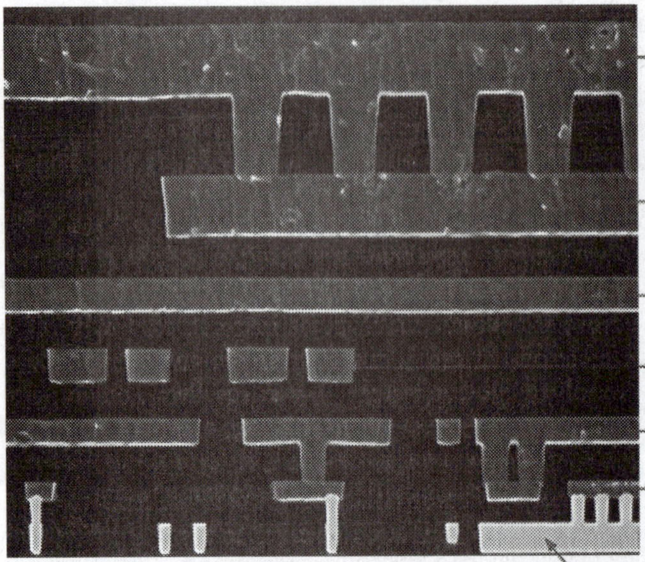

Copper layer 6
Copper layer 5
Copper layer 4
Copper layer 3
Copper layer 2
Copper layer 1
Local tungsten interconnect

Fig. 2.38 Cross sectional view of copper multi-layer interconnection structure that includes six layers of copper and an initial layer of tungsten.

Fig. 2.39 A photomicrograph of a multi-layer interconnection structure where the insulation has been removed exposing the copper plated vias and wiring lines.

Back End of Line (BEOL) Processing

The development of a new chip is an expensive undertaking with costs of tens of millions of dollars for early prototypes. To defray these costs the chips are often designed for a number of different applications. However, these different applications require modifications to the circuits contained on the basic chip. It is possible to accommodate these modifications with BEOL metal customization of one or more layers of interconnections. The concept followed in designing a chip for several different applications is shown in Fig. 2.40. Before initiating the chip development the circuit designers consider the different target applications and incorporate elements on the chip to enable the chip to serve all of these applications. Of course many of the elements on the chip are common to all applications; however, some elements are unique. The particular configuration required for each application is achieved by custom wiring. The custom wiring is performed using application unique masks for the top layers as indicated in Fig. 2.40.

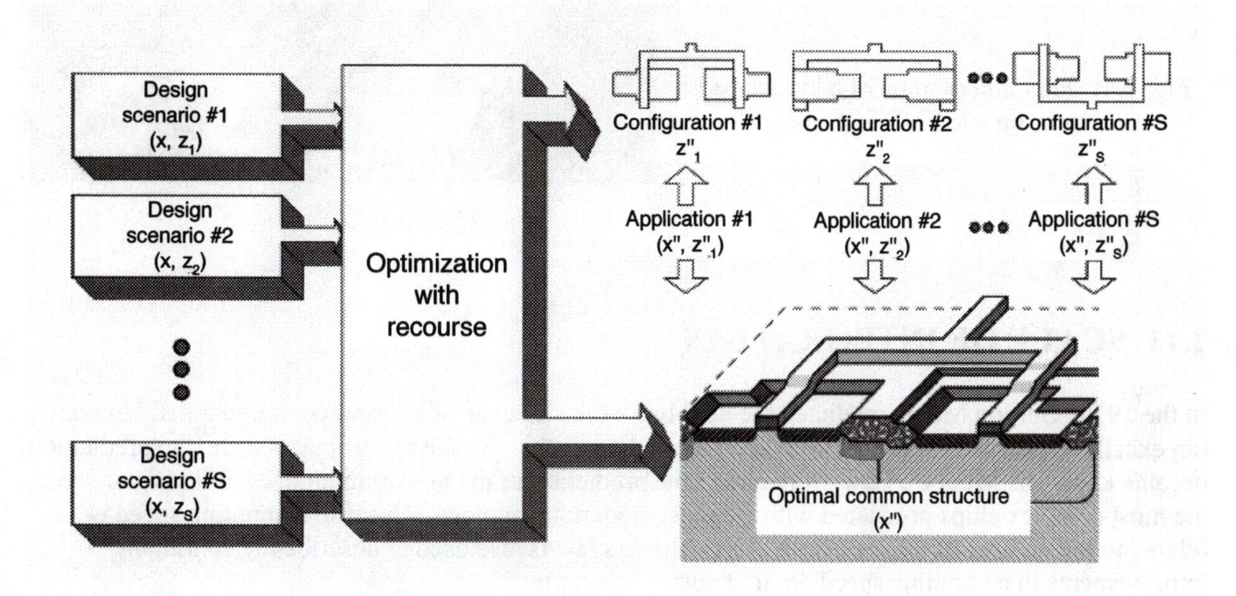

Fig. 2.40 Custom on chip connections enables a common chip design to serve several different applications.

This approach is similar in concept to using gate arrays and wiring those arrays to perform a specific function. However, with increased gate density on a chip wiring of the gate arrays is no longer feasible. The approach to day is to incorporate a number of different functional elements on a chip and then to inner connect those elements to serve a specific application. The development of the on-chip inner connection structures for multi-application chips is called Back End of the Line (BEOL) tasks.

2.10.5 Micro-Electro-Mechanical Systems (MEMS)

Micro-Electro-Mechanical Systems (MEMS) is an integration of mechanical components, sensors, actuators and electronics on a silicon substrate. MEMS are fabricated by using photoresist and selective etching methods to etch away silicon to form a wide variety of micro-mechanical devices. The etching process is precise and complex mechanical components can be fabricated as indicated by the interlocking fork set in Fig. 2.41.

The electronic devices included in a specific MEMS package are produced using standard processes that have already been described in previous sections. Because the MEMS devices are usually very small, the packaging methods used to house them are similar to those used to house chips. While the production methods are similar for MEMS and micro-electronic devices, the applications differ significantly. For this reason, we will not treat MEMS, other than this brief introduction.

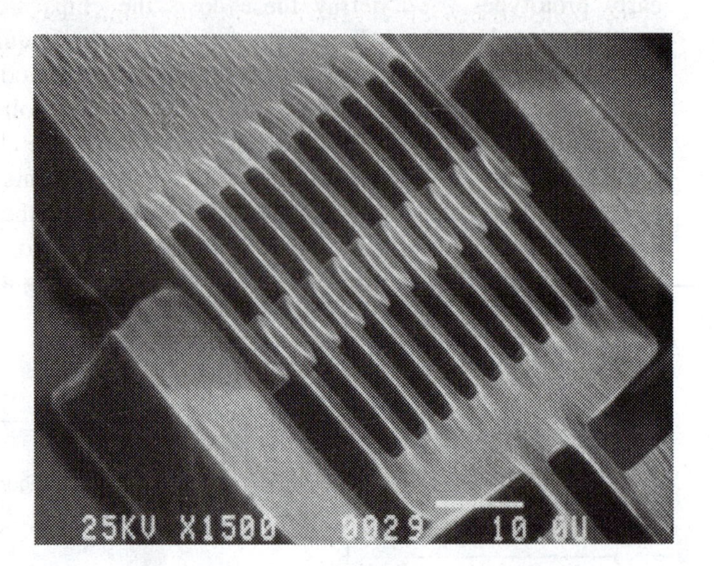

Fig. 2.41 Photomicrograph of a inner laced forks on a MEMS device.

2.11 SCALE OF INTEGRATION

In the 1970s Gordon Moore predicted the doubling of the number of transistors on integrated circuits (an excellent measure of chip performance) every 18 months. By the end of that decade, the predicition became known as **Moore's Law**. It permited the prediction of the maximim number of transistors on the most complex chips producted with the most modern technology. Because computing speed is related to the number of transistors on a chip Moore's law is also used to describe the continuing improvements in computing speed per unit cost.

Moore's law is empirical because it was formulated based on early observations. However, its global acceptance over the past 25 years has elevated its status. The law now serves as a goal for the semiconductor industry. Today semiconductor manufacturers invest significant amounts of money and engineering talent to follow Moore's Law because they know their competitors will be able to double their transistor count every 18 months and gain competitive advantage.

The implication of Moore's law for semiconductor manufacturers is immense. A typical major design project such as a new CPU takes between two to four years to reach production. Consider this developemnt time together with the implications of Moore's law. Moore's law is equivalent to a 1% per week improvement in performance for the industry. This fact implies that a short delay of a month or two in bring the new chip to market can mean the difference between a successful product one that is too large or too slow for customers to buy.

Recent semiconductor industry technology "roadmaps" predict that Moore's Law will continue for several chip generations. However the doubling time may be extended from 18 months to three years. Depending on the doubling time used in the preditions, Moore's law implies an increase in the number of transistors by a fctor from 10 to 100 in the coming decase.

2.11.1 Scaling of Feature Size

Since the introduction of integrated circuits in 1959, there has been a rapid increase in the number of components (transistors, resistors, capacitors and diodes) which can be placed on a single chip. The dramatic increase in the number of transistors, presented in Fig. 1.16, is due to a decrease in feature size and an increase in the size of the area used for the chip. Today, an advanced chip may support as many as 500 million transistors, with gate lengths of about 40 nm. Continued progress in reducing gate lengths to about 10 to 15 nm is anticipated in the coming decade unless limits on the lithographic process slow further development. The trend line for the marked reduction in feature size (gate length and on-chip wiring pitch) over the past four and a half decades is shown in Fig. 2.42.

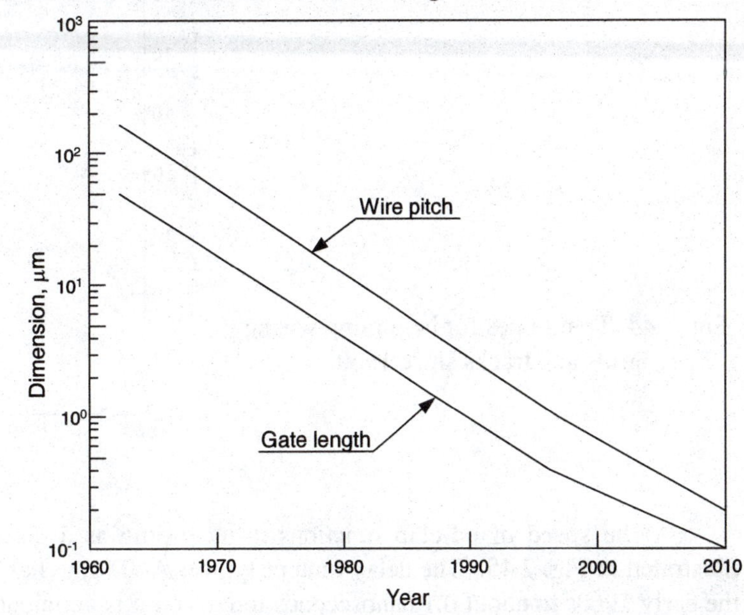

Fig. 2.42 Trend lines for scaling down the gate length and on-chip wiring pitch since 1960.

The increase in the number of components per chip reflects the gate length, the on-chip wiring pitch and the size of the silicon chip (die). The size of the silicon dies has also been increasing with continuous improvements in chip manufacturing methods that enable larger size chips while maintaining satisfactory yields. The trend line for increasing chip size with time is presented in Fig. 2.43.

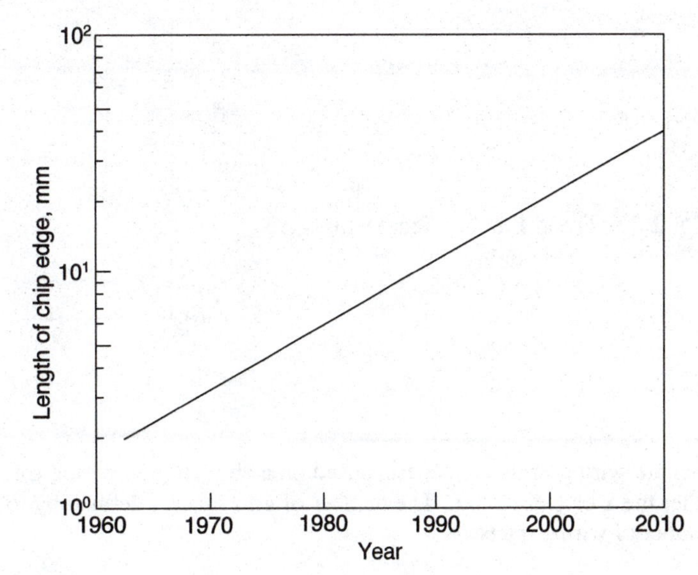

Fig. 2.43 Trend line for increasing chip size since 1960.

The trends presented in Fig. 2.42 and Fig. 2.43 yield an increase in the number of wiring tracks and the number of wiring grids[6]. The trend lines showing this increase is presented in Fig. 2.44. The number of grids on a chip is an indication of the amount of functionality that can be achieved in designing a complex integrated circuit. Modern chips have about 10^9 grid areas and by 2010 it is expected that they will have 10^{10} wiring grid areas.

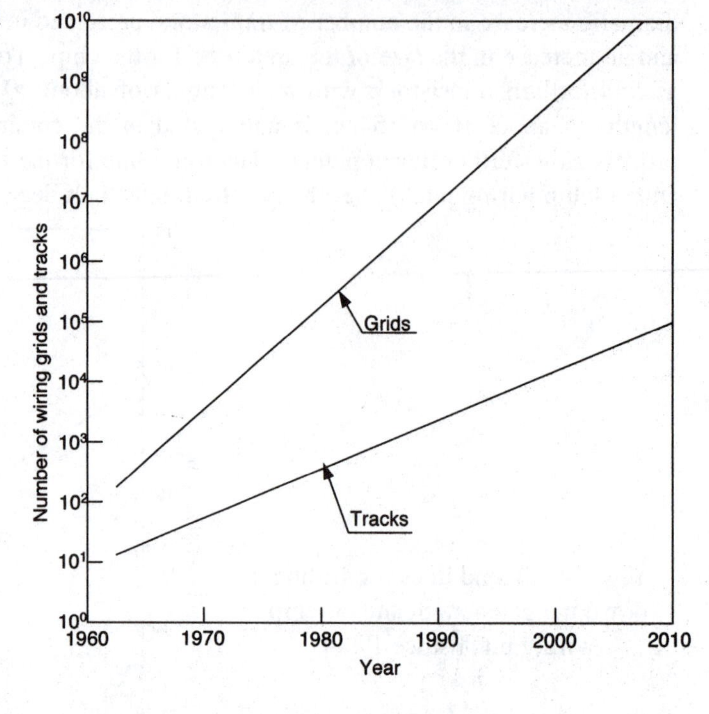

Fig. 2.44 Trend lines for increasing wiring grids and tracks since 1960.

The speed of on-chip functions is increasing as a consequence of decreasing gate length as illustrated in Fig. 2.45. The delay time of typical AND gate has decreased from tens of nanoseconds in the early 1960s to about 0.1 nanoseconds today and it is anticipated that it will decrease to a few tens of picoseconds by 2010.

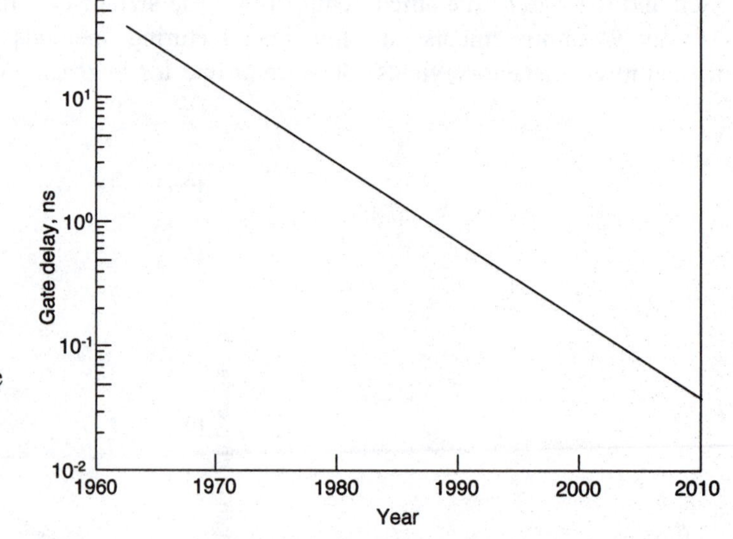

Fig. 2.45 Trend line for decreasing gate delay.

[6] Square wiring grids are superimposed on a chip and the wiring tracks extend along a sequence of these grids in either the x or y direction. The number of grid squares defined by the wiring tracks increases as the square of the number of wiring tracts.

A final graph showing the benefit of reducing feature size is presented in Fig. 2.46. This graph shows the enhanced capability of a typical silicon chip with time. The capability of a chip depends on the number of functions it carries and the speed of these functions. Hence, a measure of capability is given by the number of wiring grids divided by the gate delay. The results of Fig. 2.46 indicate that chip capability has increased from about 4 in the early 1960s to nearly 100 billion today.

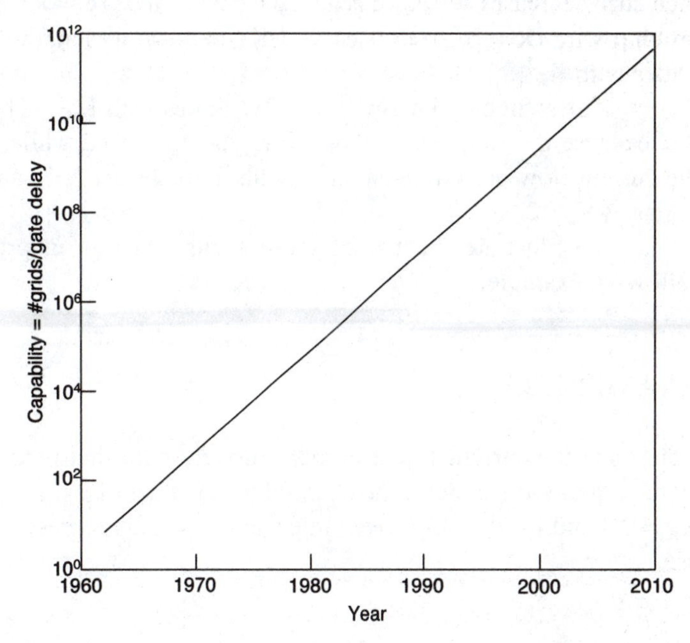

Fig. 2.46 Trend line for increasing chip capacity with time.

2.11.2 Scaling of Effects on Feature Size

The scaling down the on-chip feature size affects many electrical characteristics of transistors and other electrical components. The relationship of the change in the electrical parameters due to the scaling factor[7] K_w for the wire pitch and K_e for the chip edge dimension is given in Table 2.5.

Wire pitch and chip edge dimensions both scale linearly with K_w and K_e respectively. The chip area A scales as K_e^2 and the number of tracks and the grids on a chip scale as K_e/K_w and $(K_e/K_w)^2$, respectively. To maintain constant field strength, the voltage is also scaled by K_w.

Table 2.5
Influence of scaling factors on-chip electrical parameters.

Parameter	Symbol	Relationship	Scaling factor
Wire pitch	K_w		
Chip edge	K_e		
Area	A		K_e^2
Tracts/chip	N_T		K_e/K_w
Grids/chip	N_G	N_T^2	$(K_e/K_w)^2$
Voltage	V		K_w
Wire capacitance	C_w		K_w
Wire resistance	R_w		$1/K_w$
Switching time	t_{sw}		K_w
Switching energy	E_{sw}	$C_w V^2$	K_w^3
Current	I_w	$C_w V/t_{sw}$	K_w
	I_A	I_w/K_w^2	$1/K_w$
	I_{Chip}	AI_A	K_e^2/K_w

[7] The scale factors are approximately $K_w = 0.90$ and $K_e = 1.05$.

The area A of an on chip capacitor decreases as a function of K_w^2 but the thickness h of the dielectric decreases with K_w; hence, the capacitance ($C = kA/h$) decreases with K_w. The resistance of an on chip wire ($R = \rho L/A$) scales as $1/K_w$ because its length scales with K_w and its cross sectional area scales with K_w^2.

The switching energy $E_{sw} = CV^2$ scales with K_w^3. The current flowing along a wire I_w scales as K_w, because the frequency of switching has increased while the voltage has decreased. Note also that the current flow per unit area on the chip I_A scales as $1/K_w$ and the total current flowing into a chip I_{chip} scales as K_e^2/K_w.

To illustrate the use of these scaling factors in predicting future requirements consider the following example.

EXAMPLE 2.1

Determine the current I_{chip} required to drive a chip that is to be developed four years from today. The average today for the current supplied to this type of chip is 10 A. The scale factors for the wire pitch $K_w = 0.9$ and for the chip edge dimension $K_e = 1.05$ represent annual changes anticipated for the feature size.

$$\left[I_{chip}\right]_4 = \left[I_{chip}\right]_0 \left(\frac{K_e^2}{K_w}\right)^4 = 10\left(\frac{(1.05)^2}{0.9}\right)^4 = 22.52 \text{ A}$$

2.11.3 Effects of Scaling the Wires

As the feature sizes continue to decrease, the scaling of the wires becomes more important in terms of power, delay and density. To show the effect of the wire size on electrical parameters including power distribution and on-chip communications, consider the data presented in Table 2.6. This table shows the scaling relationship for a transistor length wire, a fixed length wire (1 μm) and a length of wire across the entire chip.

Table 2.6
Influence of scaling wire size on electrical parameters, power distribution and on-chip communication.

Parameter	Transistor length	Fixed 1 μm	Chip length
C	K_w	1	K_e
R	$1/K_w$	$1/K_w^2$	K_e/K_w^2
I	K_w	1	K_e
IR	1	$1/K_w^2$	$(K_w/K_w)^2$
IR/V	$1/K_w$	$1/K_w^3$	K_e^2/K_w^3
RC	1	$1/K_w^2$	$(K_w/K_w)^2$
RC/t_{sw}	$1/K_w$	$1/K_w^3$	K_e^2/K_w^3

The first two data rows show how the electrical parameters capacitance C and resistance R change with K_w and K_e. Note that the results differ depending on the purpose of the wire—whether it is a fixed length, device length, or if it extends across the chip.

The future difficulties in current and power distribution are characterized by the rows identified as I, IR and IR/V. The data in these three rows represent the current flowing through a single supply wire. The current I decreases with K_w if the wire is local (device length), but increases with K_e if the wire crosses the chip. The voltage drop over the length of a wire IR does not change for device length wires, but the IR loss increases with $1/K_w^2$ for fixed length wires and as $(K_e/K_w)^2$ for wires that cross the chip. The more important parameter is IR/V, which is the voltage drop relative to the supply voltage. For a chip length wire, this parameter is increasing as K_e^2/K_w^3 or 51% per year for $K_w = 0.9$ and $K_e = 1.05$. The scaling factor K_e^2/K_w^3 represents a serious problem in future developments because the voltage drop relative to the supply voltage for long on-chip wiring is increasing at an alarming rate.

The problem associated with on-chip communication is represented by the data presented in the RC and RC/t_{sw} of Table 2.6. The time delay of signal propagation is related to the RC constant which increases as $1/K_w^2$ for a fixed length wire and as $(K_e/K_w)^2$ for a cross-chip length wire. The parameter RC/t_{sw} represents the wire delay time relative to the gate delay time (switching time). Note that RC/t_{sw} is increasing as $1/K_w^3$ for fixed length wires and as K_e^2/K_w^3 for cross-chip length wires. For $K_w = 0.90$ and $K_e = 1.05$, $K_e^2/K_w^3 = 1.51$, which represents an annual increase of 51% in wire delay time relative to the gate delay. It is apparent that wiring delays will become more important than gate delay unless the chip architecture minimizes wire lengths.

2.12 I/O COUNT AND RENT'S RULE

Bonding pads are provided around the perimeter of the chip to permit input and output signals to enter or leave the chip. Bonding pads are also provided for the supply voltages and the ground lines which provide the power to the chip. These bonding pads, shown in Fig. 2.47, are terminals for the interconnections required to wire a group of chips together to provide digital logic functions or memory required in a complete electronic system. In more modern high density chips the bonding pads are distributed of the entire surface of the chip. In these cases, the chip is connected to the chip carrier by using controlled collapsible chip connections (C4), which will be described later in Chapter 4.

Fig. 2.47 Top surface of a chip showing perimeter bonding pads and leads for the first level package.

The number of I/O increases with the number of gates and it is important to estimate the number required on the chip, on the circuit board or the sub assembly which consists of several circuit boards and a back panel. A common method for estimating the I/O count is to utilize Rent's rule [1, 7], which is an empirical equation developed in the early 1960's[8]. Rent's rule relates gate count and I/O count for random logic connections as:

$$N_{I/O} = kN_G^{\,p} \tag{2.6}$$

where $N_{I/O}$ is the number of I/O, N_G is the number of gates involved, k is a constant depending on the function of the card and p the exponent, which also depends on the function of the card, varies between 0.1 and 0.80. Some values of Rent's constant k and Rent's exponent p are given in Table 2.7 for modern card designs with different functions.

Table 2.7
Values of the Rent constant p and the Rent exponent k for select functions

Design Type	Date	Rent's coefficient	Rent's exponent
SRAM	1990	6	0.12
Gate arrays	1990	1.9	0.50
Chip and Module	1990	1.4	0.63
Microprocessors	1990	0.82	0.45
Microprocessors	1995	2.09	0.36
ASIC control logic IFU*	2004	0.8	0.69
ASIC control logic FPU*	2004	2.2	0.66
ASIC control logic FXU*	2004	4.4	0.61
ASIC control logic IDU*	2004	20.5	0.30
ASIC control logic ISU*	2004	23.3	0.31
ASIC control logic LSU*	2004	7.3	0.46

*IFU is an instruction-fetch unit; FPU is a floating-point unit; FXU is a fixed point unit; IDU is an instruction-decode unit; ISU is an instruction-sequence unit; LSU is a load store unit.

Another similar relation that accounts for signal rate and clock cycle is given by:

$$B = KC^{\alpha} \tag{2.7}$$

where B is the bandwidth (bits/s) and C is the capability (number of gates × frequency).

The value of the exponent p in Rent's rule or the exponent α in Eq. (2.7) is very important as the exponent has a profound effect on $N_{I/O}$ particularly when N_G is large. The value of the exponent k

[8] The original memorandum written by E. F. Rent is shown in a recent publication by M. Y. Lanzerotti et al in reference [7].

required for a given system depends strongly on performance and the digital function. It was possible to build systems with Rent's exponent of about k = 0.7 in the 1960s soon after this relationship was developed; however, reduced feature size of the on-chip components has made it impossible to maintain this exponent. Today lower exponents are often achieved as indicated in Table 2.7.

The impressive scaling down of feature size in modern ULSI designs has acerbated the I/O problem. The number of bonding pads that can be placed on a chip increases as the perimeter of the chip multiplied by the linear density of the pads. This statement is valid even for area bonding (flip-chip) because the pads must be routed out from under the chip to its perimeter at the next level of packaging. As indicated in Fig. 2.48, the pin count on a chip is increasing at about 12% per year. This increase in pin count is due to increased perimeter dimensions (6%/year) as well as improved bonding pad densities (6% per year).

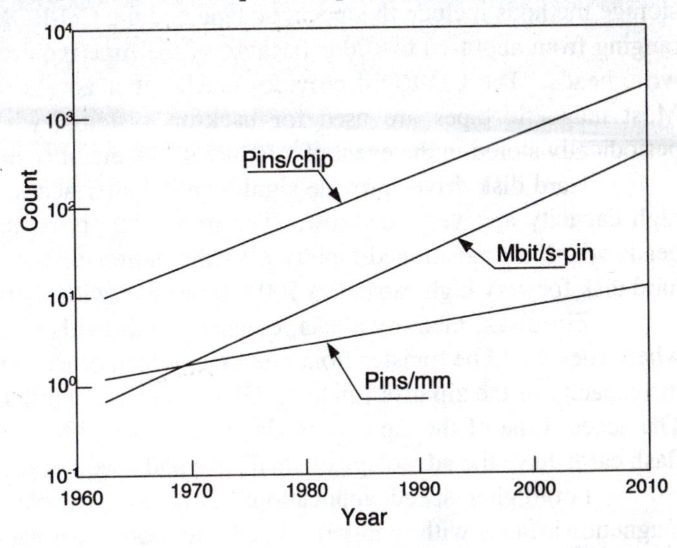

Fig. 2.48 Trend lines for increasing pin count, pin density and signal rate time.

Today large high-performance chips[9] contain more than a thousand bonding pads. It is clear that the new chip designs will have a much higher I/O count than the logic chips of the past. Design-for-performance chips will require chip carriers and circuit boards to accommodate pin counts of 1,000 to 2,000 pins at signal rates of 200 to 300 Mb/(s-pin).

2.13 MEMORY DEVICES

Digital logic circuits process extremely large amounts of information and it is necessary to store large amounts of data during and after processing to fully utilize the capabilities of a digital system. A large number of different storage devices are employed in this process including:

- Memory chips
  - RAM—random access memory
  - ROM—read only memory
  - PROM—programmable read only memory
- Magnetic disks
  - Floppy
  - Hard
- Compact disks
- Digital video disks
- Magnetic tapes

[9] The IBM Z990 processor employs a 16.8 × 16.8 mm signal control chip with 1,666 signal connections. The 16 chips that constitute the processors are mounted on a multi-chip-module (MCM) that contains a total of 5,184 pins of which 2930 are for signal connections.

Storage of data in memory requires the capability to write or to transfer information to the memory devices and to read or retrieve this information from memory at some later time. All of the memory devices listed above have the capability of reading and writing. However, the ROM and PROM are basically used only for reading as the information stored in these devices is essentially permanent and is written to memory in initializing the digital system.

The important characteristics of memory devices include storage capacity, access time, portability, reliability and cost. Access time and cost of the various devices trade off. The lower cost storage methods include the magnetic tape and the CD-ROM. The access time for the tape is very long ranging from about 10 to 100 s, because of the time required to position the tape relative to the read or write heads. The CD-ROM provides much better access time at a lower cost than the magnetic tape. Most magnetic tapes are used for back-up systems, where the entire contents of a computer are periodically stored in the event of a failure of the memory in the computer.

Hard disk drives provide significantly better access time about 9 to 10 ms while retaining very high capacity and very low cost. The improved performance is achieved by using flying read-write heads which are positioned rapidly over the entire surface of the rotating disks. Disk memories use a hard disk for very high capacities 500 GB or more with short access times.

Zip disks, memory sticks, compact flash cards and floppy disks are used for portable storage where files are to be transfer from one system to another. Zip disks are superior to floppy disks because the capacity of the zip disk (up to 35 GB) is much larger than the capacity of the floppy disk (1.44 MB). The access time of the zip disk is also less than that for the floppy disk. Memory sticks and compact flash cards have the advantage of small size and speed of recording data.

For higher speed applications, it is not possible to use electro-mechanical means (moving magnetic surfaces with read-write heads) to store information, because access time is so long that the computer's performance is degraded. For these applications, the information is stored electronically with access times which range from 10 to 100 ns. The dynamic random access memory (DRAM) is the most common way to read and write data with relatively low cost. DRAM consists of an array of storage cells as illustrated in Fig. 2.49. Each cell provides a memory location where either 0 or 1 may be stored. A particular cell is located by a row and a column address, and is activated only when both the row and column lines are high (1). The switching of a unit cell is illustrated in Fig. 2.50, where a MOS transistor is shown connected to row and column selection lines.

The data is stored in the unit cell by charging a small capacitor (10-15 fF). If the capacitor is charged the unit cell is high and storing 1, but if there is no charge the unit cell is storing 0. Because the unit cell loses charge, due either to leakage or a voltage loss which occurs during reading, it must be recharged periodically (every 1 to 2 ms the entire array of cell is replenished). The memory array is controlled by a regeneration circuit, which periodically samples each cell and replenishes the capacitors as required. After replenishment, the regeneration circuit permits reading and writing to the cell as identified by the row and column selection lines.

The capacity of dynamic RAM's has increased rapidly while its price/bit has dropped dramatically over the past years. In 2006 the price of 1 GB DDR-2-667 MHz module with a single 1 GB DRAM chip cost from about \$100 to \$110, depending on the supplier. The price of memory[10] for this popular module was slightly more than 1×10^{-6} cents per bit. The chip carrier for the DRAM is almost always a J leaded plastic chip carrier, because the number of I/O required for even large capacity DRAM is relatively small. Heat generation is also minimal as only one cell is active (replenish, read or write) at any instant. The primary requirement in packaging memory chips is to achieve high density in a location in close proximity to the logic circuits which address the DRAM chips.

---

[10] The price of DRAM chips dropped about 40% in 2005 due to over capacity of the semiconductor manufactures and relatively weak demand for this type of chip..

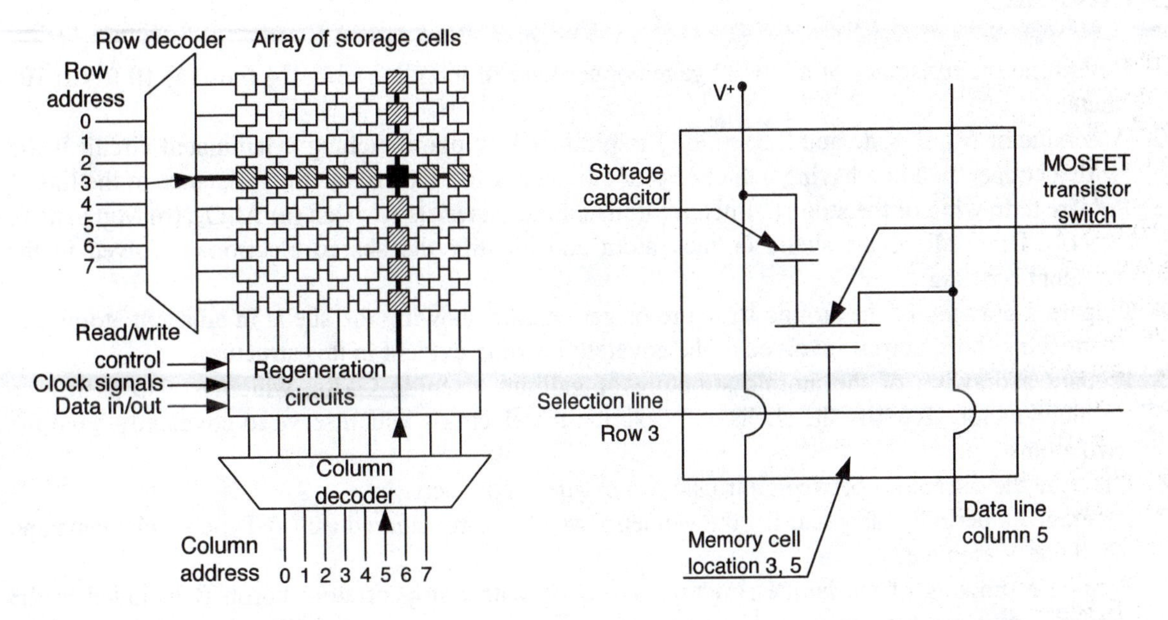

Fig. 2.49 Array of storage cells for DRAM.

Fig. 2.50 A unit storage cell for DRAM.

REFERENCES

1. Landman, B., and R. L. Russo, "On a Pin vs. Block relationship for Partitioning of Logic graphs," IEEE Transactions on Computers, Vol. C-20, No. 12, pp. 1469-1479, 1971.
2. Winkle, T.-M., et al., "First- and Second-Level Packaging of the z990 Processor Cage, IBM Journal of Research and Development, Vol. 48 No. 3/4, 2004.
3. Select chapters in Microelectronics, Scientific America, 1977 which include: R. N. Noyce, "Microelectronics" p. 2-11; J. D. Meindl, "Microelectronic Circuit Elements" p. 12-25; W. C. Holton, "The Large Scale Integration of Microelectronic Circuits" p. 26-39; D. A. Hodges, "Microelectronic Memories" p. 54-65.
4. Taylor, J., Zafiratos, C. and M. A. Dubson, Modern Physics for Scientists and Engineers, 2nd Edition, Prentice Hall, New York, NY, 2005.
5. Shackelford, J. F., Introduction to Material Science for Engineers, 6th Edition, Prentice Hall, New York, NY, 2005.
6. Hill, J. W., Petrucci, R. H., McCreary, T. W. and S. S. Perry, General Chemistry, 4th Edition, Prentice Hall, New York, NY, 2005.
7. Lanzerotti, M. Y., Fiorenza, G. and R. A. Rand, "Microminature Packaging and Integrated Circuitry: The work of E. F. Rent, with an Application to on-chip Interconnection Requirements", IBM Journal for Research and Development, Vol. 49, No. 4/5, 2005.

EXERCISES

2.1 Determine the resistance of a No. 22 gage copper wire of length (a) 3m, (b) 6 in. (c) 10 ft. (d) 100 meters.

2.2 A conductor 6 mil wide and 2.5 in. long is produced by photo-etching a laminated circuit board with a copper cladding having a thickness of 1.0 ounces/ft^2. Determine the resistance of the line.

2.3 Prepare a drawing of the atomic structure of insulating materials such as: (a) Al_2O_3, (b) MgO and (c) Si_3N_4. Draw all of the shells in each atom and identify the shared electrons involved in the covalent bonding.

2.4 Prepare a drawing of the atomic structure of germanium showing the shells in adjacent atoms and identifying the electrons involved in the covalent bonding evident in this structure.

2.5 Prepare a drawing of the atomic structure of gallium arsenide GaAs, which is also used as a semiconductor. Identify the shells and describe the electrons which serve to covalently bond the two atoms.

2.6 Describe the difference between intrinsic and extrinsic conduction.

2.7 Reference a periodic table and list the elements which are considered as: (a) Type III elements and (b) Type V elements.

2.8 Prepare a drawing of the lattice structure of silicon with a dopant atom boron B included in the lattice. Describe the conducting characteristics of this new material. Will the resistivity of the new material depend on the number of boron atoms included in the silicon lattice?

2.9 Repeat Exercise 2.8 but use arsenic as the dopant.

2.10 A batch of P type silicon is to be formulated with a resistivity of 10 Ω-cm. If the total batch has a volume 30 liters, determine the volume of silicon and boron which are mixed together prior to melting.

2.11 Repeat Exercise 2.10 but change the resistivity to 5 ohm-cm.

2.12 For a P-N junction describe the depletion layer. Indicate how the thickness of the depletion layer is adjusted in a semiconductor device.

2.13 What is the mechanical equivalent to a diode?

2.14 Reference Fig. 2.8 and note the bias voltages applied to the electrodes. Explain the direction of hole, electron and current flow shown in this illustration.

2.15 For a P-N junction forming a diode, define forward bias and reverse bias. Leakage current is observed in a diode subjected to reverse bias. Why? How large are the leakage currents in comparison to the forward currents?

2.16 Consider the transistor switch shown in Fig. 2.11 and use Ohms law to explain why the output voltage is either high or low when the switch is off or on.

2.17 A wafer 300 mm in diameter is used to fabricate chips with a die size of 10 by 12 mm. Determine the number of dice which can be processed on each wafer. If the yield from the process is 62%, estimate the number of good dies which can be obtained from each wafer. If 20 wafers are processed in a boat of wafers, find the number of good dies expected. Comment on your results.

2.18 Prepare a table showing the voltage (high or low) on the source, gate and drain and the resulting switch state for a NMOS transistor of: (a) enhancement mode and (b) depletion mode.

2.19 Repeat Exercise 2.18 for a PMOS transistor.

2.20 In Fig. 2.17 a section view of a CMOS transistor which utilizes series connected NMOS and PMOS is shown. Both of these transistors operate in the enhancement mode. Prepare an equivalent drawing for a CMOS transistor fabricated from NMOS and PMOS transistors operating in the depletion mode.

2.21 Perform a trade-off analysis citing the relative advantages and disadvantages of bipolar, NMOS and CMOS transistors.

2.22 The EXCLUSIVE-OR gate is shown in Fig. E2.22. Construct a truth table for this logic element.

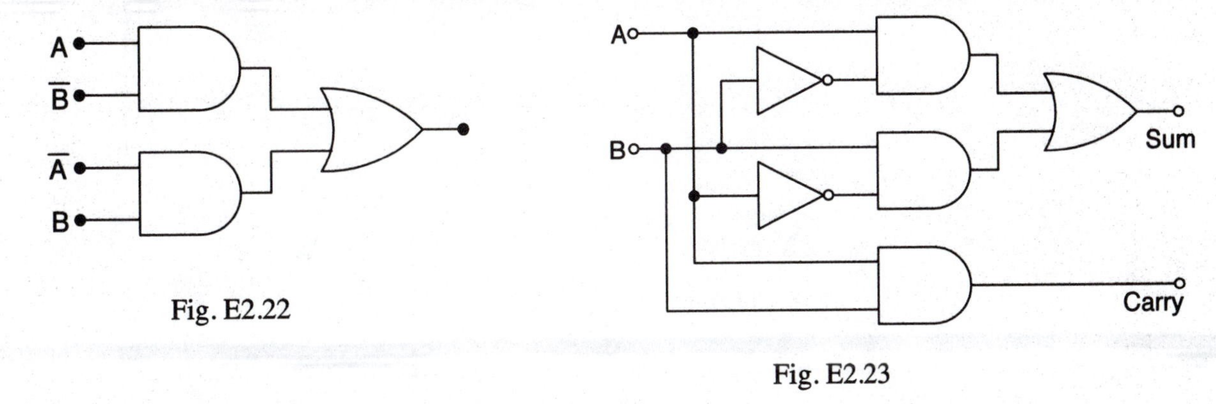

Fig. E2.22

Fig. E2.23

2.23 A logic gate half-adder is shown in Fig. E2.23. Trace inputs AB of 00, 01, 10, and 11 and determine the corresponding outputs on the sum and carry terminals.

2.24 A binary adder shown in Fig. 2.22 has three inputs A, B, and C. For inputs of 000, 001, 010, 011, 100, 101, 110 and 111 trace through the circuit and indicate the output on the sum and carry bits.

2.25 Describe the advantages and disadvantages of CMOS used in the construction of logic circuits.

2.26 Describe the process for producing silicon ingots used in semiconductor manufacturing.

2.27 Describe the process for removing wafers from the single crystal silicon ingots.

2.28 List the various steps in the lithographic process used to implant the P and N features on a chip.

2.29 Describe the preparation of the pads for wire bonding; tape automated bonding; and flip-chip bonding.

2.30 Describe the damascene plating process and indicate why it is so important in the production of high-end chips.

2.31 Explain what is meant by the term "Back End of Line" (BEOL).

2.32 What is the definition of MEMS?

2.33 Estimate when the components placed on a single chip will reach 10^{12}. What assumptions have you made in your prediction?

2.34 Determine the current to drive a chip that is to be produced four years from today. The chip that is to be replaced requires a current of 12A. The scale factors for wire pitch and edge size are 0.9 and 1.05 respectively.

2.35 Explain the problem of on-chip communication as the feature size on chips continues to decrease.

2.36 Following Rent's rule estimate the number of I/O required on a circuit card with 10,000 logic gates. Take $k = 2$ and $p = 0.45$.

2.37 For the same circuit card as described in Exercise 2.36, plot the I/O required if p is varied from 0.3 to 0.7. Retain $k = 2$ in this exercise.

CHAPTER 3

CIRCUIT ANALYSIS

3.1 INTRODUCTION

Logic circuits, used widely in electronic systems, involve gates which are switched from say a low state to a high state to produce a voltage pulse with a rise time t_r, as illustrated in Fig. 3.1. The frequency components of the pulse, which depend upon its rise time and amplitude, can be determined from a Fourier series expansion representing the pulse. The Fourier series can be written as:

$$V(t) = \sum_{n=0}^{\infty}\left(A_n \cos n\omega t + B_n \sin n\omega t\right) = \sum_{n=0}^{\infty} C_n e^{jn\omega t} \qquad (3.1)$$

where A_n, B_n and C_n are coefficients determined to fit the pulse. Exercises given at the end of the chapter illustrate the changes in frequency content of the pulse with variations in the rise time. If the gate switches with a very short rise time t_r, about 1 ns or less, then very high frequencies must be transmitted without significant distortion if the fidelity of the pulse is to be maintained. On the other hand, if the gate switches slowly with t_r = 1 μs or more the predominate frequencies are much lower. In this case, the pulse shape is easier to maintain and ringing (oscillation) with transient voltages which over and/or under shoot the steady state voltage is minimized.

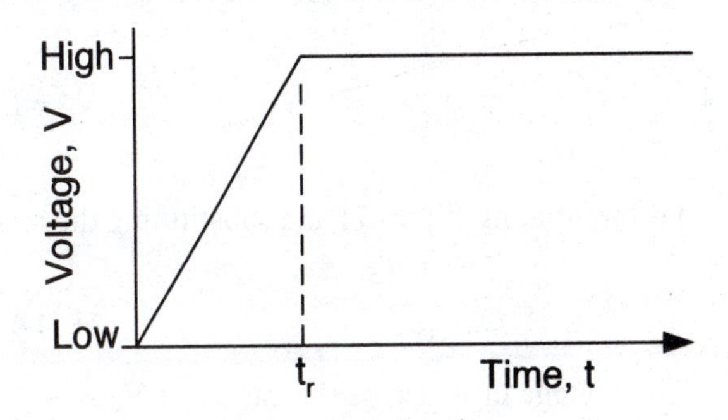

Fig. 3.1 Voltage pulse produced by switching a logic gate from the low to the high state.

Two different types of circuit analyses are employed in circuit design depending on the frequencies and the distance of propagation which are involved. For frequencies of less than about 100 MHz with relatively short propagation distances, a traditional analog circuit analysis is usually sufficient to adequately describe the circuit behavior. For frequencies above 100 MHz with longer propagation distances, the circuit wiring becomes much more critical. Distributed inductance and capacitance present in all circuit lines (wires) becomes significant at high frequencies and affect the circuit's behavior. Transmission line theory is introduced later in this chapter to account for the effects of distributed capacitance and inductance on the observed behavior of pulses propagating over extended circuits operating at high frequencies.

3.2 THE ANALOG METHOD OF ANALYSIS

The analog method of analysis is usually employed to treat sinusoidal signals with a constant frequency ω. This approach can also be used to advantage in treating pulse like signals associated with digital systems providing the predominant frequencies involved in the Fourier representation of the pulse are less than about 100 MHz. The upper limit of the analog method of analysis depends on the length of the lines, the magnitude of the distributed inductance and capacitance and the critically of the arrival times of the signal pulses involved. As an example of the analog method of analysis, consider the circuit shown in Fig. 3.2. This circuit is driven by a power supply which provides a sinusoidal input voltage V_s and current I_s which can be written as:

$$V_s = V_a e^{j\omega t} \quad \text{and} \quad I_s = I_a e^{j\omega t} \tag{3.2}$$

where V_a and I_a are amplitudes of voltage and current, and ω is the circular frequency of the signal.

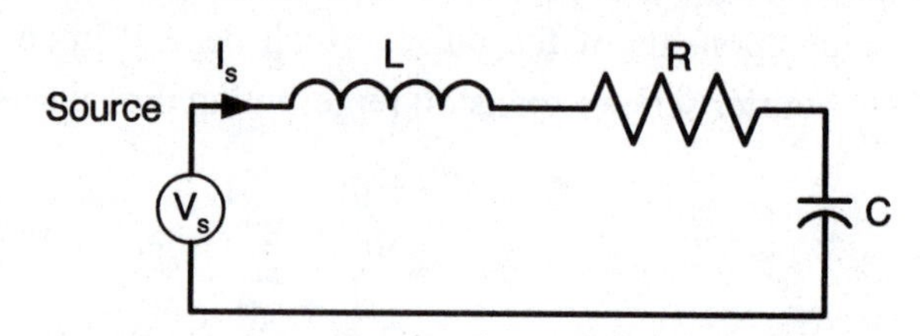

Fig. 3.2 An analog circuit with discrete components including a resistor R, an inductor L and a capacitor C.

To analyze this circuit, determine the voltage drop across each discrete component and neglect the effect of the wiring by assuming that wiring impedance is small compared to the impedance of the discrete components R, C and L. The voltage drop across the resistor V_R is then given by:

$$V_R = I_s R = I_a R e^{j\omega t} \tag{3.3}$$

The voltage drop across the inductor V_L is given by:

$$V_L = L\frac{dI_s}{dt} \tag{3.4}$$

Differentiating Eq. (3.2) and substituting the result into Eq. (3.4) gives:

$$V_L = j\omega I_a L e^{j\omega t} \tag{3.5}$$

The voltage drop across the capacitor V_C is:

$$V_C = q/C \tag{3.6}$$

where q is the charge. Because,

$$I = \frac{dq}{dt} \tag{3.7}$$

it is clear from Eqs. (3.2), (3.6) and (3.7) that:

$$V_C = \frac{1}{C}\int I dt = \left(\frac{I_a}{j\omega C}\right) e^{j\omega t} = -j\left(\frac{I_a}{\omega C}\right) e^{j\omega t} \tag{3.8}$$

If the voltage drop across the components are summed and equated to the input voltage, the following relation between the amplitudes of the voltage and current is obtained:

$$V_a = \left[R + j\left(\omega L - \frac{1}{\omega C} \right) \right] I_a \tag{3.9}$$

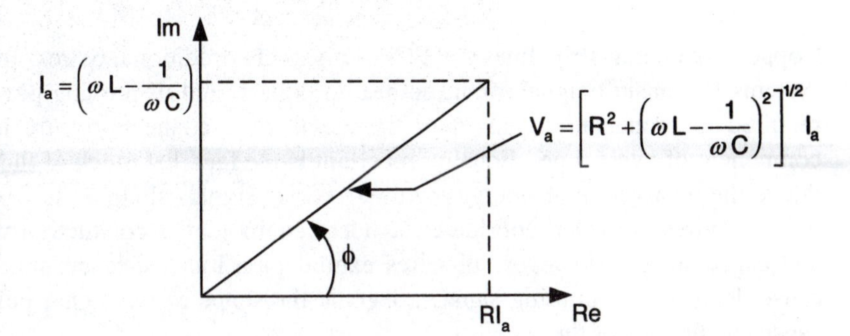

Fig. 3.3 Representation of the impedance components on the complex plane.

Examination of Eq. (3.9) shows that a complex function with real and imaginary parts is involved in the relation between V_a and I_a. If this complex function is divided into its real and imaginary parts, as illustrated in Fig. 3.3, a more functional expression for V_a in terms of I_a is obtained:

$$V_a = ZI_a \tag{3.10}$$

where the total impedance of the circuit Z is given by:

$$Z = \left[R^2 + \left(\omega L - \frac{1}{\omega C} \right)^2 \right]^{1/2} \tag{3.11}$$

While Eq. (3.10) and Eq. (3.11) define the amplitudes of the voltage and the current, the phase of the voltage relative to the current remains to be determined. Reference to Fig. 3.3 shows the phase angle ϕ, which is given by:

$$\phi = \tan^{-1}\left(\frac{\omega L - 1/\omega C}{R} \right) \tag{3.12}$$

If $\phi > 0$ as depicted in Fig. 3.3, the voltage leads the current and the complete expression for V(t) is:

$$V(t) = \left[R^2 + \left(\omega L - \frac{1}{\omega C} \right)^2 \right]^{1/2} I_a e^{j(\omega t + \phi)} \tag{3.13}$$

Equation (3.13) shows the amplitude of the voltage and the relative phase of the voltage with respect to the current. Because the frequency ω is limited, the effects of distributed inductance and capacitance due to the circuit wiring may be neglected. The results obtained are independent of the wiring parameters such as its type, length and gage. Probes for measuring voltage and current from the source with signals displayed on an oscilloscope would display I_a, V_a and the phase angle ϕ. These values are controlled by the value of L, R and C associated with discrete components, if the distributed inductance, resistance and capacitance associated with the wiring is small. For relatively low frequencies and short lines, the discrete values of R, C and L associated with individual components in the circuit are dominate and the traditional analog approach is adequate to describe circuit behavior.

3.3 ELECTRICAL PROPERTIES OF WIRES

Copper conductor strip lines on PCBs and more traditional wires and cable are used in electronic systems to transmit signals from point a to point b and to provide power and clock pulses. The time required to propagate signals over the length of a connecting wire is in many instances the major component of cycle time. Equally important is the fact that much of the power dissipated in a system is due to the relatively high energy required to drive signals along these wires.

Wires are often considered as ideal equipotential conductors with zero resistance, inductance and capacitance. However, all wires exhibit parasitic resistance, inductance and capacitance, which cause delay in transmitting signals, degrade the shape of the signal pulses and in some extreme cases cause oscillations of the signals.

Wiring strip lines on a PCB act like LC transmission lines along which signals pulses propagate at wave velocities. If this transmission line is terminated with the proper (matched) impedance, the line transmits the signal with the incident wave and the signal pulse is absorbed by the termination when it reaches its destination. It is important to note that a second pulse is not reflected from the end of a properly terminated transmission line: hence, the line is clear and free of reflecting waves.

3.3.1 Resistance, Capacitance and Inductance of Wires

Resistance

The resistance of a wire of length ***L*** is given by:

$$R = \rho\frac{\boldsymbol{L}}{A} = \begin{cases} \dfrac{\rho \boldsymbol{L}}{wh} & \text{rectangular cross section} \\ \dfrac{4\rho \boldsymbol{L}}{\pi D^2} & \text{circular cross section} \end{cases} \quad (3.14a)$$

The resistance per unit length of wire R' is given by:

$$R' = \frac{\rho}{A} = \begin{cases} \dfrac{\rho}{wh} & \text{rectangular cross section} \\ \dfrac{4\rho}{\pi D^2} & \text{circular cross section} \end{cases} \quad (3.14b)$$

where h and w are the thickness and width of a rectangular cross section, respectively.

D is the diameter of a circular cross section and ρ is the resistivity of the conductor material.

In some instances, the resistance of a rectangular conductor is specified in Ω/square when w = ***L***. In this case, the resistance per square is given by:

$$R_{Sq} = \rho/h \quad (3.15)$$

Reference to Table 2.1 shows that silver has the lowest resistivity; however, it is rarely used for wiring because of its high cost. Copper, which is nearly conductive as silver, is the most commonly employed material for a wide variety of applications including cables, ground planes, strip-lines, buss-bars, etc. Aluminum is less conductive than copper, but it offers more conductivity per unit cost and is more compatible with the fabrication of integrated circuits. Aluminum is widely used for power cables and low cost buss-bars. Tungsten and other refractor metals are used with ceramic chip carriers where high-temperature firing occurs after the metal is deposited. The noble metals, gold and silver, are used as plating materials because of their ability to resist oxidations.

Capacitance

The capacitance C of a conductor is more difficult to determine because it depends on the materials that surround it. Because adjacent conductors can screen or divert field lines, numerical methods are often employed to achieve accurate determinations of capacitance. However, it is possible to make relatively good estimates of capacitance for the different wire geometries defined in Fig. 3.4.

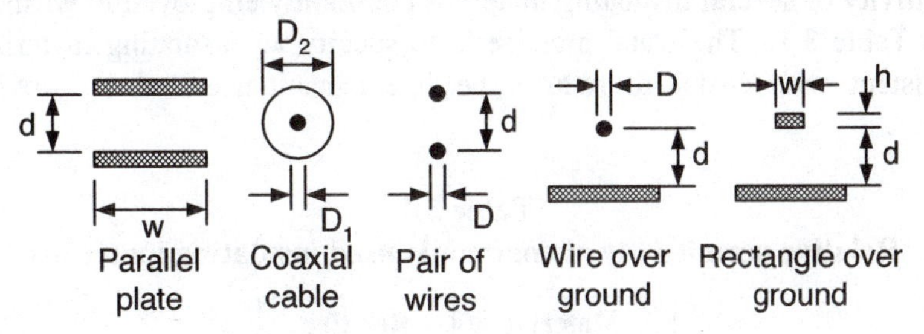

Fig. 3.4 Cross sections of common wiring.

The simplest case is the parallel plate capacitor defined in Fig. 3.4. Its capacitance C'_{pp} per unit length is given by:

$$C'_{pp} = \frac{w\varepsilon}{d} \tag{3.16}$$

where ε is the permittivity of the insulating material separating the two conductors.

The capacitance C'_{cac} per unit length of the coaxial cable defined in Fig. 3.4 is given by:

$$C'_{cac} = \frac{2\pi\varepsilon}{\log_e(D_2/D_1)} \tag{3.17}$$

The capacitance C'_{pw} per unit length of a pair of parallel wires defined in Fig. 3.4 is given by:

$$C'_{pw} = \frac{\pi\varepsilon}{\log_e(2d/D)} \tag{3.18}$$

The capacitance C'_{swg} per unit length of a single wire over a ground plane defined in Fig. 3.4 is given by:

$$C'_{swg} = \frac{2\pi\varepsilon}{\log_e(4d/D)} \tag{3.19}$$

The capacitance C'_{msl} per unit length of a microstripline over a ground plane defined in Fig. 3.4 is given by:

$$C'_{msl} = \frac{w\varepsilon}{d} + \frac{2\pi\varepsilon}{\log_e(d/h)} \tag{3.20}$$

All of the equations for estimating the capacitance per unit length of conductor depend on the permittivity of the material surrounding the conductors. The permittivity is defined as:

$$\varepsilon = \varepsilon_0 \, \varepsilon_r \tag{3.21}$$

where $\varepsilon_0 = 8.854 \times 10^{-12}$ F/m is the permittivity of free space.
ε_r is the relative permittivity of an insulating material.

The relative permittivity of several insulating materials commonly employed in wiring or printed circuit boards is given in Table 3.1. The usual practice is to specify an insulating material with the lowest permittivity (consistent with costs) to reduce the line capacitance and to provide a high signal propagation velocity.

Table 3.1
Relative permittivity of commonly used insulating materials

Material	Relative Permittivity
Air	1
Teflon	2
Polyimide	3
Silicon dioxide	3.9
Glass-epoxy	4.7
Alumina	10
Silicon	11.7

Inductance

When the conductors in a transmission line are completely surrounded by a uniform dielectric, the capacitance per unit length C' and the inductance L' per unit length are related by:

$$C'L' = \varepsilon \, \mu \tag{3.22}$$

where $\mu = 4\pi \times 10^{-7}$ H/m is the permeability of free space.

By combining Eq. (3.18) and Eq. (3.22), it is possible to write an expression for the inductance L'_{pw} per unit length of two parallel wires surrounded by a homogeneous insulator as:

$$L'_{pw} = \frac{\varepsilon\mu}{C'_{pw}} = \frac{\mu}{\pi}\log_e\left(\frac{2d}{D}\right) \tag{3.23}$$

When the conductors in a transmission line are bounded by two different insulating materials, as is the situation with strip lines on PCBs, Eq. (3.23) is not valid. In these cases, an average dielectric constant is employed to determine the distributed capacitance.

3.4 WIRES AS TRANSMISSION LINES

Unfortunately a wire is not an ideal conductor because it exhibits resistance, capacitance and inductance that are distributed along its length. To examine these distributed parameters, consider an infinitesimal length of real wire that is modeled as indicated in Fig. 3.5.

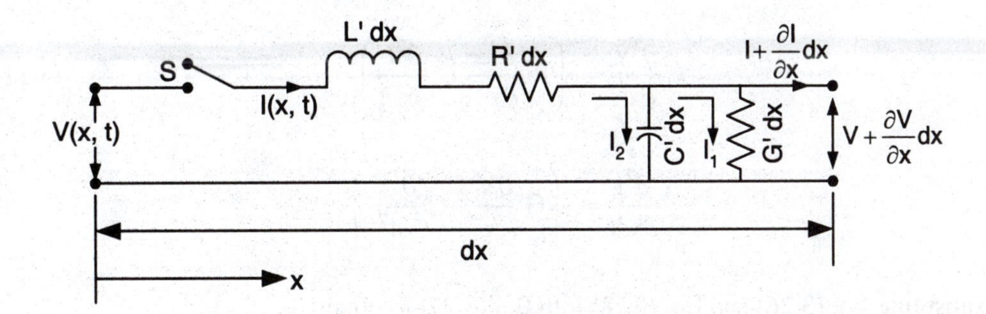

Fig. 3.5 Equivalent circuit for a transmission line of length dx.

At very high frequencies or for very long lines at more moderate frequencies, the distributed inductance and capacitance of the wiring between components affects the behavior of the circuit. To show the effect of distributed electrical parameters consider an element dx of the wiring, illustrated in Fig. 3.5, which contains a voltage source on one end and an open circuit on the other end. In this circuit L' is the distributed inductance, R' is the distributed resistance and C' is the distributed capacitance. These quantities are all specified per unit length of the line. The term G' is the inverse of the distributed resistance to ground per unit length of line.

$$G' = \frac{1}{R'_g} \tag{3.24}$$

where R'_g is the distributed resistance to ground of the insulation separating the two conductors.

To determine the voltage and the current flow in this line segment, close the switch at t = 0 and write the voltage drop over the length dx of the line as:

$$V - \left[V + \left(\frac{\partial V}{\partial x}\right) dx\right] = \left(R'I + L'\frac{\partial I}{\partial t}\right) dx$$

This expression reduces to:

$$\frac{\partial V}{\partial x} = -\left(R'I + L'\frac{\partial I}{\partial t}\right) \tag{3.25}$$

Consider next, the current flow and note that:

$$I - \left[I + \left(\frac{\partial I}{\partial x} \right) dx \right] = I_1 + I_2 = \left(G'V + C' \frac{\partial V}{\partial t} \right) dx$$

this can be simplified to give:

$$\frac{\partial I}{\partial x} = -\left(G'V + C' \frac{\partial V}{\partial t} \right) \tag{3.26}$$

Differentiate Eq. (3.25) with respect to x and Eq. (3.26) with respect to t to obtain:

$$\frac{\partial^2 V}{\partial x^2} = -\left(R' \frac{\partial I}{\partial x} + L' \frac{\partial^2 I}{\partial x \partial t} \right) \tag{3.27}$$

$$\frac{\partial^2 I}{\partial x \partial t} = -\left(G' \frac{\partial V}{\partial t} + C' \frac{\partial^2 V}{\partial t^2} \right) \tag{3.28}$$

Finally, substitute Eq. (3.26) and Eq. (3.28) into Eq. (3.27) to obtain:

$$\frac{\partial^2 V}{\partial x^2} = R'G'V + (R'C' + L'G') \frac{\partial V}{\partial t} + L'C' \frac{\partial^2 V}{\partial t^2} \tag{3.29}$$

This partial differential equation describes the relationship between voltage and current in a transmission line in terms of its distributed parameters R', G', L' and C'. It is known as the telegrapher's equation and is the basic relation governing transmission line theory. Transmission line theory is used to characterize the behavior of pulses propagating over circuit lines from one digital device or subsystem to another when the frequency components are relatively high or if the distances involved are sufficiently long.

In most applications, the distributed resistance to ground R'_g is sufficiently high to assume that the conductance G' = 0; hence, Eq. 3.29 reduces to:

$$\frac{\partial^2 V}{\partial x^2} = R'C' \frac{\partial V}{\partial t} + L'C' \frac{\partial^2 V}{\partial t^2} \tag{3.30}$$

This relation governs the propagation of on-chip signals as well as signals transmitted from chip to chip along PCB strip lines. For relatively long on-chip wires the resistance R' is the dominant parasitic parameter and Eq. (3.30) can be approximated by the diffusion equation, which is given by:

$$\frac{\partial^2 V}{\partial x^2} = R'C' \frac{\partial V}{\partial t} \tag{3.31}$$

This relation shows that the voltage level of signals decay with time as they propagate along a length of wire.

For PCB strip lines and circuit board to circuit board cables, the distributed resistance R' is negligible and the distributed inductance and capacitance are dominant. In this situation, Eq. (3.30) reduces to the well known wave equation.

$$\frac{\partial^2 V}{\partial x^2} = L'C' \frac{\partial^2 V}{\partial t^2} = \frac{1}{c^2}\frac{\partial^2 V}{\partial t^2} \qquad (3.32)$$

This relation shows that the signals propagate down the transmission line as waves with a velocity c that depends on the value of $\frac{1}{\sqrt{L'C'}}$.

3.5 SINUSOIDAL SIGNAL PROPAGATION ALONG A TRANSMISSION LINE

To explore the behavior of a transmission line with an open end consider the case of a sinusoidal signal with an input voltage and an input current, which may be written as:

$$V(x,t) = V(x)e^{j\omega t} \quad \text{and} \quad I(x,t) = I(x)e^{j\omega t} \qquad (3.33)$$

Substituting Eq. (3.33) into Eqs. (3.25) and (3.26) gives:

$$\frac{dV}{dx} = -(R' + j\omega L')I(x) \qquad (3.34a)$$

$$\frac{dI}{dx} = -(G' + j\omega C')V(x) = -j\omega C'V(x) \qquad (3.34b)$$

where it is assumed that the distributed conductance G' = 0.

Differentiate Eq. (3.34a) with respect to x and substitute Eq. (3.34b) into the resulting equation to yield:

$$\frac{d^2V}{dx^2} - (R' + j\omega L')j\omega C'V(x) = 0 \qquad (3.35)$$

This is an ordinary second order differential equation which has a solution of the form:

$$V(x) = Ae^{-\lambda x} + Be^{\lambda x} \qquad (3.36)$$

The expression for λ is given by the roots of the characteristic equation as:

$$\lambda = [(R' + j\omega L')\, j\omega C']^{1/2} \qquad (3.37)$$

Substituting Eq. (3.36) into Eq. (3.33) gives:

$$V(x,t) = \left(Ae^{-\lambda x} + Be^{-\lambda x}\right)e^{j\omega t}$$

This equation can be rewritten as:

$$V(x,t) = Ae^{(j\omega t - \lambda x)} + Be^{(j\omega t + \lambda x)} \qquad (3.38)$$

At this point, it is important to interpret the solutions given by Eqs. (3.37) and (3.38) because they describe the two most important features of transmission lines. The results of Eq. (3.38) indicate that the sinusoidal voltage divides into two parts $Ae^{(j\omega t-\lambda x)}$ and $Be^{(j\omega t+\lambda x)}$. The part associated with $Ae^{(j\omega t-\lambda x)}$ is the forward wave which propagates as a damped sinusoid along the positive x axis. The part associated with $Be^{(j\omega t+\lambda x)}$ is the backward wave that is also a damped sinusoid propagating along the negative x axis. The waves are illustrated in Fig. 3.6. The amplitude of each sinusoid is given by either $Ae^{(j\omega t-\lambda x)}$ or $Be^{(j\omega t+\lambda x)}$.

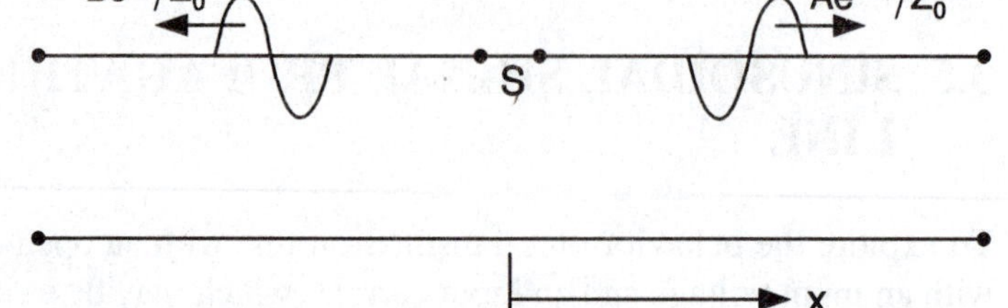

Fig. 3.6 Propagation of a sinusoidal voltage signal in the positive and negative x directions along a transmission line.

The term λ given in Eq. (3.37) shows that the amplitude of the sinusoid decreases as the signal propagates along the line. This decrease in amplitude with propagation in either the positive or negative x direction is due to two different effects—attenuation and dispersion. The term λ is often expressed as:

$$\lambda = \alpha + j\beta \qquad (3.39)$$

to describe both of these effects. In Eq. (3.39), α is the attenuation constant and β is the phase constant. It is possible to determine α and β in terms of L', R' and C' by equating Eqs. (3.37) and (3.39) to obtain:

$$\alpha^2 = \frac{1}{2}\left\{-\omega^2 L'C' \pm \left[\left(\omega^2 L'C'\right)^2 + \omega^2\left(R'C'\right)^2\right]^{1/2}\right\}$$

$$\beta^2 = \frac{1}{2}\left\{\omega^2 L'C' \pm \left[\left(\omega^2 L'C'\right)^2 + \omega^2\left(R'C'\right)^2\right]^{1/2}\right\} \qquad (3.40)$$

For a transmission line which is classified as a low loss line where ωL' >> R' and ωC' >> G', the relations for α and β reduce to:

$$\alpha = 0 \quad \text{and} \quad \beta = \omega\sqrt{L'C'} \qquad (3.41)$$

From Eq. (3.41) it is evident that the decay of the sinusoid propagating along a low loss line is due to the phase constant β because the attenuation constant α has vanished.

The current I(x) is determined by substituting Eq. (3.36) into Eq. (3.34a) and solving to obtain:

$$I(x) = \frac{Ae^{-\lambda x} - Be^{\lambda x}}{Z_0} \qquad (3.42)$$

where

$$Z_0 = \left[\frac{R' + j\omega L'}{j\omega C'}\right]^{1/2} = \left[\frac{\omega L' - jR'}{\omega C'}\right]^{1/2} \qquad (3.43)$$

The term Z_0 is known as the characteristic impedance of the transmission line. Examination of Eq. (3.42) indicates that two waves carry the current along the line, with the forward wave I_f propagating in the positive x direction with amplitude $Ae^{-\lambda x}/Z_0$ and the backward wave I_b propagating in the negative x direction with amplitude $-Be^{\lambda x}/Z_0$. The negative sign on the backward current term indicates that the sign of I_b is opposite the sign of the backward term V_b for the voltage.

For a low loss line where $R' \Rightarrow 0$, the characteristic impedance becomes:

$$Z_0 = \sqrt{\frac{L'}{C'}} \tag{3.44}$$

The characteristic impedance is purely resistive (i.e. Z_0 is a real number) for a low loss line with the attenuation constant $\alpha = 0$.

3.6 TERMINATION OF TRANSMISSION LINES

In the previous section, the behavior of a sinusoidal voltage signal propagating along an infinitely long transmission line was described. In practical applications transmission lines are of finite length and they are terminated in some manner at both ends. The method of termination markedly affects the behavior of high frequency signals because their pulses can reflect and change the state of a transmission line. To study these reflections, it is important to examine the effects of different types of line termination.

To explore the effect of the line termination, consider a transmission line with an arbitrary termination impedance Z_L placed at x = 0 and a sinusoidal voltage source at x = – ***L*** as shown in Fig. 3.7. At the termination location, we can write:

$$\frac{V(0)}{I(0)} = Z_L \tag{a}$$

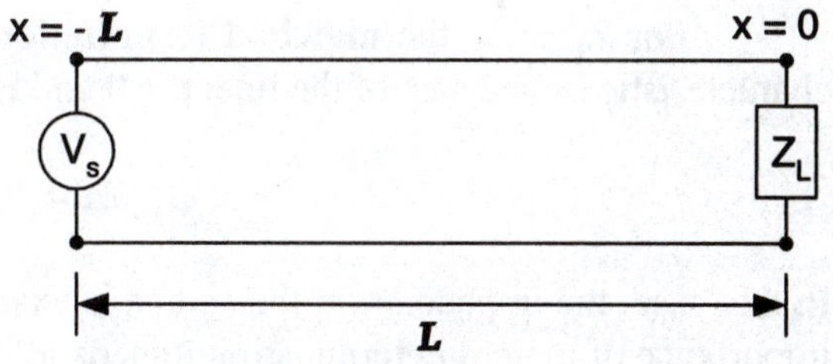

x = - ***L***   x = 0

V_s Z_L

L

Fig. 3.8 Transmission line of length ***L*** with an arbitrary termination impedance Z_L.

From Eqs. (3.36), (3.42) and (a), it is evident that:

$$Z_L = \frac{(A+B)}{(A-B)} Z_0 \tag{b}$$

Solving Eq. (b) for the ratio B/A gives:

$$\frac{B}{A} = \frac{(Z_L - Z_0)}{(Z_0 + Z_L)} = p \tag{3.45}$$

The impedance at the source Z_S due to the combined effect of the line and the termination load is given by:

$$Z_S = \frac{V(-\boldsymbol{L})}{I(-\boldsymbol{L})} \tag{c}$$

or

$$Z_S = \frac{Ae^{\lambda L} + Be^{-\lambda L}}{Ae^{\lambda L} - Be^{-\lambda L}} Z_0 \qquad \text{(d)}$$

Substituting Eq. (3.45) into Eq. (d) yields:

$$Z_S = Z_0 \frac{e^{\lambda L} + pe^{-\lambda L}}{e^{\lambda L} - pe^{-\lambda L}} \qquad (3.46)$$

This expression can also be written as:

$$Z_S = Z_0 \frac{Z_L + Z_0 \tanh \lambda L}{Z_L \tanh \lambda L + Z_0} \qquad (3.47)$$

The behavior of this line clearly depends upon the value of p given by Eq. (3.47). For $Z_L \Rightarrow \infty$, the **open circuit case**, p = 1 and Eq. (3.46) reduces to:

$$Z_{SO} = Z_0 \frac{e^{\lambda L} + e^{-\lambda L}}{e^{\lambda L} - e^{-\lambda L}} = Z_0 \coth \lambda L \qquad (3.48)$$

For $Z_L = 0$, **the short circuit case**, p = – 1 and Eq. (3.46) reduces to:

$$Z_{SS} = \frac{V(L)}{I(L)} = Z_0 \frac{e^{\lambda L} - e^{-\lambda L}}{e^{\lambda L} + e^{-\lambda L}} = Z_0 \tanh \lambda L \qquad (3.49)$$

For $Z_L = Z_0$, the **matched termination case**, where the impedance of the load is equal to the characteristic impedance of the line, p = 0 and Eq. (3.47) becomes:

$$Z_S = Z_0 = Z_L \qquad (3.50)$$

In this case, the impedance at the source is exactly the same as the load impedance, which indicates the importance of matching termination impedance to the characteristic line impedance.

3.7 PULSE PROPAGATION ALONG A LOW LOSS LINE

Next let's consider pulse propagation along transmission lines because the signal generated as a gate switches from high to low or vice versa is a low amplitude voltage pulse with an abrupt front. For a low-loss line where R' and G' may be neglected, the telegrapher's Eq. (3.29) reduces to the wave equation:

$$\frac{\partial^2 V}{\partial t^2} = c^2 \frac{\partial^2 V}{\partial x^2} \qquad (3.51)$$

where c is the velocity of propagation of the signal that is given by:

$$c = 1/\sqrt{L'C'} \qquad (3.52)$$

A solution for the wave equation, well suited for describing pulse propagation, is the sum of two functions f and g that is written as:

$$V(x, t) = f(x - ct) + g(x + ct) \tag{3.53}$$

The first function in this equation, $f(x - ct)$, represents a wave which propagates at constant amplitude in the positive x direction as t increases so that the argument $(x - ct)$ remains constant. The second function $g(x + ct)$, represents a wave propagating in the negative x direction with constant amplitude. The velocity of propagation of both of the waves is c, which is defined in Eq. (3.52).

The current $I(x, t)$ is determined by substituting Eq. (3.53) into Eq. (3.25) and utilizing the low loss conditions to obtain:

$$\frac{\partial f}{\partial x} + \frac{\partial g}{\partial x} = -L' \frac{\partial I}{\partial t} \tag{a}$$

Integrating Eq. (a) to obtain $I(x, t)$ gives:

$$I(x,t) = \left(\frac{1}{cL'}\right)\left[f(x - ct) - g(x + ct)\right] \tag{b}$$

Substituting Eqs. (3.44) and (3.52) into Eq. (b) leads to:

$$I(x,t) = \left(\frac{1}{Z_0}\right)\left[f(x - ct) - g(x + ct)\right] \tag{3.54}$$

This relation can be expressed as:

$$I = (I_f + I_b) \tag{3.55}$$

From this result it is evident that the propagation of current pulses can be described with forward and backward waves. The current I_f associated with the forward wave is:

$$I_f = \frac{f(x - ct)}{Z_0} = \frac{V_f}{Z_0} \tag{3.56}$$

where V_f is the voltage associated with the forward wave. The current in the backward wave, I_b is given by:

$$I_b = -\frac{g(x + ct)}{Z_0} = -\frac{V_b}{Z_0} \tag{3.57}$$

where V_b is the voltage of the backward wave. It should be noted in Eq. (3.57) that the sign of the current is opposite that of the voltage for the pulses propagating in the negative x direction.

The importance of this development is to recognize that the signal pulse travel as waves and that voltage and current do not appear instantaneously on the wires on either the chip or the circuit board. Voltage and current pulses must be considered separately because in some cases they are of opposite sign. The effect of line termination at both ends of a line is also important as the signal pulse reflects from the line ends and affects both the voltage and the currents with time over the length of the transmission lines. The methods for treating pulse reflect from the ends of the lines is presented in the next section.

3.8 EFFECT OF TERMINATION ON PULSE PROPAGATION

The behavior of a pulse propagating along a transmission line depends upon the termination at both ends of the line. To simplify the initial analysis consider that the line is terminated at the input end with a voltage source which exhibits an output impedance $Z_S = Z_0$ as shown in Fig. 3.8 Matching the source impedance to the characteristic impedance of the line eliminates the reflection of the pulses from the source end and reduces the complexity of the analysis. Attention will be focused on the pulse propagating down the line to the load end and the behavior of the pulse when it reaches the position $x = \boldsymbol{L}$. We will consider an end with an arbitrary resistive load R_L and then show pulse reflections from a transmission line with an $R_L > Z_0$ and $R_L < Z_0$.

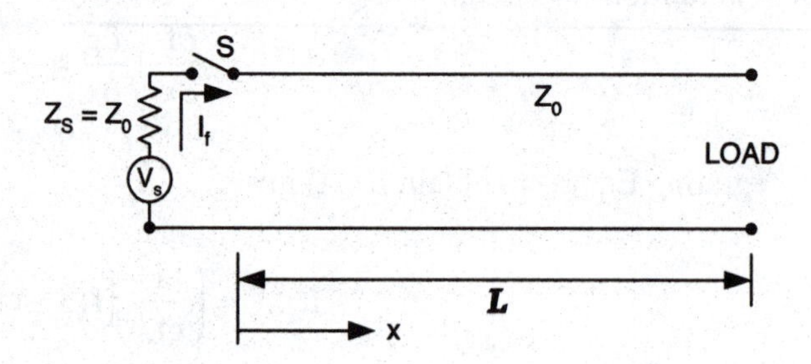

Fig. 3.8 Transmission line with a dc voltage source with an impedance $Z_S = Z_0$. and an arbitrary load resistance $Z_L = R_L$.

3.8.1 Arbitrary Resistive Load at the Transmission Line End

Examine the transmission line of length $\boldsymbol{L}$ shown in Fig. 3.8 with a source voltage V_s and an open end at $x = \boldsymbol{L}$. When the switch S is closed at t = 0, the transmission line at the position x = 0 is subjected to a dc voltage $V = V_s$. Substituting this initial condition into Eq. (3.53) gives:

$$f(x - ct) + g(x + ct) = V_s$$

or

$$V_f + V_b = V_s \tag{a}$$

From Fig. 3.10 and Eq. (3.56) it is evident that:

$$V_f = I_f Z_0 \tag{b}$$

Kirchhoff's law gives:

$$V_s = I_f Z_S + I_f Z_0 = I_f (Z_S + Z_0) \tag{c}$$

From Eqs. (b) and (c) it is clear that:

$$I_f = \frac{V_s}{(Z_s + Z_0)} \tag{d}$$

$$V_f = \frac{Z_0}{(Z_s + Z_0)} V_s \tag{e}$$

with $Z_s = Z_0$ then:

$$V_f = \frac{V_s}{2} \qquad \text{at } t = 0 \tag{f}$$

$$I_f = \frac{V_f}{Z_0} = \frac{V_s}{2Z_0} \qquad \text{at } t = 0 \tag{g}$$

For the time, $0 \leq t \leq L/c$, the voltage and the current pulses propagate down the line at a constant velocity c as indicated in Fig. 3.9b.

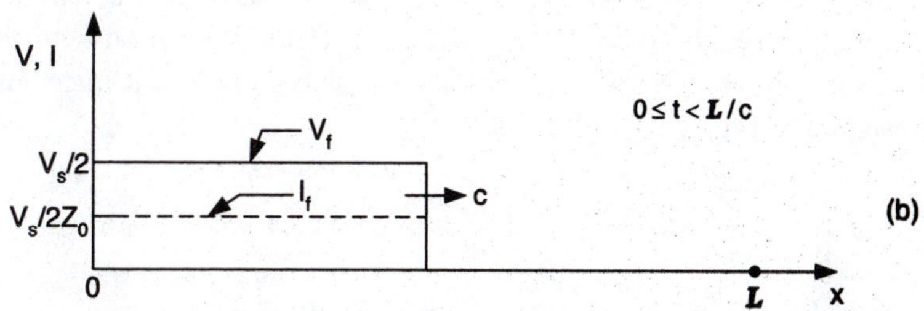

Fig. 3.9 Voltage and current pulses propagating down a transmission line.

When the pulses reach the end of the line at $t = L/c$, they encounter the terminating load $Z_L = R_L$. The boundary condition which governs the reflection process yields:

$$I = V/Z_L = (V_f + V_b)/Z_L \qquad \text{(h)}$$

Note, from Eqs. (3.56), (3.57) and (h) that:

$$V_f + V_b = Z_L(I_f + I_b) \qquad \text{(i)}$$

and

$$(V_s/2) - Z_0 I_b = Z_L[(V_s/2Z_0) + I_b] \qquad \text{(j)}$$

Solving Eq. (j) for I_b yields:

$$I_b = \frac{V_s}{2Z_0}\left(\frac{Z_0 - Z_L}{Z_L + Z_0}\right) \qquad \text{(k)}$$

and

$$V_b = -Z_0 I_b = -\frac{V_s}{2}\left(\frac{Z_0 - Z_L}{Z_L + Z_0}\right) \qquad \text{(l)}$$

It is clear that the signs of the reflected voltage and current pulses I_b and V_b depend upon the magnitude of Z_L compared to Z_0. Three cases should be considered which include (a) $Z_0 = Z_L$, (b) $Z_0 < Z_L$ and (c) $Z_0 > Z_L$. For case (a), where **the load impedance matches the characteristic line impedance**, there is no reflection and $V_b = I_b = 0$. In this case, the voltage and current for $t > L/c$ are given by:

$$V(x) = V_s/2 \quad \text{and} \quad I(x) = V_s/2Z_0 \qquad (3.58)$$

The voltage and current distributions along the line for this case are presented in Fig. 3.10a.

For case (b) where $Z_0 < Z_L$, reference to Eqs. (k) and (l) indicates that the reflected current will be negative and the reflected voltage will be positive. The combined forward and backward components for the voltage and the current during the interval $\mathbf{L}/c \leq t \leq 2\mathbf{L}/c$ are given by:

$$V = V_f + V_b = \frac{V_s Z_L}{(Z_L + Z_0)} \tag{3.59}$$

$$I = I_f + I_b = \frac{V_s}{Z_L + Z_0} \tag{3.60}$$

For case (c) where $Z_0 > Z_L$, Eqs. (k) and (l) show that I_b will be positive and V_b will be negative. The combine forward and backward pulses give the voltage and current as indicated in Eqs. (3.59) and (3.60), respectively. Graphs showing the voltage and current distributions for the cases where $Z_0 \neq Z_L$ are presented in Fig. 3.10b and 3.10c.

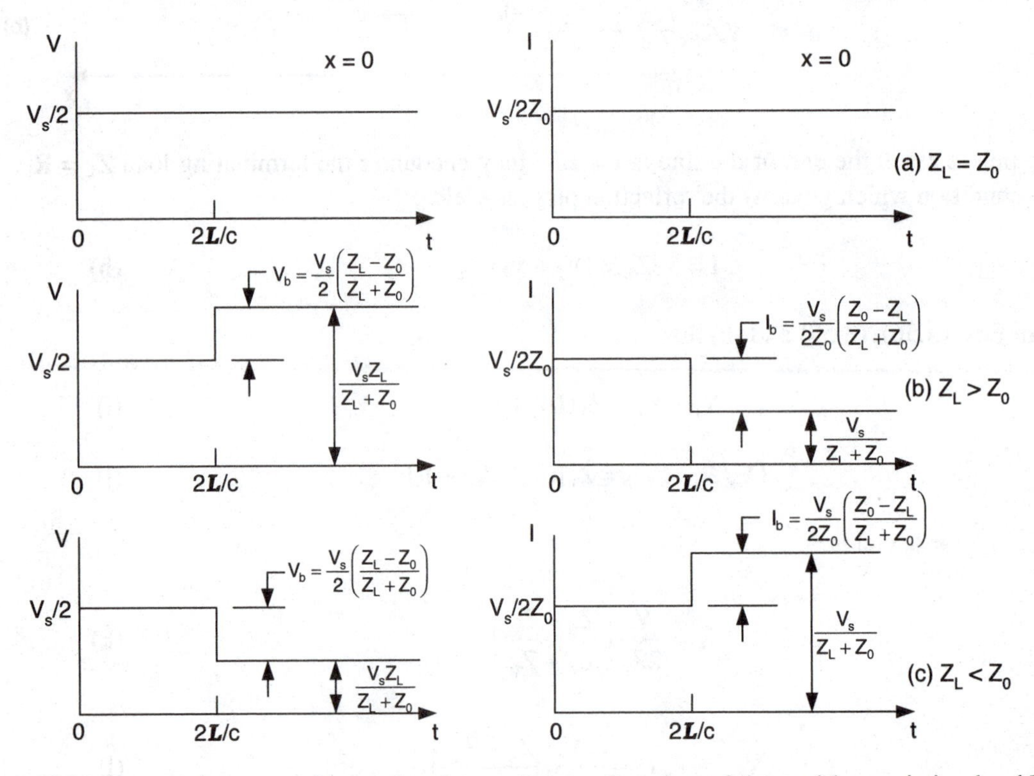

Fig. 3.10 Voltage and current observed at the source as a function of time with a resistive load R_L at the end of the transmission line where x = L.

3.8.2 Arbitrary Resistive Load at Both Ends of the Transmission Line

In the previous subsection, the source impedance was matched to the characteristic impedance of the line and reflections did not occur at x = 0. Consider here the more general case where $Z_s \neq Z_0$ and $Z_L \neq Z_0$ with the line not properly terminated at either end as indicated in Fig. 3.11.

When the switch S is closed at t = 0, the forward voltage V_f and current I_f are given from Eqs. (d) and (e) from Section 3.8.1 as:

$$I_f = \frac{V_s}{(Z_s + Z_0)} \tag{3.61}$$

$$V_f = \frac{Z_0}{(Z_s + Z_0)} V_s \tag{3.62}$$

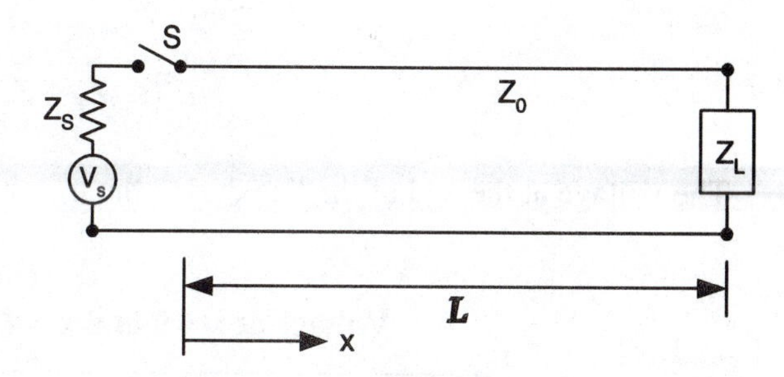

Fig. 3.11 Transmission line with a dc source voltage V_s with $Z_s \neq Z_0$ and a load impedance $Z_L \neq Z_0$.

This pulse propagates up and down the line and reflecting from both the load and the supply ends many times before steady state conditions are achieved. To show the voltage at either end of the line during this transition period it is convenient to define reflection coefficients C_V for the voltage and C_I for the current as:

$$C_V = V_r/V_i \qquad \text{and} \qquad C_I = I_r/I_i \tag{3.63}$$

where the subscripts r and i refer to the reflected and incident pulses respectively. Following the procedure described previously, it is evident that Eqs. (3.63) may be recast as:

$$C_{VL} = \frac{Z_L - Z_0}{Z_L + Z_0} \quad \text{for the load end} \tag{3.64}$$

$$C_{Vs} = \frac{Z_s - Z_0}{Z_s + Z_0} \quad \text{for the source end} \tag{3.65}$$

and

$$C_I = - C_V \tag{3.66}$$

Note, that the reflection coefficient depends upon the termination at the end of the line (source or load) where the reflection occurs. The reflection coefficients can be either positive or negative depending on the magnitude of Z_s, Z_L and Z_0.

Consider the first reflection which occurs at the load end of the line at t = ***L***/c. The reflected voltage is given by Eq. (3.63) as:

$$V_{r1} = C_{VL}V_i = C_{VL}V_f \tag{a}$$

and the total voltage at x = ***L*** is:

$$V_1 = V_{r1} + V_f = V_f(1 + C_{VL}) \tag{b}$$

The second reflection takes place at the source where $V_{r1} = C_{VL}V_f$. Hence, Eq. (3.64) gives:

$$V_{r2} = C_{Vs}V_{r1} = C_{Vs}C_{VL}V_f \tag{c}$$

and $$V_2 = V_{r2} + V_1 = V_f(1 + C_{VL} + C_{Vs}C_{VL}) \quad \text{(d)}$$

This process continues as the pulse propagates up and down the line reflecting from the ends. The reflected pulse amplitudes diminish in size and a steady state voltage V_{ss} given below is developed along the entire line after a large number of reflections.

$$V_{ss} = \frac{Z_L}{Z_L + Z_s} V_s \quad (3.67)$$

The voltage at the end of the line during the transition period is shown in Fig. 3.12 and Table 3.2.

Table 3.2
Voltage at x = 0 and x = *L* during the transition period

Reflection Number	Reflection Coefficient at x = *L*	Reflection Coefficient at x = 0	Total Voltage	Location x/*L*
1	C_{VL}	----	$V_s + V_{r1}$	1
2	----	$C_{Vs}C_{VL}$	$V_s + V_{r1} + V_{r2}$	0
3	$C_{Vs}C_{VL}^2$	----	$V_s + V_{r1} + V_{r2} + V_{r3}$	1
4	----	$C_{Vs}^2C_{VL}^2$	$V_s + \Sigma V_{rn}$	0
5	$C_{Vs}^2C_{VL}^3$	----	$V_s + \Sigma V_{rn}$	1

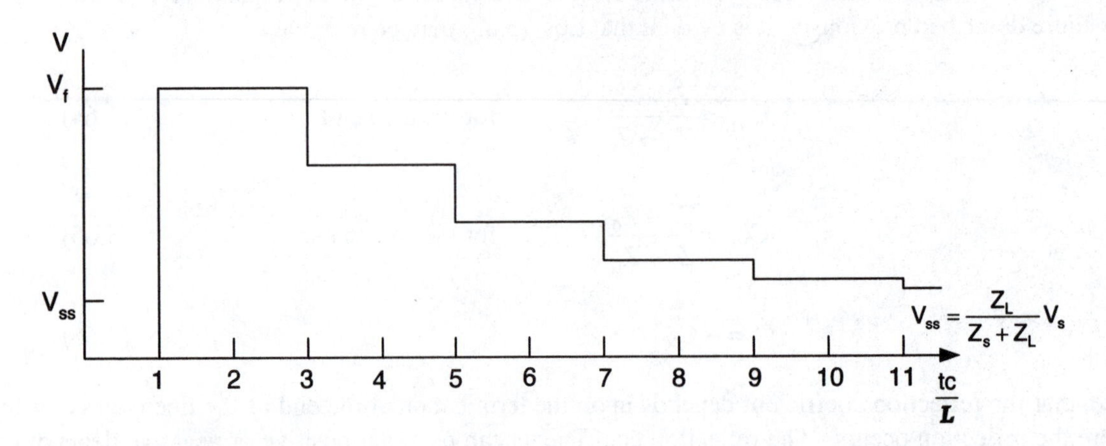

Fig. 3.12 Voltage at x = *L* as a function of normalized time during the transition period.

The voltage on the line may be approximated with an exponential relation of the form:

$$V = V_{ss} + (V_f - V_{ss})e^{t/\tau} \quad (3.68)$$

where the time constant τ is given by:

$$\tau = \frac{2\mathbf{L}/c}{\log_e(C_{Vs}C_{VL})} \quad (3.69)$$

Note, that the time constant τ will be negative since $C_{Vs}C_{VL} < 1$.

These results indicate that an improperly terminated transmission line requires a significant amount of time before the voltage stabilizes. This transition time represents a serious degradation of the line because the voltage must be stabilized at the load end in order to insure the delivery of a specified voltage pulse to the gate of a transistor. If the line were properly terminated with $Z_L = Z_s = Z_0$, then the time required for a stabilized signal to reach the load is limited to the propagation delay $t = \boldsymbol{L}/c$, which is much less than $t \geq 3\tau$.

3.9 REFLECTIONS FROM DISCONTINUITIES

In some circuits discrete components such as resistors, capacitors and inductors are placed in the line as illustrated in Fig. 3.13. These discrete components represent discontinuities in the line and affect signals propagating along it. Consider the resistor located in the line at position $x = \boldsymbol{L}_1$ as shown in Fig. 3.13 and note that the input voltage V_{in} to the resistor is:

$$V_{in} = V_f + V_b \tag{a}$$

where V_b is the pulse reflected from the resistor.

The voltage out of the resistor V_0 is given by Eq. (3.57) as:

$$V_0 = Z_0 (I_f + I_b) \tag{b}$$

and the voltage drop across the resistor V_R is given by:

$$V_R = R(I_f + I_b) \tag{c}$$

Combining Eqs. (a), (b) and (c) and noting that $V_f = V_s/2$ when $Z_s = Z_0$, leads to the relation for the reflected pulse as:

$$V_b = \frac{V_s}{2}\left(\frac{R}{2Z_0 + R}\right) \tag{3.70}$$

Adding the incident V_f and the reflected V_b pulses gives the total voltage as:

$$V = V_s\left(\frac{Z_0 + R}{2Z_0 + R}\right) \tag{3.71}$$

which appears at the source after a delay time $t_{pd} = 2\boldsymbol{L}/c$ as indicated in Fig. 3.14a.

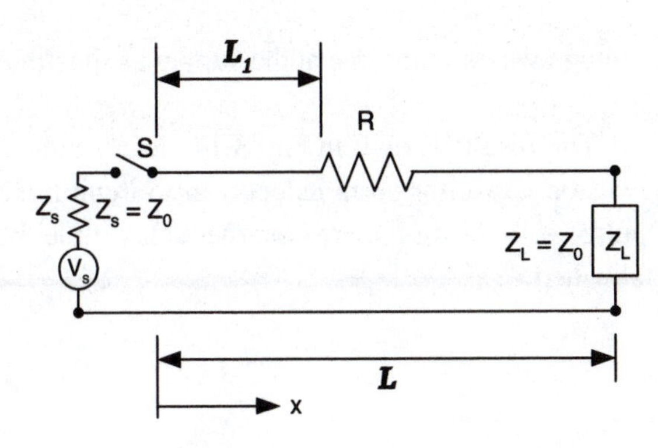

Fig. 3.13 Placement of a discrete resistor in a transmission line.

If a capacitor is placed in the line at position x = ***L***, replacing the resistor, the signal is initially grounded and the capacitor begins to develop a charge when the forward pulse encounters it. A reflected current pulse I_b is produced to provide the charging current where:

$$I_b = \frac{V_s}{2Z_0} e^{-t_1/Z_0C} \tag{d}$$

where t_1 is the time measured after the forward pulse reaches the capacitor. The sum of the forward and reflected voltage pulses give the voltage V as:

$$V = V_f - Z_0 I_b \tag{e}$$

This relation can be written as:

$$V = \frac{V_s}{2}\left(1 - e^{-t_1/(Z_0C)}\right) \tag{3.72}$$

This voltage signal arrives at the source after a delay time of $t_{pd} = 2\boldsymbol{L_1}/c$ as illustrated in Fig. 3.14b.

Finally, an inductor in the line at x = $\boldsymbol{L_1}$ produces a reflected voltage pulse:

$$V_b = L\frac{dI}{dt} \tag{f}$$

Noting that the current flow through the inductor is given by:

$$I = \frac{V_s}{2Z_0}\left(1 - e^{-(Z_0t_1)/L}\right) \tag{g}$$

Substituting Eq. (g) into Eq. (f) yields:

$$V_b = \frac{V_s}{2} e^{-(Z_0t_1)/L} \tag{h}$$

Adding the forward and reflected pulses gives the voltage as:

$$V = \frac{V_s}{2}\left(1 + e^{-(Z_0t_1)/L}\right) \tag{3.73}$$

The voltage varies with time at the supply as illustrated in Fig. 3.14c.

The results shown in Fig. 3.14 clearly indicate that the placement of a passive component such as a resistor, capacitor or an inductor in a circuit line, which acts as a transmission line, seriously affects the pulse voltage and increases the delay time before the stability of the voltage on the line is reestablished.

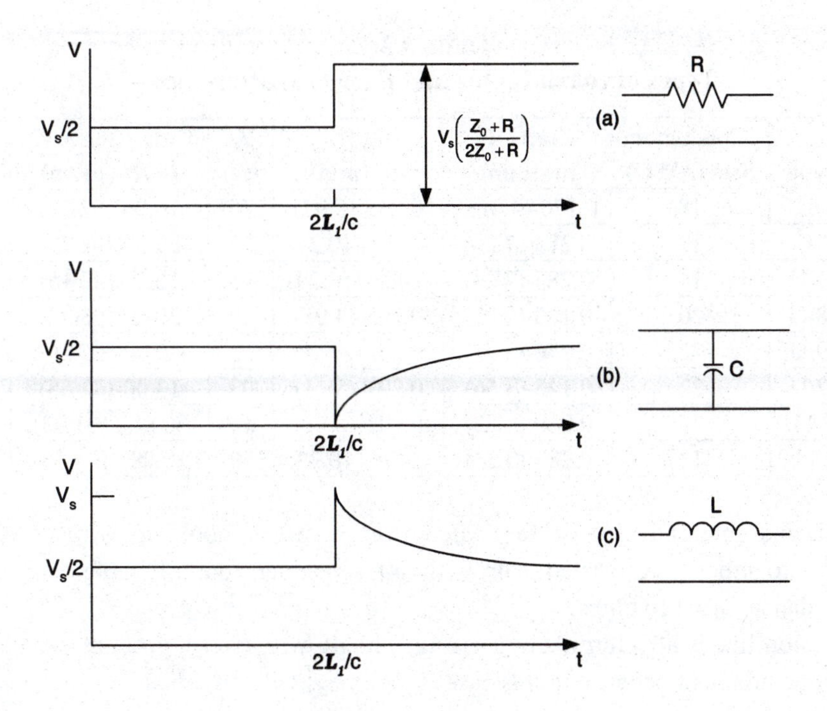

Fig. 3.14 Voltage time traces recorded at the source (x = 0) showing the influence of discrete components located at x = $\boldsymbol{L_1}$.

3.10 CHARACTERISTIC IMPEDANCE OF CONDUCTORS

A transmission line consists of two conductors separated by an insulating material. The characteristic impedance of the line depends on the geometry of the conductors, their spacing and the field properties, permittivity ε and permeability μ, of the insulating material. Coaxial cable is commonly employed as a transmission line because its construction leads to carefully controlled characteristic impedance. A coaxial cable, shown in Fig. 3.15, consists of a central circular conductor of radius a and shield of radius b. Insulating material fills the annular area between the two conductors. The characteristic impedance Z_0 of the coaxial cable is:

$$Z_0 = \frac{1}{2\pi}\sqrt{\frac{\mu}{\varepsilon}}\log_e\left(\frac{b}{a}\right) \tag{3.74}$$

A second insulating material which covers the shield does not affect the characteristic impedance. The outer insulation is use only to isolate the shield and to protect the cable assembly. Many different types of coaxial cable are commercially available and Table 3.3 provides a select listing of cables with Z_0 ranging from 50 to 125 ohms.

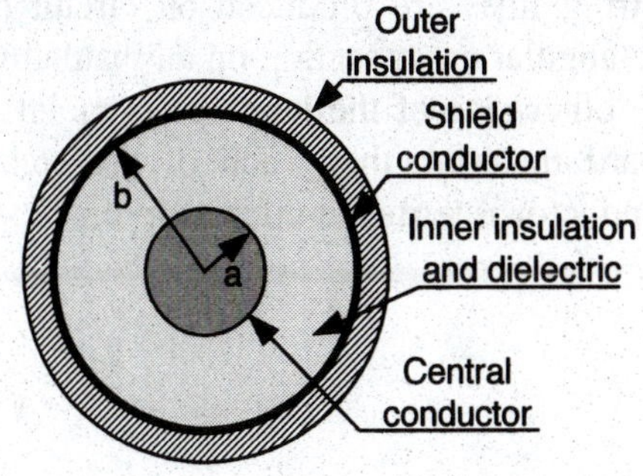

Fig. 3.15 Cross section of a coaxial cable

Table 3.3
Types of coaxial cable and their characteristics

Type	Conductor Size (AWG)	Core OD in.-(mm)	OD in.-(mm)	Z_0 (Ω)	Capacitance C' pF/ft-(pF/m)
6/U	18	0.170-(4.318)	0.233-(5.92)	75	16.00-(52.50)
8/U	13	0.285-(7.24)	0.405-(10.29)	52	29.50-(96.79)
11/U	14	0.285-(7.24)	0.405-(10.29)	75	16.20-(53.15)
58/U	20	0.166-(2.95)	0.195-(4.95)	50	30.00-(98.43)
59/U	22	0.146-(3.71)	0.242-(6.15)	73	20.50-(67.26)
62/U	22	0.145-(3.56)	0.201-(5.11)	93	13.00-(42.65)
174/U	26	0.060-(1.52)	0.103-(2.62)	50	30.08-(101.05)
213/U	13	0.285-(7.24)	0.405-(10.29)	50	30.08-(101.05)

A transmission line can also be fabricated from standard hook up wire (AWG 24-28) by twisting two wires together. A twisted pair with 30 turns per foot of cable length will exhibit characteristic impedance $Z_0 = 110$ ohms.

A transmission line is also formed by passing a single wire over a ground plane as indicated in Fig. 3.16. The characteristic impedance in this case is given by:

$$Z_0 = \frac{60}{\sqrt{\varepsilon_r}} \log_e\left(\frac{4d}{D}\right) \tag{3.75}$$

where ε_r is the effective dielectric constant of the material surrounding the wire. When wires are used for connections in a back plane, they are often placed in close proximity to a ground plane and each wire acts as a transmission line. The characteristic impedance of these wires cannot be accurately controlled and Z_0 will range from 75 to 150 ohms depending upon d and the proximity of adjacent wires.

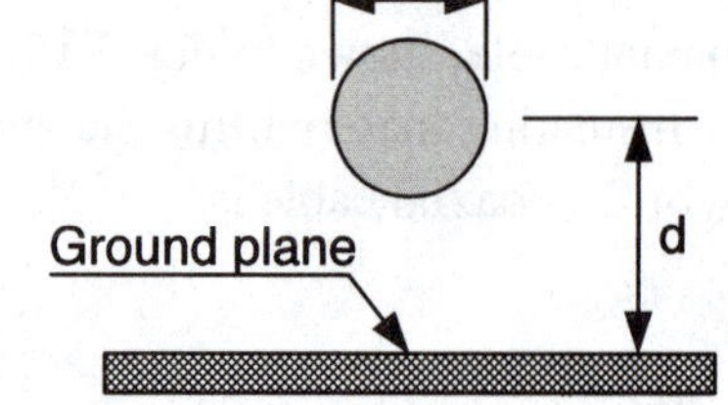

Fig. 3.16 Transmission line formed by a wire in close proximity to a ground plane.

3.11 TRANSMISSION LINES ON CIRCUIT BOARDS

Wiring lines are produced on circuit boards by an etching process which forms conductors with a rectangular cross section on an insulating laminate as illustrated in Fig. 3.17. If the copper cladding on the other side of the laminate is maintained as a ground plane separated from the signal plane by the board material, then each of the etched conductors represents a transmission line. This type of conductor is termed a microstrip line and its characteristic impedance is given by:

$$Z_0 = \frac{87}{\sqrt{\varepsilon_r + 1.41}} \log_e\left(\frac{5.98d}{0.8w + h}\right) \tag{3.76}$$

where ε_r is the relative dielectric constant of the board material (4.7 for the common glass-epoxy laminate G 10) and d, w and h[1] are dimensions defined in Fig. 3.17.

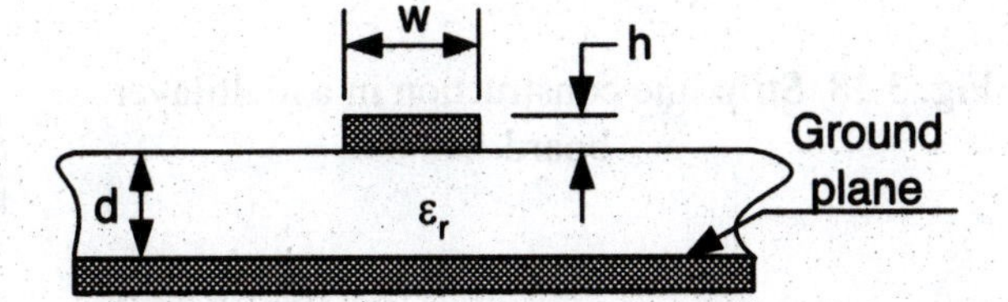

Fig. 3.17 Cross section of a microstrip line etched on a PCB with a ground plane.

The capacitance per unit length of microstrip line is a function of the line width w, the thickness d of the board and the permittivity of the board material as indicated by Eqs. (3.20) and (3.21). The capacitance per unit length for a 5 mil wide and 0.7 mil thick microstrip line formed on a glass-epoxy board with a thickness of 30 mils is given by:

$$C'_{msl} = \frac{w\varepsilon}{d} + \frac{2\pi\varepsilon}{\log_e(d/h)} = \varepsilon_0\varepsilon_r\left[\frac{w}{d} + \frac{2\pi}{\log_e(d/h)}\right] \quad \text{(a)}$$

$$C'_{msl} = (8.854)(4.7)\left[\frac{5}{30} + \frac{2\pi}{\log_e(30/0.7)}\right] \text{ in pF/m} \quad \text{(b)}$$

$$C'_{msl} = 76.51 \text{ pF/m} = 23.33 \text{ pF/ft} \quad \text{(c)}$$

The inductance per unit length L' of a microstripline is usually computed from the capacitance by using Eq. 3.44 to obtain:

$$L' = Z_0^2\, C' \quad (3.77)$$

The velocity of the pulse moving along the microstripline is given by:

$$c = \frac{1}{\sqrt{L'C'}} \quad (3.78a)$$

The propagation delay is given by:

$$t_{pd} = \mathbf{L}/c = \sqrt{L'C'} \quad (3.78b)$$

For a microstrip line the time delay can be written as:

$$t_{pd} = 1.017\sqrt{0.475\varepsilon_r + 0.67} \quad (3.79)$$

where t_{pd} is given in ns/ft.

It should be noted that the time delay depends only on the dielectric constant of the board material and is independent of the geometry of the circuit traces. For glass epoxy laminate with $\varepsilon_r = 4.7$, the time to propagate a pulse a distance of one foot is 1.73 ns.

[1] The thickness of copper cladding is expressed in ounces per square foot. One ounce copper cladding is 1.4 mil thick, 2 ounce cladding is 2.8 mil thick, etc.

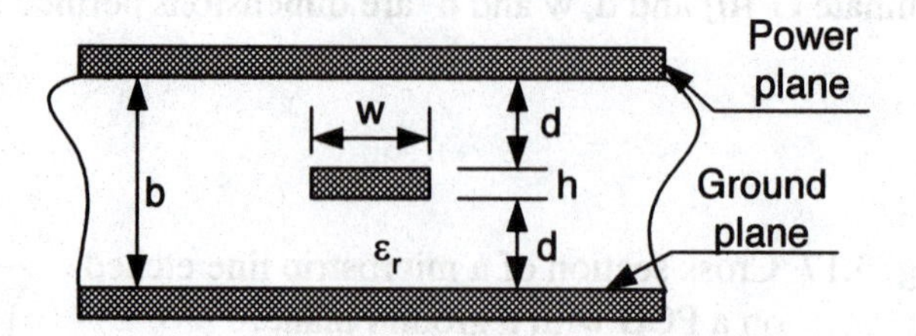

Fig. 3.18 Strip line construction in a multilayer board.

When circuit boards are fabricated with many layers, the signal planes are often sandwiched between a ground plane and a power plane as shown in Fig. 3.18. In this case, the wiring trace is called a strip line and it acts as a transmission line with Z_0 given by:

$$Z_0 = \frac{60}{\sqrt{\varepsilon_r}} \log_e \left[\frac{4b}{0.67\pi w(0.8 + h/w)} \right] \qquad (3.80)$$

where the dimensions b, w and h are defined in Fig. 3.18.

The capacitance per unit length C' for a stripline is dependent upon its line width w and spacing b between the power and ground planes, as shown in the graph of Fig. 3.19. The inductance is determined from Eq. (3.77). The delay time t_{pd} is determined in terms of ns from:

$$t_{pd} = 1.017\sqrt{\varepsilon_r} \qquad (3.81)$$

It is clear that t_{pd} depends on the relative dielectric constant of the board material. For glass-epoxy boards, t_{pd} = 2.26 ns/ft with stripline traces.

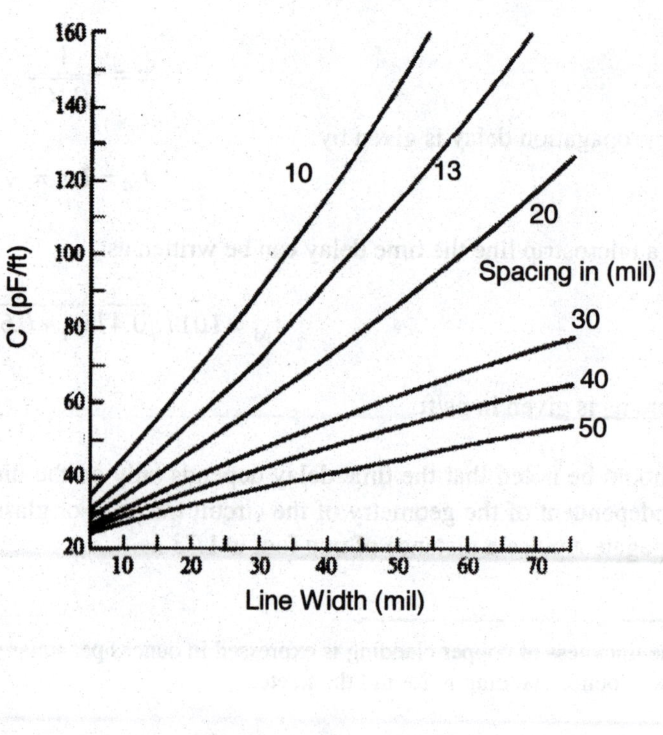

Fig. 3.19 Distributed capacitance C' for a stripline as a function of width w and the spacing between the ground and strip line.
G-10 material; 1oz Cu, ε_r = 4.7.

3.12 RESISTANCE OF PRINTED CIRCUIT LINES

The resistance R of a trace or wire of length $\boldsymbol{L}$ on a printed circuit board may be determined from Eq. (3.14) as:

$$R = \rho \frac{\boldsymbol{L}}{A} = \begin{cases} \dfrac{\rho \boldsymbol{L}}{wh} & \text{rectangular cross section} \\ \dfrac{4\rho \boldsymbol{L}}{\pi D^2} & \text{circular cross section} \end{cases} \tag{3.14a}$$

where w and h are the line width and thickness respectively and D is the wire diameter. The resistivity of copper is $\rho = 1.724 \times 10^{-6}$ ohm-cm.

The thickness of the copper cladding is usually specified in the number of ounces of copper per square foot of board area with 1 or 2 oz representing the common thickness corresponding to thicknesses of 0.00135 or 0.0027 in. respectively. Substituting these values of t into Eq. (3.70) gives:

$$R/\boldsymbol{L} = R' = 0.503/w \qquad \text{for 1 oz copper} \tag{3.82a}$$

$$R/\boldsymbol{L} = R' = 0.251/w \qquad \text{for 2 oz copper} \tag{3.82b}$$

where the line width is expressed in mils[2].

For a line width of w = 5 mils, the resistance per inch of a 1 oz copper line is about 25 milliohm/in. The line width required for digital signal transmission is not critical. In signal transmission the power required is quite small (a few mW), and the currents are of the order of a few mA. In these cases the line width is selected to ease the manufacturing process and to give higher PCB yields with w ranging from 5 to 10 mils. Circuit lines carrying power to the chips are much wider and the value of w is selected to provide adequate current capacity. The current capacity depends on the allowable temperature increase in the line. Results showing current capacity for microstrip lines as a function of line width are presented in Fig. 3.20 for conductors etched in 2 oz copper. If the lines are etched in 1 oz cladding, the current capacity is about 60% of that show in Fig. 3.20.

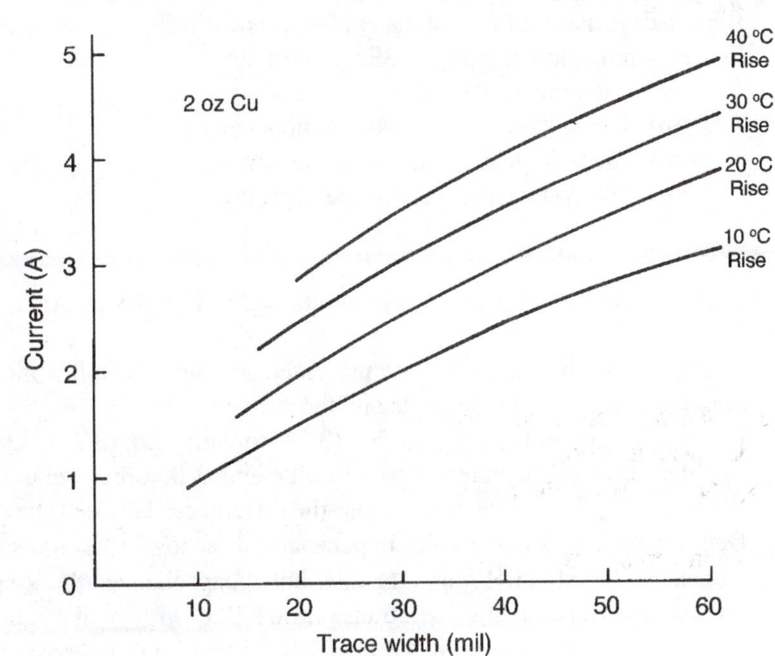

Fig. 3.20 Allowable current for surface traces as a function of their width and temperature increase.

[2] Dimensions of lines on printed circuit boards are usually expressed in mils. One mil = 1/1000 inch.

REFERENCES

1. Folland, G. B., Advanced Calculus, Prentice-Hall, Upper Saddle River, NJ, 2002
2. Haberman, R. Applied Partial Differential Equations, 4th edition, Prentice Hall, Upper Saddle River, NJ, 2004.
3. Saff, E. and A. D. Snider, Fundamentals of Complex Analysis, 3rd edition, Prentice Hall, Upper Saddle River, NJ, 2004.
4. Dally, W. J. and J. W. Poulton, Digital Systems Engineering, Cambridge University Press, New York, NY, 1998.
5. Dally, J. W., Riley W. F. and Mc Connell, Instrumentation for Engineering Measurements, 2nd edition, John Wiley, New York, 1993.
6. Nilsson, J. W. and S. A. Riedel, Introductory Circuits for Electrical and Computer Engineering, Prentice Hall, Upper Saddle River, NJ, 2002.
7. Cheng, D. K. Field and Wave Electromagnetics, 2nd edition, Prentice Hall, Upper Saddle River, NJ, 1989
8. Ramo, S., Whinnery J. R. and Van Duzer, T., Fields and Waves in Communication Electronics, John Wiley, New York, 1965.
9. Tugal, D. and Tugal, O, Data Transmission, 2nd edition, McGraw-Hill, New York, 1988.
10. Konsowski, S. G. and A. R. Helland, Electronic Packaging of high-Speed Circuitry, McGraw-Hill, New York, NY, 1997.

EXERCISES

3.1 If the high and low voltages shown in Fig. 3.1 are 1 and 0 volts respectively find the coefficients A_n and B_n in Eq. (3.1) for n = 0 to 3. Take t_r as 100, 10, 1, 0.1 and 0.01 nano-seconds. Show how the frequencies required to represent the pulse increase as the rise time decreases.

3.2 Repeat Exercise 3.1 but find the coefficients C_n in Eq. (3.1) instead of A_n and B_n.

3.3 For the circuit shown in Fig. 3.3Ex determine:

(a) The impedance.
(b) The voltage drop across the inductance.
(c) The voltage drop across the resistance.
(d) The voltage drop across the capacitor.

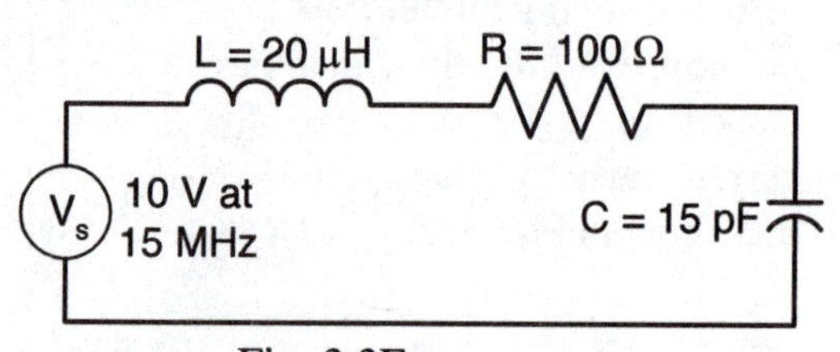

Fig. 3.3Ex

3.4 Find the phase angle ϕ for the circuit defined in Fig. 3.3Ex. Also, determine the current amplitude I_a.

3.5 Prepare a graph similar to that presented in Fig. 3.3 indicating the intercepts on the real and imaginary axes, the phase angle and the voltage V_a.

3.6 Use Eqs. (3.26) and (3.28) with Eq. (3.27) to verify Eq. (3.28), which is the telegraphers equation.

3.7 List the features illustrated in the circuit element that represents a segment of a transmission line shown in Fig. 3.5. Note in particular the differences between the circuits represented in Figs 3.2.

3.8 Determine the characteristic impedance of a low loss transmission line if the distributed inductance L' = 0.2 μH/ft and the distributed capacitance C' = 20 pF/ft.

3.9 For an open ended low loss transmission line with a distributed capacitance C' = 30 pF/ft, determine the impedance of the line if is 30 feet long and carrying a signal with a frequency of 500 MHz.

3.10 Determine the current flow in the open ended transmission line described in Exercise 3.9 if the voltage amplitude of the sinusoidal signal is 1.2 volts. What is the basic cause of current flow in an open ended line?

3.11 Consider a transmission line with a shorted end as illustrated in Fig. 3.8. If the line is 14 in. long with $L' = 5\ \mu H/ft$ and $R' = 2\ m\Omega/ft$, determine the line impedance. For a source operating at 2.2 V with a frequency of 400 MHz, find the current flow from the source to ground.

3.12 Determine the velocity of the electrical signal propagating in the transmission line described in Exercise 3.8.

3.13 Consider the transmission line illustrated in Fig. 3.10 with $Z_s = Z_0 = 50\ \Omega$ and $V_s = 3$ V dc. Determine the forward voltage and current when the switch is initially closed. Find the magnitude of the backward voltage and current pulse if the end of the line is terminated with a 20 Ω resistor.

3.14 If the transmission line of Exercise 3.13 is 28 in. long and the pulse is propagating at a reciprocal velocity of 2 ns/ft, construct a diagram similar to that presented in Fig. 3.10, which shows development of the voltage and current along the line for:

   (a) $0 \leq t \leq 2\boldsymbol{L}/c$.

   (b) $2\boldsymbol{L}/c \leq t \leq 4\boldsymbol{L}/c$.

3.15 Repeat Exercise 3.13 replacing the 20 Ω terminating resistor with a 75 Ω resistor.

3.16 Repeat Exercise 3.14 using a 75 Ω resistor at the source end of the line.

3.17 Determine the time constant τ for a transmission line with $Z_s = Z_0/3$ and $Z_L = 3Z_0$. Consider a line with $\boldsymbol{L} = 18$ in. and a reciprocal velocity $1/c = 2.5$ ns/ft.

3.18 If the supply voltage $V_s = 1.5V$ is applied to the line described in Exercise 3.17, find the voltage on the line as a function of time. Construct a voltage-normalized time diagram similar to that presented in Fig. 3.15 for this case. Compare the voltage predicted by Eq. 3.69 with that shown in your diagram.

3.19 A discrete resistor with $R = 1.0\ k\Omega$ is inserted into a transmission line with $Z_s = Z_0 = 50\ \Omega$. Determine the magnitude of the voltage pulse reflected from the resistor. Find the magnitude of the voltage which appears at the source after a delay time of $t_d = 2\boldsymbol{L}/c$.

3.20 Verify Eq. 3.72.

3.21 A capacitor with $C = 2\ \mu F$ is placed across the end of a 72 in. long transmission line. If $Z_s = Z_0 = 93\ \Omega$ with $V_s = 4V$, write an equation describing the voltage at the source after $t = 2\boldsymbol{L}/c$.

3.22 Replace the capacitor in Exercise 3.21 with a discrete in-line inductor $L = 120\ \mu H$ positioned at $x = \boldsymbol{L}$. Write an equation describing the voltage at the source after $t = 2\boldsymbol{L}/c$.

3.23 A bare copper wire is used as a telegraph line connecting two locations 100 miles apart. The wire is strung on poles and on the average the wire is 20 ft. above the ground. If the ground water is and average of 25 feet below the surface determine the characteristic impedance of the line. Note that the ground water serves as the second conductor of the transmission line.

3.24 A Teflon insulated wire with $\varepsilon_r = 2.3$ is supported by a copper ground plane. The wire is 22 gage with $d = 0.032$ in. and the diameter of the Teflon insulation is $D = 0.062$ in. Estimate the characteristic impedance of this wire.

3.25 A microstripline defined in Fig 3.17 is etched with a width $w = 6$ mils in copper cladding with a thickness of 1.0 oz/ft^2. The glass epoxy laminate separating the line from the ground plane is 30 mils thick. Determine:

   (a) Its characteristic impedance Z_0.

   (b) Its distributed capacitance C'.

   (c) Its distributed inductance L'.

   (d) The signal velocity.

3.26 A strip line defined in Fig. 3.18, with w = 5 mils is fabricated from 1 oz/ft^2 copper. This line is embedded between two ground planes with a spacing b = 20 mils. Determine the same four quantities as indicated in Exercise 3.25.
3.27 If the microstrip line in Exercise 3.25 is 6 in. long, determine its resistance R and distributed resistance R'.
3.28 If the strip line of Exercise 3.26 is 15 in. long, determine its resistance R and distributed resistance R'.
3.29 A micro-strip line 30 mils wide is used to distribute power to several chips on a PCB. If the maximum current carried by the line is 3A, estimate the increase in temperature of the line.
3.30 Describe why microstrip lines must be treated as transmission lines.
3.31 As a designer what factors do you control to establish the characteristic impedance of a strip line?
3.32 How does the electrical engineer control the source and load end impedance of a circuit line on a printed circuit card.

PART 2

PACKAGING

CHAPTER 4

FIRST LEVEL PACKAGING--- THE CHIP CARRIER

4.1 INTRODUCTION

The chip carrier is the housing for a thin and fragile silicon chip. The chip carrier serves several purposes. Firstly, it protects the chip from the detrimental effects of the environment (humidity and dust). Second it protects the chip from damage due to abusive handling in assembly. Third, it isolates the chip from the forces due to either shock or vibration that may occur in service. Fourth, it serves to facilitate the interconnection of the circuits on the chip which are extremely compact to the more widely spaced interconnecting pads or holes located on the circuit boards. Package design has a major impact on chip performance and functionality. Finally, the chip carrier serves to facilitate the assembly process by providing a rugged housing which can be handled with automatic machinery without endangering the chip. The chip carrier also provides pins, leads or pads which serve as bases for solder joints. These pins, leads and/or pads are designed so that the solder joints can be made with automatic soldering processes without danger of solder defects such as solder bridges between leads, joints with insufficient solder, solder splatter, or voids of significant size in the solder. The chip carrier is also involved in the heat transfer process. The heat generated on the circuit surface of the chip must pass through the chip and then through the carrier as the first step in the path of heat flow from the source to the heat sink.

There are several markedly different approaches to the design of a first level package. The two main stems of the different design approach are illustrated in Fig. 4.1.

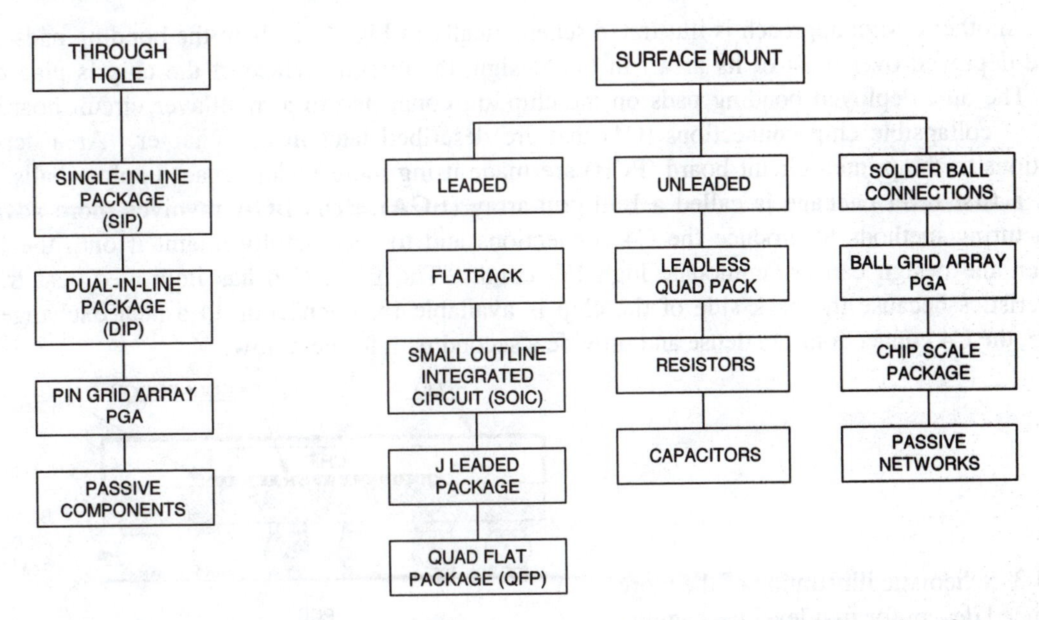

Fig. 4.1 The first level packages can be divided in two classes—through-hole and surface mount.

The materials used to fabricate the packages are also different depending on the application. Some chip carriers are made from molded plastic, others from ceramic and some modern chip carriers are constructed with laminated organic materials.

Still another distinguishing feature is the method of connecting the I/O pads on the chip to the first level package. The older method, still widely used, is wire bonding thin gold, aluminum or copper wires to connect the pads. In more modern approaches, the chip is flipped and C4 (controlled collapsible chip connections) are made between the pads deployed over the area of the chip and those on the chip carrier.

A final feature in the design of a chip carrier is its thermal resistance. Many logic type chips generate significant amount of heat (tens of watts) that must be dissipated without a large increase in the temperature of the chip. Some chip carriers are design with thermal vias and others with metal lids that contact the back of the chip and allow for attachment of a heat exchanger.

In more mature designs, the parts of the chip carrier include the chip as the central element, the case, the leads and lead frame, the chip to package bond, the bonding wires and the lid, as indicated in Fig. 4.2. In this chip carrier, the bonding wires are welded to bonding pads distributed around the perimeter of the chip. Many different types of chip carriers in use today conform closely in concept to the design shown in this figure. Unfortunately this concept, while commonly utilized, is deficient from two different view points. First, the flow of heat from the chip is severely limited as both the top and bottom surfaces are bounded by wiring planes which are thermal insulators. The dissipation of heat from this type of chip carrier is difficult and severe thermal penalties are encountered with chip carriers having wiring schemes that thermally isolate the chip. Second, the single row of bonding pads about the perimeter of the chip limit the number of I/O that this type of chip carrier can accommodate. As a consequence it is not suitable for many high performance chips with millions of transistors that require a very high I/O count.

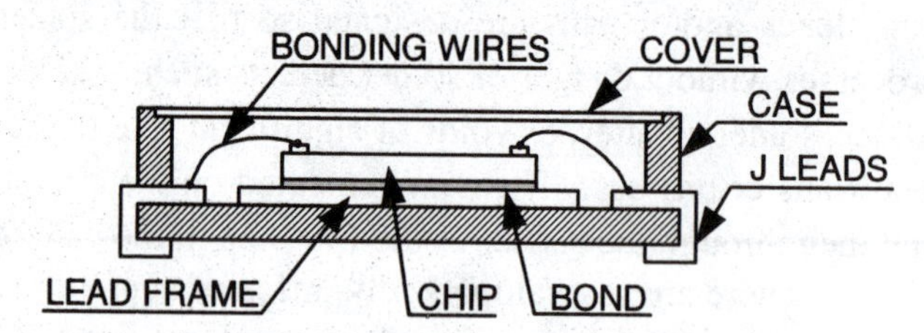

Fig. 4.2 Typical features of a traditional chip carrier with perimeter wiring connections.

Another design approach is illustrated schematically in Fig. 4.3, where the bonding pads on the chip are deployed over most of its area. In this design, the circuit surface of the chip is placed face down. The area deployed bonding pads on the chip are connected to a multilayer circuit board with controlled collapsible chip connections (C4) that are described later in this chapter. Area deployed connections to the printed circuit board (PCB) are made using more widely spaced solder balls. This type of a first level package is called a Ball-grid-array (BGA). The BGA involves more advanced manufacturing methods to produce the C4 connections and to successfully mount it onto the PCBs. However, the design can accommodate high I/O count. The BGA also has improved heat transfer characteristics because the back side of the chip is available for connection to a heat exchanger. In addition, the C4 connections are dense and provide a second path for heat flow.

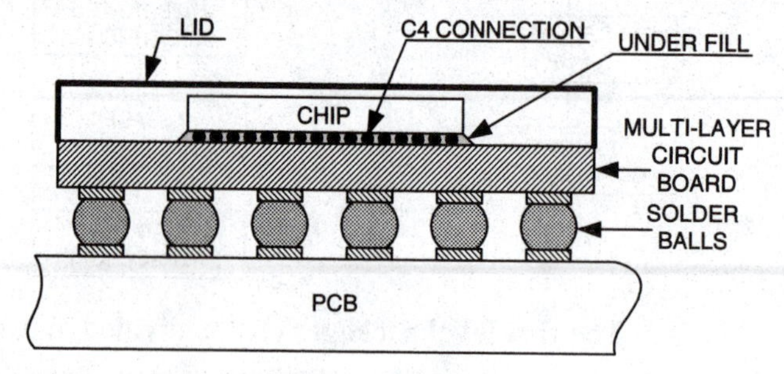

Fig. 4.3 Schematic illustration of the more advanced design for first level packaging.

Key design features for a chip carrier involve I/O count, package size (area and height), heat dissipation or thermal resistance, hermeticity and cost. I/O count is extremely important in housing modern Ultra Large Scale Integration (ULSI) chips which contain tens of millions of gates. Access to these gates through a high I/O count is required to fully utilize a high percentage of the capacity of the chip. These modern high-gate logic chips dissipate large quantities of heat and reliable operation of these dense chips requires that the chip carrier facilitate rather than impede the transfer of heat from the chip to a heat sink. Hermeticity is required to insure reliable operation of the chip over very long periods of time. Very small amounts of moisture which gain entry into the chip cavity of the carrier during assembly or by diffusion during operation permits corrosion of the metal wiring on the surface of the chip to occur. The presence of corrosion seriously degrades the chip circuits and reduces the life of the component. To eliminate the detrimental effects of moisture induced corrosion, some chip carriers are made of ceramic or glass-ceramic materials that prevent moisture entry by diffusion. Also, organic materials will outgas (release volatiles) with time; hence, they are prohibited within the chip cavity because the chemicals released when organics outgas can induce corrosion.

The common denominator of price and performance, typical of all of design philosophy for electronic systems, holds for chip carriers. Very low cost plastic chip carriers are available that are quite suitable for chips which dissipate low power, require modest I/O count to achieve maximum performance and do not require a high degree of hermeticity to insure suitable reliability for the product involved. However, for high performance systems operating at very high speed, the low cost plastic chip carrier is not adequate because it often does not have sufficient I/O count, its thermal resistance is much too high to handle the high power level required and its hermeticity is not sufficient to insure the degree of reliability required. At this time, significant research and development is underway to produce new chip carriers that are capable of housing modern high-speed, high-powered and high-I/O features found in the ultra-large-scale-integration (ULSI) chips which are currently under development.

4.2 TYPES OF CHIP CARRIERS

There are three general classifications used to describe chip carriers, which include:

1. Direct chip attachment to the printed circuit board
   - Chip and wire bonding
   - Flip-chip
2. Through-hole chip carriers
   - DIP
   - Pin-grid-array
3. Surface mount chip carriers
   - Leaded
   - Leadless
   - Ball-grid-arrays
   - Chip scale packages

Let's consider each of these classifications describing typical chip carriers that are currently employed.

4.2.1 Direct Die Attachment

Direct die attachment (DDA) sometimes called Chip-on-Board (CoB) or Wafer Level Packaging (WLP), involves an assembly process where the die is directly mounted on and electrically connected to the printed circuit board (PCB). The traditional first level package is eliminated. This process simplifies the design and manufacturing of the product, and improves its performance because of the shorter interconnection paths that result due to direct attachment.

There are two methods commonly employed to attach a chip directly to a printed wiring board—direct die attach with wire bonding and flip-chip. Direct chip attachment with these bonding techniques is described in the next subsections.

Chip on Board with Wire Bonding

One approach followed in eliminating the first level package is the Chip-on-Board (COB) that is illustrated in Fig. 4.4. In this approach, the back side of the chip is bonded directly to the printed circuit board using an adhesive. After the adhesive has cured, wire bonding, described later in this chapter, is used to make the electrical connections from the perimeter chip bonding pads to the bonding pads on the circuit board. After the chip has been tested to insure it is functioning satisfactorily, the chip and bonding wires are encapsulated with a plastic glob. This approach is low in cost and is often employed on low end products that are disposable. One difficulty is the possibility of failure due to differences in the thermal expansion of the silicon chip and the printed circuit board. To mitigate the differences in these thermal expansions, a compliant adhesive is usually employed to bonds the chip to the board.

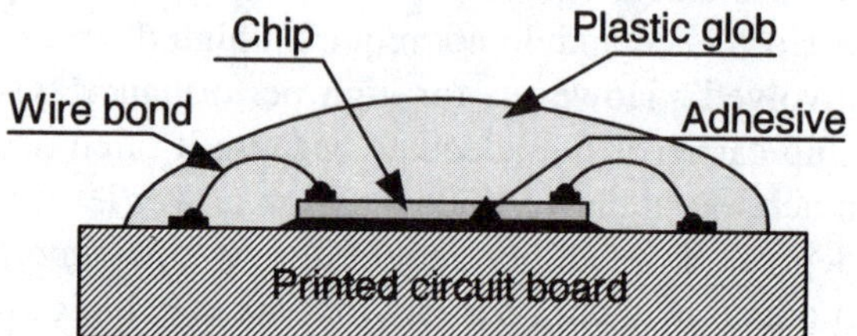

Fig. 4.4 Illustration of chip-on-board packaging that eliminates the first level chip carrier.

Flip-chip on-Board

Chips that have been processed with flip-chip bumps may be attached directly to a printed circuit board as illustrated in Fig. 4.5. The small solder balls on the back side of the chip produced by the bumping process (see Fig. 2.36) are placed over the corresponding bonding pads on the circuit board. The solder joint is made in a reflow oven that melts the solder balls forming joints that encase the solder pads. The solder mask prevents solder bridging. To mitigate the possibility of failure due to differences in the thermal expansion of the silicon chip and the printed circuit board, an underfill adhesive is placed between the chip and the board. The underfill adhesive is drawn into the small gap between the chip and the board by capillary action. After curing, the underfill adhesive also serves to seal the circuits from the environment. Hence it is not necessary to use an adhesive glob over the assembly. Because the solder bumping process enables the placement of solder balls on close centers over the entire area of the chip, higher I/O counts can be achieved than with the chip-on-board approach described above.

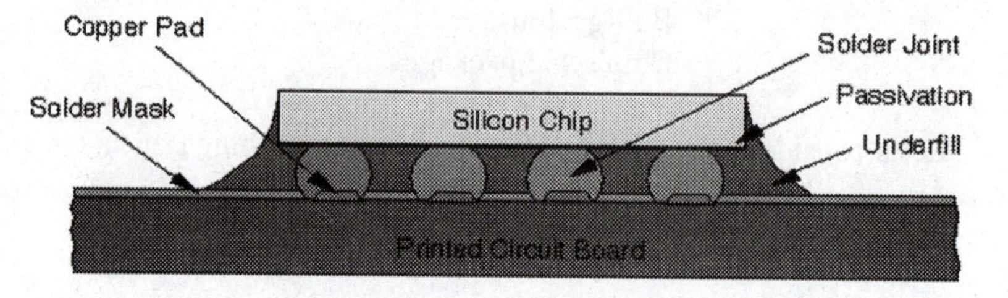

Fig. 4.5 Illustration showing the essential features used to eliminate the first level chip carrier by employing the flip-chip on-board approach.

The dies are prepared for direct attachment at the wafer level by forming small solder balls over the entire area of the chip. When the wafer fabrication process is complete, the wafers are re-passivated and a copper redistribution metal layer is patterned and sandwiched between layers of a low-dielectric passivation material. Then the wafer bumping process, illustrated in Fig. 4.6, involves the placement of small solder bumps directly to the circuit area of the dies. These solder bumps (small balls) are utilized to make solder connections to the pads on the PCB.

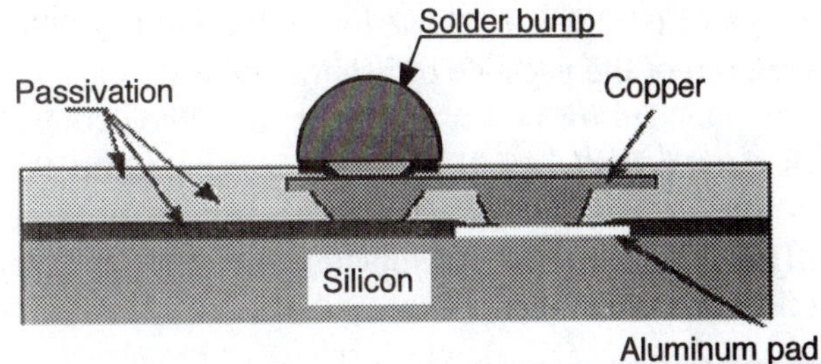

Fig. 4.6 Placement of a solder bump directly on a chip to facilitate direct attachment to a PCB.

Several different substrates are employed with direct die attachment. Ceramic and glass-ceramic substrates exhibit excellent dielectric and thermal properties. Organic substrates that weigh and cost less than the ceramic and glass-ceramic substrates are also used. These organic substrates have the advantage of lower dielectric constant. Flexible substrates that are pliable, have the ability to bend and support relatively dense circuits are used in some products.

The DDA process consists of only a few steps. For DDA with wire bonding, the chip is adhesively bonded to the PCB with its face up. The connections from the chip to the PCB are made with wire bonding. Then the die and wires are encapsulated with a plastic to protect it from the environment. For flip-chip bonding, solder bumps on the chip are aligned with the pads on the PCB and the mechanical and electrical connections are made by melting (reflowing) the solder bumps. If the PCB is made from an organic laminate, it is necessary to fill the small gap between the chip and the PCB. This process, known as underfilling, reduces the thermally induced shear strain on the solder bumps and extends the cyclic life of the product. The installation may be encapsulated to further protect the chip from mechanical and chemical damage.

4.2.2 Through-Hole Chip Carriers

The concept of through-hole chip carriers is illustrated in Fig. 4.7 where a cut away view of the chip carrier and circuit board is shown. With this approach the bonding pads on the chip are connected to leads that extend from the chip carrier. These leads are then inserted into plated-through-holes found on the circuit board. After insertion the leads are clinched and cut. The clenched leads hold the chip carriers in place on the PCB prior to soldering. The solder joints are made in a wave soldering process that is described in Chapter 6. The solder joints serve to mechanical fasten the chip carrier to the circuit board and to make the required electrical connections.

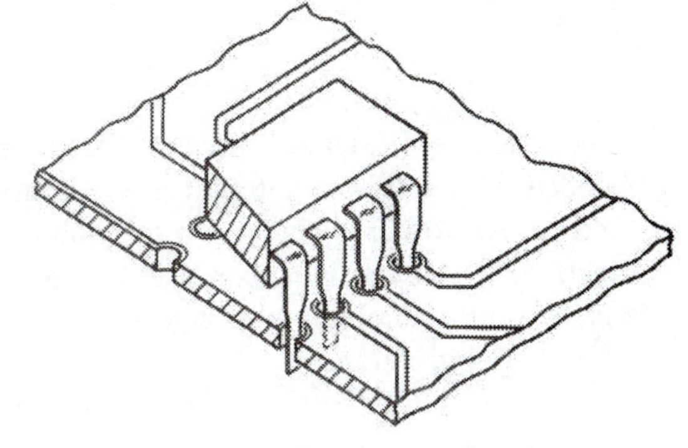

Fig. 4.7 Chip carriers with leads for through-hole mounting.

Metal Cans

The TO can is the oldest type of chip carrier and its shape reflects the transition from vacuum tubes to the design of packages to house low I/O count transistors on very small silicon chips. The pins emerge from bottom of the package on a pin circle. TO cans are still used today to house discrete devices, such as chips intended for power transistors or silicon control rectifiers where power dissipation is very high and the lead count is small. In some instances, the TO can is fitted with a flange that serves as a heat sink as shown in Fig. 4.8. The flange provides a large contact area for heat dissipation, and is bolted directly to a heat frame to complete a low resistance thermal path from the chip to the heat frame.

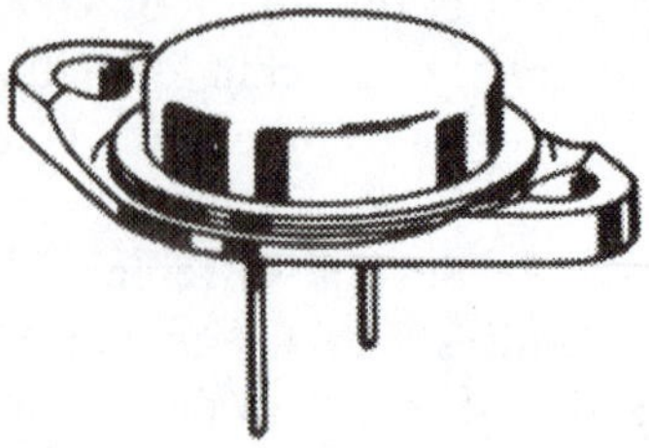

Fig. 4.8 A TO can with a heavy flange to enhance heat transfer from its base.

The DIP

One of the oldest types of chip carrier found in electronic packaging today is the DIP (dual in-line package) that is illustrated in Fig. 4.9. The DIP incorporates two rows of pins which are arranged on 100 mil (2.54 mm) centers along the longer side of the rectangular package. The standard DIP family of chip carriers is available in different sizes and accommodates I/O count from 8 to 64. The pins are short and relatively large in cross sectional dimension, which makes them stiff and robust. This robustness is an advantage in automatic assembly as the leads are inserted into holes on the circuit board by pick and place machines. The pins inserted through the circuit board are all soldered from the underside of the board by passing the assembled PCB through a standing wave of molten solder. With this wave soldering process, all of the solder joints on the board are made quickly and efficiently in a single pass through the wave. The width of the pins is increased near the body of the chip carrier to provide a shoulder that serves to seat the DIP a distance 20 to 40 mil[1] (0.51 to 1.02 mm) from the top surface of the circuit card. This stand off is necessary to provide a space under the carrier so that solder fluxes can be cleaned from the board after assembly is complete. Removal of the solder flux is important to prevent corrosion of the exposed metal on the PCB.

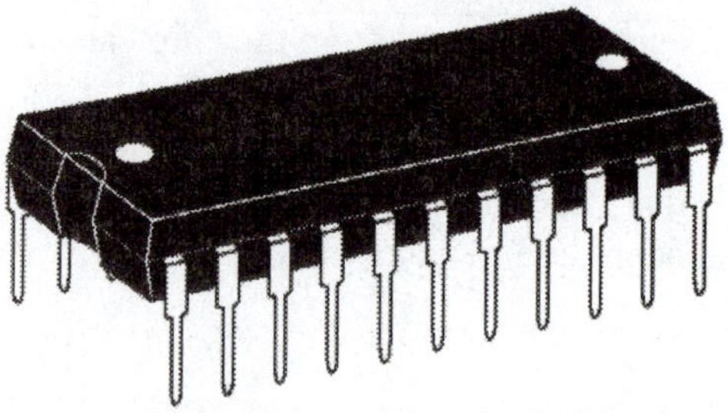

Fig. 4.9 Illustration of a 22 pin dual-in-line-package (DIP).

The internal details of the DIP are presented in the cut away view shown in Fig. 4.10. The chip is bonded, back side down, to a lead frame using an adhesive such as epoxy. The I/O pads on the chip are connected to the leads with wire bonding. The chip is then tested to insure that it is functioning correctly. The assembly is then placed in a two-side mold and a silica-filled epoxy is injection molded about the chip, bonding wires and leads. The leads are then clipped to size and plated to prevent corrosion during storage.

---

[1] Mil is still commonly used as a unit in specifying dimensions in the packaging business. One mil is equal to 0.001 in., or 0.0254 mm, or 25.4 μm.

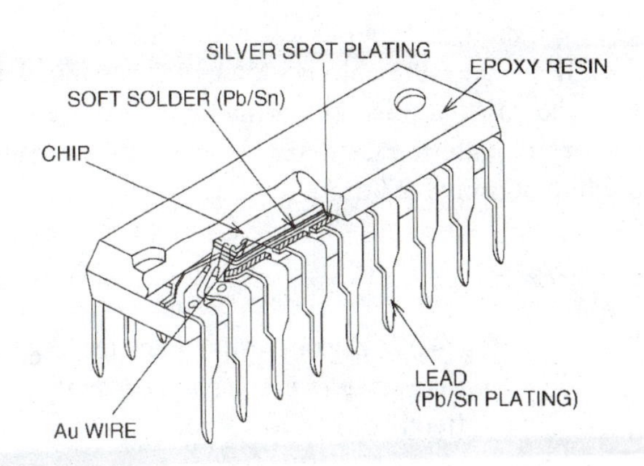

Fig. 4.10 Cut away view showing construction details of a DIP.

The standard DIPs have a lead pitch specified as 0.100 (2.54 mm) and the row spacing of 0.300 in. (7.62 mm). However, several DIP variants exist, mostly differentiated by packaging material and lead pitch.

- Ceramic Dual In-line Package (CERDIP)
- Plastic Dual In-line Package (PDIP)
- Shrink Plastic Dual In-line Package (SPDIP)—A shrink version of the PDIP with a 70 mil (1.778 mm) lead pitch

Plastic cases formed by injection molding are low cost and provide suitable housings for low and moderate cost product. Ceramic cases are used for product which has stringent hermeticity requirements. These ceramic cases are much more costly than molded plastic packages and their application is usually limited to military products that require a high degree of hermeticity to achieve the high reliabilities necessary for strategic systems. Ceramic chip carriers are also used with high performance commercial computers and signal processors where system availability is essential.

The DIP family has some disadvantages which include poor area efficiency, limited I/O count and poor wirability that limit its usefulness in housing modern high density logic chips. The area efficiency is poor because in most cases the pins are on 100 mil (2.54 mm) centers and deployed on only two sides of the package. In more area efficient chip carriers, the pins are on 50 mil (1.27 mm) centers (or less) and are deployed on all four sides of the package. The maximum I/O count on a standard DIP is limited to 64. To increase this I/O count without changing the pin pitch would require prohibitively long packages with significant lead lengths within the chip carrier. Wirability is poor because the through-hole mounting requires annular solder pads on the circuit board with an outside diameter of about 50 mil (1.27 mm). These pads leave only 50 mil (1.27 mm) for the wiring channel between pins, which cause a problem in providing for more than three wiring traces per wiring channel. A detailed description of wiring channels is presented in Chapter 5.

Other In-line Chip Carriers

The single-in-line package (SIP) is presented in Fig. 4.11. This package is similar to the DIP except it features a single row of pins on 100 mil (2.54 mm) centers. The chip in the SIP is mounted on a metal lead frame and encased with a filled plastic. It is used to house chips with relatively low I/O count.

Fig. 4.11 Single-in-line package (SIP).

Another through-hole chip carrier, known as a ZIP, is shown in Fig. 4.12. The ZIP is similar to the SIP because the chip is mounted perpendicular to the surface of the printed circuit board. However, with the ZIP the pins are narrow and staggered. While the pins are on 100 mil (2.54 mm) centers, the staggered pattern essentially reduces the projected pitch to 50 mil (1.27 mm).

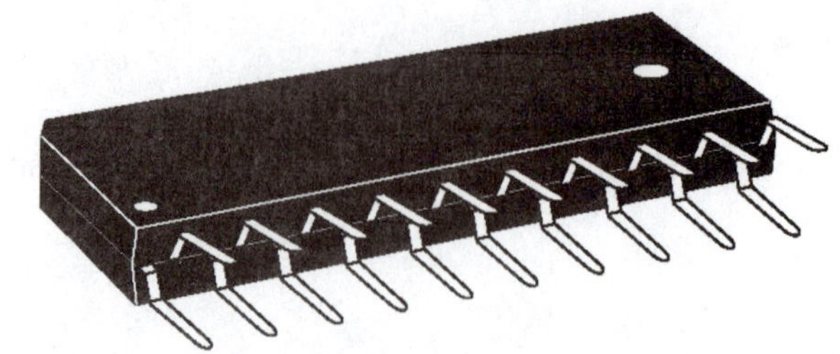

Fig. 4.12 Zigzag in-line package (ZIP) utilizes staggered pins to reduce their effective pitch to 50 mil (1.27 mm).

The Pin-Grid-Array

Pin-grid-array carriers are employed as a substitute for DIP's when additional I/O count is required or if lower thermal resistance is necessary. The pin array carrier, illustrated in Fig. 4.13, shows the general features of this design. The body of the package is fabricated from either organic laminate or ceramic (usually a multi-layer alumina) depending on hermeticity requirements. The body provides a cavity for the chip. A ledge around the cavity is used as the support surface for bonding pads that are the connections for the bonding wires which lead to the chip. Vias (holes containing conductors) lead from these pads to intermediate signal and power planes which carry the wiring traces from the vias to the pin locations. The pins which are 20 to 25 mil (0.51 to 0.64 mm) in diameter are brazed into the ceramic substrate[2]. After the chip is bonded in the package, a ceramic or Kovar (an alloy with a low coefficient of thermal expansion) lid is placed over the cavity and sealed using inorganic solders that do not outgas.

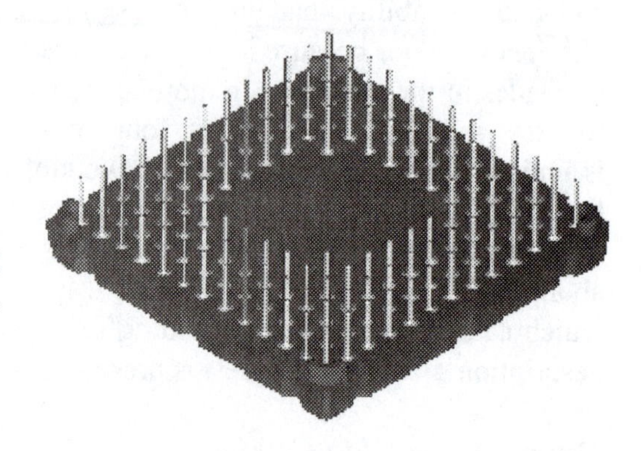

Fig. 4.13 Photograph of a pin-grid-array (PGA) showing pins deployed over the bottom surface of the chip carrier.

There are several variations in the design of the PGAs, mostly differentiated by packaging material and lead pitch.

- Ceramic Pin-grid-array (CPGA)
- Ceramic Micro Pin-grid-array (CμPGA)
- Organic Pin-grid-array (OPGA)
- Organic Micro Pin-grid-array (OμPGA)

---

[2] If the pin-grid-array (PGA) is fabricated from plastic, the pins are molded into the body of the package.

With the ceramic pin-grid-array, flip chip with C4 connections provide several advantages—increased I/O count, smaller chip size because the perimeter pads are not necessary, shorter electrical connections and improved manufacturing efficiency. The micro pin-grid-arrays are on 50 mil (1.27 mm) centers and consequently pin counts are increased by a factor of four for the same package size. The CμPGA are available with pin counts up to 940 in a package with dimensions of 7 × 40 × 40 mm.

Organic chip carriers are built with resin based materials such as the higher grade epoxy bismaleimide triazine (BT) reinforced with glass fibers, or the more traditional epoxy-glass circuit boards. The BT carriers evolved from printed wiring board technology and utilize plated-through-hole vias for their layer to layer connections. Another type of substrate for chip carriers, developed by IBM, is the called a surface laminar circuit, which uses a standard PCB to provide a high-density surface layer. This density is achieved by substituting microvias for the plated-through-hole vias thereby providing a two- to three-times reduction in the land diameter. This substrate design eliminates the relatively thick surface layer of copper foil that is used in the standard PCB design. With a lower total external copper thickness, finer lines and spaces can be used on package fan-outs as illustrated in Fig. 4.14. Organic chip carriers can achieve pin counts exceeding 1,000. Chips are attached either by thin gold wires with bond pad pitch spacings as low as 60 μm or less or by flip chip.

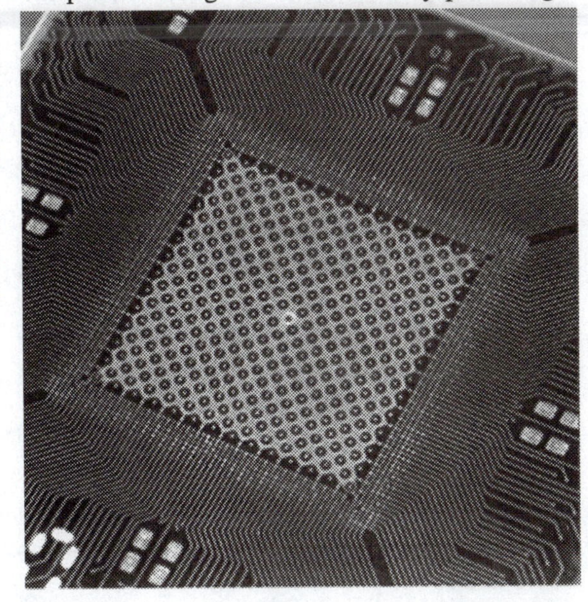

Fig. 4.14 Fine pitch circuit lines etched on organic laminate to produce the fan-out required for a OμPGA.

With pin-grid-arrays, intended for chips having a high power rating, the surface area directly below the chip is not used for pins. Instead, this area is reserved for a low resistance thermal path leading to a heat frame. The method of coupling the heat frame to the package is illustrated in Fig. 4.15. Through vias, located directly beneath the chip are filled with solder to provide a high conduction path to the bottom surface of the chip carrier. A heat sink is employed that contacts the bottom of the package and extends completely through the PCB to the heat frame. While pin-grid-arrays handle very large I/O count and can be designed to couple directly with a heat frame, they have two disadvantages. Firstly, they are relatively high in cost because the production processes for multi-layered ceramics are limited to a very few companies and the market is not competitive. The ceramic PGAs are usually employed on only high-end applications where hermeticity is important. Second, the cost of the PCB is elevated because of the need to drill and plate the large number of holes required for the pins. Lower cost chip carriers using solder balls instead of pins offer a lower cost alternative in many applications.

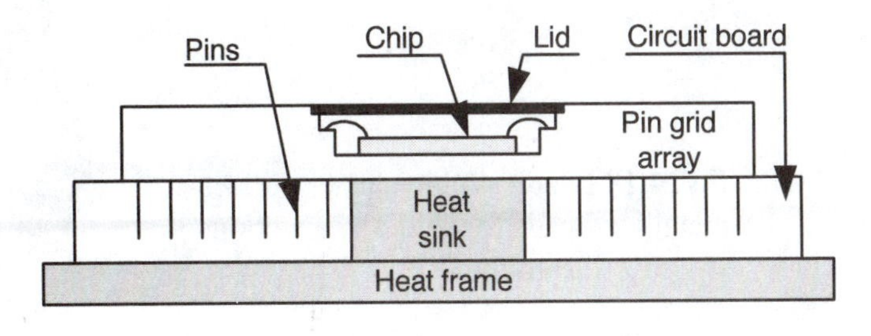

Fig. 4.15 Placing a heat sink directly below a PGA chip carrier facilitates heat flow through the circuit board to the heat frame.

Sockets

Sockets are used to accommodate either chip carriers or cable connectors. Sockets are incorporated into a circuit design when chip replacement is probable. For example, a circuit being developed for a prototype system is likely to encounter several changes. In these cases, new chips housed in pin-grid-arrays can be substituted by inserting them into an appropriate socket. Sockets are also useful when chip replacement or repair is anticipated. High I/O count chips are difficult to remove from the circuit board because of the large number of leads that must be unsoldered simultaneously. Sockets can also be used to accommodate chip carriers with pins on circuit boards with surface mounted components. The socket illustrated in Fig. 4.16 shows receptacles on one side to accommodate pins and pads on the opposite side for surface mounting to a circuit board.

Fig. 4.16 Top and bottom views of a socket to accommodate pin-grid-arrays on a circuit board with only surface mounted components.

4.2.3 Surface Mounted Chip Carriers

A major distinction between types of chip carriers pertains to the type of leads used for the I/O emerging from the carrier. In addition to the pins used in through-hole mounting of chip carriers, three other types of leads are currently employed. The first is a J-lead intended for connection to the surface of the circuit board at a solder pad as shown in Fig 4.17.

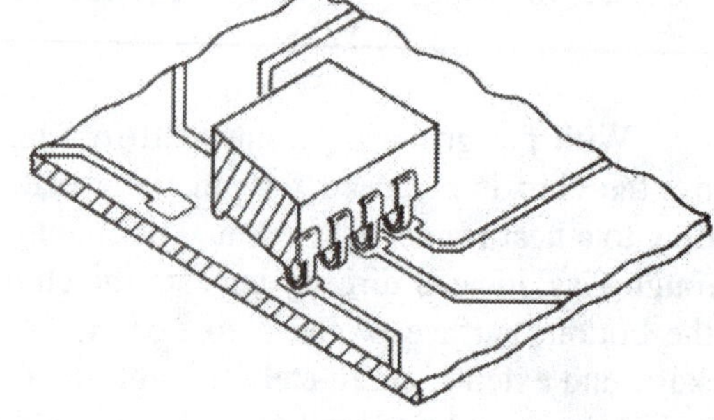

Fig. 4.17 Chip carrier with J-leads for surface mounting.

The second type is leadless, where metallized pads are provided on the chip carrier to provide soldering surfaces for connection to the circuit board. An illustration of this type of package is presented in Fig. 4.18.

Fig. 4.18 Leadless chip carrier utilizes metallized pads for solder connections to the PCB.

The third type of chip carrier is the Ball-Grid-Array (BGA), where solder balls are deployed over the area of the bottom of the chip carrier to provide the connections to the PCB as indicated Fig. 4.19.

Fig. 4.19 A ball-grid array (BGA) chip carrier showing solder balls deployed on its bottom surface.

There are a very large number of surface mounted chip carriers in use today. They have been in use for several decades and the number and size of the packages have proliferated. We will describe many of these surface mounted packages in more detail in the following subsections.

Discrete Device Packages

There are still many products developed that require small chips or passive devices such as resistors. These devices typically have a very low I/O count and are housed in Small-Outline Transistor (SOT) packages. Usually only one transistor on a small (30 × 30 mil) chip is encapsulated in an SOT. The power to be dissipated is small—typically less than 0.5 W. An example of a three lead plastic encapsulated SOT is presented in Fig. 4.20. Note in this case the leads are the gull wing type.

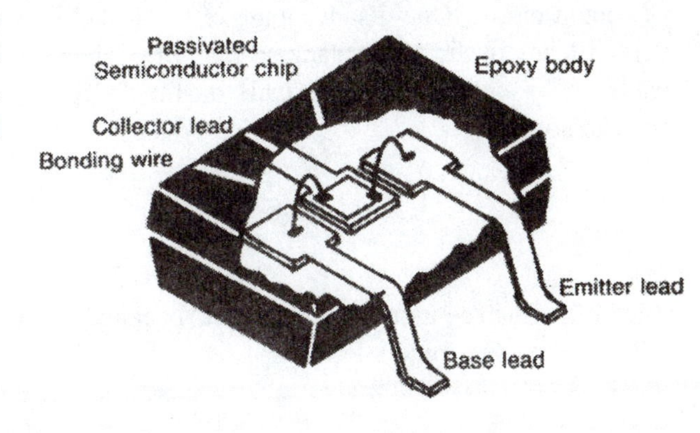

Fig. 4.20 Small outline transistor (SOT) package.

Discrete devices with higher power levels (greater than 1 W) are packaged in DPACKs. As shown in Fig. 4.21, the DPACK incorporates a heavy slug of copper that enhances the heat flow from the chip to the environment. Note a hole is provided in the copper slug so it can be bolted to a suitable heat sink. Silicon control rectifiers and power transistors, which require significant quantities of heat to be dissipated, are often housed in DPACKs.

Fig. 4.21 DPACK package for discrete devices with high heat dissipation requirements.

Small Outline Packages

One version of the small outline integrated circuit (SOIC) chip carrier is shown in Fig. 4.22. This package is similar to the DIP in that it incorporates leads deployed along both sides of the package; however, it differs in that the leads centers vary from 20 to 50 mils (0.5 to 1.27 mm) rather than the 100 mil (2.54 mm) centers found on the standard DIP. For some designs, the leads are shaped like gull wings and soldered to the pads on the PCB as indicated in Fig. 4.23. When placed on the PCB, the SOICs are held in place by the adhesion of the solder paste on the bonding pads or by a drop of epoxy that is placed under the chip carrier. All of the connections are soldered simultaneously in a soldering process. Surface tension in the melted solder assisting in aligning the gull wing leads with the bonding pads on the PCB.

Fig. 4.22 Gull wing configuration on a SOIC chip carrier.

Another surface mounted chip carrier is available which incorporates a J lead on 50 mil (1.27 mm) centers. This chip carrier, known as the small outline J (SOJ), is illustrated in Fig. 4.24. The J lead configuration, which essentially folds the lead into a small pocket on the underside of the package, reduces lead deformations during shipping and handling. The J lead is fully configured for assembly. It is possible to position the packages on the board, bond them with solder paste and make all of the solder joints simultaneously by using either a vapor phase or IR reflow soldering process. The J lead with a pocket protecting the lead and the solder joint connecting the lead to the circuit board pad is shown in Fig. 4.25. Clearly the J leaded chip carrier is superior to the conventional DIP in designs involving high I/O count chips. One disadvantage of the J-lead new package is the difficulty in maintaining all of the curved leads in the same plane so that all of the leads are in contact with the pads on the PCB prior to soldering.

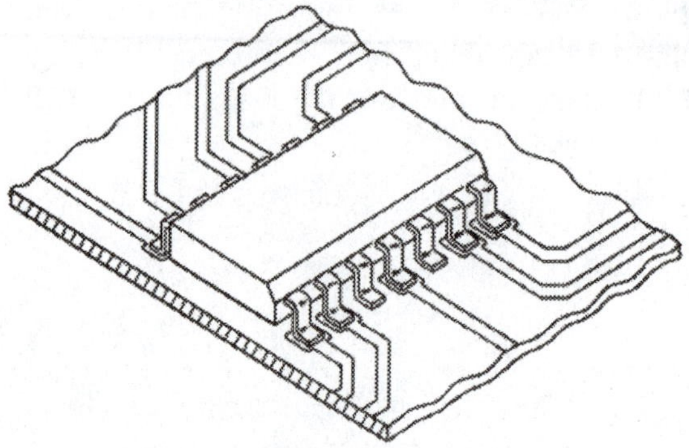

Fig. 4.23 Surface mounting of a SOIC package onto a PCB.

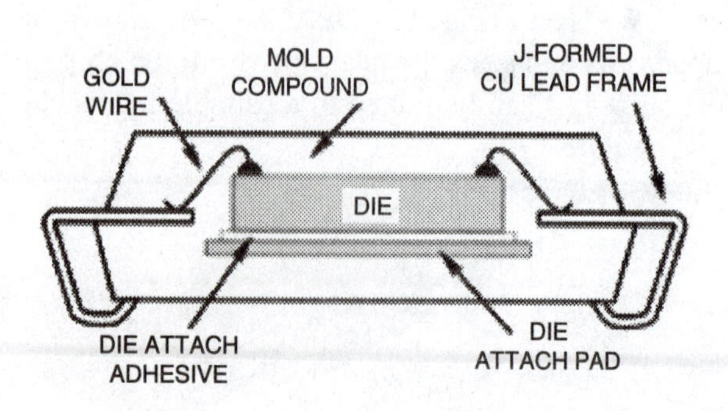

Fig. 4.24 Construction details of the SOJ—a J-leaded chip carrier.

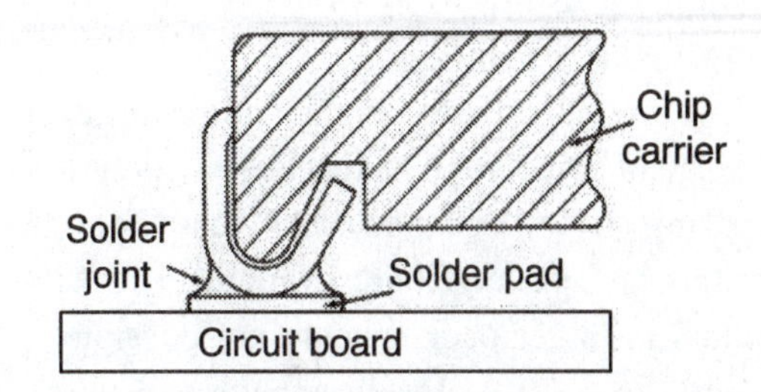

Fig. 4.25 Cross section of a J lead package showing the pocket protecting its lead, and the solder joint connecting the lead to the solder pad on the PCB.

Quad Flat Packs

With surface mounted chip carriers, the leads from the package are soldered to pads that are formed on the top surface of the circuit card as indicated in Fig. 4.17. The fact that a hole is not necessary to accommodate the mounting of the lead wire permits the leads to be placed on closer centers enabling the design of more area efficient chip carriers. With chip I/O count increasing and efficient use of circuit board area becoming more important, the advantages of surface mounting with its smaller package size has lead to many new developments in first level packaging.

The major advantage of the quad flatpack, shown in Fig. 4.26, is its relatively small size compared to DIP with the same I/O count. Because the leads are on closer centers and on all four sides, the quad flatpack is nearly four times as efficient as the DIP in utilization of circuit board real estate. Quad flatpacks have the advantage of reducing the size of second and third level packaging and improved electrical performance due to shorter lead length.

Fig. 4.26 A quad flatpack with gull wing leads on four sides.

The quad flatpack requires care in handling during the production process. Its leads are only 3 to 6 mils (0.076 to 0.15 mm) thick; hence, they are fragile and must be protected to prevent damage or misalignment during assembly. The plastic quad flatpack (PQFP) was developed to provide better protection for its leads. Its design incorporates corner bumpers that extend beyond the leads to provide some degree of protection. The PQFP is intended for low-cost, high-lead count applications. The small lead-pitch of a PQFP enables this design to accommodate higher lead count chips than possible in PDIP, PLCC, and SOIC packages. The lead pitch of the PQFP varies from 15 to 40 mil (0.4 to 1.0 mm) depending on lead count and package size. When the benefits of the PQFP package configuration were realized, the design was extended to lower lead counts. Lead counts ranging from 44 to 304 are now commercially available. In manufacturing this package, the chip is bonded to the die pad of a lead frame, and then wire bonding or TAB is used to complete the connections between the perimeter pads on the chip and the fingers on the lead frame. A silica-filled epoxy is injection-molded to encapsulate the device, bonding wires and lead frame. The leads are trimmed and formed in a gull-wing formation. A drawing of a 100 lead PQFP is presented in Fig. 4.27.

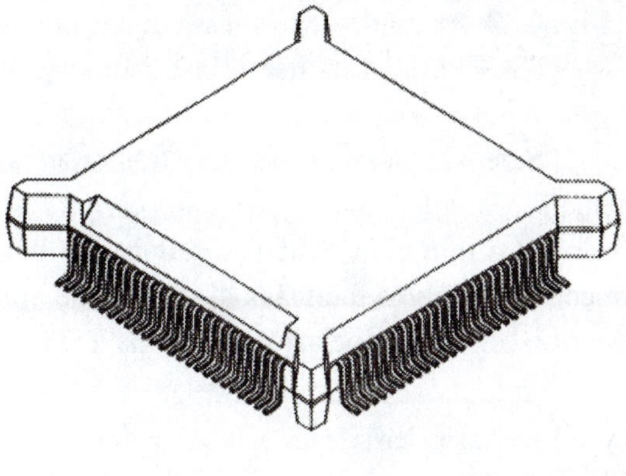

Fig. 4.27 Drawing of a 100 lead PQFP with a 25 mil (0.63 mm) lead pitch.

Quad Flat J Packages

The quad flat J type chip carrier, sometimes referred to as (PLCC), is illustrated in Fig. 4.28. This drawing shows J shaped leads on all four sides of the chip carrier. The leads are on 50 mil (1.27 mm) centers and packages are available with lead counts of 18, 20, 28, 32, 44, 52, 68 and 84. Higher lead counts in larger packages are not possible because the flatness of the packages (and the circuit boards) cannot be control closely enough to insure that all of the leads contact the solder pads. Simultaneous contact of the leads to the solder pads is essential to insure quality solder joints.

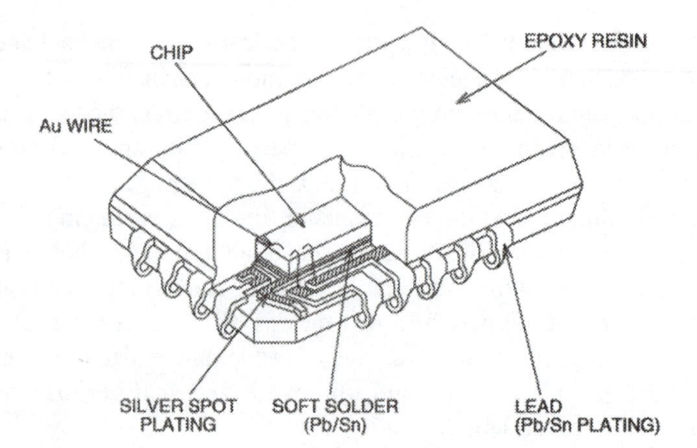

Fig. 4.28 Drawing showing fabrication details of a quad flat J chip carrier.

The J-leaded chip carriers cannot be used to house very high I/O count chips; however, they are much easier to handle than the fine pitch gull wing type chip carriers.

Thin Small Outline Package (TSOP)

Recent developments of portable electronic products such as cell phones, music recorders and cameras have driven the design of thin small outline package (TSOP) to house certain types of chips. A typical example of a Type I package[3] is presented in Fig. 4.29. These are surface mounted chip carriers with gull wing leads. The lead pitch is usually 20 mil (0.5 mm).

Fig. 4.29 Example of a thin small outline package (TSOP) used to house a flash memory chip.

Size and weight are very important design considerations in developing portable electronic products. Smaller and lighter chip carriers are demanded. To meet this demand, the manufacturers developed a thin chip carrier to reduce the height required above the printed circuit board. They also reduced the pitch of the leads to reduce the area required on the circuit board. An example of a height reduction of four to one afforded by the TSOP compared to a PQFP is shown in Fig. 4.30.

---

[3] Type I packages have leads extending from their shorter side and Type II packages have leads extending from their longer side.

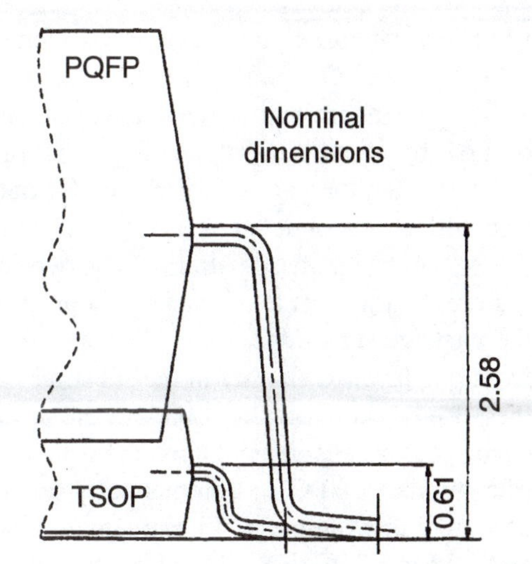

Fig. 4.30 Height reduction achieved in the dimension above a circuit board by using TSOP chip carriers instead of PQFP chip carriers.

4.2.4 Leadless Surface Mounted Chip Carriers

Leads from a chip carrier are a nuisance because they are fragile and easily deformed in shipping and handling. It is possible to eliminate the leads entirely by replacing them with metallized pads fixed directly on the chip carrier as shown in Fig. 4.18. These metallized pads are usually on 50 mil (1.27 mm) centers and provide an area efficient chip carrier. The pads on the chip carrier are connected to corresponding pads on the circuit board by using solder paste that is melted in a soldering process described more completely in Chapter 6.

The area efficiency of the leadless chip carrier is mitigated by a very serious disadvantage involving solder joint failure. The circuit board assembly is subjected to large thermally induced strains due to a mismatch of the coefficients of expansion of the chip carrier and the circuit board materials. Thermal cycling may be encountered in operation due to power on power off, or thermal cycles may be imposed as a manufacturing screening test or in qualification testing. During these thermal cycles, the solder joints are exposed to thermally induced shearing strain of relatively large magnitude. The effect of this strain is to induce fatigue failures of the solder joints. The failures begin with the solder joints located at the corners of the chip carrier and then move inward to more centrally located solder joints. The number of thermal cycles required to initiate fatigue failure depends on the temperature extremes involved in the thermal cycling, the rate of temperature change, the mismatch in the coefficients of thermal expansion and the size of the chip carrier.

4.2.5 Solder Ball Surface Mounted Chip Carriers

Ball-grid-array

The ball-grid-array (BGA), shown in Fig. 4.19, is similar to the pin-grid-array (PGA) except that solder balls have been deployed on its bottom surface instead of pins. The BGA solder balls have a distinct advantage over the PGA's pins in that they can be placed on much closer centers. The solder ball-grid-array on the underside of the package is used to connect the chip carrier to the PCB substrate. The grid spacings have been standardized under JEDEC guidelines. The most commonly used grid spacings are 32, 40 and 50 mil (0.8, 1.0, and 1.27 mm), although finer pitches are sometimes employed in BGA designs. Solder ball diameter varies from 20 to 35 mil (0.50 to 89 mm). With flip chip connections, a typical commercially available BGA with a 45 mm square body supports a 44 × 44 array of solder balls to provide an I/O count[4] of 1728.

There are many BGA packages, but the die for all types is connected to its substrate by the wire or TAB bonding, illustrated in Fig. 4.31, or by the flip–chip direct attachment that is shown in Fig. 4.32. BGA is often the package of choice for optimizing device electrical performance. They are lightweight, thin with short connections, and they minimize the use of board space.

A schematic illustration of a two-layer, plastic-encapsulated PBGA showing many of the design features of this first level package is presented in Fig. 4.31. The chip is connected to the wiring pads on the package substrate with gold wires. Wiring connections are made using vias which carry the signals from one layer to another in the substrate. Solder balls bonded to the bottom surface of the substrate are then used to connect the PBGA to the printed circuit board in a solder reflow operation. The diameter of the solder balls depend upon their pitch, with diameters that are 50 to 60% of the pitch dimension. The number of I/O in commercially available PBGAs depends on the pitch and the size of the chip carrier. For example, a 40 mm square PBGA with a 50 mil (1.27 mm) pitch is available with 564 I/O and a 35 mm square PBGA with a 40 mil (1.0 mm) pitch is available with 1156 I/O.

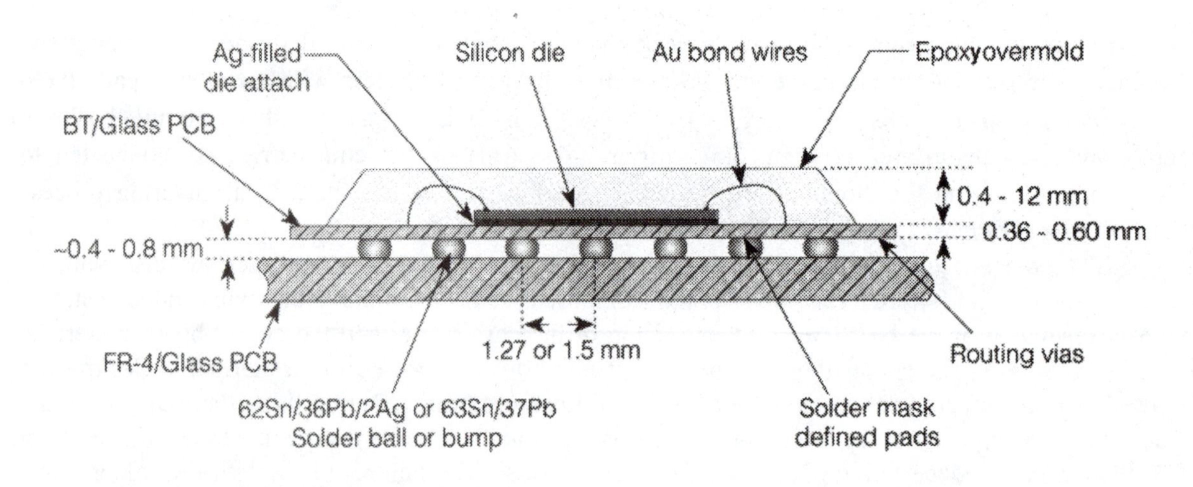

Fig. 4.31 Schematic illustration of wire bonding connections in a two-layered PBGA.

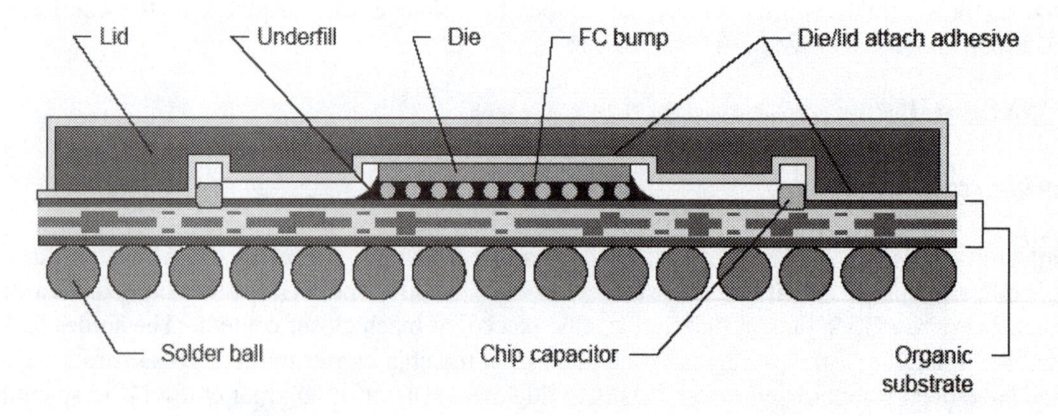

Fig. 4.32 Schematic illustration of flip chip C4 connections in a four-layered BGA.

[4] An array of 44 × 44 solder balls yields an I/O count of 1936. The lower number cited here reflects the fact that the solder balls under the chip were omitted in this design.

Ball-grid-array packages are available in ceramic, plastic and organic laminate materials. The plastic ball-grid-array (PBGA) employs a wire bond to connect the bonding pads on the chip to the organic substrate as shown in Fig, 4.31. In this design, the solder balls are a lead free composition containing tin, silver and copper. The flip chip ball-grid-array (FCBGA), shown in of Fig. 4.32 utilizes and area array of C4 connections on the chip that are bonded to an organic substrate. Two other versions of ball-grid-arrays are available; both use flip chip connections to ceramic substrates. The first is the CBGA which is superior to the PBGA because it is hermetically sealed. A higher melting point solder (90% Pb and 10% Sn) is used in fabricating the solder ball[5]. Note that a lead free solder is used to bond the ball to both substrate and the PCB. A column version (CCGA) is also available where the solder balls have been replaced by short 20 mil (0.5 mm) diameter solder columns as shown in Fig. 4.33. The column heights vary from 50 to 87 mil (1.27 to 2.2 mm). These columns are used to space the ceramic chip carrier some distance away from the PCB. The increase in this dimension enables more effective cleaning and inspection. More importantly the added height of the solder joint decreases the shearing strains due to thermal cycling thereby increases its fatigue life.

Fig. 4.33 Column grid array with columns replacing solder balls to improve solder joint fatigue life.

The ball-grid-array with flip chip C4 connections lends itself to improved thermal design. As shown in Fig. 4.34, the narrow space, between the back of the die and the copper lid that serves as a heat spreader, is filled with thermal grease. This design reduces the thermal resistance between the chip and the cooling medium passing over the chip carrier. The flip chip with its C4 connections enhances heat flow downward through the substrate. The solder balls on close centers provide a low resistance path to the PCB.

Ball-grid-arrays exhibit several advantages when compared to leaded chip carriers, which include:

1. Higher I/O capability than surface mounted packages with either gull or J leads and indicated in Fig. 4.35.
2. Higher I/O for a specified footprint results in increased circuit density.
3. Reduced weight, size and cost because smaller packages enable more chips placed on a PCB.

---

[5] An exception is provided in the RoHS directive that allows solder with a high lead content (above 85%) in chip carriers.

4. The area array of the solder balls permit larger lead pitches for high I/O packages greatly reducing soldering problems.
5. Higher speed circuits because the connections are shorter with less resistance and inductance.
6. Improved heat dissipation (see Fig. 4.34).
7. BGA manufacturing methods are easy to extend to multi-chip modules.
8. Manufacturing processes are compatible with existing stencil printing and robotic mounting equipment.
9. Manufacturing advantages include:
   - Reduced coplanarity problems
   - Self centering during reflow reduces placement requirements
   - Larger pitch reduces solder bridging during paste printing
   - Elimination of fragile leads reduces handling problems
   - Higher yields achieved with BGAs during assembly

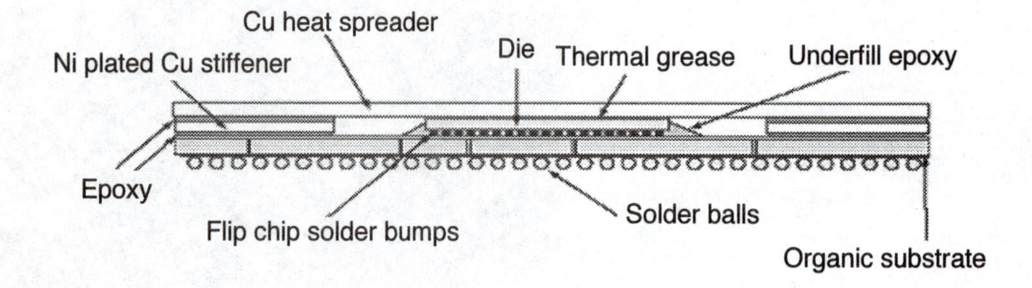

Fig. 4.34 A BGA with flip chip bonding designed to enhance heat transfer from the chip's back side.

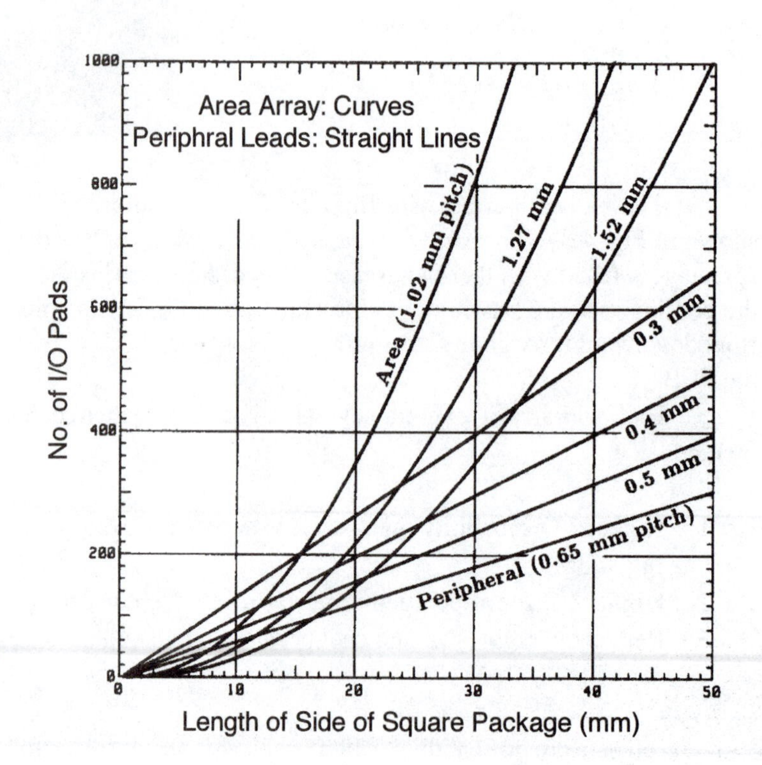

Fig. 4.35 I/O count as a function of package size for peripheral and area deployed bonding pads with different pitch.

There are some disadvantages associated with using BGA chip carriers that include

1. It is not possible to rework individual solder joints.
2. It is impossible to visually inspect the solder joints beneath to package except for CCGA column packages.
3. The rigidity of the chip carrier and the solder joints increases the possibility of solder joint failure by thermal cycling.

Chip Scale Packaging

Pressures continue to build for the chip manufactures to develop product that will enable designers to develop systems in smaller, thinner and lower weight envelops. Of particular importance is the market for portable products such as cell phones, laptop computers, music recorders and players, and video cameras. Current packaging trends are illustrated in the block diagram presented in Fig. 4.36. This diagram begins with attempts to improve the packaging density of the quad flat pack with a lead pitch of 20 mil (0.5 mm). Attempts to move to 16 mil (0.4 mm) pitch have introduced soldering difficulties that have escalated assembly costs. Direct chip attachment with the flip chip process has been hampered by the technical challenges in testing, handling and placement of bare dies. Chip scale packaging (CSP) has developed as a suitable transition step between fine pitch leaded chip carriers and direct attachment of the chip to the circuit board with flip chip.

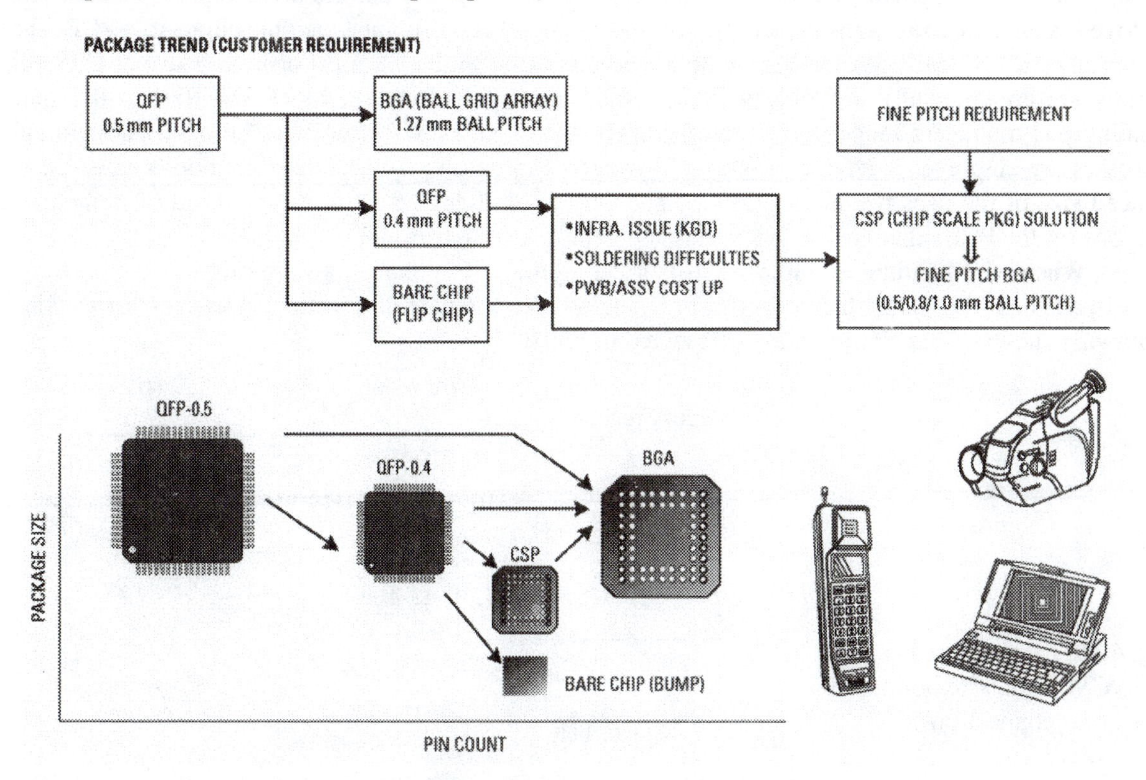

Fig. 4.36 Diagram showing trends to reduce the size and weight of chip carriers for portable electronics.

The Chip Scale Package (CSP) has evolved from the ball-grid-array, and is similar in many respects. One important distinction is the area of the chip relative to the area of the CSP. The package area of the CSP is limited to 1.2 times the chip area. This compares to a package area for either a BGA or a fine pitch quad flat pack that is four times the chip area. The photograph of a BGA presented in Fig.

4.37 shows the large area surrounding the chip that is reserved for the array of solder balls. It is clear that the development of the CSP chip carrier requires a marked reduction in the area of the chip carrier.

Fig. 4.37 The area of a BGA is about four times the area of the chip that it houses.

The CSP is a small package intended for chips with a lower I/O count. It employs solder balls deployed in an area array on its bottom surface for electrical and mechanical connections to a PCB. The definition of a CSP calls for packages with a lead pitch of 32 mil (0.8 mm) or less. Current CPS chip carriers are commercially available with lead pitches varying from 32 to 16 mil (0.8 to 0.4 mm). Bonding pad diameters range from 8 to 16 mil (0.20 to 0.40 mm). Both wire bonding and flip chip processes are employed with this package. However, flip chip is the preferred method because of the reduced size of the package that it affords. The very fine pitch CPS requires advanced manufacturing methods for the PCBs that involve microvias and very thin microstrip lines.

When wire bonding is employed in fabricating the CSP, connections are often made to a single row of pads deployed about the perimeter of a single layer laminate that forms the base of the package. A drawing showing this construction is presented in Fig. 4.38.

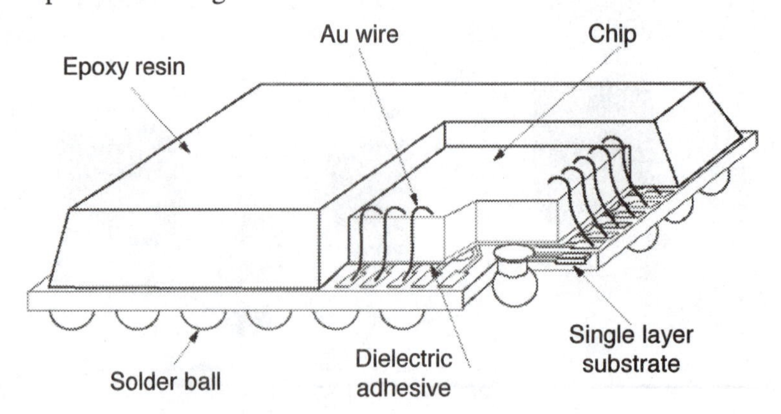

Fig. 4.38 Fabrication details for a CSP with wire bonded connections.

Fabrication details for a CSP with flip chip connections are presented in Fig. 4.39. This illustration shows that the back of the chip is exposed; however, the front of the chip with the circuits is protected by the substrate and the underfill adhesive. Flip chip bonding enables chips with higher I/O count to be placed in CSP and is superior to wire bonding for most applications.

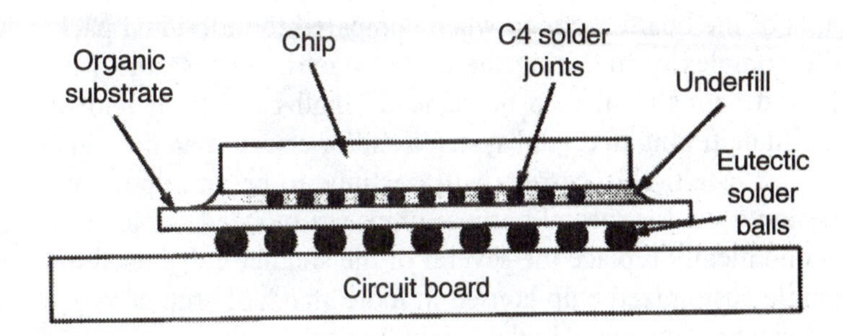

Fig. 4.39 Design of Motorola's slightly larger than IC chip package using flip chip bonding.

For chips with relatively low count I/O CSP offers several advantages when compared to direct chip attach, which include:

- Speed of testing the chip.
- Superior die protection
- Better reliability without underfill
- Much easier to rework
- Superior yield due to easier handling and assembly

4.2.6 Multi-chip Modules

In most cases, electronic systems are assembled using a large number of components that are packaged as separate items. These individual components are mounted on a circuit card or cards and connected together with PCB wiring, connectors and cabling. In some cases, it is disadvantageous to package all of the components individually because of the need for added space for the connections and the loss of performance due to wiring length induced delays. It is often advantageous to house several devices which form a circuit function in a single multi-chip package.

Hybrid packages or multi-chip modules (MCM) are first level carriers which house several chips and other passive components together and provide the wiring necessary for circuit connections. Multi-chip modules are less important today than they were a decade ago, because modern chips have higher circuit density and systems on a chip have become more common. Nevertheless, MCM offer advantages when housing more complex products that perform more than one function. An example of a MCM is shown in Fig. 4.40. This module implements a 32-bit microprocessor, an ASIC co-processor with built-in set of interfaces, 6 Mbytes of protected instruction memory and 32 Mbytes of protected data memory. It is designed in a radiation tolerant technology qualified for space applications.

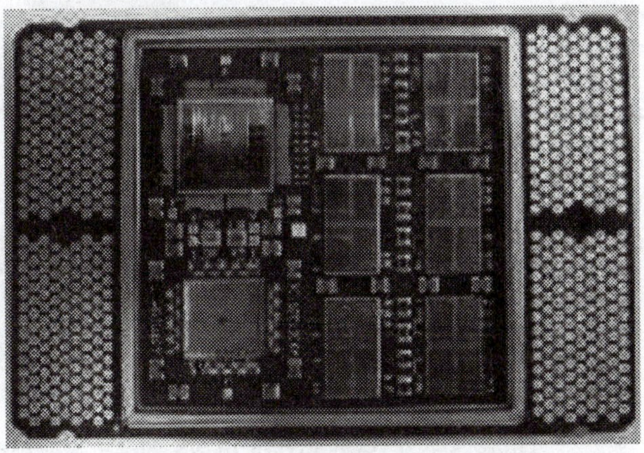

Fig. 4.40 Multi-chip module containing logic and memory chips as well as passive components.

In a sense the MCM is a combination of a first and second level package. It provides the protection function of a first level package and the wiring function of the second level package. The hybrids offer three advantages. In collecting together several relatively small chips, MCMs reduce the area of the board required when compared to individual packages for each chip. Secondly, they reduce the complexity in wiring the circuit board, because the wiring between the chips is completed in the hybrid. Finally, the on board lead lengths are short and often with closely controlled impedance to facilitate timing in digital systems and/or circuit matching in analog systems.

Multi-chip carriers will continue to be an important first level packaging concept until custom design costs for special purpose chips are reduced sufficiently for the development of custom chips that economically replace the several of the smaller chips used in a MCM. Replacement of an MCM with a single customized chip housed in more standard chip carrier will provide a lower cost assembly, a more efficient design and a higher performance circuit.

The Thermal Conduction Multi-chip Module (TCMCM)

The thermal conduction module (TCM) was a multi-chip package designed and produced by IBM for use in its high performance computers in the early 1980s. The concept was adapted and modified by other firms for the design of the CPU for main frame and other large computers. The TCM has been replaced by a more modern multi-chip carrier that embodies the basic concepts introduced 25 years ago. In this text, we will discuss the newer design of the thermal conduction multi-chip module (TCMCM) in this chapter and in Chapter 9, because this method of packaging represents a major step forward in efficient design of electronic systems. The TCMCM is a very efficient (area and wiring length) chip carrier where the functions of the first and second level packages have been combined. In Chapter 9, we will consider the merits of this package in dissipating very large quantities of heat with a small thermal penalty.

The components involved in an assembly of a TCMCM that houses four chips for a high-performance IBM server, are shown in Fig. 4.41.

Fig. 4.41 A four chip multi-chip module used in an IBM high-performance server.

There are several layers in the stack making up this module. At the top, a heavy aluminum cover serves as a heat sink and a support for heat exchangers. Below the cover, two of the four SiC heat spreaders are visible. These heat spreaders are on top of the chips and serve to provide more effective heat conduction to the heat sink. The chips are not visible because they are covered by the heat spreaders that are much larger than the chips. The chips are bonded to a glass-ceramic substrate with

C4 connections. The next layer in the stack is a land grid array that contains an array of springs that connect the pads on the glass-ceramic substrate to the circuit board. The spring connections (fuzz buttons) are required because the coplanarity of the surfaces of the large glass-ceramic substrate and the circuit board is not sufficient for C4 connections. The spring connections also enable the module to be disconnected and reconnected to the printed circuit board.

This TCMCM house four high-performance chips each with 170×10^6 transistors. Connections on the chip are made with several layers of copper traces. Each chip is attached to the glass-ceramic substrate with 7018 C4 solder connections. The substrate, which is made from 84 layers, incorporates 1.7×10^6 internal copper vias and an area array of 100 μm contact pads placed on 200 μm centers on its top surface. The glass-ceramic substrate also contains 190 meters of co-sintered copper wiring. Off module pads (5,100) on a one millimeter pitch are located on the bottom surface of the substrate to provide connections to a unique land grid array. The package 85 by 85 mm in size contains up to 32 processors that operate at frequencies from 1.1 to 1.3 GHz. The module is capable of dissipating up to 624 W of heat.

Stacked Die Memory Packaging

In recent years, efforts to reduce area required on circuit board for memory chips have lead to the development of stacked dies in a single package or stacking of dies in separate CSP packages. An example of Amkor's stacked CSP is presented in Fig. 4.42. In this arrangement, a smaller memory die is adhesively bonded to a larger one. Bonding pads about the perimeter of both chips are available for wire bonding. When wire bonding is complete, the assembly is encapsulated and solder balls are plated and reflowed over the array of pads located on the bottom side of the package.

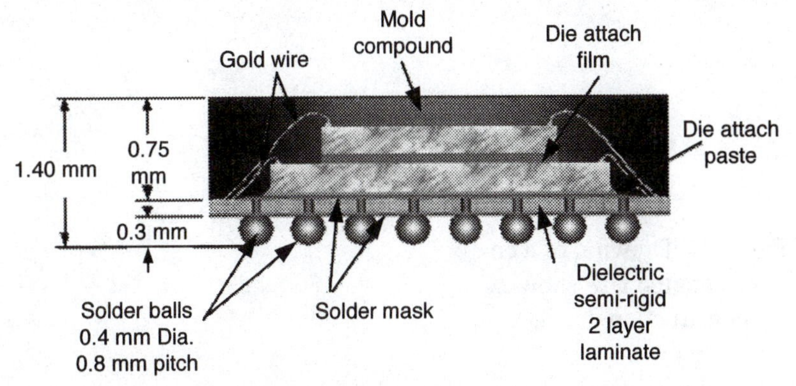

Fig. 4.42 Stacked memory dies in Amkor's CSP.

Another arrangement used to reduce circuit board area involves stacking CSP as indicated in Fig 4.43. In this photograph, three CSP are stacked to form a single package. The wiring in the substrates of the three packages must accommodate the combination of three memory chips and provide I/O on the bottom laminate, which enables connections that access to all three chips.

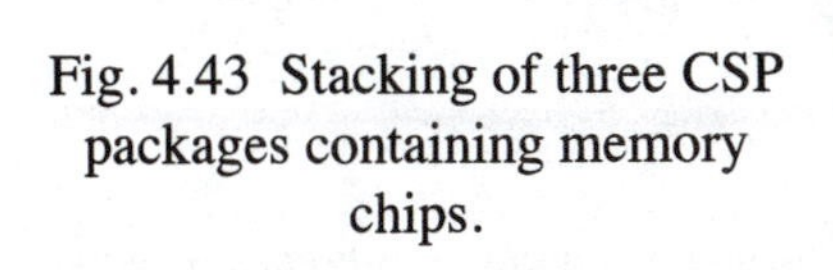

Fig. 4.43 Stacking of three CSP packages containing memory chips.

4.2.7 Ceramic Chip Carriers

The emphasis in this treatment has been on chip carriers encapsulated in plastic. Plastic chip carriers constitute by far the largest segment of the market[6]. Many of the first level packages described previously are available in ceramic versions; however, their significantly higher cost often prohibits wide spread usage. Ceramic chip carriers are usually specified for military and high performance applications where cost is less of a factor than performance. The most significant advantage that ceramic chip carriers have over plastic chip carriers is hermeticity. Moisture can in time diffuse through the plastic encapsulation and begin to corrode the circuit lines and bonding pads on the chip; thus, degrading the reliability of an electronic system. Ceramic is impervious. Moisture does not diffuse through the ceramic, glass or metal used in constructing the ceramic chip carriers.

There are two methods used in producing ceramic chip carriers—co-fired multi-layered construction and pressed two-layer construction. An example of a 14 pin DIP fabricated from ceramic is presented in Fig. 4.44. The body is made from three layers of ceramic green sheets. Circuit lines are printed onto the middle sheet with a glass-refractory metal mixture. The three sheets are stacked together in alignment and fired. During the firing process the ceramic green sheets are sintered and the volatiles are driven off the refractory metal ink forming a conducting line. The chip is brazed into the cavity of the ceramic body with gold silicon solder. The chip is wire bonded to the refractory metal pads deployed about the cavity. A metal lid is soldered in place under a vacuum to seal the chip within the ceramic package.

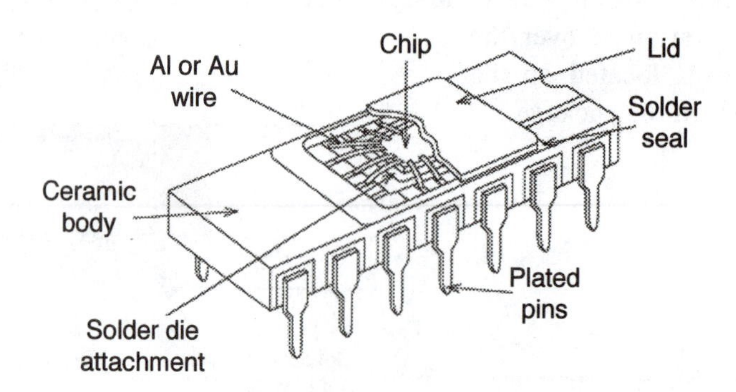

Fig. 4.44 Drawing of a co-fired ceramic DIP showing construction details.

An example of pressed two layer construction of a ceramic surface mounted chip carrier is depicted in Fig. 4.45. In this approach, the base and the top layers are prepared by pressing a mixture of ceramic powder blended with a binder into molds. The two pieces are then sintered individually to form high density ceramic parts. A sandwich is then formed with the two ceramic layers, a lead frame and two layers of low melting point glass. The sandwich is then heated until the glass fuzes, welding all of the components together. The chip is brazed into the cavity of the package and connected with wire bonding. Finally a metal lid is solder to form a hermetically sealed package.

The pressed two-layer ceramic chip carriers are suitable for low pin count packages; however, for chips with higher I/O count two layers are not sufficient to accommodate all of the wiring that is required. In these cases, co-fired multilayer construction is employed where additional layers are added to accommodate vias and wiring in both the x and y directions. An example of a pin-grid-array with several layers of co-fired ceramics is presented in Fig. 4.46.

[6] When the market is measured in numbers of packages used plastic chip carriers are dominant; however, when measured in dollars, ceramic packages constitute about 2/3 of the market.

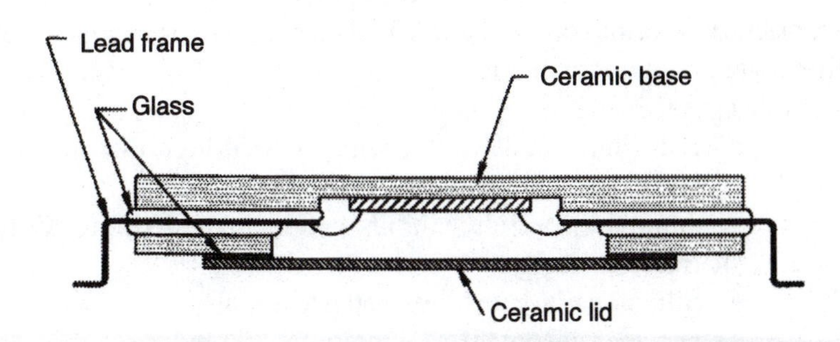

Fig. 4.45 Construction details of a surface mounted ceramic chip carrier with two-layers.

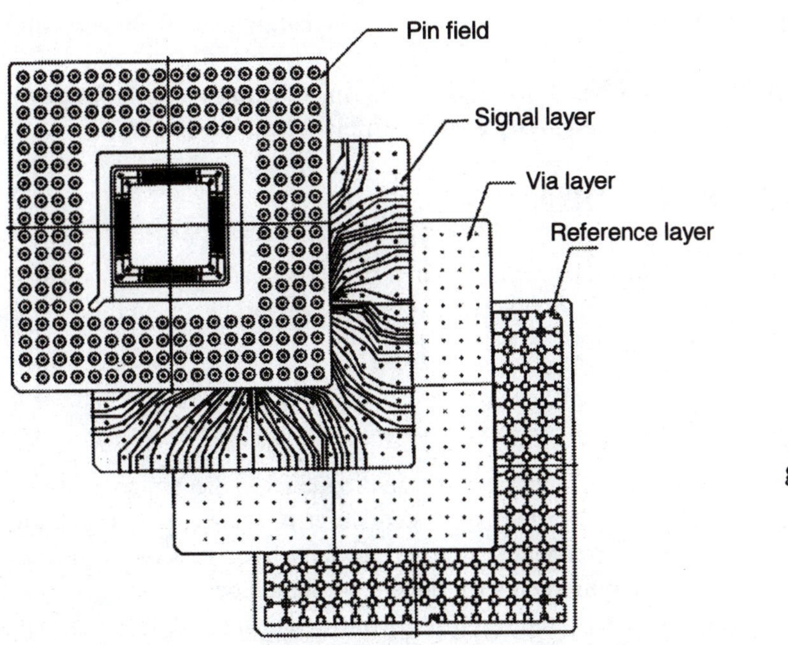

Fig. 4.46 Layers of co-fired green sheets form the substrate for a pin array chip carrier.

Co-fired green sheets are also used to form multi-layered substrates for multi-chip-modules, as illustrated in Fig. 4.47. These substrates are made up of tens of layers. The top layer has the metal pads to which the flip chips are bonded. Next several layers are devoted to redistribution where the wiring lines are fanned out to provide additional width and spacing. Many center layers are used in transmitting signal pulses between the chips and the I/O pads on the module. Finally the bottom layers are used to distribute power to the chips and to connect with the I/O pins or pads.

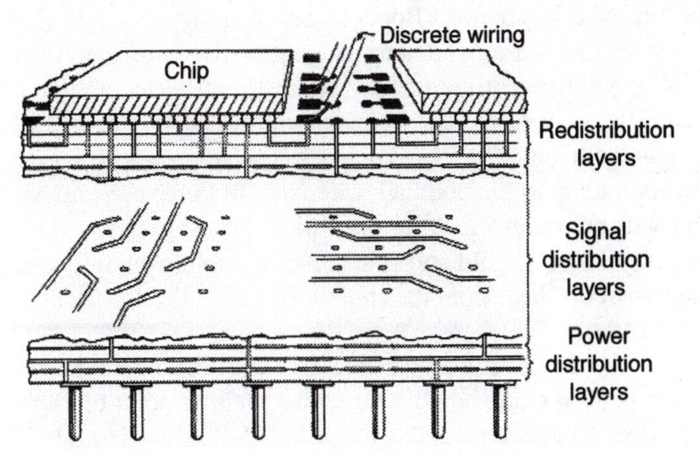

Fig. 4.47 Multi-layer co-fired ceramic substrate for a multi-chip module.

Most ceramic chip carriers are fabricated from alumina (Al_2O_3) with small amounts of additives such as glass, magnesium oxide and calcium oxide. The exact composition is considered proprietary information by the manufacturers. Other ceramic are often considered for applications in first level packaging including:

- Beryllium oxide (BeO) with its very high thermal conductivity; however, it is expensive and extremely toxic.
- Aluminum nitride exhibits better thermal conductivity than alumina, but it is brittle and fractures easily.
- Silicon carbide exhibits better thermal conductivity but is a conductor. It is used as a heat spreader in some IBM's multi-chip modules.
- Tungsten carbide is extremely hard and is used in fabricating cutting tools.
- Silicon nitride (Si_3N_4) is a high-temperature, high-toughness ceramic used in very high temperature applications.
- Boron nitride (BN), in its cubic crystal form, is used in substrates for high power components where its high thermal conductivity enables efficient heat dissipation.

4.3 CHIP CARRIER MANUFACTURING

Many manufacturing processes are employed in enclosing a chip in a first level package with its leads ready to be attached to a second level package. We will describe several of the processes in this section, and introduce some properties of the material employed.

4.3.1 Chip to Chip Carrier Mounting

The term 'die bonding' describes the operation of attaching the semiconductor die either to its chip carrier or to some substrate. The die is first picked from a separated wafer or waffle tray, aligned to a target pad on the carrier or substrate, and then permanently attached, usually by means of an inorganic solder or a particle filled epoxy adhesive. The requirements for the die bond include:

- It must not transmit significant stresses to the fragile chip during curing or during thermal cycling.
- It must provide excellent adhesion to both the chip and substrate materials without voids.
- It must withstand operational temperature without degrading or outgassing.
- It should exhibit good thermal conductivity to enhance heat transfer from the chip

The chip is bonded to the chip carrier to prevent any movement of the chip relative to its housing during the life of the product. Bonding methods vary considerably and depend on the type of chip carrier. Consider first a high-performance chip carrier fabricated from multi-layer ceramic. This carrier usually must meet stringent hermeticity requirements. Hence, it is essential that the bonding material used to attach the chip to the carrier maintain this hermeticity. In these ceramic housings, eutectic solder of gold and silicon (97.15% gold and 2.85% silicon) with a melting point of 363 °C or a silver filled glass are commonly as the bonding materials. Both the eutectic solder and the silver filled glass are inorganic and will not outgas making them ideal bonding agents in view of the hermeticity requirements. The eutectic solder of gold and silicon also exhibits a high thermal conductivity of 296 W/m-°C and aids in the transfer of heat from the chip to the case. The tensile strength of this solder is greater than 7,000 psi (48.3 MPa).

Other eutectic solders are also used for die attach to ceramic substrates such as a gold-tin alloy which is 80% gold and 20% tin with a melting point of 280°C (556°F). This alloy is lower in cost and its reflow temperature is also lower at 315-320°C. This alloy provides excellent wetting characteristics, high joint strength, excellent resistance to corrosion and high thermal conductivity. Its reliability has been proven in for die-attach and for solder lid sealing in applications for over 30 years.

The bonding of the chip to plastic carriers is entirely different because hermeticity usually is not a major concern. Instead, the problem encountered is with the difference in the thermal coefficient of expansion (TCE) of the silicon chip and the plastic housing. The TCE of silicon is 4 PPM/°C and that of plastic about 80 PPM/°C. Bonding the chip directly to a plastic substrate would result in large thermal stresses induced in the chip due to temperature changes. This problem is circumvented by introducing a lead frame into the housing. The lead frame serves several purposes. Firstly, it is fabricated from an alloy which has a thermal coefficient of expansion much more closely matched to silicon. Second, it serves as a bonding surface. Third, it provides the I/O leads that fan-out from the package. Finally, the lead frame provides a surface to which the small bonding wires from the chip can be welded. A typical lead wire frame for a 40 lead DIP is shown in Fig. 4.48.

Properties of lead frame materials that are important include: density, modulus of elasticity, tensile strength, thermal conductivity and thermal coefficient of expansion (TCE). Physical properties of three common lead frame materials are presented in Table 4.1.

Adhesive die attach materials are suspensions of metal particles in a polymeric carrier. The particles are several μm in size, usually in the form of thin flakes of silver. The polymer provides adhesion and cohesion to make a bond with the adequate mechanical strength, while the metal particles provide electrical and thermal conductivity. Note that conductive resins are often used even though an electrical connection is not required, in order to enhance thermal performance. The polymer most commonly used is a solvent-free, high purity epoxy resin. Epoxy adhesives reduce costs, shorten cure cycle time and produce stress-free bonds. When selecting a particle filled epoxy it is important to specify a solvent-free product to reduce the formation of voids in the adhesive joint and to improve heat transfer across the bond line. Purity of the adhesive is also an important consideration. Aluminum corrosion failures of the bonds occur when hydrolysable ions react with water vapor to form organic acids. In more recent years, adhesives that contain lower and lower levels of hydrolysable ions, in particular chlorine, sodium and ammonium are specified for this application.

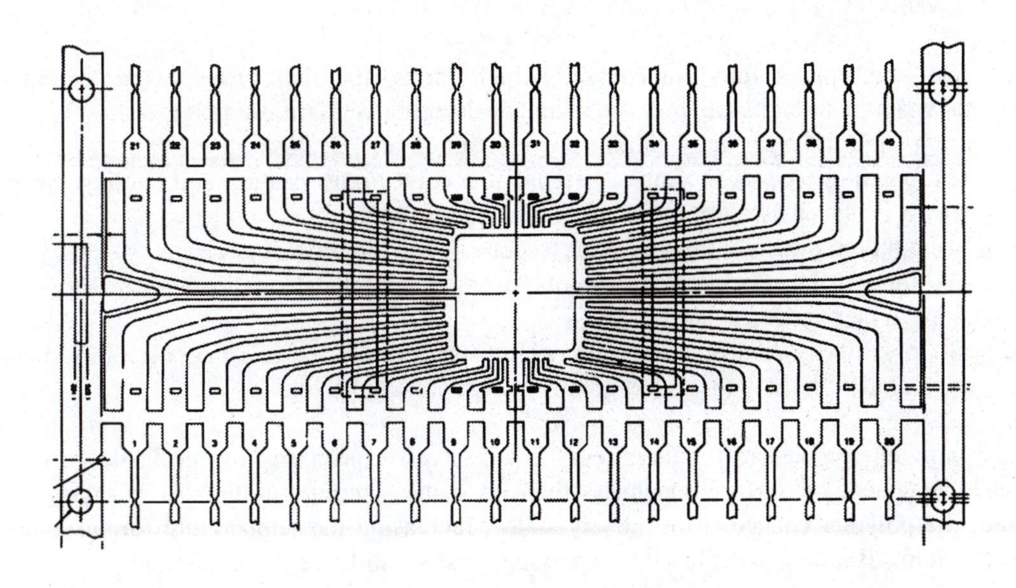

Fig. 4.48 A typical lead frame used for chip bonding in a 40 lead DIP.

Table 4.1
Physical properties of common lead frame materials

Properties	Units	Copper Alloy MF 202	Alloy 42	Kovar
Density	kg/m^3 (g/cc)	8880 (8.8)	8100 (8.1)	8400 (8.4)
Modulus of Elasticity	GPa	113	145	138
Tensile Strength	MPa	540	615	627
Thermal Conductivity	W/m°K	160	15.7	17.5
Coefficient of Thermal Expansion	ppm/°K	17.0	4.5	5.3

Polyimide die attach resins are specified when higher temperatures are involved in subsequent processes because they withstand higher temperatures than epoxies. Polyimide adhesives are thixotropic pastes containing approximately 70% silver powder in a polyimide resin that has been dissolved in a solvent with a high boiling point. To minimize the quantities of solvents and other vapors released during cure, the base polyimide used is a low molecular weight resin which is cure by an addition reaction. Although in use since the mid-1980s, the main objection to polyimide is the difficulty in removing all the organics that may outgas later in service, causing premature failure of the device.

Epoxies have extremely large molecules which are cross-linked in three dimensions to give relatively high-modulus polymers with good adhesion, low shrinkage and high-strength. This strength and rigidity, which was ideal when chips were much smaller, can cause problems when bonding large dies. The requirement for the adhesive to bond these large chips involves mechanically decoupling the larger die with a low TCE from a substrate with a higher TCE. To accomplish this decoupling requires a more flexible, lower-modulus adhesive. Thermoplastics are inherently lower-modulus adhesives because they are made from linear molecules that do not cross link. There are a number of commercially available filled thermoplastic adhesives suitable for die bonding that can be processed in the range from 200°C to 400°C.

The advantages of thermoplastic adhesives are presented below:

- They are supplied fully polymerized, which implies that their properties are determined by the manufacturer; hence, they are more consistent than thermoset polymers.
- Their shelf life is virtually unlimited even without refrigeration.
- The bonding process is simple, because it involves only heating and cooling the polymer after it is placed in contact with the surfaces to be bonded.
- The bonding time is extremely short (seconds), and the process is clean.
- The die can be reworked, disassembled or repositioned by reheating allowing defective chips to be removed and replaced.
- The polymer is used in a dry state, which significantly reduces the possibility of voids as compared with those thermoset adhesives that contain solvents.

The bonding mechanism for thermoplastic adhesives is primarily by mechanical interlocking and is more dependent on surface roughness than on material composition. At its glass transition temperature, the polymer changes to a rubbery state. Increasing the temperature further causes it to become semi-fluid. Bonding is achieved by applying pressure to force the semi-fluid polymer into the scratches and pits on the surface of the lead frame. Sufficient time under pressure is permitted for heat to distribute at the bond line and for the polymer to penetrate the surface roughness. Cooling the assembly causes the polymer to solidify making a mechanical bond.

4.3.2 Chip to Chip Carrier Connections

The I/O from the chip consists of a number of bonding pads usually arranged in a straight line around the four edges edge of the chip. The bonding pads are very small, usually 40 to 50 μm square on 60 μm centers. On very dense chips, two staggered rows of bonding pads are placed around the chip's perimeter. When flip chip bonding is employed solder bumps are deployed over the area of the chip. In this case, the solder bumps are placed on pitches of about 150 μm. Connections are made between these bonding pads on the chip and the metal fingers on the lead frame for leaded chip carriers. For flip chip, the solder bumps are aligned with the pads on the chip carrier and connections are made by reflowing the solder bumps. Three different methods are used in making these connections which include: automatic wire bonding, tape automated bonding and flip-chip with solder bump reflow. Each of these methods will be described in the following subsections.

Automated Wire Bonding

Wire bonding is performed using either a thermo-compression or an ultrasonic bonding process. The wire materials are gold, aluminum and more recently copper. With thermo-compression bonding[7], a gold ball is first formed by melting the end of the wire, which is held by a capillary bonding tool, with an electronic flame-off (EFO) procedure. The free-air ball, shown in Fig. 4.49a, has a diameter that is about twice the size of the wire diameter. The size of the free air ball size is controlled by the current employed in the EFO procedure and the tail length of the wire extending from the capillary tool.

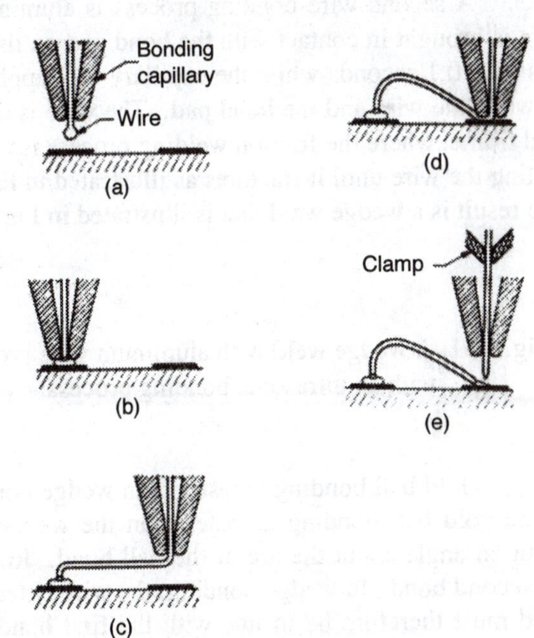

Fig. 4.49 Processing steps involved in producing ball-wedge joints with a thermal compression welding process.

[7] The thermal ultrasonic process is also employed where the metallurgical bond is made using heat, pressure and ultrasonic forces. In ultrasonic bonding, the wire is forced against the bonding pad with a head that is vibrated with a shearing motion at a frequency of about 60 kHz. The local slip of the wire produces intimate contact of the two materials and friction welding occurs producing the joint.

The free-air ball is then brought into contact with the bond pad. Pressure and heat are applied to the ball with the capillary tool for less than 0.1 second, as indicated in Fig. 4.49b, to form the metallurgical weld between the ball and the bond pad. The weld is formed by the diffusion of the atoms from the wire and pad materials into each other. The combination of temperature and pressure promotes diffusion producing a strong welded joint between the bonding wire and the pad. The shape of the end of the gold wire after attachment is illustrated in Fig. 4.50.

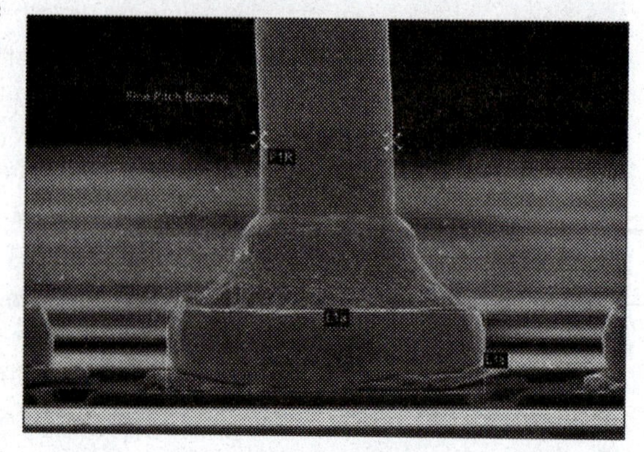

Fig. 4.50 Photomicrograph of a 20 μm diameter gold wire bonded to a chip pad.

After the ball is bonded to the die, the capillary pulls wire from the supply spool to free a sufficient length to make a connection to the lead frame of the chip carrier (see Fig. 4.49c). Then the capillary lays the wire down in a parabolic or elliptical curvature to form a loop between the chip and lead frame pads as shown in Fig. 4.49d. The looping profile is controlled by the wire bonder's software. Achieving low-loop, long lead bonding is not difficult because a programmable looping algorithm optimizes its formation for the different connection lengths. When the capillary reaches the second bond position, a wedge shaped bond is formed on the lead frame by applying heat and pressure with the capillary tube as shown in Fig. 4.49d.

A second wire-bonding process is aluminum wedge bonding. During this process, a clamped wire is brought in contact with the bond pad on the chip. Ultrasonic forces are applied to the wire for a less than 0.1 seconds while the capillary tool applies a compression force. A friction weld is produced between the wire and the bond pad. The wire is then extended to the specified bonding location on the lead frame, where the friction welding process is repeated. The wire is then broken off by clamping and pulling the wire until it fractures as illustrated in Fig. 4.49e. The result is a wedge weld that is illustrated in Fig. 4.51.

Fig. 4.51 A wedge weld with aluminum wire produced with an ultrasonic bonding process.

Gold ball bonding is faster than wedge bonding with about 18,000 welds per hour. The speed of the gold ball bonding is faster than the wedge bonding because the second bond can be formed about an angle about the arc of the ball bond. Rotation of the capillary tool is not necessary to form the second bond. In wedge bonding, the wire is fed under the bottom of the capillary tool. The second bond must therefore be in line with the first bond. To achieve alignment the capillary tool must be rotated. This added motion slows the process and as a consequence wedge bonding often takes longer to form the second weld than gold ball bonding.

Aluminum wedge bonding can be used with slightly smaller pitches than gold ball bonding because the size of bonding pad on the chip is only 25 to 30% larger than the diameter of the bonding wire. Because the deformation of the gold ball is larger, the bonding pad size must be 60 to 80% larger than the diameter of the gold wire. Aluminum wire is also used when higher temperatures are required in subsequent manufacturing processes, because higher temperatures degrade the gold wire weld by producing intermetallics at the gold/aluminum joint.

The two most common metals used for the bonding wire are gold or aluminum with wire diameters varying from about 20 to 50 μm depending on the pitch of the bonding pads on the chip. Aluminum is used primarily in making wedge type joints with ultrasonic bonding on chips with very small bonding pad pitch (< 45 μm). Gold is easily bonded, using the ball-wedge method illustrated in Fig. 4.49. The ductile properties of gold aid in the welding process and high production rates can be achieved. Also, the gold is a noble metal and resists corrosion that can occur with plastic chip carriers.

Copper has been introduced as a bonding wire in recent years. Copper is more economical than gold, has superior electrical properties and is more compatible with copper metallization on chips. Because copper metallization results in significant reductions in the power dissipated by logic chips, its usage has grown in the past decade. When making wire bonds with copper, a nitrogen blanket is used to prevent the formation of copper oxide during the bonding process. It is also necessary to maintain tight control of processing parameters on the wire bonding machines because copper is harder than gold, and higher forces are required to effect the weld. These high forces if not closely controlled may damage the die.

Tape Automated Bonding (TAB)

In tape automated bonding (TAB), the small diameter wires described previously are replaced with a precisely etch series of foil leads supported on a plastic carrier. An example of typical TAB tape is shown in Fig. 4.52. The plastic carrier is in the form of a roll of film with sprocket holes so that the TAB lead pattern can be automatically positioned over a chip. The lead pattern is etched in very thin copper foil and then plated with gold to protect the foil from oxidation during storage and to facilitate welding to the chip bonding pads. The pattern is positioned over the chip bonding pads and a heated pressure head is lowered over the assembly to form a thermo compression bond to all of the chip pads. After this inner lead bond is complete, the chip may be tested to insure that it is totally functional before completing the assembly of the chip carrier. The final step in the TAB process is making the connections to the pads on the chip carrier. This series of bonds is made in the same way, with a larger pressure head designed to conform to the pad array on the chip carrier.

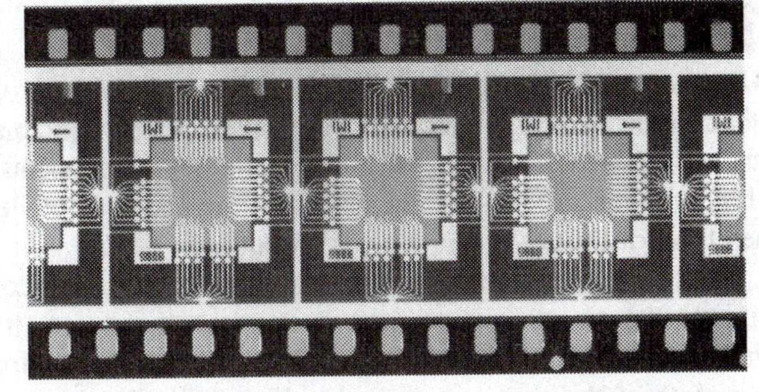

Fig. 4.52 Sprocket holes in the polyimide tape are used to position the tape over the die.

TAB is typically a single sided polyimide based circuit, although tape with metal on both of its sides is available. Copper is either electro-deposited to the tape or rolled copper is attached to the tape using an adhesive. The circuit traces are then produced using a photolithography process that will be described in Chapter 5. The main advantage of TAB is the small pitch of the bonding pads that can be accommodated. Currently, line pitches of 45 μm can be achieved (22.5 micron lines/spaces), which enables the used of TAB with high-density circuits that require a high pin count. The polyimide film is available in three widths, 35 mm, 48 mm and 70 mm, allowing the process to accommodate different size chips. The other critical dimension is length of the part, which is typically measured in sprocket pitches (see Fig. 4.52). Each sprocket pitch is equal to 4.75 mm, and most lithographic systems are capable of an image length of 16 pitches (76 mm). The base polyimide film is typically between 50 to 100 microns thick.

The critical feature for TAB processing is cutting holes in the tape to expose the circuit traces that are to be bonded to the chip. The leads on the tape are cantilevered over the position of each gold bump on the die. Similarly, other holes are cut in the tape to allow for connections to the lead frame or directly to the PCB.

There are two common methods of welding the gold bump on a die to the lead on a TAB trace—a single-point process with heat and ultrasonic forces applied to the bonding pad on the chip or a gang thermal compression process. In single-point bonding, the weld is made to each of the die's bond sites individually. This single-point process is more time-consuming than the gang bonding. Gang bonding employs a specially designed bonding head to apply pressure and temperature to simultaneously make thermal compression (diffusion) welds between the TAB leads and gold bumps on the die. The gang bonding process has a high throughput rate, and is preferred to the single-point bonding process.

TAB has several advantages over wire bonding which include:

- TAB permits use of smaller bond pads and smaller bonding pitch
- Bonding pads can be deployed over the entire area of the chip to increase the I/O count
- The quantity of gold required for bonding is decreased.
- Less time is required in production if gang bonding is employed.
- TAB connections provide better electrical performance (reduced noise and higher frequency.
- TAB connections can be made on flexible circuit materials.
- TAB enables the chip to be tested after completing only the inner lead connections.

The disadvantages of TAB include:

- The additional time and cost due to fabricating and patterning the tape.
- The capital expense for TAB equipment that is needed in addition to the machines used for wire bonding.

TAB is a better alternative to conventional wire bonding if very fine bond pitch, reduced die size and higher circuit density are required. It is also the only suitable manufacturing method when the circuits must be flexible (i.e. printers, flex cables, folding components, etc.) TAB is usually more cost-effective in high-volume production, because the time and cost of developing the tape become less important when large quantities of a product are produced.

The welded joints formed with TAB require additional processing of the chip bonding pads. This added processing is performed on the wafer before it is separated into chips. The aluminum bonding pads are plated with chromium and copper to develop a bump which raises the pad well above the surface of the chip as indicated in Fig. 4.53. The bumps are then gold plated to facilitate thermo-compression welding. The chromium barrier layer serves to limit migration of the gold and aluminum atoms and inhibits the formation of Al-Au intermetallics which can degrade the joint.

Flip Chip Connections to a Chip Carrier

In standard chip carriers, the chip is positioned on its back surface and electrical connections are made to the bonding pads located about the perimeter on its top surface. In chip carriers using flip chip technology, this standard orientation of the chip is reversed. The chip is placed face downward and the back side of the chip faces upwards. This flip chip orientation has a significant advantage in that it concentrates the electrical functions on the underside of the chip leaving the top side free for use in developing a highly efficient method for heat dissipation. In addition, today's trend toward higher device clock speeds, smaller devices for mobile and portable applications and higher I/O count make flip-chip the technology of choice. Flip chip technology offers several other advantages including: superior electrical performance due to the shorter electrical connections between the chip and substrate that result in smaller package size. Also attachment of high I/O devices becomes very cost competitive in flip-chip, compared to wire bonding, because hundreds of I/Os can be connected in a single process step.

Chromium barrier
Copper
Photo resist
Aluminum
Aluminum bonding pad
Passivation oxide
Copper bump
Gold plating

Fig. 4.53 Processing steps followed in placing gold plated solder bumps on a chip in preparation for TAB bonding.

The flip chip method of connecting the bonding pads to the chip carrier was originally developed by IBM in 1964 and was used exclusively by that company for many years in its high performance computers and signal processors. The original process was identified as C4 by IBM to identify a controlled collapsible chip connection. In this process, the chip bonding pads are deployed in an area array over the surface of chip in contrast to the deployment of these pads around its perimeter. By using these area arrays of bonding pads a higher I/O count is possible.

To insure a low and stable contact resistance at the bump-bond pad interface, the aluminum bond pads are metallized to eliminate non-conductive aluminum oxide. The under bump metallization involves evaporation of layers of chrome, copper and gold over the entire surface of the wafer. This layered structure acts as a hermetic seal, provides an electrically conductive diffusion barrier, and establishes a good mechanical base for the solder bump. Solder bumps are placed on these pads by either an evaporative process that is the preferred method for lead based solders, or by electroplating, which can be used for the application of any metal. After the deposition, a reflow process forms the spherical solder bumps as shown in Fig. 4.54.

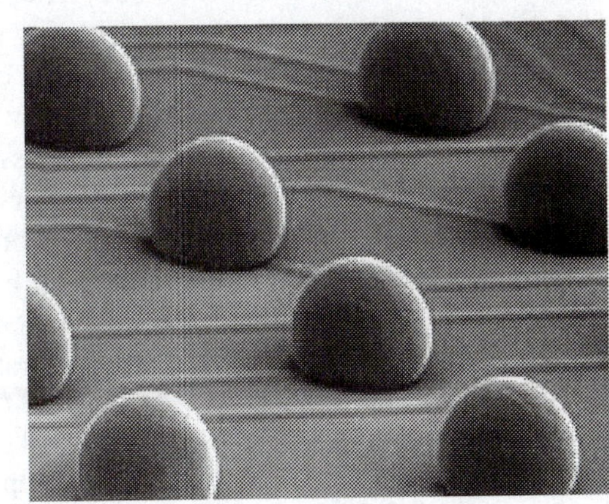

Fig. 4.54 Solder bumps deployed over the active area of the chip.

The spacing on the bonding pads vary with the application, but high performance processors employ pads on 200 μm centers that yields 2500 solder bumps per square centimeter. The solder bumps are 100 μm in diameter. The composition of the solder bumps is usually 97% lead and 3% tin. Matching bonding pads are produced on the substrate[8] so that the pads on the chip and the ceramic coincide. The chip with its solder bumps is placed over the substrate with matching pads aligned. The assembly is heated until the solder spheres begin to soften and a controlled collapse of the sphere takes place as the solder simultaneously wets both pads. Surface tension of the solder controls the geometry of the solder joints until the solder solidifies. Solder flow from the joints along the circuit traces on either the chip or the substrate is prevented by plating these traces with metals like chromium which inhibit solder flow. This treatment controls the exact quantity of solder at the joint and provides for solder pillars about 75 to 100 μm high between the chip and the substrate. Precise control of the quantity of solder also prevents the formation of solder bridges between the closely spaced leads during the reflow process. After the completion of the reflow process, the space between the chip and the substrate is washed to remove any flux residues. When organic substrate materials are used, a significant mismatch exists in the thermal coefficient of expansion of the organic laminate and the silicon of the chip. In these cases, the gap between the chip and the substrate is filled with an adhesive. This adhesive, known as an underfill, serves to reduce the shear strains that result from temperature cycling that occurs in service.

4.3.3 Interposer Substrates

Designers of first, second and third level packaging have recognized for many decades that several layers of conductors are needed to make all of the interconnections necessary to wire a chip, PCB or mother board. However, there are several approaches to connecting the layers. The oldest method for connecting the conducting layers in organic boards was by plated-through-holes. Circuits boards are manufactured, bonded together, several boards are stacked and then drilled. The trough holes are plated with copper to form vertical conductors. With ceramic, multi-layer substrates are fabricated from green sheets that have printed circuits and filled via holes. These multi-layered substrates are co-fired to form in-plane circuits as well as conductors through the filled vias to provide z direction connections.

A circuit interposer is defined as either a material or construction that electrically and mechanically connects a pair of conducting layers in the z direction without interfering with x-y plane wiring. Perhaps the simplest interposer is an anisotropic conductive adhesive. In the 1980s, a process was introduced for making z direction connections using an adhesive that incorporated small solder balls in a B-staged epoxy film. The concept was to join pairs of double-sided circuit boards with this anisotropic adhesive film. The circuit boards were stacked and placed in a laminating press where heat cured the adhesive and pressure forced the solder powder to contact and connect the opposing pair of copper pads as shown in Fig. 4.55. The solder particles formed small but robust connections that remained intact during subsequent soldering operations.

One difficult with the anisotropic conductive adhesive was the random pattern of the small solder balls that limited its application to low density circuit boards. To extend the application to circuit boards with increased density it was necessary to develop a patterned anisotropic conductor. In the 1990s designers began to develop interposers using flexible circuits because of the thinness, compliancy, and ability to support high-density circuits on polyimide films. Microvias were laser drilled through the polyimide film with dry film adhesive on both sides. These microvias were filled with a conducting adhesive and B-staged. Double-sided circuit boards are then bonded together with these patterned interposer films to make high-density substrates.

---

[8] Different substrate materials are employed depending on the application. Ceramic, glass-ceramic and organic laminates with BT epoxies are the most commonly used materials.

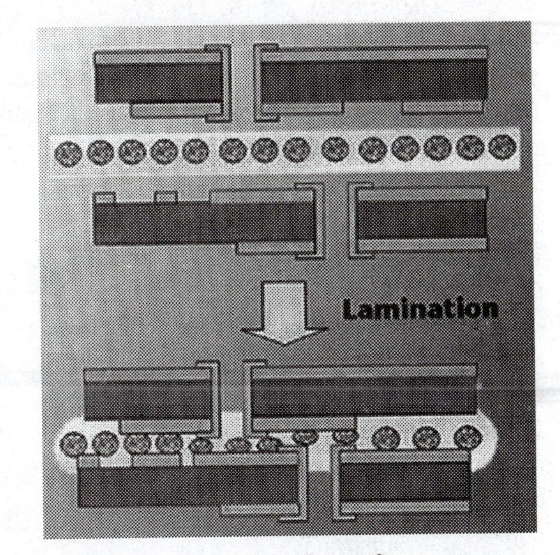

Fig. 4.55 An adhesive interposer to make connections between two double-sided circuit boards.

In both ball-grid-arrays and CSPs, the chips are bonded to a substrate that often serves as interposer because it is positioned between the chip and the I/O leads. The interposer has several functions in addition to supporting the chip and sometimes enclosing the package. The substrate contains the wiring pads for either wire bonding of flip chip connections. It may have vias and several layers of wiring to distribute the signals to the array of solder balls. The interposer provides a test bed for verifying the electrical integrity of the chip. Finally it provides a surface to support the solder mask required to prevent solder bridging during reflow operations. The coefficient of thermal expansion of the materials used in fabricating the interposer is very important because they markedly influence the thermal strains imposed on the chip or the C4 joints if the chip is bonded with flip-chip processes.

There are four general types of interposers—lead frame, rigid, flexible and wafer. We will describe each below:

Lead Frame Interposer

The lead frame interposer is employed on memory chips with a low I/O count. These chips have a small number of gull wing leads extending from two sides of the package. The construction is similar to that found in plastic encapsulated packages except the lead frame chip extends over the chip. Connections are made to lead wires along the length of the chip with wire bonding.

Rigid Interposer

Rigid interposers are made from either organic or ceramic substrates. As the name implies, they are rigid (high elastic modulus). The illustration, presented in Fig 4.38, shows a single layer laminate with wire bonding; however, rigid interposers often have several layers with microvias, blind vias and wiring planes with flip chip bonding. Materials for the ceramic interposers are usually mixtures of alumina and glass that are sintered together to form a material with a low coefficient of thermal expansion (4 to 6 × 10^{-6}/°C). Circuit traces and vias are formed from refractory metals. Matching the coefficient of thermal expansion of the interposer and the chip minimizes the thermal strains induced in the chip during operation.

Organic substrates are often employed with ball-grid-array assemblies as illustrated in Fig. 4.56. The BGA package is sometimes fabricated by wire-bonding a chip on a substrate made of a two-metal layer copper clad bismaleimide triazine (BT) laminate. The BT laminate is used in place of the standard and multi-functional FR4 laminates because of its high glass transition temperature of 170 – 215° C and heat resistance (exposure to 230° C for 30 minutes with no degradation). The standard core thickness of this two-layer substrate is typically 0.2 mm. The copper cladding is typically ½ ounce (18 μm thick) with foils bonded on each side. Four-metal layer substrate designs are sometimes specified to provide additional power and/or ground planes that improve electrical and thermal performance.

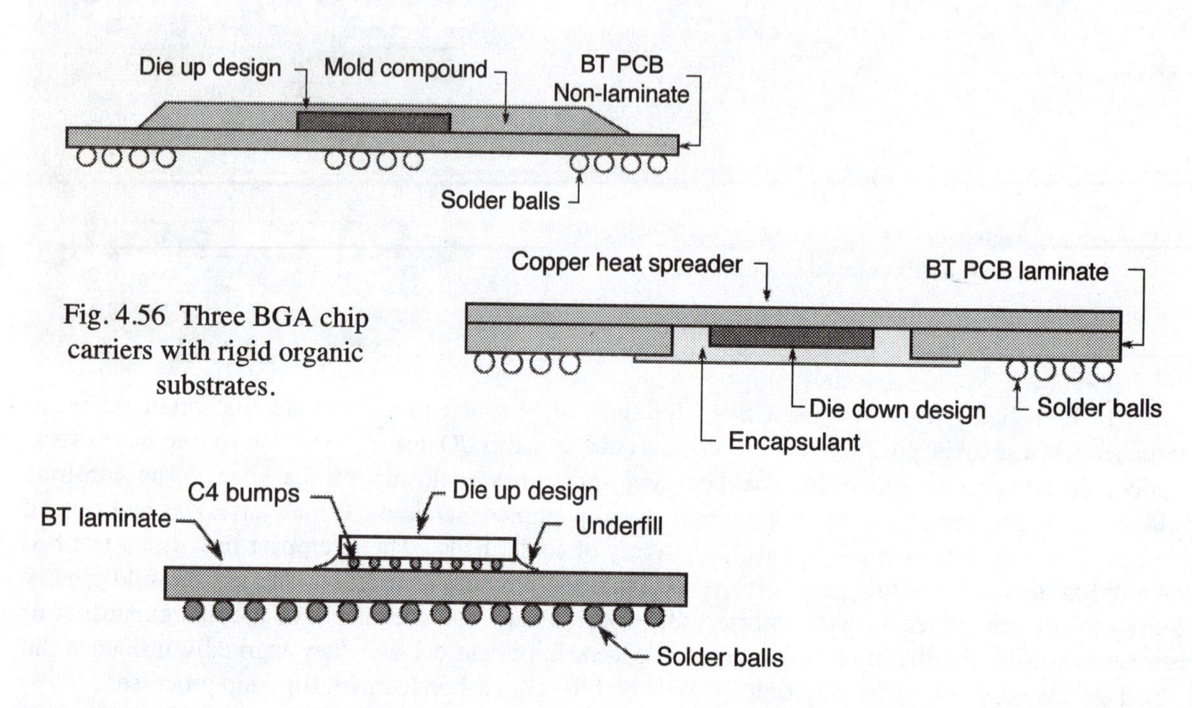

Fig. 4.56 Three BGA chip carriers with rigid organic substrates.

Flexible Interposer

Flexible interposers use a polyimide flexible circuit to connect the chip to the substrate similar to TAB tape interconnection. The most popular CSP of this type is the μBGA package. Interconnecting the chip to the interposer may include flip-chip bonding, wire bonding or TAB bonding. The concept of using flexible circuits as interposers is illustrated in Fig. 4.57. The flexible circuit, which may be multi-layered, contains the wiring and microvias needed to connect the chip I/O to the solder balls which serve as the package I/O. Connections from chip pads to the pads on the flexible circuit are wire bonded. Spacers are added to separate the chips and to provide the clearance required for the wire bonding. Solder balls are formed on the flexible circuit using procedures already described.

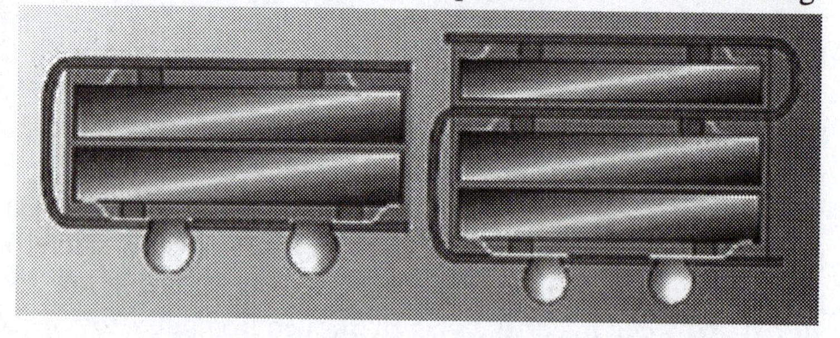

Fig. 4.57 Flexible circuits used as interposers for connecting chip I/O to package I/O.

Some molded interposers are considered flexible, because they also use a flexible circuit interconnection, as shown in Fig. 4.58. However, this type of package is encapsulated with conventional molded technology to protect the chip and its interconnections. Low modulus elastomers are employed in connecting the chip to the polyimide tape that serves as the flexible circuit. The ease of laser drilling microvias in thin film polyimide compared to laser drilling in glass reinforced epoxy is a major advantage in the growing use of flexible interposers.

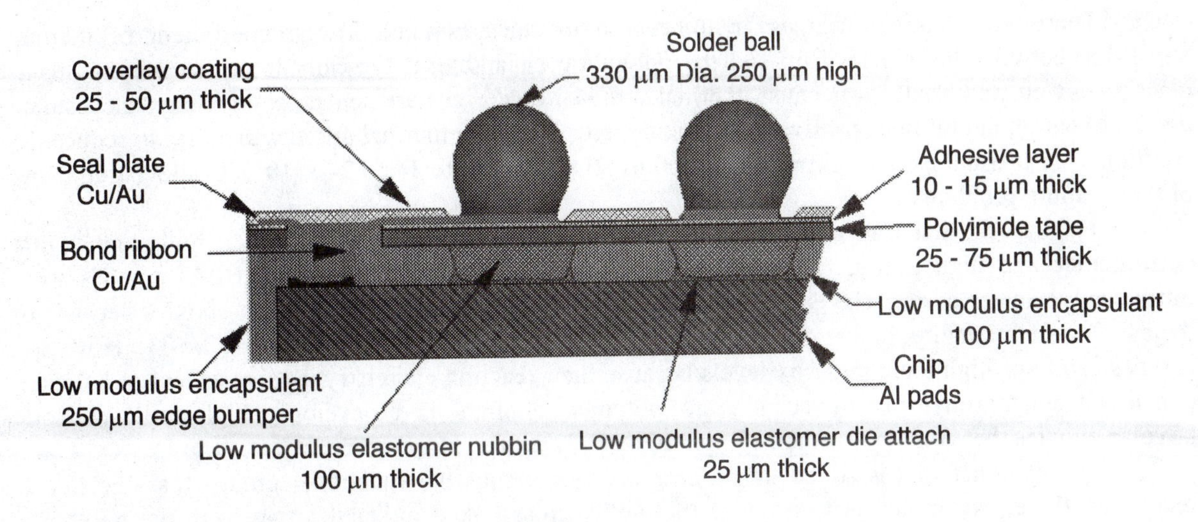

Fig. 4.58 Construction details of a molded flexible interposer.

Wafer Level Assembly

Some manufacturers are producing CSP using a process known as wafer level assembly. In this type of CSP, a chip is processed and assembled on the wafer before it is diced to form a single chip. A thin-film metallization process is used to place conducting lines that extended into the normal wafer scribe areas to redistribute the chip bond pads to a standard grid array footprint. The redistribution of IC interconnections to create other pad configurations allows the integrated circuit to be converted to a CSP. A diagram showing the detail of the transformation from an IC pad to a solder bump on a CSP is presented in Fig. 4.59. Note that polyimide dielectric polymer layers are used to isolate redistribution routings and to cover the entire area of the chip face in redistributing the I/O.

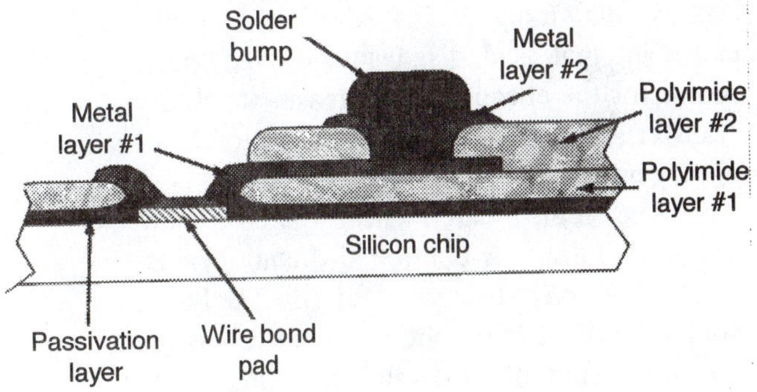

Solder bump
Metal layer #2
Polyimide layer #2
Metal layer #1
Polyimide layer #1
Silicon chip
Passivation layer
Wire bond pad

Fig. 4.59 Section view of a CSP solder bump produced on the wafer before dicing.

4.3.4 Molding

Plastic encapsulation of microelectronic devices is preferred because of costs. Plastic compounds are used to encapsulate chip on board, first level packages and as underfill and encapsulation of flip-chip assemblies. In this discussion, we will deal with those plastic materials used to encapsulation first level chip carriers. The primary purpose of the plastic chip carrier is the same as with the ceramic packages—to provide environmental and mechanical protection of the microelectronic device.

Environmental protection has been a concern because of the ability of water to diffuse through the plastic and causing corrosion of the aluminum pads and chip metals. This problem has been mitigated by formulating relatively pure encapsulating polymers with low ionic content. The mechanical protection is provided by employing polymers with a relatively high modulus of elasticity (rigid).

Thermo-mechanical stresses arise because of the mismatch in the coefficient of thermal expansion between the silicon chip and the polymeric encapsulant. Temperature changes that occur during start-up and shut-down cause thermal stresses that can result in chip cracking, encapsulant cracking, wire bond failures, passivation cracking, etc. The polymers are filled with silica to reduce the coefficient of thermal expansion from about 60 to 80 × 10^{-6} /°C to 14 to 24 × 10^{-6}/°C alleviating many of these failure problems.

The earliest materials used for encapsulation were silicones because of their high-temperature performance and high purity. However, silicones are costly and bisphenol-A (BPA) epoxies were introduced to lower costs. The BPA epoxies were replaced with phenolic novolac epoxies because of their higher functionality, higher cross-linking and increased glass transition temperature T_g. However, epoxies can have high ionic impurity levels because their reaction chemistry uses an excess of halogen-containing epichlorohydrin. In recent years, polymer suppliers have developed high-purity phenolic novolac epoxies that contain less than 25 ppm of hydrolyzeable chlorine.

The hardener that is added to the base epoxy contains the functional groups that chemically react with the epoxy molecule to produce the highly cross-linked molecules that form the polymeric encapsulants. The epoxy resin and hardeners are mixed with fillers and other additives then the composition is partially cured (B staged). This B staged compound is solid and can be ground into powder, formed into pellets and stored until it is ready to use. Because the epoxy resin and its hardener are reactive, the pellets are stored at low temperatures to prolong their shelf life.

The filler material, usually fused silica, serves to reinforce the epoxy enhancing its strength. To improve bonding between epoxy and silica, epoxy silanes and amino silanes are used as coupling agents. The addition of the silica (65 to 70% by weight) reduces the coefficient of thermal expansion and increases its resistance to thermal stresses. The added silica also increases the thermal conductivity of the polymer to 0.6 – 2.0 W/(m-°C) thereby improving the ability to dissipate heat through the encapsulant. Other additives include elastomeric modifiers to improve the epoxy's toughness and carbon black for its coloring. The filler also increases the modulus of elasticity of the encapsulant that is a disadvantage. A higher modulus for the encapsulant increases its ability to transfer stresses to the chip, passivation layer and the wire bonds.

The chips are encapsulated using injection molding. A controlled quantity of B-staged pellets is heated until the polymer becomes fluid. A hydraulic ram then drives the resin into a split mold as shown in Fig. 4.60. The lead frame, chip and bonding wires are in place before the mold is closed. Heaters are provided in both sides of the mold to maintain the temperature of the compound as it fills the cavities. The encapsulant is held at temperature in the molds until it is cured (C staged).

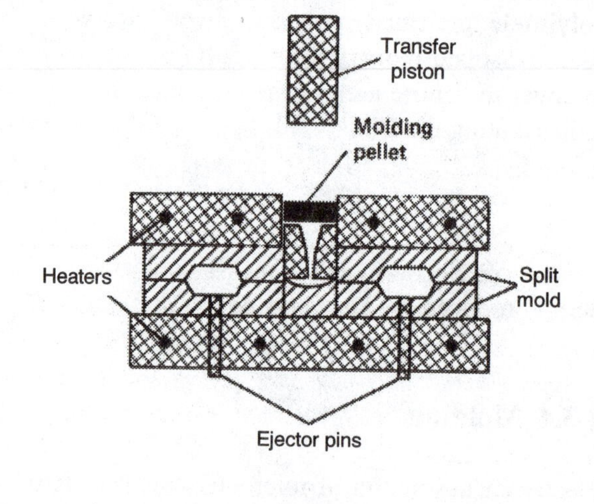

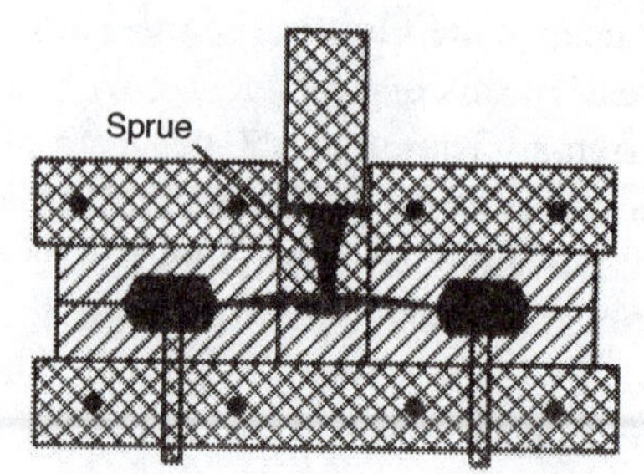

Fig. 4.60 Injection molding plastic encapsulated chips.

After the curing process is complete, cold water is circulated in the mold to cool the encapsulated chips. The die is opened and ejector pins are engaged to push the product from the lower mold. The flashing and sprue are cut from the molded chip carriers to complete the injection molding process.

The properties of molding compound used for encapsulating chips are presented in Table 4.2.

Table 4.2
Properties of an encapsulating molding compound

Property	Units	Molding Compound
Density	g/cm^3	1.9
Modulus of Elasticity	GPa	15-20
Tensile Strength	MPa	95-150
Thermal Conductivity	W/(m-°C)	0.7-0.9
Glass Transition Temperature	°C	180-225
Coefficient of Thermal Expansion	ppm/°C	α_1 = 12-18 α_2 = 41-65
Volume Resistivity	Ω-cm	>10^{14}

4.3.5 Transport Packing

Safe transportation of the bare die is important to maintain overall quality and ensure higher yields when they are assembled into a product. It is essential to select the appropriate transport carrier for the bare dies. There are a number of different methods used for transporting chips from one location to another prior to their encapsulation (and protection) in a first level package. These include Waffle Pack, Gel-Pak, Tape and Reel, Wafer Jar or Vial, Wafer Cassette. The transport carrier is designed to move individual dies or complete wafers from one location to another without incurring damage, maintaining clean room cleanliness levels and retaining position and orientation so that the dies can be acquired with pick and place robots. Also important is prevention of damage due to electro-static-discharge (ESD) during transport and subsequent storage. Finally, protection must be provided to prevent corrosion of the sub micron size features on the chip during extended storage periods. Illustrations showing the essential features of each of these transport methods will be presented below:

Waffle Pack

Waffle packs have been the traditional carrier for dies that have been cut from their wafer, especially those that has not been functionally tested. The waffle pack, shown in Fig. 4.61, is made of conductive, polypropylene that provides the grounding needed to avoid electrostatic discharge. It is available in either 2 × 2 or 4 × 4 inch sizes. Indented pockets are formed in an array across the tray for housing each chip. The number of pockets per waffle pack varies depending on the die size. A chip is placed in each pocket with its orientated consistent with all other chips in the tray. A sheet of lint-free, glassine paper is placed over the array of chips, and then another waffle tray is stacked on top. Several waffle packs are uniformly stacked and locked in place. Several of these stacks of waffle packs are sealed in a dry pack bag for extra cushioning protection during shipment. A humidity indicator card and a desiccant are included in the bag when it is packed under vacuum.

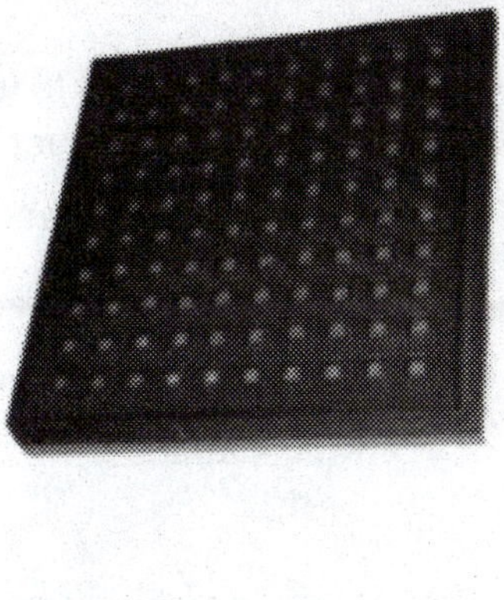

Fig. 4.61 A waffle pack holds an array of chips in a rectangular array of pockets formed in a conducting polymer tray.

Gel-Pak

In a Gel-Pak, the chips are arranged in a rectangular array in a manner similar to that used for the waffle pack except for the absence of pockets. The Gel-Pak is essentially a flat tray that has a non-adhesive but tacky membrane over which the chips are arranged. This tacky membrane minimizes movement of the chips in transport. The trays are available in 2 × 2 and 4 × 4 inch sizes. They are vacuum packed in a foil bag that provides protection from ESD and from the corrosive effects of the environment. A photograph of a Gel Pak holding a large number of small chips is presented in Fig. 4.62.

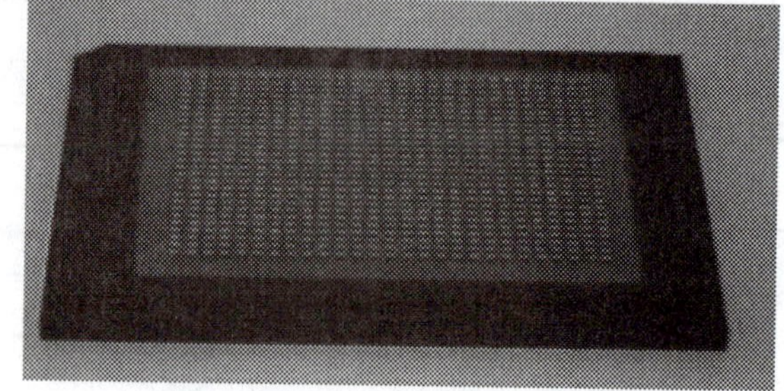

Fig. 4.62 A Gel-Pak supports a large number of chips on a tacky membrane during transport.

Surf Tape and Reel

The surf tape and reel carrier[9], shown in Fig. 4.63, is designed for shipping large numbers of chips in a format compatible with a high-volume, automated board assembly operation. The tape and reel is preferred over waffle packs because it eliminates movement of the die during shipment, reducing the risk of damage. In addition, many different chip sizes can be accommodated with only a few standard tape sizes.

The design for surf tape and reel is similar to the standard tape and-reel carriers that are commonly employed for conventional surface-mount packages. The system consists of a surf tape wrapped on a standard 7 in. (178 mm) diameter reel. Two sizes of surf tape, which fit on the same standard reel, are available to accommodate a wide range of chip sizes.

The surf tape, shown in Fig. 4.63, consists of a conductive, polystyrene tape with rectangular depressions formed in the tape to provide a slightly recessed pocket for each chip. Two strips of sticky tape are applied along the edges of the window. This tape has a pressure-sensitive adhesive coating that holds each chip in place eliminating the risk of damage to the chips due to movement during transit. The sticky tape allows for easy and safe removal of each chip by the pick-and-place robot. Each chip is loaded into the surf tape with its topside exposed and chip orientation is consistent throughout the tape. Because the pockets are recessed, each chip is protected from exposure by the next layer of surf tape. The tape is indexed into the pick and place robot using the sprocket holes to advance the tape for each placement cycle.

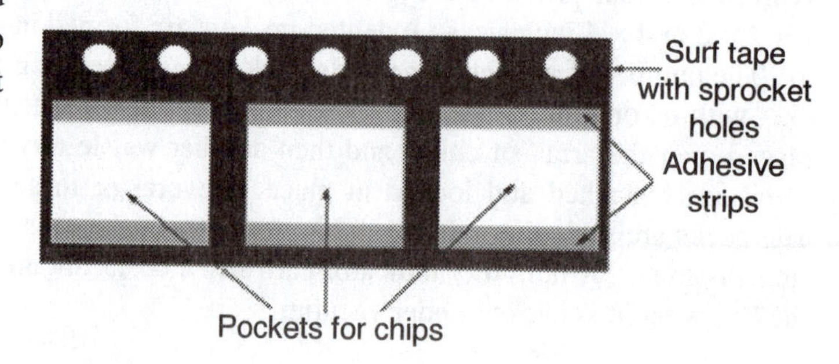

Fig. 4.63 Surf tape used to transport chips in a format suitable for high speed assembly operations.

[9] The surf tape and reel system and the tape and reel system are so similar that only the surf tape and reel system is described here.

Wafer Jar or Vial

Wafer jars are the standard carrier for fabricated wafers that have not been functionally tested. The wafers are packed in a wafer jar that is made of conductive polypropylene. A piece of antistatic, polyurethane foam is placed on the bottom of the jar to cushion any impact that might occur during transit. The wafers are stacked in the jar with a piece of lint-free filter paper between each wafer. Another layer of foam is used to fill the jar. The jar lid is also fabricated conductive polypropylene.

Wafer Cassette

The wafer cassette is used to transport complete wafers in a boat-like-box. The boat is fitted with slots that accommodate the wafers as shown in Fig. 4.64. The slots are arranged to provide spaces between adjacent wafers. Both the top and the bottom of the boat are fabricated from conductive polypropylene to provide protection from ESD. The cassettes can hold up to 25 wafers with diameters of 200 and 300 mm. The boats are filled with an inert gas before they are sealed to protect the wafers from the environment.

Fig. 4.64 Wafer cassette for transporting and storing wafers prior to dicing testing and assembly in chip carriers.

4.4 FIRST LEVEL PACKAGING FOR DISCRETE COMPONENTS

While the emphasis in this chapter has been on chip carriers, it should be recognized that first level packages are also important in protecting passive devices like resistors, capacitors and diodes. The packaging of these passive devices involves both through hole and surface mount. Because each of these components different to some degree with regard to their packaging, they will be covered individually.

4.4.1 Resistors

Resistors are available in a very wide assortment of package types as illustrated in Fig. 4.65. There are several reasons for the lack of standardization in the types of packages employed. Firstly, the range of product where resistors are employed is quite large and the amount of power dissipated varies considerably. Secondly, the importance of circuit density changes significantly with the type of product. Next, the importance of resistor temperature differs markedly with the application. Finally, the assembly process used for the board may influence the choice of package. To avoid having to pass the board through another line to assemble the resistors, the package for the resistors often is selected to match the carrier for the chips.

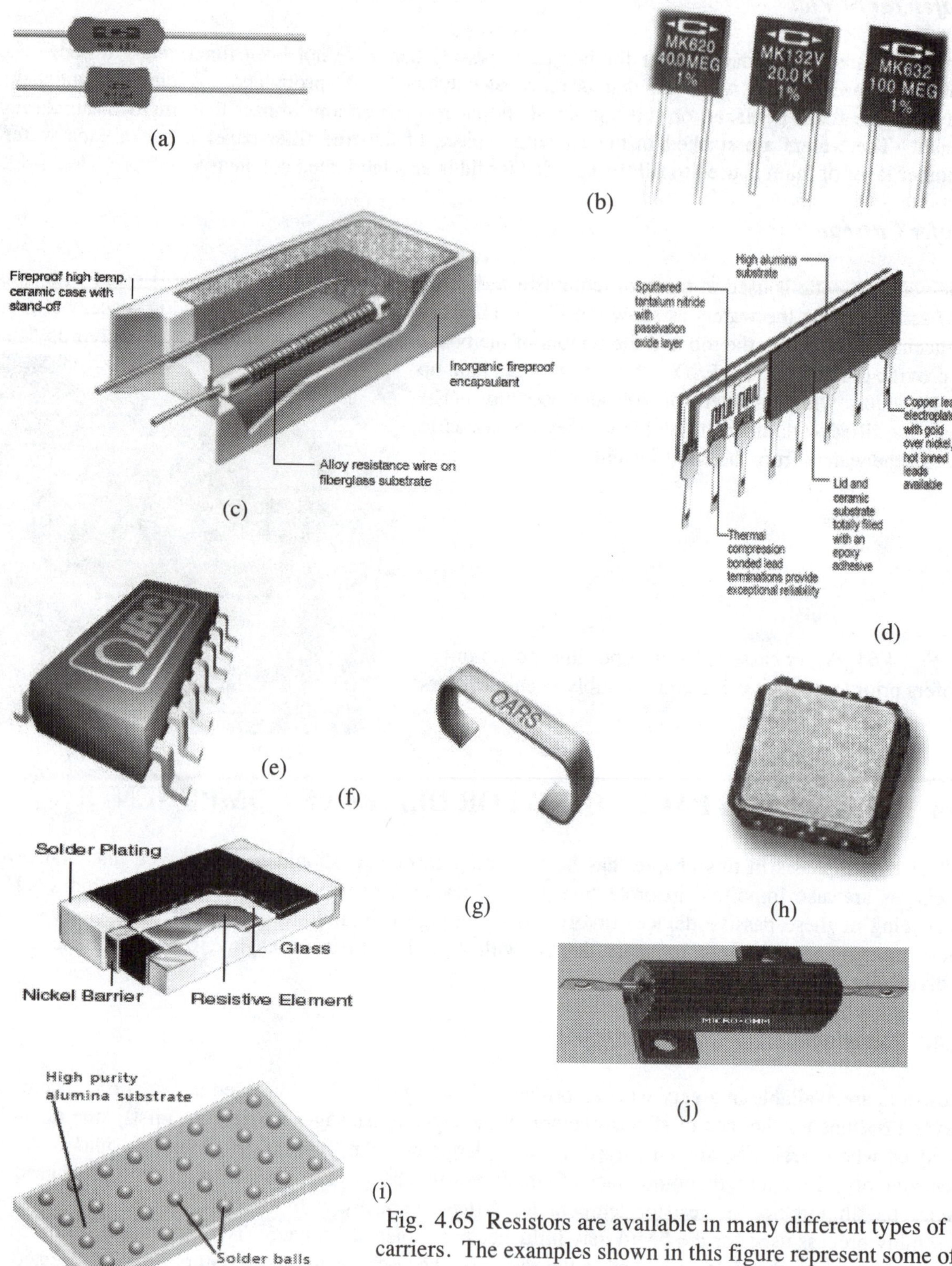

Fig. 4.65 Resistors are available in many different types of carriers. The examples shown in this figure represent some of the different carriers available.

A common resistor package is the axial lead type (See Fig. 4.65a) because it is easy to assemble into through-hole type circuit boards. The radial lead package shown in Fig. 4.65b utilizes less board space and dissipates heat better than the cylindrical axial leaded package. A power resistor with radial leads, shown in Fig. 4.65c, is encased in a ceramic case to withstand high temperatures. The SIP, shown in Fig. 4.65d, houses a resistor network. SIPs save board space and conform to assembly procedures involving boards containing only through-hole components. All of these resistor packages are the through-hole type and can be assembled with DIPs connected to the PCB's with a wave soldering process.

The SOIC carrier and the leadless packages, shown in Fig. 4.65e, f, g and h, are surface mounted carriers that are used to conform to the surface mount chip carriers, which house the IC's placed on the circuit board. The component, illustrated in Fig. 4.65f, is a zero ohm resistor that provides a method to connect one circuit to another. The leadless resistor shown in Fig. 4.65g is one of the most common surface mounted resistors in use today. It comes in a wide variety of sizes with all of the standard resistance values in use today. A network of resistors is housed in the leadless quad pack shown in Fig. 4.65h. Networks of resistors are also housed in ball-grid-array carriers as illustrated in Fig. 4.65i. The power resistor shown in Fig. 4.65j incorporates an aluminum heat sink. This heat sink is fabricated with fins to improve the amount of heat dissipation from the resistor.

Temperature plays an important role in the life and performance of resistors. The resistors are rated for specific power dissipation, and this rating is based on an ambient temperature T_a of 25 °C. If the resistors are used at a temperature higher than T_a, they must be derated (used at a power level lower than the rated value) to avoid early failures or a change in the value of their resistance with time. Typical derating curves for two types of resistors are shown in Fig. 4.66a and Fig. 4.66b.

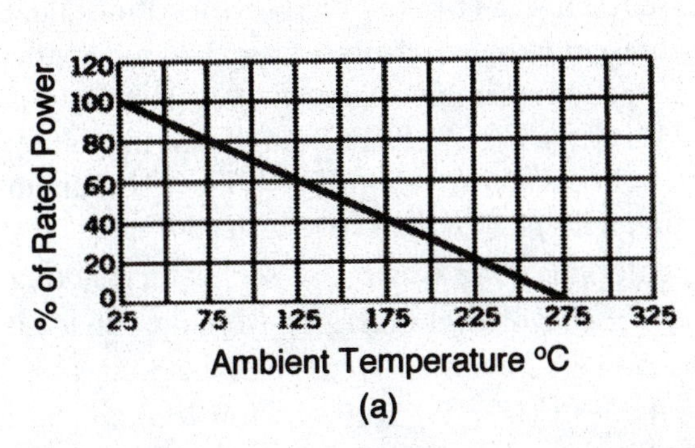

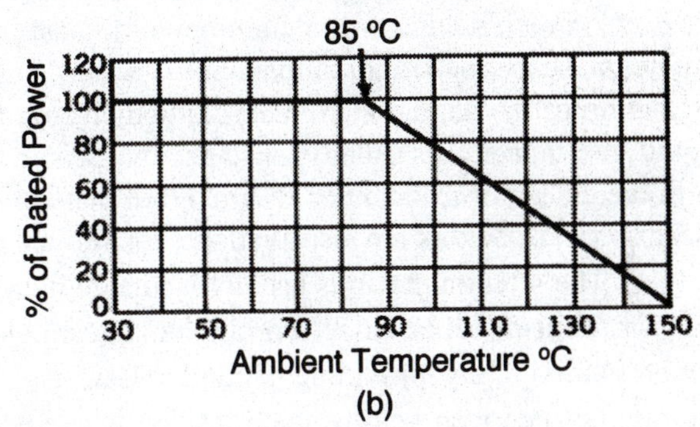

Fig. 4.66 Derating curves for two types of resistors. (a) Wire wound power resistor. (b) Precision thin film resistor.

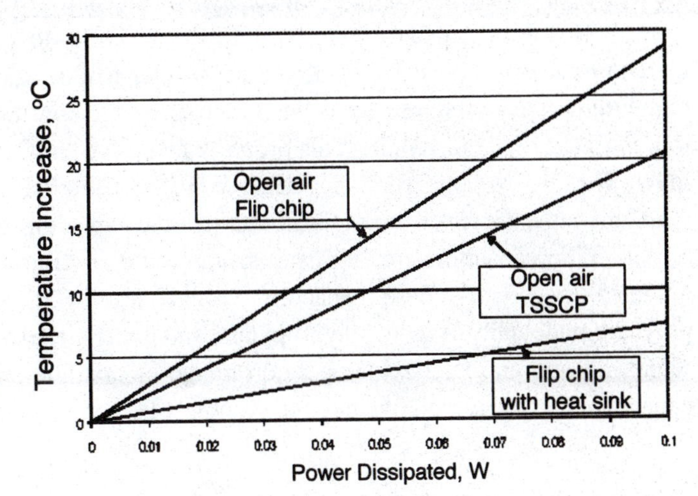

Fig. 4.67 Temperature increase due to self heating of resistors housed in flip chip packages.

Self heating must also be considered in determining the temperature of a resistor. As the power dissipated by a resistor increases, its temperature increases relative to the ambient temperature. The thermal resistance which controls the slope of the temperature-power relation is a function of the size of the resistor and the materials used in its construction. In this case, the resistors were packaged using flip chip technology with and without a heat sink. Significant temperature increases were noted with in the flip-chip package without the heat sink. Cylindrical resistors with axial leads also undergo appreciable increases in temperature in order to dissipate power. Often resistors are packaged in an epoxy case that is molded about the core to provide protection against moisture. It should be noted that the thermal resistance penalty associated with this type of encapsulation is quite high with values ranging from 135 to 195 °C/W. A more detailed description of thermal resistance and its effect on component temperature is presented in Chapter 9.

4.4.2 Capacitors

Capacitors like resistors are packaged in a wide variety of carriers that are usually selected to be consistent with the packaging of the other devices on the circuit board. The PCB usually provides support and wiring for the capacitor; however, in some instances the capacitor is too large and it must be mounted on the chassis. Capacitors are generally divided into four categories, namely aluminum electrolytic, tantalum, film and ceramic. The aluminum electrolytic utilizes aluminum foil as the conductor and aluminum oxide as the dielectric. Electrolytes used in aluminum electrolytic capacitors have near-neutral pH and no aggressive species that might attack aluminum or its oxide. Typical solvents include ethylene glycol, dimethylformamide and g-butyrolactone. Mixtures of these are often used. Common solutes include ammonium, quaternary ammonium, and tertiary amine for cations, and borate or dicarboxylate for anions. Some water is usually present in the electrolyte (1 to 3%) to support formation of the anodic aluminum oxide dielectric after capacitor assembly. Because the electrolyte is liquid, the capacitor is usually packaged in a sealed can as shown in Fig. 4.68a. Circuit board mounting, with the capacitor leads being inserted into plated through holes, is illustrated in Fig. 4.68b. Aluminum electrolytic capacitors are also available for surface mounting with pads at their base instead of pins.

The aluminum can is usually insulated by a vinyl chloride sleeve which also serves for marking the part number, capacitance, working voltage and polarity. The lead wires emerging from the can are a copper based alloy providing a material which can be soldered with conventional methods. An aluminum to copper weld is made inside the can because the lead emerging from the foil is aluminum. In connecting electrolytic capacitors to the PCB, it is important that polarity be established and confirmed. If the capacitor is used in reverse polarity, its life is reduced in the best of circumstances or the capacitor is destroyed in a worst case.

Tantalum capacitors are also of the electrolytic type and if the electrolyte is a liquid, the tantalum capacitor is sealed in a manner similar to that shown in Fig. 4.68. The advantage of using tantalum in place of aluminum is due to the dielectric constant of tantalum oxide ($\varepsilon = 11.6$) as compared to aluminum oxide ($\varepsilon = 4.5$). This property gives the tantalum capacitor a much higher capacitance at a given working voltage for a specified volume. Because capacitor size is extremely important, the tantalum capacitor has gained a large share of the capacitor market.

Film capacitors are fabricated from polymeric films either metallized directly or between layers of metallic foils. These capacitors utilize several different film materials including polyester, polycarbonate, polystyrene, polypropylene and paper. These capacitors have the advantage of very high reliability and an extended temperature range of operation – 55 to 125 °C. However, for a specified capacitance, they are relative large in volume when compared to the electrolytic type capacitors describe previously.

Ceramic capacitors are available in single or multiple layers using materials such as titanium acid barium for the dielectric. Manufacturers of multilayer ceramic capacitors (MLCC) have increased the capacitance per unit volume of the ceramic carrier. In a multilayer capacitor, increased capacity is accomplished by increasing the number of electrodes, and by using thinner ceramic dielectric and electrode layers. The multilayer construction of a ceramic capacitor, presented in Fig. 4.69, shows the internal electrodes, which are typically made from palladium or silver palladium alloys. One factor that complicates the effort to produce thinner electrodes is the trade off between metal electrode thickness and electrical performance. Thicker electrodes provide more continuous, higher conductivity paths for capacitor charging and discharging, but thicker electrodes also increase the cost of manufacturing due to the use of additional quantities of costly metals (silver/palladium).

Capacitors like resistors are affected detrimentally by increasing temperature. Increasing the operational temperature decreases life and decreases the maximum allowable working voltage.

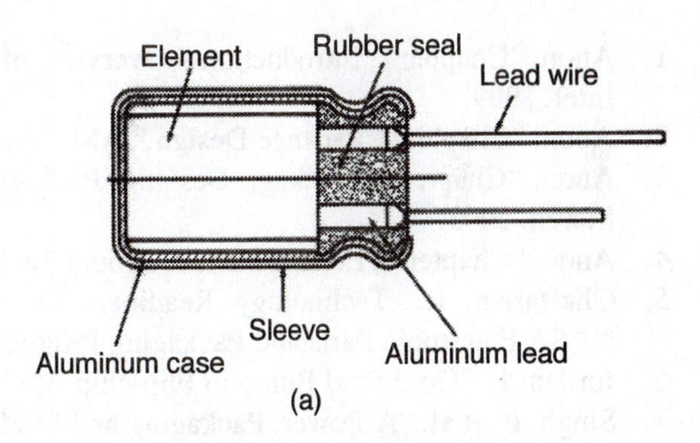

Fig. 4.68 (a) Construction details for an aluminum electrolytic capacitor. (b) Mounting methods for axial and radial leaded capacitors.

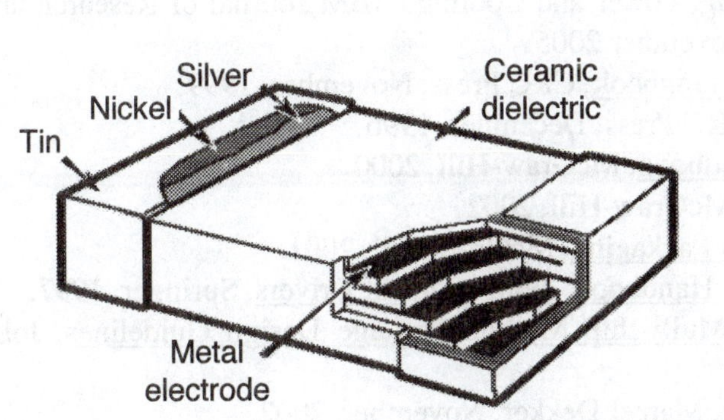

Fig. 4.69 A small high-capacity ceramic capacitor that is surface mounted on a printed circuit board.

4.4.3 Diodes

Packaging for diodes depends largely on their purpose. Signal diodes either pass or block signals representing low voltages and low currents. These diodes are housed either in small diameter glass or plastic cylinders with axial leads for through-hole PCBs or in a variety of surface mounted packages. Surface mounted packages include two terminal leadless carriers, SOICs for diode arrays, two leaded L plastic encased carriers and even CSP with solder ball connections. The diodes with axial leads are mounted to a circuit board using through-hole technology the same way that an axial lead resistor is mounted to a board. The only difference is that the diode has a polarity and must be inserted in the correct direction for the circuit to function properly. The diodes housed in surface mounted packages are also fastened to the board using the same solder reflow operations as other leaded chip carriers.

Diodes used in power supplies to rectify ac voltages are packaged in a wide assortment of housings. As the lead count is very low, the emphasis in packaging the diode is on dissipating large amounts of heat from the device to the heat sink with a small thermal penalty. Most of these power diode packages have a base or a flange which can be fastened directly to a heat sink to minimize the thermal resistance of the diode package.

REFERENCES

1. Anon, "Chapter 1, Introduction—Overview of Intel Packaging Technology," Packaging Data Book, Intel, 1999.
2. Anon, "Chapter 2, Package Design," AMD Packages and Packing Publication, AMD, June 2004.
3. Anon, "Chapter 2, Package Design," Packages and Packing Methodologies Handbook, Spansion, March, 2005.
4. Anon, "Chapter 7, Leaded Surface Mount Technology (SMT)," Packaging Data Book, Intel, 2000.
5. Ghaffarian, R., Technology Readiness Overview: Ball-grid-array and Chip Scale Packaging," NASA Electronic Parts and Packaging Program, January 2003.
6. Jordan, J., "Gold Stud Bump in Flip-chip Applications," Palomar Technologies, 2002.
7. Singh, P. et al, "A Power, Packaging and Cooling Overview of the IBM eServer z900, IBM Journal of Research and Development, Vol. 46, No. 6, pp 711-738, November 2002.
8. Knickerbocker, J. U., et al, "An Advanced Multi-chip Module (MCM) for High-performance UNIX Servers," IBM Journal of Research and Development, Vol. 46, No. 6, pp 779-804, November 2002.
9. Goodman, T. W. and Vardaman, E. J., "FCIP and Expanding Markets for Flip Chip," Tech Search International, Inc., July 1997 Report.
10. Gilleo, K., "Flip or Flop", Circuits Assembly, Feb. 1997.
11. Crippen, M. J., et al, "BladeCenter Packaging, Power and Cooling," IBM Journal of Research and Development, Vol. 49, No. 6, pp 905-920, November 2005.
12. Blackwell, G. R., The Electronic Packaging Handbook, CRC Press, November, 1999.
13. Witaker, J. C., The Electronics Handbook, CRC Press, December, 1996.
14. Walsh, R. A., Electromechanical Design Handbook, McGraw-Hill, 2000.
15. Lau, J. H., et al, Electronics Manufacturing, McGraw-Hill, 2002.
16. Chapman, S., Fundamentals of Microsystems Packaging, McGraw-Hill, 2001.
17. Tummala, R. R., Microelectronics Packaging Handbook: Technological Drivers, Springer, 1997.
18. Pecht, M., Integrated Circuit, Hybrid and Multi-chip Module Package Design Guidelines, John Wiley, 1994.
19. Rice, R. W. Ceramic Fabrication Technology, Marcel Dekker, November, 2002.
20. Pecht, M., Handbook of Electronic Package Design, Marcel Dekker, 1991.
21. Lyman, J. Microelectronics Interconnection and Packaging McGraw-Hill, New York (1980).
22. Blodgett, A. J. and D. R. Barbour, "Thermal Conduction Module: A High Performance Multi layer Ceramic Package", IBM Journal Research Development, Vol. 26, No. 1, pp. 30-36, 1982.
23. Burger, W. G. and C. W. Weigel, "Multi-Layer Ceramics Manufacturing", IBM Journal Research Development, vol. 27, No. 1, pp. 11-19, 1983.

EXERCISES

4.1 What is a chip carrier and what purposes does it serve?

4.2 What are the two main approaches in the design of chip carriers?

4.3 Describe the three different types of surface mount chip carriers.

4.4 Prepare a sketch showing the arrangement of components in a DIP type chip carrier.

4.5 Prepare a sketch showing the arrangement of components in a ball-grid-array. What assumption did you make in preparing this sketch?

4.6 Ceramic chip carriers are very expensive. Why are they used when plastic chip carriers are much less costly?

4.7 Describe the general features of through-hole technology.

4.8 Describe the general features of surface mount technology with leaded chip carriers.

4.9 Describe the general features of surface mount technology with leadless chip carriers.

4.10 Describe the general features of surface mount technology with ball-grid-array packaging.

4.11 Describe the general features of direct chip attachment considering both wire bonding and flip chip connections.

4.12 What are the advantages and disadvantages of the DIP family of chip carriers?

4.13 Determine the area efficiency of a DIP which is used to house a chip which is 4 by 6 mm in size. Consider the number of pins on the DIP to be an open variable and graph the area efficiency as a function of I/O count.

4.14 What are some variants in the design of DIPs?

4.15 What is a SIP and what devices are housed in this package?

4.16 What is a ZIP? How does it differ from a SIP?

4.17 How is the pin-grid-array related to the DIP? Why do some designers specify the pin-grid-array in the design of new product?

4.18 Develop an equation showing the area of a pin-grid-array as a function of pin count. Assume the chip size is 12 × 12 mm and that pins are not placed under the chip cavity. Consider a square array on 100 mil (2.54 mm) centers. Also consider as a second case a staggered pin arrangement with the same pitch. Prepare a graph showing this relation.

4.19 What are some variants in the design of pin-grid-arrays?

4.20 Describe the two materials used to fabricate the substrates for pin-grid-arrays.

4.21 Describe the two techniques used with pin-grid-arrays for enhancing it heat dissipation capably.

4.22 What is a socket and why is it used to support pin-grid-arrays?

4.23 What is the oldest type of chip carrier and why is it still in use after more than 50 years?

4.24 Describe the advantages and disadvantages of surface mounted chip carriers relative to the DIP.

4.25 Prepare a list of the various types of surface mounted chip carriers.

4.26 Describe the purpose of the solder balls located at the center of the BGA illustrated in Fig. 4.19.

4.27 Describe the general characteristics of discrete device packages.

4.28 Go to the Internet and find a supplier of plastic quad flat packs (PQFP). Describe their product line. Are drawings of the package footprint available to aid you in the subsequent design of the printed circuit board?

4.29 Describe the general features of a SOIC type chip carrier

4.30 Describe the general features of a J leaded chip carrier. Why is this package limited to a lead pitch of 50 mil (1.27 mm)?

4.31 Write an equation describing the area of a J leaded chip carrier as a function of I/O count and prepare a graph showing this relation.

4.32 Describe the general features of a leadless chip carrier. What are the two most important disadvantages of this type of chip carrier?

4.33 Describe the general features of a thin-small-outline-package (TSOP). What type of devices is it used to house? In what type of product is it found?
4.34 The ball-grid-array is a relatively new chip carrier design. Describe its general features.
4.35 What are the principle advantages of the ball-grid-array? What are its disadvantages?
4.36 Write an equation describing the area of a ball-grid-array as a function of I/O count and prepare a graph showing this relation.
4.37 Sketch the features on a ball-grid-array that enhances its heat transfer capability.
4.38 What is the purpose of the column grid array illustrated in Fig. 4.33? Explain why the columns are more resistant to thermal fatigue that the typical solder ball found on a BGA.
4.39 Describe the difference between chip scale packaging and ball-grid-arrays.
4.40 Cite the primary applications of CSP.
4.41 Prepare a sketch of a CSP using flip-chip connections. Include in the sketch the connections to the PCB.
4.42 What is a multi-chip module? What are its advantages? When is it used?
4.43 Describe the features of the IBM's TCMCM.
4.44 Why are multi-chip modules still in use today when chips can be design with a complete system on a single chip?
4.45 What is meant by die stacking and why is this practice employed?
4.46 Cite the advantages and disadvantageous of ceramic chip carriers.
4.47 Describe the method of producing multi-layer ceramic substrates using the co-fired process.
4.48 Describe the process of solder bumping on a chip so that it may be bonded directly to a printed circuit board.
4.49 How is the mismatch between the temperature coefficient of expansion of the chip and the printed circuit board managed for wire bonding applications?
4.50 How is the mismatch between the temperature coefficient of expansion of the chip and the printed circuit board managed for flip-chip applications?
4.51 Describe the process for bonding the chip to a ceramic substrate.
4.52 What is eutectic solder? Why is it important to consider eutectic solder in many manufacturing processes?
4.53 Would you consider the gold-tin eutectic as a replacement solder for the gold-silicon eutectic? Why?
4.54 Describe the process of bonding the chip to the frame of a leaded chip carrier.
4.55 Discuss the engineering trade-off between using an epoxy adhesive and a thermoplastic for bonding a chip to a lead frame.
4.56 In what applications is polyimide used as an adhesive for bonding chips to lead frames?
4.57 Why are the thermosetting adhesives filled when used in chip bonding processes? What is a common filler material?
4.58 Describe the physical properties of materials used to fabricate lead frames.
4.59 The physical properties of three alloys used for lead frames are given in Table 4.1. Copper alloy MF 202 is less expensive and has better heat transfer characteristics than the other two alloys (Alloy 42 and Kovar). Why are the other alloys used in some designs?
4.60 Describe the automated wire bonding process for gold wire.
4.61 Write an equation for the number of wire bonding pads which can be placed about the perimeter of a chip. Prepare a graph of the I/O count as a function of perimeter length for three different pad sizes and spacings.
4.62 How does the automated wire bonding process change when aluminum wire is used to replace gold wire.
4.63 Is it possible to use copper instead of gold or aluminum in wire bonding? Cite the advantages and disadvantage of wire bond with copper.

4.64 Calculate the cost of the gold used in the bonding wires for a 100 pin chip carrier. Take the cost of gold wire at $500 per troy ounce. Consider at least two common wire diameters. Comment on the cost trade-off between gold and aluminum bonding wires.

4.65 Describe the tape automated wire process for making connections between the chip and the chip carrier.

4.66 Cite the advantages and disadvantages of the TAB process.

4.67 Describe the process for producing solder bumps on chips that are to be connected using flip-chip technology.

4.68 Cite the advantages and disadvantages of the flip-chip process process.

4.69 How is it possible to use flip chip technology with organic substrates which exhibit a much higher temperature coefficient of expansion than a silicon chip?

4.70 What is the purpose of an interposer?

4.71 Sketch the design of a chip carrier with a rigid interposer using flip-chip connections. Include the PCB in this sketch.

4.72 Sketch the design of a chip carrier with a flexible interposer that uses a flexible circuit for making the required connections between the chip and the PCB.

4.73 Sketch the features of a interposer manufactured at the wafer. Cite the advantages of this process and indicate why it may become the dominate process for producing CSP in the future.

4.74 Describe the process and the materials used to encapsulate chips and their connections with plastic.

4.75 Why is important to carefully package bare dies for shipment from one facility to another? Describe three of the methods for shipping individual dies.

4.76 Describe two methods for shipping wafers after they have been processes but before the chips have been cut from the wafer.

4.77 Examine a circuit board from an available PC and identify the passive components such as the capacitors and resistors. Describe some features of the board and the first level packages worth noting as design techniques.

4.78 Describe at least four through-hole packages for resistors.

4.79 Why would a design use a zero ohm resistor?

4.80 Resistors can be housed in leadless carriers without concern for solder joint failures due to cyclic thermal fatigue. Why?

4.81 Reference the power resistor shown in Fig. 4.65j. Where will this resistor be mounted. Cite the reasons for your answer.

4.82 Is it necessary to derate resistors when they are required to operate at elevated temperatures? Why?

4.83 Is self heating a consideration in derating a resistor? Why?

4.84 Explain why it was not necessary to consider self heating in the discussion pertaining to the packaging of capacitors. Consider capacitors in both ac and dc circuits in the explanation.

4.85 There are four types of capacitors. Describe the criteria for selection of the type used in a specific design.

4.86 What are the characteristics of a diode that is important in the selection of a first level package?

4.87 Prepare a caption for the figure shown in Fig. Ex4.87.

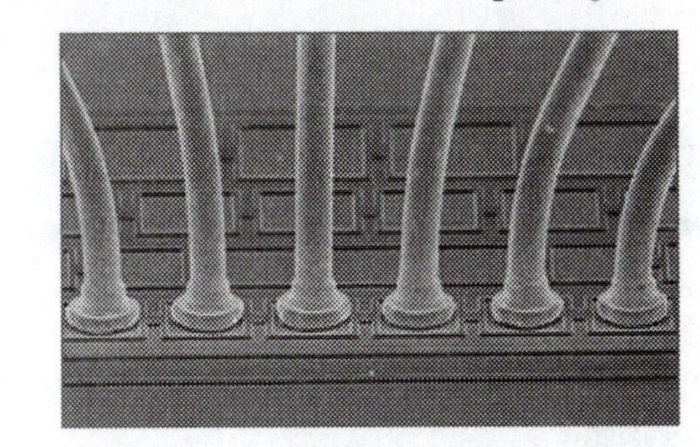

Fig. Ex4.87

CHAPTER 5

SECOND LEVEL PACKAGING SUBSTRATES AND PRINTED CIRCUIT BOARDS

5.1 INTRODUCTION

The printed circuit board (PCB) is a major element in the mechanical design of an electronic system because it is the primary field replaceable unit. The PCB embodies many important functions that include:

1. A mounting surface for most of the components involved in the design of a product.
2. Soldering pads to facilitate 1st to 2nd level and 2nd to 3rd level electrical connections. These pads also provide the mechanical anchor points for attaching the components to the PCB.
3. Wiring channels to provide conduits for chip to chip connections.
4. A test bed to provide accessible and organized points that are probed in testing the circuits.
5. A marking surface to assist in identification of the components to be placed on the board in assembly. Other board markings are also added to support manufacturing operations.
6. A controlled profile on one side of the card with additional markings to support field maintenance and service.

Each of these functions of the circuit board will be discussed individually. The circuit board serves as a mounting plane for the various components such as integrated circuits (IC's), resistors, capacitors and diodes. The board provides a method for component attachment, either through-hole, leaded surface mount or ball grid arrays to the PCB. This attachment essentially fixes the position of each component for the life of the product. The board acting as a planar mounting structure also mitigates shock and vibration forces that are transmitted to the components through the board. In this sense, the PCB, like the chip carrier, acts to protect the components. Because the circuit boards are usually relatively thin planes with large lateral dimensions, the weight of the components that can be supported is limited. Usually heavy components, like transformers and larger displays, are mounted directly on the chassis because they can impart forces large enough to damage the board in a shock or vibratory environment.

The circuit board is usually a laminate fabricated from glass cloth and an epoxy matrix with a thin copper cladding on one or both sides. The cladding is configured by a lithographic process to any shape required to provide land patterns and the wiring connections to the electronic components. Surface mount land patterns, also called footprints or pads, define the sites where components are to be soldered to the PCB. The design of the land patterns is critical, because the design affects the solder joint strength and hence the reliability of the PCB. The design of the land patterns also affects solder defects, the ability to properly clean the PCB, the ability to test each component, and the ability to rework or repair a component.

A typical land pattern etched from the copper cladding, illustrated in Fig. 5.1, provides soldering surfaces that are precisely aligned to match the corresponding leads or pads from the components. The size and shape of these solder land patterns control to a large degree the formation and geometry of the solder joints connecting the chip carrier leads to the board. Other solder pads are provided for attachment of the leads for the I/O cables or connectors that carry power or signals to other PCB's in the system.

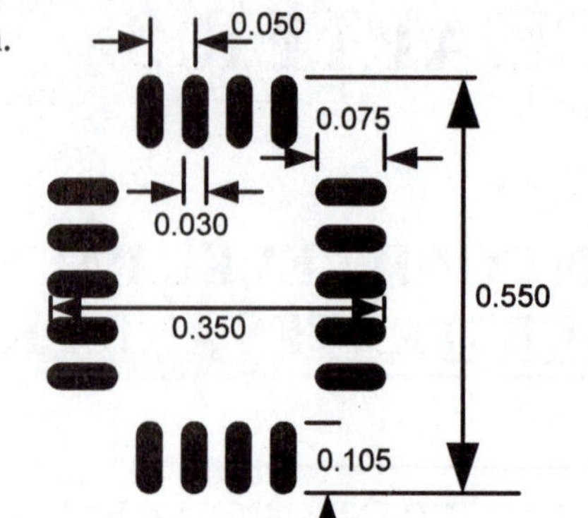

Fig. 5.1 The land pattern etched in copper to correspond with the leads on a J leaded package with 18 leads on 50 mil centers. Dimensions are in inches.

The circuit board also supports the wiring traces, either microstrip lines or strip lines that make the required connections among the components located on the board. These microstrip lines are arranged in wiring channels as illustrated in Fig. 5.2. The wiring traces are etched from the copper cladding and a typical circuit board may support 1000 to 10,000 individual traces leading from one wiring point to another. Wiring traces include conductors for the signal that are small in cross sectional area because of the low currents involved in switching the logic gates. However, traces for the conducting power to the components are much larger in width to accommodate the higher currents required to operate each device. Most circuit boards contain several layers with the wiring planes embedded on interior layers. The power and ground wiring (V^+ and V^-), in multilayer construction, is often provided with nearly solid copper planes dedicated to supplying the required current. The power and ground planes also serve to control the characteristic impedance of the strip lines located on adjacent signal planes.

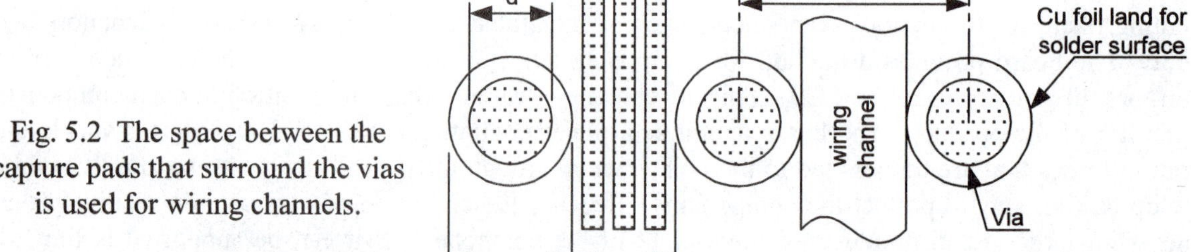

Fig. 5.2 The space between the capture pads that surround the vias is used for wiring channels.

When the circuit board is completely assembled, as shown in Fig. 5.3, it contains many different types of components and perhaps 100 or more devices. Because failure of at least one of these devices over the life of the product is probable, it is important that the circuit board be field replaceable. When a PCB is replaced in the field, it is usually repaired at a service center that is equipped to remove and replace different types of individual components.

When repairing a circuit board, it is necessary to probe many different locations with a voltmeter or an oscilloscope to ascertain that if any of the components have failed. To facilitate probing and subsequent repair, the devices should have land patterns that are accessible to a voltage probe[1]. Also, the leads should be accessible for specialized soldering tools that can melt the solder at all of the

[1] For ball grid arrays probing is difficult if not impossible because a large fraction of the connections located under the chip carrier are inaccessible.

chip carrier joints simultaneously to permit rapid removal of the devices without damage to the board and the remaining devices. To facilitate testing and repair, a clear spacing between the pads of through-hole or surface mount components is usually 100 mil (2.54 mm). Clear spaces of 50 mil (1.25 mm) are also specified around the edges of the PCB to allow clearance for test probes.

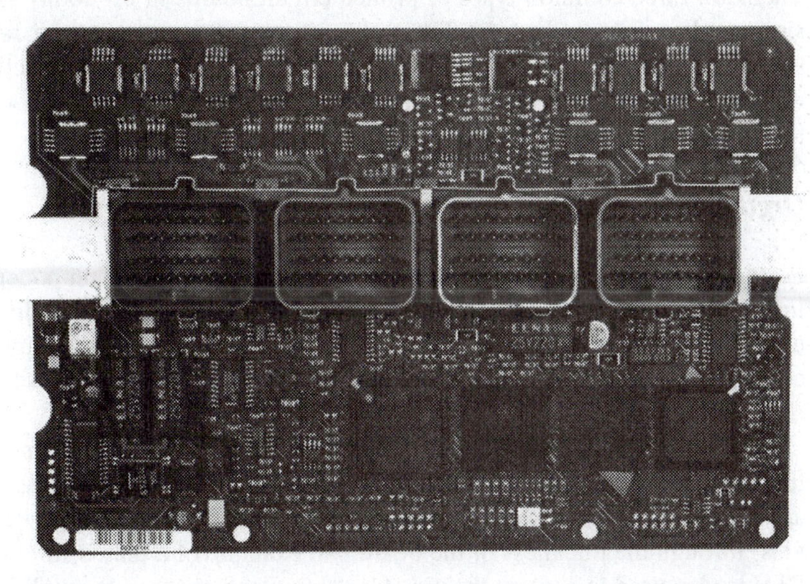

Fig. 5.3 Photograph of an assembled PCB.

Because the copper cladding on the circuit board material can be etched to any configuration, it is easy to mark the circuit card to facilitate the manufacturing process[2]. The example presented in Fig. 5.4 shows markings identifying the location various IC's employed in assembling this board. Other markings can be added to indicate the part number, revisions and the date of circuit board production. It is clearly easy to mark the board with component outlines, part numbers, production dates, lot numbers or any other information that will assist in production control, component assembly and inventory control.

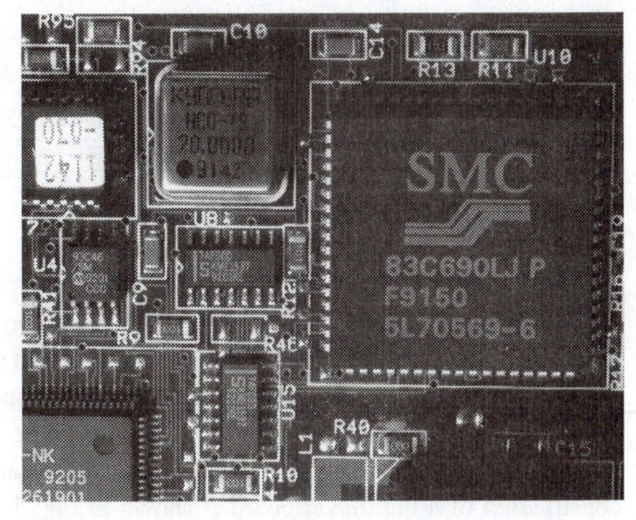

Fig. 5.4 Markings identify the location of components that are mounted onto a PCB.

To facilitate maintenance, the shape of the card is modified to prevent improper substitution in when the PCB is replaced in the field. The profile of the circuit card along the edge where the connector pads are located is usually slotted or notched. These slots act as keys and permit the correct board to be inserted into the edge connector but prohibit the incorrect substitution of the wrong part number. Markings of the part number and the probe sites with reference voltages can also be placed on the card to assist maintenance personnel in servicing the card either in the field or at the repair facility.

---

[2] Markings can also be applied to the PCB at very low cost using a screening process.

5.2 TYPES OF PRINTED CIRCUIT BOARDS (PCB)

There are three common types of printed circuit boards in use today including the single-sided, double-sided and multilayer board. These three types of boards can be fabricated from organic laminates (polymeric composites), flexible organics or ceramics. The selection of a specific type of board depends to a large degree on circuit density and the number of connections that must be made between the components mounted on the board.

Organic Laminates

For very low-density, low-cost product where relatively few components with limited I/O are involved, low cost, polymeric composite, single sided boards that have circuit lines on only one side are often used. Holes are drilled or punched into these single sided laminates to support pins from the components; however, these holes are usually not plated and the solder joint is limited to the region between the pin and the top surface soldering pad.

For low to moderate density product, double sided boards are employed that incorporate circuit lines on both the top and bottom surface of the laminate. The circuits on the opposite sides of the board are connected as necessary by plated-through-holes (PTH) that serve as transverse conductors (called vias) through the thickness of the board. An example of a plated-through-hole is shown in Fig. 5.5. The PTH is also used to contain the pins for components that use through-hole mounting. In these cases, the pin and the copper plating on the wall of the PTH are soldered together and the solder joint encompasses the top and bottom capture pads and fills the entire hole.

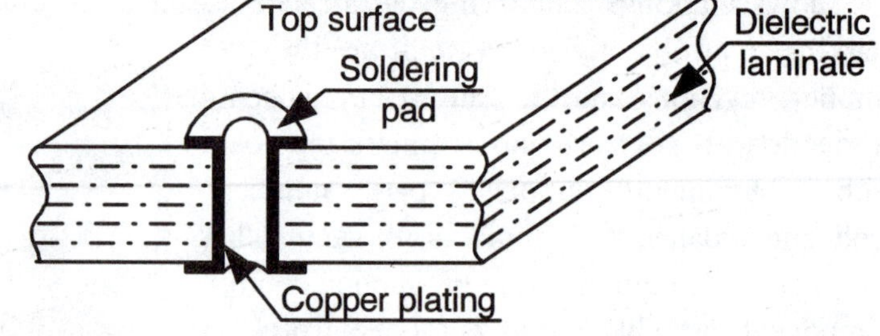

Fig. 5.5 Plated-through-hole connecting top and bottom surfaces of a double-side circuit board.

For two-sided PCBs with surface mounted components, the leads from the chip carriers are soldered to surface land pads. PTHs on PCBs, incorporating surface mounted chip carriers, are filled with solder and only serve to connect circuit traces on one side of the board with traces on the other side.

The most common PCB in use today is the multilayer board (MLB) because these boards accommodate the high I/O count chips used in most products. Multilayer boards are fabricated from polymeric composites containing several layers of laminates—each with two wiring planes. Typical construction of a multilayer board is shown in Fig. 5.6. The layers are prepared in a sandwich with a pre-impregnated epoxy-glass-cloth sheet, known as pre-preg, positioned between each layer. The epoxy in the pre-preg at this stage is only partially polymerized. After a stack of laminates and prepreg are assembled, it is placed in a hot platen press or an autoclave to complete polymerization of the prepreg layers. This lamination process permanently bonds all of the layers together. It is important that the registration of all n layers be maintained for subsequent drilling of through thickness vias. The MLB consists of a single multilayer laminate with 2n wiring planes. Plated-through-holes act as vias to connect wiring in the interior planes with the leads from the surface mounted chip carriers. Other plated-through-holes are used to support through-hole chip carriers and they become part of the soldered joints as described above. The plated-through-holes may also be filled with solder and used for thermal vias to enhance the conduction of heat through the thickness of the laminate.

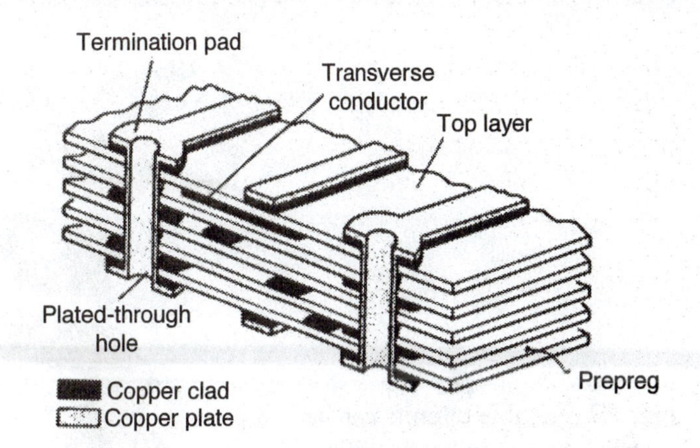

Fig. 5.6 Multilayered circuit board using plated-through-holes to connect wiring traces on the top and bottom surfaces with wiring traces on interior wiring planes.

Each of these three types of circuit boards has its advantages and disadvantages. The single-sided board is the lowest cost and is easy to maintain. However, it supports only low density circuits, control of electrical parameters is not possible and crossovers of wiring traces must be accommodated with jumpers. Double-sided boards permit much higher density, some degree of impedance control is possible and crossovers can be eliminated with through thickness vias. These advantages are gained with a modest increase in cost. Multilayer boards permit very high density circuits to be connected, but at a significant increase in cost. MLBs also have the advantages of excellent impedance control of the signal lines and high power switching through the use of ground and power planes capable of conducting very high currents. Disadvantages, in addition to cost, include lower yields in manufacturing with an increasing number of layers and difficulty in repairing shorts or opens that may develop at interior wiring planes.

Organic Flexible Circuits

Organic flexible circuits are fabricated from polymeric films that are bonded to thin copper foils with adhesives. They can be one-sided, double-sided or multi-ply. Flexible circuits have many applications including, interposers for first level packaging, circuit boards, and as a substitute for wire harnesses. The thin polymer films employed in their fabrication offers considerable weight and space savings of thicker more rigid organic laminates. Flexible circuits can be bent, twisted, and folded to utilize all three dimensions of an enclosure.

Although flexible circuits are thin, they can be used as substrates for surface mounted chip carriers. For dense circuits several layers of copper clad polyimide film can be bonded together with film adhesives. Vias are drilled in the thin polymeric film with a laser and filled with a conducting adhesive to form z direction connections. Circuit wiring is formed using photolithographic techniques similar to those used for processing the rigid organic laminates. The minimum line width and spacing that can be achieved is a function of the thickness of the copper foil and is not influenced by the thickness of the polyimide films.

Flexible circuits are often used to replace cables and wiring harnesses. The flat conductors found on a flexible circuit have a much greater surface to volume ratio than round wire—a geometry that facilitates the dissipation of heat. Flexible circuits are also used almost exclusively on cables that move with the traveling bed of electronic products such as a scanner. If properly designed with an adequate radius to thickness ratio, their fatigue life exceeds 100,000 cycles.

Examples of several flexible circuits are presented in Fig 5.7.

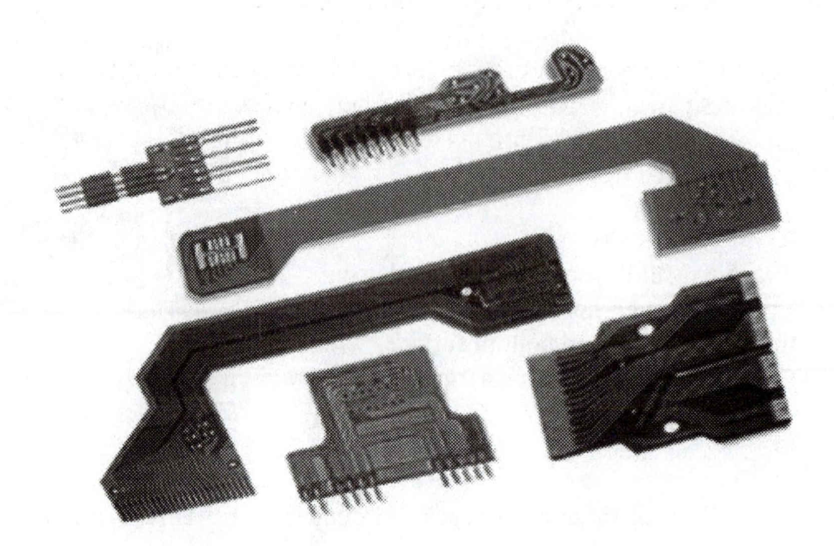

Fig. 5.7 Flexible circuits can be twisted or bent to better utilize space in an electronic enclosure.

Ceramic Circuit Boards or Substrates

Most ceramic circuit boards or substrates for multi-chip-modules are fabricated from green sheets. These green sheets, described in Section 4.2.7, are paper thin layers of glass-ceramic powders that are bonded together with a resin binder. The paper-like sheets are punched with small holes to form visa and printed with refractory metal inks where circuit wiring is required. The vias are also filled with refractory metal ink. The sheets are stacked in registration, placed under pressure and fired at high temperature. The resin binder burns off, the ceramic sheets shrink and the ceramic particles are sintered together. The resulting substrate is employed on high-end product where reliability is a major factor in the design. The low temperature of coefficient of expansion of the glass-ceramic material closely matches that of silicon; consequently, chips can be direct attached with a flip-chip process to the ceramic circuit boards without significant danger of solder ball failures due to relatively large cyclic temperature changes.

5.3 CIRCUIT BOARD MATERIALS

A large number of materials are available for use in fabricating circuit boards. Three factors are involved in the material selection namely cost, electrical characteristics and mechanical properties. Cost factors are extremely important because the market for circuit boards is extremely large. While the overall market in 2000 was about $10 billion in revenue, a Frost & Sullivan industry analyst forecasts the market to reach revenues of nearly $17 billion in 2006. The analysis also finds that most single- and double-layer rigid printed circuit board production has shifted to the Pacific Rim; however, North America remains the technological leader in multilayer rigid boards. The immense volume of the market implies that cost will be a major consideration. Mechanical properties of importance include coefficient of thermal expansion, dimensional stability, flexural strength, modulus of elasticity, thermal resistance and resistance to chemicals and solvents. The electrical characteristics of concern include surface and volume resistivity, dielectric constant, dielectric strength, and dissipation and loss factors. These electrical properties are measured using the test methods described below.

5.3.1 Test Methods for Electrical Properties

Surface resistivity is measured by using the specimen shown in Fig. 5.8. A voltage of 500 V is applied across the center and ring electrodes and the bottom electrode is grounded to provide a guard that eliminates the effects of currents that pass through the thickness of the laminate. The surface resistivity ρ_s is determined from:

$$\rho_s = \frac{R_s p}{s} \tag{5.1}$$

where R_s is the surface resistance, p is the perimeter of the ring electrode and s is the spacing between the center and ring electrodes.

The units for ρ_s are given as ohms per square and usually range from 10^7 to 10^{10} Ω/square for polymers. Note, that units are not given to the word square. This word is inserted as a reminder that we are dealing with surface resistivity.

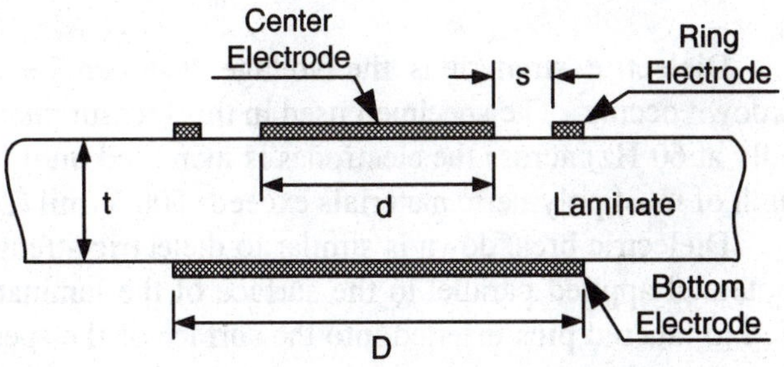

Fig. 5.8 Test specimen used to measure surface resistivity and volume resistivity (ASTM D257).

Volume resistivity ρ_v is also measured with the specimen illustrated in Fig. 5.8; however, the voltage is applied between the center and bottom electrodes and the ring electrode is grounded to serve as a guard. The relation for ρ_v is given by:

$$\rho_v = \frac{R_v A}{t} \tag{5.2}$$

where R_v is the volume resistance, A is the area of the center electrode and t is the thickness of the laminate.

Most polymers exhibit $\rho_v > 10^{15}$ Ω-cm.

The dielectric constant ε_r of a material is defined by:

$$\varepsilon_r = \frac{C_r}{C_v} \tag{5.3}$$

where C_r is the capacitance of a capacitor that uses the material being measured as the dielectric and C_v is the capacitance of the same capacitor with vacuum as the dielectric.

A low dielectric constant is necessary for circuit boards used in high-speed signal processing where distributed capacitance is an important property. Most polymers exhibit ε_r that range from 2 to 4; however, when these polymers are reinforced with glass fibers that have $\varepsilon_r \approx 6$, the effective dielectric constant for the laminate increases to $4 < \varepsilon_r < 5$. The relative dielectric constant depends on the percent of glass and resin. A common laminate, known as FR-4, with 50 to 55% epoxy as the matrix exhibits a relative dielectric constant of 4.6 to 4.8 as indicated in Fig. 5.9. This data was measured at a frequency of 1 MHz; however, at high frequencies the dielectric constants of both glass and epoxy are lower. Clearly frequency must be considered in calculating capacitance and characteristic impedance of the circuit lines (traces) on the laminate.

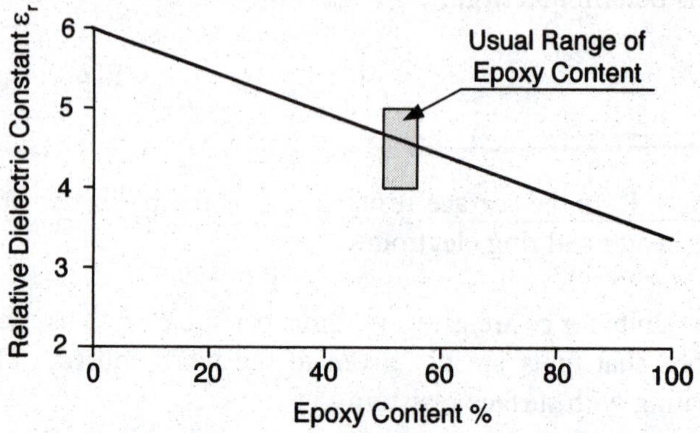

Fig. 5.9 Relative dielectric constant as a function of epoxy content for a FR-4 laminate.

Dielectric strength is the voltage that can be applied per unit thickness before electrical breakdown occurs. The specimen used in this measurement is shown in Fig. 5.10a. The voltage applied (usually at 60 Hz) across the electrodes is increased until failure of the insulation occurs. The dielectric strength of most polymeric materials exceeds 500 V/mil (20,000 V/mm).

Dielectric breakdown is similar to dielectric strength except that the measurement is made with the potential applied parallel to the surface of the laminate. As indicated in Fig. 5.10b, the voltage is applied to tapered pins inserted into the surface of the specimen on 25.4 mm centers. To avoid surface conduction, which occurs at high voltages, the specimen is immersed in oil. Typical results for dielectric break down for polymers usually exceed 10 kV.

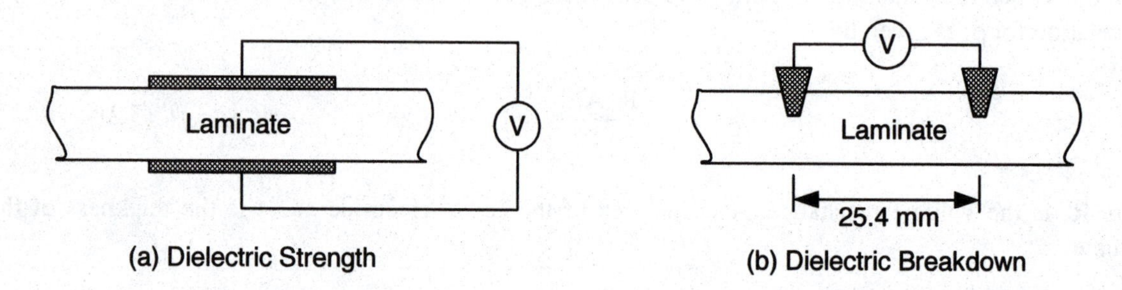

Fig. 5.10 Test arrangement for measuring the dielectric strength and breakdown voltage (ASTM D149). Both specimens are immersed in silicone oil to eliminate surface conduction.

The dissipation factor is a measure of the loss of power due to the insulating material. This quantity is defined as the ratio of the reactance to the resistance that is related to the loss angle δ by:

$$\tan\delta = \frac{1}{R}\left[\omega L - \left(\frac{1}{\omega C}\right)\right] \tag{5.4}$$

The loss factor F_L is related to the dissipation factor because:

$$F_L = \varepsilon_r \tan\delta \tag{5.5}$$

The dissipation factor for polymers decreases with increasing frequency. This property, usually reported for a frequency of 1 MHz, ranges from 10^{-2} to 10^{-4}. To minimize power losses it is important to select materials with low dissipation losses.

The type of laminate selected in design depends upon the application. If the product involves high frequency analog circuits, it is essential that the signal integrity (voltage amplitude and frequency) be maintained. Line losses must be minimized because these circuits process voltage signals with small amplitudes. Fortunately these circuits usually are not dense and the PCBs do not involve more than a few signal layers. Thus the multilayer boards are relatively thin and manufacturing is not a major concern. Maintaining a low and constant dielectric constant to insure low loss lines on the board is the objective of the design.

If the PCB is intended for a digital product, then the circuit is probably very dense and many signal planes are required to connect all of the leads or pads on the chip carriers. As a consequence the PCB is relatively thick and manufacturing becomes more difficult. Solder processes at elevated temperature results in higher stresses on the copper plated vias. To avoid failure of these vias, the glass transition temperature T_g of the polymeric matrix must be higher than the soldering or reworking temperatures imposed on the PCB.

5.3.2 Mechanical Properties

Mechanical properties such as flexural strength and modulus of elasticity are well understood and will not be described here. However, the dimensional stability and the temperature coefficient of expansion (TCE) require further explanation. Dimension stability is a major concern at elevated temperatures associated with soldering the circuit boards. If the laminate material under goes large dimensional changes at soldering temperatures, product yield is adversely affected. The temperature coefficient of expansion is an excellent indicator of the dimensional stability of the polymers used in circuit board fabrication. Consider the z direction expansion temperature curves for several epoxies, two polyimides and cyanate ester shown in Fig. 5.11. The slopes of these curves give the temperature coefficient of expansion α_z. Note, that there is a sharp discontinuity in each of these curves at the glass transition temperature T_g. The glass transition temperature is where the polymer undergoes a phase transition and changes from a glass-like to a rubbery material. For temperatures higher than T_g, the material is too soft to maintain dimensional stability if forces are applied in processing. Also, the coefficient of expansion in the thickness direction of the laminate α_z is markedly affected when the temperature of the laminate exceeds the transition temperature. A typical example showing α_z as a function of temperature for an epoxy glass laminate is presented in Fig. 5.12. As α_z increases the expansion of the thickness of the laminate increases that produces very high stresses on the copper plated-through-holes. The result, particularly if the temperature of the PCB is cycled during operation, is cracking of the copper plating in the through-holes due to fatigue. Clearly, the polyimides with their higher T_g exhibit superior dimensional stability at elevated temperatures than the epoxies.

The operating environment is also important in selecting materials. Humidity resistance is critical because the water absorption by the polymers affects their electrical properties and can seriously degrade performance. The maximum operating temperature is also important as both mechanical and electrical properties degrade rapidly when the glass transition temperature T_g for the polymer is exceeded.

The majority of circuit board laminates are fabricated using polymers as a matrix for either a glass-cloth, cotton-fabric or kraft-paper reinforcement. These laminates are often known as rigid PCBs and are fabricated using thermosetting polymers such as phenolic, epoxy or polyimide. Flexible circuits are also produced; these circuits employ thin plastic films without reinforcement to support the circuit traces.

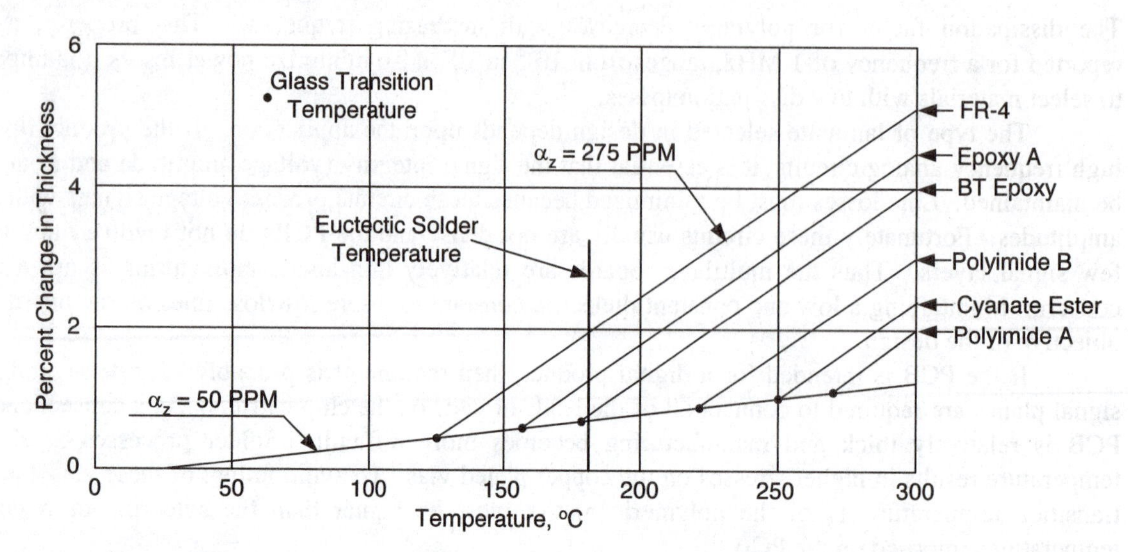

Fig. 5.11 Expansion of several circuit board materials with temperature. The sharp discontinuity defines the glass-transition temperature T_g. (The slope α_z is the temperature coefficient of expansion).

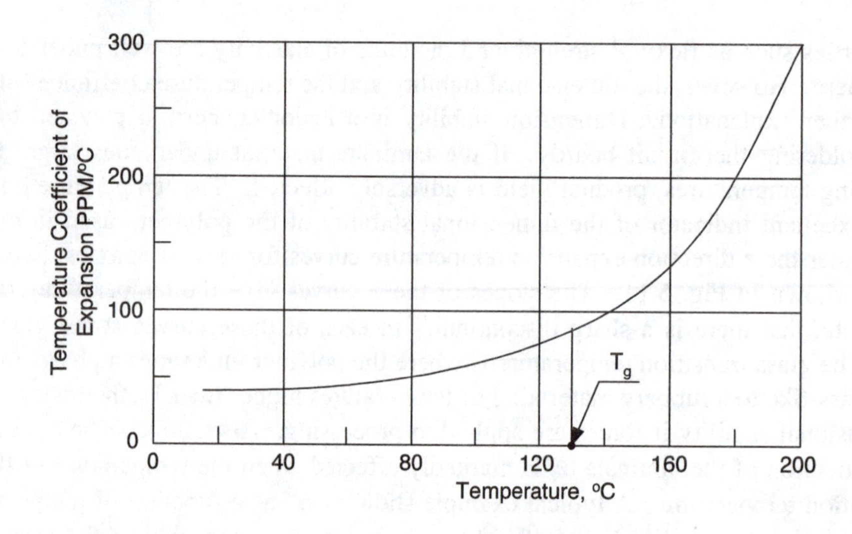

Fig. 5.12 Temperature coefficient of expansion α_z in the thickness direction of a glass-epoxy laminate.

5.3.3 Polymeric Reinforced Laminates

Reinforcing Materials

Let's first consider the reinforcing materials used in fabricating circuit boards. Cotton cloth or paper is used with phenolic resins to produce insulating boards that are used in a number of electrical application and for simple circuit boards. These materials can be punched (instead of drilled) to produce rough holes used for mounting components. The surface finish of the holes is not sufficiently smooth to enable plating, so the boards are restricted to single side mounting.

NEMA grades X, XX and XXX are paper reinforced phenolic with a natural color that is typically light tan to brown. Phenolic polymers are the oldest, best-known general purpose thermoset resins. They are among the lowest in cost and easiest to process. Phenolics are quite adequate for some electrical applications. However, they are not equivalent to epoxies in resistance to moisture, humidity, dimensional stability, shrinkage and retention of electrical properties in extreme environments. The paper reinforced phenolics have good electric strength properties with moderate mechanical strength and are often used for template material and/or back-up material. A similar laminate XXXPC affords the opportunity to form the holes by cold punching that saves on drilling costs for single sided boards. Other products FR-2 and FR-3 are similar to the XX grades except that the phenolic matrix, in the FR grades, has additives that make it flame resistive.

Glass cloth is used as the reinforcing material of choice for the majority of laminates to provide strength and structural integrity. Most glass cloth is woven from yarns made from E-glass, S-glass, or NE-glass. E-glass is the most common of the different types of glass fibers used in the yarns that are employed in weaving fiberglass fabrics. Because E-glass is manufactured in huge volumes, it is inexpensive. S-glass exhibits higher strength, but it is usually used in structural applications. Finally, NE-glass exhibits improved electrical and mechanical performance when compared to E-glass and it is used in high performance laminate product lines.

Table 5.1
Properties of fiberglass materials

Property	E-glass	NE-glass
Coefficient of Thermal Expansion ppm/°C	5.5	3.4
Dielectric Constant @1 MHz	6.6	4.4
Dissipation Factor @1 MHz	0.0012	0.0006

Table 5.1 shows a comparison of the properties of the two different types of glass used in electronic applications. NE-glass exhibits better temperature stability, lower relative dielectric constant, and lower dissipation factor than the equivalent E-glass. Lower dielectric constant and dissipation factor are important properties in high-frequency, high-performance signaling applications.

To better understand the electrical properties of epoxy laminate substrates, it is necessary to develop an understanding of the construction of the glass cloth used as reinforcement. Fibers of various diameters (5 to 13 μm) are combined to form strands. A typical strand contains from 102 to 816 filaments, and a yarn usually contains a single strand. The number of filaments in a strand determines the thickness of the yarn, the coarseness of the weave and the thickness of the cloth. Examples of three different cloths used as reinforcing materials in PCB production are illustrated in Fig. 5.13.

A laminate is manufactured with several layers these fabrics. Each layer is filled with a polymer (usually epoxy). The laminates are cured to form core sheets of various thicknesses. In other cases, the cure is incomplete (B staged) to produce prepreg sheets. The electrical characteristics of epoxy fiberglass composite changes with the weave pattern of the glass cloth. The loose, open weave of 1080 glass cloth allows significant interstitial epoxy fill. For example, the pitch of 1080 style weave is 16.7 mil in the warp direction and 21.3 mil in the fill direction. The fact that the pitch of the fabric weave pattern is larger than pitch of the circuit traces on the circuit board has electrical consequences. The circuit traces can be positioned directly over either warp or fill yarns or centered between two strands of either warp or fill yarns. For each of these configurations, the local electrical and mechanical properties differ. Local changes in the relative dielectric constant affect the propagation velocity of the signal and the time of flight for a specific wire length. For short rise times, corresponding to 3Gb signals, matching the length of the signal wires does not insure a time of flight match. Time of flight variations for wires with identical length in close proximity on the same surface may reach 5% in common cases corresponding to changes in the signal propagation velocity of ± 4 to ± 7 ps/in.

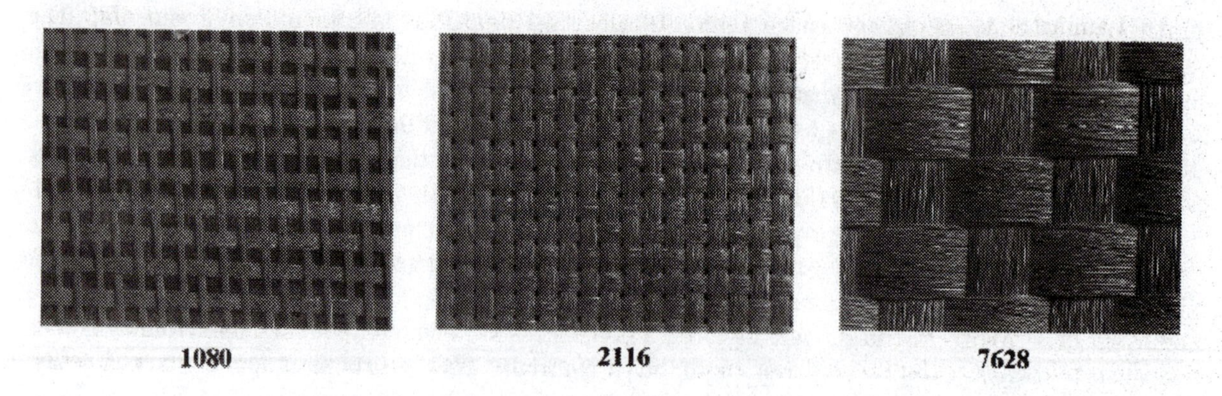

Fig. 5.13 Examples of different size strands and different weaving patterns used to produce glass cloth.

Recently, DuPont Advanced Fiber Systems introduced a proprietary organic reinforcement for laminate materials known under the trade name of Thermount. Its main component is an aramid polymer fiber called Kevlar, which has been woven into cloth to reinforcement circuit boards used in military electronics for several decades. The Kevlar fiber in Thermount is chopped and the short fibers are randomly oriented in a production process similar to that used to make paper. Compared with glass fiber reinforcements, it exhibits a high modulus, low density and low dielectric constant. The short fibers, have a negative axial coefficient of thermal expansion due to their anisotropic properties. Its primary disadvantage is its high cost.

Polymeric Materials

A large number of polymeric materials are used in producing laminates with different mechanical and electrical characteristics. The most common polymers in use today include polyester, epoxy, multifunctional epoxy, polyphenylene oxide (PPO) and polyimide. The exact chemical composition of these matrix materials is proprietary and the polymers are usually identified by trade names. Properties of significant importance depend upon the application; however, the glass transition temperature is usually of serious concern because it markedly affects the laminate's coefficient of thermal expansion.

FR-4 materials are still the predominant materials[3] used for printed circuit boards. They are available in a wide array of single and double sided laminates with different thickness and copper cladding weights. FR-404 and FR-406 are used in many applications. FR-404 is a multifunctional epoxy system with a glass transition temperature T_g of 150 °C and a dielectric constant D_k of 4.6. FR-406 is a tetra functional epoxy system that has a T_g of 170 °C and a D_k of 4.3. FR-406 is employed in applications where exposure to higher temperature is encountered either during soldering or subsequent operations. FR-406 is also recommended for multilayer boards when layer counts exceed 8. Its higher T_g minimizes z-axis expansion and the potential for through-hole barrel cracking and solder pad lifting.

The standard FR-4 laminates described above are used for most applications; however, polymers with enhanced properties are employed in high performance applications that include:

1. Printed circuit boards with special thermal requirements, because they will operate in high-temperature environments or because they must accommodate high power densities.
2. High-count, multilayer boards are difficult to manufacture, because they require extremely precise registration.

---

[3] A similar product—G-10—is produced without a fire retardant chemical. This laminate is used in most military applications where the out gassing of fire retardant chemicals in confined spaces is considered more dangerous than the possibility of circuit board fires.

3. Laminates requiring controlled (low) thermal expansion required for flip-chip and chip-on-board mounting or enhanced via reliability in thick MLBs.
4. Circuit boards with special electrical requirements including:
   - A requirement for a low dielectric constant for short propagation delays.
   - Lower crosstalk and higher clock rates.
   - A low dissipation factor to reduce signal attenuation and provide improved signal integrity.
5. Higher density, low thickness and lower power consumption in portable electronics.

FR-4 is the most widely used laminate because its properties satisfy the electrical and mechanical needs of many applications. The thermal and electrical properties of FR-4 and several laminates with other polymeric matrix materials are presented in Table 5.2.

Table 5.2
Properties of FR-4 and modified epoxy based high performance laminates.

Property	Unit	ED130 FR-4	GE GETEK	ISOLA FR-408	ARLON 45N	Nelco N4000-13
Cu Peel Strength	N/mm	1.76-2.00	1.35-1.48	1.45-1.49	1.45-1.49	1.52-1.60
T_g by DMA	°C	145	181	219	NA	245
TCE x axis	ppm/°C	14	NA	13	15	NA
y axis	ppm/°C	13	NA	13	15	NA
z axis	ppm/°C	175	3.80%	70	55	55
Dielectric Constant Dk @ 1 MHz	---	4.61	3.75	4.57	4.5	4.50
Dissipation Factor Df @ 1 MHz	---	0.01642	0.009	0.019	0.025	0.007
Volume Resistivity	MΩ	8×10^7	1×10^6	1×10^8	2.6×10^7	---
Surface Resistance	MΩ	2×10^5	$>1\times10^4$	1×10^6	2.9×10^7	---

Epoxy-based resins have been modified to improve their properties to meet the demands of higher circuit densities and high operating frequencies. Increasing the glass transition temperature T_g results in improved through-hole reliability because the coefficient of thermal expansion at temperatures below T_g is much lower than that above it. The attempt is to minimized z-axis thermal expansion in the range up to the reflow soldering temperature. The increase in T_g is achieved by increasing cross-linking of the polymer by adding a higher content of tetra- or multifunctional epoxy within the resin mix. This modification of the polymer reduces the copper peel strength and makes manufacturing (drilling small diameter holes) more difficult because the modified epoxies are harder and more brittle.

The modified epoxies with higher T_g materials have better dimensional stability than the conventional FR-4 laminate. This fact is important when producing dense circuit boards, because these boards are not exposed to operating or solder reflow temperatures much higher than the glass transition temperatures.

The thermal and electrical properties of laminates fabricated with alternative resins are presented in Table 5.3. From this table it is clear that cyanate ester resin has a low dielectric constant and a low dissipation factor; hence, it is suitable for high-frequency applications. Also cyanate ester exhibits a relatively high T_g that results in a very low z-axis expansion in the range up to soldering temperature, enhancing through-hole reliability. There are a few disadvantages associated with cyanate ester, which include relatively low copper adhesion, separation of the copper barrel from though-hole walls, delamination and a rate of high moisture absorption.

BT, developed by Mitsubishi Gas Chemical, is a blend of bismaleimide/triazine and epoxy resin that provides enhanced thermal, mechanical and electrical performance when compared to standard epoxy systems. BT epoxy exhibits several advantages such as a high T_g (185 °C), a relatively low coefficient of thermal expansion and excellent electrical insulation in high humidity and at higher temperatures.

High frequency applications, above 500 MHz, require laminates with special properties. Electrical characteristics, not as critical for DC and high speed digital circuitry, are essential when selecting materials for radio frequency and microwave systems. Control of the constant and layer thickness and providing a low dissipation factor are essential in these high-frequency applications. In these applications, the laminate is more than a support for the conductor. It forms part of the circuit because a signal line becomes an element of the circuit with distributed resistance, capacitance and inductance. At high-frequencies, dimensions of the traces and spaces influence the system performance.

Table 5.3
Properties of laminates with alternative resins

Property	Unit	Cyanate Ester	BT	Rogers RO4003	Asahi A-PPE	Polyimide
Cu Peel Strength	N/mm	1.51			1.28-1.38	
T_g by DMA	°C	286	185	280	232	270
TCE x axis	ppm/°C			11		
y axis	ppm/°C			14		
z axis	ppm/°C	43-52	50	46	49-50	65
Dielectric Constant Dk @ 1 MHz	---	3.44	4.1	3.38	3.71	4.0
Dissipation Factor Df @ 1 MHz	---	0.005	0.013	0.0027	0.003	0.0043

Reference to Table 5.3 shows the excellent properties of the Rogers' RO4003 laminate. This laminate in fabricated with a hydrocarbon resin filled with fine-grained ceramics. The enhanced electrical properties of this laminate, that make it ideal for high-frequency applications, are achieved by controlling the filler content.

A-PPE is a relatively new polymer, developed for circuit board applications by Asahi Kasei Chemical. It is a modified polyphenylene ester. The laminate has excellent properties including a low dielectric constant and a low dissipation factor, a high glass transition temperature and a low coefficient of thermal expansion in the z direction. For high-frequency applications, A-PPE outperforms all the epoxy based laminates. In comparison with cyanate ester, it exhibits a very low water absorption 0.05% compared to 0.5% for cyanate ester.

Laminates with a polyimide matrix also have excellent high temperature properties. Thermal stability makes polyimide laminates attractive in applications with stringent high temperature requirements. A low coefficient of thermal expansion in the z direction enables plated-through-hole reliability in very thick multilayer-boards. However, polyimide laminates are expensive compared to epoxy based FR4 laminates. Processing is also more difficult that adds to costs. The usual application for polyimide laminates are military where costs are less important for limited production runs and reliability at elevated temperatures that are encountered in qualification testing is much more important.

PTFE (polytetraflouroethylene) reinforced with glass cloth was the first high-frequency laminate developed for high performance printed circuit boards. To fabricate these laminates, PTFE resin is saturated on a woven glass cloth. By varying the ratio of glass to resin, it is possible to control the dielectric constant and dissipation factor. The dielectric constant varies from 2.17 to 3.2, depending on the style of glass cloth and the volume fraction of the glass. The dissipation factor will also vary from 0.0009 to 0.0035, with high volume fractions of glass producing higher dielectric constants and higher loss factors. The lower dielectric constant laminates can be used at frequencies up to 30 GHz.

PTFE laminates exhibit controlled dielectric constant and low dissipation factor; however they are not suitable for some applications that require control of the dielectric constant as a function of temperature or require low z-axis expansion. By replacing the glass cloth reinforcement with ceramic filler, inherent changes in the PTFE matrix due to temperature can be negated. With ceramic fillers stable properties over a wide range of operating temperatures can be achieved. A slightly different PTFE/glass cloth laminate is obtained by adding ceramic filler. In this modification, the intent is not to improve temperature stability, but to increase dielectric constant without sacrificing dissipation factor. With improved control of the dielectric constant and the loss factor these ceramic/glass PTFE laminates can be employed to frequencies as high as 40 GHz.

Mechanical properties of some of the more common glass-fiber reinforced laminates are presented in Table 5.4.

Table 5.4
Mechanical and electrical properties of laminates used for circuit boards

Property	FR-4 Epoxy	Isola P25 & P95 polyimide	FR-2 Phenolic	Teflon PTFE Glass
Flexural Strength (ksi) (MPa)	80 552	76 524	20 138	15 103
Flexural Modulus (ksi) (GPa)	2.5×10^3 17.2	4×10^3 27.5	1.2×10^3 8.3	
Interlaminar Bond Strength (ksi) (MPa)	2.5 17.2	2.1 14.5	1.0 6.9	0.8 5.5
Dielectric Constant (1 MHz, 25 °C)	4.3	4.4	4.0	2.4
Dissipation Factor (1 MHz, 25 °C)	0.019	0.016	0.03	0.001
Volume Resistivity ($\times 10^{15}$ Ω-cm)	0.2	0.24	0.01	0.2
Electric Strength (kV/mm)	22	47	25	36
Glass Transition Temperature (°C)	120	260	60	327
Temperature Coefficient of Expansion (PPM/°C) In-plane	12	15	18	20
Water Absorption Percent in 24 h	0.1	0.54	0.6	0.02
Maximum Operating Temperature (°C)	150	280	105	300

5.3.4 Polyimide Films

Polyimide films are used without reinforcement for flexible circuits. Flexible circuits are fabricated from foils of ductile copper that is adhesively bonded to a thin insulating film. The polyimide film known as Kapton (a trade name for a DuPont product) is widely used as the insulating carrier in thickness ranging from 25 to 125 μm (1 to 5 mil). High temperature properties of polyimide described in the previously make it the most suitable of the polymeric films for the flexible circuit applications. Three types of polyimide films are available from DuPont that include the type HN film, the VN film and the FN laminated film. The HN film can be laminated, metallized, punched, formed, or adhesive coated. The VN, is similar to the HN film with all of its properties plus superior dimensional stability. The FN, is a HN film coated on one or both sides with Teflon® FEP fluoropolymer resin. This coating imparts heat sealability, provides a moisture barrier, and enhances its chemical resistance. The coated FN film is available in a number of combinations of polyimide and Teflon® FEP thicknesses.

Properties of the polyimide films used in flexible circuits are given in the Table 5.5.

Table 5.5
Properties of polyimide (Kapton) films

Film Designation	Type 100 HN 25 μm (1 mil)	Type VN 25 μm (1 mil)	Type* 120FN616
Mechanical Properties			
Ultimate Tensile Strength MPa (psi)	231 (33,500)	231 (33,500)	207 (30,000)
Yield Strength (3%) MPa (psi)	69 (10,000)	69 (10,000)	61 (9,000)
Tensile Modulus GPa (psi)	2.5 (370,000)	2.5 (370,000)	2.48 (360,000)
Initial Tear Strength N (lb)	7.2 (1.6)	7.2 (1.6)	11.8 (2.6)
Propagating Tear Strength N (lb)	0.07 (0.02)	0.07 (0.02)	0.07 (0.02)
Ultimate Elongation, %	72	72	75
Thermal Properties			
TCE	20 ppm/°C (11 ppm/°F)	20 ppm/°C (11 ppm/°F)	
Thermal Conductivity W/(m-°K)	0.12	0.12	
Specific Heat J(g-°K)	1.09	1.09	
Glass Transition Temperature, °C	360 to 410	360 to 410	
Shrinkage	0.17 30 min at 150 °C	0.17 30 min at 150 °C	
Electrical Properties			
Dielectric Strength V/μm (V/mil)	303 (7,700)	303 (7,700)	272 (6,900)
Dielectric Constant	3.4	3.4	3.1
Dissipation Factor	0.0018	0.0018	0.0015
Volume Resistivity Ω-cm	0.5×10^{17}	0.5×10^{17}	1.4×10^{17}
Chemical Properties			
Chemical Resistance	Excellent except for strong bases		
Moisture Absorption at (23 °C and 50% RH	1.8	1.8	1.3

A lower cost polyester film (Mylar) can be employed for flexible circuits; however, the polyester films are limited to a maximum temperature of 150 °C. This temperature is too low for soldering using normal materials and processes. As a result, the utilization of polyester films is limited to applications where connections can be made without soldering.

Flexible printed circuits are sometimes employed instead of cables to save weight and space. They are also used to make electrical connections to moving parts. For example the read/write head in a hard disk drive is connected with a flexible circuit to the stationary I/O circuit for the drive. It must withstand of the order of 10^8 cycles over the life of the disk. The flex circuits are usually four layer laminates that include a support, adhesive, conductor and cover layers. The support and cover layers are usually made from polyimide films with thickness ranging from 13 to 125 μm. Polyimides are used for the base dielectric film because they are dimensionally stable, non-flammable and can withstand soldering temperatures. The polyimide films also exhibit high initiation tear strength. The characteristics that differentiate the various types of polyimide films are stiffness, thermal conductivity, oxygen and water vapor permeability, and the coefficient of thermal expansion.

Metal foil, usually copper, provides traces to conduct signals and power. Copper contributes to the mechanical structure of the circuit. The foil selected is based on factors such as cost, tensile strength, ductility, conductivity, solderability, fatigue strength and surface finish. The copper foil employed is formed by repeated rolling of billets and sheets to produce a foil with a thickness of 1/3 to 2 oz/ft^2. Grain controlled electrodeposited foil, discussed in Section 5.3.5, is also used to conduct signals and power. The copper foil is bonded to the polyimide supporting and cover layers with an adhesive film.

Various adhesive are used to provide the flexible laminates with mechanical strength, resistance to solder temperatures and chemical resistance. They are an integral part of the dielectric packaging of the signal, power and ground circuit lines, and as such, affect the characteristic impedance of the circuit. Several properties affect the performance of an adhesive in specific applications, which include flammability, coefficient of thermal expansion, bond strength, higher temperature properties, elastic modulus, moisture and insulation resistance, and fatigue life. Two adhesives commonly used are acrylic or phenolic butyral. For dynamic applications, where a large number of cycles are anticipated, the fatigue life of the flexible laminate is improved by a factor of about 100 by using the phenol butyral adhesive.

5.3.5 Copper Foil

Copper foil is commonly used as a cladding material on one or both sides of the laminate employed in fabricating a printed circuit board. Because copper sheets were used in the construction industry for roofing applications many decades prior to their application on PCB's, the thickness of the copper foils is measured in ounces per square foot. Converting 1 ounce/square foot gives a thickness of 35 μm (1.4 mil) for the 1 ounce cladding and 70 μm (2.8 mil) for the 2 ounce cladding. Copper foil is produced either by an electrodeposition process[4] or by rolling. The rolled copper foil is superior in terms of the uniformity of the surface profile between the front and back sides and its fatigue strength, while the electrodeposited copper foil can easily be manufactured to meet the requirements for wider or thinner products.

Electrodeposited copper is manufactured in a specially configured electrolytic plating tank as shown in Fig. 5.14. The copper from anodes is deposited onto the cathode that is in the form of a highly polished stainless steel drum two to three meters in diameter and about one meter wide. By controlling the speed of rotation of the drum and the plating cell current, the thickness of the copper deposit can be closely controlled. As the drum rotates, the foil is removed to provide a product that is very smooth on the side in contact with the drum and rough on the other side. Handling problems limit the minimum thickness of electrodeposited copper to 3/8 oz/ft^2, or 0.5 mil (13 μm). Thinner foils can be produced if the copper is deposited on an aluminum foil carrier, instead of the stainless steel drum, and it is subsequently stripped from the carrier. This process is capable of producing electrodeposited copper as thin as 1/8 oz/ft^2, or 0.18 mil (5 μm).

The roughness of the foil on the anode side is considered an advantage, because this rough surface provides a mechanical attachment when it is adhesively bonded to the circuit board laminates.

---

[4] A new electrodeposition process that enables grain orientation provides enhanced strength and tests show it has advantages over rolled and annealed foils..

The electrodeposited copper exhibits a grain structure with elongated grains in the direction perpendicular to the plating surface. This grain structure accounts for the rough surface and while the rough structure improves adhesion it is not conducive to high strength or flexibility of the foils. In normal rigid board applications, this lack of flexibility is not a problem as the stiffness of the circuit board limits the strain imposed on the copper foil. However, in flexible circuit applications, the conductors are often subjected to repeat flexing that impose many cycles of relatively high strain on the copper foil. In these applications, electrodeposited foils are not satisfactory and it is necessary to specify rolled and annealed foils.

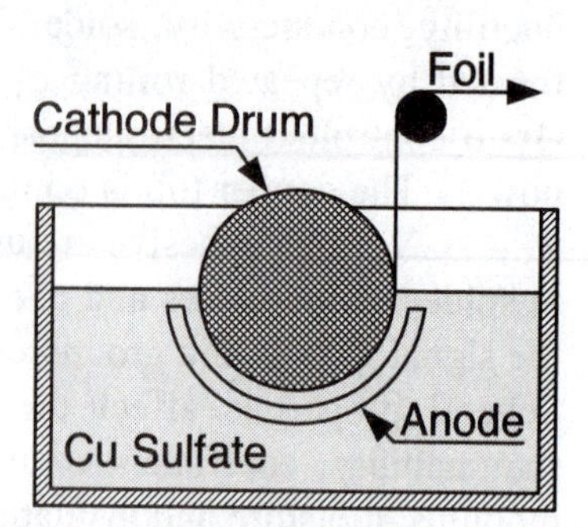

Fig. 5.14 Process for producing electrolytic copper foil.

Rolled copper foil is manufactured by successively rolling, with intermediate annealing, a copper ingot cast from electrolytic copper. The final reduction in thickness is usually performed in a rolling mill that is capable of producing foils of uniform thickness down to 12 μm (0.5 mil). The rolled copper foil increases in cost as the foil becomes thinner because more passes and annealing cycles are required to gradually reduce its thickness. The strength and elongation of rolled copper foil is superior to the standard electrodeposited foils because of its grain structure. In rolled foils, the grains are elongated and parallel to the surface of the laminate providing a structure resistant to failure due to cyclic strains.

The difference in the manufacture of the foils creates different foil properties that are important to design engineers. For rolled copper foil, the grain structure is flattened and oriented in longitudinal plane of the foil as the copper is squeezed between the rollers. The surface finish of both sides of the foil is as smooth as the rollers. Hence, the surface roughness of rolled foil is much less than that of electrodeposited copper. Surface roughness of the copper foil controls the bond strength and the loss (dissipation) factor. Bond strength is directly related to surface roughness with rougher surfaces producing stronger bond. For example, 1.0 oz/ft^2 electrodeposited copper foil with a surface roughness of 2.5 μm has typical copper peel strength of 2.5 N/mm that compares to peel strength of 1.6 N/mm for rolled copper foil with a roughness of 1.5 μm on the same substrate material.

With the standard electrodeposited process, copper crystals initiate on the stainless steel drum and grow with their long axes at right angles to the plane of the foil. The drum side surface finish is as smooth as the drum, but the outer side surface finish is rougher due to the protruding ends of the crystals. This rough side is exposed to a secondary plating treatment that adds to this roughness.

Another important property affected by surface roughness is loss factor for a high-frequency signal in a copper conductor. The loss factor is a measure of the loss of signal power as it travels along a conductor. For relatively low frequencies, one or two GHz, the loss factor in the conductor is small compared to the losses caused by the power dissipation factor due to the dielectric (insulating) material. However, at higher frequencies of 10 or more GHz, the conductor losses become more significant. In these applications, rolled copper foil with its smoother surfaces is superior. Rolled and annealed copper foil, exhibits lower losses by about 15% than standard electrodeposited copper foil.

Another important distinction between rolled and the standard electrodeposited copper foil is in their fatigue strength. Rolled copper foil with its grain structure lying in the plane of the foil exhibits much higher fatigue strength than standard electrodeposited copper foil. The standard electrodeposited foil with its grains oriented perpendicular to the plane of the foil is prone to crack initiation and is particularly weak in fatigue. Hence, it should not be used in dynamic flexible circuits that are intended for high cycle service.

Two new copper foils have recently been developed that are finding application in flexible circuits. The first foil consists of a thin film of copper sputtered directly onto either flexible or rigid polyimide without adhesive. These films are used in many applications because the very thin copper layer permits higher density traces. Sputtered copper film is more expensive than copper foil that is adhesively bonded to laminates; consequently, it is employed only when high density traces (three mil lines and spaces) are required.

Second, a new manufacturing process[5] to produce electrodeposited copper foil with superior properties has recently been developed. This new proprietary plating process eliminates the detrimental perpendicular grain structure responsible for the poor fatigue resistance of standard electrodeposited copper foil. A comparison of the grain structure for rolled foil, standard electrodeposited foil and grain controlled electrodeposited copper foil is presented in Fig. 5.15. The grain controlled electrodeposited copper foils have a unique, fine-grained structure that eliminates the columnar grains known to reduce fatigue performance in standard electrodeposited copper foils.

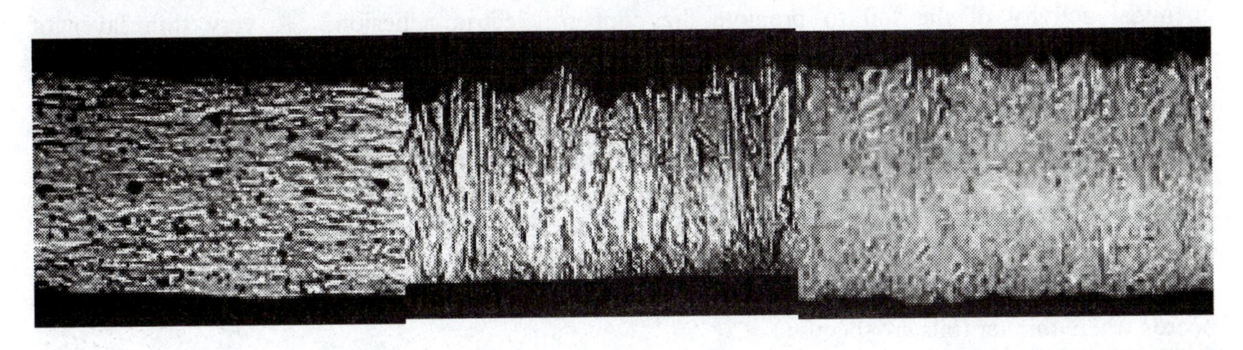

(a) Rolled foil (b) Standard electrodeposited foil (c) Grain controlled electrodeposited foil.

Fig. 5.15 Grain structure of three types of copper foil produced with different manufacturing processes.

Laminates based on this grain controlled electrodeposited foil have been shown to outperform rolled and annealed foil laminates in the three different types of flexible circuits. The grain controlled foil is currently being employed in periodically-flexing or flex-to-install applications, such as those found in many cell phone applications. Laboratory test results indicate that it should also perform well in dynamic applications where the number of rolling-flexing-cycles approaches a million or more.

Grain controlled copper foils have higher tensile and yield strength than rolled and annealed foils. This difference in strength is due to the grain size that is produced by the manufacturing methods for the two types of foils. Grain controlled electrodeposited foils have smaller grain size by a factor of nearly ten than the rolled and annealed foil and higher strength. Also, the annealing process that gives the rolled foil its ductility reduces its tensile strength.

The most significant advantage of the grain controlled foils is their availability in thickness of only 0.375 oz/ft^2 or (12 μm). Thinner foils permit finer trace/space and higher density in circuits. Thinner foils usually enable faster throughput in the etching process and higher quality trace walls. The thinner rolled and annealed foil require more time to produce and their costs increase significantly as their thickness decreases. On the other hand, the grain controlled plating process requires less time as the deposited layer becomes thinner. The lower product yields from handling such very thin foils

[5] This process has been developed by Gould Electronics.

diminish this advantage, but the price for thinner rolled annealed copper foils is about 30% higher than the cost for thinner electrodeposited, grain controlled foils.

Properties of grain controlled and rolled and annealed foils are presented in Table 5.6.

Table 5.6
Properties of electrodeposited, grain controlled and rolled and annealed foils

Property	Grain Controlled Foil		Rolled and Annealed Foil	
	18 μm (0.5 oz/ft^2)	35 μm (1.0 oz/ft^2)	18 μm (0.5 oz/ft^2)	35 μm (1.0 oz/ft^2)
Tensile Strength	65-70 ksi 449-483 MPa	65-70 ksi 449-483 MPa	20 ksi 138 MPa	22 ksi 152 MPa
Elongation	13-17%	15-20%	8%	13%
Volume Resistivity	1.82 MΩ-cm	1.78 MΩ-cm	1.78 MΩ-cm	1.74 MΩ-cm

Copper foils are usually treated before they are laminated to a substrate. This treatment may include up to three different additives on one surface or the other to improve the performance of the foil in subsequent steps in the manufacturing process. A bonding treatment on the rough side enhances the chemical activity of the foil to promote the copper/dielectric adhesion. A very thin layer of electroplated and stable alpha brass may be applied to provide a thermal barrier that prevents degradation of the copper/epoxy bond during processing at high temperatures. Foil stabilizers are added to the smooth surface passivating the foil to prevent it from oxidizing or staining during storage.

Because the requirements for circuit board density continue to increase to accommodate more modern high gate count IC's, circuit trace width has decrease and thinner foils are commonly employed in today's products. Electrodeposited copper foil has an advantage in cost with decreasing thickness. The time required for deposition is reduced with thickness permitting faster rotation of the drum and increased productivity. For this reason electrodeposited copper foil is commonly used in printed circuit boards where thinner foils are specified.

5.3.6 Ceramics

Ceramic substrates have the advantage that their coefficient of thermal expansion is closely matched to that of silicon; thus, the thermal stresses produced in the solder balls connecting the chip to the substrate are reduced significantly. Organic substrates have higher temperature coefficients of expansion and underfill adhesives used to mitigate the thermal stresses are mandatory; however, the organic substrates are much more economical. Both ceramic and organic substrates are multi-layered to provide the required wiring planes necessary for high I/O devices.

There are two types of ceramic substrates that are commonly employed today: a low temperature co-fired ceramic (LTCC) that is fabricated from a glass-alumina mixture and the high temperature co-fired ceramic (HTCC) that is fabricated from alumina powder. The glass-alumina is sintered at about 900 °C and the alumina is sintered at about 1500 °C. Both substrate types are fabricated using the green sheet process. This process involves the production of paper thin sheets of ceramic powders that are held together with resin binders. Vias are punched in the green sheets and filled with a conducting metal. The selection of this metal depends on the firing temperature[6] of the green sheets. A pattern of conductor traces is then screen-printed onto each green sheet. A number of sheets are stacked together for firing at a suitable sintering temperature. The glass/ceramic has an advantage that is fired at a much lower temperature and its shrinkage is less than that of sintered

[6] For the low firing temperatures glass/ceramics, both silver and copper may be employed as conductors. However, for the high sintering temperatures associated with alumina, refractory metals such as cobalt and tungsten are required.

alumina. During firing of both types of materials, pressure is applied to the stack normal to the plane of the sheets to prevent warpage.

Both glass-alumina and alumina can be fabricated into relatively large size substrates. These substrates can be used for bonding single flip-chip or for bonding many chips on larger modules (100 × 100 mm in size). Large substrates such as the 76 × 76 mm substrate used in the IBM multi-chip-module (MCM) supports four 20 × 20 mm chip sites each with 7018 solder bumps. Connections to 70 different layers in this module are made with 90 μm vias filled with co-fired copper.

Resistors can be printed on these ceramic substrates. The placement of printed resistors below select components and on the back surfaces of substrates is a significant factor in reducing module size and in increasing component density. In addition, ceramic substrates can be plated using the conventional Ni/Au electro-less plating process that is compatible with either gold, aluminum or copper wire-bonding.

5.4 FOOTPRINT DESIGN

The media for the design of a footprint is a sheet of copper clad laminate that is large compared to typical circuit board dimensions. The laminate is cut or contoured to size as specified for the product. The circuit traces and the pads used for chip mounting are produced by photolithography and an etching process. This process removes most of the traditional constraints from design that are due to the difficulties of machining contours. Essentially we are free to design the soldering surfaces and the circuit traces that provide the required mounting surfaces and connectivity in the smallest area possible. The primary constraint is the quality of the etching process and the thickness of the copper cladding that limits line width and line spacing.

The design of a PCB is usually performed with the aid of a computer program that provides: (1) footprint layout, (2) the number of layers, (3) component placement and (4) routing of the circuit traces. We will cover footprint layout in this section and defer the coverage of placement, layer count and routing to Sections 5.5 to 5.7.

The footprint layout is arranged to accommodate different types of chip carrier or some other component. As an example, the footprint for a 16 pin DIP is shown in Fig. 5.16. The footprint drawing shows the location of the plated-through-holes for each of the 16 pins together with their tolerances. Note, that the tolerances along the length of the chip carrier are not cumulative. The pins on the DIP are very rigid in the longitudinal direction of the chip carrier and, because all the pins are inserted together, it is necessary that close tolerances be maintained over the chip carrier's length. The tolerances in the transverse direction are relatively large (10 mil). The pins are flexible in this direction and the insertion device adjusts the pins inwardly to fit the row to row spacing while maintaining alignment of the row of pins. The dimensions of the hole diameter d and the solder pad diameter D in Fig 5.16 are not specified. These dimensions depend to a large degree on the precision that can be maintained in several of the manufacturing processes.

A large scale drawing of the maximum sized pin from a DIP in a plated-through-hole is shown in Fig. 5.17. In this drawing d = 30 mil and D = 50 mil and the clearance and concentricity of the hole and solder pad is evident. A large diameter solder pad is useful in accommodating registration errors; however, large pads utilize valuable board area and reduce the width of the wiring channel.

The influence of registration/drilling errors with solder pads having D = 50 mil with a plated-through-hole 30 mil in diameter is shown in Fig. 5.18. The center of the hole and the center of the pad are located in two different manufacturing processes. Differences in registration between these two processes produce errors in locating the centers for both d and D. With D = 50 mil (1.27 mm) it is possible to accommodate errors $e_x = 0$ and $e_y = 5$ mil (0.13 mm) while maintaining an annular region suitable for soldering. However, when $e_x = e_y = 5$ mil (0.13 mm), the annular region is seriously distorted and effective soldering is impaired. Indeed, when $e_x = e_y = 7$ mil 0.18 mm), the hole breaks the edge of the pad and the board is considered a reject.

The footprint for a 12 lead flat pack type chip carrier is shown in Fig. 5.19. While the leads for the flat pack are on 50 mil (1.27 mm) centers, the footprint is designed with holes on a 100 mil (2.54 mm) grid. This practice is often followed and is to accommodate drilling machines that are established to drill holes in rectangular arrays on 100 mil (2.54 mm) centers. Because the flat pack is a surface mounted package, the holes shown in Fig. 5.19 serve only as vias used to connect flat pack leads to internal wiring planes. The holes may also serve as thermal vias to enhance heat transfer through the board. The diameter of the plated-through-holes can be reduced providing that manufacturing has the ability to drill and plate the smaller diameter holes.

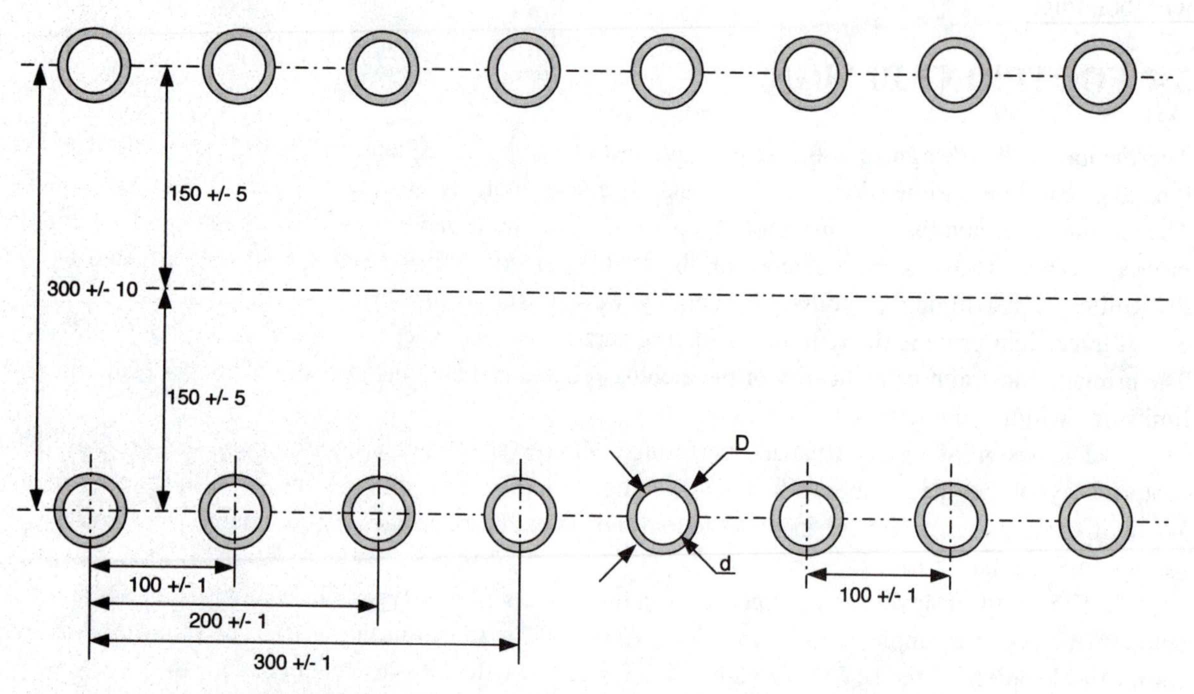

Fig. 5.16 Footprint for a 16 pin DIP. Dimensions are in mil.

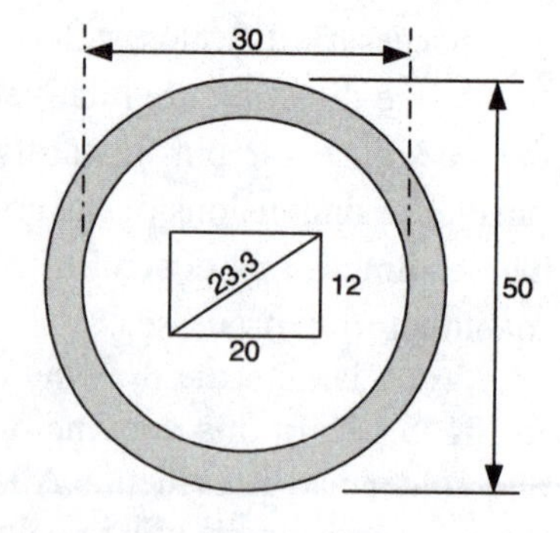

Fig. 5.17 Scale drawing showing the dimensions of a DIP pin in a plated-through-hole. Dimensions are in mil.

The footprint presented in Fig. 5.1 is for the J leaded chip carrier. In this design, the solder pads are on 50 mil (1.27 mm) centers with a width of 30 mil (0.76 mm) to accommodate an 18 mil (0.46 mm) wide J lead (see Fig. 4.7). The length 75 to 105 mil (1.9 to 2.7 mm) for the solder pad is to provide a long base for generous solder fillets with a large radius of curvature at the J lead as shown in Fig. 5.20.

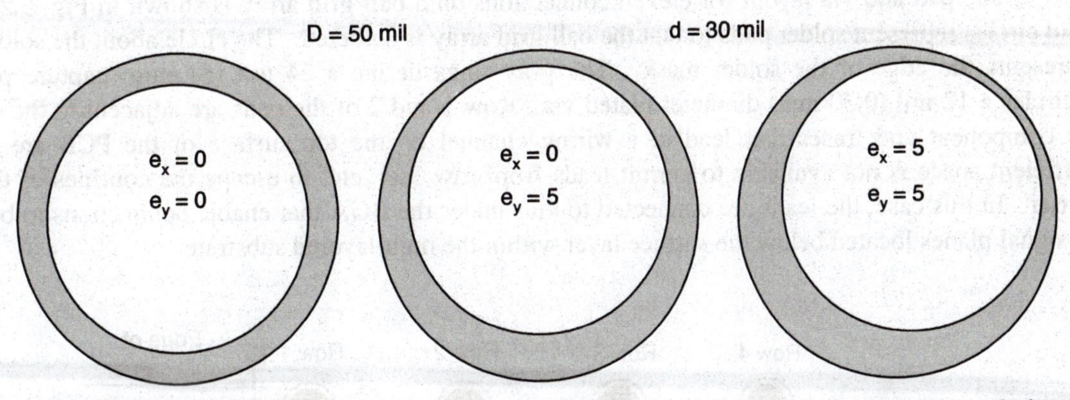

Fig. 5.18 Influence of drilling registration errors in placement of the plated-through-hole relative to the solder pad.

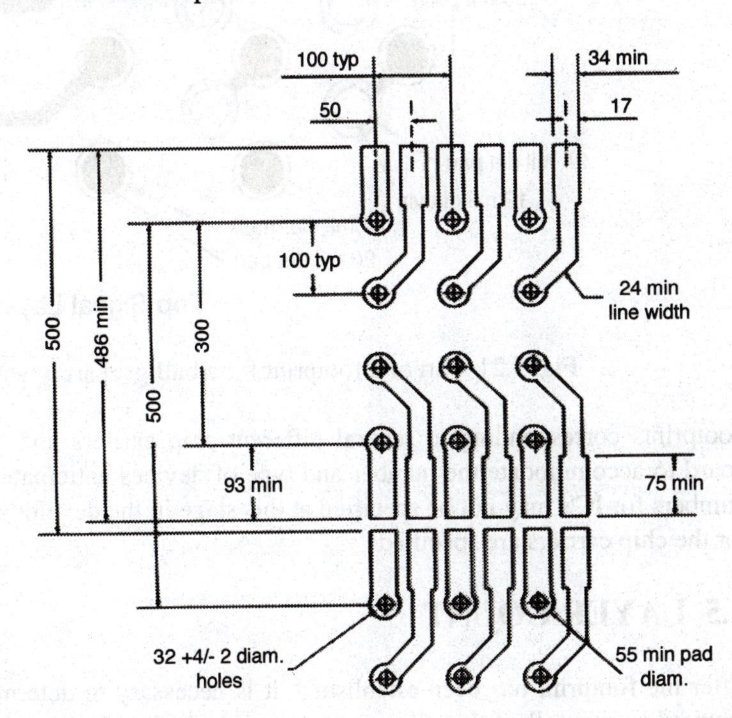

Fig. 5.19 Footprint for a 12 lead flat pack chip carrier. Dimensions are in mil.

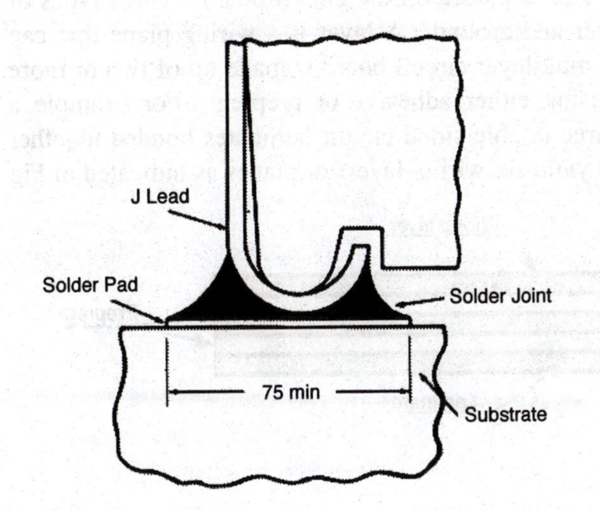

Fig. 5.20 Length of a solder pad to accommodate the relatively long solder joint used with a J leaded chip carrier.

The pad and via layout for eleven connections on a ball grid array is shown in Fig. 5.21. The solid circles represent solder pads to that the ball grid array is soldered. The circle about the solder pads represents the edge of the solder mask. The gray rings define a 24 mil (61 mm) capture pad that encircles a 12 mil (0.30 mm) diameter plated via. Row 1 and 2 of the pads are adjacent to the edge of the component and traces that lead to a wiring channel on the top surface of the PCB are shown. Sufficient space is not available to permit leads from row 3, 4, etc. to escape the confines of the chip carrier. In this case, the leads are connected to vias under the BGA that enable connections to be made on signal planes located below the surface layer within the multilayered substrate.

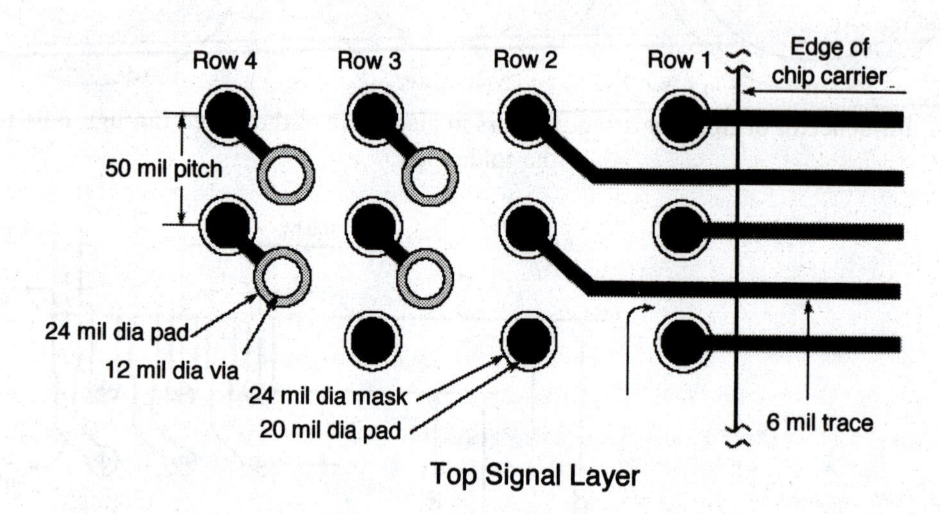

Fig. 5.21 Part of a footprint for a ball grid array with pads on 50 mil pitch.

Footprints corresponding to several different chip carriers and components are arranged on a circuit board to accommodate the number and type of devices anticipated in the development. Note, that part numbers for IC's may not be specified at this stage in the development, but the type and number of leads for the chip carriers are specified.

5.5 LAYER COUNT

After the footprint has been established, it is necessary to determine the number of layers that will be required to wire all of the chips and passive devices to be placed on the circuit board. Three types of wiring are employed—wires for signal pulses, power and ground. A layer is a wiring plane that can carry signal pulses or currents for power. A typical multilayer circuit board is made up of two or more double sided laminates that are bonded together using either adhesive or prepreg. For example a multilayer circuit board could be fabricated from three double sided circuit laminates bonded together with two layers of prepreg. This construction would yield six wiring layers or planes as indicated in Fig. 5.22.

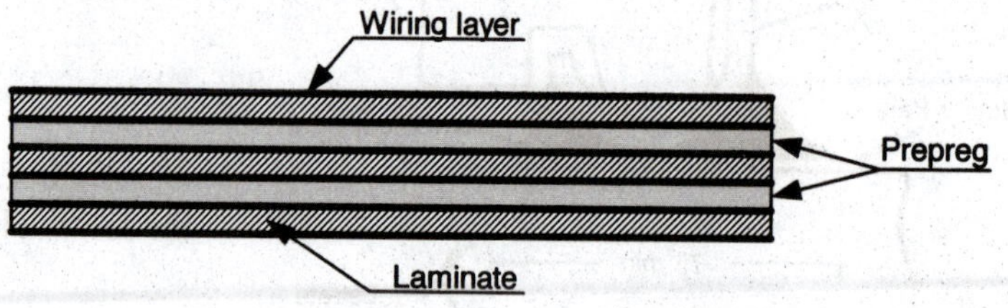

Fig. 5.22 Multilayer circuit board construction that yields six wiring layers.

The top layer is reserved for the foot print and the bottom layer is often used as the ground (one of the two layers utilized to supply the power). Both the power and ground planes are made with nearly solid copper cladding or with lines that are wide and thick to support the currents required to supply power and voltage to various components. The remaining layers would be used for signal wiring. The signal lines on a given wiring plane are usually oriented in either the x or y directions. Connections linking wiring across a given laminate are made with vias. The vias may extend through all of the layers and the prepreg, they may be blind vias and extend though only the top or bottom laminates or they may be buried and extend through one or more interior layers and not be visible from either exposed surface.

Usually an estimate of the number of signal layers is made after the components are placed, but before the routing is attempted. The final determination of the number of signal layers required is made after the routing is complete and the line width, line spacing and visa distribution has been established.

Factors that control the number of signal layers needed in a multilayered circuit board include:

1. Routable substrate area
2. Component type
3. Line width and space width (wiring density)
4. Percent signal connections
5. Routing efficiency

Routable substrate area A_r includes all of the area on a wiring plane that can be used for lines, via holes, capture pads and chip pads. Areas around the edges, areas used for clamping the boards, areas for tooling holes, and areas of cut-outs are excluded from the total surface area. It is relatively easy to determine this area A_r by subtracting the restricted area from the total surface area of a layer. The component area A_c is also easy to determine by noting the projected surface area of each component and summing them. It is clear that:

$$A_c < A_r \text{ for single sided placement}$$
$$A_c < 2A_r \text{ for single sided placement} \qquad (5.6)$$

for the components to fit on the circuit board. The larger A_r is when compared to A_c, the more area that is available for wiring.

The component type is also an important factor. Trough-hole chip carriers consume more area on a wiring plane because they extend through all of the layers and have relatively large capture pads on each layer. Surface mount packages are more efficient as they often can be wired using blind and/or buried vias. Ball grid arrays are even more efficient because the wiring pitch is small and the area under the package is effectively employed for pads and lines.

Line width and space width are also a significant factor. High density wiring (3 mil (75 μm) lines and 3 mil (75 μm) spaces enables many more connections to be placed in a single wiring channel; hence fewer layers would be required with fine lines and spaces.

The percent of the I/O on a chip that must be connected on signal planes also affect the number of layers. Some leads are used for voltage and ground connections and others are for timing and still others may not be used. The number of signal connections for chips varies from about 30 to 80% of the I/O on a chip.

Routing efficiency pertains to the ability of the routing program to utilize all of the area on the routing channels on a wiring layer. This efficiency depends on the wiring algorithms used in the routing program and the number and type of vias that are permitted in wiring the multilayer circuit board. Typical routing efficiencies vary from about 50 to 75%.

There are a two different methods for estimating the number of signal layers required in the design of a multilayered circuit board, which include the density and connectivity methods. We will briefly describe the density method because it is the simpler of the two approaches. The connectivity method that is much more involved is described in reference [3].

Density Method

In the application of the density method, a reference component is selected to initially represent the devices that are to be placed on the circuit board. Then the number of signal planes required is determined based on the routing area available for the reference component. The number of signal planes N_L required is estimated from a density function as:

$$N_L = f\left(\frac{1}{D_s}\right) \tag{5.7}$$

where the density of the substrate D_s is given by:

$$D_s = \frac{A_r}{N_r} \tag{5.8}$$

where N_r is the total number of reference devices to be placed on the circuit board.

The number of layers or signal planes depends on the type of laminate and the type of reference device. Clearly as the routing area available for each reference device increases, the density increases and the number of layers required decreases.

To account for the fact that different types of components are included on a circuit board, it is necessary to introduce a correlation factor that accounts for the differences in the wiring demand for various types of devices. It is defined by:

$$F_i = \frac{WD_i}{WD_r} \tag{5.9}$$

where WD_i and WD_r are the wiring demands for the ith and reference components, respectively.

The total number of reference components to be placed on the circuit board is given by:

$$N_r = \sum_{i=1}^{N_n} F_i \tag{5.10}$$

Clearly substituting Eq. (5.10) into Eq. (5.8) gives the substrate density. Substituting that result into Eq. (5.7) yields a first estimate of the layer count. The problem with this approach is determining the wiring demand for all of the components mounted on the board as well as the reference component. One method for determining wring demand is based on the number of I/O of the active devices on the board relative to the I/O count on the reference device. Using this method, Eq. (5.9) may be rewritten as:

$$F_i = \frac{(N_{I/O})_i}{(N_{I/O})_r} \tag{5.11}$$

For passive devices like resistors and capacitors with two leads F_i is usually taken as a number in the range of 0.10 to 0.25. For board mounted connectors, F_i may be determined from the connector lead count divided 20.

Summary

The number of layers required depends on the number of chips placed on the board, the I/O of these chips, the number and location of the vias and the density of the lines and spaces on each signal plane. Circuit boards with several high performance chips that contain many hundreds of I/O often require ten or more wiring layers even with dense lines and spaces and an abundance of vias. The importance of circuit density is illustrated in Fig. 5.23.

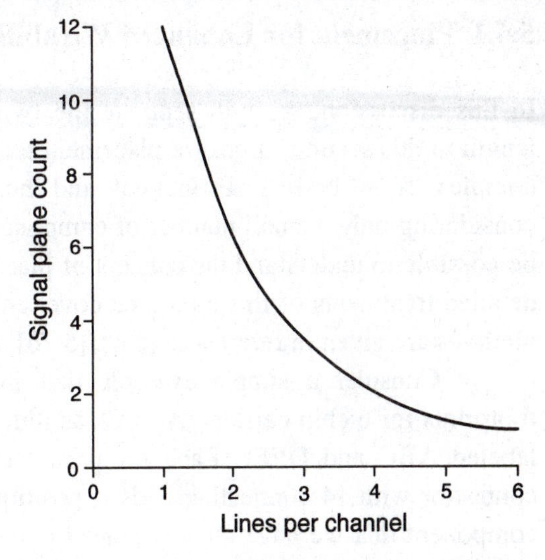

Fig. 5.23 Typical signal plane count with circuit lines per wiring channel showing the importance of fine lines and spaces.

The cost of a multilayer board increases with layer count so the determination of the exact number of layers is important. In practice, electrical and mechanical engineers with considerable experience make the layer count determination for the collection of components to be placed on a specific circuit board. If this count is too low, it will be impossible to wire all of the devices on the available wiring layers and additional layers will be required in a redesign. If the number of layers is excessive, the wiring task is easier, but the cost of the added layer penalizes the design.

5.6 PLACEMENT FOR ENHANCED WIRABILITY

A typical circuit board supports many active and passive components and it is often necessary to locate 30 to 100 or more components over the available area of a board. The placement (selecting the location) of each of these components on the board is not arbitrary. There are two different design goals influencing the location where each device is placed. The first goal is wirability, which pertains to minimizing the total length of wire required to connect the devices. Minimizing wire length reduces transit time of the signal pulses and alleviates congestion of the circuit traces that results in crossovers and crosstalk. Wirability also involves clustering functionally related components, designing signal lines to meet electrical specifications for resistance, current and capacitance, and shielding analog and digital signals to prevent cross talk.

The second design goal is to place the devices at locations that lead to improved reliability of the system. This goal is achieved by placing the more critical (less reliable) components at locations on the circuit board where their temperatures will be minimized. The reliability and wirability design goals often conflict because a placement scheme that ranks well from a wiring view point may rank poorly when reliability is considered.

Regardless of which design goal that is given preference, placement is a difficult problem because of the very large number of unique placement possibilities that exist. The number k of placement options for a board with N components is k = N! Of course, this fact implies that a very simple board with say only 8 devices has k = 8! = 40,320 combinations to consider if placement is to be performed rigorously. With circuit boards populated by a large number of devices, the placement options involved tax large computers with placement codes. These codes attempt to optimize circuit board performance with an best possible selection for the location of each of the components.

5.6.1 Placement for Enhanced Wirability

In this discussion, we will separate the two design goals and consider placement to minimize wire length in this section. Because placement techniques that handle very large numbers of components are complex from both mathematical and numerical viewpoints, we will simplify the problem by considering only a small number of components with limited I/O. With this simplified approach it will be possible to understand the concept of placement to enhance circuit board wiring. More rigorous and detailed treatments of this topic that cover cost functions, constructive methods and iterative placement methods are given in references [5.3], [5.16] and [5.17].

Consider a simple example that involves circuit board with an edge card connector and footprints for 6 chip carriers (A to F), as illustrated in Fig. 5.24. The footprints are aligned in two rows labeled ABC and DEF. Each footprint has 8 pads where connections are to be made. The edge connector with 14 connecting pads is positioned along the lower edge of the card. The connector is a component that we have already placed (along the edge of the card); hence, we consider it seeded and permanently located. An x-y coordinate system for the board is established with its origin at pad no. 1 of the edge connector. The pads are arranged on a grid so that wire length can be determined as grid units. In determining the wire length s between two points, the Manhattan distance s is employed where:

$$s = x + y \tag{5.12}$$

The concept of rectilinear measurement of the length between two points that is illustrated in Fig. 5.25 is an accurate measure of distance in circuit boards because the circuit traces usually run in either the x or y directions. Diagonals, that yield shorter lengths between two points, are rarely used in routing the circuit lines because diagonal traces block wiring channels and can lead to crossovers[7].

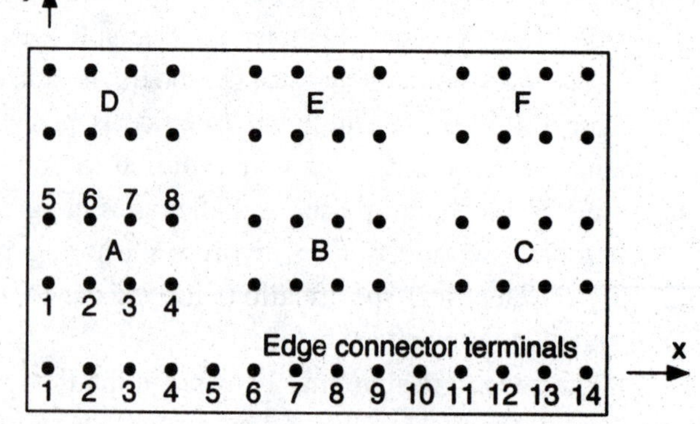

Fig. 5.24 Circuit board with footprints for six chip carriers each with 8 leads and a 14 pin edge connector.

[7] In some cases when traces are turned through an angle of 90°, a short diagonal is used to eliminate the sharp corner. This practice reduces the wiring length slightly.

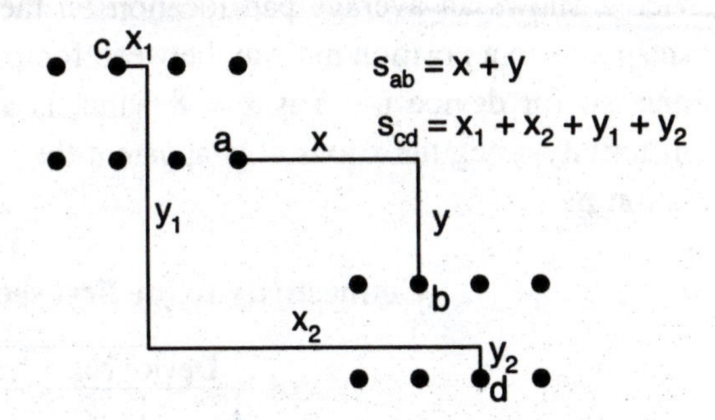

Fig. 5.25 Manhattan measure of distance between two pairs of points, a-b and c-d.

Table 5.7
Wire list for the PCB illustrated in Fig. 5.24

From Device		To Device		From Device		To Device	
Device No.	**Pin No.**	**Device No.***	**Pin No.**	**Device No.**	**Pin No.**	**Device No.***	**Pin No.**
1	1	2	5	4	1	3	6
	2	2	6		2	3	7
	3	5	5		3	6	6
	4	ec	1		4	ec	2
	5	5	6		5	6	7
	6	3	3		6	6	1
	7	3	5		7	6	2
	8	ec	13		8	ec	14
2	1	ec	3	5	1	ec	8
	2	ec	4		2	ec	9
	3	ec	5		3	ec	10
	4	ec	1		4	ec	2
	5	1	1		5	1	3
	6	1	2		6	1	5
	7	3	1		7	3	2
	8	ec	13		8	ec	14
3	1	2	7	6	1	4	6
	2	5	7		2	4	7
	3	1	6		3	6	5
	4	ec	1		4	ec	2
	5	1	7		5	6	3
	6	4	1		6	4	3
	7	4	2		7	4	5
	8	ec	13		8	ec	14

- ec designates the edge connector

Returning to the placement problem, we are to determine, from a list of six devices (nos. 1 to 6), which device is to be placed at the locations A to F on the circuit board. To make this determination, we refer to a wire list that gives the connectivity between pins on each device and to the edge connector. An example wire list for this illustration is given in Table 5.7. The wire list is examined to determine the connectivity with the first seeded component (the edge connector). This connectivity, listed in Table 5.8, shows that devices nos. 2 and 5 each have five connections to the edge connector and all of the other devices have only two leads to supply the power. The high connectivity of devices 2 and 5 indicates that they should be placed in the lower row at locations A, B or C. The equal connectivity of these two devices with the edge connector permits us to seed them together. Examining the wire list for device 2 shows an average pad location on the edge card connector of x = 5.2 units. This location corresponds to a position midway between footprints A and B. The average pad location along the edge connector for device no. 5 is x = 8.6 that is adjacent to the B footprint. With this information on connectivity along the x axis, it is apparent that device no. 2 be seeded at location A and device no. 5 at location B.

Table 5.8
Connectivity to the first seeded device (edge card connector)

Device No.	Number of Connections
1	2
2	5
3	2
4	2
5	5
6	2

After placing devices 1 and 2, the placement problem reduces to positioning the remaining four devices. The opening at location C is adjacent the edge connector and device no. 5. The listing of remaining connectivity, shown in Table 5.9, indicates that device no. 1 has four connections, device no. 3 has three connections and devices no. 4 and 6 have only two connections to the seeded components positioned adjacent to location C. Maximizing the connectivity count places device no. 1 in location C.

Table 5.9
Connectivity to the seeded devices adjacent to location C
(Edge card connector and device No. 5)

	Number of connections		
Device No.	To Device No. 5	To edge connector	Total No.
1	2	2	4
3	1	2	3
4	0	2	2
6	0	2	2

To position the remaining three components the connectivity matrix presented in Table 5.10 is prepared. This matrix shows a large connectivity count between devices 4 and 6 and indicates that devices 4 and 6 should be placed side by side. Positioning device no. 4 in position E and device no. 6 in position D, gives position F to the remaining device no.3 where it connects well to devices 1 and 4. The diagram shown in Fig. 5.26 indicates the final placement and the connectivity to adjacent locations.

Table 5.10
Connectivity matrix for devices 3, 4 and 6

Device No.	1	2	3	4	5	6
3	2	1	0	2	1	0
4	0	0	2	0	0	4
6	0	0	0	4	0	2

This simple example indicates a process for determining the optimum placement of components to enhance the wirability of the board. The importance of this topic will become more evident in Section 5.7 where board routing is described. In real design situations, where a large number of components increases complexity, manual placement techniques require excessive personnel time. The placement process is automated by using extensive computer simulations with placement algorithms in an attempt to arrange the devices at locations that reduces the total length of the circuit traces required to connect the devices.

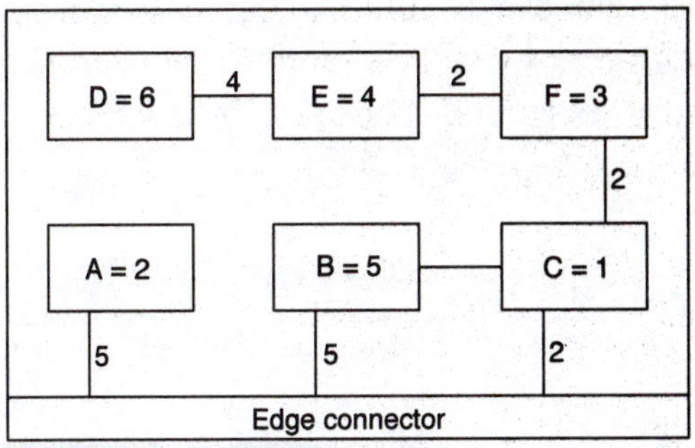

D = 6 — 4 — E = 4 — 2 — F = 3
2
A = 2   B = 5 — C = 1
5   5   2
Edge connector

Fig. 5.26 Placement of six devices showing connectivity to adjacent components.

5.6.2 Placement for Enhanced Reliability

The failure rate λ of an integrated circuit depends strongly on the junction temperature T_j of each chip. To achieve maximum reliability for a given board, it is necessary to consider the sum of the individual failure rates for each component placed on the board. The failure rate for a board supporting N components is given by:

$$\lambda_T = \lambda_1 + \lambda_2 + \ldots\ldots + \lambda_N \quad (5.13$$

To optimize board reliability the components must be placed to minimize the total failure rate λ_T. This is a difficult problem to solve in the exact sense because of the number of computations required. There are N! placement combinations to consider with a different λ_T for each combination. We could consider each combination, compute each junction temperature, determine the individual values of λ_i , then λ_T and finally select the combination corresponding to minimum value of λ_T. However, with a large number of components, the factorial N! becomes excessive. Clearly, this procedure is too involved, and approximate methods are usually employed.

Osterman and Pecht [5.17] introduced an effective approximate method for placement of IC's on conduction cooled circuit boards of the type shown in Fig. 5.27. Their approach was to first divide the circuit board into rows as indicated in Fig. 5.27 and then to consider each row independently. This approach assumes that heat flows in only the ± x directions, which is not exactly true. However, this assumption permits each row of IC's to be placed without considering the effect of adjacent rows. This simplification reduces the number of placement possibilities from N! to (N/r)!, where r is the number of rows. Because (N/r)! << N!, the complexity of the placement procedure for maximum reliability is reduced significantly.

The failure rate of a component clearly depends on its junction temperature and this temperature depends on the location of the device on the board. For conduction cooled boards, like the one illustrated in Fig. 5.27, the locations adjacent to the heat sink give minimum junction temperatures and locations near the center of the board results in maximum junction temperatures. (Procedures for determining junction temperatures for conduction cooled systems are described in Chapter 10. In this development, we assume the reader to be familiar with analysis methods of heat transfer by conduction.)

The approach is to examine the IC's to be placed in a single row and to select from this group the most critical IC, namely, the component with the highest failure rate. The most critical IC is placed adjacent to the heat sink at the location marked no. 1 in Fig. 5.27.

Osterman and Pecht have defined a priority ranking number PRN for each component i given by:

$$PRN_i = [d\lambda_i/dT_i^j]q_i \quad (5.14)$$

where q_i is the rate of heat dissipation from the ith component.

The subscript i = 1 to (N/r)! is used to identify the components to be placed. The PRN is determined for each component positioned in the row under consideration. The maximum value of PRN identifies the most critical component. This component is positioned at the lowest temperature location 1 that is adjacent to the heat sink.

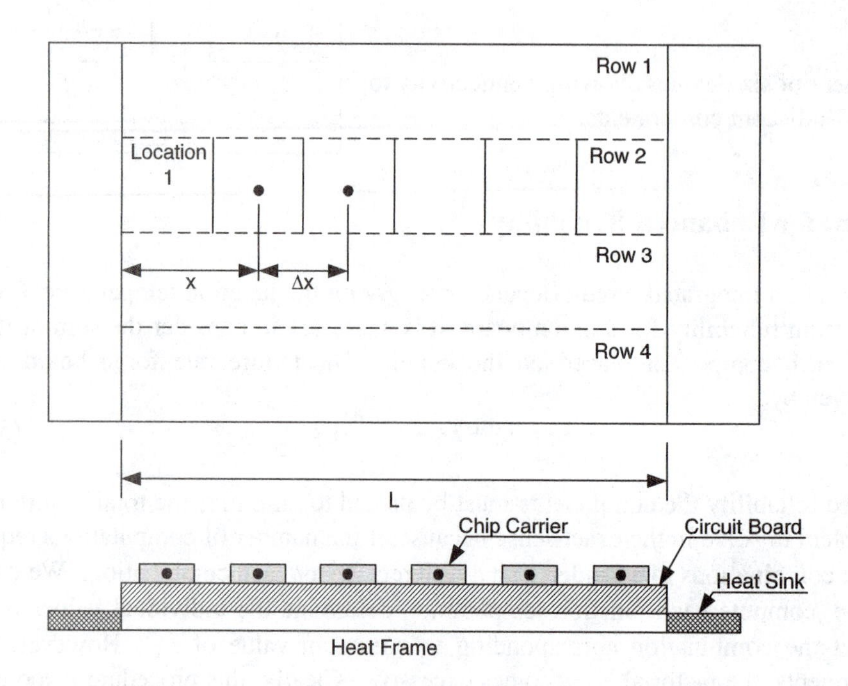

Fig. 5.27 Placement on a conduction cooled circuit board to enhance reliability.

The value of PRN_i depends on the slope of the failure rate temperature relation, and is obtained by differentiating Eq (1.1) to give:

$$\frac{d\lambda_i}{dT_i^j} = B_i C_i (T_i^j)^{-2} e^{C_i\left[(T_i^j - 298)/298T_i^j\right]} \tag{5.15}$$

The junction temperature T_i^j that is an unknown in Eq (5.15) depends upon the location being considered for placement. In this approach, we start with position located at $x = x_1$ as shown in Fig. 5.28. The junction temperature is determined using the electrical analog illustrated in Fig. 5.29 that permits us to write:

$$T_i^j = T_i^f + q_i R_i^{jf} \tag{5.16}$$

where T_i^f is the temperature of the heat frame at $x = x_1$ and R_i^{jf} is the combined thermal resistance including the chip, chip carrier and circuit board referenced to the heat frame.

It should be noted that placement controls T_i^f and hence T_i^j. The term $q_i R_i^{jf}$ is important but this term is a constant independent of the placement location. The resistor network shown in Fig. 5.30a could be used to precisely solve for T_i^f; however, for a large number of components the computations become extensive. It is possible to reduce the computational complexities by considering the approximate model shown in Fig. 5.30b.

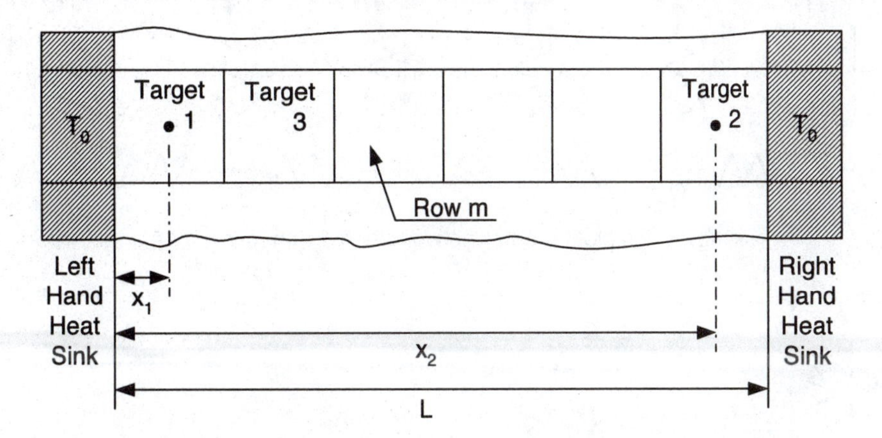

Fig. 5.28 Locations of target 1 and 2 for placement along row m.

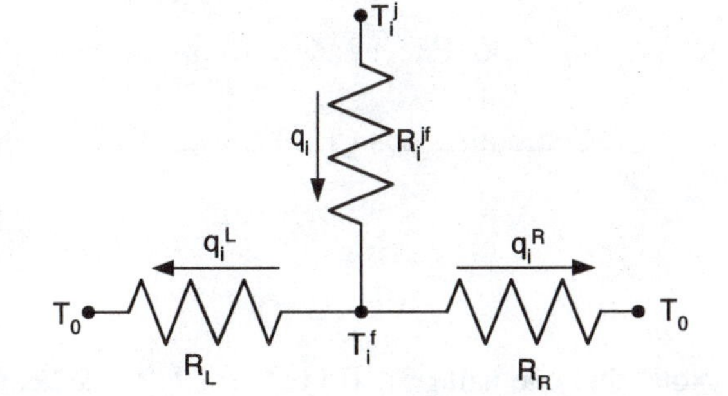

Fig. 5.29 Resistor network showing the heat flow to the heat sinks and the relation between T_i^f and T_i^j.

In this approximate model, the ith component is placed over the target location and the remainder of the heat q_c is considered to be dissipated at the center line of the heat frame. We write q_c as:

$$q_c = q_t - q_i \tag{5.17}$$

where q_t is the sum of the heat dissipated by all the components to be placed in the row. To account for the branching of the flow of heat at node A in Fig. 5.30b, we write:

$$q_i = q_i^L + q_i^R \tag{5.18}$$

The heat flow to the left and to the right may be expressed as:

$$\begin{aligned} q_i^L &= \frac{\left(T_i^f - T_0\right)}{R_L} = \frac{\left(T_i^f - T_0\right)kA}{x_1} \\ q_i^R &= \frac{\left(T_i^f - T_0\right)}{R_R} = \frac{\left(T_i^f - T_0\right)kA}{(L - x_1)} \end{aligned} \tag{5.19}$$

where k is the coefficient of thermal conductivity for the heat frame material. A is the cross-sectional area of the heat frame corresponding to the row width. L and x_1 are dimensions defined in Fig 5.28. R_L and R_R are thermal resistances of the heat frame.

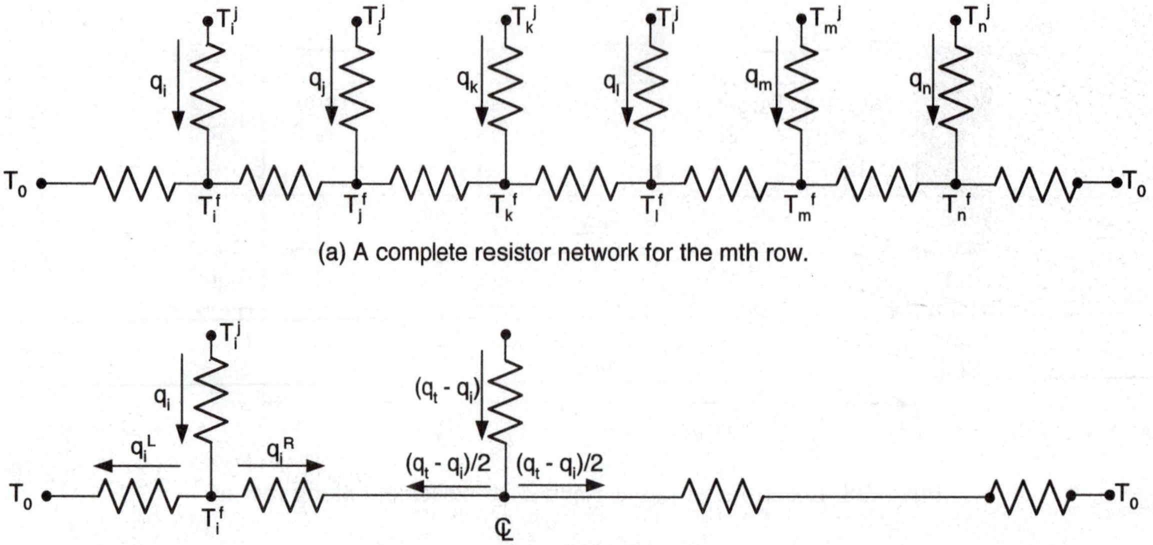

(a) A complete resistor network for the mth row.

(b) An approximate thermal model for the m th row.

Fig. 5.30 Exact and approximate thermal resistance models used to determine T_i^f and T_i^j.

Combining Eqs (5.18) and (5.19) gives the temperature T_i^f due only to q_i as:

$$T_i^f - T_0 = \frac{x_1(L - x_1)q_i}{kAL} \tag{5.20}$$

Note, that one half of q_i flows to the left side heat sink and produces an increase in T_i^f due to R_L, that is:

$$T_i^f - T_0 = \frac{x_1(q_t - q_i)}{2kA} \tag{5.21}$$

Using superposition and combining Eqs (5.20) and (5.21) gives the frame temperature as:

$$T_i^f = \frac{x_1(L - x_1)q_i}{kAL} + \frac{x_1(q_t - q_i)}{2kA} + T_0 \tag{5.22}$$

Next, use Eq (5.16) and (5.22) to obtain the junction temperature:

$$T_i^j = \frac{x_1(L - x_1)q_i}{kAL} + \frac{x_1(q_t - q_i)}{2kA} + T_0 + q_i R_i^{jf} \tag{5.23}$$

Finally, substitution of numerical results from Eq (5.23) into Eq (5.15) gives $d\lambda_i/dT_i^j$, which is then substituted into Eq (5.14) to obtain the priority ranking number PRN_i. This process is repeated for each of the components to be placed in the subject row. Selection of the maximum value of PRN_i permits the placement of the first component at the most favorable location $x = x_1$.

The procedure is repeated with the remaining components and the target for placement of the second component is location 2 at $x = x_2$ (see Fig. 5.28) adjacent to the right hand heat sink. Determination of T_i^j follows the form of Eq (5.23) except that the heat flow is in the opposite direction to the right hand heat sink. We can accommodate this mirror image flow by letting:

$$x_1 = L - x_2 \quad \text{and} \quad L - x_1 = x_2 \tag{5.24}$$

Substitution of Eq (5.24) into (5.23) enables T_i^j to be determined at location 2. Again PRN_i is determined for the remaining components, the maximum value is selected and the second most critical component is identified and placed at $x = x_2$.

Placement of the remaining components follows the same process with T_i^j determined at location $x = x_3$ from the resistance network presented in Fig. 5.31. Examples of this placement procedure to maximize reliability are given in the exercises following Chapter 10, because of the conduction method of heat transfer used in this placement approach.

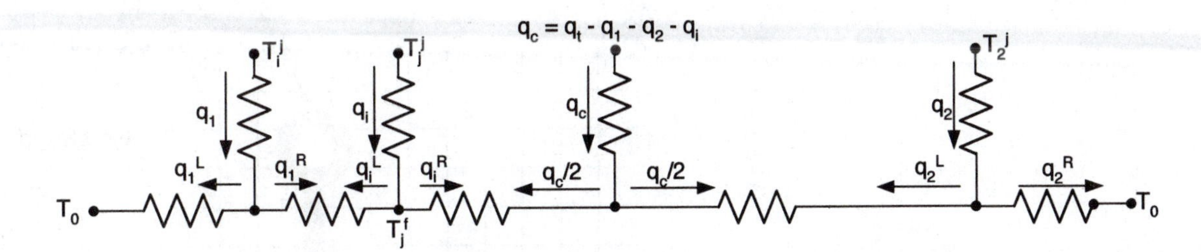

Fig. 5.31 Resistor network used to determine the junction temperatures required to place the third and fourth components.

5.7 ROUTING METHODS

Routing a circuit board pertains to the layout of the traces from one point to another on the circuit board. This layout is based on a wire list, as illustrated in Table 5.4, which describes the required connectivity between the I/O for the components placed on a given board. Routing appears simple. We need to connect prescribed pins or pads together according to the wire list, without crossovers that produce connections, which are unspecified and detrimental. Also, we must avoid long, parallel and closely spaced paths that produce cross talk between signal traces. While simple in concept, actual routing is quite complex, because of the density of the components placed on a board and the large number of pads that must be connected together within a limited circuit board area. Indeed, manual placement of the traces on a dense board with 1000 or more pads requires many tedious hours of work and is prohibitively expensive. A more effective approach is to employ one or more automatic routing programs that utilize wiring algorithms to route most of the traces. Manual intervention at the end of the automatic routing may be necessary to clear the crossovers and to place the wires that could not be placed automatically with the controlling algorithms. The combination of the automatic and manual approaches is usually employed because it reduces the time required in routing the board.

In this description of routing, we will cover four important topics; namely, surface organization, design rules that constrain routing options, automatic routing programs and a few classical wiring algorithms.

5.7.1 Surface Organization

To layout circuit traces in a logical and systematic process, the signal planes are organized to provide defined pathways (channels) for the traces. This process is analogous to providing right of ways for the streets in planning and allocating space in the design of a new city. There are four common methods used for surface organization that include grids, gridless, plastic grid and channel. The grid method divides the entire surface into a uniform grid with a fixed pitch p in both the x and y directions. Traces and half spaces may be placed on any grid band that does not interfere with a pad, component lead, via or reserved area. Vias may be placed at intersections of the x and y grid bands. The grid method of surface organization is illustrated in Fig. 5.32a. Reference to this illustration shows the grid lines forming the grid bands. In each grid band, one or more circuit traces each of width w may be etched. A space s is required between traces and from the geometry shown in Fig. 5.32b it is evident that:

$$w + s = p \tag{5.25}$$

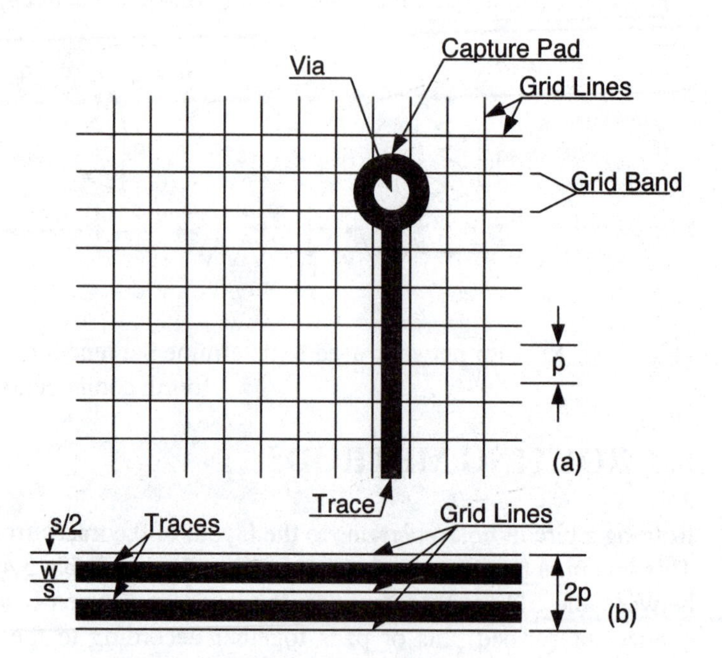

Fig. 5.32 Grid technique for surface organization. (a) Grid layout showing pitch, capture pad, via and trace. (b) definition of space s, half-space s/2, trace width w and grid pitch p.

Commonly employed dimensions for p, s and w are listed in Table 5.11. The particular pitch selected depends upon the center-to-center dimension of the leads on the chip carriers and the capability of manufacturing to produce high quality boards with fine line etching. For example the pitch p = 25 mil (0.635 mm) is effective with through-hole chip carriers where the center-to-center dimension is 100 mil (2.54 mm). Two traces may be fit into the space between the 50 mil diameter pads. A pitch p = 10 mil (0.25 mm) would be better because five traces could be placed between the pads; however, the trace width would only be 5 mil. This width is small enough to be considered as a fine line. Most firms producing circuit boards today control their etching process well enough to insure adequate yields with fine line traces [4 to 5 mil (0.100 to 0.125 mm) wide]. Trace widths of 3 mil (75 μm) or less are considered leading edge or state of the art manufacturing capability. The grid method is excellent approach to surface organization when all of the components on the board are similar and regular patterns can be extended over the entire surface of the board.

The gridless surface permits the use of traces where w, s and p can be varied at select regions on the board in order to avoid interference with obstacles such as vias or pins. This approach permits better utilization of board space, but it increases the complexity of the automatic routing process significantly. It also requires fine line etching capabilities in production, because in crowded areas of the board the pitch actually used is small. In some cases, where a two-sided board is used in place of a multilayered board, the added cost of gridless routing is justified.

Table 5.11
Dimensions in mil for trace geometries with grid surface organization

Pitch, p	Width, w	Space, s
25	13	12
20	10	10
15	8	7
12	6	6
10	5	5
8	4	4
6	3	3
4	2	2

The plastic grid surface organization is similar to the gridless method, because this technique permits each routing grid to have a different pitch. This feature assists in routing those odd components that do not fit the fixed grid, which is suitable for a majority of the components. The plastic grid method is a compromise between the grid and gridless methods. It usually reduces the need for manual routing of many of the off grid components that occurs when the fixed grid method is employed. It reduces the time required for routing when compared with the gridless technique.

The channel method organizes the area of the board into channels that pass between the pads of the components as illustrated in Fig. 5.33. Vias are placed between the channels using the same centers as the component's I/O. This procedure reserves space for both the vias and the traces so that the two functions do not compete for board area. This advanced reservation of board area is important in routing very dense boards that require several layers to resolve all of the wiring assignments.

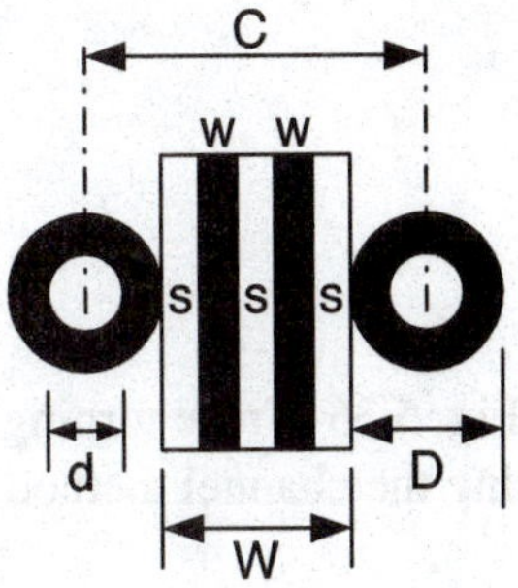

Fig. 5.33 Two-trace wiring channel between through-hole pads with controlling dimensions shown.

The width W of the wiring channel is illustrated in Figs. 5.33 and 5.34 for through-hole and surface mount boards, respectively. The channel width W is determined by the footprint as:

$$W = C - D \qquad \text{or} \qquad W = C - b \tag{5.26}$$

A wiring channel usually contains one to five wiring traces depending upon the available width W and the degree of control of the etching process that dictates the minimum trace and space width w and s, respectively. Note, that the dimensions of the traces in the wiring channel are given by:

$$W = w + 2s \tag{5.27a}$$

$$W = 2w + 3s \tag{5.27b}$$

$$W = 3w + 4s \tag{5.27c}$$

$$W = 4w + 5s \tag{5.27d}$$

$$W = 5w + 6s \tag{5.27e}$$

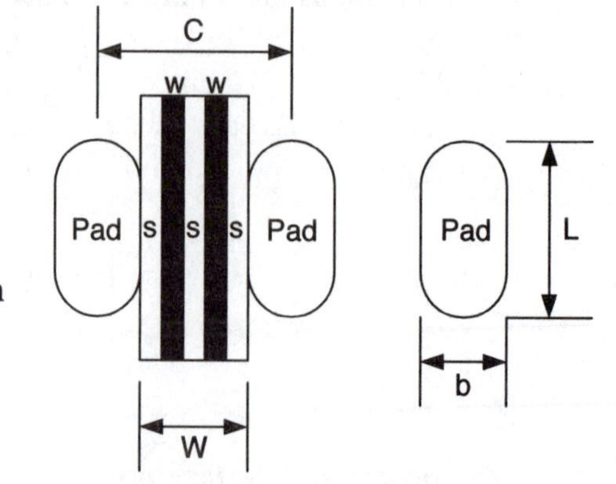

Fig. 5.34 Two-trace wiring channel between SMT pads with controlling dimensions shown.

Wiring channels extend in both the x and y directions and intersections of x and y channels allow for the traces to be turned on a given signal plane. Trace turning is also accomplished by intersecting say the x trace on one signal plane with a via that connects to the y trace on another signal plane. An example of trace turning with the channel method of surface organization on a multilayered board is presented in Fig. 5.35.

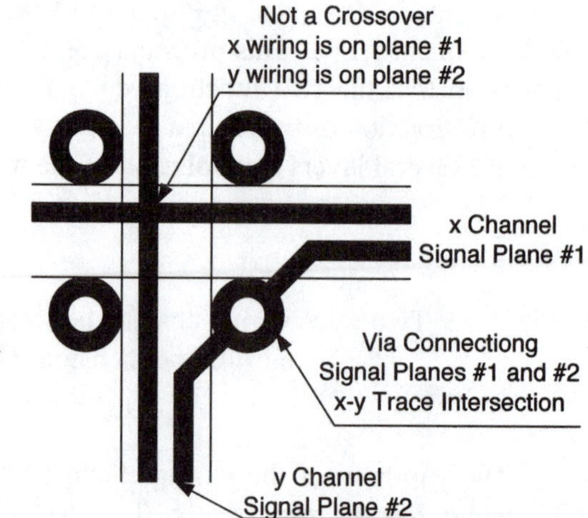

Fig. 5.35 Trace turning on a multilayer circuit board using the channel method of organization for each signal plane.

Surface organization for the ball grid array (BGA) chip carrier differs from the layout commonly employed for packages with leads or pads deployed about their perimeter. The BGA utilizes the area under the chip carrier for an area array of bonding pads. Moreover, the solder balls used for connecting the chip carrier to the circuit board are on very close centers. This arrangement requires an arrangement of bonding pads and vias to provide escape routes for the connections at locations under the chip carrier.

A common layout for the surface of a printed circuit board to accommodate a BGA is illustrated in Fig. 5.36. Note that the two outside columns of the BGA pads have traces on the top surface of the PCB. However, with the close spacing of the surface mounting pads, sufficient space does not exist to route the traces outward from the internal rows and columns. These internal connections are made by using blind vias that connect the surface pads to interior signal planes below the BGA.

The dimensions of the vias, pads and traces used in accommodating a BGA vary to some degree, but in general, feature sizes are small requiring excellent control in manufacturing methods. For example, various feature dimensions for a 0.80 mm pitch BGA are listed below:

- 0.45 mm (11 mil) diameter surface mounting pad
- 0.60 mm (15 mil) diameter solder mask opening
- 0.40 mm (10 mil) diameter via capture pad
- 0.20 mm (5 mil) diameter via
- 0.15 mm (4 mil) trace width
- 0.10 mm (2.5 mil) minimum spacing

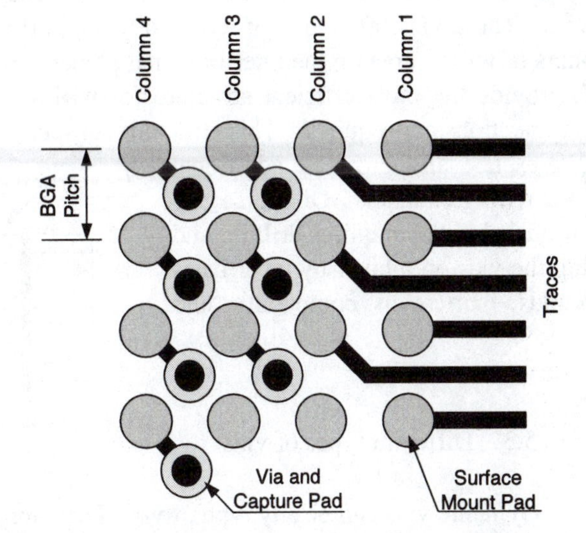

Fig. 5.36 Surface organization used to provide escape routing under a BGA.

5.7.2 Design Rules for Routing

Design rules are developed by mutual agreement between the design and manufacturing functions. These rules are imposed to force the designer to layout circuit board employing only those features that can be produced with processes and equipment that are available for manufacturing the product. The firms with the best equipment, the most advanced processes, and quality control methods that insure high yields can establish design rules that permit the development of the high density circuit boards. Several design features including: via hole diameter, capture pad diameter, trace and spacing dimensions, layered vias, dynamic vias, blind vias, buss structure, and buried resistors should be considered in writing the design rules that govern the board layout and routing.

Via hole diameter is extremely important in routing because it is a significant factor in controlling the real estate available for wiring. With PCB's for through-hole devices, there is little choice in specifying this diameter because the via serves to hold the pin as well as to connect one signal layer to another. Reference to Figs. 5.17 and 5.18 indicates that d = 30 to 34 mil (0.76 to 86 mm) and D = 50 or 60 mil (1.27 to 1.52 mm) are common design rules for through-hole layout. For surface mount devices, a via hole diameter is dependent upon the fabrication processes because electrical function can be accomplished with holes that are only a few mil in diameter. Today many fabricators[8] prefer a 14 mil (0.35 mm) hole drilled in a 22 mil (0.55 mm) pad for SMT. However, with leads spaced at 20 mil (0.50 mm) centers on SMT chip carriers, the available space for a wiring grid or channel located between the pads vanishes. Improved drilling methods permits the via hole diameter and the via capture pad diameter to be reduced allowing for additional wiring channels on all layers of a multilayer circuit board.

Trace and space width are also specified as design rules, which are extremely important in organizing the board surface areas. The copper cladding is etched away to produce a space of width s and a circuit trace of width w. The photo lithography and etching processes, described later in Chapter

[8] Leading edge manufactures can drill 9 mil (0225 mm) diameter via holes in the center of 16 mil (040 mm) diameter capture pads.

6, control the precision and smoothness of the traces and the yield of the boards being processed. Fabricators today are producing lines and spaces that usually range from s = w = 3 to 4 mil (75 to 100 μm) with 4 mil (100 μm) considered standard and 2 to 3 mil (50 to 75 μm) considered as fine line technology. When fine line technology becomes more common place, the density of the circuit boards that can be wired will increase and the number of signal layers required to wire the boards will decrease.

Via structure must also be considered in design rules particularly with multilayered boards. Many fabricators restrict design and permit only through-hole vias on a fixed grid like those shown in Fig. 5.5. These vias often are not required to support the chip carrier and they effectively destroy large amounts of wiring area on the interior signal planes. Layered or buried vias, which are illustrated in Fig. 5.37, provide the most efficient structure for wiring. A via is employed on only those layers where interconnections are required. This structure conserves area on all the wiring planes not affected by the via's presence and provides additional area for wiring channels. Of course, the use of buried layers requires drilling and plating the vias on a layer by layer basis that adds fabrication complexity and cost.

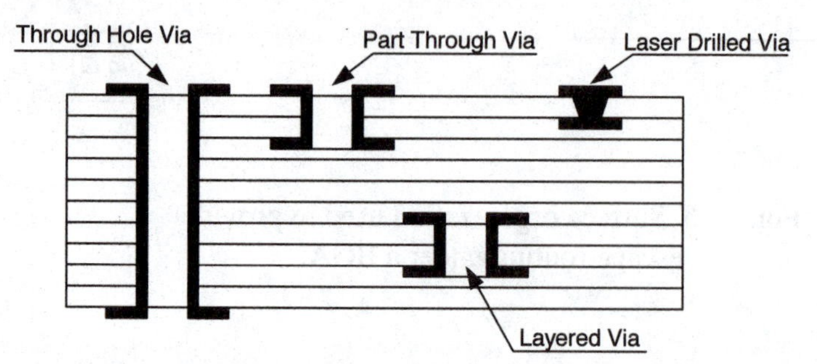

Fig. 5.37 Different types of vias.

Dynamic vias can be any type of via. They are called dynamic because they are used only when it is necessary to make a connection with one or more interior layers. This is in contrast to a fixed via structure commonly employed where the vias are drilled on fixed centers whether they are required or not. Dynamic vias are placed when routing SMT boards and only those vias required for connectivity are inserted into the laminate. This procedure conserves area on the interior layers and permits more dense circuits to be wired on fewer layers.

Buss structure refers to power distribution that often utilizes significant amount of wiring area. Signal traces can be narrow with a small cross sectional area because of the small currents associated with the signal pulses; however, wiring to supply power usually requires conductors with large cross sectional areas for the high power chips. There are three commonly utilized buss structures depending upon circuit density. The first is the power trace that is identical the signal trace except that the width w is about 60 mil to provide the cross sectional area required to handle higher currents. The power trace is commonly used in lower density circuits mounted on two-sided boards.

For moderate density circuits with high switching speeds, small buss bars mounted on the circuit board provide the conductors to distribute the power. These buss bars, illustrated in Fig. 5.38, carry the V^+ and the V^- to a row of IC's as indicated in Fig. 5.39. The bus bars conserve board space, add rigidity to the board and provide a conductor with low inductance and low resistance consistent with requirements to limit noise generation in high-speed switching. This approach for distributing the power is most useful in two-sided boards where high inductance in the power distribution conductors can cause large voltage oscillations when 32 or 64 drivers are switched simultaneously in a nanosecond or less.

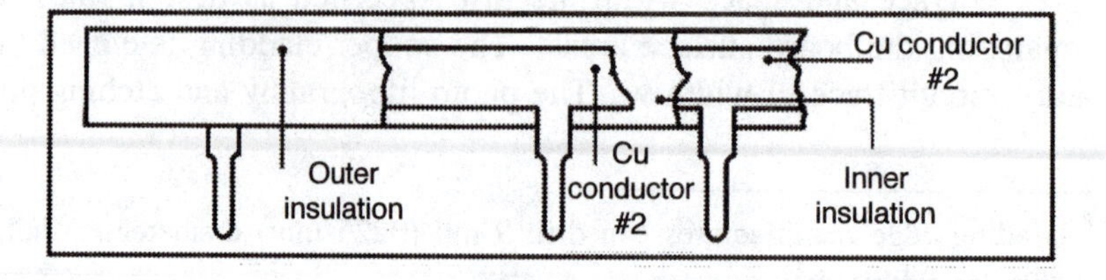

Fig. 5.38 Bus bar for through-hole mounting on a PCB.

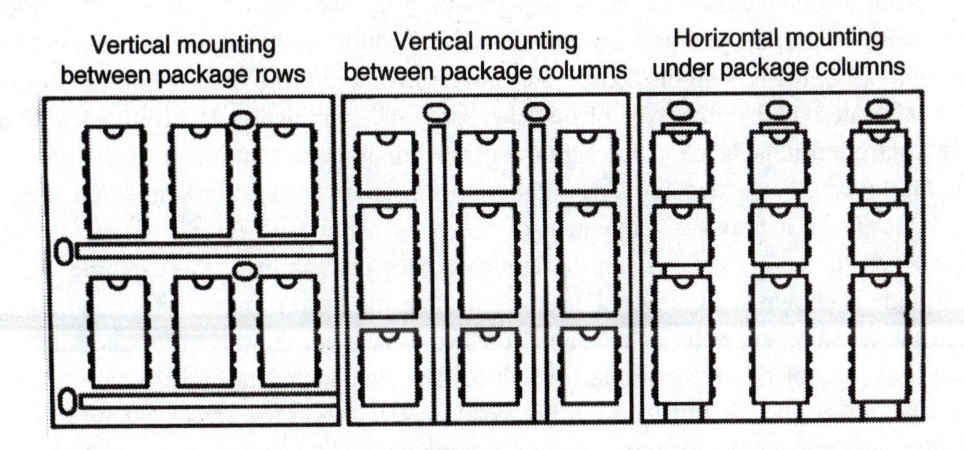

Fig. 5.39 Power distribution to rows and columns of through-hole chip carriers with bus bars.

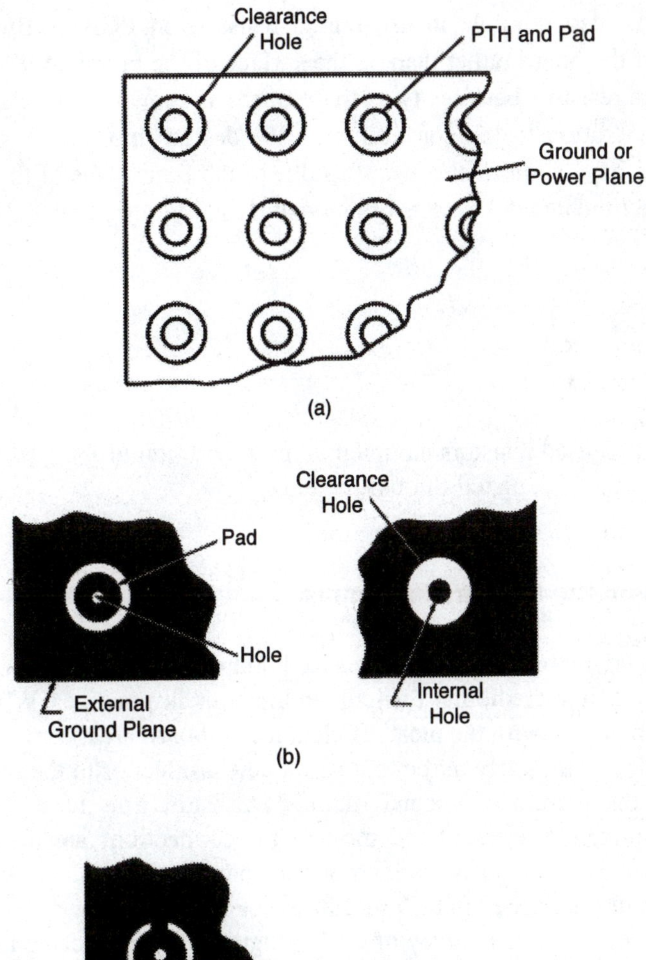

Fig. 5.40 Power plane design for a multilayered circuit board. (a) Patterned copper cladding forms the power planes.
(b) Detail showing pad isolation.
(c) Detail showing connection of a via capture pad to a power plane.

The third method for power distribution is used in multilayer boards (MLB). With MLBs the V^+ power is usually supplied over an entire plane making up one side of a layer. This approach has several advantages. First, the power plane provides a low inductance and low resistance conductor to supply voltage to all of the IC's on the board. Secondly, the power plane in close proximity to the signal plane, which is located on the other side of the layer, permits construction of strip lines and micro strip lines to give controlled impedance of the signal wiring. Finally, the continuity of the power plane on one side of the layer gives added in-plane rigidity to the layer during the lamination process. This rigidity is important as it prevents movement of the pads located on interior layers at the elevated temperatures when the epoxy softens and the laminate becomes viscous. An example of a power or ground (V^-) plane is shown in Fig. 5.40. Observe here that the power plane is not continuous as holes are etched in the cladding either to isolate signal pins or to connect the power/ground pins. Details of the clearance hole for isolation are given in Fig 5.40b. The clearance detail for the case when the pin is connected to the power plane is illustrated in Fig. 5.40c. Note, two short traces extend from the power plane to the pad, and the annular clearance area is maintained. This practice prevents high rates of heat transfer from the PTH to the copper power plane during the soldering operation. If the PTH is connected directly to the power plane, adequate temperature of the plated-though-hole is difficult to maintain during the soldering operation and a cold soldered joint may result.

It is also possible to use buried resistors in PCB's. Buried resistors are mounted in the z direction of the board rather than on the surface of the board. Vias may be placed in the board to house low wattage resistors between two circuit traces as shown in Fig. 5.41.

In addition to the constraints on the design imposed by the limitations in the manufacturing process, we have additional constraints due to the limitations of the automatic routing programs. Some of the more fundamental aspects of automatic routing programs will be discussed in the next sub-section.

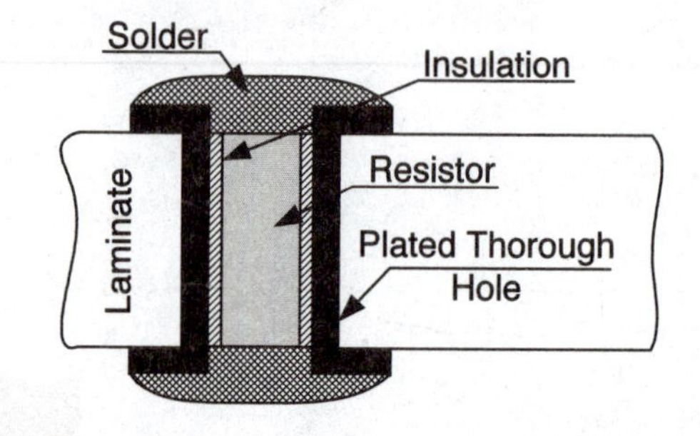

Fig. 5.41 Buried resistors mounted in the z direction in a plated-through-hole.

5.7.3 Automatic Routing Programs

The task of laying out all of the routes for hundreds or thousands of wires on a circuit board is simple in concept, but it is incredibility difficult and tedious in practice. While circuit boards can still be routed by manual methods with the most efficient use of board area, the time required to accomplish this task is not conducive to the early introduction of a new product or to the maintenance of a satisfied work force. To reduce the tedium of manual routing, software firms have developed automatic wiring routines where component placement and most of the connections are made with computer automated design (CAD) methods. The software incorporates one or more of the wiring algorithms and accommodates the design rules described in the preceding section.

A recent internet survey of CAD programs for PCB design indicated availability of 53 different products ranging in cost from about $200 to $10,000 depending on program capability and the number of designers the program supports. In most cases, these programs run on personal computers and the cost of computer hardware usually is modest. In addition to vigorous developments by software firms, the larger electronic companies have developed their own CAD programs that are usually proprietary.

Some of these CAD tools include a sophisticated engineering database management system that allows the electronic designer to easily create and revise electronic circuit designs (digital or analog) directly on a personal computer. The programs can access thousands of standard or custom on-line libraries in seconds, retrieving the necessary information to support the current design task. The programs are essentially on-line systems that construct databases as the design is developed. In other CAD tools, the designer creates a complete schematic and then use post-processing utilities to describe the circuit. All of the CAD tools incorporate an auto-interconnect routing engine that lays out schematic interconnects. Many programs generate a variety of output formats, which include a standard SPICE list output, schematic drawings and input data for developing phototools employed in the photolithographic process used to create the various layers that are laminated together to form the multilayer board.

While it is not within the scope of this text to describe these automatic routing programs, it is possible to list features that should be included in CAD programs as indicated below:

1. Automatic routing for boards containing through-hole, surface mounted devices and ball-grid arrays.
2. Variable grid routing.
3. Daisy chain connection of nets.
4. Routing the shortest connection first.
5. Routing high-speed time critical circuits before the non critical circuits.
6. Automatic placement and connection of line termination resistors.
7. Preliminary assignment of critical pads for priority routing.
8. Report of traces that exceed a specified length.
9. Report of parallel length for traces.
10. Assignment of traces to specific layers.
11. Assignment of trace width and space width by layer.
12. Control of via usage per net or circuit.
13. Control of stub length.
14. Matching trace length on time critical circuits.
15. Listing of crossovers and incomplete circuits.

Most of these features are available in CAD system available today. The more expensive programs include more features, and they handle circuit designs with more signal layers. As software continues improves more of these features will be available at lower cost permitting better circuit board design in less time.

In addition to the features listed above, the CAD system should interface with other facilities necessary in the production of the board. Most important is the interface between the CAD system and the photoplotter that is used to transmit the master files for the phototools. This interface can be a modem that allows design information to be transmitted to the photoplotter over a high bandwidth connection. Of course, the format of the design data must be consistent with the format required by the photoplotter. Next, the CAD system should produce a drilling file that is used with the numerically controlled drills to provide the coordinates of all via and tooling holes. Finally, the CAD system should provide the design data necessary for the solder mask. The form of this data depends on the solder mask process, screened or dry film, and includes the locations, geometry and sizes of the areas to be exposed to the solder.

5.7.4 Wiring Algorithms

The task of routing a circuit board can be divided into four parts that include:

1. Preparing the wire list
2. Assigning traces to specified layers
3. Assigning the order for making the connections
4. Laying out wiring traces

In this treatment, we will not attempt to completely describe all of the methods to accomplish these four tasks because to do so would require a separate textbook. Instead, we will introduce each task with one or two techniques that illustrate the concepts involved.

The wire list (see Table 5.4) shows the connections between the leads of the various devices on the circuit board and is related to the circuits being placed on the circuit board. With digital circuits, many common points occur in the circuit, where the voltage will be either high or low at the same instant. A connection of these common points is called a net and the wire list should connect these points together while minimizing the length of the traces used to make the specified connections. One approach to determine the most suitable wire list comes from the classic traveling salesman problem. This problem involves a salesman who is to travel to N cities and return to the starting point (home) following the shortest route.

To illustrate this approach consider the four points shown in Fig. 5.42 and labeled A to D. The scheme is to start from home and travel to a different city and determine the minimum travel distance. In this simple example, there are only three unique routes (remember we are not concerned with the direction of travel), as illustrated in Fig. 5.42 b, c and d. The result in Fig. 5.42b shows that the minimum distance traveled is 10.5 units as compared to 12.1 and 12.4 for the other two unique routes. The wire list for this net would specify the sequence A, B, D, C, A to define the wiring route.

This example is quite simple, because we selected N = 4; however, the complexity increases rapidly for greater values of N. In general, the number of routes k to consider in traveling to N cities is given by:

$$k = (N - 1)!/2 \qquad (5.28)$$

For N = 8 there are 2520 unique routes, that should be considered in minimizing trace length. Clearly, algorithms and computers are required for this task if wire length is to be minimized with large nets.

Layering involves the assignment of specified traces to one of the signal planes in a multilayered circuit board. Of course, in one-sided circuit boards layering is not possible. In two-sided circuit boards, layering is not necessary because one side is essentially reserved for the x traces and the other side for y traces. The need for layering and the use of a multilayered board to facilitate routing becomes evident if all of the traces specified on the wire list cannot be placed on a two-sided board. Additional wiring planes are added to accommodate traces that cause the most crossovers. The addition of wiring planes adds significant cost to the PCB's and consequently each interior wiring plane that is added should significantly enhance wirability.

In commercial products, where wiring placed on external planes is common practice, the first step in layering is to remove the power and ground traces from the external planes and place them on a single internal plane. This clears area on the exterior planes allowing for additional signal traces. If additional wiring planes are required, more layers are added to the circuit board as illustrated in Fig. 5.43. Assignment of the circuits or nets to the various layers is based primarily on those connections producing crossovers. These crossover connections are assigned to an interior plane until crossovers become excessive on these planes. At this point, additional planes are added or the remaining traces are manually routed.

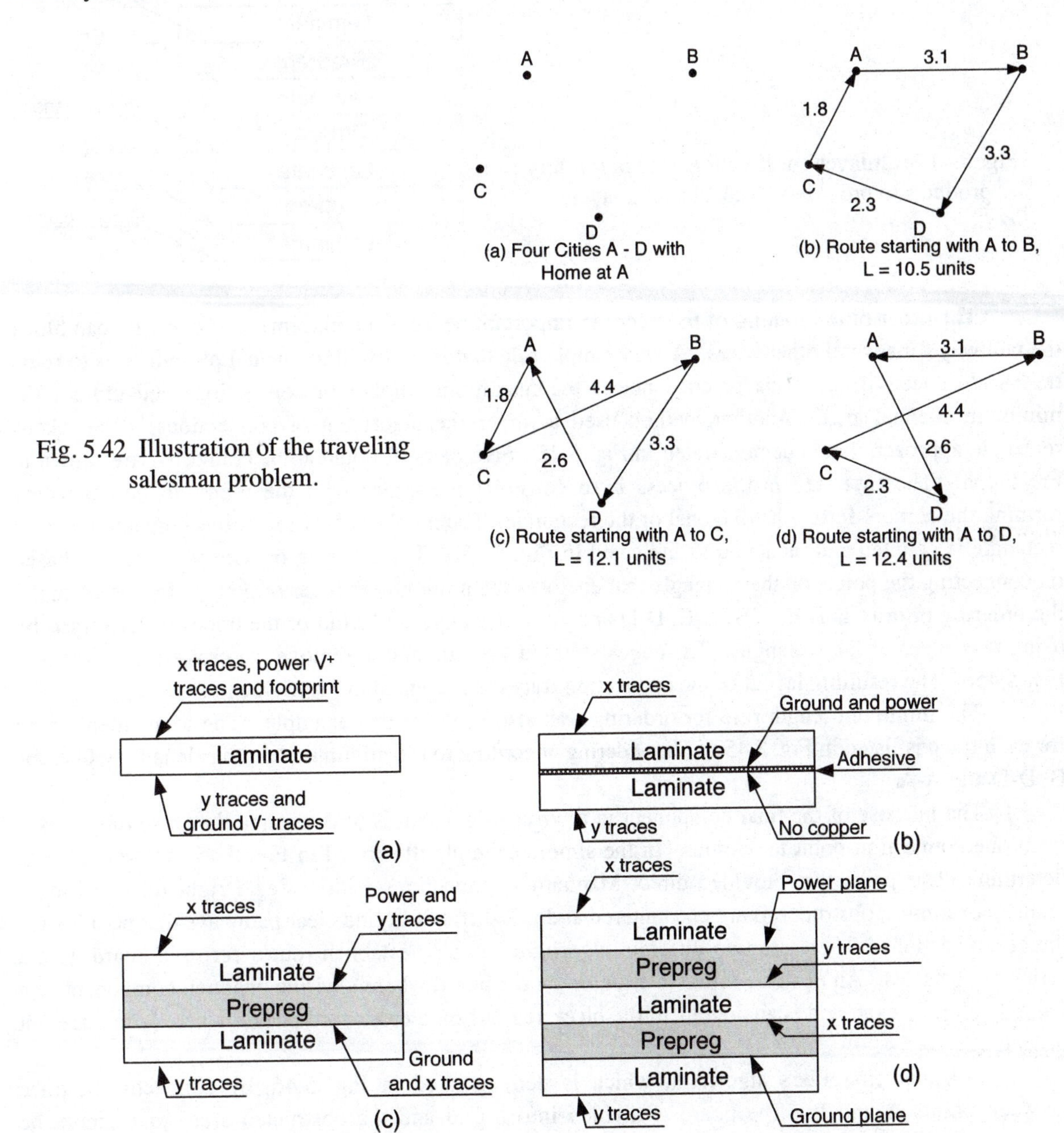

Fig. 5.42 Illustration of the traveling salesman problem.

Fig. 5.43 Multilayer construction used with commercial products providing additional wiring planes. (a) Two-side circuit board. (b) Two layer adhesively bonded three-plane circuit board. (c) Two-layer prepreg bonded four-plane circuit board. (d) Three-layer prepreg bonded six-plane circuit board.

With military systems, signal traces on the exterior planes are often not permitted. All signal wiring is placed on interior planes. It is common practice to alternate between signal and power or ground planes as indicated in Fig. 5.44. Assignment of connections to the signal planes is again based on crossovers. All of the traces are initially placed on the outermost signal planes and those routes producing a large number of crossovers are progressively moved to the more interior layers.

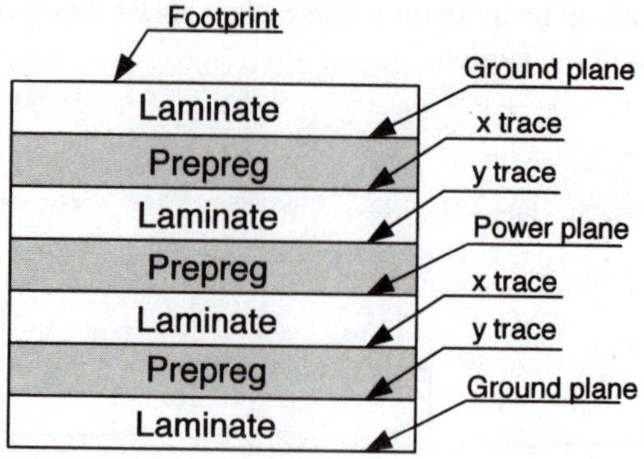

Fig. 5.44 Multilayer construction used in military products to provide additional signal layers.

The order of the routing of the traces is important because the placement of one wire can block the pathway for several other wires. A very simple rule that is followed in manual placement is to route the shortest leads first. This practice places the maximum number of connections and utilizes the minimum channel area. Another method used to order the placement of connections is the Aker's rectangle approach that is demonstrated in Fig. 5.45. Five pairs of points to be connected are shown in Fig 5.45a. The first step in the process is to construct rectangles with the points to be connected forming the corners across the diagonal of the rectangle. The number of wiring points contained in each rectangle is counted and tabulated as indicated in Fig. 5.45b. The ordering of wire placement is based on connecting the points on the rectangle that encloses the minimum number of points. In this example, the ordering priority is E-E, B-B, C-C, D-D and A-A. Finally, the layout of the traces is performed by using two edges of the rectangles. The edges selected for elimination remove the crossovers evident in Fig. 5.45b. The resulting layout of the connecting traces is presented in Fig. 5.45c.

The minimum length rule for ordering was also used for this example. The Manhattan length for each trace is listed in Fig. 5.45c. The ordering according to the minimum length rule is E-E, C-C, B-B, D-D and A-A.

The purpose of the final component in a wiring algorithm is to determine the route of the trace from one connection point to another. In the simple example illustrated in Fig. 5.45, we were able to determine clear paths that provided direct Manhattan connections with a single right turn. In more realistic examples, obstructions are encountered and it is difficult to find clear pathways that permit such direct connections. There are two different algorithms used to establish routes across a board with a series of obstacles. An obstacle refers to a region on the board where a wiring channel, component or a via may already exist and is analogous to the black regions on a crossword puzzle where letters are not permitted.

Consider first Lee's algorithm, which is demonstrated in Fig. 5.46, by connecting a trace between points A and B. The board is divided into a grid and the obstructed areas that cannot be crossed are blackened. Beginning at point A, we number adjacent grid squares 1 as indicated in Fig. 5.46a. Because Manhattan wiring is employed the grid squares on the diagonal cannot be numbered. The numbering process is continued by placing 2 adjacent to each 1 and then 3 adjacent to each 2 as illustrated in Fig. 5.46b. The numbering process is completed in Fig 5.46c, when the number 18 is at grid square adjacent to the B location. The final step is the construction of the route from A to B, which is shown in Fig. 5.46d. It should be noted that there are many other routes each 18 units long that connect A and B. All of these routes are evident from an inspection of Fig. 5.46d. Other routes longer than 18 units can also be determined.

The graphic illustration of Lee's algorithm is used only to describe the method. In practice, the method is implemented with a computer that stores the locations of the entire grid structure. The numbering advances automatically from a single point. A route is established when the square adjacent to the second point is filled with a number. The advantage of Lee's algorithm is that it will always find a route between two points if one exists. Moreover, this route will be of minimum length. The disadvantage is that method requires a very large amount of memory and significant computer time when the number of grid squares on the signal layer is large.

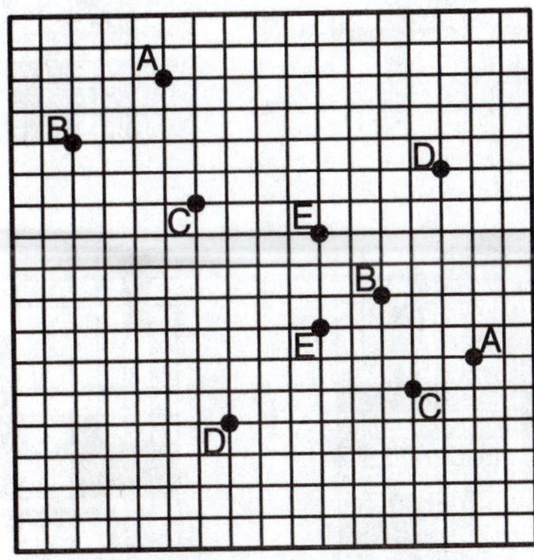

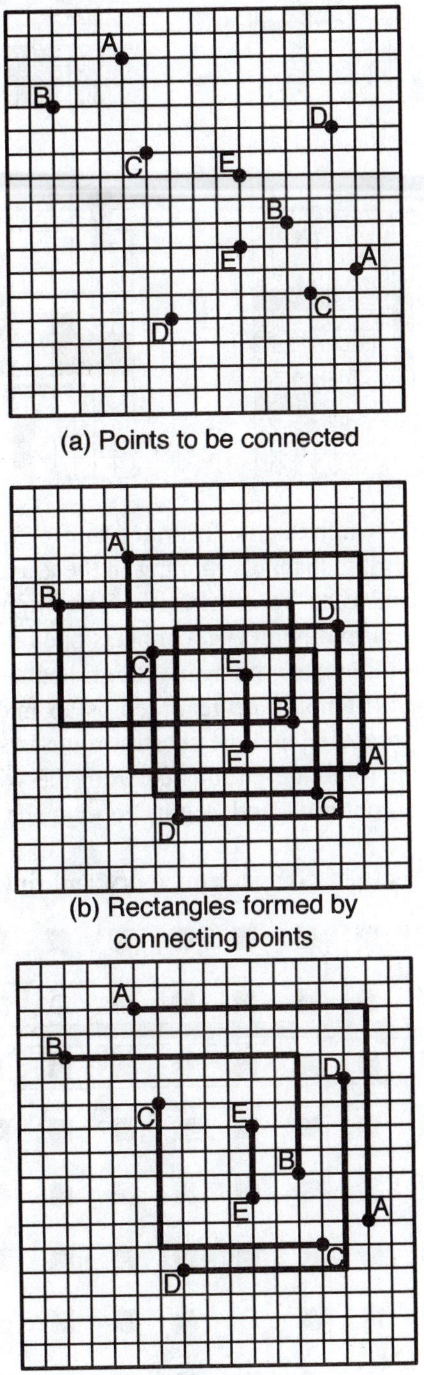

(a) Points to be connected

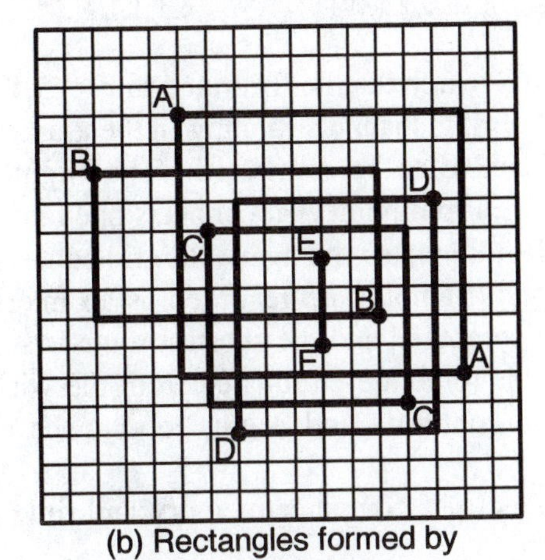

(b) Rectangles formed by connecting points

Rectangle	Points enclosed	Ordering
A	5	5 th
B	2	2 nd
C	3	3 rd
D	4	4 th
E	0	1 st

(c) Connections

Connections	Trace length	Ordering
A-A	19	5 th
B-B	15	3 rd
C-C	13	2 nd
D-D	15	4 th
E-E	3	1 st

Fig. 5.45 Aker's method for ordering placement of connections.

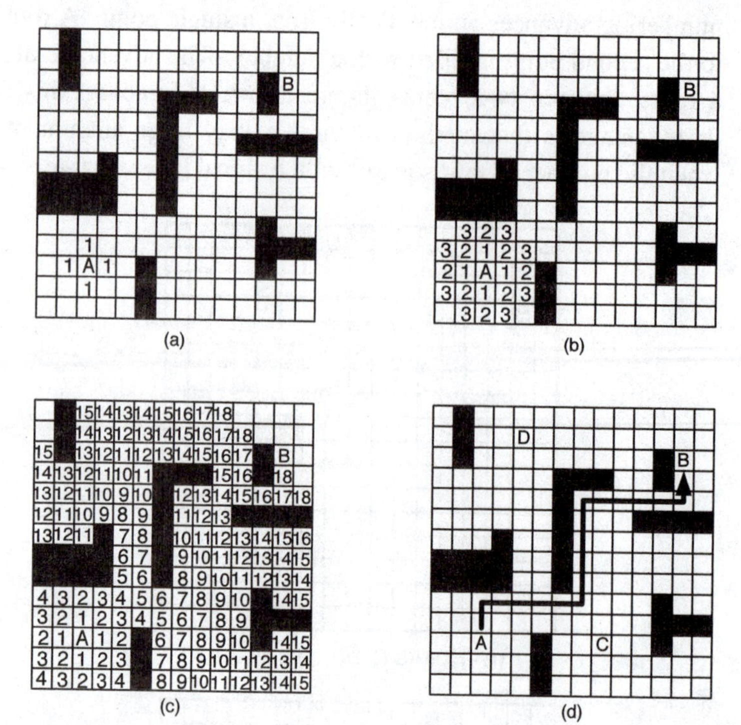

Fig. 5.46 Lee's algorithm for determining minimum length connection between point A and point B.

Another approach is the Hightower [20] fast line search that is illustrated in Fig. 5.47. The field of the circuit board is arranged with wiring points and wiring channels running in the x and y directions between the connection points. We locate points A and B to be connected. The Hightower search routine begins simultaneously at both connection points. From point A it connects open grid squares in the ± y direction of the wiring channel. From point B it connects the available grid squares in the ± x direction. The locations of the intersections (16 in this example) are recorded as an intersection that identifies a possible route. The length of each route is computed and the shortest route is selected from the solution set. The grid track used in establishing this route are eliminated from the wiring channel (this area now becomes an obstacle) and the program continues and begins routing the next pair of wiring points.

The advantage of the Hightower approach is reduced computer memory and time required to place most of the wires on a board. Also, it is usually possible to place more wires per layer with this approach than with Lee's algorithm. The disadvantages are that the routes are not always minimum length and that manual routing of some of the last connections may be necessary.

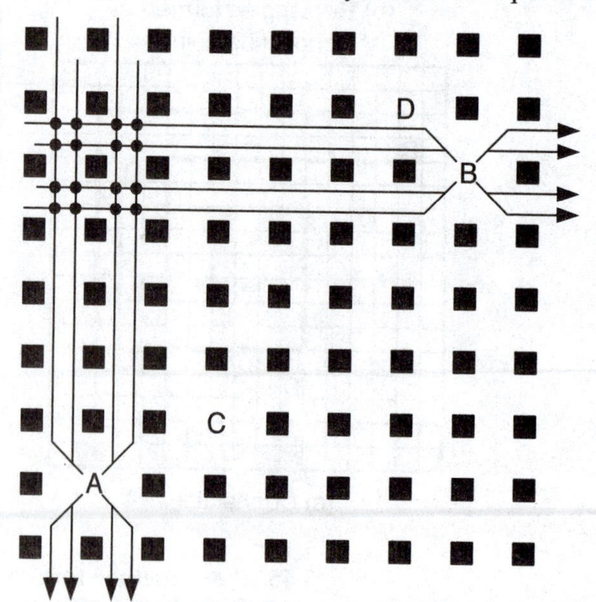

Fig. 5.47 Graphic illustration of Hightower's fast line search.

REFERENCES

1. Dally, W. J. and J. W. Poulton, Digital System Engineering, Cambridge University Press, New York, NY, 1998.
2. Pecht, M. (ed.), Handbook of Electronic Package Design, Marcel Dekker, New York, NY, 1991.
3. Pecht, M. (ed.) Placement and Routing of Electronic Modules, Marcel Dekker, New York, NY 1993.
4. Anon, "Surface Mount Requirements for Advanced Packaging Solutions," Amkor Technology, Chandler, AZ, 2004.
5. Anon, "Designing with Fineline BGA Packages," Application Note 114, Version 3.0, Altera Corporation, San Jose, CA, 2005.
6. Department of Defense, "MIL-HDBK-217F-2 Reliability Prediction for Electronic Equipment", Rome Air Development Center , Defense Printing Service, Philadelphia, PA.
7. Blackwell, G. R., The Electronic Packaging Handbook, CRC Press, November, 1999.
8. Witaker, J. C., The Electronics Handbook, CRC Press, December, 1996.
9. Walsh, R. A., Electromechanical Design Handbook, McGraw-Hill, 2000.
10. Lau, J. H., et al, Electronics Manufacturing, McGraw-Hill, 2002.
11. Chapman, S., Fundamentals of Microsystems Packaging, McGraw-Hill, 2001.
12. Tummala, R. R., Microelectronics Packaging Handbook: Technological Drivers, Springer, 1997.
13. Prasad, R. P., Surface Mount Technology, Springer, New York, NY, 1997.
14. Lyman, J. Microelectronics Interconnection and Packaging, McGraw-Hill, New York (1980).
15. Harper, C. A., High Performance Printed Circuit Boards, McGraw-Hill, New York, NY, 2000.
16. Hanan, M. and J. Kurtzberg, "Placement Techniques" in Design Automation of Digital Systems: Theory and Techniques, Editor, M. A. Breuer, New York, Prentice-Hall, Chpt. 5, 1972.
17. Osterman, M. D. and M. Pecht, "Placement of Integrated Circuits for Reliability on Conductively Cooled Printed Wiring Boards", ASME paper 87-WA/EEP-8, Dec. 1987.
18. Aranoff, S. and Y. Abulaffio, "Routing Printed Circuit Boards", IEEE 18th Design Automation Conference: p 130-136, 1981.
19. Lee, C. Y. "An Algorithm for Path Connections and Its Applications", IRE Trans. on Electronic Computers, Vol. EC-10, No. 3, p 346-365, 1961.
20. Hightower, D. W. "A Solution to Line Routing Problems on the Continuous Plane," Proceedings 6th Design Automation Workshop, 1969.

EXERCISES

5.1 Consider a J leaded chip carrier soldered to a circuit board with the attachment areas described in the footprint shown in Fig. 5.1. Determine the force necessary to shear the chip carrier from the board if the shear strength of the solder used to attach the carrier to the board is 2,500 psi. Comment on the implications of your result. State the assumptions made in your analysis.

5.2 A small transformer with a mass of 0.5 kg is attached in the center of a circuit board with dimensions of 2 × 125 × 300 mm. The short edges of the board are simply supported and the long edges are free. If the product is subjected to a shock load of 9 G's, determine the stresses imposed on the PCB by the transformer. Comment on the significance of your results.

5.3 Prepare a list of information that can be etched onto a PCB to assist in the manufacturing process. What is the cost of providing this information on the circuit card?

5.4 Examine a circuit card from an available PC and classify if the board is one-sided, two-sided or a multilayered. Describe the features you examined in your classification process.

5.5 For the circuit card of Exercise 5.4, identify the plated-through-holes and describe (sketch) the solder joint that is formed when a connection is made to their capture pad. Describe the size of the pads and indicate the manner in that these pads affected the shape of the solder joint.

5.6 For the circuit card of Exercise 5.4, identify the pads for the surface mounted components and describe (sketch) the solder joint that is formed. Describe the shape of the solder pads and indicate the manner in that these pads affected the shape of the solder joint.

5.7 For the circuit card of Exercise 5.4, determine how it is connected to the rest of the electronic system. Examine this connection and describe the key code used to avoid accidental insertion of this card into the wrong slot.

5.8 Suppose an electronic system has 8 different types of cards located in a single row all connected to the same mother board (back panel). Design a key code for the PCB's that will insure insertion of the cards in one and only one slot. How many part numbers will you have in implementing your design? Discuss the implication of increasing significantly the number of part numbers involved in the construction of a system.

5.9 List the three common circuit boards in use today. Describe the advantages and disadvantages associated with each type of board. Classify by product type the usual application of these three types of circuit boards.

5.10 A multilayer circuit board is fabricated from 6 laminate layers. How many wiring planes could be incorporated in this board? What method is used to connect a wire on one signal plane to a wire on another plane? Some wiring planes are reserved for the power V and the ground. Describe the features of copper conductors used for these planes. Describe the features usually found on the wiring planes used to transmit signals.

5.11 Describe organic laminates, flexible organics and ceramic substrates.

5.12 Cite the advantages and disadvantages of organic laminates.

5.13 Cite the advantages and disadvantages of flexible organic circuit connections.

5.14 Cite the advantages and disadvantages of ceramic substrates for second level packaging.

5.15 Define the following terms as used to describe wiring.

a. Crossovers

b. Shorts

c. Opens

5.16 The surface resistivity of a laminate with two parallel conductors is 10^8 Ω/square. If the two conductors are 200 mm long and spaced 125 μm mil apart, determine the current flow along the surface from one conductor to the other. The voltage difference between the traces is 2.5 V.

5.17 Determine the resistance of a laminate 1.5 mm thick, if the conductor on the top surface is 350 mm long and 0.10 mm wide. A ground plane covers the bottom surface of the laminate. Note, that $\rho_v = 10^{15}$ Ω-cm.

5.18 For the conductor and the laminate described in Exercise 5.17, find the total and distributed capacitance of the line if the dielectric constant $\varepsilon_r = 4.7$. Find the time required to charge this line if it is open at one end. Find the time to charge the line if it is terminated with a resistor matching the characteristic impedance of the line.

5.19 Write the specifications for a flexure test designed to measure the strength and modulus of a typical circuit board laminate. In the specification, provide the specimen dimensions, points of loading, points to measure either the strain or the displacement, transducer requirements and the formulas to be used in the data analysis.

5.20 From the expansion-temperature curves shown in Fig. 5.11, determine the glass transition temperatures for epoxy A, BT epoxy and polyimide A.

5.21 Estimate the thermal stress imposed on a copper PTH 2.5 mm long that is in a circuit board made from a FR-4 laminate. The circuit board is temperature cycled from 20 to 160 °C. Assume the glass transition temperature of this laminate is 130 °C.

5.22 List the reinforcing materials commonly employed with organic laminates.

5.23 Explain why the weaving pattern of the glass cloth used as a reinforcing material may be important in some applications.

5.24 List the polymers commonly used as matrix materials for circuit boards.

5.25 Compare FR-2 and FR-4 circuit board materials for use in a very high volume competitively priced product intended for a normal office environment.

5.26 Compare the properties of organic laminates fabricated with FR-4 and FR-408 as their matrix polymers.

5.27 Compare the properties of organic laminates fabricated with BT epoxy and GETEK as their matrix polymers.

5.28 Compare the properties of organic laminates fabricated with cyanate ester and polyimide as their matrix polymers.

5.29 Discuss the two polymers commonly used in the design of flexible circuits. Cite the advantages and disadvantages of each polymer.

5.30 Describe the three types of applications that employ flexible circuits.

5.31 Construct a graph showing the thickness of copper foil in units of in. and μm as a function of ounces/ft^2.

5.32 Discuss the advantages of copper foils produced by the two different electrolytic plating processes. Compare this type of foil with the copper foils produced by rolling.

5.33 A 1.0 oz. copper foil is bonded to a polyimide film 75 μm thick and etched to form conductors for a flexible circuit. The circuit is flexed repeatedly forming a loop with a 40 mm radius of curvature. Determine the strain and stresses in both the polyimide film and in the copper.

5.34 Describe the two types of ceramic substrates. In this description, indicate their composition and the method of processing each type.

5.35 Use a CAD program to layout the footprint for a standard 8 pin DIP showing dimensions of the holes. Include in the layout, some of the holes associated with the four neighboring DIPs. Define the pad diameters D = 50 mil (1.27 mm) and determine the width of the wiring channels between the pins. What controls the spacing of the neighboring DIPs.

5.36 Repeat Exercise 5.35 changing D to 40 mil (1.0 mm). If the conductor traces and the spaces are both 4 mil (100 μm) wide, how many wires can be placed in a wiring channel?

5.37 Repeat Exercise 5.35 changing D to 30 mil (0.75 mm). If the traces are the spaces are both 3 mil (75 μm) wide, how many wires can be placed in a wiring channel?

5.38 Write an equation that relates pin spacing, pad diameter, hole diameter, width of the annulus, trace width, number of traces, and space width that can be used to analyze a foot print for a through-hole chip carrier.

5.39 A 0.5 mm diameter hole is to be drilled in a 0.75 mm pad. Prepare a drawing showing the effect of registration errors in drilling. Consider registration errors of 25, 50, and 75 μm in the x direction, in the y direction and simultaneously in both directions. If each error has a 5% probability of occurring, what is the probability of drilling a satisfactory hole through the land?

5.40 A circuit board has 600 holes of the type described in Exercise 5.39. For the same drilling errors and probabilities, determine the percentage yield in producing the board. What can the designer do to improve the yield of the board in production? What can the manufacturing engineer do to improve the yield?

5.41 A new chip carrier is designed with a U lead intended for surface mount solder attachment. The thickness of the U lead is 0.125 mm, and the inside radius of curvature of the U is 0.25 mm.

Design a solder pad that will provide a solder joint with generous fillets. Show a scale drawing of the solder joint.

5.42 Describe the terms crossover and cross talk as used to describe problems in routing traces when performing circuit board layout.

5.43 Layout four rows of a footprint for a pin grid array chip carrier with 20 solder balls in each of the four rows. The solder balls are 10 mil (0.25 mm) in diameter on 40 mil (1.00 mm) centers. In your scale drawing, specify the diameters of the solder mask holes, solder pads, capture pads, and via holes.

5.44 Describe the approach and list the equations required for determining an estimate of layer count using the density method.

5.45 For a given routable area, develop a mathematical relation that describes the curve for signal plane count as a function of lines per wiring channel as shown in Fig. 5.23.

5.46 Using the wire list shown in Table 5.Ex.46 prepare a connectivity table to the first seeded component, that is the edge connector.

5.47 Place the bottom row of components if the board format shown in Fig. 5.24 is used with the wiring list of Table 5.Ex.46.

Table Ex 5.46
Wire list for the circuit board shown in Fig. 5.24

Device No.	Pin No.	Device No.	Pin No.	Device No.	Pin No.	Device No.	Pin No.
1	1	3	6	4	1	ec	5
	2	5	1		2	ec	6
	3	ec	11		3	ec	9
	4	ec	7		4	ec	7
	5	ec	12		5	ec	10
	6	ec	13		6	6	6
	7	ec	14		7	6	7
	8	ec	8		8	ec	8
2	1	6	3	5	1	1	2
	2	6	5		2	3	7
	3	2	7		3	2	5
	4	ec	7		4	ec	7
	5	5	3		5	6	1
	6	5	6		6	2	6
	7	2	3		7	6	2
	8	ec	8		8	ec	8
3	1	ec	1	6	1	5	5
	2	ec	2		2	5	7
	3	ec	3		3	2	1
	4	ec	7		4	ec	7
	5	ec	4		5	2	2
	6	1	1		6	4	6
	7	5	2		7	4	7
	8	ec	8		8	ec	8

5.48 Prepare a connectivity matrix similar to the one shown in Table 5.9 for the components remaining after the solution of Exercise 5.47. Then place the remaining three components to minimize wire length.

5.49 Using the results from Exercise 5.48 prepare a drawing showing the placement of the six components and the edge connector. On this drawing draw lines connecting the pins according to the wiring list of Table 5.Ex.46.

5.50 Fifteen components are to be placed on a circuit board that uses conduction cooling similar to that shown in Fig. 5.27. The component locations are to be selected to minimize the failure rate of the circuit board. Determine the number of possible placement patterns for the components. If the circuit board is arranged with three rows of five components and each row is placed separately, determine the number of possible placement patterns. Estimate the savings in computational effort afforded by the placement by rows approach.

5.51 Consider the grid and channel methods of surface organization and sketch the intersection of a circuit trace with via pads for both methods.

5.52 For the circuit board shown in Fig. 5.24 layout the surface of the board using the grid method. Use a grid pitch of 20 mil (0.50 mm) and take 50 mil (1.25 mm) for the pad centers. Note, that the layout should be based on using a two-sided board and include both the top and bottom layers.

5.53 Repeat Exercise 5.52 but using the channel method for surface organization.

5.54 Using the wire list shown in Table 5.7, the placement described in the text for this wire list, and the surface organization of Exercise 5.52, manually route the six component board.

5.55 Repeat Exercise 5.54 but use the surface organization method (channel) that was used in Exercise 5.53.

5.56 Use the grid method to organize the surface of a board, which is to be used with a flat pack with surface mounted leads on 50 mil (1.25 mm) centers. Layout foot prints showing four 8 leaded chip carriers. On the layout show the wiring dimensions indicating p, s, w and W.

5.57 Repeat Exercise 5.56 but use the channel method of surface organization instead of the grid method.

5.58 If d is the diameter of a via hole and D is the diameter of its capture pad, prepare a table showing the number of conductors in a wiring channel between the pads as a function of d and the trace width w. Let D = 2d and s = w. Consider pad spacings of 100, 50, 40, and 20 mil (2.54, 1.25, 1.00 and 0.50 mm). Comment on your results.

5.59 Prepare a scale drawing showing the footprint, vias and the wiring associated with a ball grid array chip carrier if it has solder balls on 40 mil (1.00 mm) centers. The ball grid array has its solder ball arranged in a square array with 15 × 15 balls.

5.60 Describe four different types of vias. Sketch each type.

5.61 Locate several circuit boards from different products. Examine these boards and identify the method used to supply the power to the various components. Identify the power buss structure.

5.62 In routing a PCB, why should the shortest connections be made first? Explain the implication of making a daisy chain connection of logic nets.

5.63 Why is it important in automatic routing to have a program that lists the crossovers and incomplete circuits at the conclusion of the run? What do you do with this list?

5.64 Why should a circuit board designer give priority to routing the I/O of critical components?

5.65 Why should we be concerned with matching the trace length on time critical components?

5.66 Why should we be concerned with via usage per net or circuit?

5.67 For the four points shown in Fig 5.Ex.67, use the traveling salesman approach to determine the minimum length for the trace connecting the points.

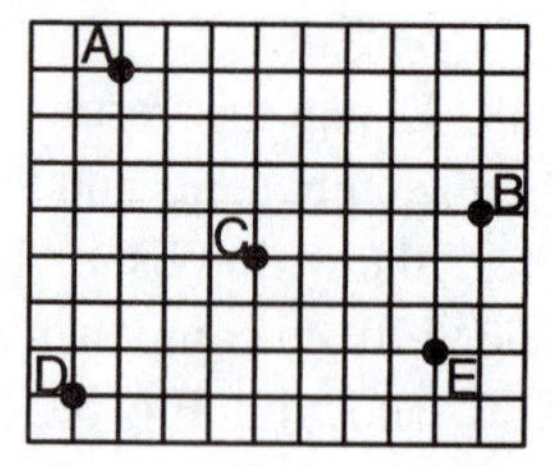

Fig 5.Ex.67

5.68 Prepare a table showing the number of possibilities that must be considered in using the traveling salesman approach to determining the minimum route of a net as the number of points in the net varies from 3 to 10. Comment on these results.

5.69 Determine the minimum connection length for the 5 points shown in Fig 5.Ex69.

Fig 5.Ex.69

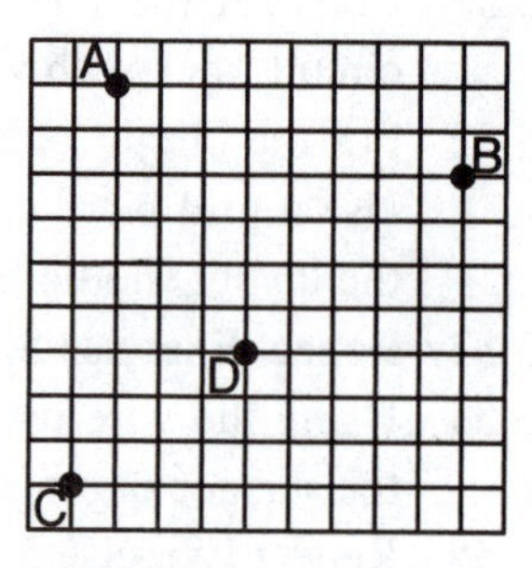

5.70 Use Aker's rectangle method to determine the wiring order for the six pairs of points shown in Fig 5.Ex70.

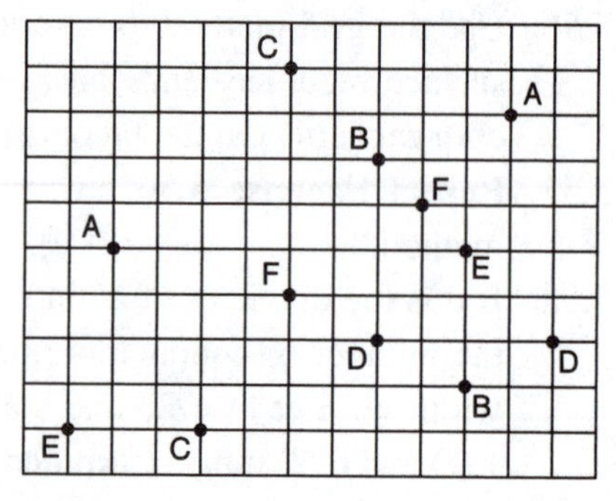

Fig 5.Ex.70

5.71 Use the minimum length rule to determine the order of wiring for the six pairs of points shown in Fig. 5.Ex70.

5.72 Describe the difference between Manhattan distance and Euclidean distance. Write the equations for both of the distances. Give two reasons for the use of Manhattan distances in design of PCB's.

5.73 Use Lee's algorithm to determine the route from point A to B in Fig 5.Ex73.

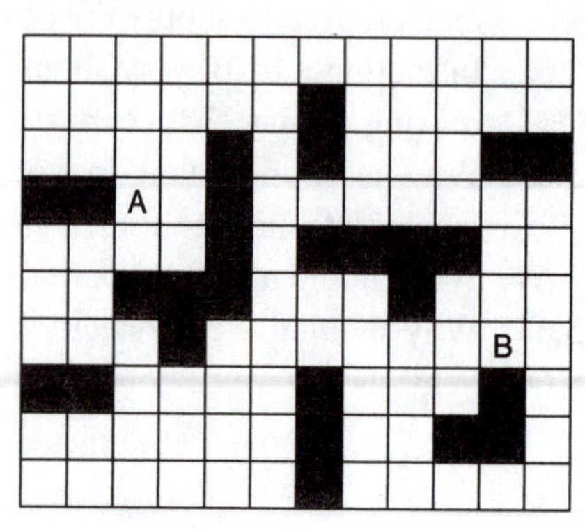

Fig 5.Ex.73

5.74 Use Lee's algorithm to determine the route from point A to B in Fig 5.Ex74.

Fig 5.Ex.74

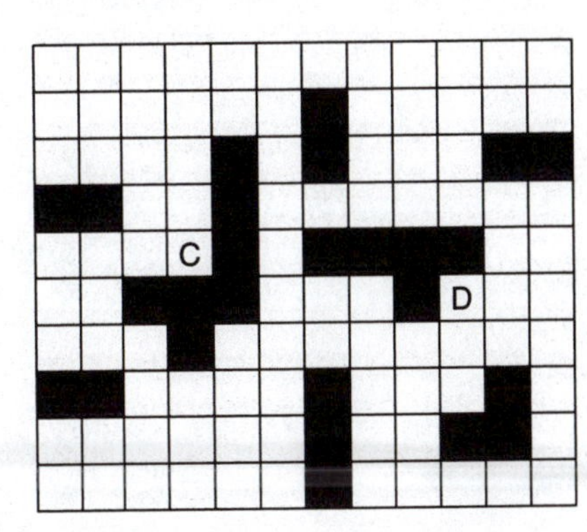

5.75 For the example illustrated in Fig. 5.46, determine the number of unique routes between A and B that are 18 units long. Layout these routes on a grid drawing.

5.76 Suppose that the 18 unit routes to point B shown in Fig. 5.46 are excluded. Find the next shortest route.

5.77 Use Lee's algorithm to wire point pairs C and D that are shown in Fig. 5.46. Hint, use the minimum length rule to establish the order of the routing.

5.78 Consider the 16 intersections obtained with the Hightower line search illustrated in Fig. 5.47. Determine the minimum line length and block out the grid squares associated with this route.

5.79 After completing Exercise 5.78 and blocking the grid squares for the route from point A to point B, determine the route from point C to point D. List the minimum length of the available paths and block the grid squares used for this route. Next place a pair of points on the diagram of Fig. 5.47 that are impossible to wire.

5.80 Suppose you have a only few wires that cannot be wired without adding an interior wiring plane. Because of significant costs you clearly do not want to add another wiring plane. How can you resolve the problem?

CHAPTER 6

PRODUCTION OF PRINTED CIRCUIT BOARDS

6.1 INTRODUCTION

There are three classifications for printed circuit boards—organic rigid, organic flexible and ceramic. We have already discussed many characteristics of these circuit boards in Chapter 5 and have described briefly some of the manufacturing methods employed in their production. In this chapter, the manufacturing methods will be described in more detail. Of particular importance are the lithographic processes that are used to create the conductors and surface features for all three types of circuit boards. Much of the technology associated with the printing industry was transferred to the electronics industry where it has been improved and adapted to meet the growing demand for higher density associated with conducting lines and smaller features sizes on closer centers.

Drilling is also an important process because holes are required for registration pins, plated-though-holes and other via structures. Moreover, the number of holes drilled in a typical PCB is huge, and their diameter varies from a few mil (75 μm) to 0.25 in. (6.35 mm). Several processes used to drill these holes including mechanical methods, laser drilling, photo forming and plasma etching are described.

Laminates of different types are used in various applications. The copper clad fiber-glass, reinforced circuit board is a laminate. Multilayer boards are laminates formed with prepreg and processed double sided printed circuit boards. High-density printed circuit boards (HDI-PCBs) are a different type of laminate with different via structures. The lamination processes used in fabricating these various products are described.

Plating processes are used to form vias and plated through holes and to provide corrosion and wear resistant coatings over the copper cladding remaining on the PCB after etching. Three manufacturing processes are described for applying these plated surfaces. These processes include electrolytic, electrolysis and immersion plating.

The functions of solder masks are described as well as the methods used in applying these coatings to select surfaces on the printed circuit board.

Flexible circuits used primarily to replace collated flat wire cable and wiring harnesses are described in considerable detail. The methods used to manufacture flexible circuits are similar to those used to produce PCBs. The steps used in the manufacturing process are outlined and the materials used for the base and cover films, adhesives and copper cladding are described.

Finally processes used to produce ceramic and glass-ceramic substrates are described. These processes included methods to produce dense ceramic disks and "green" sheets. Thin film, thick film and co-fired methods of applying metallization to the ceramic substrates are covered. The processes to produce multilayered substrates with green sheets and copper metallization are described.

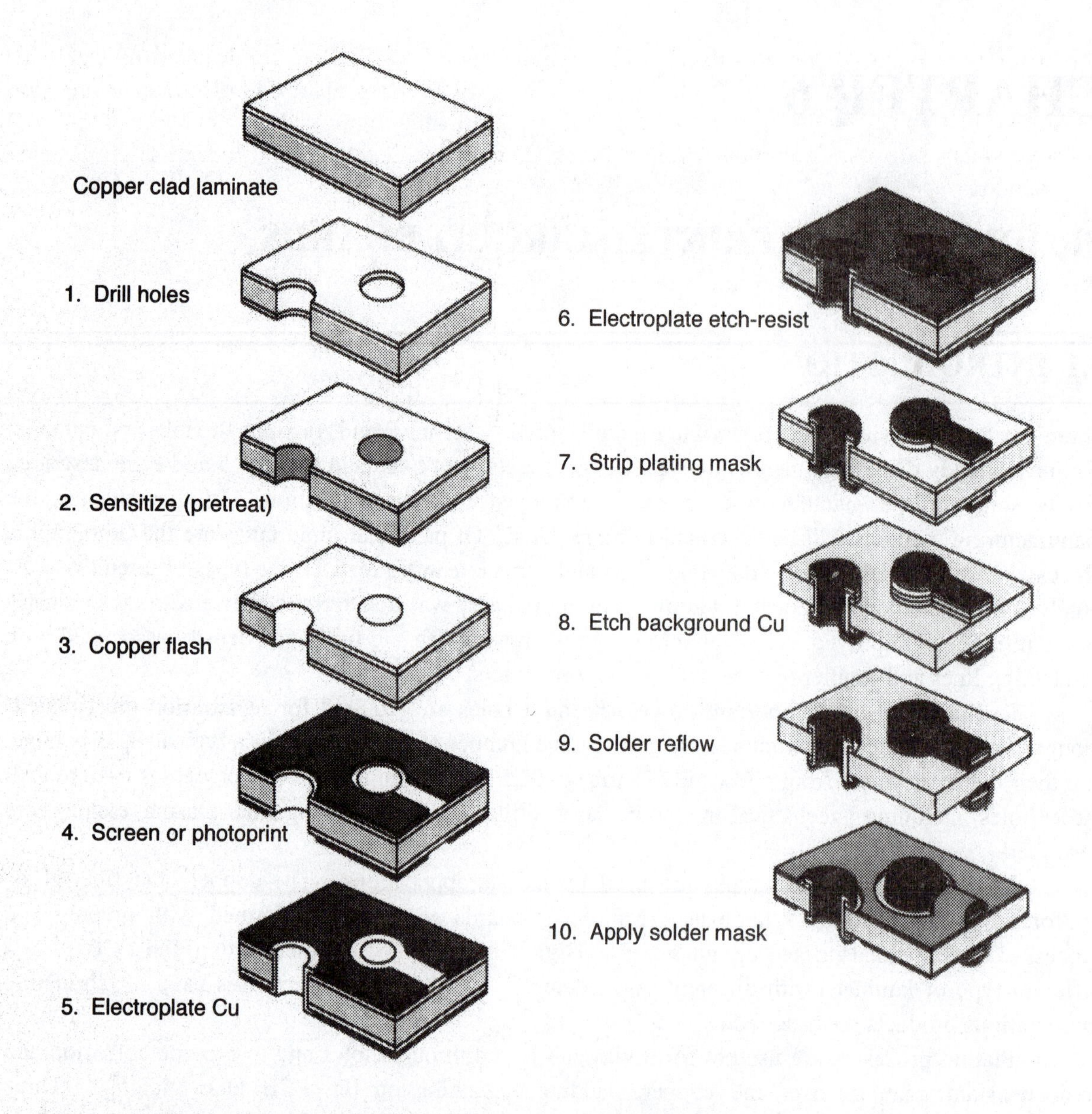

Fig. 6.1 Illustration depicting manufacturing steps employed in fabricating printed circuit boards.

6.2 OVERVIEW OF THE PRODUCTION PROCESS

A printed circuit board (PCB), as described in Chapter 5, fulfills many functions in the basic art of packaging electronic systems. Indeed, the design of the PCB is often considered the most important level of the packaging task as its development involves a significant portion of mechanical engineering time when developing a new electronic product. Also, the production of PCB's is a large and important business. Over several hundred million square feet of reinforced polymeric circuit board are produced annually worldwide, and the market continues to grow.

Due to the technical and economic importance of the PCB, it is critical that this component be designed not only to accommodate the electrical and mechanical requirements, but also be designed so that it can be manufactured with high quality at competitive costs. Good design implies an understanding of the manufacturing processes so that design requirements are in close conformity with production capabilities. We will describe here basic methods, procedures and processes that are used to produce circuit boards. We will then cover assembly operations that are followed in mounting the components to the board and the soldering processes used to make all of the electrical connections separately in Chapter 7. As the name "printed" circuit boards implies, the production of the PCB involves many processes common to the printing or lithography industries. Indeed, it is the close correspondence of the two production systems that has lead to the rapid development of a highly efficient processes for producing specialized circuit boards rapidly at remarkably low cost. A sequence of processes, illustrated in Fig. 6.1, shows many of the manufacturing steps involved. The interface between design and manufacturing functions is the layout of the circuit board. This layout is represented by a master copy of the detail design that is used to prepare a phototool. The phototool is a pattern that is imaged onto each copper clad board with a printing operation that employs a photo sensitive coating called photoresist. After the photoresist is developed, the surface features of the board are formed with an etching operation where the copper foil that is not required for the pads or traces is chemically removed. Holes used for vias and mounting are drilled through the board and then plated with copper to form plated-through-holes (PTHs), buried and/or surface vias. Solder plating or an alternative finish is employed to protect the exposed copper surfaces from oxidation and to pre-tin select soldering surfaces. Solder masks are applied to the board to control the flow and deposition of the solder during automatic soldering processes. Components are assembled at the proper locations on the board and the leads are soldered to the mounting pads. Following soldering the operations, the board is cleaned and then stored in inventory until it is used in a new product or as a field replaceable unit in the repair of an existing product.

We will cover each of these steps in detail in this chapter or in Chapter 7. We will also introduce a brief description of the manufacturing processes employed to produce flexible circuits and ceramic substrates that have more limited, but important, applications in commercial electronic systems.

6.3 PREPARATION OF MASTER CIRCUIT LAYOUTS

The interface between the design and manufacturing functions is in preparing the design layout that shows the features of each plane (layer) of the circuit board. The layout of these features is called "artwork" that is usually prepared by the design engineers and accepted by the manufacturing engineers. It is still possible to prepare artwork manually with tape and decals that are placed on an oversize sheet of Mylar. The final master is made by photographing this oversized sheet of Mylar using a scale factor that produces accurate feature sizes. However, today almost all master circuit layouts are produced using computer driven automated equipment. Before describing this equipment, it should be understood that artwork is the master copy of a detailed design. It is used in a series of photographic processes to produce the detail features of each plane (layer) of the PCB. The requirement for high precision and quality in the preparation of the artwork is self evident, because each master layout represents the first step in a long and intricate series of production processes used in fabricating the circuit board.

6.3.1 Automated Artwork by Photoplotting

Manual layout of circuit board features has been replaced with automated procedures that enable the board outlines and chip carrier footprints to be designed using a computer aided design (CAD) system. The CAD system provides a digital record of all of the feature sizes, locations and dimensions. The circuits are then wired using a computer aided engineering (CAE) system that automatically routes the circuit traces and places and records all the required vias. A description of the use of CAD/CAE in the PCB design process was given previously in Chapter 5.

The digital output of the CAD/CAE system is usually in the form of a number of different files maintained in several storage locations including the hard drive of the designer and the hard drive residing in engineering records. One of these files, called the drill file, contains a record of the coordinates of via holes, and it is used as input for numerically controlled drilling machines. Another file, that contains the feature information, is used to drive a photoplotter. Photoplotters are used to produce the master "artwork" that consists of a photographic image of the conducting lines and board features with a 1/1 scale on either photographic film or glass.

A photoplotter is what the name implies—a plotting instrument that writes on a photographic emulsion using a beam of light. The beam of light is used to draw accurate lines and shapes on film or glass plates following data instructions to generate what is called "artwork". Thus, the photoplotter replaces the very time consuming task of taping lines on a Mylar master by manual methods.

The Gerber Scientific Instrument Company developed a vector based photoplotter in the late 1960's, and for about 20 years it was the industry standard. The mechanism used in this early photoprinter is illustrated in Fig. 6.2. An ultraviolet light that could be turned on and off was used as the source. This light was focused on one of a series of holes drilled in an aperture wheel. The light transmitted through the hole aligned with the light source exposed UV sensitive photographic film located below the aperture plate on a precision x-y table. These holes, known as apertures, controlled the size and shape of the light beam impinging on the film. Two basic operations were performed in preparing the master image used in PCB production—flash and draw. Flash involved of turning on and off the light so it was transmitted through a specific aperture to produce a feature such as a pad. Draw was used to produce a line (trace). The appropriate aperture was rotated into the light path to provide the correct line width. Then the light would be turned on and remain on while the x-y table moved in the preprogrammed x and y directions to draw the line on a film attached to the table. These were time consuming tasks and many plots required an hour or more to complete.

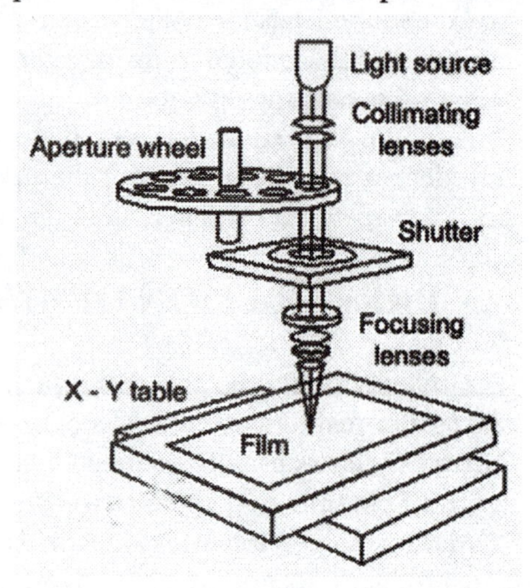

Fig. 6.2 Schematic illustration of the optics and the mechanism associated with Gerber's vector plotter.

This company also developed Gerber data as a language to drive its photoplotter. This type of data is often referred to as numerically controlled (NC) data. This data language enabled the production of artwork from data rather than preparing it with tape and decals. This development was a major step towards modern photoplotting and permitted Gerber's to dominate the PCB industry for two decades. During this period, their data format became an industry standard. Today almost all CAD/CAE systems generate Gerber data because photoplotters of all types can operate from this data format.

Raster photoplotters form images by connecting pixels together to develop patterns representing pads and circuit traces. As the size of the pixel decreases, the quality of the image increases. Generally feature dimensions must exceed the pixel size by a factor of six to eight to produce images with acceptable quality. Raster photoplotters move a flat bed or a rotating drum that transports the film under a beam of light. However, a raster photoplotter does not image a single feature at a time. Instead, a laser light beam sweeps across the film switching on or off at each pixel location to form small dark or light elements of all of the features located along the sweep line. The table or drum is indexed and the line sweep is repeated until the entire area of the plane is plotted[1]. Because of the raster process, the

---

[1] The action of a raster type photoplotter is identical to a laser printer except that the pixel size is smaller. The typical laser printer used in the office is capable of 1,200 points per inch (ppi). Indeed laser printers could be used

time required to cover the area of the plane is independent of the number and the complexity of the features. The computer that interfaces with the plotter must read and process the CAD/CAE file before initiating the plotting and the computer processing time is dependent on the signal plane complexity, but the plotting time is independent of the circuit density or complexity.

Because of the speed and cost of the raster photoplotters, they have replaced the vector plotters. Also with time and the improvement of laser light sources, the quality of the raster photoplotters has improved as their costs have decreased. The price of most quality photoplotters today is in the range of $50,000 to $200,000. For small companies, this is a significant investment and a significant utilization factor is required to justify the capital expenditure. For those companies with a limited number of applications, the use of service bureaus that offer quality photoplotting services at fixed prices with very short turn around times is usually advisable.

The output from a photoplotter is a negative or positive image, with a 1/1 scale that serves as the master for the phototools used in production. This master image can be written on film or a glass plate (on a flat bed photoplotter) depending on the precision required. The master is always safely stored and used only to make phototools. Other outputs from photoplotters are images showing the features of the solder masks.

Resolution, accuracy and productivity are three important characteristics of a photoplotter. Of course, resolution is related to the number of pixels per inch (ppi) used in producing the photo-master. In a higher performance photoplotters[2], the resolution is adjustable ranging from 8,000 to 40,000 ppi (0.32 to 1.6 ppμm). At 40,000 ppi the size of a pixel is 25 μin. (0.625 μm). With the photoplotter operating at 40,000 ppi, the minimum line width that can be generated on the film master is specified as 4 μm, which implies a scale factor of 6.4 between pixel size and minimum feature size.

The productivity of the unit referenced in the footnote for an 18 in. by 24 in. master is dependent the selection of the number of pixels per inch. Clearly the photoplotter can operate at higher speed if the number of pixels it must generate per inch is reduced. At 8,000, 20,000 and 40,000 ppi, the productivity measured in master films per hour is 21.5, 10.0 and 2.7, respectively.

High-performance photoplotters are large and relatively expensive. The equipment shown in Fig. 6.3 weighs slightly more than 1.5 tons and requires a 30 ft^2 footprint. The light source in this unit is a 5 mW HeNe laser that generates a coherent beam of light at 632.8 nm. The accuracy of this photoplotter operating at 40,000 ppi is shown in Table 6.1.

Fig. 6.3 Photograph of a modern high performance photoplotter capable of high resolution photo masters at high production rates.

to produce masters from which some PCBs could be produced; however, the image would be of poor quality and PCBs with small features could not be produced with sufficient accuracy.

[2] The specifications described here correspond to the Calibr8tor 600 & 6500 series photoplotters marketed by Mania Technologie AG, Belgium.

Table 6.1
Accuracy and repeatability of a high-performance photoplotter

Parameter	(mil)	(μm)
Geometric plotter accuracy	0.16	4.0
Geometric plotter repeatability	0.08	2.0
Global positioning accuracy	0.01	0.25
Geometric accuracy on film	0.50	12.5
Geometric repeatability on film	0.50	12.5
Line width variation	0.08	2.0

The films used in the photoplotters are also important. Both Kodak and Fuji have developed specialized films for photoplotters. The Kodak emulsions are on a 7 mil (0.175 mm) ESTAR base that is dimensionally stable. The film's thermal coefficient of linear expansion is 0.0018%/°C. Its humidity coefficient of linear expansion is 0.0009% per %RH after the film is processed. The emulsions used exhibit very high contrast to produce sharp line-edge quality. They have anti-static properties and resist attracting dust. They are also provided with a scratch resistant overcoat. The emulsion exhibits a matte finish that permits better vacuum draw down, which in turn enables closer contact of the phototool with the photoresist. Photoplotters from different manufactures use different light sources with HeNe and Argon being the most popular. Clearly, the emulsion on the film employed must match the wavelength of the laser.

A new technology called laser direct imaging is emerging. In this approach, the phototool is eliminated. Individual circuit boards, coated with a high sensitivity photoresist, are placed on a flat bed photoplotter. The patterns for the pads and traces are written directly into the photoresist using a high powered Argon laser. The feasibility of this approach depends entirely on the speed and resolution of the photoplotter. For large production quantities, it is difficult to match the speed of contact printing where the circuit features are transferred onto the photoresist. With contact printing, a single flash of light is sufficient to expose the entire pattern. For small production quantities, some savings with direct imaging is achieved by eliminating the photo master and the phototools used in production. Some improvement in quality might also be possible by eliminating the phototools and contact printing. However, these savings in costs are small and the improvement in quality is modest. It appears difficult for laser direct imaging to overcome the cost benefits of the production rates that can be achieved with contact printing except for very small production quantities.

6.4 LITHOGRAPHY

Lithography and intaglio printing have been employed by artists and publishers for more than a century for making reproductions of their artworks or publications. The lithographic processes developed in the printing industries for making printing plates were adapted to the manufacture of printed circuit boards in the early 1940's when printed circuit boards were first introduced. Two specific processes were adapted from the photolithographic industry. First, the process to transfer of the image of the circuit board features from the phototool to the copper clad laminate using photoresist techniques. Second, the etching of the copper cladding to chemically machine all of the individual features required on a specific signal or power plane. These two processes are described in the following subsections.

6.4.1 Photoresist Printing

Photoresists are polymeric coatings that are sensitive to light and are similar in many respects to black and white photographic emulsions used in standard photography. In addition, the photopolymers must be resistant to select chemicals, adhere well to metallic surfaces particularly copper, be water soluble and form thin films. These photopolymers are applied to the copper clad laminates to produce a continuous, light sensitive film over the copper surface. After the photoresist has dried, a phototool is placed over this film and brought into close contact with the laminate. A vacuum frame is employed to remove the layer of air separating the phototool from the photoresist surface. The photoresist is then subjected to an ultraviolet (UV) light source to expose those regions of the coating under the transparent regions of the phototool. The exposed regions of the photoresist undergo chemical and physical changes that make it insoluble to many chemical solutions. The regions not exposed to light are not affected and the coating remains soluble. After exposure, the laminate is placed in a developing solution that strips those regions of the photoresist that have not been exposed. This developing action does not affect the exposed regions where the coating is insoluble. The resulting image, which is the negative of the phototool, can be hardened by baking. A graphic describing the steps in the negative photoresist process is shown in Fig. 6.4.

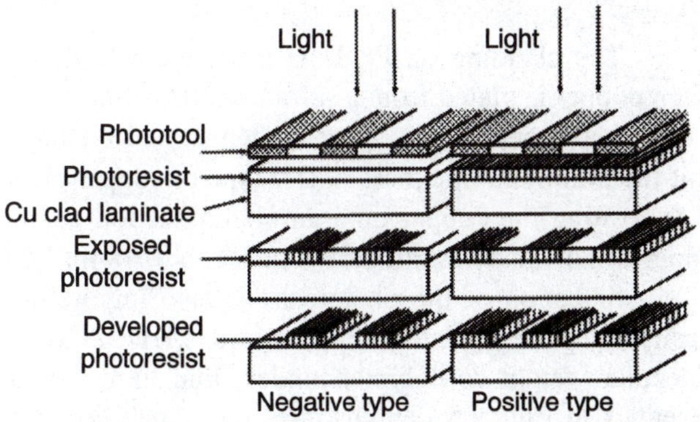

Fig. 6.4 Exposure and development in image transfer with photoresist coatings.

Positive type photopolymers are also available that act in the opposite sense than the negative resists. Positive type resists give coatings that are insoluble prior to exposure. Those areas exposed to the light are changed chemically and physically and become soluble.

Commercial photoresists are available as liquid or dry film products. Liquid resists may be applied to the copper clad laminates by spraying, rolling or dipping. Dry films are applied by laminating in a rubber roller laminating press. Dry films are preferred[3] because they are ready to be exposed immediately after lamination. Liquid resists must be dried and baked and this requirement delays the process and adds to the floor space necessary to accommodate the production line.

The exposure of photoresist becomes critical as the line width decreases. For wider lines with $w < 8$ to 10 mil (0.20 to 0.25 mm), mercury arc lights, arranged in a reflector to give a uniform intensity of light over the field of the phototool, provide an exposure adequate to resolve the lines. For fine lines with $w < 5$ mil (0.125 mm), the definition of the line becomes more difficult to achieve. It is necessary to employ collimated light in the plate maker, thin resist films and closer control of the exposure and developing times. When the proper plate maker is used with precise process control, the fidelity of the reproduction of the lines and spaces in the photoresist is excellent. An example of the quality of the photoresist pattern that can be achieved is indicated by the SEM photomicrograph presented in Fig. 6.5.

For fine line products with circuit trace widths of 75 μm, several procedures must be implemented to insure high product yields in manufacturing. First, in selecting the laminate, the weave of the glass cloth reinforcement must be considered. It is critical that the surface of the laminate be

[3] Currently more than 90% of inner layers are produced using dry film resists.

plane and free of surface discontinuities. Moreover, the copper cladding thickness must be uniform and as thin as possible consistent with current carrying requirements.

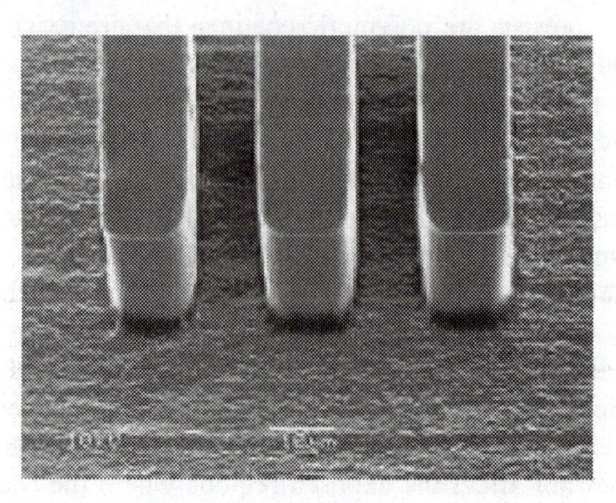

Fig. 6.5 SEM image showing the resolution capability of dry film resist. The lines and spaces in this photomicrograph are 15 μm wide and the resist is 30 μm thick.

The cladding on PCBs is usually electrodeposited copper made by an electrolytic process in which copper is plated from a sulfate solution onto a slowly rotating stainless steel drum. The thin layer of electrically deposited copper is then peeled off the drum in a continuous sheet. Handling problems limit the minimum thickness of this type of copper cladding to 3/8 ounce, or 0.5 mil (13 μm). However, by depositing the copper onto an aluminum foil carrier instead of the stainless steel drum, the plated copper can be made as thin as 1/8 ounce, or 0.18 mil (5 μm).

Selection of the photoresist is also important in insuring higher yields in production of high density integrated printed circuit boards (HDI-PCB). Newer high resolution photoresists utilize low molecular weight co-polymer binders that limit swelling during development. The photomicrograph presented in Fig. 6.5 demonstrates the capabilities of these new photoresists. They are capable of yielding an aspect ratio of photoresist thickness to space width equal to 2:1.

Two different processes are used to laminate the dry film photoresists to the copper clad laminates. The more traditional process involves passing the board and two sheets of dry film between two rubber coated lamination rollers. To improve the flow of the photoresist films into the surface roughness of the copper cladding, the lamination rollers are heated to enable the temperature of the photoresist film to be increased. Also photoresists with low viscosity at lamination temperatures are selected.

The second process involves wet lamination of the dry films. In this process, a small amount of water is applied to the copper cladding prior to lamination. This water, which is absorbed by the dry film during lamination, acts to expel any air from the interface between the cladding and the resist and prevents the formation of interfacial voids.

Transfer of the image from the phototool to the photoresist coated laminate is accomplished with an imaging system. Imaging systems may be characterized as contact printers or non-contact printers. In the contact printing processes, the phototool is brought into hard contact with the photoresist by applying a vacuum between the two. In soft contact imaging, mechanical forces hold the two surfaces together but a vacuum is not created. Non-collimated illumination sources require hard contact between the phototool and the resist. Highly collimated sources are required for soft-contact imaging, but they are recommended for hard-contact systems when fine lines are required.

Non-contact printers include proximity printers, projection printers and laser direct imaging systems. Proximity printers are used for wet or tacky photoresists so that a separation between the photoresist and phototool can be maintained. Proximity printers require a highly collimated light source to prevent the image from blurring. In projection printers, a lens is placed between the phototool and the resist, keeping the artwork a great distance from the surface of the PCB. Laser direct imaging systems do not use a phototool because CAD data are used in a flat bed photoplotter to sweep a laser beam over the photoresist to generate the image.

Illumination intensity, wavelength(s), exposure dose, uniformity and collimation are among the many factors that influence the quality of the image formed in the resist by the imaging system. The spectral output from the illumination source must include sufficient energy at wavelengths that are absorbed by the initiator in the resist to complete polymerization of the polymer. Most UV-sensitive photoresists have initiators that are tailored to absorb the 365 nm peak found in mercury and mercury/xenon arc lamps. When collimation is necessary, as in soft contact, proximity, or off-contact printing, the additional optics that are necessary reduce the intensity and increase the exposure time for an equivalent exposure dose.

The printing environment must be carefully controlled and maintained for high yield output. Cleanliness is extremely importance. A well-maintained laminar flow clean room with submicron particle filters is a requirement. The operators, maintenance and engineering personnel must wear appropriate clean room garments and follow clean room procedures to minimize defects caused by dust, fibers, hair, and other particulate contaminates. Sources of contamination, which are carried onto the boards being processed, include glass fibers, copper flakes and epoxy particles. Hair and flakes of skin are due to personnel found in the clean room.

Yellow lighting in the printing room is usually required to prevent unintentional exposure of the UV-sensitive photoresist, while tight temperature and humidity controls are necessary to minimize film and substrate distortion that can cause registration errors.

Transfer of images by the photoresist method is easier for single sided boards than for double side or multilayer boards. With single sided boards, the registration issue does not arise because only one independent pattern is employed for wiring. With double sided boards, both sides of the laminate must be imaged requiring the PCB to be exposed in two separate operations. Because there are many vias connecting one side with the other, it is imperative that these two images be in precise registration. Registration is usually accomplished by drilling three tooling holes in the laminate and positioning the laminate in the printing frame over registration pins that pass through these alignment holes. The use of tooling holes insures precise registration providing the holes are placed accurately and the pins have a close fit in these holes.

Another process to insure registration of the top and bottom layers of a laminate is to mount the top and bottom phototools into a two-sided glass frame. Then the two phototools for the top and bottom surfaces are adjusted until they are in perfect registration. Next, a laminate coated on both sides with photoresist is placed in the glass frame. The frame is closed, a vacuum is drawn and both sides of the laminate are exposed simultaneously.

6.4.2 Etching and Cleaning

Etching is a chemical machining process that strips the copper cladding away from those regions of the board not protected by the photoresist, which is impervious to the attack of the etchant. The copper that remains after the etching process serves to define the board features such as circuit lines, capture pads and solder pads. There are two common processes in use today in etching the circuit features. The first employs a photoresist coating to protect the copper cladding from the etchant. The second utilizes a solder or tin plate 0.3 to 1 mil (7.5 to 2.5 μm) thick as the protective coating. These two processes are illustrated in Fig. 6.6.

Etching is performed in automated etching lines, illustrated in Fig. 6.7, which incorporate sprays that deliver the etchant to both sides of the board at high velocity. Alkaline ammonia (NH_4OH) that can be used in continuously operated lines with automatic solution replenishment, is the most commonly employed etchant. Ammonium hydroxide and ammonium salts combine with copper ions to form cupric ammonium complex ions $[Cu(NH_3)_4^{2}]$ that hold the dissolved copper in solution. The concentration of dissolved copper is usually in the range from 18 to 30 oz per gallon. The spent etchant with high concentrations of dissolved copper is recycled and the copper recovered. This recycling is very important as it eliminates the need to dispose of large amounts of toxic chemicals. After etching, the board is washed to thoroughly remove the etchant; then it is treated with a mild acid to neutralize its surfaces, and washed again before air drying and storage.

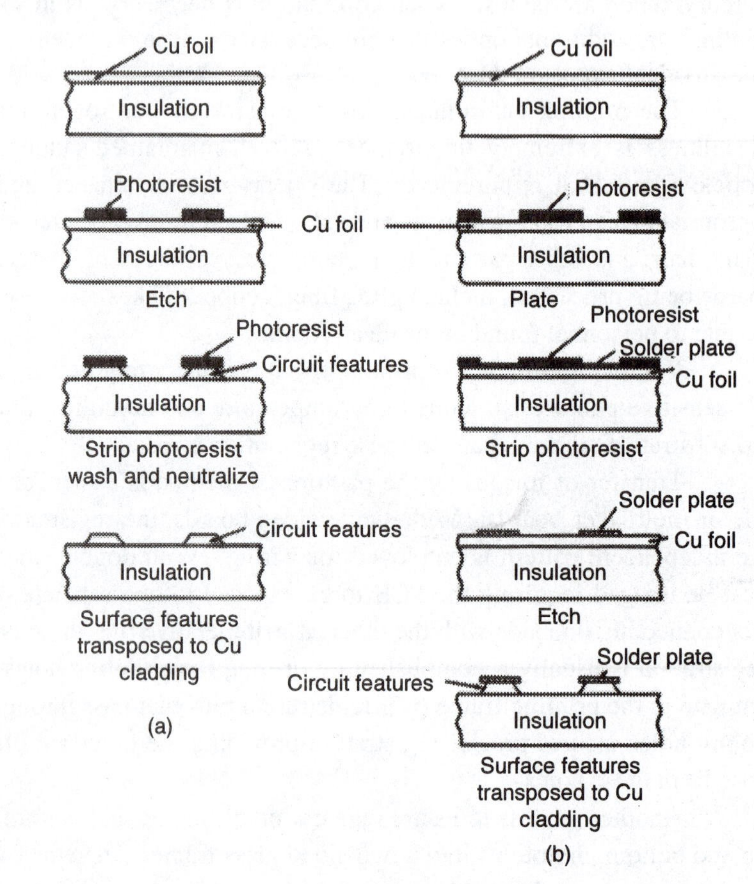

Fig. 6.6 Two techniques for using etch resistant coatings.
(a) Direct photoresist coating to protect surface features.
(b) Tin or solder plating to protect surface features.

The etching process is not selective. This fact implies that the exposed copper cladding will be etched away in all directions at nearly the same rate. Spraying the etchant onto the surface of the board introduces a modest degree of selectivity into the process because fresh etchant is always being delivered to the bottom of the channels where the copper is being dissolved. However, the lack of selectivity in etching causes under cutting the copper below the photoresist coating as shown in Fig. 6.8.

This under cutting is extremely important. The board features, as etched, are a different dimension than the features on the phototool and imaged on the photoresist. The top surface of the copper cladding is under sized on each edge by a distance u. The magnitude of the under cut dimension u is determined by the etch factor F_e that is defined as:

$$F_e = u/t \qquad (6.1)$$

where t is the thickness of the copper cladding.

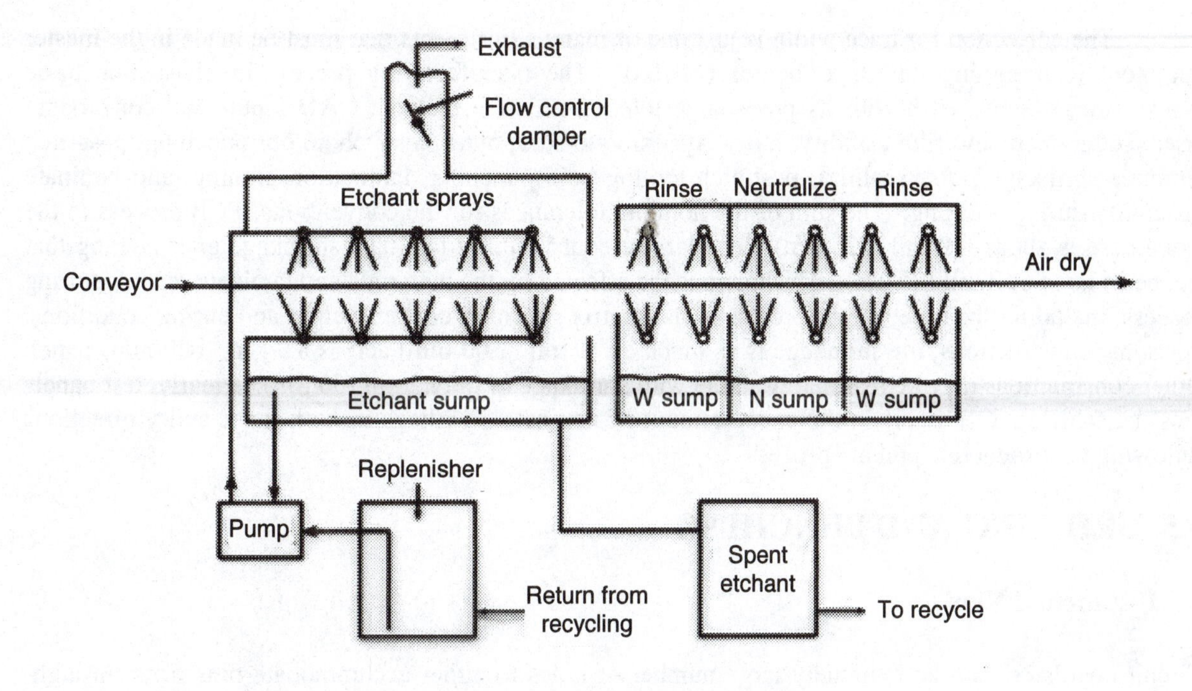

Fig. 6.7 An automated etching and cleaning process.

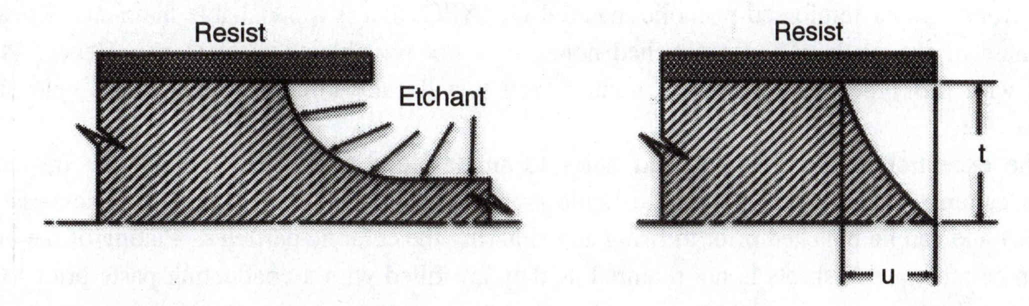

Fig. 6.8 Illustration showing the etchant undercutting the copper below the layer of resist.

It is important in the layout of the circuit features to take into account the etch factor that requires the designer to increase the line width on the layout. For example, consider a circuit trace is to be 4 mil (0.100 mm) wide and etched into a foil with a thickness equivalent to 0.5 oz of copper with an etch factor of 0.7. The circuit trace on the layout should have a width of

$$w = 4 + (0.7 \times 0.675)2 = 4.95 \text{ mil} \qquad \text{(a)}$$

to accommodate for the dimensional changes anticipated by under etching.

The relation used to account for under cutting the resist is:

$$w_L = w_f + 2F_e\, t \qquad (6.2)$$

where subscripts L and f refer to the width at layout and the final width after etching.

The correction for trace width is just one of many adjustments that must be made in the master phototool in designing multilayer-boards (MLBs). The manufacturing process involves nine basic image process steps, each with its predictable tolerance. These include: CAD inputs and conversion, laser plotter accuracy, film stability, inner layer side-to-side, outer layer phototool punching, post-etch laminate shrinkage (after scaling), post-etch tooling hole punching, lamination pinning, and laminate instability during pressing. The sum of the nominal tolerances for the conventional PCB process in the worst case is about ± 9 mil (225 μm). The largest contributor is laminate shrinkage after etching that can be as large as 3 mil (75 μm). This tolerance is affected by the materials and laminate manufacturing process, including the type of reinforcement and matrix polymer, copper weight and curing conditions. For some constructions, the shrinkage is as much as 40 mil (1.00 mm) across a 24 in. (600 mm) panel. Other constructions may be reasonably stable with shrinkage of only 2 mil (50 μm). Clearly, test panels must be fabricated to ascertain the exact amount of laminate shrinkage in both the x and y directions following the production etching process.

6.5 DRILLING AND PUNCHING

6.5.1 Punched Vias

Circuit boards contain an extremely large number of holes to either accommodate pins from through-hole mounted components or to serve as vias used for connections between layers and for heat conduction. These holes are formed either by drilling or by punching depending upon the type of laminate used in constructing the board. Forming holes by punching is limited to single sided PCB's fabricated from a paper reinforced phenolic material (XXXPC) that is a punchable laminate. Because the roughness of the walls of these punched holes, it is not possible to plate them. Hence, PCBs fabricated with this type of laminate are limited very low density circuits found with single sided applications.

The exception to limiting punched holes to single sided substrates is when the insulating material is ceramic green sheet. Because ceramic green sheets are similar to paper in thickness and texture, they and can be punched prior to firing and sintering the ceramic particles. Plating of the holes punched in ceramic green sheets is not required as they are filled with a conducting paste prior to the sintering operation.

6.5.2 Mechanically Drilled Vias

Mechanical drilling is performed with multiple-head, numerically-controlled drilling machines, using small diameter twist-drill-bits fabricated from carbide as shown in Fig. 6.9. Manual drilling procedures are not usually employed. Location of the center of the holes to within ± 1 mil (25 μm) and the need to reduce labor costs dictates the use of multiple spindles and numerically controlled drilling machines. With modern drilling stations equipped with four drill spindles, it is possible to drill 200 to 400 holes per minute with positional accuracy of ± 0.2 mil (5 μm) and diameter tolerance of ± 1 mil (25 μm).

Two different types of holes are usually drilled in the copper clad laminates. The first, tooling holes are of relatively large diameter, 125 to 250 mil (3.125 to 6.25 mm). The second, via holes are small diameter holes that accommodate chip carrier pins and circuit connections either through or within layers of the MLB. Via holes can be divided into several categories that include:

1. Through-hole chip carriers
2. Leaded surface mount chip carriers

3. Ball grid array chip carriers
4. Top surface blind vias
5. Inner-layer vias.

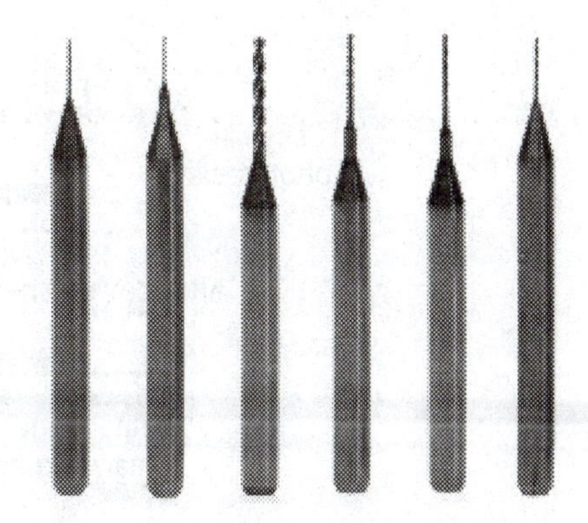

Fig. 6.9 Solid carbide drills for drilling holes in copper clad laminates.

To accept pins, holes are usually drilled with a diameter of about 40 mil (1.0 mm) so that clearance for the pin exists after the hole is plated and its diameter is reduced to about 32 mil (0.8 mm). The via holes for the other categories may be much smaller in diameter, because these holes are used only to connect one signal layer to another. Indeed, the minimum diameter for vias used for circuit connection is established by drilling capabilities and cost. If via hole diameters are reduced below 8 mil (250 μm), drill breakage increases and mechanically drilling is not the most cost effective method of producing via holes. Currently, most companies will not attempt to drill holes less than 4 to 6 mil (100 to 150 μm) in diameter in production, although smaller drill diameters are available.

6.5.3 Laser Vias

About 95% of the holes drilled in circuit board are formed using mechanical means. However, for the smallest diameter holes, ranging from about 25 to 250 μm, other methods such as laser drilling, plasma etching and photo formation of vias are employed. Laser drilling is performed with small diameter coherent beams of light from high-wattage CO_2 and UV lasers. The laser beams literally burn through thin layers of polymeric materials to form vias about 1 to 2 mil (25 to 50 μm) in diameter. Pulsed lasers produce a better defined hole than continuous lasers and the UV type lasers can produce smaller diameter beams; hence, they are capable of smaller diameter vias. Laser micro-vias in polymeric films are drilled with relative ease; however, for laminates reinforced with glass cloth, the marked difference in the melting temperatures of polymers and glass fibers cause significant difficulties. The resulting laser drilled holes are often very rough and irregular in shape with fiber ends protruding.

6.5.4 Photo-vias

A relatively new epoxy photo-dielectric dry film has been developed that may be used to add sequential layers to PCBs providing high I/O redistribution. Photo-vias are formed in layers of a photo-dielectric film that is photo-sensitive like the photoresist coatings described previously. Dry photo-sensitive films are usually preferred instead of liquids because they can be applied by lamination, which insures a more uniform thickness than that obtained when spraying liquid photo-sensitive coatings. With dry film photo-dielectrics, film thickness in the range of 1.0 to 2.5 mil (25 to 62 μm) is usually employed. The uniform thickness of dry photo-sensitive film enables predictable circuit performance.

Vias are formed by imaging using the same phototools and equipment employed to expose photoresist coatings, as shown in Fig. 6.10a. After exposure, the film is developed in a solvent-free, caustic aqueous solution. Consequently, no equipment is needed to recapture solvents in the developer. This development process reduces operating cost and eliminates regulatory concerns. After development and curing the photo-sensitive film, the vias are catalyzed and plated to connect them with their capture pads.

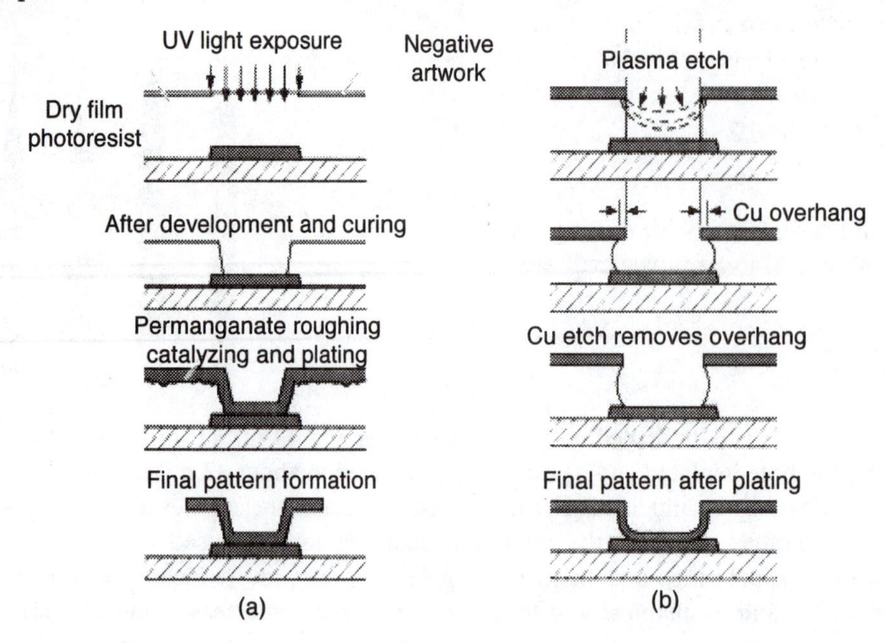

Fig. 6.10 Schematic illustration of via formation with photo-dielectric coatings and plasma etching.

6.5.5 Plasma Etched Vias

Plasma etched vias are formed using a mask to protect circuit features surrounding the via locations. The surface is then exposed to a plasma that attacks a polymer film in which vias are to be formed as shown in Fig. 6.10b. As with any etching process, undercutting occurs leaving a copper over hang that is subsequently removed with chemical etching. After catalyzing, the vias are plated to connect them to their respective pads. Plasma etching does not become cost effective until the density of holes increases to about 150 hole/in.^{2} (0.25 hole/mm^2). Photo-vias, described above, are usually more cost effective than any of the other methods for generating vias if their diameter is 4 mil (100 μm) or less.

6.5.6 Registration

With reduction in the diameter of via holes, registration of the hole with respect to the capture pad becomes more critical. A common dimension for the width of the annular ring formed by drilling a hole through a circular pad is 1/3 times the diameter of the PTH. This rule implies that the capture pad diameter D is given by:

$$D = 1.6\ d \qquad (6.3)$$

where d is the via hole diameter.

For example with this sizing rule, a 24 mil (0.6 mm) via has a capture pad with a diameter of 1.6 × 24 = 36.4 mil (0.96 mm) with an annular ring 7.2 mil (0.18 mm) wide. These tolerances give high yields in production even when the circuit board contains many hundreds of holes.

For surface mount chip carriers, via diameters are usually smaller. This fact implies that the tolerances on the position of the holes and on the diameter will be tighter and more precision in drilling will be required. The scale drawing presented in Fig. 6.11 shows the problem of maintaining concentricity in pad positioning and drilling. Remember that the position of the pad and the position of the center of the hole are established in two independent manufacturing processes. Errors in either one or both can result in pad break out and rejection of the board.

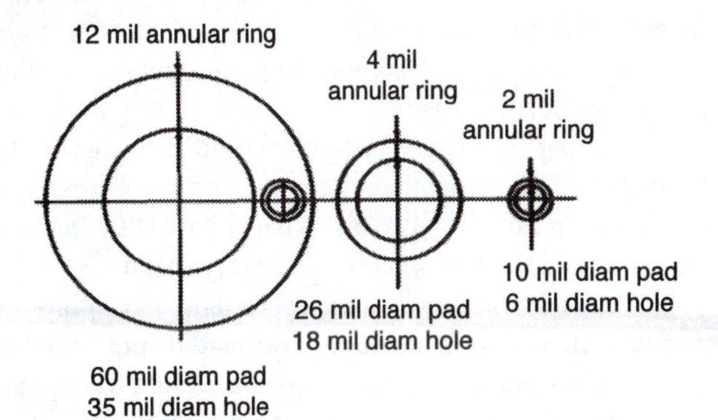

Fig. 6.11 Scale drawing indicating the importance of registration, drilling position and diameter drilling tolerances as the diameters of vias decrease.

Usually laminates are stacked in a pile, with registration maintained by the tooling holes. Then several boards are drilled together. The number of boards that can be drilled in a stack depends on the board thickness and the aspect ratio that can be handled at the drilling station without excessive drill breakage. The aspect ratio, the depth of the stack of boards divided by the diameter of the drill that can be achieved usually ranges from about 4 to 8, depending on the diameter of the drill. To avoid drill deflection upon entry and the formation of burrs upon exit, the stack of circuit boards is usually sandwiched between entry and backup layers. These layers are fabricated from materials that facilitate a low-force entry and a burr-free exit. An illustration of drilling a two high stack of multilayer boards, presented in Fig. 6.12, shows entry and exit layers.

The entry material is soft metal clad board material. Its function is to reduce drill bit wander on PCB drilling machines equipped with solenoid or pneumatic driven Z-axis control. These systems tend to hammer the bit into the board's surface before the machine begins to drill. The soft metal clad entry material helps absorb this impact allowing the bit to begin cutting before it encounters the harder copper cladding of the board material. PCB drilling systems with a stepper motor driven Z-axis usually do not require entry sheets because the rate of decent of the drill is be accurately controlled and the hammer effect is avoided.

Fig. 6.12 Cross section of a two-high stack of multilayer boards with an entry overlay and a foil board back-up to aid in drill burr-free holes.

The ability of a drilling machine to support a number of different board materials is determined by the spindle motor (the faster the better) and the mechanism used to hold down the stack. On basic machines, a DC spindle motor with a 5,000 to 40,000 RPM variable speed controller is adequate to drill hard clad board material like FR-4. In some cases, a higher speed spindle equipped with a speed controller (up to 100,000 RPM) is recommended. This system can drill harder substrate boards at lower spindle speeds while higher speeds are used to drill softer circuit board materials.

The board hold down system for rigid circuit boards require a pressure clamp that exerts sufficient force to flatten a stack of PCBs to conform to the machine's rigid drilling bed. Vacuum hold down systems are employed with lower modulus, flexible board types that conform more readily to the machine's flat drilling bed.

Drilling generates a cloud of fine dust that must be removed from the milling surface as the holes in the board are drilled. PCB drilling machines, designed for debris collection, equip each spindle with a vacuum nozzle. The air exhaust from the vacuum system is filtered and the debris collected is disposed of in an environmentally friendly manner.

The selection of z axis speed and the spindle speed is a critical step in the production of correctly drilled circuit boards. Fortunately the more modern PCB drilling machines are automatic. There is a prompt for the board type and weight of copper cladding before beginning a drilling sequence. At this point, the controlling software automatically sets the correct z axis speed and spindle speeds for each drill bit. In systems that do not prompt for board type, manual adjustment of the settings for each bit must be made to produce the best possible milling results. Setting incorrect drilling parameters usually results in poorly machined boards and broken drill bits.

6.6 LAMINATION

Lamination is the first step in the production of a typical single or double sided board. Thin copper foils that form the clad surfaces are spread over a polished steel sheet that serves to control the flatness of the foil. Next, several sheets of prepreg are placed over the copper foil until the specified thickness of the laminate is achieved. Prepreg is a sheet material formed by impregnating glass fabric with liquid epoxy (stage A). The epoxy resin is partially cured and becomes rubbery (stage B) permitting the prepreg sheets to be cut and handled easily. A second foil of copper is then placed on top of the prepreg lay up. Finally, another polished steel sheet (called a press pan) is placed over the assembly. Mold release is used on the press pans to insure that the laminates will separate easily after the polymerization of the epoxy is completed. This process is repeated until a stack of about 100 boards is developed. The sheets are pressed together in a large hydraulic platen press, which develops about 1,000 psi (6.895 MPa) pressure over the area of the sheets, typically 36 by 72 in. (0.914 by 1.828 m) in size. In addition, the platens of the press are heated to increase the temperature of the stack. When the temperature of the partially cured (B staged) epoxy increases, the resin becomes more fluid and the voids are eliminated as it flows. With time and temperature, the polymerization of the epoxy continues until the it cures and becomes rigid (C stage). The laminates are separated from the stack and are heated in an oven at a temperature slightly above the glass transition temperature of the epoxy to relieve the residual stresses developed in the laminate as the epoxy cured. A photomicrograph of a section of a laminate showing bundles of glass fibers (yarns) embedded in the epoxy matrix is presented in Fig. 6.13.

6.6.1 Standard Multilayered Boards

Lamination is also used in producing multilayer-boards (MLBs). In this instance, several layers, each representing a two sided circuit board with a previously etched power and signal planes, are laminated together. The process is similar to that described above in that prepreg is used to separate the two-sided boards. The thickness of the prepreg is controlled to properly space the signal and power planes. However, the process is more difficult because registration of the layers is very important in the build-up of a multilayer-board. These layers are connected electrically after lamination by drilling and plating through holes. The registration in lamination is achieved by using fixtures that incorporate two or more registration pins that pass through tooling holes drilled in each layer. Because of the need for close control over registration, the size of the multilayer panel is smaller—20 by 20 in. (500 by 500 mm) is common. Also, the stack height is usually limited to less than an inch to better control heat conduction through the stack. A typical laminate lay-up with a laminating fixture is shown in Fig. 6.14.

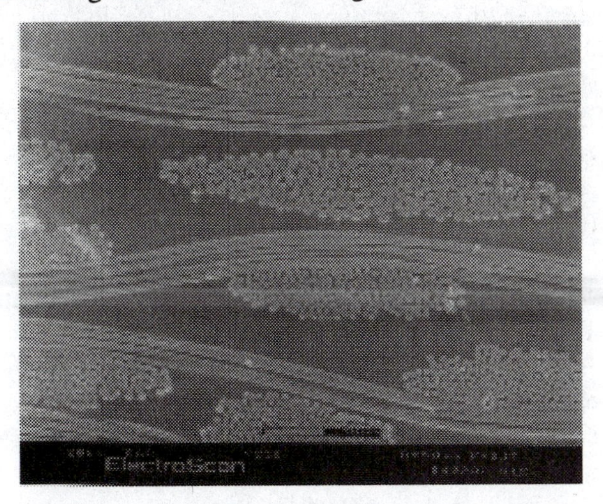

Fig. 6.13 Photomicrograph of a glass-cloth reinforced epoxy laminate.

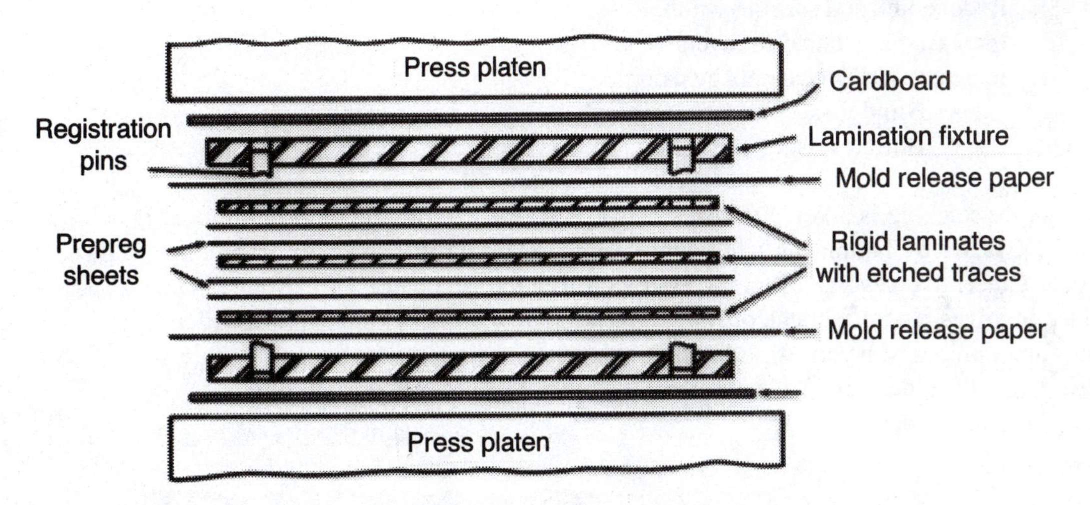

Fig. 6.14 Lay-up of a six layer circuit board in a platen press prior to lamination.

A vacuum laminating press is shown in Fig. 6.15. The press applies pressure to the laminate stack while heating the stack under a vacuum. The vacuum reduces the amount of air, moisture and volatile entrapment. It also enables lower laminating pressure, improves registration and reduces residual stresses that often cause the MLBs to warp.

6.6.2 High Density Integrated Printed Circuit Boards

High density integrated printed circuit boards, (HDI-PCB) are employed in products where space and weight are the primary factors driving the design of the circuit cards. High density interconnects (HDI) are defined as boards with a higher wiring density per unit area than conventional PCBs. They have finer lines and spaces (less than 75 μm), smaller vias (less than 150 μm) and capture pads (less than 400 μm), and higher connection pad density (greater than 20 pads/cm^2) than in conventional PCBs. In addition to reducing size and weight, HDI, enhances electrical performance. The terms HDI, build-up board (BUB), sequential build-up (SBU) and microvia technology are used interchangeably.

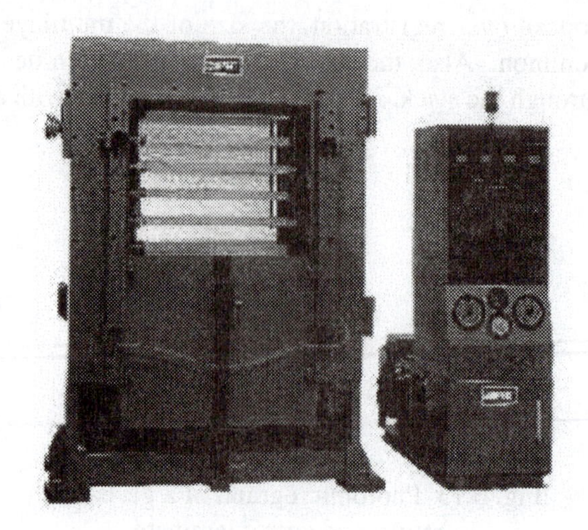

Fig. 6.15 A modern vacuum press used to laminate multilayer-boards.

There are four basic design guides for HDI-PCBs as indicated below:

1. Reduce hole diameter
2. Reduce line and spacing width
3. Increase the number of layers
4. Increase wiring channels by using
    - Blind vias
    - Buried vias

Layer stacking is a key difference between HDI-PCB and the more traditional PCB. Engineers design HDI-PCBs by laminating additional layers containing microvias on traditional PCBs that serve as cores. Currently, several different HDI construction sequences are used. Type I construction sequence involves a core fabricated from conventional rigid or flexible two-sided laminates. This core contains any number of layers using through-hole vias. A single layer containing microvias is laminated to one or both sides of the core, as presented in Fig. 6.16.

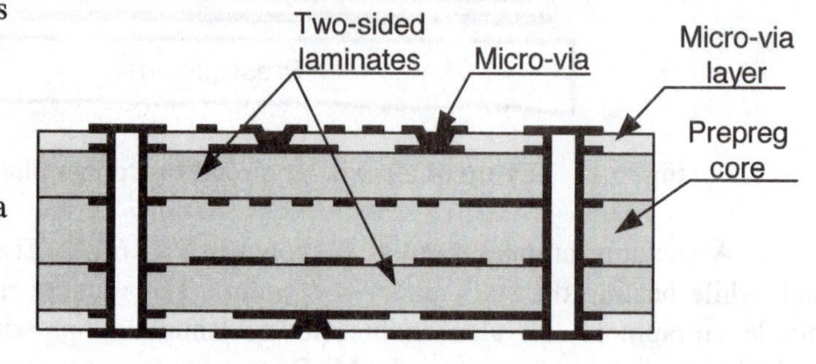

Fig. 6.16 Type I construction sequence for a HDI-PCB contains a single micro-via layer on each side of the core and two-sided laminates.

Type II construction is similar, but the vias in the core are fabricated before adding the microvia layers as shown in Fig. 6.17.

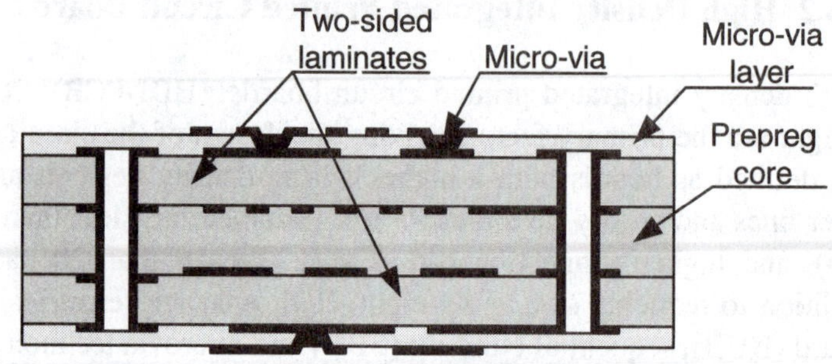

Fig. 6.17 Type II construction sequence with microvia layer laminated over previously drilled and PTHs.

The sequence of construction in Type III, HDI-PCBs has at least two microvia layers on at least one of the core's surfaces as indicated in Fig. 6.18. Several other construction sequences are used by circuit designers. One type employs a passive core that functions as a thermal buffer. Another type known as coreless construction utilizes two PCBs bonded together with adhesive. Micro-layers are laminated to this sandwich of PCBs.

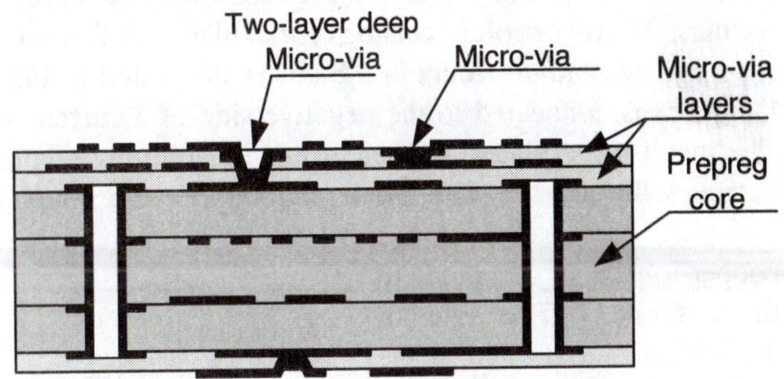

Fig. 6.18 Type III construction sequence with two microvia layer laminated over previously drilled and PTHs. Note the one and two layer deep micro-vias.

The microvia layers in the HDI-PCBs are not fabricated using conventional glass-cloth-reinforced epoxy. Instead non-reinforced polymers are used. The rigidity of the HDI-PCB is due to the prepreg core and two-sided laminates located in the center of the stack. Two materials often employed in laminating the microvia layers are resin coated copper (RCC) and polyimide film with copper coating. The resin coated copper consists of a 60 μm thick layer of resin with an 18 μm copper foil. The polyimide film is the same as that used in flexible circuits that was described previously in Chapter 5.

The design rules followed in laying out pads, vias and circuit traces are presented in Table 6.2. The symbols defining the dimensions of the PCB features are given in Fig. 6.19.

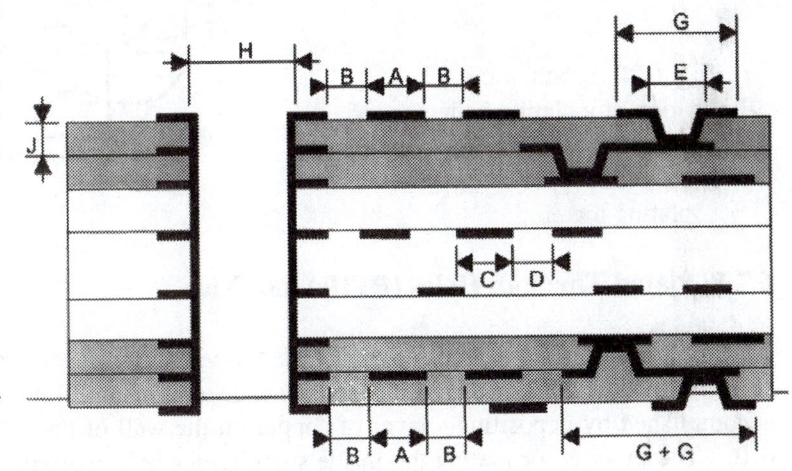

Fig. 6.19 A drawing defining the feature dimensions for HDI-PCBs.

Table 6.2
Dimensions of features found on HDI PCBs.

Symbol	Feature	Advanced Manufacturing	State of the Art Manufacturing
A	Line width outer layer	125 μm	100 μm
B	Spacing width outer layer	125 μm	100 μm
C	Line width inner layer	100 μm	75 μm
D	Spacing width inner layer	100 μm	75 μm
E	Microvia hole diameter	125 μm	100 μm
G	Microvia pad diameter	300 μm	250 μm
J	Aspect ratio E/J	2/1	1/1

6.7 PLATING

Plating is the electro-chemical deposition of thin layers of various metals on a PCB's metallic surfaces. Electro-plating is employed in several steps in manufacturing a circuit board, including: finishing PTHs and other vias, providing an etch resistant surface, applying a gold coating for connector contacts and forming thin tin or solder coatings for oxidation resistance.

The plating occurs in a bath, as illustrated in Fig 6.20, where the work piece (the copper clad laminate) is connected to the negative side of a current supply making it the cathode in a DC circuit. The metal to be plated, copper in this illustration, is connected to the positive side of the supply and becomes the anode. The anode and cathode are coupled together with an electrolyte that serves to transport positively charge copper ions from the anode to the cathode. The rate of deposition is controlled by the cell voltage that is adjusted to give current densities consistent with anode to cathode spacing and the specified deposition rates.

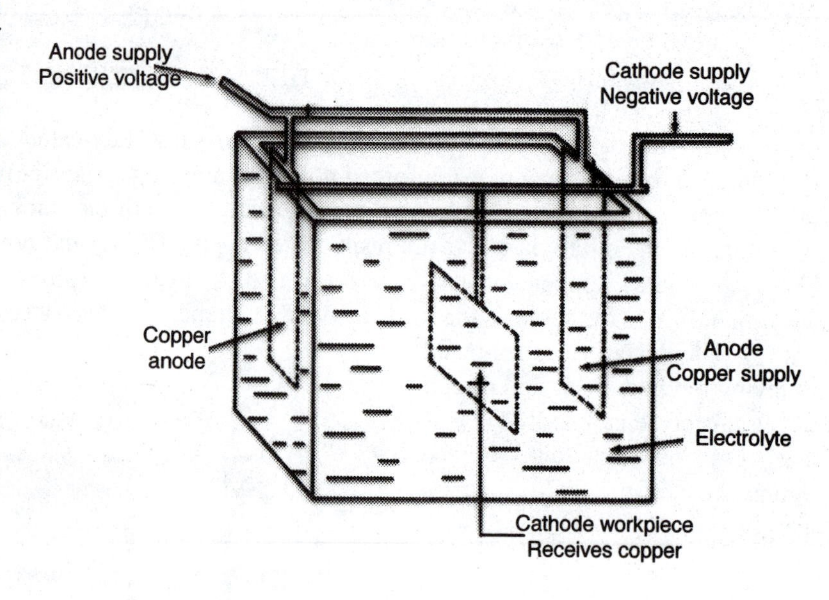

Fig. 6.20 Schematic illustration of a plating bath showing the anode and cathode with copper as the plating material.

6.7.1 Plated Through Holes (PTHs) and Vias

After holes have been drilled for the PTHs and vias, it is necessary to connect the circuits on the two sides of the PCB or between outer and inner layers on multilayer-boards. The connection is usually accomplished by depositing a layer of copper on the wall of these holes to form vias, as shown in Fig. 6.16. The process for plating the inside surface of a hole requires two steps. First, the formation of a very thin conducting layer is plated on the wall of the hole using an electroless process to deposit a very thin layer of copper. This process is followed with the deposition of a thicker layer of electrolytic plated copper.

Electroless plating of copper is a chemical deposition process that makes use of special activators to sensitize the walls of the holes. These activators are colloidal solutions of stannous and palladium ions. When the board is immersed in these colloidal solutions, the stannous ions are adsorbed onto the surface to be catalyzed. Subsequently the palladium ions are converted to a metallic state. This series of reactions creates precious metal sites that are necessary to initiate an electroless copper reaction. The board is then transferred to an electroless copper bath where copper ions are reduced to metallic copper, which in turn is deposited onto the palladium sites. The deposition rates are quite slow 1.5 μm/h. To reduce process time only a very thin layer of copper is deposited in this manner. The board is then transferred to the electrolytic plating bath where 0.2 to 0.5 mil (5 to 12.5 μm) of copper is flash plated on the walls of the holes to give a stronger and thicker film of copper. Additional copper is added in subsequent plating operations to achieve a wall thickness of 1 to 3 mil (25 to 75 μm).

6.7.2 Panel and Pattern Plating

Two different approaches are followed in developing the full thickness of the copper plate in the through hole—panel and pattern plating. In panel plating, the entire surface of the copper cladding including the inside surface of all of the holes is plated immediately after drilling the holes and applying the electroless copper. This panel plating operation precedes the image transfer process.

Pattern plating, on the other hand, is selective because the copper is deposited only on the walls of the holes, pads and the circuit traces. Of course, pattern plating requires completion of the image transfer process before plating begins. The presence of the resist patterns influences the distribution of the current density making it difficult to achieve copper coatings that are uniform in thickness over the entire area of the board.

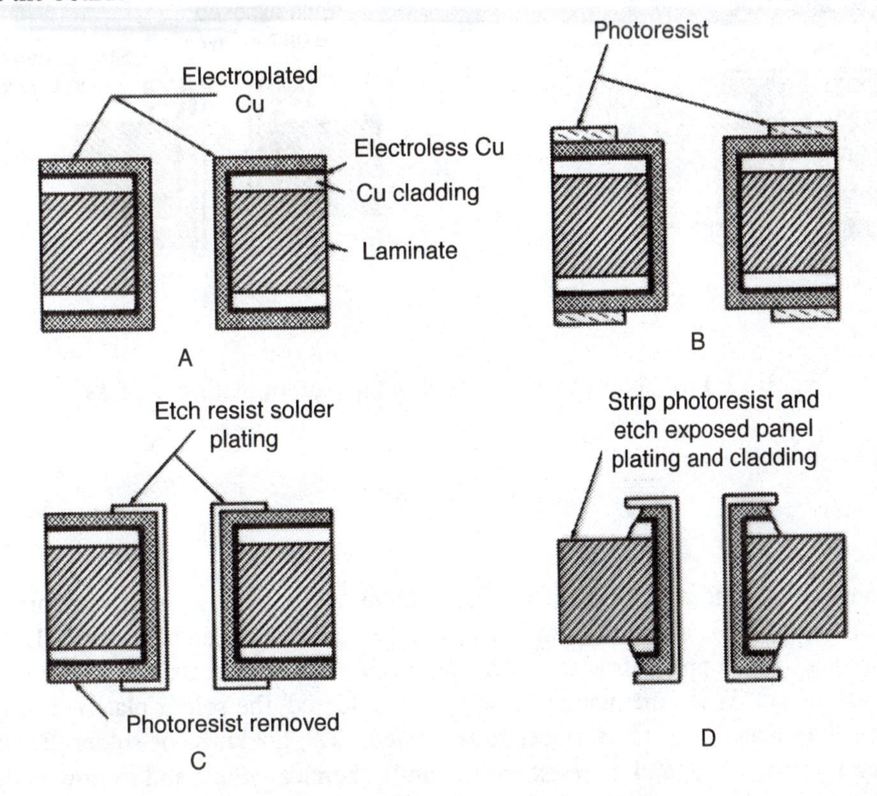

Fig. 6.21 Production steps in producing PTHs or vias using the panel plating process.

The panel plating process is illustrated in a four step sequence in Fig. 6.21. In the first step, a 1.0 mil (25 μm) thick layer of copper is deposited over the wall of the hole and the entire surface of the cladding. In the second step, photoresist is applied and the image is transferred using techniques described in Section 6.4.1. In the third step, a thin layer of solder is plated over the areas of the board that are free of resist. Finally the resist is stripped exposing copper that is subsequently removed by etching. The solder plating serves as etch resistant coating that protects the copper from the chemical attack of the etchant.

The pattern plating approach involves a three step sequence of operations illustrated in Fig. 6.22. First the photoresist is applied and an image of the layer's features is transferred. Next, copper and then solder plating are deposited over those board features not protected by the resist. Finally, the resist is stripped and the exposed copper is etched away to produce the complete pattern on the surface of the PCB.

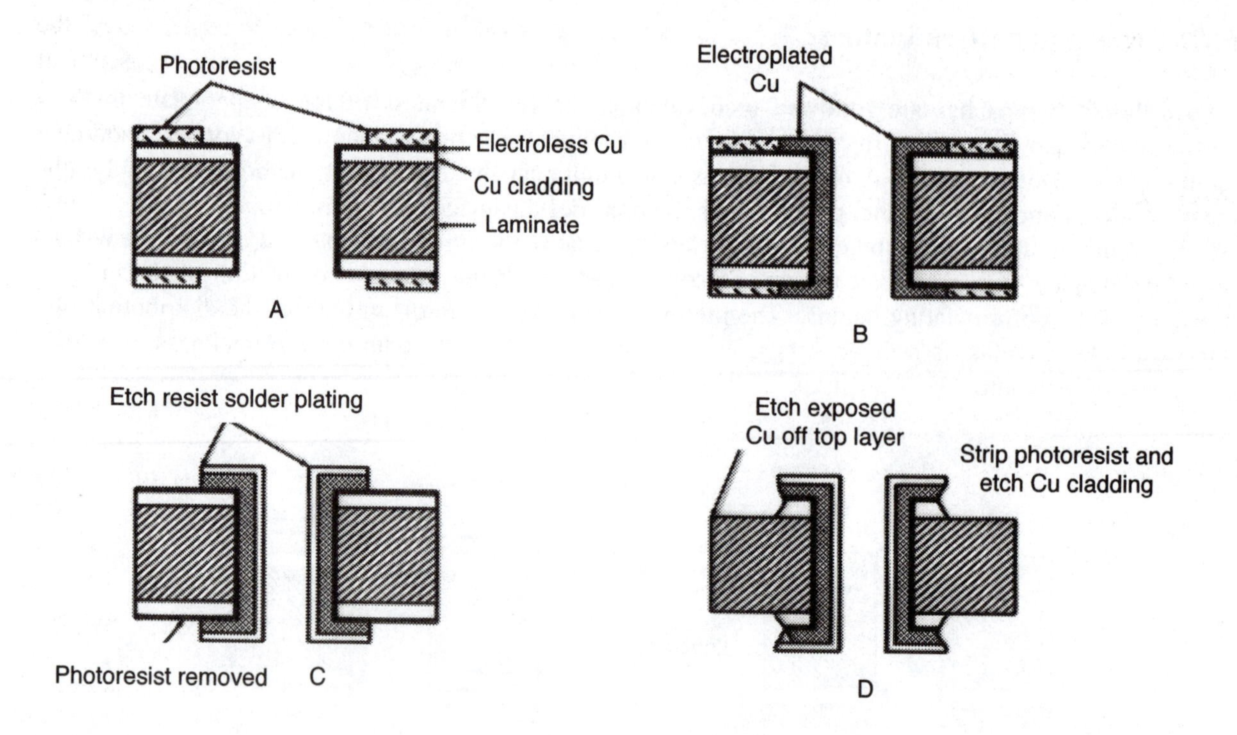

Fig. 6.22 Production sequence used in the pattern plating process.

6.7.3 Plating Finishes

Solder Plate

This plating consists of a composition of lead-free solder and serves as etch resistant coating that protects circuit features, PTHs and vias as the excess copper cladding is etched away. The solder plate also serves to protect the copper surfaces on the pads and traces from oxidation during the storage period prior to assembly. When the board is ready to be soldered, the solder plate is reflowed. In the liquid state, the solder acts to wet the surfaces to be joined. The thickness of solder plate is about 0.3 mil (7.5 μm) that is sufficiently thick to resist the etchant's chemical attack and to flow under the action of surface tensions when heated above its melting temperature. However, the recent adoption of lead-free solder for plating and copper surfaces and for soldering connections on chip carriers elevates the processing temperatures. This topic will be discussed in more detail in Chapter 7.

HASL Hot Air Solder Leveling

Hot air solder leveling (HASL) is a technique for ensuring printed circuit board solderability during fabrication and assembly. HASL protects coated surfaces from corrosion and contamination with the application of lead-free solders to the exposed copper clad surfaces on a PCB. The solder coatings provide protection for either the entire copper surface of a laminate or selective areas not covered with a solder mask. The solder mask covers those copper surfaces not requiring solder connections during assembly and leaves solder sites exposed.

The process consists of pre-cleaning, preheating, flux coating, solder coating, leveling with hot air knives, cooling, and post-cleaning. Pre-cleaning employs a mild etch, water rinse and drying in hot air. The mild etch removes the oxides and organic surface contaminants from the copper surface. Pre-cleaning is performed on an in-line conveyorized system, where the mild etchant is sprayed onto the laminate and it is transported at a controlled rate through the sprays to achieve a uniform depth of etching. Preheating serves two functions: it reduces thermal shock when the PCB is immersed in molten solder and helps prevent blocked vias. The most common preheating system uses an infrared (IR) conveyorized pre-heater with variable intensity and speed to control the amount of heating. The exit temperature, which is dependent on the thickness of the PCB, is monitored to control the process. Applying sufficient preheat allows for reduced dwell times in the molten solder. Also, solder is cleared from the vias more easily with pressurized air if the panel temperatures are higher.

Flux is applied either by immersion or with brush-rolls. Many different fluxes are available and the selection of their viscosity and acidity depend on the process and equipment used to apply the solder coating. Reduced dwell time in the molten solder requires the flux to disperse quickly to allow for adequate solder wetting. Flux selection is based on ease of coverage, solder mask type, base materials such as nickel/gold, and lead-free solder alloys.

Solder coating is applied by immersing the PCB in molten solder. A glycol-based oil blanket that is compatible to the flux limits dross formation on the solder. In most common process, the PCBs are feed by a conveyor through the molten solder with a dwell time of about two seconds. The flux is displaced by the solder and an intermetallic bond (Cu_6Sn_5) is formed between the copper cladding and the solder. Some types of copper contamination can be displaced with additional time in the molten solder; however, increasing dwell times also increases the thickness of the intermetallic layer. This layer is not solderable. Hence, the surface and must have an overcoat of solder with sufficient thickness, after hot air leveling, to ensure solderability.

Leveling is accomplished immediately after the PCB has emerged from the molten solder bath. The PCB is passed through pressurized hot air knives to remove excess molten solder and to level the solder on the copper cladding. In addition, pressurized hot air clears excess solder from the vias. The pressurized air passes through heaters and then to the air knife. Typical air temperatures are 400°to 500°F (204°C to 260°C), with pressures ranging from 12 to 30 psi. Leveling is controlled by the air knife configuration and operational parameters, which include the distance from PCB to air knife, air knife angle, air pressure and speed of the PCB past the air knife. After the molten solder is leveled, it solidifies in a few seconds. The temperature of the PCB is reduced in a controlled manner to prevent panel warpage and thermal shock when the laminate is subjected to the liquids used in the post-cleaning operation.

Post-cleaning is the final step in the HASL process. Depending on the product and its application, post-cleaning can involve removal of flux residuals to complete removal of essentially all of the ionic contaminants. The most common variables involved in the post-cleaning operation are water quality and water additives such as detergent, cleaners and solvents. Other parameters that affect cleanliness are flow rates, water pressure and the use of brushes to facilitate removal of the flux residuals.

HASL is the predominant final finish for PCBs in manufacturing facilities worldwide. The process is predictable, the solder coat is known and the industry produces billions of solder joints daily. However, three factors are motivating the electronics industry to consider HASL alternatives: cost, new technology and the uncertainty associated with lead-free solders.

6.7.4 Alternative Plating Finishes

There are several final finishes that are alternatives to the HASL process. These finishes include organic solderability preservatives (OSPs), electroless nickel/ immersion gold (ENIG) or new metallic immersion metals such as silver (IAg) and tin (ISn). These processes enable the production of a lead-free PCB and provide flat coplanar surfaces to allow more effective placement of BGAs. The alternative processes provide coatings more suitable for finer pitch and area array devices. Finally, most alternatives improve assembly operations and long term reliability while reducing cost. Alternatives like OSPs, immersion silver and immersion tin can provide a 20 to 30 percent reduction in final finishing costs.

The use of alternatives is expected to increase and eventually replace HASL as the final finish of choice. Alternatives ENIG, OSPs, immersion tin and immersion silver all provide lead-free, solderable, coplanar surfaces that provide significant improvement in assembly yields. Each of these alternatives will be described below.

Electroless Nickel Immersion Gold (ENIG)

The use of electroless-nickel-immersion-gold (ENIG), a metallization finish for PCBs, has grown significantly over the past decade due to the introduction of fine-pitch surface mount chip carriers and ball grid arrays. Electroless nickel is applied after copper plating, imaging and etching operations, but usually before the solder mask is applied. The nickel coating protects all exposed copper surfaces. It is applied in a process employing a palladium catalyst. The nickel is chemically deposited in processing tanks to a nominal thickness of 4 to 5 μm. The coating consists of approximately 92% nickel with about 8% phosphorous that is co-deposited with the nickel. The phosphorus reduces the ductility of the nickel significantly, but it is an inherent part of the chemical deposition process. The thin nickel coating acts as barrier layer between the copper and the gold, preventing undesirable intermetallics from forming. The nickel layer also adds strength to plated-through-holes and vias.

Immersion gold is applied after the electroless nickel process and covers all of the nickel coated surfaces. It is a molecular replacement process that involves replacing the nickel molecules with gold molecules in a chemical processing tank. The gold that is deposited is 99.99% pure, and its thickness is allowed to build up to 0.1 to 0.2 μm. Although this very thin coating is somewhat porous, it provides excellent corrosion resistance and solderability properties, and it retards oxidation of the nickel layer.

Electrolytic gold can also be applied after the electrolytic nickel coating. However it must be applied before etching and the application of the solder mask. Thus this gold finish covers only the top surfaces of all exposed nickel traces and features. After the final etch operation, the trace and feature sidewalls of both the copper and the nickel are exposed. Electrolytic gold, also known as 'flash gold' is plated to a nominal thickness of 0.1 to 0.25 μm. It provides excellent corrosion resistance and solderability properties as well as sealing the nickel surface to eliminate top surface oxidation. Thin electrolytic gold, 0.1 to 0.25 μm thick, is not used for wire bonding. Thicker electrolytic gold (0.75 μm) are used when wire bonding is employed.

For coatings that provide corrosion resistance, extended shelf life, nominal metal thickness, metal purity, and solderability, both electroless-nickel-immersion-gold and electrolytic-nickel-electrolytic-gold provide good to excellent performance. Either metallization is suitable for standard or extra-fine pitch (i.e. 1.27 to 0.5 mm) surface mounted chip carriers or BGAs. For ultra-fine pitch less that 0.4 mm, the electroless-nickel-immersion-gold coatings appear to perform better.

Organic Solderability Preservative (OSP)

With the growing use of ball grid arrays and finer pitch chip carriers, an extremely flat circuit board surface is critical for the simultaneous contact of all of the leads or solder balls. In addition the tin-lead compositions that have in the past been used for the HASL process are not compatible with lead-free soldering. An organic solderability preservative (OSP) is one of the alternative surface finishes. The OSP is an anti-tarnish coating of an organic-based compound that is applied over the copper surfaces on a PCB to prevent them from oxidizing. This water-based organic compound selectively bonds with copper to provide an organic-metallic layer that protects the copper. Coatings 0.7 to 1.0 μm thick are applied after solder mask application. OSP's are protective coatings that must be removed during the soldering process with active fluxes. The flux must come in direct contact with the coating and penetrate it in order for the solder to bond to the copper on the PCB surface. Another important concern is potential incompatibility with some flux chemistries and solvents. Solderability can be degraded when the PCB is subjected to multiple thermal cycles during the soldering operations. Also contact with the coated surface as the PCB is handled degrades the organic coating impacting solder process integrity

The exact formulation of OSPs is proprietary; however, the coatings are formulated from organic compounds selected from the group consisting of benzimidazoles, alkylimidazoles, benzotriazozles and alkyltriazoles and metal particles of solder-wettable metals. OSPs are the lowest cost alternative finish. They are compatible with many multi-metal surfaces, and provide the highest bond strength.

Immersion Metal Coatings

Coating by electroplating is fast and relatively inexpensive, but it suffers from a serious problem when attempts are made to apply coatings to PCBs where the pads and holes have already been imaged and etched. It is necessary to make an electrical connection to every single feature that is to be plated. This is not the case for either electroless or immersion plating. Characteristic that differentiates the two processes—electroless and immersion—has to do with the plating process itself.

When a metal is dissolved in a plating bath, it becomes an ion and carries a positive electrical charge. When the ion is deposited on the workpiece, electrons are added to the metal and the charge is neutralized. In an electroplating bath, the electrons are provided by a power supply that is essentially an electron pump. In an electroless plating bath, the electrons are provide by a chemical in the bath that is called a reducing agent. In an electroless nickel bath, the reducing agent it is either hypophosphite or dimethyl amine borane (a very expensive reducing agent).

In an immersion plating bath, the electrons are supplied by the base metal and a reducing agent is not required. Because the base metal provides electrons to the bath, the base metal itself goes into solution, and the process is known as a replacement reaction with more noble metals (tin and silver) replacing the base metal (copper). The important feature of an immersion plating process is that it is self limiting. When the base metal, copper for a PCB, is covered with more noble, corrosion resistant metal, the plating stops. Immersion plating baths are usually formulations of metal salts, alkalis and complex agents such as lactic, glycolic, or malic acid salts). This process yields a very thin, dense, non-porous coating that is very economical.

Immersion White Tin and Immersion Silver

Two immersion finishes that are being employed more frequently are immersion white tin (ISn) and immersion silver (IAg). Both finishes tend to provide thinner coatings than ENIG or traditional HASL. They typically contain an organic component that retards the surface oxidation preventing one of the most common soldering problems. The finishes are extremely flat, and are ideal for use with fine pitch devices and BGAs. These surface finishes are already used in high-volume production and research to optimize their properties and further development continues.

The process for applying immersion white tin is very simple. The exposed copper on the PCBs is first cleaned with an acid cleaner and micro-etch. The laminates then go through a pre-dip for about one minute, an immersion tin bath at room temperature for about one minute, and then into a heated immersion tin bath at 60-65° C for six to eight minutes. The PCBs are then rinsed and dried in warm air.

The areas where copper was exposed on the PCBs is coated with a flat, uniform coating of white tin that is 0.7 to 1.0 μm thick. The chemicals in the immersion bath are formulated to create a fine and sufficiently dense and stable grain structure that is capable of suppressing the growth of an intermetallic layer. This grain structure distinguishes it from traditional tins, which have a porous structure that is unstable and insufficient to suppress the intermetallic layer. Moreover, this fine grain structure allows immersion white tin to resist the dendritic growth or tin whiskers that form under certain environmental conditions with ordinary tin.

A board coated with immersion white tin has a shelf life of one year before soldering problems develop. This finish is excellent for fine pitch applications, and it can withstand multiple heat cycles. The chemistry in an immersion white tin baths is controlled by measuring specific gravity and tin content. A concentrated replenisher is added to the bath when the tin content drops below a specified level.

Immersion silver is still a relatively new process when compared to OSP and ENIG. However, in recent years, extensive testing and high volume production have established the reliability of this protective coating. The solder wetting characteristics of this finish make it adaptable to existing no-clean wave soldering processes. This surface finish is a potential alternative for most applications, including shielding, aluminum wire bonding, connector contacts and soldering. The contact resistance of this coating remains low after aging or reflow processes. Also, as a metallic coating, immersion silver is easy to inspect with low or no magnification.

Conclusions

OSP, immersion silver and immersion white tin all provide high, first-pass assembly yields with both water soluble and no-clean assembly technologies. The proper application for each coating is dependent on the PCB design requirements. A comparison of the five surface finishes used to protect the copper on the PCBs during the storage interval before soldering is shown in Table 6.3.

6.7.5 Gold Plating

Gold is a noble metal and as such it is very resistant to oxidation and corrosion. It is also a very good conductor with a resistivity of only 2.44×10^{-6} Ω-cm. These two properties indicate that gold plate is an ideal coating for connector parts that come into sliding contact. The cost of gold is quite high, currently in the neighborhood of $600 per troy ounce, depending upon the fragility of the financial community. Because of its high cost, it is clearly important to minimize the thickness of the gold coating that is required. MIL SPECS call for gold coatings with a minimum thickness of 80 μin. (2 μm); however, in commercial applications thinner coatings are usually employed. Most gold plate is deposited on top of a layer of 0.2 mil (5 μm) thick nickel plate. The nickel plate serves to smooth the surface producing an improved base for applying gold plate that is free of pinholes and porosity. The nickel plate is chemically stable and acts together with the gold to produce a corrosion resistant surface.

In connector applications, where two contacts are forced together and designed to wipe as the contacts are seated, wear of the surfaces is important. When wear is a criterion, the two surfaces (the pin and the contact) are usually plated with gold of differing hardness. Hard gold, an alloy of gold and nickel, is used on the pin and pure soft gold is used on the female contacts. With proper plating procedures the life of connectors often exceeds 500 cycles before wear becomes excessive.

Table 6.3
Comparison of the characteristics of five different surface finishes

Properties	White Tin	HASL	Ni/Au	OSP	Silver
Fine Pitch	Yes	Problem	Yes	Yes	Yes
Flatness of Pad	Yes	No	Yes	Yes	Yes
Multiple Solder Cycles	Yes	Yes	Yes	Problem	Yes
Bare Board Testing	Yes	Yes	Yes	Problem	?
Dimensional Stress	None	High	None	None	None
Controllability	High	Low	Med	High	Med
Cost Factor	Med	Med	High	Med	High
Compatible with All Flux	Yes	Yes	No	No	Yes
Solder Pot Contamination	No	No	Yes	No	No
Solder Mask Compatibility	Yes	Yes	Problem	Yes	Yes
Rework Board	Yes	Yes	Problem	Problem	Problem
Shelf Life	Long	Long	Long	Med	Med

Very thin gold plate is sometimes used on expensive components to protect the leads of the chip carrier from oxidation during storage. When these components are soldered to the circuit board, the gold plate is stripped by the molten solder and the solder joint is essentially composed of an lead-free alloy rich in tin. Problems can arise if the gold plate is too thick and is not removed by the molten solder. In these cases, the gold becomes an important alloying element in the solder contained in the joint. Indeed, if the concentration of gold is excessive, a thin layer of a gold-tin intermetallic forms at the bond line of the solder joint. This intermetallic is very brittle and prone to failure and it impairs the strength of the solder joint.

6.8 SOLDER MASKS

After the circuit board has been drilled, plated, imaged, etched, neutralized and cleaned, it is ready for the solder mask to be applied to its exposed surfaces. The role of the solder mask is to control the placement of the solder during an automated soldering process. The mask also provides a protective barrier against surface contamination during the operating life of the PCB. The solder mask is a thin polymeric coating that is applied over all areas of the board that will not be soldered during assembly. The solder mask can be applied with two different processes—by screening or by film lamination.

The film lamination process is almost identical to the photoresist process because a photopolymer film, which serves as the mask, is roller laminated to the board. The photopolymer is then imaged using a phototool that permits exposure of those areas where solder joints are to be made. The exposed area is removed by stripping and the remaining film forms the solder mask. The mask is placed with the precision of the original photoresist process. The resolution of the solder masks formed by film lamination is excellent; however, the cost is higher than that incurred in the screen printing process that is described next.

The screening process, illustrated in Fig. 6.23, utilizes a fine mesh screen usually fabricated from a thin sheet of stainless steel with a very dense array of small holes. A stencil film covering only those areas on the PCB to be soldered is bonded to the screen to block the holes in these areas. The screen is mounted in a frame that keeps it tightly stretched. The screen is lowered in registration over the circuit board and liquid polymer is forced through the screen with a wiper tool called a squeegee. A modern manually operated stencil machine with a stainless steel screen is illustrated in Fig. 6.24.

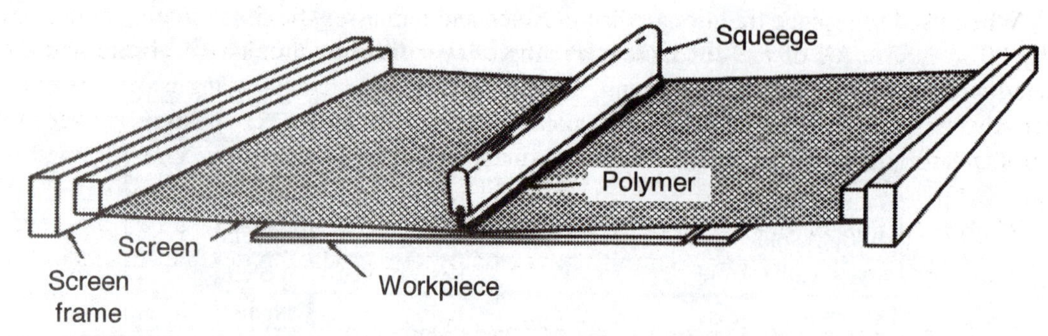

Fig. 6.23 Illustration of the screening process used to transfer images from a stencil to a PCB.

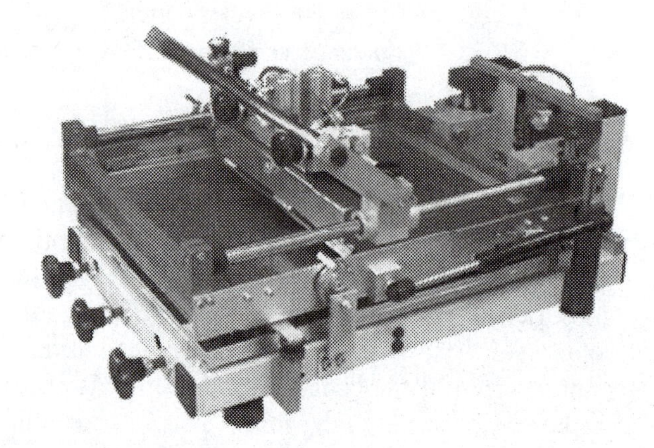

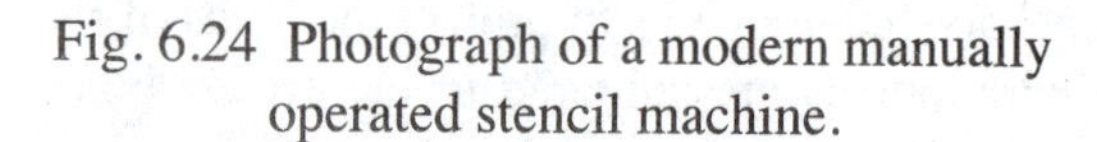
Fig. 6.24 Photograph of a modern manually operated stencil machine.

The area pattern of the polymer transferred to the board is controlled by the openings in the stencil and the thickness of the coating is controlled by the combined thickness of the stencil and the screen. After screen printing, the thin coating of polymer is permitted to dry forming a solder mask. While lower in cost, registration and resolution of the solder mask applied with screen printing may not be sufficient for dense circuits with pads on very close centers.

In addition to solder control during assembly and corrosion control over the life of the product, the solder mask serves a third important function. The mask provides an insulating coating that encases the circuit traces over major areas of the board. This insulation prevents voltage induced dendritic growth of copper crystals that may occur with time between closely spaced traces. Dendritic crystals grow on free surfaces and will produce short circuits when they extend from one wiring trace to an adjacent one. Finally, the mask protects the surface of the board from detrimental effects of moisture over extended periods of time.

6.9 PRODUCTION OF FLEXIBLE SUBSTRATES

A flexible printed circuit is similar in many respects to a rigid PCB, although it usually serves a different function. In most applications it is used to replace either flat collated cables or a wiring harness. Its flexibility distinguishes it from a rigid PCB. The fact that it must bend, sometimes thousands of times, requires its copper cladding, base and cover films and adhesives to be flexible and to be able to endure repeated straining. Most flexible circuits are less than 10 mil (0.254 mm) thick. Thin dielectric films are used for repeated flexing applications, and thicker films are used for intermittent flexing or for bending only during assembly.

When used to replace traditional wiring cables and harnesses, flexible circuits reduce assembly costs by 20 to 50%. All of the conductors on single layer flexible circuits are visible and the work associated with measuring, cutting, stripping, tinning, routing and lacing a wring harness is eliminated. Another advantage is in the production of via holes. Because the base or cover films are very thin, via holes of standard size can be punch. Very small diameter holes for microvias can be formed by laser drilling.

The production of flexible circuits utilizes many of the same processes that are used in the production of rigid PCBs. The manufacturing steps employed in producing flexible circuits includes:

1. Clean and degrease the copper foil used as a cladding.
2. Adhesively bond copper foil to the base polymeric film.
3. Apply either liquid or dry film photoresist coatings.
4. Image the conductor lines and other features.
5. Develop the photoresist pattern.
6. Etch away the copper cladding from areas not required to for the conductors and other features.
7. Clean and neutralize the copper lines and base polymeric film.
8. Coat copper with a finish if required.
9. Punch the required holes in the protective cover layer.
10. Apply the protective cover film as an overcoat in a lamination process.
11. Cut laminate to finals shape and size.

The materials employed for the dielectric substrate include polyimide films, polyester films, aramids, reinforced epoxies, and fluorocarbons. These films are tough and provide the mechanical strength required to resist handling stresses and the strains imposed by repeated bending. Of course, the dielectric substrates provide the required insulation, surface resistance and high breakdown voltage.

The strain ε produced when bending a flexible circuit depends on the radius of curvature R of the arc through which the flexible circuit is bent. This strain is given by:

$$\varepsilon = \frac{t}{2R + t} \tag{6.4}$$

where t is the thickness of the flexible laminate.

In continuous flexing applications the base film and the cover film are the same material and thickness. This design places the copper cladding on the neutral axis and minimizes the imposed bending strains. The copper cladding used in flexible circuits is grain size controlled electrolytic copper or rolled and annealed copper foil. The characteristics of these copper foils were discussed previously in Section 5.3.5.

Several different polymers are employed to bond the copper foils to the substrate films. These include polyester, acrylic, modified epoxy, polyimide, fluorocarbon, and butyral phenolic. The choice of the adhesive depends on the polymeric film used as the substrate and the protective cover, because the adhesive must be compatible with these films. All of these polymers provide adequate adhesion, electrical properties, chemical resistance and flexibility. The distinguishing properties among the adhesives are temperature resistance, moisture absorption and cost as indicated in Table 6.4.

The protective cover is usually applied as a layer of polymeric film identical to the film serving as the substrate. This protective film is adhesively bonded over the surface of the copper conductors. Holes to access pads and hole for registration are punched into the film before it is laminated to the substrate. The protective film is positioned in registration over the conductor pattern and the lamination is performed under heat and pressure.

Table 6.4
Comparison of adhesives for flexible circuits

Type	Temperature Resistance	Moisture Absorption	Cost
Polyester	Fair	Fair	Low
Acrylic	Very Good	Poor	Moderate
Modified Epoxy	Good	Good	High
Polyimide	Excellent	Poor	Very high
Fluorocarbon	Very good	Excellent	Moderate
Butyral Phenolic	Good	Fair	Moderate

In some flexible circuits liquid polymers such as acrylated polyurethane are screen printed over the conductors. These polymers are cured with IR heating or with UV radiation to form thin but tough coatings that protect the copper conductors.

6.10 CERAMIC CIRCUIT BOARDS

Ceramics are employed in the production of chip carriers, hybrid circuit substrates, which support a relatively small number of components, and larger more traditional circuit boards. In chip carrier applications, ceramics have the advantage of superior hermeticity and relatively low thermal resistance. In circuit board applications, both hybrid and conventional, the ceramics have the advantage of a temperature coefficient of expansion that is closely matched to a silicon chip for direct chip attachment or a ceramic chip carrier for more traditional connections. A second advantage, particularly important for hybrids, is the ability to print resistors directly onto the ceramic substrates. Another advantage is the stability of the ceramic with both time and temperature in harsh environments.

Circuit boards fabricated from ceramics are comprised of three different components that include: the ceramic or glass-ceramic substrate, sintered metal circuit traces and pads and glass insulation. We will describe the components individually in subsequent sub sections.

6.10.1 Ceramic Substrates, Materials and Processes

Some of the ceramic materials employed as substrates are listed in Table 6.5. Alumina based ceramics, containing 80 to 90% Al_2O_3, are the most commonly used ceramic because of their strength, high electrical resistivity and economy. Proprietary mixtures of glass and alumina are also employed because they have lower dielectric constant and a lower sintering temperature. The substrates are produced in sheet form by dry pressing or sheet casting, followed by sintering. For substrate applications, where heat transfer is the most important consideration, aluminum nitride (AlN) and beryllium oxide (BeO) can be employed because they have higher thermal conductivities than alumina. Beryllium oxide has the highest thermal conductivity of any of the ceramic oxides; however, it toxicity in powder form and its high cost severely limit its applications. Aluminum nitride, which is lower in cost and not toxic, is frequently used when high thermal conductivity is a serious issue.

In recent years, nitrides and carbides have been considered for packaging applications. These materials are very strong and offer interesting combinations of properties. For example, the thermal conductivity of both aluminum nitrate and silicon carbide is high and in addition their temperature coefficient of expansion closely matches that of silicon. Because silicon carbide is conductive, it cannot be used in applications where insulation is required. Hence, it is used primarily in a first level package as a heat spreading device attached to the back side of a chip. Aluminum nitride has excellent electrical resistivity and is used instead of alumina in both substrates and first level packages when a reduction in thermal resistance is required.

Table 6.5
Select properties of ceramic substrate materials

Material	Thermal expansion Coefficient × 10^{-6}/°K	Thermal Conductivity (W/m°K)	Dielectric Constant
Aluminum Oxide	6.7	25-35	9.0-9.5
Silicon	2.7	150-160	
Aluminum Nitride	3.3	140-170	6.5-10.0
Silicon Nitride	2.3	25-35	6.0-10.0
Silicon Carbide	3.5	120-250	20-40
Beryllium Oxide	9.0	220-240	6-7

The dry pressing and sintering method of producing thin flat plates (disks) for ceramic substrates is illustrated in Fig. 6.25. The fabrication process involves mixing ceramic powders together with a binder and a solvent. This mixture is spray dried and granulated to produce a coarse grain powder containing a binder. The granulated powder is pressed into thin "green" disks. These are stacked together with graphite spacers to prevent them from bonding together. The stack is placed in a vacuum furnace and heated under pressure to sintering temperatures. Dense ceramic disks with very low porosity are form with this process.

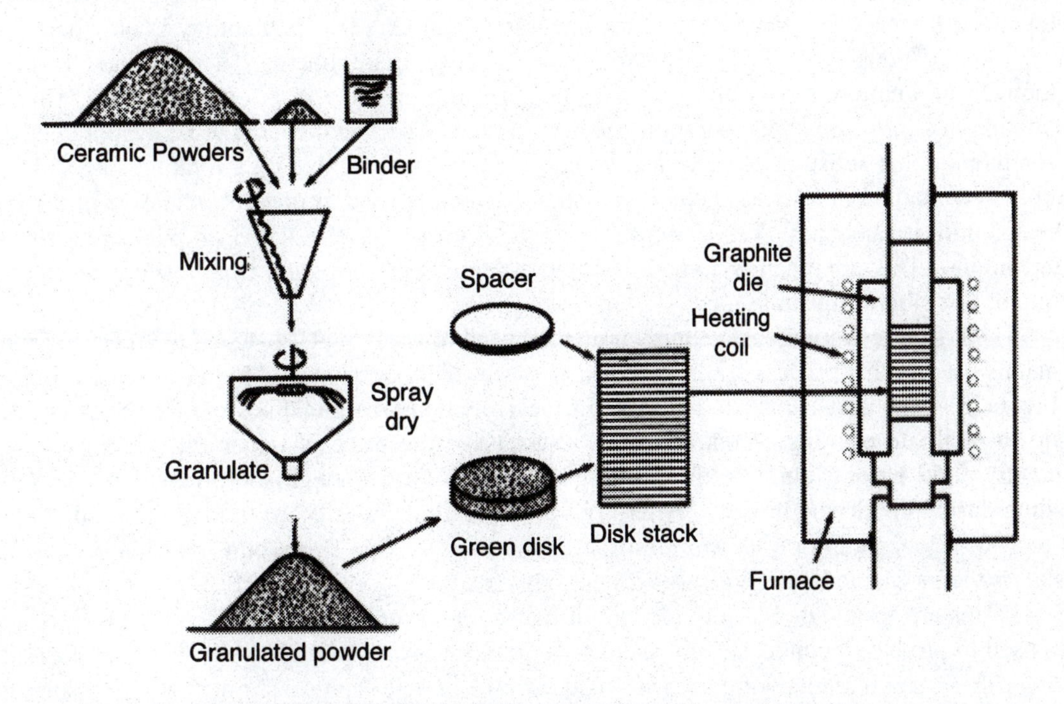

Fig. 6.25 Method of producing high density ceramic disks.

A second method for producing much thinner and larger ceramic plates involves the development of a "green" sheet. To form a green sheet, ceramic and glass powders are blended together and then milled in a ball mill to ensure proper particle size. The milled powders are mixed with organic binders and a solvent to form a thick slurry. This slurry is formed into a uniform thickness sheet by casting the mixture onto a continuous plastic belt, as shown in Fig. 6.26, which supports and transports the green sheet as the solvents in the mixture are removed by evaporation. Upon drying, the green sheet is similar to a piece of paper. The resin binder holds the particles of glass and ceramic together, providing a reasonable degree of flexibility and strength. These properties enable unfired sheets to be handled during subsequent processing operations. In the green state, prior to sintering, the sheets can be cut and holes punched. In addition, circuit lines and vias can be printed onto the green sheet with metallized inks. The green sheets become thin ceramics only after they are fired. The firing drives off the organic binders and fuse (sinter) together the glass and ceramic particles. During the sintering process the sheet shrinks by about 15 to 20 per cent. This change in dimensions must be accommodated in the design layout where the sizes of the different features are increased to account for this shrinkage.

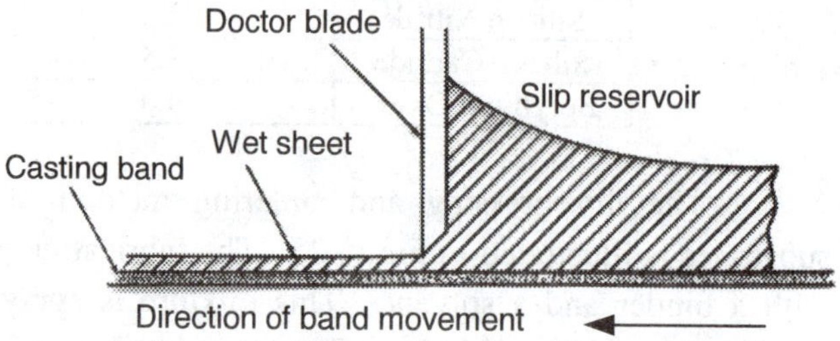

Doctor blade
Slip reservoir
Wet sheet
Casting band
Direction of band movement

Fig. 6.26 Continuous casting of green sheets of ceramic and glass powders.

6.10.2 Metallization

There are three methods commonly employed to apply the metal required to form pads and circuit traces on ceramic substrates, namely thin film, thick film and co-fired. Thin film technologies have been developed for the past fifty years to produce integrated circuits. This technology is now being applied in a few applications of high density ceramic or silicon circuit boards. The method involves the development of a pattern representing the pads and circuit traces in a layer of photoresist. The metal is then applied in a thin film by evaporation under a vacuum. The metal charge to be evaporated is placed in a crucible and the substrate to be coated is placed above this crucible on a rotating stage. The entire assembly is contained in a vacuum chamber. High vacuum is used to prevent oxidation of the metal as it is heated until it vaporizes. Very thin pads and circuit traces, usually 1 to 3μm thick are formed using this technique. This approach is sometimes used when silicon chips are flip-chip bonded with direct attachment to a silicon substrate.

Thick film technology is a common method used to apply metal features to a ceramic substrate. The metals used in this process are in the form of pastes that are deposited by a screen printing process (see Fig. 6.23). The pastes contain a binder to adhesively bond the metal powders to the substrate and a vehicle to facilitate printing. Control of the viscosity of the paste is important. The paste must be sufficiently fluid to pass through the printing screen yet firm enough to form sharp edges without spreading during the drying process. After drying, the printed deposits are fired at elevated temperature (800 to 900 °C depending on the composition of the paste). This firing burns off the organic binders and sinters the inorganic binders (glasses) to the substrate.

Pastes are formulated to provide circuit traces representing both conductors and resistors. For pastes used to produce a conductor, the metal is in the form of fine particles with an equivalent diameter of 1 to 3 μm. A small amount of glass frit is added to the paste to control shrinkage during sintering and to improve bonding of the metal to the substrate. Gold particles perform well as a metallized conductor and would be more widely employed except for its cost. Gold I/O pads on chips can be wire bonded and the gold coating is resistant to oxidation at the sintering temperatures. The properties of gold can be improved with the addition of either platinum or palladium that enhance its solderability and increase its resistance to leaching in solder. Silver is sometimes used because of its reduced cost, but it tends to migrate under voltage with time. Additions of platinum and/or palladium reduce the tendency of silver to migrate and improve its resistance to leaching. Copper can be employed to advantage in pastes because it is resistant to migration, does not leach in solder, has a high conductivity and is available at a much lower cost. The difficulty in using copper based pastes is in the firing operation. Copper will oxidize in air and it is necessary to use furnaces with nitrogen or argon cover gases. Of course, the absence of oxygen in the furnace makes it impossible to burn out the organic binders. To alleviate this problem, the binders are oxidized with air as the cover gas in the furnace at relatively low temperature. Then the furnace is flushed to remove the air and any hydrocarbons that remain in the cover gas from the binder. Finally, the furnace is flooded with nitrogen or argon prior to elevating its temperature to sinter the remaining copper particles.

Resistors are formed directly on the substrate by sintering a mixture of metal and oxides together to form a high resistivity conductor. The geometry of the line (width, length and thickness) controls the resistance of the element. Ruthenium mixed with glass frit is used for high resistivity films and silver palladium alloys are compounded with glass for the lower resistivity films. The actual size of the resistor depends upon the power dissipation, but minimum sizes are usually about 10 mil (0.25 mm) wide by 15 mil (38 mm) long. Accuracy tolerances on the resistance are usually ± 20%. Laser trimming of the resistors is often employed to modify their values to provide improved accuracies.

Co-fired metal conductors and resistors are produced using methods nearly identical to those used with the thick film. More fluid pastes much like printing inks are either screened or printed onto the ceramic to form the surface features required. The primary difference is the fact that the ceramic substrate is in the form of a "green" sheet. The green sheet and the pastes are fired together and both are sintered in the same operation. The use of the co-fired process places special requirements on the pastes used for metallization. If the green sheets are made from alumina, the firing temperatures that are necessary for sintering the ceramics are quite high and the alloys described previously are not suitable at these higher temperatures. Instead molybdenum, tungsten and other refractory metal alloys are employed that do not chemically react with the ceramic at these elevated temperatures. However, if the green sheets are fabricated from a glass-alumina mixture, the firing temperatures are lower and more metals may be employed as conductors. If the firing furnace is programmed to operate at a lower temperature with air as a cover gas, the resin binders can be oxidized and the ceramic sheets consolidated. Then the cover gas is flushed and changed to nitrogen or argon and the temperature increased to completely sinter the glass ceramic sheets and the metallization. The thermal conduction multi-chip module (TCMCM) that utilizes a multilayer ceramic substrate for the substrate, illustrated in Fig. 4.33, was produced using a co-fired process than enabled 70 layers of copper interconnects.

6.10.3 Glasses as Dielectrics and Seals

Multilayered circuit boards are produced with metallization on ceramic substrates. With the co-fired process, layers of green sheet with printing on one side only are stacked together in registration. The stack of sheets is sintered as a single unit and each sheet remains in registration even with shrinkage of 15 to 20%. With the thick film process, the production of multilayer boards differs appreciably as several printings and firings are used to fabricate the required number of signal and power planes. The first firing is for the metal paste screened onto a ceramic substrate as described previously. The next step is to apply, by printing, a layer of an insulating dielectric over this layer of metallization. The dielectric is a glass paste that is screen printed with via holes aligned to the first layer of metallization. The glass paste is dried and fired to form a continuous layer of insulation of uniform thickness with via holes at select locations to provide signal plane connections. This layer of glass provides the surface for the application of a second layer of metal paste. The printing and firing processes are repeated many times to fabricate a multilayered ceramic circuit board.

Glass is also widely used for seals in a wide range of packaging applications. Glass-to-metal seals are vacuum tight assemblies of glasses with a metal enclosure used to convey conductors through the wall of a hermetically sealed package. A typical glass-to-metal seal consists of a metal enclosure in which a pre-formed sintered glass element is sealed. The sintered glass element in turn encloses one or more leads that it seals. Because of the different expansion coefficients of various glasses and metals, thermal stresses occur during the sintering process. By designing the metal enclosures and selecting the glasses correctly, it is possible to obtain glass-to metal seals in which these stresses do not cause failure of the assembly.

There are two design approaches used in creating glass-to-metal seals—matched and compression seals. For matched glass-to-metal seals, the thermal expansion coefficient α of both the glass and the metal must closely coincide at any temperature between ambient and the sintering temperature of the glass. The matching of the thermal coefficient of expansion of the glass and the metal insures that the thermal stresses generated in the glass never become large enough to cause failure as the package cools from the sintering temperature to ambient temperature. The outer electrodes of matched glass-to-metal seals are usually fabricated from thin sheet metal (0.2 – 0.5 mm thick) in the shape of shells, caps and tubes. Under optimal processing conditions, these glass-to-metal seals are essentially stress-free at ambient temperature. An advantage of matched glass-to-metal seals is their relatively low weight. However, their sensitivity to mechanical stresses is a disadvantage. High thermal shock loads due to severe localized heating of the external metal element, which may result during assembly, can cause failure of the glass seals to occur.

The compressive strength of glass is much higher than its tensile strength, which varies from 20 to 80 MPa depending on the composition of the glass. This fact is utilized in designing glass-to-metal seals that apply uniform compressive stresses to the glass element. These compressive stresses prevent the development of tensile stresses in the seal, even when it is exposed to relatively severe mechanical and thermal loads. To design a compression glass-to-metals seal, a metal for the external enclosure is selected with a thermal expansion coefficient that is greater than that of the sealing glass and the internal leads. After the glass is sintered, the external enclosure shrinks more than the glass seal forming a compression ring around the sealing element. The compressive stresses exerted on the glass must be balanced by tensile stresses developed in the compression ring. To ensure that the stress applied to the external enclosure does not exceed its elastic limit, compression rings with sufficient wall thicknesses and appropriate material combinations are used. A few typical designs illustrating the use of heavy wall enclosures for compression glass-to-metal seals are presented in Fig. 6.27.

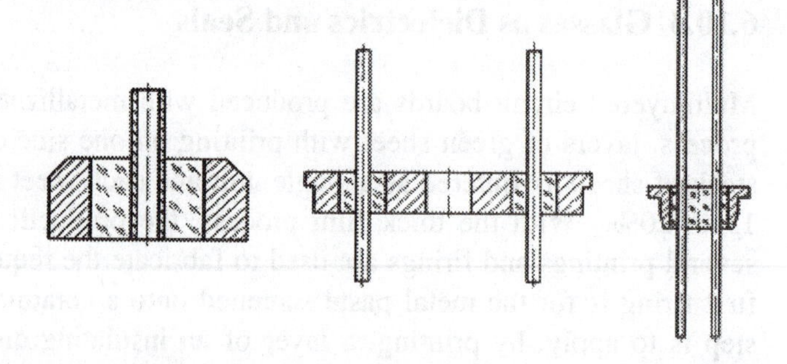

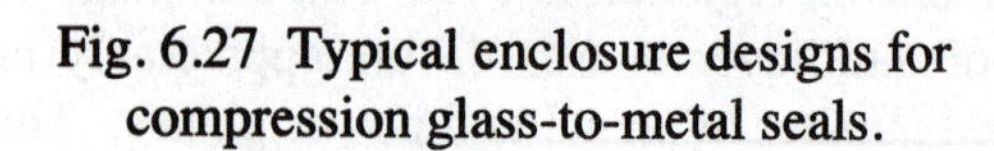

Fig. 6.27 Typical enclosure designs for compression glass-to-metal seals.

Due to their high mechanical and thermal stability, compression glass-to-metal seals are robust and can be processed with minimal risk of damage. The glasses commonly used for sealing are listed in Table 6.6 together with their thermal coefficient of expansion and glass transformation temperature.

Table 6.6
Characteristics of glasses used for seals

Glass No.[1]	Applications	Thermal Coefficient of expansion × 10^{-6} /°K	Transformation temperature T_g (°C)
S-8061	Compression seals Austenitic steels and NiFe alloys	9.3	467
8250	Matched seals with NiCo 29-18 and NiFe 42	5.0	492
8350	Compression seals with steels and NiFe alloys	9.0	520
8421	Compression seals with steels and NiFe alloys	9.6	525
8422	Compression seals with steels and NiFe alloys	6.6	540
8537	Battery seals Seals with steels and NiFe alloys	9.1	480
8629	Compression seals with steels and NiCo 29-18	7.6	529
8630	Compression seals with steels and NiFe alloys	9.1	440

[1] The glass numbers refer to compositions commercial available from Schott North America, Inc.

REFERENCES

1 Ducas, C., "Electrical Apparatus and Methods of Manufacturing the Same" US Patent No. 1,563,731, Dec. 1, 1925.
2 Blackwell, G. R., The Electronic Packaging Handbook, CRC Press, November, 1999.
3 Witaker, J. C., The Electronics Handbook, CRC Press, December, 1996.
4 Walsh, R. A., Electromechanical Design Handbook, McGraw-Hill, 2000.
5 Lau, J. H., et al, Electronics Manufacturing, McGraw-Hill, 2003.
6 Chapman, S., Fundamentals of Microsystems Packaging, McGraw-Hill, 2001.
7 Tummala, R. R., Microelectronics Packaging Handbook: Technological Drivers, Springer, 1997.
8 Rice, R. W. Ceramic Fabrication Technology, Marcel Dekker, November, 2002.
9 Pecht, M., Handbook of Electronic Package Design, Marcel Dekker, 1991.
10 DeForest, W. S., Photoresists-Materials and Processes, McGraw Hill, 1975.
11 Mullen, J., "How to Use Surface Mount Technology", Texas Instruments, Inc. 1984.
12 Ikegami, A. and T. Yasuda, "High Thermal Conductive SiC Substrate and Its Applications", 5th European Hybrid Microelectronics Conference, ISHM, Reston Va., pp. 465-471, 1985.
13 Tummala, R. R. and R. B. Shaw, "Ceramics in Microelectronics," Ceramics International, Vol. 13, pp. 1-11, 1987.

14 Holmes, P. J. and R. G. Loasby, Handbook of Thick Film Technology, Electrochemical Publications Ltd., Glasgow, Scotland, 1975.
15 Robertson, C. T., Printed Circuit Board: Designer's Reference Basics, Prentice Hall, Upper Saddle River, NJ, 2004.
16 Harper, C., Electronic Materials and Processes Handbook, 3rd edition, McGraw Hill, New York, NY 2004.
17 Lee, Y. C. and W. T. Chen, Manufacturing Challenges in Electronic Packaging, Chapman Hall, London, UK, 1996.
18 Rosato, D. V., et al, Plastics Engineering Manufacturing and Data Handbook, Kluwer Academic Publishers, Norwell, MA 2001.
19 Schroder, C., Printed Circuit Design Using AutoCAD, Butterworth-Heinemann, London, UK, 1996.
20 Blackshear, E. D. et al, "The Evolution of Build-up Package Technology and Its Design Challenges," IBM Journal of Research and Development, Vol. 49, No. 4/5, pp. 641-659, 2005.

EXERCISES

6.1 Prepare a block diagram showing the key steps in the production of a printed circuit board. Begin with the design engineering layout and track the product flow to inventory.

6.2 Plot a line using the principles of raster graphics and comment on the quality of the line if it is along the x axis, along the y axis and along a 45 degree diagonal. Comment on the quality of the image produced as a function of the points per unit length used in the plot.

6.3 Plot an equilateral triangle using the principles of raster graphics. Comment on the quality of the image produced as a function of the points per unit length used in the plot.

6.4 Plot a circle using the principles of raster. Comment on the quality of the image produced as a function of the points per unit length used in the plot.

6.5 Prepare a cost analysis showing the cost trade off between buying and operating a photoplotter and using a service bureau to prepare the required artwork. Go to the Internet to ascertain the costs as required to support your analysis. Also state all of the assumptions used in your cost analysis.

6.6 A master film was produced for a fine line circuit. The film was produced at a temperature of 20 °C and 45% relative humidity. A new employee stores the master in a room where the temperature reaches 24 °C and the relative humidity increases to 74%. If a circuit layout is 150 by 200 mm in size, discuss the errors that would occur if this master is used in its present state to produce a phototool to be used in production.

6.7 Describe the differences between negative type photoresist and positive type photoresist.

6.8 Design a frame to hold the laminate in a photoresist printing operation. The frame should incorporate some feature that will permit registration of the image on both sides of the laminate. Discuss the principle of operation for your registration technique.

6.9 Describe the difference between dry film and liquid photoresist. Cite the advantages and disadvantages of each product.

6.10 Describe the precautions used in developing photoresist with an exposed image of the circuit.

6.11 Discuss the implications of the lack of selectivity in the etching process.

6.12 An etching process is yielding product with an etch factor $F_e = 0.6$. Determine the width of lines to be used in the phototool if the line width on the board is to be 5 mil (0.125 mm). Consider copper cladding of weights 1/8, 1/4 and 1/2 oz/ft^2.

6.13 What are the factors that determine the minimum line width and spacing that can be achieved with the photoresist process as it is used in the production of PCBs.

6.14 What is often the largest source of error in producing accurate and correctly placed surface features on a PCB? Discuss all of the sources of error.

6.15 Describe the types of circuit boards that are produced with punched holes. Give some characteristics of the products likely to use this type of PCB.

6.16 Prepare a graph showing the diameter of the capture pad and the thickness of the annular ring on a PCB as a function of the via hole diameter d. The design rule for the layout indicates that the thickness of the annular ring is (a) 1/2 d, (b) 1/3 d and (c) 1/4 d.

6.17 Write the equations needed to perform a tolerance analysis for hole break out in a PCB. Before beginning, list all of the factors that contribute to the error accumulation.

6.18 A drilling station is capable of operation at an aspect ratio of 8 without excessive drill breakage. Suppose you are planning to drill 12 mil (0.30 mm) diameter holes in circuit boards that are 25 mil (0.625 mm) thick. How many boards do you place in the stack to be drilled simultaneously?

6.19 What is a microvia? Describe how the holes are made for these small diameter vias. Discuss the advantages and disadvantages of the three methods used to produce these small diameter holes.

6.20 Describe the process used to produce the two sided epoxy-glass laminate.

6.21 Describe the lamination process used to fabricate multilayer circuit boards. Indicate how registration between the layers in each board is maintained. Discuss control of the stack during the elevated temperature curing cycle.

6.22 What are the advantages of using a vacuum press in the lamination of multilayered circuit boards?

6.23 What is a high density integrated printed circuit board (HDI-PCB)? Describe several different constructions using this design approach.

6.24 Write a engineering paper based on Reference [20] that describes the sequential build-up technology. Cite the advantages and disadvantages of this design approach.

6.25 What materials are used in the outer layers of a HDI-PCB? What is unique about these materials?

6.26 Describe at least three types of HDI-PCBs.

6.27 Why is the plating process used so frequently in the production of PCB's?

6.28 What factors control the deposition rates in a plating process?

6.29 Describe the electroless plating process showing the essential differences with the electrolytic plating process. What is the role of the stannous and palladium ions in this process?

6.30 Describe the immersion plating process used to coat copper surfaces with tin and gold.

6.31 Describe the panel and pattern plating processes. List the advantages and disadvantages of both processes.

6.32 Solder is an alloy containing tin, lead silver, and other elements. How can two or more of these elements be plated to form a coating that is composed of the proper amounts of each element?

6.33 How will the new EU regulations affect the use of lead in the design and manufacture of electronic systems?

6.34 Describe the hot air soldering level process.

6.35 Show the price of gold with respect to time beginning with 1932. What are the most important factors that affect the price of this material?

6.36 What are the two most important properties of gold that warrant its use in the design of PCBs and connectors?

6.37 Describe the advantages of using photo polymer films for solder masks.

6.38 Describe the advantages of using liquid polymers and the screening operation for the application of solder masks.

6.39 List the defects that could occur if a PCB was soldered using the wave soldering process without the benefit of a solder mask.

6.40 Describe the screen printing process. What materials are usually applied to a PCB using this process?

6.41 What are the typical applications of flexible circuits? List at least three.

6.42 How are flexible circuits employed in producing stacked first level chip carriers.

6.43 Describe the processing steps used in manufacturing a typical flexible circuit.

6.44 Compare the advantages and disadvantages of printed circuit boards of the organic (glass epoxy) and ceramic types.

6.45 Discuss the common substrate materials for ceramic PCB's. Include in your discussion the nitrides and carbides.

6.46 What are the three common methods used to form surface features on ceramic or silicon substrates? Briefly describe each of these methods.

6.47 Describe the process for manufacturing "green" sheets of ceramic. What constituent gives the sheets strength and flexibility?

6.48 Describe the manufacturing process used to produce thicker plates of ceramic materials.

6.49 Describe the primary difference in the firing process for conductor pastes made from the noble metals and those made from copper.

6.50 Why would you prefer to place your resistances on a circuit board by direct firing rather that by placement of discrete components?

6.51 Describe the manufacturing process for producing a multilayer ceramic circuit board using thick film technology.

6.52 Describe the manufacturing process for producing a multilayer ceramic circuit board using co-fired technology.

6.53 Describe the manufacturing process for producing a multilayer silicon circuit board using thin film technology.

6.54 Describe the two design approaches for producing glass-to-metal seals. Which is the preferred approach and why?

CHAPTER 7

ELECTRONICS MANUFACTURING CHIP CARRIER TO SUBSTRATE

7.1 INTRODUCTION

The manufacture of an electronic product using chip carriers, passive devices and printed circuit boards as components requires the careful control of a large number of sophisticated processes. To begin the assembly process, all of the components to be included in a specific product or products are inspected to insure that they meet the specifications necessary for the successful operation of the product over its lifetime. The printed circuit boards are readied for assembly with the application of solder paste over those pads that will receive components with surface mount leads or solder balls. At this stage of the process, the PCBs are inspected to insure that the solder paste has been applied to all of the pads and that it is the correct thickness and consistency. Next the components are placed at their prescribed location using pick and place machines that operate with high placement rates and with high accuracy. The components are then soldered in place using one or more soldering processes that will be described in detail later in this chapter. After the soldering operations, the PCB is cleaned and an adhesive is inserted under those components requiring underfill. Finally the circuit board is either conformal coated or potted to protect it from corrosion and other abuses if it is to operate in harsh environments.

Quality assurance is a significant factor in ensuring satisfactory performance of the PCBs and the entire electronic assembly. There are many aspects to well managed quality assurance programs. Quality assurance requires inspection of incoming components and frequent inspections of the processes used as the board moves along the assembly line. Statistical quality control is employed to monitor and adjust the operating parameters of most manufacturing processes. Finally, screen tests at the board and system level are conducted to insure that the individual subsystems and the entire system are performing in accordance with the specifications. The objective is to ship a defect free product that performs satisfactorily in the field over its specified life.

7.2 MOUNTING TECHNOLOGIES

There are three different methods used to mount the components, both active and passive, to the printed circuit board. These mounting concepts were defined in previous chapters, but they will be reviewed in this section to focus on the three markedly different typed of leads encountered in assembly processes. The difference in the chip carrier leads often result in mixed package types on a PCB, which introduces difficulties in the soldering processes used to attach components and make the electrical connections. The first type of attachment involves through-hole chip carriers and passive devices with wire leads either axial or radial. Attachments of this type are made by inserting the leads into plated through holes, cutting and clinching them as shown in Fig. 7.1. The attachment is made with solder joints that fill the plated through holes and form fillets on both sides of the circuit board. A wave soldering process, which will be described later in this chapter, is used to form through hole solder joints. While through-hole mount is still common, it is being replaced by leaded surface mounted chip carriers and leadless components including ball grid arrays.

Fig. 7.1 Through-hole mounting of a chip carrier to a printed circuit board.

The second type of attachment is with surface mounted, leaded chip carriers as illustrated in Fig. 7.2. In this design, chip carriers with either gull wing or J leads are placed over copper pads that have been coated with solder paste. When the paste dries it serves as a weak adhesive and holds the chip carriers in place as the solder joints are made in a reflow soldering operation. In some cases, a wave soldering process is used to form the solder joints with leaded surface mounted chip carriers, but the chip carriers must be bonded in place with a strong adhesive to withstand the force of the solder wave. The solder joints form generous fillets between the solder pads and the leads.

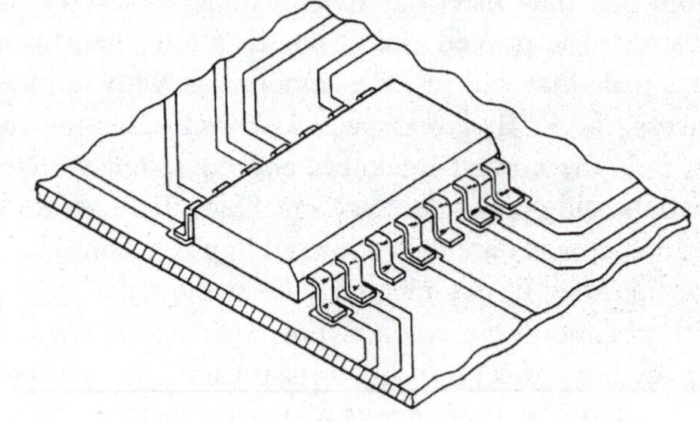

Fig 7.2 Surface mounting a gull wing leaded chip carrier to a PCB.

Passive devices, resistors, capacitors and inductors are often packaged without leads. These leadless packages have metallized ends to provide soldering surfaces. They are placed over copper pads that have been coated with solder paste. The solder joints are made in a reflow soldering operation at the same time as the leaded components are soldered. An example of the generous fillet at the connection to the PCB that is formed in the reflow solder process is shown in Fig. 7.3.

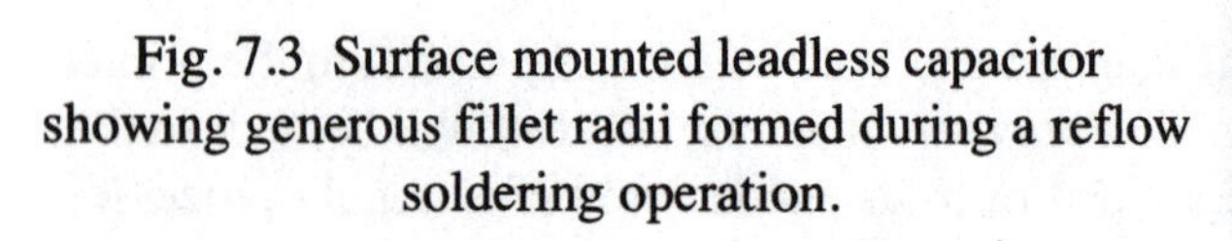

Fig. 7.3 Surface mounted leadless capacitor showing generous fillet radii formed during a reflow soldering operation.

The ball grid array (BGA) and its cousin the chip scale package (CSP) are also surface mounted. The solder balls are deployed in an area array under the surface of the chip carrier. While the solder balls exhibit some compliance, they are not nearly as flexible as the gull or J leads. However, they are more compliant than the leadless packages. An illustration of the solder balls on a BGA, presented in Fig. 7.4, shows the solder balls are made from high leaded solder (95 to 97% Pb). The solder balls are attached to the chip carrier with a solder that melts at a much lower temperature than the high lead solder. To attach the BGA, it is positioned over the array of copper pads that have been coated with solder paste. The soldering is preformed in a reflow operation. The temperatures achieved in this soldering operation cause the lower melting point solder to melt, flow and to form fillets about the solder balls. The solder balls remain solid and serve to separate the chip carrier from the printed circuit board. When the lower melting point solder melts, surface tension forces are generated that tend to center the solder balls over the solder pads correcting small placement errors.

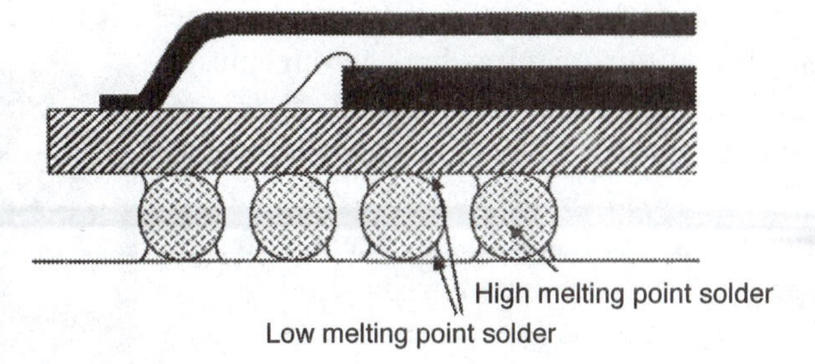

Fig. 7.4 Attachment of a BGA with solder balls to a printed circuit board with low melting point solder.

7.2.1 Solder Screen Printing

The screening process, illustrated previously in Fig. 6.23, utilizes a fine mesh screen usually fabricated from a thin sheet of stainless steel with a very dense array of small diameter holes. A stencil film covering all of the holes on this fine mesh screen, except those over the pads on the PCB to be soldered, is bonded to the screen. The screen with stencil attached is mounted in a frame that keeps it tightly stretched. The screen is lowered, in registration, over the circuit board and solder paste is forced through the screen with a wiper tool called a squeegee. The openings in the stencil are in alignment with the copper pads and the solder paste is transferred in a controlled manner to cover these pads with a thin layer of solder paste. The solder paste is transferred to the pads when the screen is lifted from the PCB.

The area of solder paste deposited on the pads is controlled by the size and shape of openings in the stencil and the solder paste thickness is controlled by the combined thickness of the stencil and the screen as well as the viscosity of the solder paste. After screen printing, the solder paste is usually not permitted to dry, because it serves as a weak adhesive to hold the chip carriers and passive devices in place prior to reflow soldering.

The alignment of the paste with the pads, its thickness and its flow when the screen is lifted is checked periodically either by a well trained operator or by an inspector. The tackiness of the paste is also checked to determine if it is sufficient to bond the chip carriers that will be placed in the next processing step.

Solder paste can also be applied with pick and place machines that are equipped with a syringe to dispense a drop of solder paste over the solder pads as illustrated in Fig. 7.5. While pick and place machines are fast (several thousand executions per hour), they cannot compete with the throughput of a well designed and operated solder screening machine.

Fig. 7.5 Pick and place machine used to dispense solder paste or adhesive at specified coordinates.

7.2.2 Component Placement

As the name implies, the task of component placement involves selecting the correct chip carrier or other device, orienting it correctly and placing it onto the circuit board so that its leads cover the solder coated pads. This is a simple task, but it is much more difficult to execute because of the large number of different types of components to be placed, their density and the relatively tight tolerances required in placement of many of the components. Speed is also essential because competition drives lower costs and that in turn requires high throughput at placement rates up to several thousand components per hour. An example of the different types of components, their orientations and sizes, and their different type of leads is presented in Fig. 7.6.

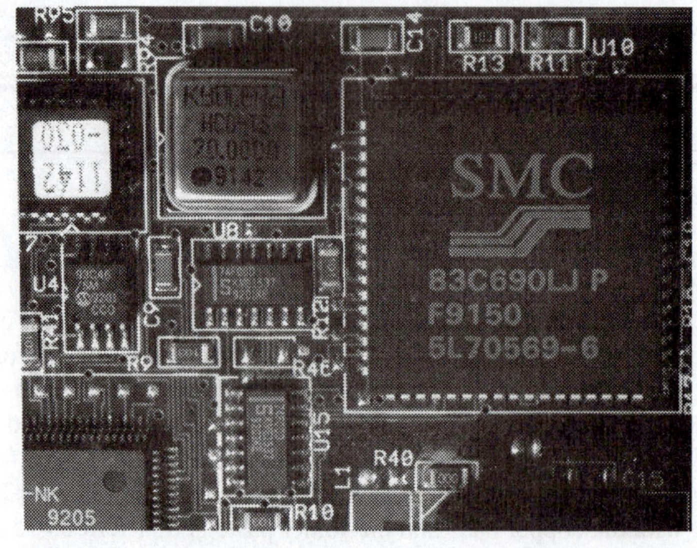

Fig. 7.6 Placement of components using a mixture of packaging technologies on a single PCB.

Placement of components on PCBs is performed for almost all production runs by pick and place machines. These machines vary considerably from model to model and manufacturer to manufacturer and their cost and placement rates vary accordingly. The simplest machine consist of an automated head equipped with an end effecter to acquire and place the components, and a table with a system to hold the printed circuit board at a well defined location. The automated head is equipped to move with four degrees of freedom, permitting linear motion in the x, y and z directions and rotary motion θ about the z axis. The printed circuit board is positioned at a target location on the table by using registration pins or using an imagining system that locates the printed circuit board using fiduciary marks printed on the board. An example of a set of fiduciary marks located near the edge of a PCB is presented in Fig. 7.7. Note the cross hairs used to in the alignment. When the board is located on the table, its footprint is known and the locations of all of the solder pads and plated through holes is specified in a digital format. This positioning data is stored in computer memory.

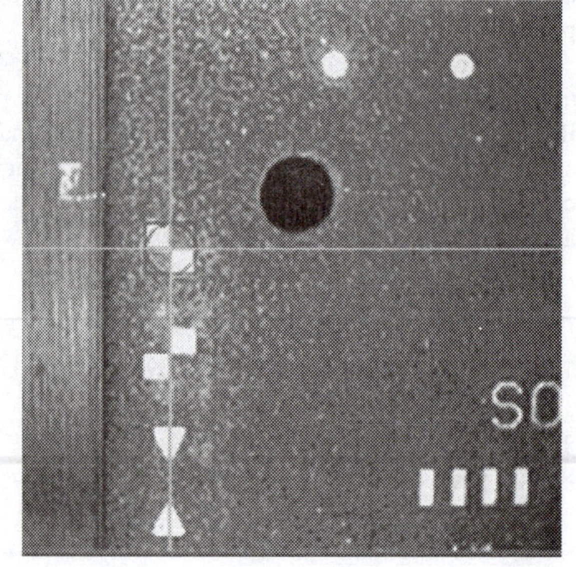

Fig. 7.7 Fiduciary marks on a PCB used to position and align it with the pick and place system.

The pick and place machine is equipped with a number of different feeders containing the various component that are placed on the board. Each component is held in a feeder, which presents the component with the correct orientation at a known location $[(x_0, y_0, z_0, \theta_0)_i]$ around the perimeter of the table. Various component feeder mechanisms used to supply the pick and place machine include:

1. Tubes (called sticks)
2. Reels
3. Standard JEDEC palettes
4. Vibration feeders

The pick and place machine, illustrated in Fig. 7.8, shows several reel feeders located in front and below the table top. In this position, the reels can be easily replaced with a new supply of components. The tape, upon which the components are mounted, is sprocket driven to the component target location $[(x_0, y_0, z_0, \theta_0)_i]$ at the perimeter of the table.

Table 7.8 Pick and place machine with reel feeders in the lower foreground.

The automated head is programmed to move to a feeder point $[(x_0, y_0, z_0, \theta_0)_i]$ to acquire a specific component. The acquisition is usually made with a small vacuum cup attached to the end the z rod. Linear motors are used to move the head to position $[(x_1, y_1, z_1, \theta_1)_i]$, which defines the component's placement location. If necessary, the z rod is rotated by a specified angle θ to align the leads with the solder pads. When alignment is achieved, the z rod is moved downward to position z_2, the leads contact the solder pads, vacuum is released and the component is placed. An example of chip carrier placement with a pick and place machine equipped with a vision system is presented in Fig. 7.9.

Fig. 7.9 Placement of a chip carrier using a vision system.

Some more complex pick and place machine are equipped with multiple heads and multiple spindles for each head. An example of a two head machine with four spindles per head is presented in Fig. 7.10. The additional heads and the multiple spindles permit placement of several components simultaneously. One head can be placing (four components in this case) while the other head is acquiring the next four components scheduled to be placed. However, effective use of such machine requires considerable uniformity in the layout of the footprint to enable simultaneous placement of several components with a single head, because the four z rods are at fixed distances from one another. The advantage of the machine complexity is the significant increase in throughput in the assembly process. The added cost of the machine can be offset by savings in assembly time of a specific PCB if the production volumes are sufficiently large.

Fig. 7.10 A two head, four spindle machine with four at a time placement and four at a time acquisition.

The speed of the pick and place machine is essentially controlled by the time to move the heads and positioning rods. The computer system speeds are sufficiently high to permit the use of fast image processing software to check all connections before placement. For example, suppose one of the solder balls on a BGA is missing. This defect is detected and the component is rejected and a new component acquired. Many pick and place machines are equipped with a laser system that measures component movement from its acquisition position to its place position to insure accurate placement. The computer systems usually employ Widows as an operating system.

Software employed in operating the pick and place machine is critical to successful operation. The software usually contain of three elements. First, the CAD files showing the locations of all the features on the PCB must be converted into a CAM files that are used to drive the pick and place machine. Software that converts these files must also be sufficiently flexible to accommodate the hundred or more (ASCII) data formats used in the assembly of the CAD files.

Second, information must be provided or generated that identifies component acquisition sites and placement sites. In addition information on the sequencing of the placement operation must be provided. On many machines equipped with vision systems, operators locate these positions with cross hairs viewed with magnification on a display panel.

The third software element, included in some more advanced pick and place systems, deals with management information. This software tracks the quantity at each component acquisition position, enables programming with a bar code reader, and maintains records for component traceability.

The placement rate varies with the design of the machine, the number of heads, the number of z rods per head, the component mix, the pitch of the solder pads, the uniformity of the PCB's footprint, etc. However, placement rates of 1,000 to 10,000 components per hour are quoted by the machine manufacturers. The size of the components that can be placed is again machine dependent, but range from 0201 resistors to 75 × 75 mm or 100 × 50 mm. All surface mounted components including leaded and unleaded chip carriers, BGAs, CSPs, and bare dies with flip chip bumps can be placed.

7.3 SOLDER PROCESSES

Soldering is a fastening operation where two metallic pieces are joined together by adding a third metal. It differs from welding where two pieces are joined by heating them until they soften and flow together to make a joint. Soldering and brazing are similar as both operations require a third metal. With brazing, the third metal is usually an alloy containing mostly copper and zinc, with smaller amounts of either phosphor or silver. With soldering, the third metal is an alloy with a lower melting temperature containing metals such as tin, lead, bismuth and silver. Solder is used for plating and plumbing and electrical connections in many applications. The emphasis in this chapter will be on the use of solder for electrical connections in the assembly of PCBs and cable connectors.

The number of solder joints produced annually in the U. S. and abroad is so large it is difficult to comprehend the number. With a production operation that is repeated this often each year, it is imperative to maximize the efficiency of the process and minimize defects which occur. The impact of the recent increase in complexity of the design of a circuit board, with ball-grid-arrays and pitches of less than 1.0 mm, on yield during production is alarming. At a nominal defect rate of 100 parts per million joints (PPM), an increase in the number of solder joints from 5,000 to 10,000 per board can reduce yield from 65% to 38%. Manufacturers of very complex boards with more than 30,000 joints have reported six defects per board, even with their manufacturing processes operating under control.

Defective joints must be identified during quality control inspections and repaired before the board is incorporated in a product and shipped. The cost of a defective solder joint escalates as a function of time after the soldering process is complete. After the board is placed in the system, the defective board must be found, and then the defective joint must be located before repair can be initiated. After the product is shipped, the customer must deal with a product that does not function properly, which is clearly not acceptable under any circumstances. Depending upon the type of product, failure rates have a direct impact upon end costs and a manufacturer's reputation. For example, a manufacturer of cheap products that are not safety related can more easily deal with releasing products with defects. However, a manufacture shipping safety related products, for a high volume product such as an automobile, must insure defect-free product. If just one in 10,000 boards failed in the field resulting in injury or loss of life, a vehicle recall might be mandated at a cost of hundreds of millions of dollars. Expensive litigation is even more probable.

A sharp reduction in the number of defective solder joints can be achieved with markedly improved process control. Defect rates of 10 to 30 PPM are possible. However, achieving these rates requires diligent attention to the details of cleaning prior to soldering, elimination of holes and voids in the solder plate and rigorous control of every aspect of the soldering process. There are several soldering methods employed in the industry depending on the type of solder joint being fabricated or the type of solder coat being applied. We will describe some of the more common methods used in soldering electronic components in the following sub-sections.

Solder alloys for electrical connections should exhibit high strength, high thermal conductivity, low electrical resistivity and corrosion resistance. For the past six decades, the most common alloy used for soldering was a eutectic consisting of 63% lead and 37% tin. This alloy has a low melting point (183 °C), is relatively low cost and can be employed in soldering machines where thousands of solder joints can be formed nearly instantly. However, the use of this alloy is being severely restricted due to the Restriction of Hazardous Substances (RoHS) Directive 2002/95/EC[1], which limits the use of lead, mercury, cadmium, hexavalent chromium, polybrominated biphenyls or polybrominated biphenyl esters

[1] This directive completed on January 27, 2003 effectively eliminates the commonly used solders containing lead from commercial electronic equipment sold in the European Union (EU) as of July 1, 2006. Some exceptions are permitted such as high melting point solders containing more than 85% lead, lead in solders for servers, networking equipment and piezoelectric devices. Electronic products for military applications are also excluded from this directive.

in electronic equipment. Today most manufactures of electronic components and systems with global markets have already switched to lead-free solders.

7.3.1 Reflow Soldering

Reflow soldering is a high-speed high-volume method for making solder joints. With the reflow process, all of the solder joints on the board are made as essentially the same time, although there are may be small time differences due to temperature gradients and the time required for different size components to reach an equilibrium temperature. With this process, the solder is applied to the copper pads on the PCB by plating or by screen printing as previously described. The leads on the components to be placed are also tinned with solder. In effect all of the solder is in place and additional solder is not required when forming solder joints in a reflow process. Flux is employed either directly in the solder paste or as a coating over the solder plated pads.

In the solder reflow process, the PCB is subjected to a thermal cycle with a time temperature profile that includes a pre-heat ramp, a pre-heat dwell time, another ramp to reach soldering temperatures where the solder melts and flows, followed by a rapid cool-down to prevent grain growth during solidification. An example of such a time temperature profile is presented in Fig. 7.11.

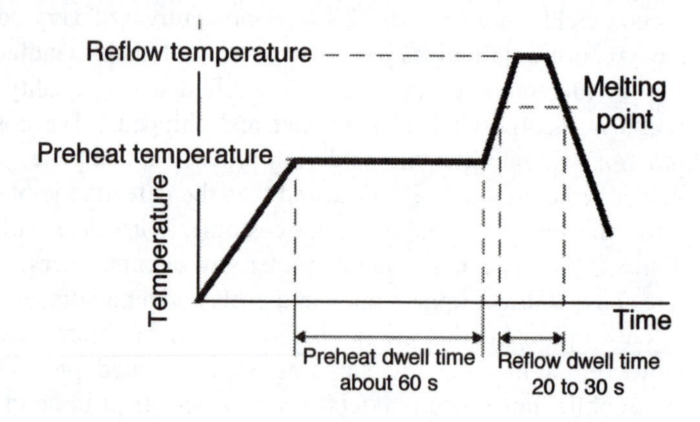

Fig. 7.11 Typical time-temperature profile for a reflow process. Note the temperatures will vary with the solder alloy used for the joints.

7.3.2 Infrared (IR) Reflow

Infrared (IR) soldering is a reflow process that employs an infrared heat source to raise the temperature of the solder until it melts, flows and wets the surfaces to be joined. The IR reflow process is similar to the vapor phase process described in the next section except for the method of transferring heat to the solder joints. The IR process has the advantage of flexibility as its temperature profile can be adjusted simply by controlling the power to the IR heat sources. It has the disadvantage that the temperature distribution over the surface of the board is not as uniform as that achieved with the vapor phase reflow process. The costs tradeoff between IR and vapor phase is nearly equal—capital costs of the IR unit is less than the vapor phase unit, but the operating costs for the IR system are higher on a per board basis than the vapor phase system.

There are three different IR systems in use: lamp IR, panel IR and panel IR with forced air. We will describe only the panel IR systems because they are the most cost effective. The emitting element for a panel IR heater is depicted in Fig. 7.12. Electrical heating elements are located between a radiation plate, which emits IR, and another plate that reflects the IR back into the heating chamber. A thick layer of insulation inhibits heat transfer to the building in which the IR unit is housed. Panel heaters are located above and below the heating chamber. With panel IR units (without forced air) using heaters similar to that shown in Fig. 7.12, about 40% of the heat is transferred to the circuit boards by radiation and 60% by natural convection.

The forced convection IR system is similar to the natural convection panel system except for the addition of duct work and fans enabling more air or cover gas to circulate through the heating chamber. The availability of additional air lowers the temperature of the chamber allowing lower conveyor speeds, more uniform temperature distribution and better control over the process. Also if nitrogen is used as a cover gas, oxidation of the solder and the metallic surfaces is reduced improving the quality of the solder joints. A schematic drawing of an IR system with forced convection is presented in Fig. 7.13. A photograph of a commercial system is shown in Fig. 7.14.

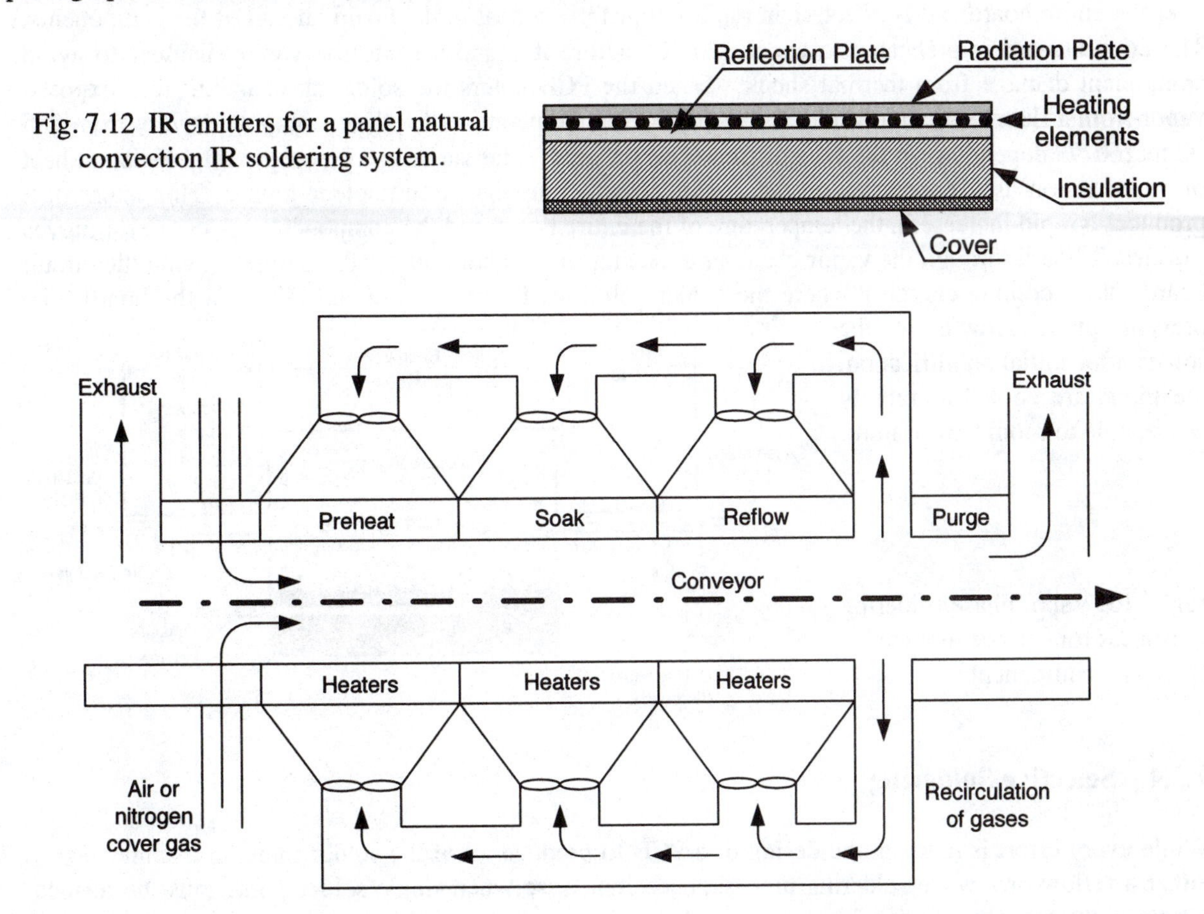

Fig. 7.12 IR emitters for a panel natural convection IR soldering system.

Fig. 7.13 Schematic of an IR system using both radiant heaters and forced convection with a cover gas.

Fig. 7.14 Commercial IR system for soldering surface mounted components using reflow methods.

7.3.3 Vapor Phase

Vapor phase soldering, illustrated in Fig. 7.15, is an excellent process employed to produce high-quality solder joints with surface mounted components. This process achieves excellent temperature control over the entire board and is effective in rapidly transferring heat to the board and all of the components. The circuit board is preheated to about 120 °C, before it is moved into the vapor chamber to avoid component damage from thermal shock. When the PCB enters the soldering chamber, it is exposed vapor from a fluorinert compound. The condensation temperatures for these compounds vary from 155 °C to 260 °C depending on the type of fluorocarbon used. The vaporized Freon gives up its latent heat of vaporization as it condenses on the circuit board, components and solder joints. This latent heat produces a rapid increase in the temperature of the entire board and the components. A conveyor moves the circuit board through the vapor chamber exposing[2] it for about 40 to 60 s before moving the circuit board into a cooling chamber where the solder solidifies forming all of the joints on the board. To prevent grain growth in the solder after initial solidification, the joints are cooled as rapidly as possible to room temperature.

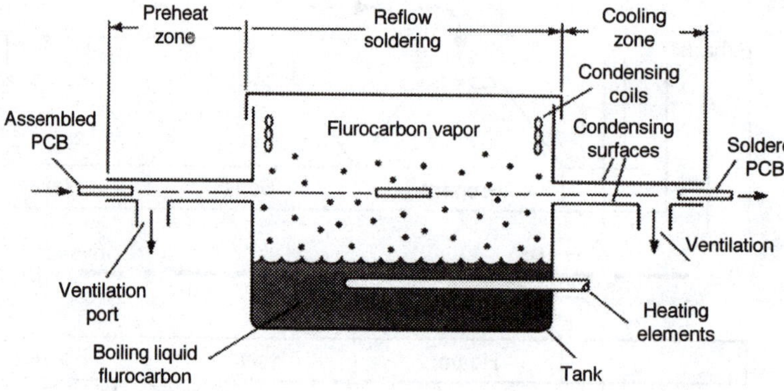

Fig. 7.15 Vapor phase soldering process for surface mounted components.

7.3.4 Selective Soldering

While every effort is made in the design of a PCB to produce all of the solder joints in a single step in either a reflow or a wave soldering process, occasions arise when single solder joints must be formed. Single joints are often made with a small soldering iron equipped with a temperature controlled tip. A line of solder joints, often needed for connecting flat cables or flexible circuits to connectors, is usually made with a hot bar. In some special situations automated machines are used to control either a fine pointed soldering iron or a flame to heat the joints. In all of these selective soldering processes, flux and solder are usually added as the joint is heated.

Hot Bar

Hot bar soldering as the name implies utilizes a soldering iron with a bar-like tip. It is used to simultaneously apply heat to a row of leads or terminals. Its use is limited to soldering leaded chip carriers where the leads are so flexible that they must be constrained during the soldering process. It is also an effective method for soldering flat flexible cables to connectors with pitches as small as 20 mil (0.5 mm). In operation, the leads and pads are pretinned so that additional solder is not required in the joining operation. The purpose of the hot bar is to apply the heat uniformly to a row of terminals in a controlled manner.

[2] The time of exposure to the vapor depends on the type of components on the circuit board. The vapor phase process reduces the time the components are exposed to high-temperatures because the heat transfer coefficient of vapor condensation is about ten times faster than hot air and eight times faster than infrared radiation heating.

In most cases, the hot bar soldering iron is a part of a soldering station that contains a low power optical microscope, and an x-y table. If the soldering is performed on circuit boards, registration pins inserted into tooling holes insure alignment. Feed mechanisms for components to be soldered are sometimes included at the soldering stations.

Point to Point Soldering Methods

For limited production and for repair work, point to point soldering methods are an option. The primary advantage of these methods is the control of the application of heat to local non-critical regions; thus, avoiding damage to temperature sensitive devices. The point to point equipment include induction or resistive soldering irons, micro-flame soldering heads, and laser soldering heads. A photograph of a micro flame soldering head is illustrated in Fig. 7.16. The small torch that produces the micro-flame is on the right side. The solder feed, which is to the left, supplies solder wire through a capillary tube. The head is mounted in a soldering station equipped with an optical microscope and a video camera, and a positioning mechanism to control head motion in the x, y, z and θ directions, as shown in Fig. 7.17.

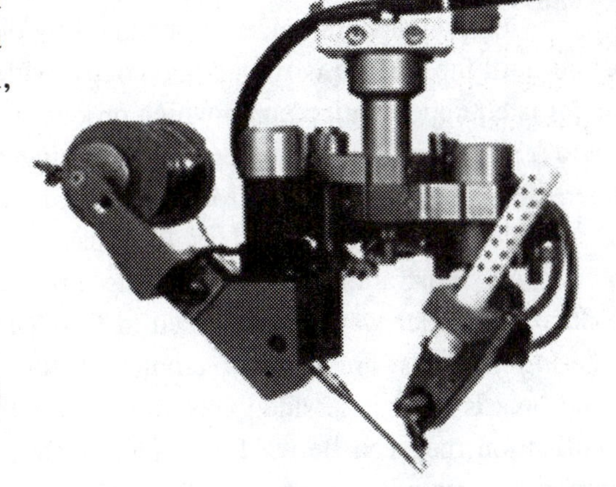

Fig. 7.16 A point-to-point soldering head that utilizes a micro flame torch and a solder wire feed.

Fig. 7.17 A point-to-point soldering station with computer control of the x, y, z and θ position of its head.

7.3.5 Solder Pots and Wave Soldering

A solder pot is used in the simplest form of soldering—pretinning leads. The pot is normally fabricated from cast iron and is heated with electrical elements arranged over the bottom and up the sides to maintain a uniform temperature distribution in the molten solder bath. The pots range in size and may contain only a few pounds of solder for a pretinning operation to several hundred pounds of solder for larger circuit board applications involving wave soldering. The heaters are thermostatically controlled to maintain the solder's temperature somewhat higher than its melting point.

The molten solder will oxidize with time in the pot and form a film called dross that floats on the surface. The dross is detrimental to the soldering process, and to inhibit its formation, rosin flux is added to the pot. The flux forms a liquid film that acts as a surface blanket reducing the exposure of the molten solder to air. Another approach is to remove the dross as it accumulates to restore the surface quality of the molten solder.

Components are soldered or tinned by dipping them into a solder pot and holding them for an instant until the surface is wetted and coated with solder. For producing solder joints on circuit boards, the pot is equipped with a pump which produces a solder wave. The solder wave is adjusted to produce a wide stream of liquid solder that impinges on the underside of the circuit board. Molten solder is supplied to the solder joints on the board as they pass through the wave. An illustration of the pumping arrangement to produce the solder wave is presented in Fig. 7.18. Wave soldering systems are usually employed for soldering circuit boards with underside joints, although with suitable modification they can be used to solder surface mounted components providing they are adhesively bonded to the PCB.

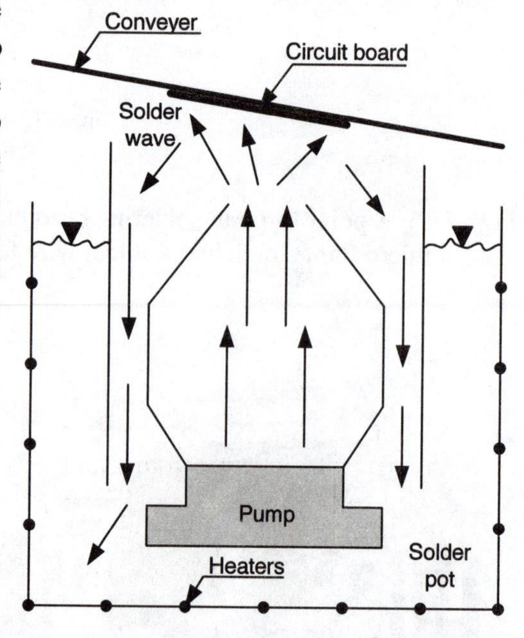

Fig. 7.18 Schematic illustration of a solder wave used to produce solder joints in a single pass.

In the wave soldering operation, a conveyor belt carries the circuit board over the wave at a slight angle to the horizontal axis (3 to 7°) to create an adjustable asymmetric wave. The angle controls the degree of asymmetry and the hydraulic forces acting on the molten solder. The solder leaving the nozzle forms a wave that strikes the underside of the board and supplies a stream of solder to form the joints before the excess solder falls back into the pot. The velocity of the wave impinging on the board should be sufficient to clear the board of flux, wiping the joints clean. As the board continues through the wave, the solder wets the component pins and fills the plated-through-hole. Finally, the board exits the wave and the solder is "peeled back" from the board. The release of the wave is as important as the initial impingement, because the release controls the amount of solder remaining in the joints. Excess solder results in a defect known as bridging where solder covers the space between two capture pads.

The formation of solder joints by transfer of liquid solder from the wave to the through hole type circuit boards is shown in Fig. 7.19.

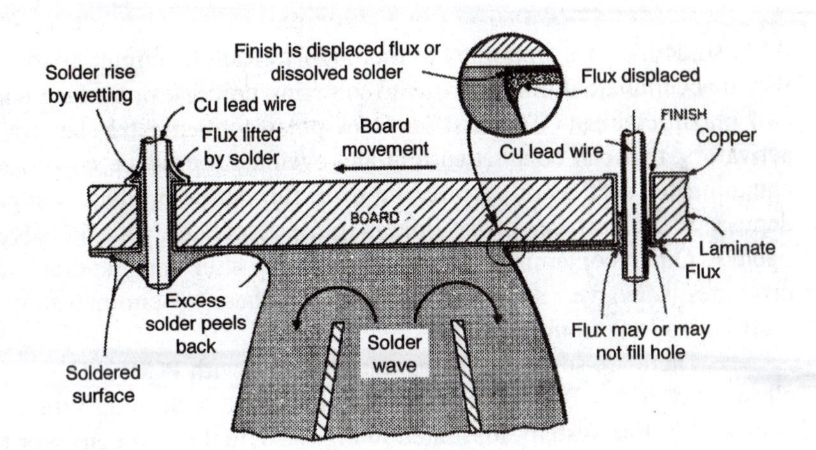

Fig. 7.19 Interaction of a solder wave to form solder joints on circuit boards with components using through hole mounting.

7.3.6 Summary of Soldering Precautions

The soldering process creates thermal stresses on all of the components mounted on a circuit board. In addition, the melting temperature of solder is often higher than the maximum rated operating temperature of the device. Hence, the amount of time the device is at peak soldering temperature should be minimized to reduce the temperature inside the chip carrier thereby improving the chip reliability. The rules listed below should be followed to minimize the thermal stresses to which devices are subjected.

1. Always preheat the device and the PCB. Failure to preheat the device and the circuit board can result in excessive thermal stresses damaging the components.
2. Maintain a temperature difference between the preheating phase and reflow phase of less than 100 °C.
3. The maximum rate of temperature change should be less than 5 °C/s when increasing the temperature from the preheating phase to soldering phase.
5. The maximum soldering temperature and time for wave soldering must not exceed 260 °C for 5 seconds.
6. The maximum temperature in a soldering process should be at least 30 °C higher than the melting point of the solder being used.
7. Rapid cooling after soldering reflow improves the grain size in the solder joints, but if the cooling is too rapid, large temperature differences occur generating high thermal stresses that may lead to solder joint failures.

7.4 POST SOLDERING OPERATIONS

After the soldering operations are complete and the board is subjected to a thorough inspection, it is prepared for assembly. There are three common processes included in post soldering operations that include cleaning the circuit board, applying underfill for BGAs or direct chip attachment and applying a conformal coating. Each of these processes will be described in the follow subsections.

7.4.1 Cleaning Methods

After soldering, it is important to thoroughly clean the circuit board to remove residual surface residues which accumulate during automated soldering processes. Surface contamination is usually divided into two classifications. The first involves polar contaminates, such as plating residues and solder flux activators, that can conduct current and eventually produce corrosion. The second includes non-polar contaminates such as oil, grease and rosin fluxes that produce non-conducting films. These films may deposit on the surfaces of the fingers (pins) associated with the edge connector producing intermittent opens. Other contaminates include body oil, skin and dandruff from workers, oil from placement machines, adhesives, solder balls, etc. All of these contaminants must be removed to achieve a quality PCB with a long trouble free life.

There are several cleaning systems used with PCBs. The most popular are solvent and aqueous cleaning systems. Both methods are similar in that a cleaning fluid is applied with high pressure sprays. A typical in-line system, illustrated in Fig. 7.20, includes a conveyor to carry the PCBs into the cleaning chambers.

Vapor collector
Solvent
Rinse water
Forced air
Spray heads
Conveyor
Sump
Sump
Heaters
Air knife

Fig. 7.20 A three chamber in-line solvent cleaning system.

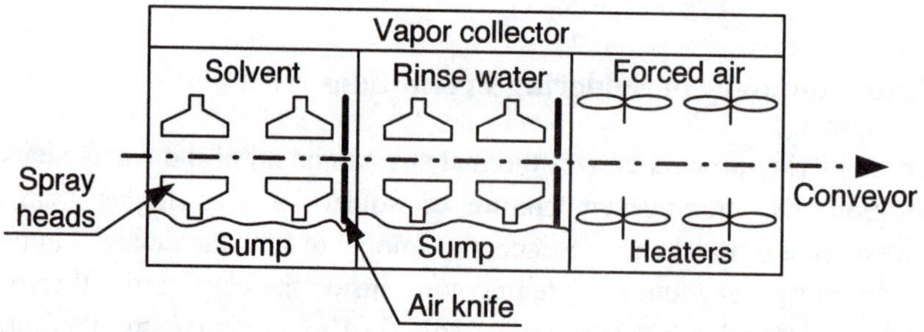

In the first chamber, a solvent is applied with relatively low pressure sprays operating at 20 to 50 psi (172 to 345 kPa). This solvent chemically attacks the flux on the circuit board and the scrubbing action of the sprays dislodges the remaining debris. The solvent falls into a sump and is recycled. The circuit board passes through a pair of air knives that remove the remaining solvent from the surfaces of the board. The solvent cleaning systems are sealed and the vapors are collected, condensed and recycled. The emissions from a solvent cleaning line must be closely controlled to be environmentally compliant.

The second chamber is also equipped with spray nozzles that wash the top and bottom surfaces of the circuit boards with rinse water. The rinse water is also collected in a separate sump and is recycled. Excess water is removed from the board with a pair of air knives located at the exit side of this rinse water chamber. The third chamber is equipped with heater elements and fans that dry both sides of the board.

The aqueous systems are similar in concept to the solvent system shown in Fig. 7.20, except the aqueous line usually has four or more chambers. The first three chambers are for pre-wash, high pressure wash and finally rinse. The fourth chamber is for drying the circuit boards. The pre-washing and high pressure washing are usually performed with saponfier solutions that are very alkaline. Because the pH of the water used for removal of the solder flux is from 10 to 12, it is imperative that the boards be rinse thoroughly and then dried. The effectiveness of the aqueous systems depends on the strength of the surfactants, the design of the spray heads, the size of water droplets delivered, the pressure of the delivery and the variation in the direction of the spray as the board moves through the chamber.

In operation, the systems controls are adjusted to select the temperature of the wash and rinse waters, the dryer oven temperature and the conveyor speed. As the water is recycled, its resistivity is monitored to ascertain if the aqueous cleaners are within the control limits. An example of a commercial cleaning system for printed circuit boards is presented in Fig. 7.21.

Fig. 7.21 A commercial in-line aqueous cleaning system.

While solvent cleaning is usually more effective than aqueous cleaning, environmental controls on the release of the solvent in either liquid or vapor form into the environment complicate its operation.

No-clean Solder Flux

No-clean solder fluxes were developed after the Montreal Protocol went into effect and the use of solvent cleaning of PCBs became more difficult because of environmental regulations. The use of no-clean solder fluxes has been used successfully by a number of manufacturers in full-scale production.

IBM successfully converted from rosin-based flux to a no-clean flux for the flip-chip process that enabled the elimination of perchloroethylene and xylene in the process. The no-clean flux was combined with a mild activator. The reflow soldering was performed in a hydrogen atmosphere, because the no-clean flux and hydrogen both clean solder surfaces and reduce oxidation in this system. The flux almost decomposes completely in the hydrogen atmosphere, leaving minimal residue. This hydrogen-flux combination system was used because the standard no-clean fluxes do not function properly at the high temperature required for flip-chip reflow soldering on a ceramic assembly.

Researchers at three laboratories evaluated a no-clean flux for potential application in military electronic assemblies as a replacement for conventional rosin-based flux. A dilute adipic acid flux coupled with a formic acid vapor with a nitrogen cover blanket and a wave soldering process was used in all tests. Flux was applied to the boards using an ultrasonic spray process. The no-clean process was comparable to the rosin-based control with regards to number of solder defects, ionic cleanliness, solder joint contact resistance, surface insulation resistance, solder joint mechanical strength and long-term storage. Some important issues exist with regard to no-clean fluxes include:

1. Visible residues occur when using many of the no-clean fluxes
2. Cleaning in the soldering area is not eliminated.
3. Using cover gasses, such as nitrogen or hydrogen, improves the quality of the board surface.

Case studies indicate that visible residues that remain on the PCBs after soldering were not detrimental to performance as these PCB's passed reliability tests. It appears that the manufacturer or customer may consider changing their acceptance criterion to adapt to harmless but visible residues. In those situations where visible residues are not acceptable, manufacturers will be required to use cover gasses such as nitrogen or hydrogen.

In some situations, cleaning is still required in the soldering process area. Cleaning is still required at the screening stage for the stencil and for poorly screened boards. Localized solvent cleaning operations are often completed after manual rework to improve the electronic assembly's appearance and to enhance automated contact testing. Also for those companies using wave soldering, cleaning of the flux spray or foam applicators is still needed. In situations where many solder balls are adhered to the PCB, cleaning the board by washing may be required. It is clear then that no-clean solders will reduce but not eliminate cleaning equipment from the soldering operation.

7.4.2 Dispensing Underfill

After the circuit board is clean, underfill is applied to some BGA and CSPs to reduce the thermal strains imposed on the solder ball joints due to temperature cycling. The underfill is an adhesive, usually and epoxy, that has been thinned so that it can flow by capillary action into the narrow space under the chip carrier. The underfill is dispensed with a syringe in a thin line about the perimeter of the chip carrier. Often the underfill is applied by robotic dispensing. The diameter of the bead of adhesive is adjusted to provide sufficient volume of underfill to completely fill the space under the chip carrier. When the capillary action is complete, the underfill adhesive is cured at elevated temperature.

7.4.3 Conformal Coating and Potting

The final step in the post soldering operation is the application of a conformal coating over all or part of the surface of the circuit board. The purpose of this coating is to minimize the degradation in electrical properties of the circuit board due to the long term effects of humidity. Recall that water vapor absorbed by the polymer in the circuit boards lowers its insulation resistance, reduces its high voltage breakdown, increases its loss factor and promotes corrosion. Conformal coatings are blends of polymers, solvents and plasticizers. Acrylics, polyurethane and epoxy are commonly used as the base polymer because of their ease in application and relatively low cost. Silicone and polyimide are also used in higher temperature applications; however, they are more difficult to apply and are much higher in price.

Conformal coatings are applied in thin layers, typically a fraction of a millimeter, onto the printed circuit board and its components. The environmental and mechanical protection they provide extends the life of the components and circuitry. Conformal coatings are usually applied by dipping, spraying and increasingly by robotic dispensing.

Conformal coatings protect electronic printed circuit boards from moisture and contaminants, preventing short circuits and corrosion of conductors and solder joints. They also minimize dendritic growth and electro-migration of metal between conductors. In addition, conformal coatings protect circuits and components from abrasion and solvents. Stress relief is provided by the coating, and it serves to maintain the board's insulation resistance.

7.5 SOLDER MATERIALS

There are a large number of different solder alloys employed in the production of electronic products. The most common until recently was the eutectic tin-lead (63% Sn and 37% Pb) solder. However in recent years most manufactures of electronic components and systems have converted to lead-free solders to meet the requirements imposed by the RoHS directive. The companies producing solders have made available new solder alloys in bar, wire and paste forms that meet the new directive. In the subsections below, the eutectic tin lead alloy will be described in detail because its characteristics and properties are well known. A number of other solder alloys will then be introduced that are compliant with the RoHS directive.

7.5.1 Eutectic Tin-Lead Solder

Solder is an alloy containing some combination of tin, lead, silver, bismuth, copper, antimony, etc. Until recently the most common solder employed was the eutectic alloy containing 63% tin (Sn) and 37% lead (Pb). This alloy has many favorable characteristics that includes its relatively low melting point (183 °C) and its suitability for use in wave soldering machines and reflow ovens. These manufacturing processes enable the simultaneous formation of thousands of solder joints at extremely low cost. The phase diagram for tin-lead solder, shown in Fig. 7.22, shows its melting temperature for the entire composition range of tin and lead.

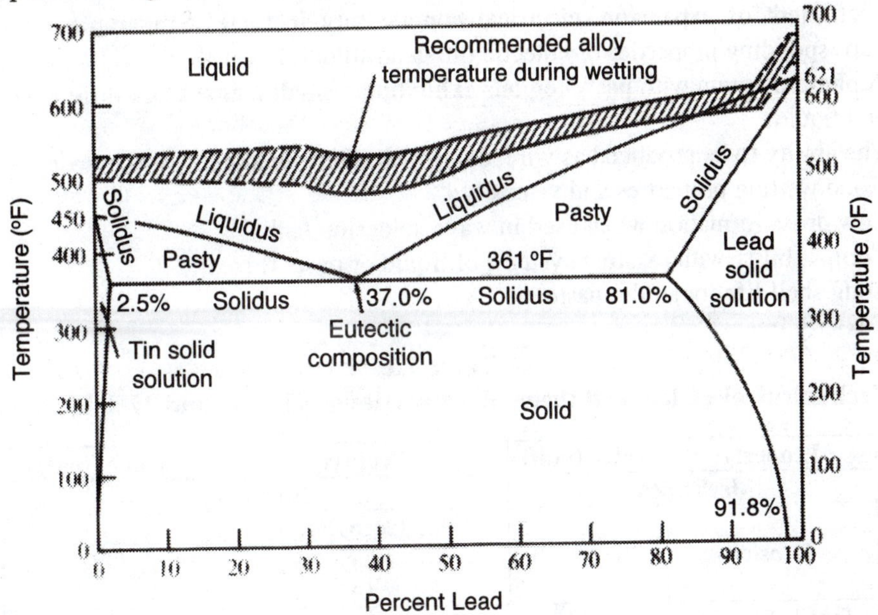

Fig. 7.22 Tin-lead phase diagram.

Let's examine the central regions of this diagram and ignore the solid solutions of tin or lead at its two extremes. For temperatures less than 361 °F (183 °C), the solder is solid. For temperatures slightly greater than this value, the solder is pasty (some solids mixed with some liquid) except at the eutectic composition of 37% Pb and 63% Sn. When the temperature is increased sufficiently, the liquidus line is reached the paste disappears and the solder becomes liquid. In production, the soldering temperatures used are always higher than that indicated by the liquidus line because of the need to compensate for the slight variations in the composition of the solder, the temperature gradients in the solder bath and in the equipment used in performing soldering operations.

High processing temperatures lead to problems in manufacturing. Circuit boards warp, thermal stresses are imposed on the chip and chip carriers, newly formed solder joints are stressed upon cooling, the copper lining in plated-through-holes and other vias are highly stressed, etc. To mitigate these problems, processing temperatures are held to a minimum. The use of eutectic tin-lead solder provided a cost effective means of minimizing soldering temperatures. The changes to lead-free solder will require higher soldering temperatures for most of the lead-free alloys.

Mechanical, electrical and thermal properties of eutectic tin-lead solder are presented in Table 7.1.

7.5.2 Lead-free Solders

In the past few years most companies producing electronic components and systems have engaged in studies of lead-free solders to meet the requirements imposed by the RoHS directive. The companies producing and supplying solders have up dated their product lines making available lead-free alloys in bar, wire and paste forms that meet the new directive.

Several factors to consider when selecting a lead-free solder alloy include:

- Melting and solidus temperatures similar to that exhibited by eutectic tin-lead solder.
- Physical properties such as ductility, tensile strength, thermal conductivity and temperature coefficient of expansion, electrical conductivity that are comparable or better than the corresponding properties of eutectic tin –lead solder.
- A phase diagram with pasty regions exhibiting a small temperature difference between solid and liquid.
- The ability to be produced as wire, controlled diameter spheres and powder as well as paste.
- Good wetting properties and viscosity.
- Low dross formation when used in wave soldering tanks.
- Compatibility with existing systems of liquid or paste fluxes.
- Long shelf life for solder paste.
- Toxicity free.

Table 7.1
Mechanical, electrical and thermal properties of 63% Sn and 37% Pb solder

Property	Value (unit)	Property	Value (unit)
Mechanical		**Electrical**	
Density	7.4 (g/cm^3)	Electrical Conductivity % IACS	11.9
Tensile Strength	30.6 (MPa)	Electrical Resistivity	14.5 (μΩ-cm)
Yield Strength	(27.2 MPa)	**Thermal**	
Total Elongation	48%	Thermal Coeff. Expansion	24-25 × 10^{-6} /°K
Elastic Modulus	34-40 (GPa)	Thermal Conductivity	50.9 [W/(m-°K)]
Poisson's Ratio	0.4	Melting Temperature	183 °C
Shear Strength	27.4 (MPa)	Wetting Angle	14 – 16°
Fatigue Strength* at 20 °C at 100 °C	16.2 (MPa) 10.2 (MPa)		
Creep Strength at 20 °C at 100 °C	3.3 (MPa) 1.0 (MPa)		
Hardening Exponent	0.033		

* This composition was 60% Sn and 40% Pb.

It is widely recognized that a single lead-free alloy to replace eutectic tin-lead solder does not exist. However, there are a number of alloys commercially available that can be employed depending on the application. Studies have been made to select the most suitable alloy for specific applications. As a result, there has been an increasing consensus in the industry for using Sn-Ag-Cu alloys to make electrical connections with surface mounted chip carriers. The large product-oriented telecommunications industry has indicated a preference for this alloy. In other applications, factors such as temperature compatibility and/or cost may dictate the selection of other alloys. For example, the lower melting point Sn-Ag-Bi alloys may be used for surface mount consumer products if high processing temperatures are detrimental to one or more components mounted on the PCB. Sn-Cu solders, without the addition of silver that adds markedly to the cost, are suitable for wave soldering applications when alloy economy is a major issue. Patent issues remain another important consideration because several Sn-Ag-Cu alloys are patented—Castin[3] solder is an example. It is possible to employ these patented alloys by licensing, but this requirement adds to the cost. A slightly different Sn-Ag-Cu

[3] This solder composition of 96.2% Sn, 2.5% Ag, 0.8% Cu and 0.5%Sb is covered by Aim's Patent No. 5,525,557.

alloy[4] can be employed that appears to be patent-free. While the movement to lead-free solders is relatively new, early data from the field suggests that lead-free solders will provide reliable connections. The strength and durability of Sn-Ag-Cu alloys appear to be equivalent to that of tin-lead alloys if the processing is controlled within specified limits.

A brief description of several of the alternative lead-free solder alloys is given in the paragraphs below:

Sn/Ag ***96.5% Sn and 3.5% Ag*** ***Melting Temperature 221°C***

This is a eutectic alloy of tin and silver that exhibits good strength and reasonable wetting of copper surfaces. It is used in connecting copper pipes in plumbing applications and as a solder in electronics. When compared to the tin-lead eutectic solder it exhibits a relatively long life in cyclic fatigue tests. Fatigue damage in tin-lead solder alloys is accelerated at elevated temperatures. In tin-lead solders, high solid solubility of lead in tin causes micro-structural defects due to grain growth with time at operating temperatures. Regions of inhomogeneous grain growth generate crack initiation sites. After crack initiation, fatigue cracks grow under the action of cyclic thermal fatigue producing solder joint failures.

The tin-silver solder alloys exhibit limited solid solubility of silver in tin increasing its resistance to grain growth. Consequently this tin-silver eutectic forms a more stable and reliable grain structure than tin-lead solders. Although the grain size in the tin-silver eutectic is stable, it is subject to diffusion of copper from the circuit board into the high tin content of this alloy. When sufficient copper is diffused into a solder joint, a brittle Cu_6Sn_5 intermetallic compound forms. A crack often initiates in the solder joint in these brittle intermetallic compounds. To reduce the diffusion rate of copper into the tin, the copper pads on the circuit board are plated with immersion gold (Au over Ni over Cu). The nickel in the immersion gold plating serves as an effective diffusion barrier preventing the migration of the copper.

Sn/Ag/Cu ***95.5% Sn, 4.0% Ag and 0.5% Cu*** ***Melting Temperature 217-219°C***

The mechanical stability of all solder joints is degraded at elevated temperatures as the solder's melting point is approached. Elevated temperature cycling produces more damage for tin-lead eutectic solder (with a melting point of 183°C) as compared to higher melting point tin-silver solders (with melting points of 216 to 227 °C). The higher melting temperatures of tin-silver solders improve their performance when operating at temperatures up to 175 °C.

The tin-silver solders do not wet copper as well as tin-lead solders using commercial fluxes. However, excellent fillets can be produced when fluxes developed for the higher temperature soldering processes are employed. Copper dissolution and the formation of a brittle intermetallic is a concern and immersion gold is often required to limit diffusion of the copper into all of the tin rich solder alloys.

Sn/Cu ***99.3% Sn and 0.7% Cu*** ***Melting Temperature 227 °C***

This lower cost solder alloy is suitable for high temperature applications required by the automotive industry and others. It is a solder alloy without either lead or silver content. Elimination of the lead meets the RoHS directive and elimination of silver decreases the cost of this alloy relative to others. As expected, early testing conducted with this alloy has shown significant improvements in creep/fatigue strength over standard tin-lead solders.

---

[4] The Engelhard and Oatey patent on the solder composition with 0.05 to 3.0% Ag and 0.5 to 6.0% Cu with Sn in remaining portion has expired.

Sn/Ag/Cu/Sb 96.2% Sn, 2.5% Ag, 0.8% Cu and 0.5% Sb Melting Temperature 217-220 °C

This alloy has similar mechanical properties and reliability characteristics to the tin-silver-copper alloy described above. However, there is some concern regarding the toxicity of the antimony present in the alloy.

Sn/Ag/Bi 91.8% Sn, 3.4% Ag and 4.8% Bi Melting Temperature 200-216 °C

Bismuth is added to the tin-silver solder alloys to lower their melting point. Another benefit of bismuth is greater joint strength indicated by ring and plug testing. Tests at Sandia National Laboratories have found no electrical failures on surface mount devices following 10,000 thermal cycles using many components mounted on standard FR-4 circuit boards and cycled at temperatures from 0 to 100°C. However, if the pads on the circuit board or the leads on the components are plated with lead, a tin-lead-bismuth ternary eutectic compound forms at 96°C that may degrade the cyclic thermal fatigue strength of the solder joints.

Sn/Sb 95% Sn and 5% Sb Melting Temperature 232-240°C

This solder is a solid solution of antimony in a tin matrix that has a relatively high melting point, making it suitable for high temperature applications. The antimony imparts strength and hardness to the tin. In comparing the yield strengths of several solder alloys, the strength of the tin-antimony solid solution was 37.2 MPa, which is comparable to the tin-silver eutectic solder. The formation of a brittle intermetallic compound Sb/Sn in this alloy is possible. The wetting behavior is significantly less than the tin-lead and tin-silver eutectic solders. Also the toxicity of Sb may become an issue in the future.

In/Sn 52% In and 48% Sn Melting Temperature 118°C

The low melting point of this indium-tin alloy is attractive if the applications do not involve higher temperatures. Indium displays good oxidation resistance, but it is susceptible to corrosion in a humid environment. It is also a very soft metal and has a tendency to cold weld. As expected this alloy exhibits poor high temperature fatigue behavior, because of its low melting point. The high indium content with its high cost limits the use of this alloy.

Au/Sn 80% Sn and 20% Au Melting Temperature 280°C

This gold-tin eutectic solder is a very strong, high-modulus material because of the formation of brittle intermetallic compounds. It is usually employed in attaching dies to lead frames or ceramic chip carriers. Problems of cracked dies have been observed with this solder when the die size is large and when a large mismatch in the temperature coefficient of expansion exists between the die and the lead frame. The high price and high quantity of gold in this alloy restricts its use in applications where cost is a factor.

7.5.3 Selecting a Lead-free Solder

From the many lead-free solder alloys available, industries in the U. S., Japan and Europe have moved toward compositions containing tin-silver-copper. In the U. S., NEMI[5] has recommended a lead-free solder composition consisting of 95.5% tin, 3.9 % silver and 0.6% copper. This composition differs slightly from the tin-silver-copper eutectic that contains 95.6% tin, 3.5% silver and 0.9% copper. The NEMI recommended alloy contains slightly more silver than the eutectic composition to account for the dilution of the solder joints from the tin rich plating on component leads that probably will be used by manufactures after RoHS is fully implemented. Other lead-free solder alloys that merit consideration for different applications are listed with their compositions and melting temperature in Table 7.2.

Table 7.2
Lead-free solder compositions.

Alloy Name	Castin	Tin-Silver Eutectic	Tin-Copper Eutectic	Tin-Silver-Copper Eutectic	NEMI Composition
Composition %	96.2 Sn, 2.5 Ag 0.8 Cu, 0.5 Sb	96.5 Sn, 3.5 Ag	99.3, 0.7 Cu	95.6 Sn, 3.5 Ag 0.9 Cu	95.5 Sn, 3.9 Ag 0.6 Cu
Melting Temp. (°C)	216	221	227	217	218

The melting temperatures of the four tin-silver solder alloys are nearly the same with a range from 216 to 221 °C. The tin-copper eutectic has a higher melting temperature (227 °C), but it is less expensive because of the absence of silver. The tin-copper eutectic is often used for wave soldering, where the chips are shielded from the solder by the packaging and the time of their exposure to the high temperature solder is limited to a few seconds.

The melting temperatures of the different tin-silver-copper solders, listed in Table 7.2, are similar because the change in their composition from one to the other is relatively small. Hence, the mechanical properties listed for the tin-silver eutectic solder in Table 7.3 may be considered representative of this class of lead-free solders.

Table 7.3
Properties of tin-silver eutectic solder (96.5% Sn and 3.5% Ag)[6].

Property	Value
Density	7.36 g/cm^3
Yield Strength	19 MPa
Tensile Strength	28 MPa
Total Elongation	16%
Elastic Modulus	26.2 GPa
Shear Strength	
at 20 °C	39.0 MPa
at 100 °C	23.5 MPa
Fatigue Strength	
at 20 °C	17.6 MPa
at 100 °C	10.5 MPa
Electrical Resistivity	7.7 μΩ-cm

[5] The National Electronics Manufacturing Initiative (NEMI) is a consortium of companies conducting research on the use of lead-free solders in the U. S. and elsewhere.

[6] Yield and tensile strength as well as total elongation are strongly dependent on the heat treatment of the test coupons used in the measurement of these properties. Annealed specimens give lower strength and higher elongation that as quenched specimens.

Important Implementation Issues

As the industry transitions from eutectic tin-lead solder to lead-free solder, there are many concerns. Some of these issues are listed below:

- Selection of lead-free solder alloys for wave and reflow soldering operations.
- Higher process temperatures (by about 30 °C) for lead-free solders may cause circuit boards to delaminate and plastic chip carriers to blister.
- Equipment and processes will require modifications to accommodate the higher soldering temperatures.
- Better process control will be required for defect free soldering with lead-free solders.
- The appearance of lead-free solder joints will require new criteria for visual and X-ray inspection methods.
- The availability of components and boards with lead-free finishes may cause delays in production.
- Training will be required to develop rework and repair capabilities.

The most significant concern is with regard to component availability. Many component companies have been slow to react to the impending change to lead-free soldering, probably due to low level demand for lead-free product. However, this situation has improved and RoHS certified announcements are now common.

Another issue is the compatibility with the higher soldering temperatures for many components including: plastic encapsulated chips, capacitors, LEDs and connectors. In many cases, the combination of soldering time and temperature for typical lead-free reflow profiles exceeds that specified on component manufacturers' datasheets. The higher reflow temperatures and faster ramps may result in increased board delamination if the laminates have high moisture content. Experience to date indicates that the number of failures is small; however, the use of lead-free soldering in the industry has been limited. Wide spread use of lead-free solders will require component manufactures to re-qualify their components to 250 or 260 °C in the near future.

A few companies[7] have been supplying components with lead-free terminations for several years using nickel/palladium plating. However, the combinations of lead-free solders and new plating materials on the component terminals will require process development to provide the solderability expected from conventional soldering. Similar wetting performance is obtained only when the lead-free soldering temperature is in excess of 250 °C. In recent years, renewed interest has been shown in developing plating systems using pure tin or high tin-based compositions that inhibit tin whiskers from initiating and growing with time and temperature cycling.

7.5.4 High Lead Solders

There are several exceptions to the RoHS directive that prevents the presence of lead in solder alloys by July 1, 2006. One exception is for the presence of lead in high melting temperature solders containing more than 85% lead. Solder balls used in ball-grid-arrays are often made from 95% lead and 5% tin or 97% lead and 3% tin to provide balls that remain solid at temperatures exceeding 550 °C. These balls are connected to the ball-grid array with lower melting temperature solder by the component manufacturer. When the ball-grid-arrays are placed on a circuit board, solder paste with an even lower melting temperature alloy is employed to make the mechanical and electrical connections.

High lead solders are also used for the bumps found on flip chip connections to both ceramic and organic chip carriers. Ceramics can withstand the high reflow temperature necessary to make the

---

[7] Texas Instruments, Inc. has been plating their lead frames with Ni-Pd coating for more than ten years.

connections of the high-lead solder bumps to the ceramic substrate. However, this is not the case for organic substrates that are temperature limited. With organic substrates, a lead-free, lower-temperature-melting solder paste is necessary to make the connection between the solder balls and the pads on the substrate.

7.5.5 Solder Paste

Solder paste is a mixture of solder alloy powder, flux and vehicle (fluid) that is used to form high quality mechanical and electrical connections at specified soldering conditions in reflow ovens. Solder pastes must have several characteristics to be effective in soldering surface mounted components.

- The paste must be sufficiently fluid to permit it to be applied by screen printing or syringe dispensing.
- After deposition on a circuit board pad, the paste must remain on the pad without spreading.
- The paste must act as an adhesive when components are pressed into it holding them in position while the board is transported by conveyor into a reflow oven.
- The paste must have sufficient fluxing action to clean the surfaces to be joined, facilitating the wetting and flow of solder when it melts.
- The solder particles in the paste must melt when heated to form a single mass of solder that is void free when cooled.

Solder paste is a suspension of solder particles (90% by weight) mixed with a flux and a vehicle. The particles are spherical and their size varies according to the application. When heated, the flux acts to remove oxide coatings from the surfaces of the metals to be joined and to prevent further oxidation during the soldering operation. The flux is comprised of two chemicals: a base and an activator to increase the fluxing action of the system. Until recently, rosin was the most widely-used base. The weakest flux base is the standard rosin (R) that contains only natural rosin without activators. A stronger flux base is rosin mildly activated (RMA), which is a system containing both rosin and a small amount of an activator. Other products employ different flux bases that are synthetic resins or water-soluble resins that enable the circuit boards to be cleaned by washing.

The flux base and vehicle (solvent and additives) determine the viscosity of the solder paste and the allowable processing time prior to placement. Additives such as thickeners and lubricants give the paste its flow characteristics.

The solder particles when manufactured are spherical, as shown in Fig. 7.23, because this shape has the lowest surface area-to-volume ratio. Therefore, for a given oxide thickness, the sphere will support the smallest amount of oxide that might possibly contaminate the solder joint. Another advantage of the spherical geometry is that the spheres have a lower tendency to jam the printing screen or clog the dispensing needle when the paste is applied.

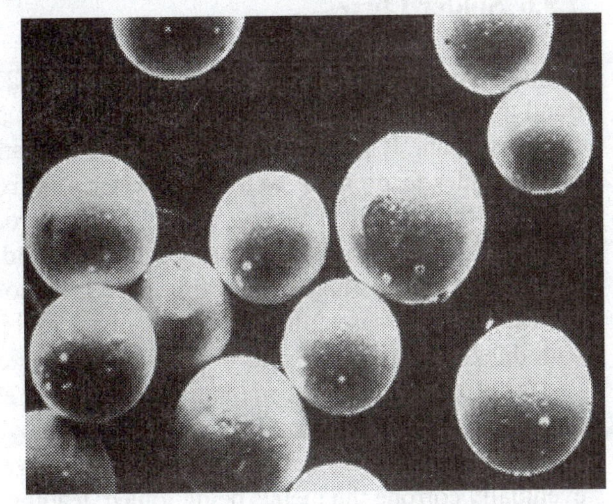

Fig. 7.23 Spherical solder balls constitute about 90% by weight of solder paste.

The particle size used in a paste is a balance between easy application and clear definition of the paste dots versus the problems caused by surface oxide and solder balling. The smaller diameter particles pass easily through the printing screen or syringe permitting the use of finer screens. The finer printing screens or nozzle diameters enable improved definition of the solder dot. However, the small diameter particles increase the amount of oxide that must be accommodated in forming the solder joint. Smaller particles also result in a more solder balling because of spattering in the reflow ovens.

Manufacturers of solder pastes have standardized on a six particle size distributions, which are defined in Table 7.4. The type of powder used in a specific application is determined by printing requirements. In the usual case, the largest particles should be 2.5 to 5 times smaller than the screen/stencil apertures. These apertures in turn are a function of the pitch of the pads on the circuit board. Type I is now obsolete because the particle size is too large for the small diameter holes used in printing screens and syringe nozzles. Type II is still in use circuit boards with pads having a pitch of 30 mil (0.75 mm) or larger. Type III and IV are used with smaller pitches 15 mil (0.375 mm) and 8 mil (0.20 mm), respectively.. Type V and VI, which have even smaller diameter particles, are intended for extreme fine pitch work.

Table 7.4
Powder size in solder paste

Powder Type	Size µm	Dot Diameter	
I	150-75	1.00 mm	40 mil
II	75-45	0.80 mm	32 mil
III	45-25	0.50 mm	20 mil
IV	38-25	0.30 mm	12 mil
V	25-15	0.25 mm	10 mil.
VI	15-5	0.10 mm	4 mil

Solder powders are produced using an atomization process that involves melting the alloy under inert gas to avoid evaporating its components. The liquid solder is fed into an atomization tower where the melt is disintegrated into a cloud of droplets by a high velocity stream of inert gas. The droplets solidify as they drop the height of the tower and are collected with a vacuum system. The solidification process is conducted in an inert atmosphere to prevent oxidation of the powder. After screening to classify the powder according to particle size, it is stored in sealed containers with nitrogen cover gas.

7.5.6 Solder Fluxes

A flux is necessary to chemically prepare the surfaces so that the melted solder will form a metallurgical bond at the solder joint. The flux removes surface contaminates that are usually oxides and leaves the surfaces clean. It momentarily prevents the formation of new oxide films and lowers the surface tension of the solder to promote wetting. The term "flux" is used fairly loosely to describe a wide range of products that exhibit a fluxing action. These products include: thin solvent-rich liquid fluxes, highly viscous fluxes used in formulating pastes and solid rosin fluxes. Either natural or synthetic rosins serve as the base for all fluxes. The flux base is dissolved in a solvent with activators to form thin liquid fluxes. In some cases, the base is dissolved in a small amount of solvent with activators and blended with thickeners and lubricants to form fluxing pastes.

Rosin fluxes are commonly employed in electronic applications. Usually fluxes which contain only rosin or rosin with mild activators are employed on circuit boards. Flux residues are removed after the soldering process because they are corrosive. If the flux residues are not removed, they will eventually corrode the metals being joined, leading to solder joint failures.

The fluxes with mild activators are more easily removed by cleaning than the more active ones. The activators employed are usually halogenated organic compounds that dissociate above their activation temperature to produce agents like HCl (hydrochloric acid), which effectively remove surface oxides. For environmental reasons, the least aggressive flux should be selected to facilitate cleaning the surface of the PCB following the soldering operation.

Traditional fluxes used with tin-lead alloys may not be adequate to overcome the slower wetting of lead-free alloys and the higher temperatures normally associated with lead-free solder alloys. Flux systems specifically formulated for lead-free soldering will require new activator packages and heat stable gelling and wetting agents to avoid solder defects. Due to the slower wetting and higher surface tension of many lead-free alloys, selecting the correct flux for lead-free soldering will prevent the increase of solder defects and greatly assist in maintaining production yields.

7.6 QUALITY ASSURANCE

The quality of each solder joint on a PCB is vitally important. If a single joint fails in service, the board fails and in all probability the systems fails. The result is customer irritation, cost of the repair (if the product is within the warranty period) and loss of prestige as a supplier of high quality, reliable products. As a consequence, manufactures have quality assurance programs in place to insure that the entire system is defect free prior to shipping. These programs differ from manufacturer to manufacture but they all have several common elements including:

1. Inspection of incoming components and materials to insure that they meet specifications and will not cause problems in the assembly process.
2. Inspection of the board after the application of solder paste to insure that pads are covered and the thickness of the paste is within specification.
3. Measuring and monitoring process control parameters such as solder temperature, dwell times, the rate of temperature change with time and cool down time.
4. Monitoring the amount of gold that diffusing into solder joints. A tin-gold intermetallic may form in the solder joint that is brittle and increase the probability of solder joint failure with time in service.
5. Inspection of the board prior to soldering to insure that the placement of the components is with specified dimensional limits.
6. Inspection of the boards after the soldering operation is complete using the appropriate equipment and software to locate any defective solder joints.
7. Statistically monitor the control of the soldering, cleaning and conformal coating processes.
8. Conduct tests on each board to insure its electrical functions.
9. Conduct tests on each system to insure it performs to all specifications.
10. Plan and execute accelerated screening test to determine the expected life of the product before failure by one or more wear out mechanisms.

7.6.1 Inspecting Components and Materials

Most manufactures order hundreds of components and scores of materials that are used in fabricating a circuit board. The vendors (suppliers) usually provide data sheets that specify the parameters which accurately define their product. For example, the chemical composition of the solder is given together with the allowable deviation in each constituent. If the vendor is well known, this data is accepted; however, if the supplier is new, the product is often tested to verify the vendor's data. Also, the company is frequently visited by the manufacture's representatives to ascertain if the supplier has a rigorous quality control program in place.

Mechanical inspection of components is usually focused on the leads and the coatings applied to the leads. This inspection also includes the tubes, tapes or carriers that are used to hold the components prior to assembly. The inspector's role is to reject out-of-specification materials and components that can result in a product defect during or after assembly.

Circuit boards are inspected to determine if they are flat, if the copper cladding thickness is within specified limits and if the cladding is well bonded to the laminate. The copper clad laminates may be subjected to the complete etching process to ascertain if undergo dimensional changes or if they warp to any significant degree. Electrical tests are also conducted to determine the electrical properties of the laminates. These tests are critical because a warped board will prevent uniform contact of the leads on surface mounted components. Changes in dimensions of the board due to the etching process will result in misregistration and cause misalignment of the leads with the solder pads.

7.6.2 Measuring Process Control Parameters

Soldering processes, such as wave and reflow soldering, are well understood and most manufactures have established temperature-time profiles for each process that insures high quality solder joints. In many cases, the control of these processes is automatic. Temperature sensors are placed at critical locations in the equipment, the temperature measured and feedback signals are used to control the heaters so as to vary the temperature of the process according to the established temperature-time profile.

A graph of this temperature time profile is displayed on a monitor. An inspector checks this display periodically to insure that the process remains under control. If for some reason, the process begins to show defective joints and the time-temperature profile is within specification, the soldering operation is halted. An investigation is initiated to ascertain the reason for the production problem. These reasons could include impurities in the solder or flux, aging of the copper cladding, poor cleaning, equipment malfunction, etc.

7.6.3 Monitoring Gold in the Solder Joint

Gold plating of component leads is common because gold plate prevents oxidation of the leads and it provides a reliable electrical contact if the component is assembled into a socket. Pure gold (24 carat) and low alloy metal gold (99%) can be soldered with relative ease. However, gold is highly reactive both with tin and lead at normal soldering temperatures. Reaction of gold with tin causes the formation of several intermetallic compounds, one of which is $AuSn_4$ that melts at about 200 °C. If the gold content in the solder joint is under 20% by weight (which is the usual case) and at temperature above 177 °C, the liquid solder is a ternary Sn-Pb-$AuSn_2$ system. Upon cooling this system transforms into a Sn-Pb-$AuSn_4$.

In wave soldering at 250 °C, gold coatings are dissolved quickly by molten solder. A coating of 2-3 μm may be dissolved completely with a contact time of 1 second in the molten solder. The tin-gold intermetallic compounds that are dispersed into the solder matrix have a needle like shape that is detrimental to solder joints because it forms a preferential cleavage plane.

The preferred approach is to minimize the amount of intermetallics forming by soldering at the minimum possible temperature and the minimum contact time. However, a flash of gold with a thickness of only 0.5 to 1.0 μm will dissolve completely regardless of the soldering temperature or time. For these very thin coatings, a high soldering temperature with long soak time is recommended. This temperature-time profile dissolves the entire gold layer, and disperses the intermetallic compound in the bulk of the solder fillet. Dispersion of the intermetallic throughout the solder joint reduces the probability of failure along an intermetallic cleave plane.

It is well-established that the $AuSn_4$ intermetallic compound is brittle and joints containing these compounds have separated completely when under tension or shear but not compression. It is difficult to predict exactly when a solder joint will be subjected to tension or shear forces; however, when the joint is subjected to either of these forces the failure is immediate and permanent. It is more reliable to use soldering processes that do not allow a layer of brittle intermetallic compound to form.

Any solder joint found with a large presence of $AuSn_4$ should be considered a reliability risk particularly if long life at elevated temperature is required. The types of intermetallic layers that will be susceptible to failure require diffusion and time to diffuse to critical regions. Time is also required for phase transition to occur as the gold interacts with the tin. In these cases, inspection and/or non-destructive screening testing may not detect the problem. An understanding of the factors influencing intermetallic growth and employing soldering processes that ensure that the joint surfaces are not coated with gold is the best approach for avoiding intermetallic failures. The vendor's process should be examined and verified for each lot of components or boards. Destructive physical analysis of samples drawn from the production lot can be used to determine if gold-tin interfaces exist and for the presence of intermetallics forming at the lead-pad interface.

7.6.4 Inspection of the Circuit Board Prior to Soldering

Inspection of sample lots of the circuit boards prior to soldering often identifies production problems before expensive components are soldered in place. After the circuit patterns are etched, the board should be checked for warping and for dimensional stability. Have the pads moved from their intended locations? Do the tooling holes or notches remain an effective means for registration of the board?

The board should also be inspected after the solder paste is applied to the pads and prior to component placement. Is the solder paste in registration with the pads? Is the paste the proper consistency? Is the paste sufficiently thick? Does the paste have sufficient tack to hold the components in place prior to the reflow operation? Well-trained operators should insure that all of the process parameters are within specifications at every step of the process; however, it is the role of the inspector to double check samples of the boards moving through production. If problems occur, the defective boards should be identified early in the process and corrected immediately. Any delay markedly increases the probability of finding more serious defects later in the process when corrections will require more time and be much more costly.

7.6.5 Inspection of the Board after Soldering

The use of smaller components, lead-free soldering, and higher component density requires higher performance from several different types of inspection systems. Each of these three factors has driven the requirement for higher resolution and accuracy from inspection processes. Smaller components such as 0201-size resistors are barely visible with the naked eye. A 10-μm pixel size is required to obtain the magnification necessary for accurate inspection. Currently the 0201 resistor is the smallest component available, but in the future the 01005 size resistor, which is to be 4 times smaller, probably will be incorporated into the design of high density circuit boards. Higher resolution in automated optical inspection (AOI) systems will require optical zoom lenses and a higher pixel-count cameras and monitors to meet future inspection demands.

Lead-free soldering also affects X-ray inspection because of the different absorption characteristics of these new solder alloys. Measurement of solder-paste volume at solder joints must account for slightly different fillet curvatures and a rougher surface finishes that are characteristics of lead-free solders. Higher temperatures are used in the reflow process that increases the probability of board warping and component failures. To alleviate some of the inspection problems, new pad shapes are being designed to aid in measuring soldered-joint integrity.

Increased circuit density results from the trend to use more area-array packages in each assembly. In the past, typical assemblies contained a few BGAs, but now assemblies are designed with tens of these area array packages. In addition, assemblies are becoming more complex with double-sided reflow used to mount mirrored BGAs. The increase in circuit and component densities is affecting throughput of the automatic inspection machines. Inspection of a single PCB involves capturing many optical or X-ray images. Then a large number of computations are performed on each image to ascertain the quality of each solder joint. The speed of the computers and the efficiency of the image analysis software employed are critical in determining the throughput.

Automated Optical Inspection (AOI) Systems

A typical automated optical inspection (AOI) system consists of a digital camera, zoom lens, lighting, computer, software and conveyor (or x-y table), as illustrated in Fig. 7.24. The camera is a critical component in the system because its resolution must be sufficient to clearly view the small features found on high density circuit boards. Moreover, its dynamic range must be sufficient to capture either gray scale or color images that are rich in the content necessary for image analysis. A combination of high magnification lenses and small pixel size provides the resolution. Zoom lenses are often employed because their magnification can be varied to accommodate several different regions of the board. Today CCD digital cameras are almost always the technology of choice. With a CCD sensor, a pixel size of less than 10 μm with a fill factor of 50% or more are required to image features currently found on high density boards.

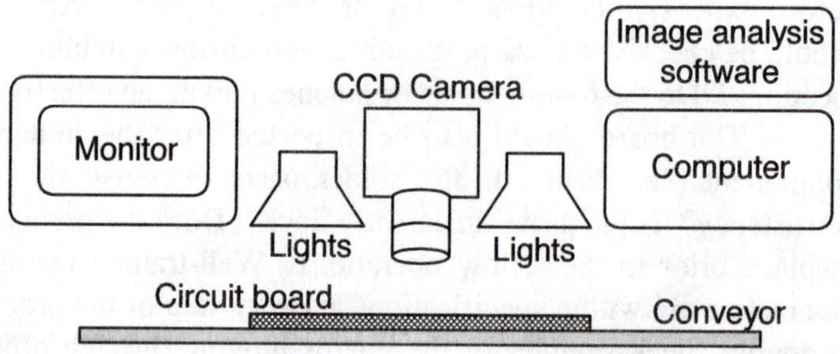

Image analysis software
CCD Camera
Monitor
Computer
Lights
Lights
Circuit board
Conveyor

Fig. 7.24 Components in a typical automated optical inspection system.

Imaging circuit board features requires uniform lighting over the region of the board subjected to image analysis. Otherwise variable lighting intensity will affect the gray scale or color measurements from one or more pixels complicating the interpretation of the image analysis. The computer speed and software efficiency are important in establishing the throughput of the inspection station. Image analysis, performed to ascertain if a particular step in the soldering process is under control, is computationally intensive. High speed computers are required to insure timely analysis and reasonable sample sizes. The monitor is used by the inspector to insure that the camera is focused correctly and that the lighting is uniform over the field of view. The monitor may also be used to view the results of an image analysis. A commercial optical inspection system showing two monitors and optical controls is illustrated in Fig. 7.25.

Software is another critical component in any inspection system. The cameras capture either gray scale or color images with both orthogonal and oblique viewing. Image analysis software is used to determine if the image is showing a production process that is under control or if the process is yielding solder joints with defects. Most image analysis software contains the classic algorithms that are listed below:

1. **Enhancing the image**: These routines involve using filtering tools to sharpen edges, remove noise or extract frequency information. Also image calibration tools are included to remove nonlinear and perspective errors caused by lens distortion and camera placement. Image calibration tools to convert pixel measurements to standard units such as μm are included in these standard routines.

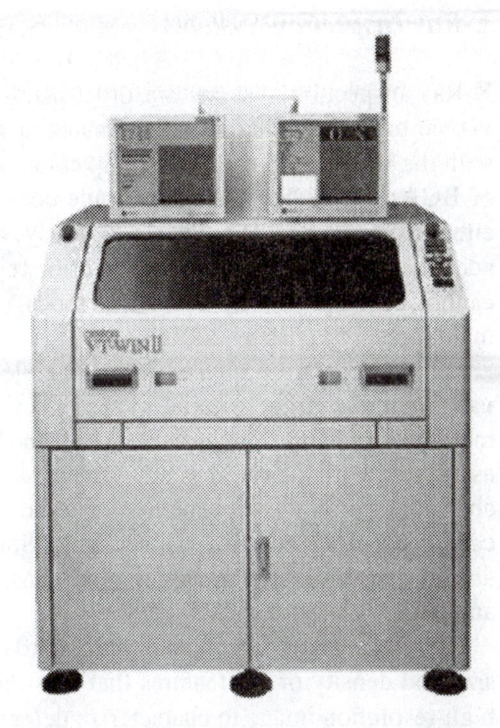

Fig. 7.25 A self contained color vision system for inspecting PCBs after either wave or reflow soldering. Courtesy of Omron Corporation.

2. **Checking for presence**: To inspect for part or feature presence, any of the color, pattern-matching or histogram tools can be employed. The result of a feature check always provides a pass/fail indication.
3. **Locating features**: Locating features is important when aligning objects or determining exact object placement. Locating features serves as a basis for all subsequent inspections. Edge detection, grey-scale pattern matching, shape matching, geometric matching, and color pattern matching are all subroutines embedded in an image analysis program to accurately locate features. These algorithms provide the feature position in rectangular x, y coordinates and a rotation angle θ to within less than 1/10 pixel.
4. **Measuring features**: After locating features, their dimensions are often measured. Typically, edge detection, particle analysis, and geometric function routines are useful in measuring distance, diameter, angles and determining feature area. These tools provide a numerical result for the dimensions or other measurements.
5. **Identifying parts**: Part identification routines are used for part compliance, tracking and verification. Routine identification methods include reading a barcode or a data code. Newer algorithms use trainable object classification techniques.

Automatic optical inspection systems are designed specifically for certain inspection functions. The commercial system illustrated in Fig. 7.25 is designed for post wave and reflow solder inspection. Its hardware and software are coordinated to address this narrow but vitally important inspection task. The system uses a color camera and a mirror arrangement that is incorporated into the optics to permit both orthogonal and angled viewing. Solder defects such as bridging, lack of wetting, absence of solder, excess solder, blow holes and solder balls can be detected. Component placement can also be ascertained and problems due to shifting or skewing, lifting, orientation, marking, and bent leads can be identified. The software in these commercial systems is proprietary and cannot be described here in detail. However, it is clear that with the small size of the features on high density circuit boards manual inspection systems are no longer adequate.

X-Ray Inspection Systems

X-Ray inspection systems are of growing importance because the number of solder connections not visible to optical systems is increasing at a rapid rate. The connections are at the chip-package level with the widespread use of flip chip connections, and at the package-board level with the growing usage of BGAs and CSPs. At the package-board level, only the outer row of balls around the perimeter of either a BGA or CSP is visible. Clearly, optical inspection methods are not adequate to ascertain the adequacy of the connections at interior regions of the package. The percentage of solder joints that cannot be inspected by optical methods is projected to grow to 50% of the solder joints on newly manufactured PCBs by 2007.

X-rays are produced when free electrons (usually from a heated filament) are accelerated in a vacuum into a target material. These electrons impart their energy into the atomic structure of the target material producing an unstable state and a subsequent release of x-rays. Because of the dangers associated with x-rays, it is advisable to avoid direct exposure to this form of radiation. However, the ability of the x-rays to penetrate materials enables them to be used effectively to inspect features that cannot be observed with visible light. Some of the applications of x-ray inspection include medical, airport baggage, radiography of welds, inspection of critical metal parts and inspection of food in cans and jars.

Inspection of printed circuit boards differs from these more routine applications, because of the size and density of the features that must be imaged. This application requires a highly magnified and high-resolution image to characterize defects that may only be a few μm in size. The resolution of an x-ray detector is limited; consequently electronic magnification of the image does not reveal additional detail. The only approach to observing micro defects is to geometrically magnify the x-ray pattern before converting it to a visible image. Geometric magnification of an X-ray image is accomplished with a micro-focused x-ray source that is illustrated in Fig. 7.26.

In a micro-focused X-ray system source, the electrons are generated at the filament and accelerated down a vacuum tube (or chamber). Near the target the beam is focused by a magnetic lens to a spot on the target about 2 to 5 μm in diameter. This spot becomes the focal point of the beam of X-rays produced by the impact of the high energy electrons onto the target material. The beam of X-rays expands and passes through a port provided in the vacuum chamber. Geometric magnification is achieved by positioning a portion of the PCB between the X-ray port and a detector that is an essential component in the imaging system. The amount of magnification depends on the relative spacing of the PCB and the detector from the focal point on the target.

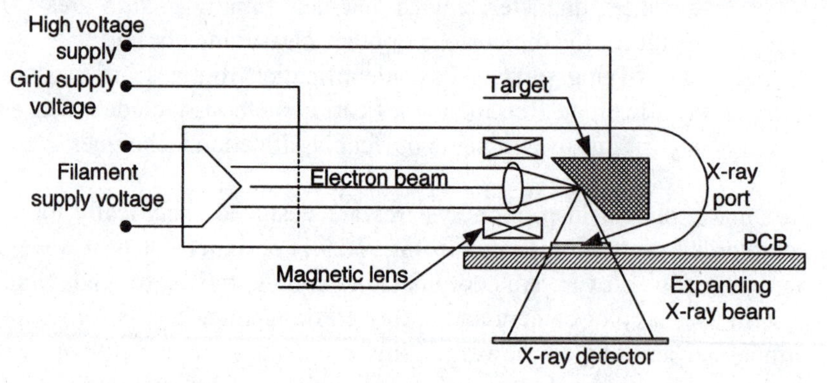

Fig. 7.26 Illustration of a micro-focused X-ray system.

There are three techniques to achieve magnification in an x-ray system. The first and most important is the geometric magnification, which is increased by reducing the distance from the PCB to the focal point, which is the preferred approach. A extension of this technique is to increase the distance between the PCB and the detector, but with large distances a low-contrast, noisy image is observed.

A second technique is to magnify the image when the signal from the detector is transferred to a display screen. If the detector is small and the display screen is large, a significant magnification will result. However, the detector must be larger than the field of view required to inspect specific components on a PCB. For example, when inspecting a BGA package, a 100 mm detector is usually the smallest recommended. Display screens are usually 350 mm in size because screen resolution is optimized at about this size. For these dimensions, the second magnification factor is about 3.5.

The third technique for magnifying the image involves pixel manipulation by image processing techniques. An X-ray inspection system with a magnification of 200X is suitable for BGA solder balls about 1 mm in diameter. However, with CSP and flip-chip solder connections that are smaller, a resolution of 2 μm is required to inspect for defects such as cracks or voids. Achieving this resolution requires a magnification of more than 1,000X[8].

High resolution is achieved with sharp images and adequate penetration of the X-rays. The image sharpness depends on the resolution of the detector and the size of the focal point on the target. Clearly, the smaller the size of focal spot the more closely it approximates a focal point. The thickness that an x-ray can penetrate an electronic component depends on the energy of the X-ray. For most applications 30 kV for the accelerating voltage provides X-rays with sufficient energy to penetrate plastics found on a populated PCB. However, voltages of 100 kV are required to provide X-rays with sufficient energy to pass through a 1 mm solder ball under a BGA.

The intensity of X-rays at a specific accelerating voltage is controlled by changing the filament current. High filament currents improve the contrast of the image; however, there are two detrimental side effects to increasing this current. First, high filament currents increase the space charge in the electron beam expanding the focal spot, which in turn reduces resolution. Second, high filament currents require higher power dissipation in the tungsten target that may melt unless it is water cooled.

One of the most useful applications in electronics for X-ray systems is the inspection of PCB assemblies. These assemblies are composed of different materials with widely different atomic numbers. Because of this diversity an x-ray image of a solder joint on a glass-fiber substrate contains high contrast and considerable detail. Where X-rays pass through areas of glass-fiber substrate, their attenuation is low, and the pixels in the imaging system are very bright. Where X-rays pass through the solder joints the attenuation is large and these silhouettes of the solder joints appear dark. Because the outline of the solder joints is an obvious transition from dark to light, solder bridging and misshaped joints are easy to detect. The detection of a missing ball under a BGA package or flip chip is also possible.

By increasing the accelerating voltage so that about 10% of the x-rays pass through a solder joint, it becomes somewhat transparent, and the voids may be observed as light circles in a dark background. Software is included on most systems to detect and analyze shorts, voids, shapes and missing balls for BGAs.

The most challenging defect to locate is a non-wetted or open joint under a BGA array. Collapsible BGA solder balls form a barrel-shaped joint with the ends of the barrel matching the size of the pad on the substrate. By obtaining an angled view of the balls, non-wetted balls will appear small spheres resting on a flat pad. If the pad diameter is much smaller than the ball diameter, then the wetted ball and the non-wetted ball become similar in shape and are difficult to differentiate. Some designers add wetting indicators to the solder pads to aid X-ray inspection. These are pads with larger diameters than the solder balls so that an orthogonal X-ray view will show a dark extension of the joint when reflowed correctly. If the wetting indicator is not visible after reflow, then the solder has not flowed onto the pad extension indicating a defective joint.

---

[8] Commercial X-rays systems are available with a 2 μm focal spot. These systems utilize a thin beryllium window to provide brighter images, improved resolution and a geometric magnification up to 2,400X.

Misaligned devices do not usually cause a problem, because surface tension of the joints pulls the component into alignment during the reflow operation. Uneven collapse of the solder balls may occur if the temperature profile across the device is not uniform. In this case, the device is cocked sitting high on one side or corner. This condition can be detected by measuring the height of the corner balls in a fairly steep-angled x-ray view. A correctly soldered component exhibits the same height on all four corner joints. A warped device or PCB will show difference height ratios between center and edge solder joints.

The integrity of flip-chip underfill is difficult to ascertain, because the silica-filled epoxy is relatively transparent to X-rays. The use of lead-free solder will not pose significant problems to X-ray inspection capabilities, because the absorption of the X-rays in tin is sufficiently high to produce the contrast required to identify the edges of the solder joints.

7.6.6 Statistical Process Control

Statistical process control is a major component of any quality assurance program. Inspection helps to eliminate defective parts, but it cannot build quality into a product. Carefully controlled processes, which are established with the correct processing parameters, are the most cost effective approach to assuring a consistent high-quality product. As an example of statistical process control, let's consider the time-temperature profile for a reflow process for a Sn-Ag-Cu solder alloy, presented in Fig. 7.27. This particular profile was established by testing a specific assembly containing ball grid arrays and other surface mounted chip carriers. It contains two temperature ramps, a soak period and a cool down ramp. The rate of temperature change is specified as well as the end points of the ramps and the peak temperature.

Modern reflow equipment is under computer control; consequently temperature-time profiles such as the one presented in Fig. 7.27 can programmed and monitored. Times, rates of temperature change and end point temperatures are measured and stored as process data. Each of these measurements[9] typically shows some variation because of small external changes that affect the reflow heaters. Usually the soldering process is sufficiently robust to accommodate small temperature variations, but larger variations can lead to solder defects. For instance, consider the soak temperature of 160 °C, as shown in Fig. 7.27. If the soak temperature were to exceed 170 °C, the solder paste activator would be destroyed and the fluxing action would be impaired. Statistical process control can avoid operating at temperature extremes that could result in defective solder joints.

The mathematics employed in statistical process control is not involved and all of the tedious computations required are performed with computers. As an example of the approach, consider the data collected during each reflow cycle, and determine the mean temperature and the standard deviation for the soak temperature. The mean $\overline{x}$ of some quantity x is determined from:

$$\overline{x} = \sum_{i=1}^{n} \frac{x_i}{n} \tag{7.1}$$

where x_i is the i th value of the quantity being measured and n is the total number of measurements.

The standard deviation S_x is a measure of the process variation and is determined from:

[9] The variation in the control times is small enough to be neglected because the electronic clocks are incredibly accurate.

$$S_x \left[\sum_{i=1}^{n} \frac{(x_i - \bar{x})^2}{n-1} \right]^{1/2} \qquad (7.2)$$

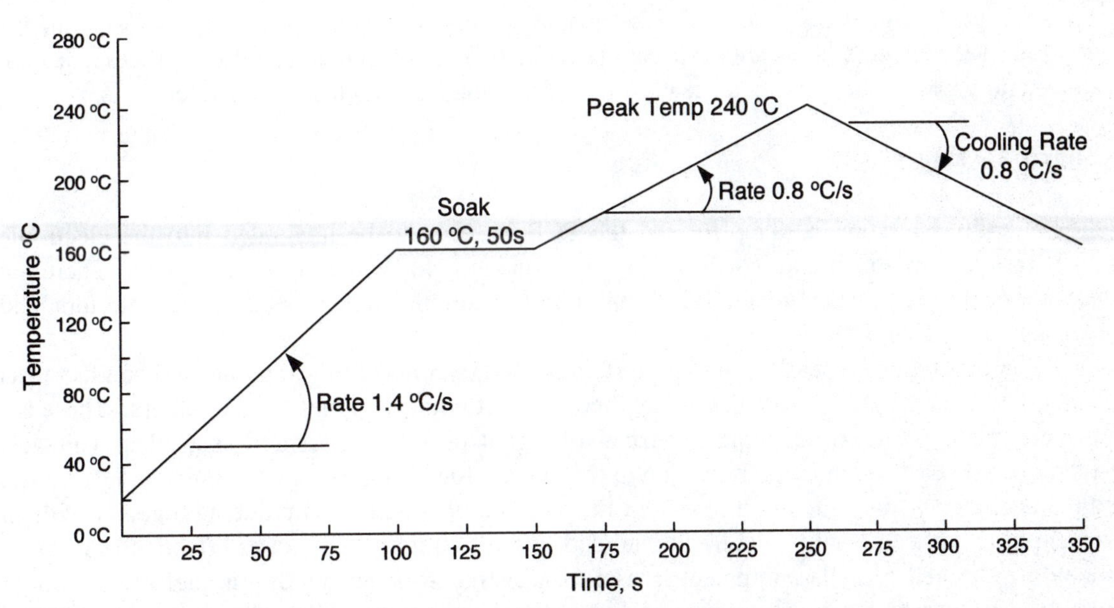

Fig. 7.27 Temperature-time profile for reflow soldering a specific assembly with lead-free solder.

If the sample size n is small, the standard deviation S_x of the sample represents an estimate of the true standard deviation σ of the population. Computation of S_x and $\bar{x}$ from a data sample is easily performed with most scientific type calculators. In most cases, the software associated with computer control of the reflow process enables the periodic determination of the mean and standard deviation of each process parameter.

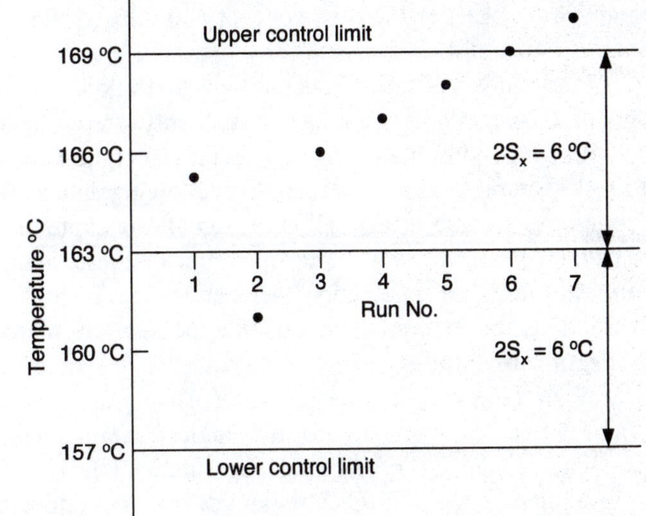

Fig. 7.28 Example of a control chart used to monitor soak temperature in a reflow soldering operation.

Determining the mean and standard deviation of a process parameter provides sufficient information to construct a control chart. Suppose the mean soak temperature is $\bar{x}$ = 163 °C and it standard deviation S_x = 3 °C. We construct the process control chart as indicated in Fig. 7.28. The upper control limit UCL is drawn at a temperature of UCL= $\bar{x} + 2S_x$ = 163 + 6 = 169 °C. The lower control limit LCL is drawn at LCL= $\bar{x} - 2S_x$ = 163 – 6 = 157 °C. The soak temperature is plotted on the control chart for each cycle of the reflow equipment. In this example, we observe that the soak temperature remained inside the control limits for the first five runs. On the sixth run the temperature coincided with the upper control limit indicating the process might be out of control. On the seventh run, the soak temperature exceeded the upper control limit indicating that the process was out of control. This statistical probability of the accuracy of this prediction is 95.4%.

When a process goes out of control it is mandatory that it be shut down and the cause of the large variation in the process be established then remedied. For the soldering reflow process, several temperatures and times are important; hence each would be monitor and charted separately.

7.6.7 Board Testing

When the soldering operation is complete and the PCB is ready to be integrated into the electronic system, it is tested to establish that it will perform the functions for which it was designed. There are many test methods—in-circuit testers, functional testers, combinational/boundary scan testers and manufacturing defect analyzers.

In some cases a bed-of-nails, in-circuit testers is used to connect to test points and conduct tests on selected circuits on the PCB. Voltages are applied at select points and measured at others. These are dynamic measurements, where the time of arrival of signal pulses is measured as well as voltages. However, in recent years a significant reduction in the defects found in integrated circuits and PCBs has changed the testing environment. With the overall defect rate of components reducing together with an improvement in component quality, the traditional bed-of-nails equipment is not as effective in locating defects. While the bed of nails equipment is useful in testing complex PCBs, the defects or timing problems, where these test systems are effective, occur less frequently. This limitation restricts the tester's utility. A more important limitation is the accessibility of test points in high-density fine-pitch chip carriers such as BGAs.

In the 1980s, a group of companies formed a Joint Test Action Group (JTAG) to address the problem of testing PCBs with advanced packaging. The approach recommended was to incorporate test hardware into standard components that could be controlled with software. This approach eliminates the need for costly in-circuit testing equipment. In 1990, the Institute of Electrical and Electronic Engineers (IEEE) refined the concept and created the IEEE Standard Test Access Port and Boundary Scan Architecture.

Boundary scan is a technology that allows complete control and monitoring of the boundary pins of a JTAG compatible device with software commands. This capability enables in-circuit testing without the need of bed-of-nail in-circuit test equipment. During standard operations, boundary cells of a JTAG compliant device are inactive, allowing data to flow through the device under normal operating conditions. During testing, all input signals are captured for analysis and all output signals are preset to test down-string devices. The operation of these scan cells is controlled through a test access port controller and an instruction register. The instruction register provides three different sets of instructions for testing the circuits that includes an interconnect test, a sample preload instruction and a by-pass instruction.

In a circuit board design, there usually are several JTAG compliant devices. All these devices can be connected together to form a single scanning chain as shown in Fig. 7.29. Alternatively, multiple scan chains can be established for simultaneously testing of components. The TAP controllers incorporated in each JTAG compliant device are connected to an external TAP control device, such as a personal computer, through an access connector. The external TAP driver performs different tests after the PCB is produced without the need of bed-of-nail in-circuit testing equipment.

A sequence of tests, listed below, is performed to insure that the circuits function properly.

- The infra-structure test that determines if all of the components are installed correctly.
- The board level interconnect test is used to check if there are opens or shorts on the PCB.

- Additional tests are often conducted when a microSPARC-IIep chip has been incorporated into the circuit board design. This chip is integrated with other JTAG-compliant devices to perform advanced board level testing.

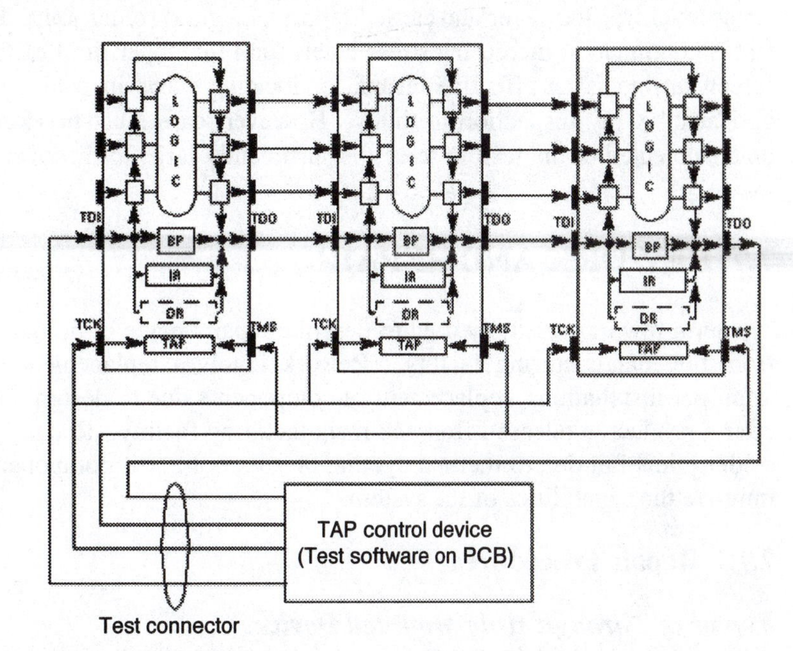

Fig. 7.29 Connections for a single boundary scanning chain test.

Board level testing has become more complex with the increasing use of fine-pitch, high-pin-count chip carriers. However, the boundary scan method permits board level testing to be performed more efficiently and at lower cost.

7.6.8 System Testing

When an electronic system is completely assembled, it is subjected to a sequence of operating tests to ascertain if it performs all of its functions in accordance with the design specifications. For the early system prototypes, these tests determine the adequacy of the electrical design as well as the mechanical assembly. However, when system prototype testing proves that the product meets the design specifications, the battery of tests is simplified and the systems usually are tested only to determine the adequacy of the mechanical assembly.

7.6.9 Stress Screening Tests

Screening tests are often employed by quality assurance personnel to eliminate latent defects in solder joints that cause premature failures. In some cases, the solder defects can be found by visual or X-ray inspection and their cause established. In other cases, the defects are more subtle and stress screening is employed to induce failure of inadequate solder joints. Stress screening involves the application of relatively high cyclic stresses to relatively small samples of the product to precipitate defects. Because the stress levels applied in the screening tests are higher than those encountered in service, the test results are considered to be accelerated.

The level of stress applied to the assembly is critical. The stress level should be high enough to induce failure in solder joints with small flaws, but not so sever to cause failure of good joints. Two different types of stress screens are usually employed—vibration and temperature cycling.

Step-stress procedures are usually employed to establish the level of a stress screen. A screening test is initiated at low stress levels and then the stress level is progressively increased in steps until a failure of a solder joint occurs. This solder joint is examined to determine if its cause was due to a defect such as voids, dewetting or misregistration. If no defect is observed, it is concluded that the stress level was too sever and caused failure of a good solder joint. If the failure was due to a defect, the test is continued at increasing stress levels until the upper limit of the stress level is established. Stress screen testing is an effective means for locating flaws in solder joints that cannot be discovered by optical or X-ray inspection methods. However, care must be exercised to insure that the number of units subjected to the testing screen is sufficiently large to discover all of the potential defective solder joints.

7.7 REWORK AND REPAIR

The term rework is usually used to describe repairs made in production prior to the release of a product from the manufacturing facility. Rework involves replacement of defective components, repair of improper installations, replacement of components due to design changes, etc. The term repair is used after a product is released from the manufacturing facility. Repairs are required if a component fails, if solder joints fail due to thermal cycling or vibration, or if components are updated from time to time to improve the capabilities of the system.

7.7.1 Repair Procedures

Repair of Through Hole Mounted Devices

The repair process for through-hole mounted components involves the following six steps:

1. Desolder the old device.
2. Remove this device.
3. Clean the plated-through-holes removing excess solder.
4. Insert the replacement device.
5. Solder the new device.
6. Clean PCB.

The difficulty in repairing through-hole mounted devices is removing the solder from all of the holes. It is not sufficient to simply heat the solder simultaneously in all of the holes until it melts and to pull the device from the circuit board. The holes must be free of excess solder before a new component can be inserted. This problem is circumvented by using a continuous vacuum desoldering iron and sucking the solder from each hole. Soldering the new device may be performed manually with a hand held soldering iron or by sending the board through a wave soldering machine.

Leaded Surface Mount Repair

The repair process for leaded mounted devices involves the following seven steps:

1. Unsolder the old device.
2. Remove this device.
3. Clean mounting pads removing excess solder.
4. Tin leads of the new device and apply solder paste to the pads.
5. Place and align the new device.
6. Reflow solder the new device.
7. Clean PCB.

The surface mounted devices are removed by applying heat to all of the leads simultaneously until the solder melts. The device is then lifted from the circuit board and the excess solder is removed from the solder pads. In most cases, a continuous vacuum soldering iron is used for this purpose. The new devices are usually placed by hand with the solder paste providing sufficient adhesion to hold the component in place until the solder joints are made in a reflow oven.

Repair of Ball Grid Arrays

The repair of ball grid arrays is very exacting but possible with specialized equipment that provides for uniform heating of the package, cleaning of the array of pads, application of controlled amounts of solder paste and placement of the new component. Repair and rework is often performed on semiautomatic placement and rework stations equipped with split vision systems, a placement head, forced convection heating on the top and bottom of the PCB and computer controlled processes.

Removal of the ball grid arrays involves preheating the circuit board and the ball grid array. Additional heat is applied to the top of the ball grid array by forced convection. The temperature is monitored with thermocouple probes placed adjacent to the BGA and on the circuit board under it. The temperatures required to melt the solder depend on its composition, but temperatures of 230 to 240 °C are required for lead-free solders. These higher temperatures are maintained for 45 to 90 s to insure that all of the solder joints have melted. A vacuum-actuated computer-controlled probe is employed to lift the BGA from the board.

After removal of the BGA, the excess solder from the solder pads is removed with a vacuum probe while the temperature of the chip site is maintained. The circuit board is then cooled to room temperature. Flux is applied to the spheres of a BGA with transfer plates (small stencils designed for the specific BGA being repaired) that allow the deposition of a controlled amount of flux.

The new BGA is accurately placed with a vacuum probe that is capable of x, y and θ motion control. After placement, the installation is heated with forced convection until the solder paste reflows. After the solder joints have been formed the assembly is cooled to room temperature. A commercially available rework/repair station for BGAs is presented in Fig. 7.30.

Fig. 7.30 Automated rework station for BGAs with a split view vision system and computer control.

REFERENCES

1. Kang, S. K.. et al, "Microstructural and Mechanical Properties of Lead-free Solders and Solder Joints Used in Microelectronic Applications," IBM Journal of Research and Development, Vol. 49, No.4/5, 2005.
2. Hwang, J. S., Modern Solder Technology for Competitive Electronics Manufacturing, McGraw-Hill, New York, 1996.
3. Puttlitz, K. J. and K. A. Stalter, Handbook of Lead-Free Solder Technology for Microelectronics Applications, Marcel Dekker, New York, NY, 2004
4. Totta, P. and K. J. Puttlitz, Area Array Interconnection Handbook, Kluwer Academic Publishers, Norwell, MA, 2001.
5. Lee, N. C., Reflow Soldering Processes and Troubleshooting, Elsevier, New York, NY, 2001.
6. Judd, M. and K. Brindley, Soldering in Electronic Assembly, 2nd Edition, Newnes, Oxford, England, 1999.
7. Frear, D. R. and J. H. Lau, The Mechanics of Solder Alloy Interconnects, Chapman and Hall, New York, NY, 1994.
8. Pecht, M. G., Solder Processes and Equipment, John Willey and Sons, New York, NY, 1993.

EXERCISES

7.1 Describe the three mounting technologies in use today and discuss the use of mixed technologies affects the soldering circuit board assembly processes.

7.2 Describe the solder screening printing process. Discuss the process variables that affect its application to high density circuit boards.

7.3 Prepare an engineering brief describing methods used to place components of various types on printed circuit boards.

7.4 Using information from the Internet, describe the features you would specify if you were writing the specification for the purchase of a new pick and place machine.

7.5 What are the factors that affect the throughput on a pick and place machine?

7.6 Prepare a brief explanation of why a simple solder joint is so important to the electronics industry.

7.7 Explain why a defect rate of 100 PPM is objectionable.

7.8 Describe the reflow soldering process.

7.9 Compare vapor phase and IR reflow processes. Cite advantages and disadvantages of each process.

7.10 Describe the IR reflow equipment.

7.11 Describe the vapor phase reflow equipment.

7.12 When would it be necessary to use point to point soldering methods?

7.13 Describe hot bar soldering and give examples of applications where it is the preferred method for soldering.

7.14 Describe a solder pot and give examples of its uses in the electronics industry.

7.15 Describe a wave soldering operation and indicate its primary function in an printed circuit board production facility.

7.16 Prepare a listing of precautions that should be considered in soldering operations.

7.17 Prepare a specification that is to be used to purchase an aqueous cleaning system for populated printed circuit cards.

7.18 Write a brief describing no-clean solder fluxes and indicate and test results showing either their effectiveness or lack thereof.

7.19 Why are certain circuit board coated with a conformal coating.

7.20 What are certain circuit board potted.

7.21 Write an engineering brief explaining the differences among welding, brazing and soldering.

7.22 Why is important to use a solder alloy with a composition that is eutectic or close to eutectic?

7.23 Provide the eutectic compositions for: (a) tin-lead, (b) tin-silver and (c) tin-copper.

7.24 Why are high processing temperatures detrimental in the production of printed circuit boards?

7.25 Why is the electronics industry converting to lead-free solder after more than 60 years experience with a less expensive and easier to use solder alloy.

7.26 Prepare a bar graph showing the cost of the materials to produce a pound of the following solder alloys:

(a) 63% Sn and 37% Pb.
(b) 96.5% Sn and 3.5% Ag
(c) 97.3% Sn and 0.7 % Cu
(d) 52% In and 48% Sn
(e) 95% Sn and 5% Sb

Recent costs[10] of these elements are: Sb = $1.85/lb, Cu = $2.50/lb, Pb= $0.81/lb, Sn = $5.35/lb, Ag = $10.95/troy ounce, In = $600/kg

7.27 Prepare an engineering brief making a recommendation for the composition of a lead-free solder for a manufacturing company that produces circuit boards populated with plated-through-hole mounted components. Justify your recommendation based on cost, yield and quality.

7.28 Repeat Exercise 7.27 but the recommendation is for a manufacturing company that produces circuit boards populated with J leaded components. Justify your recommendation based on cost, yield and quality.

7.29 Repeat Exercise 7.27 but the recommendation is for a manufacturing company that produces circuit boards populated with BGAs and other leadless components. Justify your recommendation based on cost, yield and quality.

7.30 Why are high lead solders used in BGAs, CSPs and in flip chip bonding?

7.31 What is solder paste? Why is it important in producing solder joints.

7.32 Write a specification for the solder paste to be employed in mounting a BGA with pads on 40 mil (1.0 mm) centers. Explain the rational for each aspect of the specification.

7.33 What is the purpose of solder flux?

7.34 Describe the action of the flux as a dot of solder paste printed on a copper mounting pad is heated until the solder melts, flows and wets the two surfaces to be joined.

7.35 Explain why it is important to clean the solder paste from the circuit board after the soldering operation is completed.

7.36 In some cases it is not necessary to clean the residual flux from the circuit board after the soldering operation is complete. Explain the circumstances and the chemistry involved.

7.37 A production supervisor boasts that his/her solder joint defect rate is 80 PPM somewhat better than the nominal 100 PPM defect rate. If a new circuit board design is to be produced with 8,000 solder joints compute the yield of the soldering process. Discuss the implications of your analysis on production costs and make recommendations for improvement.

---

[10] Except for Indium the price of these elements (commodities) are reported daily in the Wall Street Journal.

7.38 Prepare an engineering brief for operating personnel that summarizes soldering procedures for a wave soldering operation. Note that half of the personnel are functionally illiterate or speak Spanish and not English.

7.39 Prepare an engineering brief for operating personnel that summarizes soldering procedures for a vapor phase reflow soldering operation. Note that half of the personnel are functionally illiterate or speak Spanish and not English.

7.40 Prepare an engineering brief for operating personnel that summarizes soldering procedures for an IR reflow soldering operation. Note that half of the personnel are functionally illiterate or speak Spanish and not English.

7.41 Describe the advantages and disadvantages of a solvent system for cleaning PCBs after the soldering operation is complete.

7.42 Describe the advantages and disadvantages of an aqueous system for cleaning PCBs after the soldering operation is complete.

7.43 Describe some of the problems associated with no-clean solder fluxes and the procedures for circumventing these difficulties.

7.44 Cite the advantages and disadvantages of using conformal coatings on PCBs.

7.45 You have been assigned as a quality assurance manager in a small electronics company that produces several hundred PCBs each day for a number of different customers. Prepare an over view of a quality assurance plan that you are to present to the vice president of engineering.

7.46 Prepare a detailed plan for assuring the quality of incoming components and materials.

7.47 Specify the locations of the thermocouples used to measure temperature in a wave soldering operation. Prepare a diagram of the wave solder equipment showing these locations.

7.48 Specify the locations of the thermocouples used to measure temperature in an IR reflow soldering operation. Prepare a diagram of the wave solder equipment showing these locations.

7.49 Describe how you would monitor the gold content in a solder joint. Why is the gold content important?

7.50 Prepare instruction for an inspector who is to examine circuit board prior to soldering the components. State all of your assumptions.

7.51 Prepare instruction for an inspector who is to examine circuit board after the soldering the components. State all of your assumptions.

7.52 Write a specification to be used by the purchasing department for procuring an automated optical inspection system.

7.53 Describe the classic algorithms that are included in the image analysis program for optical inspection systems.

7.54 Write a specification to be used by the purchasing department for procuring an automated X-ray inspection system.

7.55 Explain why lens systems cannot be used to magnify the X-ray images.

7.56 Explain geometric magnification and write equation for the magnification factor as a function of the geometric parameters associated with an X-ray inspection system.

7.57 Cite the advantages and disadvantages of both the optical inspection systems and the X-ray inspection systems.

7.58 Prepare an inspection plan to find defects in BGAs with a 1.0 mm pitch. Prepare sketches for each defect that the inspector may find.

7.59 Prepare an engineering brief for the vice president of engineering explaining why you plan to implement statistical process control on each step in the process for populating and soldering components onto PCBs.

7.60 For the data presented in the table below determine an estimate of the mean and standard deviation of the peak solder temperature as measured by a thermocouple located adjacent to the PCB is the reflow oven.

Reading No.	Temperature °C	Reading No.	Temperature °C
1	242	11	241
2	244	12	244
3	246	13	246
4	244	14	245
5	242	15	244
6	240	16	241
7	238	17	239
8	237	18	240
9	238	19	242
10	239	20	243

7.61 For the data presented in Exercise 7.61, prepare a control chart with upper and lower control limits on the peak temperature. The limits should be set to provide an indication of control with a probability of error of only 4.6%.

7.62 Describe bed-of-nails in-circuit testers and their function in a quality assurance program. Also indicate their limitations.

7.63 Explain boundary scan technology and indicate its importance today and in the future for testing PCBs.

7.64 Why do companies perform system tests? After a series of system tests with prototypes have demonstrated that the product meet specifications, is the system test program modified? Why?

7.65 Why do manufactures perform stress screening tests?

7.66 Prepare a plan to establish the level of vibrations in a stress screening test to verify the reliability of solder joints in an electronic system. How will you determine sample size?

7.67 Cite reasons for using temperature cycling in a stress screening test.

7.68 Cite reasons for using vibration in a stress screening test.

7.69 Define the difference between rework and repair. Are the methods used in rework and repair essentially the same?

7.70 Prepare instruction for personnel operating a station to repair components mounted with clenched leads soldered in plated-through-holes. Specify tools to be used together with temperatures and time at temperature for removal and replacement of DIPs and leaded capacitors.

7.71 Prepare instruction for personnel operating a station to repair J leaded components that are surface mounted. Specify tools to be used together with temperatures and time at temperature for removal and replacement of these components.

7.72 Prepare instruction for personnel operating a station to repair BGAs with pads on 1.0 mm centers. Specify tools to be used together with temperatures and time at temperature for removal and replacement of the BGAs.

7.73 Write a specification to be used by the purchasing department for procuring an automated rework station for repairing BGAs with 0.8 mm pitch.

CHAPTER 8

THIRD LEVEL PACKAGING – CONNECTORS, CABLES, MODULES, CARD CAGES AND CABINETS

8.1 INTRODUCTION

Third level packaging includes all of the hardware required to house the circuit boards and peripheral equipment necessary for a complete electronic system. Large systems, with large numbers of circuit boards, are housed in one or more electronic enclosures. These enclosures support the circuit boards, provide a cooling medium, protect the components from the environment and contain the back panels, buss bars and cabling necessary for the higher level connections. In large systems, the design of the third level packaging is a significant effort as many different mechanical components are involved. Consumer-products may often be smaller systems, in which the third-level may represent the highest level of packaging, and the product-housing provides the enclosure for the printed circuit board assemblies. In several consumer products, the cabinet to cabinet connections may not be required. However, competitive pricing pressures in consumer products complicate the design effort because cost limitations often eliminate many design approaches.

Examples of complex systems with third-level of packaging involving several cabinets include ground-based defense systems, avionics, and web-servers. In addition to housing the entire electronic system, the third level packaging must permit easy access for service personnel so that the components located within the enclosure can be identified, tested and replaced. For large high performance networks, system availability is critical and rapid diagnostics and repair is essential to ensure timely recovery of a malfunctioning system. Design for accessibility is an essential part of maintenance strategy. At the same time, the third level packaging must inhibit access to prevent accidental-damage to personnel or the system. It is clear that these conflicting goals for accessibility require ingenious design of fasteners, keys and tools involved in opening and closing the enclosures.

Safety is always a paramount design requirement. One may believe that the voltages associated with a logic circuit are always low and the dangers of shock and electrocution are non existent. It is true that low voltages (less than ± 5V) are common in digital circuits; however, these low voltages are often associated with very high currents. Buss bars servicing a dense back panel can carry electrical currents of several hundred amperes. These bars must be shielded to prevent accidental shorting by service personnel. If the power buss is shorted, severe arcs will result and the possibility of blinding and burning service personnel exists. Higher voltages are found in the power supply compartments where the ac supply is converted to the dc voltages required for digital circuits. Shielding of the higher ac supply terminals is essential in meeting safety requirements.

A final factor to consider in the design of the third level package is the operator/system interface. A typical interface includes switches and indicator lights for turning the system on and off, a display for messages and/or graphics and a keyboard or keypad for entering commands. It is essential that this input/output panel be positioned to minimize operator effort in monitoring the system function and in providing commands to control system functions.

We will cover third level packaging beginning with the edge card connectors, which serves as the entrance to third connection level in the connection hierarchy, and then proceed to describe back panels and cables that are used for still higher levels of connections. We will also describe subassemblies and enclosures and then discuss cooling devices such as fans, cold plates and cold rails that are commonly required to dissipate the heat generated in operating the electronic systems.

8.2 CIRCUIT CARD CONNECTORS

On a typical PCB, we frequently connect 1,000 or more pads in forming an element of a logic function. The complete logic function requires many more components and connections and it is often necessary to employ many circuit boards to support a large number of chips involved in the complete system. These circuit boards are connected together using edge card connectors. In consumer products, multiple board assemblies may be connected with rigid board-to-board connectors or using flexible-circuits. The pin-and-socket type connectors may include the header (the male side with pins or blades) and the receptacle (the female with windows to accept the pins). Other connector types include sliding-contact connectors, and zero-insertion force connectors. Example applications of sliding-contact connectors include, multi-media card connectors, PC card bus, SD memory cards, mini-SD memory cards, SIM cards, and IO smart cards. A typical pin-and-socket connector, illustrated in Fig. 8.1, shows six features common to almost all connectors.

1. A series of pins or gull wing leads which are soldered to the circuit board to make connections to the I/O traces from the board.
2. A second series of pins which are inserted into a receptacle to carry the signals to the next connection level.
3. A core usually fabricated from an injection molded thermoplastic that spaces the pins on closely controlled centers and insulates the connector assembly.
4. Two larger diameter guide pins that serve to align the connector pins with the holes in the receptacle. The two guide pins can also be arranged to serve as a key to prevent insertion of an incorrect header into a receptacle.
5. A geometric feature to serve as a key for a unique connector/receptacle combination.
6. A case or shroud to prevent damage to the exposed pins.

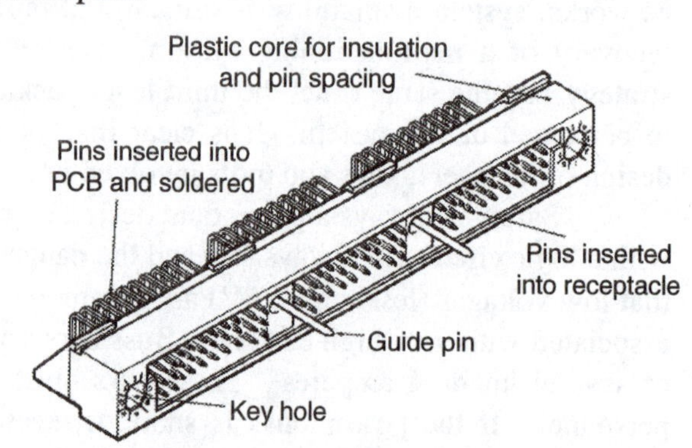

Fig. 8.1 The header on a typical edge card connector showing several common design features of this class of connectors.

There are many different types of edge card connectors ranging from the simplest header (male), presented in Fig. 8.2, to a dense connector with pins on 50 mil (1.27 mm) centers as shown in Fig. 8.3. The simple header (male connector) of Fig. 8.2 is made by etching blade surfaces from the copper cladding on one or both sides of a printed circuit board. These finger-like surfaces are plated with gold to provide a suitable surface for electrical contact. The fingers serve as blades that are inserted into a receptacle that in turn is connected to a back panel (mother board) or a cable connector.

The connector (receptacle), presented in Fig. 8.3, is capable of a high pin count because its holes (windows) are on 50 mil (1.27 mm) centers. Higher pin counts are achieved by increasing the number of rows and the number of windows per row[1]. In the end view, shown in Fig. 8.3, the gull wing leads used for surface mounting are evident. Small pitch connectors are often required to handle large I/O requirements typical of a high performance circuit board; however, higher costs per pin are incurred with these connectors because of the close tolerances necessary in manufacturing.

Fig. 8.2 An inexpensive arrangement of gold-plated fingers on the edge of a PCB serves as pins for one type of edge card connector. Note the slot in the line of fingers that acts as a key to insure compatibility with a specific receptacle.

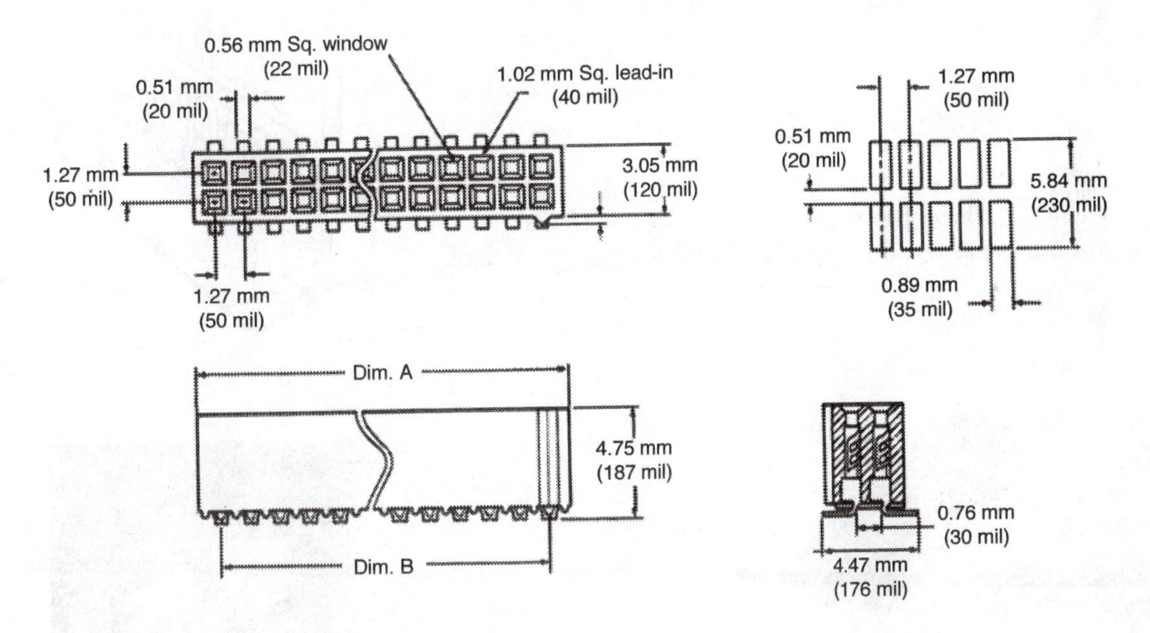

Fig. 8.3 Drawing of a surface mounted connector with a pitch of 50 mil (1.27 mm).

The receptacles often contain beam or tuning fork contacts to accept pins or blades from the edge card connector. The receptacles may be mounted on a back panel, a wire wrap board, a cable or a chassis. An example of a receptacle that accepts the gold plated blades on a PCB is illustrated in Fig. 8.4. The beam contacts are exposed in this illustration; however, in practice they are enclosed to protect them from damage.

[1] Higher density connectors are commercially available with pins on either 32 or 40 mil (0.8 or 1.0 mm) centers. These connectors are surface mounted and found in products that must be packaged in a minimum volume such as cell phones, pagers, digital cameras and mobile computers.

Other connectors mate with flat ribbon cables as shown in Fig. 8.5. When the I/O count is small and only a few PCBs are involved in a system, cable connections from board to board are often feasible. Cable connections among PCBs eliminate the back panel, which is an expensive component, and results in a considerable savings in product cost. Board-to-board connections may also be accomplished by use of flex-circuit assemblies. Flex-circuits may be single-sided, double-sided, multi-layer, or rigid flex. Single-sided flex circuit assemblies consist of one conductor layer, which offers a low cost solution. Double-sided flex-circuit assemblies have two conductor layers, and offer greater routing density per unit area compared to single-sided flex circuits. Multi-layer flex circuits may have up to 20 conductor layers. Rigid flex circuit assemblies use combination of traditional printed circuit board and flex circuit constructed as one continuous piece, with up to 30 layers on printed-circuit board. Typical flex-circuit materials include polyimide film 0.0005 inch or greater in thickness. Conductor layers may be electrodeposited copper ¼ ounces to 2 ounces or rolled-annealed copper ½ ounce to 5 ounces in thickness. Use of flex-circuit has several advantages including space and weight reduction achieved from the flex's ability to occupy all three dimensions resulting from which allows the flex to be bent around packaging and even over itself in order to fit in a much smaller enclosure. Flex circuits are much more chemically resistant than traditional printed circuit boards.

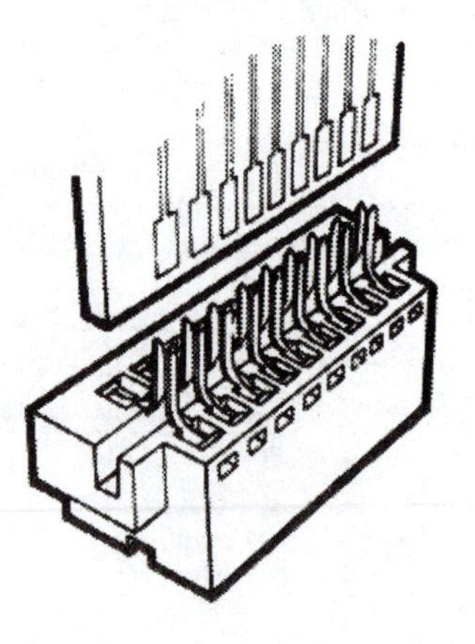

Fig. 8.4 Cantilever beam fingers in a receptacle (female) connector used to accept the gold plated blades along the edge of a PCB.

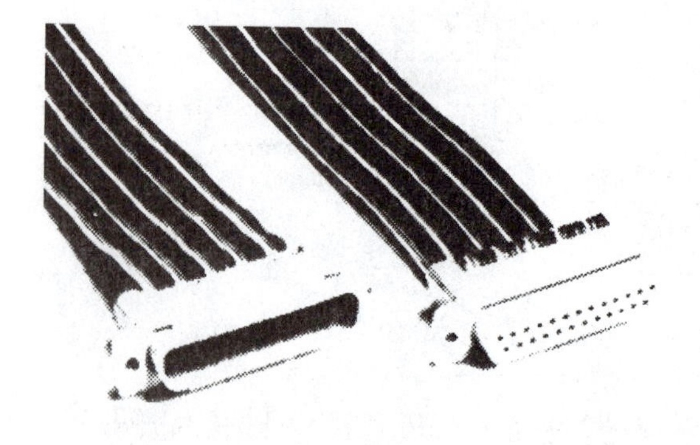

Fig. 8.5 (a) Header and receptacle connecting to a flat ribbon cable. (b) High Density Flex-Circuit Connector Assembly (Courtesy of Molex, Inc.)

The common D-shell connector found on the personal computers and many other products is presented in Fig. 8.6. This illustration shows a cut away view that reveals many of the features used in designing the header and the receptacle of the D-shell connector.

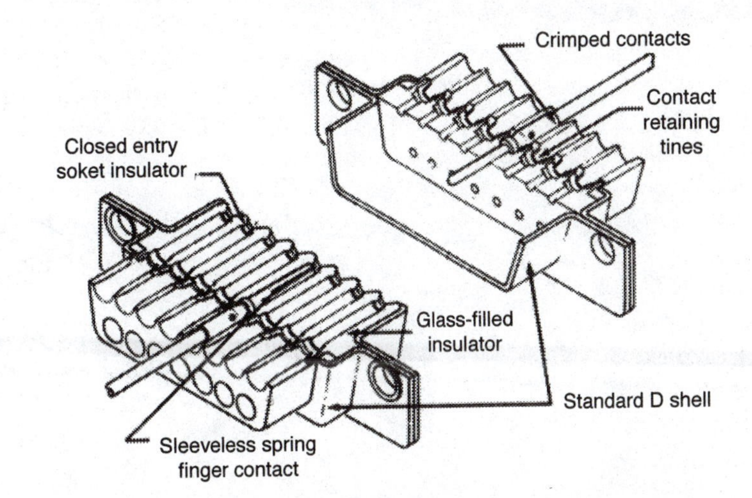

Fig. 8.6 D-shell connector used for mounting on a chassis panel.

More rugged dust proof connectors may be required for industrial and military equipment. In such cases, the left side (receptacle) of the round cable connector, shown in Fig. 8.7, is bolted to the equipment. Wiring from the interior of the chassis is connected to its terminals. A seal encases the contacts providing protection from the elements. The right side (header) of the round cable connector carries a cable that includes the wiring which connects one chassis to another. The extended shell on the header provides strain relief for the cable. A clamp is provided to fix the cable to the strain relief shell.

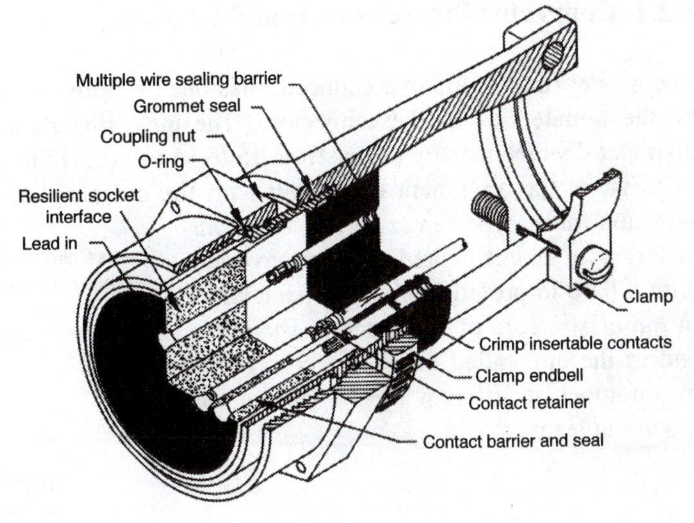

Fig. 8.7 A round cable connector used for rugged applications where dust and debris may cause malfunctions.

A typical example of PCBs (daughter boards) connect with a back panel (mother board) is illustrated in Fig. 8.8. In this illustration, pins are inserted and soldered into the mother board to provide the mating half for the receptacles. In many applications, the receptacles are surface mounted to solder pads on the back panel, eliminating the need for pins to accommodate the daughter boards.

There are many different designs of connectors; however, the basic principals used in these designs are relatively few in number. We will consider the design of the individual components including the pin, contact and insert in the next subsection.

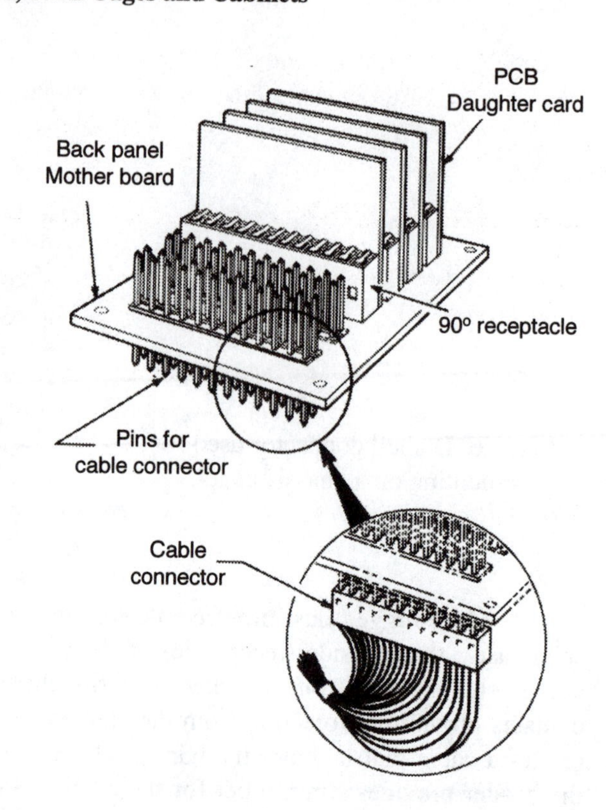

Fig. 8.8 Daughter boards mounted to a back panel with circuit card connectors.

8.2.1 Connector Pins, Contacts and Inserts

The header (male) side of a connector has one or more pins or blades that are inserted into the contacts on the female side of the connector. The pins, illustrated in Fig. 8.9, are often cylindrical with a diameter d which usually ranges from 15 to 85 mil (0.375 to 2.125 mm). The nose of the pin is rounded to assist in the alignment of the pin with the contact during engagement and tapered to reduce the insertion force. The center region of the pin is larger in diameter and usually provided with a barb to provide for positive retention in the plastic core of the header. The other end of the pin, called the tail, is configured in a number of different ways to accept the incoming leads.

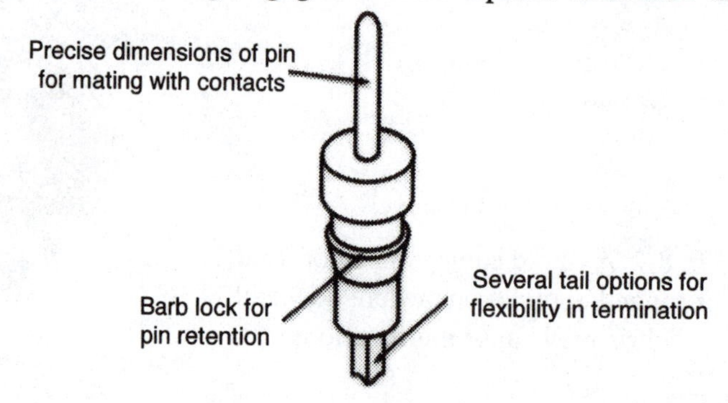

Fig. 8.9 Example of pin used in a header.

The maximum force which can be exerted on a circular pin during insertion is controlled by buckling[2]. Reference to Euler buckling theory gives the critical load associated with the collapse of the pin as:

$$P_{cr} = \frac{C\pi^2 EI}{L^2} \quad (8.1)$$

[2] The critical buckling force is usually much larger than the actual insertion force, which is usually about 1 N/pin.

where C = 1/4 for one end of the pin fixed and the other end free.
C = 2 for one end of the pin fixed and the other end guided.
$I = \pi d^4/64$ for a pin with a circular cross section.
L is the length of the pin.

Blades with a rectangular or square cross section are often used in place of circular pins. The moment of inertia for the blade, $I = b^4/12$ (square) or the smaller of $I = bh^3/12$ or $I = b^3h/12$ (rectangle), is substituted in Eq (8.1) to determine the buckling force.

Contacts are geometrically more complex than the pins because of the need for the contact to first accept and then to clamp the pin. The contact acts as a small spring and exerts a normal force on the pin. This normal force is essential in maintaining a low contact resistance across the connection. An example of a tuning fork contact used in a receptacle or pin plate is presented in Fig. 8.10.

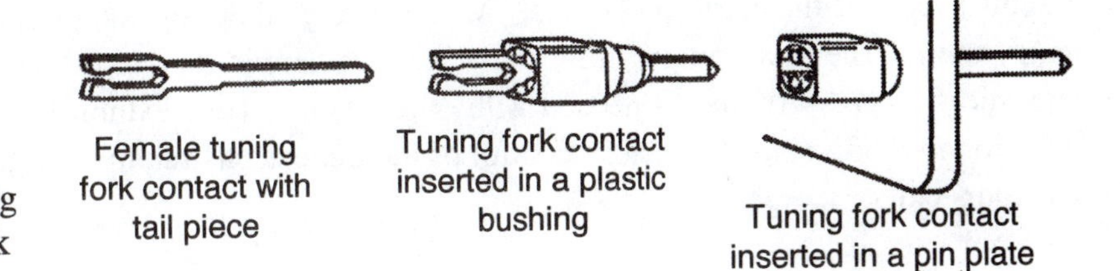

Fig. 8.10 Illustration showing the assembly of a tuning fork contact in a pin plate.

The right side of a double beam contact is illustrated in Fig. 8.11. The geometry of this contact provides several advantages. The ramp on the first cantilever beam facilitates the alignment of the pin as it is inserted into the pair of contacts. As the pin is inserted, it is wiped clean by the crown at the contact point. Both beams flex when the pin is inserted insuring a relatively low insertion force per contact. Accidental overloading of the contacts is prevented by the end of the first beam that acts as a stop when it contacts the barb located near the base of the contact. The barb also serves to lock the contact in the plastic insulation of the receptacle. A terminal lead is provided to make connections to the PCB.

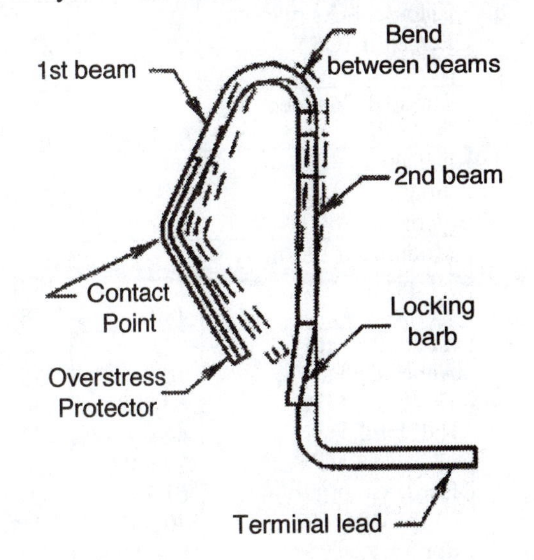

Fig. 8.11 The right side of a double beam contact used in a high density receptacle.

The materials used in fabricating the pins and the contacts must exhibit low resistivity, high strength, high modulus of elasticity and excellent wear characteristics. These materials must also be sufficiently ductile to permit easy fabrication of various connector parts by bending and stamping processes. Copper based alloys including brass, beryllium copper, phosphor-bronze and copper-nickel are commonly employed for both pins and contacts. Electrical and mechanical properties of these materials are described in Table 8.1. After the contacts and pins have been fabricated, they are plated with relatively soft metals (lead-free solder or tin) to increase the true contact areas and to further reduce the electrical resistance across the connection.

The plastic core or the insert of the header holds the pins in position with close dimensional tolerances and serves to insulate one pin from another. Plastic cores are also used with the receptacle to hold the contacts in position. In most connectors, thermoplastics are used in fabricating the cores and inserts used with both the headers and the receptacles. Thermoplastics are characterized by the ability to be softened several times without degrading their properties. In addition, they can be injection molded in high volume production at relatively low costs.

Two thermoplastics for connector applications include polyphenyleneoxide and polyethylene terephthalte (PET) polyester. Both of these thermoplastics have low water absorption rates (0.1% in 24 h); hence, they maintain their properties and are dimensionally stable. Both of these thermoplastics exhibit excellent dielectric properties. Mechanically, they are tough and rigid over a wide range of temperature. Their outstanding flow characteristics permit them to fill complex, thin-walled molds with less injection pressure than is needed with other resins. They exhibit low shrinkage upon cooling after injection molding permitting close control of the dimensions required in fabricating high-density multi-pin connectors.

Table 8.1
Copper based alloys used in fabricating connector pins and contacts

Composition, Percent	**Alloy 260 Brass 70 Cu; 30 Zn**	**Alloy 172 Beryllium-Copper 98.1 Cu; 1.9 Be**	**Alloy 510 Phosphor Bronze 94.81 Cu 5.0 Sn 0.19 P**	**Alloy 638 95.0 Cu 2.8 Al 1.8 Si 0.4 Co**	**Alloy 7.25 88.2 Cu 9.5 Ni 2.3 Sn**	**Alloy 762 Nickel Silver 59.25 Cu 28.75 Zn 12.0 Ni**
Electrical Conductivity						
Percent IACS*	28	22	15	10	11	9
μΩ-cm	0.163	0.128	0.087	0.058	0.064	0.050
Thermal Conductivity W/(m-°C)	121	107-130	69	40	54	42
Density						
lb/in^3	0.308	0.298	0.320	0.299	0.321	0.314
g/cm^3	8.54	8.26	8.86	8.29	8.89	8.70
Modulus of Elasticity						
10^6 psi	18.0	18.5	18.0	18.7	19.0	18.0
GPa	110	128	110	115	131	124
Yield Strength						
Annealed, ksi	10-33	155	22	58-67	25	29
MPa	69-228	1069	152	400-462	172	200
Half hard, ksi	42-60	175	47-68	76-89	57-73	58-82
MPa	290-414	1207	324-469	524-614	393-503	400-565
Hard, ksi	67-78	180	74-88	91-103	74-79	82-97
MPa	462-538	1241	510-607	627-710	510-545	565-669
Spring, ksi	82-91	N. A.	92-108	100-112	78-93	101-110
MPa	565-627		634-745	690-772	538-641	696-758
Extra spring, ksi	86-98	N. A.	98-110	107 min	89-102	102 min
MPa	593-676		676-758	738 min	614-703	713 min

* % International Annealed Copper Standard (IACS) = 172.41/resistivity in μΩ-cm

8.2.2 Connector Resistance and Plating Materials

The connection made between the pin and the contact is totally mechanical. The connection does not have the metallurgical bond of a soldered joint or the chemical bond of an adhesive joint. However, the electrical resistance across the connection between the pin and the contact must be reduced to a minimum, usually 10 to 15 mΩ. The connection resistance, R_c, is due to two factors, namely a resistance due to constriction effects, R_{ct}, and a resistance due to oxide films, R_f, that develop on the surfaces of both the pin and the contact.

Consider the engagement of a blade in a tuning fork contact that appears to give a large area A of contact as shown in Fig. 8.12. However, the surface finish over this area of contact is far from perfect. The surface imperfections, called asperities, produce a large number of true contact spots each of which is quite small. When the total area of true contact is determined by summing the areas a_c of the many contact spots, we find that the true contact area is often less than 1% of the apparent contact area. Increasing the normal force N on the contact enlarges the asperity contact area a_c of the existing contact spots and produces additional contact spots. However, even with excessively large normal forces the true contact area remains a very small fraction of the apparent contact area.

A better approach for increasing the true contact area and decreasing the connector resistance is to plate the surfaces of both the blade and the contact with a relatively soft material. The diameter d of the contact spot formed by bring the two plated surfaces together is shown in Fig 8.13.

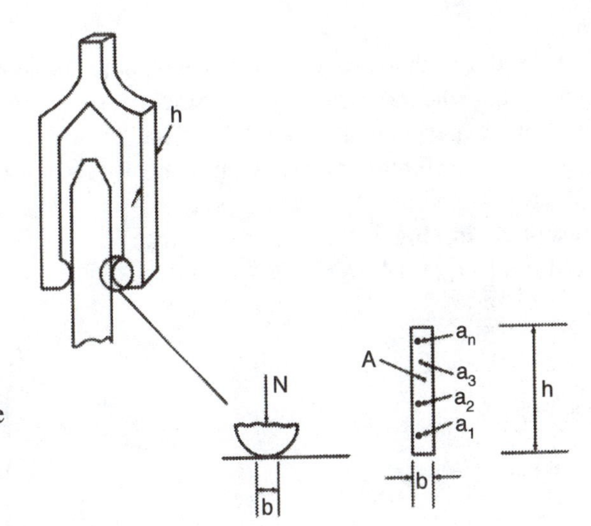

Fig. 8.12 Apparent contact area A = bh and the true contact area given by the summation of the spot areas a_1, a_2, a_3 and a_n.

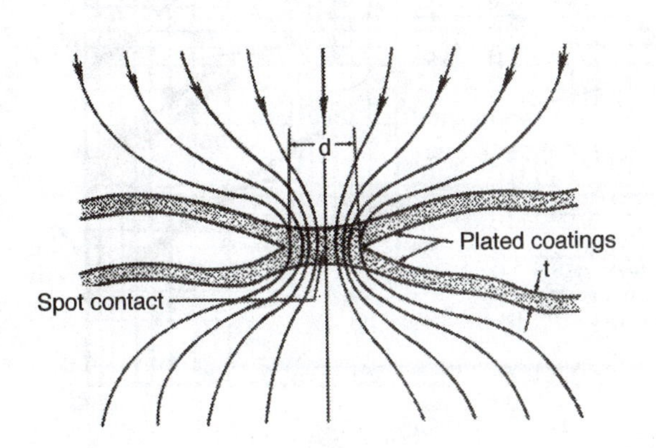

Fig. 8.13 Schematic illustration of current flow through a contact spot in a plated pair of conductors.

The diameter d of the contact area is related to the hardness H and the normal force N by:

$$d \propto \sqrt{\frac{N}{H}} \tag{8.2}$$

where the hardness is often approximated as three times the yield strength.

Next, note that the constriction resistance R_{ct} can be written as:

$$R_{ct} = \frac{\kappa\rho}{d} \tag{8.3}$$

where ρ is the resistivity of the plating material and κ is a parameter depending on geometry and type of contact.

Combining Eqs (8.2) and (8.3) gives:

$$R_{ct} \propto \kappa\rho\sqrt{\frac{H}{N}} \tag{8.4}$$

The result of this proportionality indicates the importance of both hardness and normal forces in minimizing the constriction resistance. Typical experimental results for a number of different materials showing R_{ct} as a function of normal force N, presented in Fig. 8.14, confirm the validity of Eq (8.4).

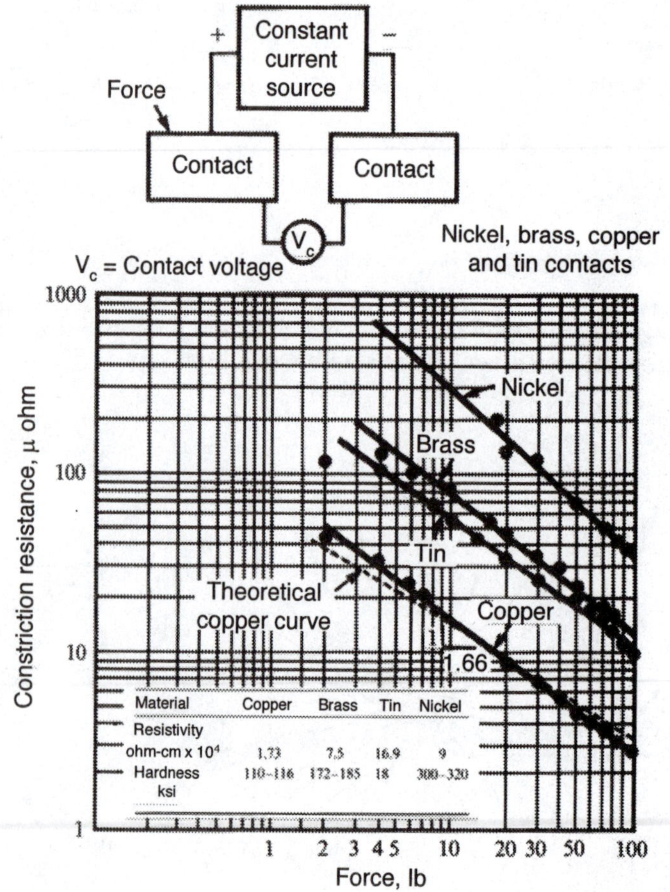

Fig. 8.14 Constriction resistance R_{ct} as a function of normal force N for several different materials.

The effect of the film resistance on connector resistance is often more important than the contact resistance. There are four factors which must be considered in evaluating the effectiveness of a connector to resist the development of detrimental film resistance. The first is the quality of the plating as indicated by the porosity of the coating. Porosity is a measure of the number of pores that occur in the coating. Clearly the pores are detrimental because they permit migration of base metals that oxidize to form surface films. Porosity may be controlled by coating thickness as indicated in Fig. 8.15; however, costs of noble metal plating at several hundreds of dollars per ounce[3] dictate extremely thin coatings. A more suitable approach for controlling the effects of porosity is to provide an under plate coating of nickel 1 to 2 μm thick. The nickel serves to form a passive, very thin, oxide film at the base of the pore sites. It also prevents migration of corrosion products from the sharp edges of the contacts. The use of the nickel under plate reduces the required thickness of the precious metals to 0.5 to 0.75 μm. A popular commercial plating system[4] consists of a nickel under-plate 1.3 μm thick, a palladium-nickel alloy with thickness from 0.4 to 1.2 μm thick depending upon the application and a thin over-flash of gold at least 0.05 μm thick.

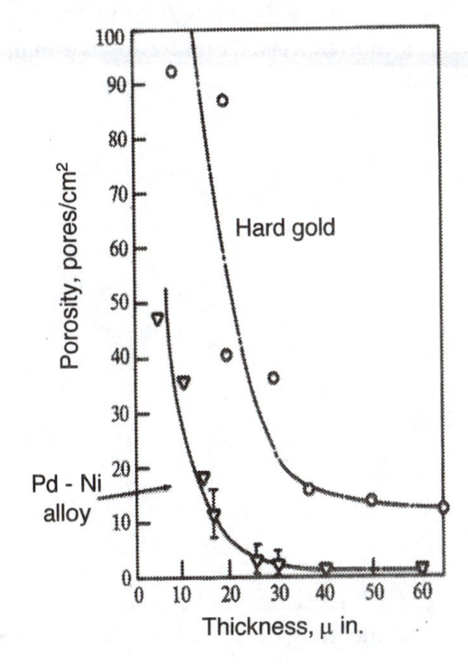

Fig. 8.15 Porosity of plated coatings as a function of coating thickness.

A second factor to consider is wear of the plating material at the contact area as the connector is cycled in service. The cycling can occur due to either the need to repeatedly mating and removal of the connector-system, or due to small motions, called fretting, produced by vibration or thermal cycling of the connector. There are several approaches for extending the life of a connector. One of the most common design methods is to use either hard gold plate or a palladium nickel alloy plate over a nickel under plating. The results of wear test with these two plating systems, presented in Fig. 8.16, indicate long life for both coatings. However the superiority of the palladium nickel alloy is evident for extended life applications.

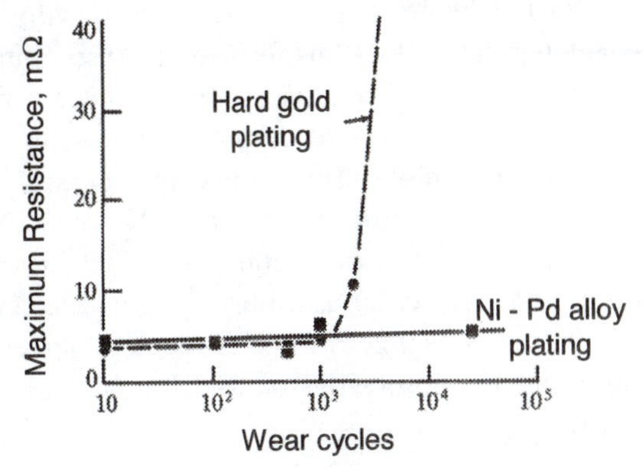

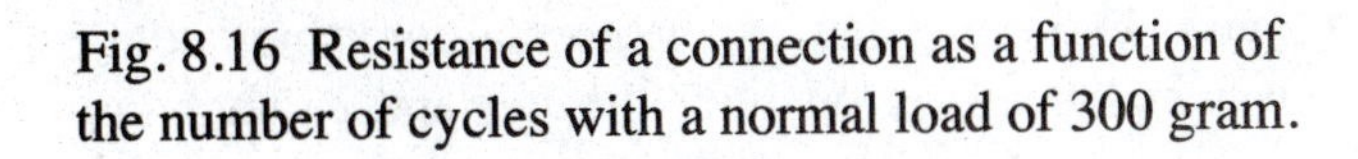
Fig. 8.16 Resistance of a connection as a function of the number of cycles with a normal load of 300 gram.

[3] The price of a troy ounce of gold, palladium and silver in early 2006 was $550, $280 and $9.20, respectively.

[4] The GTX plating system was developed by Berg Electronics in 1979 to replace hard gold with a more cost effective plating system.

In some applications, tin is used as the plating material in order to reduce costs. Tin oxidizes easily and forms a film which is thin, hard and brittle. The normal forces acting at the contact rupture this film and the metallic tin extrudes through the cracks to form a low resistance connection. Tin may also form metallic dendrites in the presence of high-temperature and moisture, resulting in migration induced resistive shorts. With repeated usage or fretting, the contact is repeatedly broken and new oxide films are generated on each cycle. With time, tin/tin oxide debris forms significantly increasing the contact resistance as shown in Fig 8.17. The life of the tin plated contact can be extended considerably by lubricating the assembly. The lubricants can be passive, such as mineral oil, or active proprietary fluids which attack oxide films. The passive lubricant reduces wear, flushes debris from the contacts, cleans the contact area and limits the amount of oxygen reaching the exposed metal at the contact. The active lubricants are more effective than the passive lubricants because they strip the oxide films from the tin.

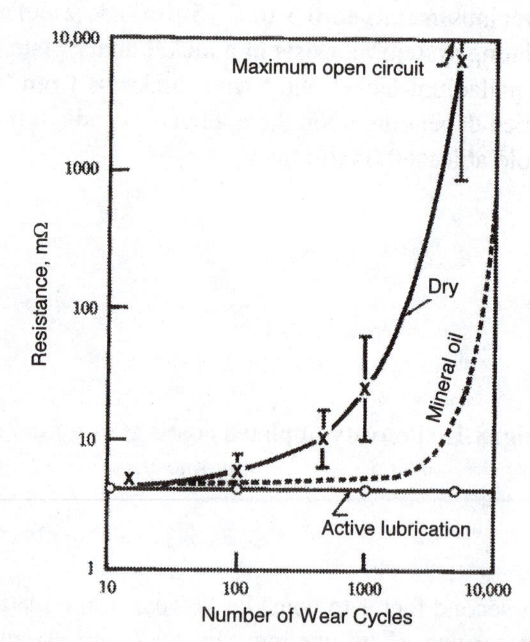

Fig. 8.17 Resistance of a connection as a function of the number of wear cycles for tin plate.

The third factor that markedly affects connector performance is the normal force acting on the contact area. High normal forces produce lower connector resistance because they reduce both the contact and film resistances. The effect of the normal force on contact resistance is clearly evident in Eq (8.4). The effect of normal forces on film resistance is less clear as film resistance depends on the characteristic of the films formed over the contact area. These films are insulators and must be ruptured during mating and removal of the connector system and the accompanying sliding of the pin over the contact that takes place during engagement. A high normal force insures that the film is ruptured; however, this high normal force increases wear rates and increases both the insertion and extraction forces. Normal forces of 2 to 3 N are commonly encountered during insertion in many types of connectors. With newer connectors that have a higher pin counts, it is necessary to reduce insertion forces to permit easer assembly. The insertion forces may be reduced if the normal forces are decreased to 1 N or less or if zero-force connectors[5] are employed. In these instances, the normal force may not be sufficient to rupture the oxide films and precious metal coatings on both the pin and contacts are mandatory.

[5] Zero force connectors employ a cam mechanism to separate the contact pair prior to inserting the pin. After insertion, the cam is rotated and the contact force is developed. The zero force connectors are larger in size than the standard connector because of the cam mechanism requires additional space in the receptacle.

The final factor to consider in the analysis of film resistance is the power transmitted through the connection. The connector should always be engaged without power (i.e. a dry circuit). When the circuit is activated, current flows through the connector and is concentrated at the contact spots shown in Fig. 8.12. The voltage drop across the contact area breaks down the films and heat from the I^2R losses melts the local high spots to establish a larger contact area. The level of power involved significantly affects behavior of the connection. For very high power, with voltages greater than 100V and currents greater than 50A, the conducting characteristic of the base metal is extremely important and plain copper is the preferred material. If the connection is to be made under power, as in a switch, then silver plate is essential to prevent fusing of the contacts due to arcing as the switch closes.

For high power applications with the currents exceeding 1A at 10V, the constriction resistance is the important factor. High normal forces and effective wiping are more important in reducing the constriction resistance than plating materials. For intermediate power levels with current less than 1A at 10V, both film and constriction resistances are important. In this power range, it is essential to seek a balanced design that includes effective plating to inhibit film development and moderate normal forces to lower constriction resistance. The balanced design provides a connection with wear resistance, a long life and stable connection resistance. At the low power levels with the current less than 0.1A, film resistance becomes the dominant factor and precious metal plating and/or lubrication are design approaches followed in providing reliable connections with a long life.

8.2.3 Connector Forces

There are three different forces which are considered in the engagement and extraction of a connector—the normal force, the insertion force and the extraction force. We have already discussed the normal force and have noted its importance in minimizing connector resistance and its influence on wear of the pin and contacts. Insertion and extraction forces are also important because they control the ease of assembly of a circuit board or a set of cables in an electronic system. To determine the insertion force, consider the free body diagram of the blade and contact shown in Fig 8.18. When the blade is inserted into the contact, the taper on the blade encounters the contact and the normal forces developed are not perpendicular to the axis of the blade. Equilibrium relations give the insertion force F_i as:

$$F_i = 2N_i\,(\sin\theta + \mu\cos\theta) \tag{8.5}$$

where θ is the taper angle, μ is the coefficient of friction and N_i is the normal force.

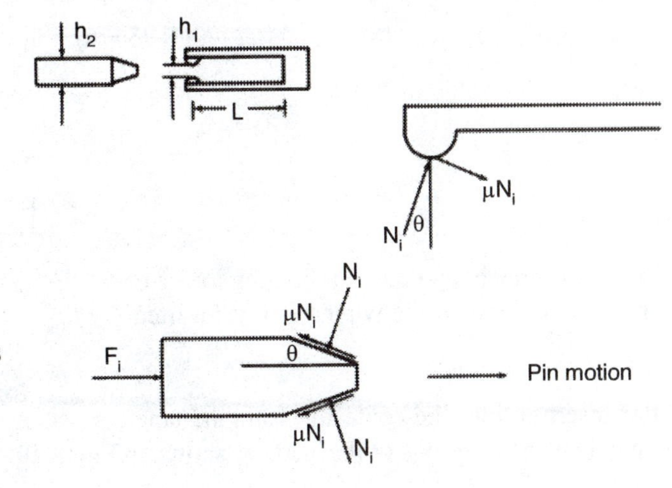

Fig 8.18 Free body diagram showing forces on the header blade during insertion.

The normal force N_i is controlled by the stiffness of the contact and the amount of interference Δh between the blade thickness and contact opening. If the contact is treated as a cantilever beam, we may write the force deflection equation as:

$$P = \frac{3EI\delta}{L^3} \tag{8.6}$$

where P is the transverse force and δ is the deflection of the free end of the cantilever beam.
The forces acting on the pressure point at the entrance to the contact, shown in Fig. 8.18, may be resolved to give the force P transverse to the beam as:

$$P = N_i (\cos\theta - \mu \sin\theta) \tag{8.7}$$

Combining Eqs (8.6) and (8.7) gives:

$$N_i = \frac{3EI\delta}{\left[L^3(\cos\theta - \mu \sin\theta)\right]} \tag{8.8}$$

Examination of this equation shows that the normal force on the blade is a function of the stiffness of the contact (EI/L^3), the deflection imposed on the contact during insertion, the taper angle and the coefficient of friction.

Next, substitute Eq (8.8) into Eq (8.5) to obtain the insertion force during engagement of the blade in the contact as:

$$\begin{aligned} F_i &= 2N_i(\sin\theta + \mu\cos\theta) \\ &= 2\left(\frac{3EI\delta}{L^3(\cos\theta - \mu\sin\theta)}\right)(\sin\theta + \mu\cos\theta) \\ &= \left(\frac{6EI\delta(\tan\theta + \mu)}{L^3(1 - \mu\tan\theta)}\right) = \left(\frac{3EI(\Delta h)(\tan\theta + \mu)}{L^3(1 - \mu\tan\theta)}\right) \end{aligned} \tag{8.9}$$

where $\Delta h = 2\delta$, is the interference given by:

$$\Delta h = 2\delta = h_2 - h_1 \tag{8.10}$$

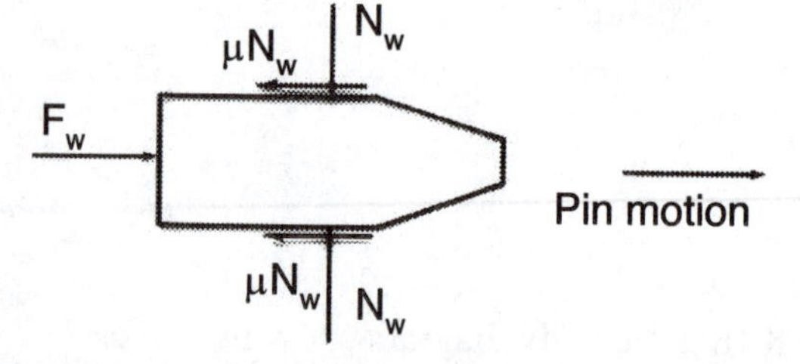

Fig. 8.19 Free body diagram showing forces on the header blade during wiping as it is inserted.

After insertion into the contact beyond the taper, the blade is wiped and the normal force N_w becomes perpendicular to the axis of the blade as shown in Fig. 8.19. The force N_w decreases to:

$$N_w = P = \frac{3EI\delta}{L^3} \tag{8.11}$$

The insertion force F_w during the wiping part of the engagement reduces to:

$$F_w = \frac{3\mu EI\Delta h}{L^3} \tag{8.12}$$

Upon extraction, the direction of the friction forces reverses, as indicated in Fig. 8.20, and the extraction force F_e becomes:

$$F_e = F_w \tag{8.13}$$

The relations given above describe the forces required to engage and disengage a single pin in a connector. In most cases, the connectors have many pins and it is necessary to multiply the single pin forces by the number of the pins being engaged. Single pin insertion forces usually range from 0.20 to 0.5 lb (0.89 to 2.22 N) and multiple pin connectors often require relatively large insertion and extraction forces. When the total insertion force exceeds about 35 lb (156 N), it is necessary to provide mechanical aids to facilitate assembly of the connector. The mechanical aids may be as simple as two screws that are used to draw the connector together or levers that are hinged to the chassis to give a mechanical advantage to service personnel making the connection.

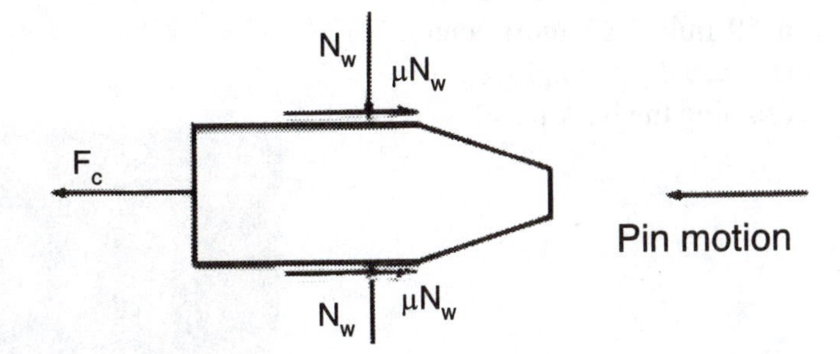

Fig. 8.20 Free body diagram showing forces on the header blade during extraction.

8.3 BACK PANEL, WIRE WRAP BOARDS AND CABLE CONNECTORS

Edge connectors on printed circuit boards permit their insertion into a back panel, a wire wrap board or a cable arrangement. All three of these components serve the same function, namely to carry power to all of the PCBs and to carry signals from one to another. The choice from among the three alternatives depends largely on complexity and on the volume of the product produced. We will describe these three methods for connecting the PCB's in the following subsections.

8.3.1 Rigid and Flex Back Panels

A back panel, often called a mother board, may be similar to a multi-layer printed circuit board in that it is laminated with alternating layers of glass-cloth reinforced plastic, each providing two planes of copper cladding for power, signal and ground. Back-panels may also be made from flex-circuits instead of rigid glass-epoxy printed circuit board (Figure 8.21b). A back panel is different from a PCB in that is larger in size, contains more layers, and the foot print is usually designed to accept the contact side of a connector. An example of a relatively small back plane, illustrated in Fig. 8.21, shows the plated through holes which accommodate the pins from the connector. In this design, the pins on the reverse side of the receptacle are press fit in the plated through holes to make connections with the interior wiring planes in the back panel. The nine receptacles shown in this figure each accommodate 96 pins from the edge connectors giving an I/O capacity of 854. This particular back panel is a six layer design with three layers or six planes devoted to signal. The two voltage planes are independent of each other and distribute the required voltages. The ground planes are connected together with several plated through holes to avoid ground loops and to ensure that the ground acts as a single plane. Power blocks are fitted to the back side of the panel to provide for high current connections to the power buss.

Back panels vary in size and complexity depending on the application. Leading edge manufacturing technology enables line and spacing widths of 3 mil (75 μm) and drill diameters of 9 mil (225 μm). Drill capture pads are 16 mil (0.40 mm) in diameter. The high density circuits possible with these feature sizes reduce the area of the back panel. Those back panels that support many high density circuit cards with high I/O count utilize many signal and power layers. In these cases, surface mount edge connectors attached to pads on 50 mil (1.27 mm) centers are usually employed in designing the back panel.

Fig. 8.21a Back panel with 11 DIN receptacles each with 96 contacts.

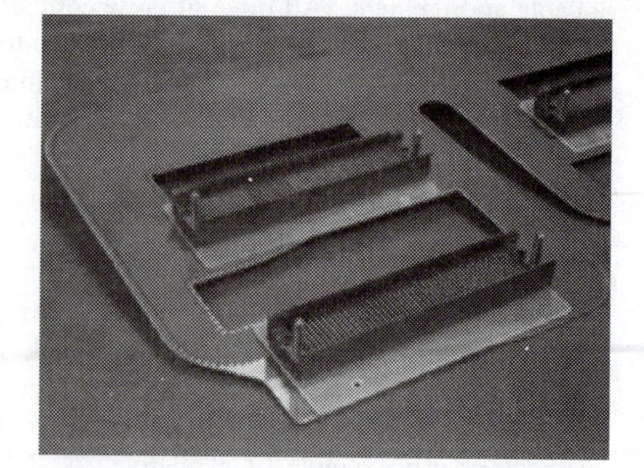

Fig. 8.21b: Flex-Circuit Back-Plane (Courtesy of Molex, Inc.)

For very high performance products, such as the IBM's BladeCenter® eServer™, unique packaging is employed to handle the very high-density, high-frequency chips used in these products. Each BladeCenter supports 14 processor blade servers. At the center of the class e Server is a four chip multi-chip module (MCM) with 170×10^6 transistors on each chip. These chips are attached to a ceramic substrate with 1,780 flip chip connections. The ceramic substrate for the four-chip module has top surface pads 100 μm in diameter on 200 μm centers. Interconnections within the substrate are accomplished with 1.7×10^6 copper vias and 190 m of co-sintered copper wiring. Another 5,100 module connections on 1.0 mm pitch are placed on the bottom surface of the substrate to allow for connections to the back panel with a land grid array.

Back panels are expensive and are usually used only in large, higher performance processors with moderate to large volume production. For smaller processors with very limited production, wire wrap panels may be more cost effective than back panels.

8.3.2 Wire Wrap Panels

Wire wrap panels, illustrated in Fig. 8.22, are similar to back panels in that they are multi-layer boards with an array of plated through holes to accommodate pins. Wire wrap panels contain layers for power and the ground with 2 or 4 oz. copper, but signal layers are not provided. On the connector side of the board, holes are available to accept component pins, connector pins or socket pins. On the wire wrap side of the board, wrapping posts extend beyond its bottom surface. The point to point connections are made by wrapping the bare ends of an insulated wire in a tight helix about a four cornered square post. A typical wire wrapped connection is illustrated in Fig. 8.23.

Fig. 8.22 A wire wrapped circuit board showing posts and wiring.

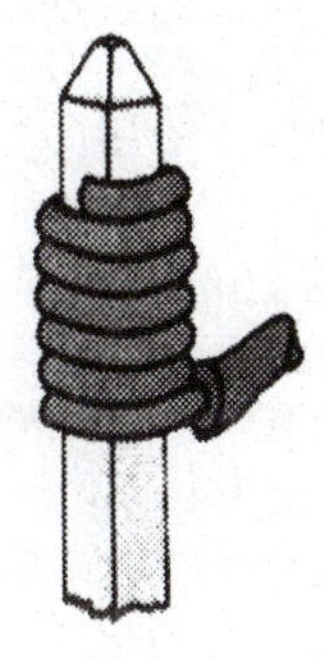

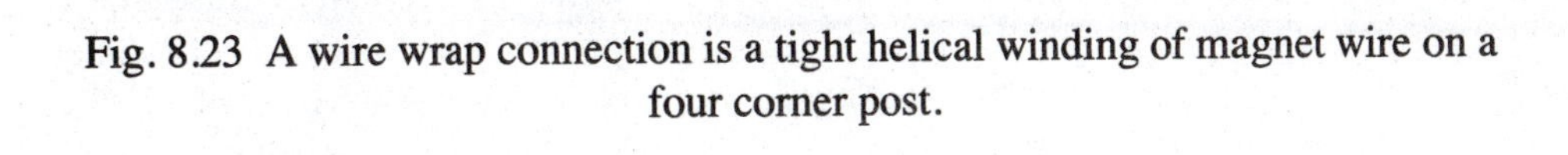

Fig. 8.23 A wire wrap connection is a tight helical winding of magnet wire on a four corner post.

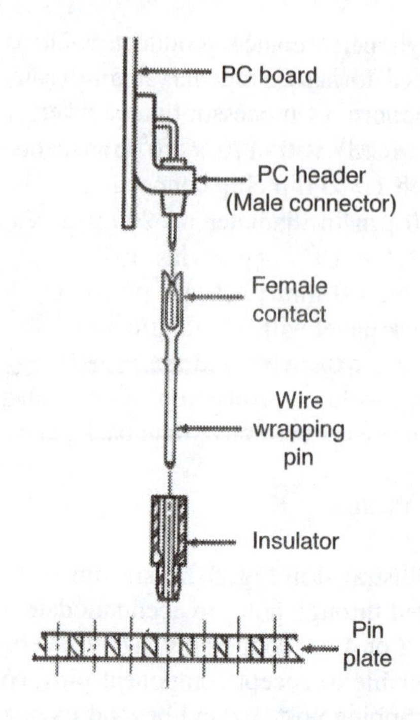

Fig. 8.24 Assembly sequence in preparing a pin plate showing the fabrication of the contact insertion of the PCB.

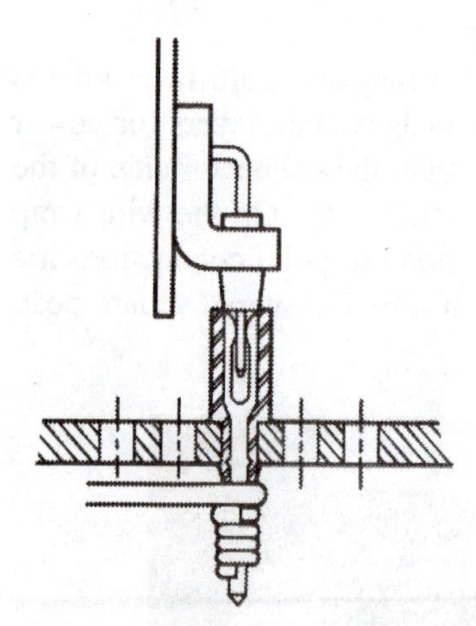

Fig. 8.25 Completed assembly of a pin plate after wire wrapping and insertion.

A pin plate provides another method of connection which is similar to the wire wrap panel. The pin plate is usually a sheet of aluminum or plastic with holes drilled in a rectangular array. Insulators and female contacts are inserted into these holes to form the receptacle portion of a connector as shown in Fig. 8.24. The tail on the contact extends beyond the lower surface of the pin plate and serves as the four cornered post used in wire wrapping. The circuit card is mated with the female contacts in the pin plate as indicated in Fig. 8.25.

Wire wrapping a pin plate or a panel provides a method of connecting together circuit boards with moderate density and size. However, if the volume of the production is large the labor cost for making the large number of wire wrapped joints becomes prohibitive and the use of a back panel becomes more cost effective. With small volume production and semi automatic operations, the initial tooling costs are moderate and wire wrapping is often the preferred approach in countries with low labor costs. Wire wrapping and pin plate connections are also used in developing prototypes of electronic systems where a large number of changes in the wire list are anticipated.

8.3.3 Cable Connected Circuit Boards

A common method for connecting a few circuit boards together that avoids the complexities and costs of a back panel or a wire wrapped board uses flat cables or flexible circuits. A flat cable, illustrated in Fig. 8.26, consists of a number of adjacent wires lying in a common plane. The wires are connected together by a web of insulation so that the multi conductor assembly resembles a sheet of parallel conductors. Connectors either male or female can be attached to the cables at one or more locations along the length of the flat cable. The conductors in the cable serve the same purpose as the back panel or wire wrap panel in connecting together two or more circuit boards. It is interesting to note, in Fig. 8.26, that the connector shown incorporates a contact that acts as a knife to cut through the cable insulation and then the contact clamps the wire to make an air tight connection. A barb on the top of the contact snaps into the connector cap and permanently locks the assembly in place without fasteners. This type of design facilitates rapid assembly which significantly reduces labor costs in preparing cable connections.

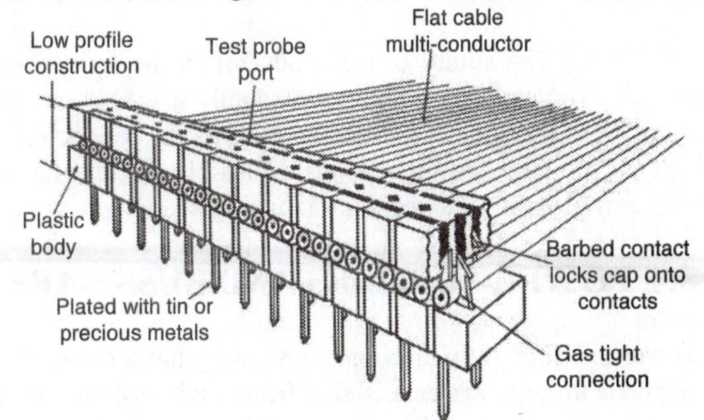

Fig. 8.26 Flat cable connections.

A similar design concept involves the use of flexible circuits instead of the flat cable as shown in Fig. 8.27. The flexible circuit is the same as a printed circuit board except that a polyimide film is used as the substrate instead of the rigid circuit board materials. The conductors are etched from rolled and annealed copper cladding that is bonded to the flexible substrate. A second layer of polyimide film is bonded over the copper conductors to form a sandwich structure which completely insulates the flexible circuit. They can be mass terminated with a connector, as shown in Fig. 8.27, or they can be soldered directly to the circuit boards to act as jumpers connecting one board to another as shown in Fig. 8.28.

Fig. 8.27 A flexible circuit used to provide connections for three receptacles.

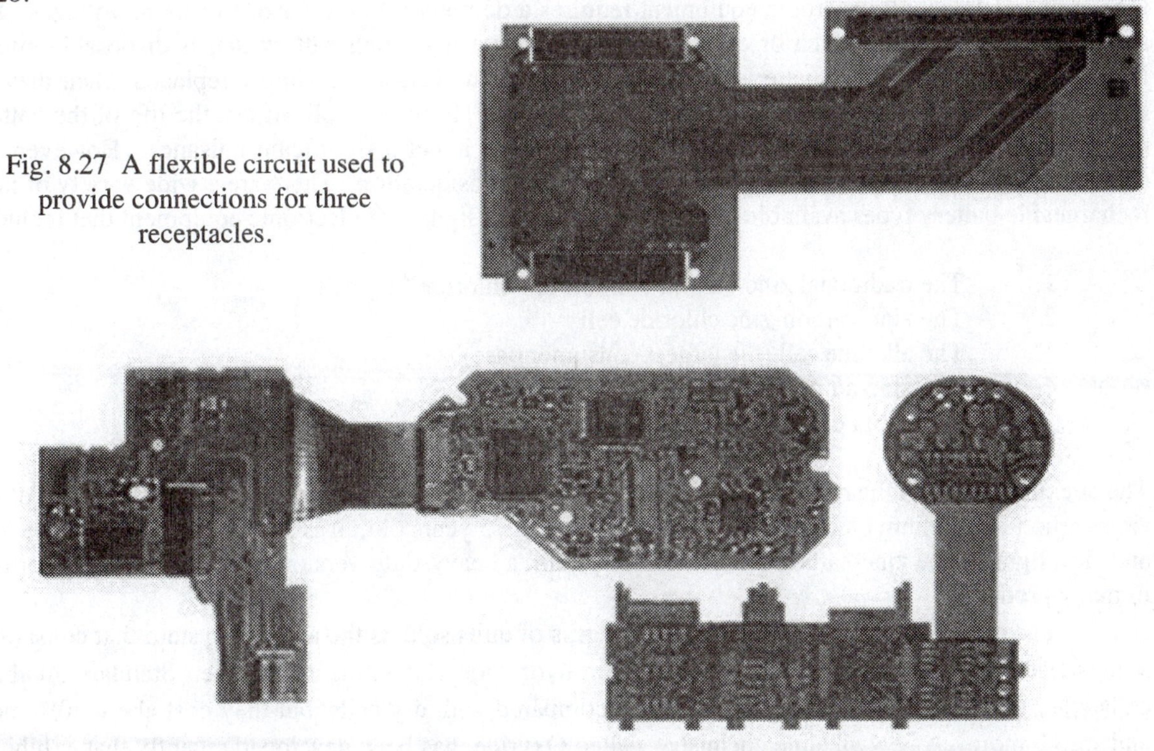

Fig. 8.28 Flexible circuits used as jumpers to connect circuit cards.

Either flat cable or flexible circuit connections between PCB's are employed when a relatively small number of connections are required. The method provides for cost effective solution to a limited class of connection problems because either flat cables or flexible circuits are much lower in cost than the back panels. The flexible circuits have additional advantages which include:

1. The ability of the conductors to be formed into almost any imaginable configuration and then used to connect randomly spaced points on a given plane.
2. The flexible substrate supports the conductors and provides a connection system capable of large cyclic motions. Flexible circuits fabricated with rolled and annealed copper foil exhibits a very high fatigue life.

8.4 POWER SUPPLIES AND BUSS BARS

Power distribution in electronic systems is not a difficult task, but it is an important topic. The power required in large systems is significant and is often specified in tens of kW. On the other hand, the power required in small systems, say your watch, is measured in microwatts. The production of the dc voltages and the distribution of the power are topics of major importance in the design of a wide variety of electronic systems that may range from less than a dollar for the circuits in a watch to several hundred thousand dollars for a large processor or server[6].

8.4.1 Batteries as Power Supplies

Non-Rechargeable

Operation of almost all electronic equipment requires a dc power source for one or more voltages. For small systems, like your watch or calculator, the dc source is a small battery that is disposable after a life of several years. These batteries are relatively inexpensive and are simply replaced when they no longer provide sufficient voltage to operate the equipment. In these applications, the life of the battery must be sufficiently long so that the need for replacement is not a significant nuisance. However, the size and weight of the batteries used is a very important consideration. There are a wide variety of non-rechargeable battery types available to meet the needs of designers of electronic equipment that include:

1. The traditional zinc-carbon-ammonium-chloride (dry cell)
2. The zinc-carbon-zinc chloride cell
3. The alkaline cell (the largest consumer base)
4. The zinc-air cell
5. A number of lithium-based chemistries

The chemistry used in the many of these batteries has been around for many years. The chemistry of the zinc-carbon-ammonium chloride "dry-cell" is more than 75 years old. It is still used in inexpensive toys and flashlights. The zinc-carbon-zinc chloride system, a heavy-duty version of the dry cell is also used in many products.

The most popular battery chemistry, in terms of units sold, is the alkaline system that consists of manganese dioxide, zinc with a caustic potassium hydroxide-zinc oxide electrolyte. Standard alkaline-cells offer greater capacity for a given cell size compared with dry cells, but they cost about 50% more and weigh more. A new alkaline chemistry, called Oxyride, has been developed recently that exhibits a

---

[6] When six BladeCenter chassis are placed in a 42U rack the power required exceeds 30kW. Customers provide the 12 VDC power supplies used in this blade server with three phase 240 VAC at 60 A.

50 to 100% longer life than standard alkaline cells. This is an important advantage for users of digital cameras and portable audio players. The Oxyride battery uses newly adopted cathode materials (Oxy nickel hydroxide and manganese dioxide and graphite). Improved manufacturing methods, which include a controlled mixing ratio of the chemicals and a new vacuum pouring technology, enables the quantity of electrolyte in the battery to be increased, resulting in longer life.

Batteries using mercury cells, once widely used in applications requiring small (coin) size and relatively low currents, have been replaced with zinc-air batteries. Compared with both mercury cells and alkaline cells of the same physical size, zinc-air cells have much higher energy density. The zinc-air cells are energized only when atmospheric oxygen is absorbed into the electrolyte through a gas-permeable, liquid-tight membrane. When a sealing tab is removed from the battery, oxygen from the air enters the cell energizing it within 5 s. While sealed, this battery has excellent shelf life with only a 2% self-discharge rate.

Of all the replaceable battery chemistries, lithium is the most interesting. Lithium is an ideal material for battery anodes because its intrinsic negative potential exceeds that of all other metals and because it is the lightest non-gaseous metal. Batteries based on lithium have the highest specific energy and energy density of all the available cell chemistries. The high energy density is a result of lithium's high potential and the fact that lithium reacts strongly with water.

When size, weight and long life at low current drain are important, the lithium batteries are clearly superior over the alkaline and carbon-zinc batteries. There are several different chemistries used in the production of lithium batteries. We will provide brief description of different types of lithium batteries below:

1. Poly-carbon monoflouride cells have an output of 2.8 V and moderately high energy density as indicated in Table 8.2. Those packaged as cylinders are manufactured with a spiral-shaped cathode and crimped polymeric seals. Though generally safe, under extreme conditions the seals can fail before the case and reactive contents in the cell escape.
2. Manganese-dioxide lithium cells are comparable to poly-carbon monoflouride cells in terms of construction, energy density, safety, and output voltage, but they exhibit about half the service life. They are preferred in applications with relatively high continuous-current or pulse-current requirements because the internal impedance is lower than that found in other lithium chemistries. They are also less expensive than poly-carbon monoflouride batteries.
3. Lithium-iodine cells do not contain liquids and spillage due to seal failure is not a problem. Its major disadvantage is high internal impedance limiting its applications to products which draw very low currents.
4. Sulfur dioxide-lithium cells are used almost exclusively in military applications. They exhibit lower energy density than manganese dioxide-lithium or carbon monoflouride lithium chemistry. Their service life and energy density are less than half that of lithium thionyl chloride batteries.
5. Lithium thionyl chloride cells have the highest energy density of all the different lithium chemistries with typical service life of 15 to 20 years. These batteries are preferred for applications with low continuous currents and low to moderate pulses in the current requirements. Their long service life and low self-discharge rate make them ideally suited for many different products. They feature an operating temperature range of – 55 to 150 °C, high capacity, small size, and an ability to withstand broad fluctuations in pressure, temperature and shock.

Table 8.2
Properties of lithium cells with different chemistries

Chemistry	Cathode Material	Specific Energy W/kg	Voltage	Operating Temp. Range (°C)	Maximum Service Life (years)	Packaging	Applications
Li/$SOCl_2$	Thionyl Chloride	700	3.6	−55 to 150	15 to 20	Bobbin welded	Industrial Consumer
Li/SO_2	Sulfur Dioxide	260	2.8	−55 to 70	5	Spiral welded vented	Aerospace Military
Li/MnO_2	Manganese Dioxide	330	3.0	−20 to 60	5	Spiral elastomeric seal	Consumer
Li/$(CF)_x$	Poly-Carbon monoflouride	310	2.8	−20 to 60	5	Spiral elastomeric seal	Consumer
Li/I_2	Iodine	230	2.7	0 to 70	10	Welded	Medical devices

Rechargeable Batteries

Other battery types such as lithium-ion, nickel-cadmium, nickel-metal-hydride and lead-acid are rechargeable so that larger power requirements can be accommodated with periodic recharges. Recharging does not eliminate the need for replacement, but replacing a rechargeable battery should be a relatively rare event. We will briefly discus each type of battery with emphasis on those used in higher performance products.

Lead Acid

The lead acid battery, found in your automobile, provides high currents and is very cost effective; however, it is large and heavy. As such, it is only used in applications when its size and weight is not a primary consideration and where high currents are often required. Examples of such applications include power for emergency lighting and for computer backup in the event of power failure.

Nickel-Cadmium Batteries

The nickel-cadmium cell with an output voltage of 1.2 V, found on most of your battery powered hand tools, is low in cost and can be recharged 500 times before the end of its life. It is favored for power tool applications because it retains its voltage well as it is discharged and it has a low life cycle cost. Another advantage is that it can be used over a wide temperature range – 50 to 60 °C. It does not compare well with nickel-metal-hydride for energy density and specific energy; consequently, it is not used in higher performance equipment where minimizing weight and size are important design considerations. A particularly serious problem with Ni-Cad cells is the toxicity of the cadmium used in its construction. This chemical has been banned in many countries.

Nickel-Metal-Hydride Batteries

Nickel-metal-hydride (Ni-MH) batteries employ nickel hydroxide for the positive electrode and alloys[7] capable of absorbing and releasing hydrogen from the negative electrode. Because Ni-MH batteries have approximately twice the energy density as Ni-Cad batteries they have taken over a large part of the market for rechargeable batteries. As with any battery, there are five main characteristics to consider in judging the adequacy of a battery for a specific application—charge, discharge, storage life, cyclic life and safety.

Both the charge and discharge characteristics of Ni-MH batteries are affected by the charging current, time and temperature. It is recommended that the Ni-MH battery be charged with a constant current of 1.0 A or less. Charging should be accomplished at temperatures from 10 to 30 °C. Overcharging should be avoided because it degrades performance. The discharge characteristics are flat at 1.2 V. Compared to Ni-Cad batteries the Ni-MH batteries have inferior high-rate discharge characteristics; consequently, they are not recommended for applications requiring high discharge currents.

The charge on a battery stored for some extended period of time depends on their self-discharge characteristics. Most batteries have a finite shelf life because of this characteristic. When stored for four weeks at a temperature of 20 °C, a Ni-MH battery retains about 85% of its capacity compared to about 80% for a Ni-Cad battery. Both types of batteries can be restored to their full capacity by recharging.

The cyclic life of both the Ni-Cad and the Ni-MH batteries is same with a mean life of 500 cycles. The capacity of both batteries degrades to some degree with life. After 500 cycles the Ni-MH battery retains 90 to 95% of its original capacity. When the internal pressure in either the Ni-Cad or the Ni-MH batteries increases due to over charging, short circuiting or reversing the charge, a safety vent opens releasing the pressure. Because this vent is self sealing, it closes when the pressure level decreases to a safe value restoring the integrity of the battery.

Lithium Ion Batteries

Lithium-ion batteries and lithium batteries are different. Lithium-ion batteries are rechargeable whereas the lithium batteries are not. Lithium-ion batteries have a very high specific energy and energy density; hence, they are specified for products where size and weight are important. A comparison of the specific energy and energy density among Lithium-ion, Ni-Cad and Ni-MH batteries is shown in Fig. 8.29. It is clear in this illustration that below average Lithium-ion batteries are superior to the better Ni-MH batteries and much better than the better Ni-Cad batteries.

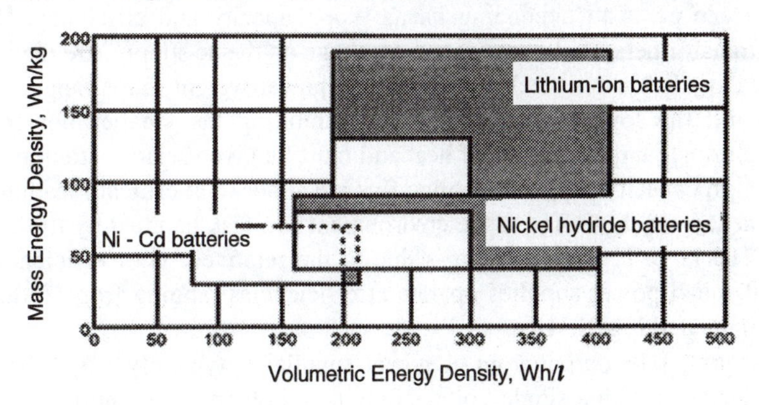

Fig. 8.29 A comparison of specific energy (mass energy density) and specific energy density (volume energy density) for three common rechargeable batteries.

---

[7] These alloys include TiFe, $ZnMn_2$, $LaNi_5$, and Mn_2Ni.

Lithium-ion cells produce 3.6 volts which is three time that produced by Ni-Cad or Ni-MH cells. This characteristic indicates that higher voltages can be achieved without stacking cells resulting in lower weight and smaller volume packaging of the Lithium-ion batteries. Lithium-ion batteries do not exhibit the memory effect observed in Ni-Cad batteries where the apparent discharge capacity is reduced when it is repeatedly discharged incompletely and then recharged. Lithium-ion batteries exhibit a relatively flat discharge voltage.

The batteries disadvantages include high cost, some evidence suggesting their performance gradually degrades with time and a concern over safety. These batteries can exhibit spontaneous combustion[8] if internal hot spots exceeding 130 °C develop.

8.4.2 Power Supplies

Power supplies are used to convert ac line voltage at 115 or 230V to a dc source with one or more fixed voltages. The power supplies may be packaged in a number of different ways. In some applications, the supply is totally enclosed and can be used as a free standing unit. In other cases, as shown in Fig. 8.30, the unit is enclosed in a safety screen and then housed in another electronic enclosure together with other parts of the system. Other power supplies are open and are assembled on a simple sheet metal platform. These open supplies must be enclosed and shielded in the enclosure to insure the safety of service personnel.

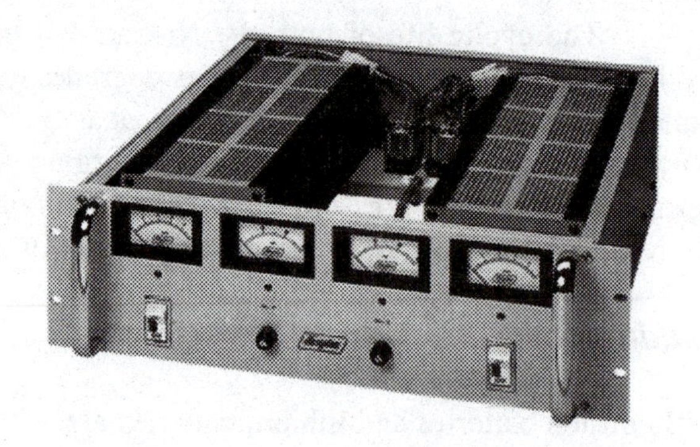

Fig. 8.30 A dual, rack-mounted power supply equipped with fuses, switches, voltmeters and ammeters.

There are several features of a power supply that must be considered before incorporating it into an enclosure. First, power supplies are large and they require from 1 to 3 in^3 (16,400 to 49,200 mm^3) of space per watt depending on its type, capacity and efficiency. Second, they are heavy, because the transformers used to convert the voltage on the ac supply, are made with large copper coils wound about heavy laminated steel cores. Typical power weight ratios range from 40 to 100 W/lb (178 to 445 N/W) with the lower capacity supplies exhibiting the smaller power weight ratios. Third, the supplies dissipate large amounts of heat and must be given serious attention in the design of a cooling system. In high capacity supplies, cooling fins are attached at critical locations to dissipate the heat. Managing the heat from these fins to the environment is a task handled by those designing the total system enclosure. The large heat dissipation is due to the relatively poor efficiencies of the power conversion process. Typical power supplies operate at efficiencies ranging from 70 to 90% depending on power output as indicated Fig. 8.31.

The output from dc power supplies is typically ± 5, ± 10, ± 12 and ± 24V. Some supplies are designed with a single voltage output and others have multiple voltages outputs. There is no difficulty

[8] In August of 2006, Dell replaced the batteries in several million of its laptop computers because of the danger due to fires that occurred from over heated batteries.

in specifying any voltage required for the system if custom designed power supplies are procured. With relatively low voltages and high power output, it is important to recognize that the supplies provide large currents. For example, a 1.0 kW supply at 5 V has a current output of 200 A. Distribution of these high currents and the termination of the distribution conductors require careful consideration to minimize voltage drops (IR) along the distribution line and power losses (I^2R_c) at the terminations.

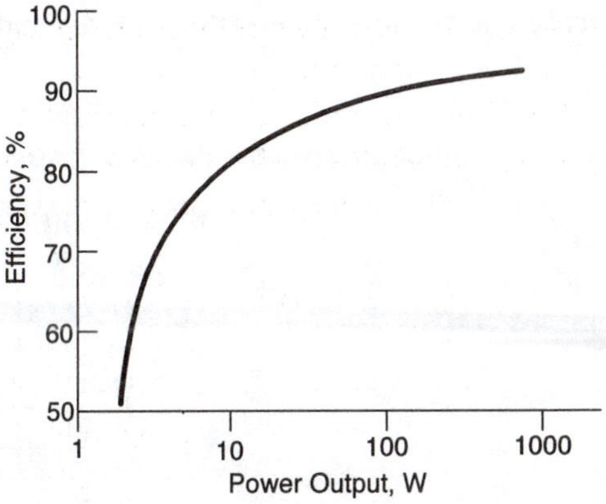

Fig. 8.31 Efficiency as a function of power output for typical power supplies.

8.4.3 Buss Bars

Buss bars are conductors in the same sense that wire is a conductor and in some instances the buss bar is simply a tin plated wire that is not insulated. We refer to bus bars here as conductors used to distribute high amperage currents from the power supplies to the back panels.

Buss bars can be heavy gage copper wire with a circular cross section. Copper wire sizes are commonly designated in the American wire gage (AWG) standard that employs numbers that extend from #0000 for the largest conductor to #40 for the smallest. The capacity of solid and stranded annealed copper wire of different gage based on the AWG number is given in Table 8.3. The current carrying capacity of a specific wire number depends upon the insulation covering the wire. The current capacity ranges from 343A for the 0000 wire to 7A for #18 wire with insulation rated at 75 °C. For buss bar applications, heavy conductors are common with wire size of #8 or less employed in normal designs. Current densities of 0.001A/circular mil are usually employed in large distribution circuits. Note, that 1.0 circular mil is the area of a circle inscribed in a 1.0 mil square.

Table 8.3
Current capacity in amperes for single insulated copper wires in free air at an ambient temperature of 40 °C.

Wire Size	Temperature rating of insulation		
AWG	60 °C	75 °C	90 °C
18			16*
16			22*
14	24*	30*	35*
12	30*	39*	45*
10	41*	51*	61*
8	55	71	83
6	73	94	109
4	96	124	145
3	112	145	169
2	128	165	192
1	148	191	223
0	171	221	258
00	198	255	298
000	229	295	345
0000	266	343	400

Note the overprotection for wires marked with the * symbol in Table 8.2 shall not exceed 7 A for #18, 10 A for # 16, 15 A for #14, 20 A for #12 and 30 A for #10 copper wire. Current carrying capacities increase as the ambient temperature decreases and decreases when the ambient temperature increases. The scale factors shown in Table 8.4 are employed to adjust the values in Table 8.3 accommodating ambient temperatures that differ from 40 °C.

Table 8.4
Scale factors for adjusting current capacity to account for ambient temperatures

Ambient	Temperature rating of insulation		
Temp. °C	60 °C	75 °C	90 °C
21-25	1.32	1.20	1.14
26-30	1.22	1.13	1.10
31-35	1.12	1.07	1.05
36-40	1.00	1.00	1.00
41-45	0.87	0.93	0.95
46-50	0.71	0.85	0.89
51-55	0.50	0.76	0.84
56-60	-----	0.65	0.77
61-70	-----	0.38	0.63
71-80	-----	-----	0.45

For very high currents bus bars are often fabricated from sheet copper using several layers of copper insulated with plastic film. The sheet copper is easy to form into long, thin but wide conductors that can be attached with a simple screw clamp to the power and ground planes on the back panel. Low resistance terminations at the junctions to the bus bar are critical to avoid excessive heating of the joints. To illustrate the possibility of heating, consider a termination that conducts 100A with a connection resistance of 10 mΩ. The power dissipated at this joint is 100W, which is sufficient to cause the connection to fail due to oxidation and eventually melting. The connection resistance must be reduced to less than 10 to 100 μΩ in high current applications to limit power dissipation at the joints to less than 0.1 to 1 W. Low resistance connections are made by designing joints with large areas and clamping them together with a bolt that is sufficiently large to apply contact forces at the joint of about 100 lb (445 N) or more. These high contact forces fracture any oxide films and produce a true contact area that is a larger fraction of the apparent contact area.

8.5 CARD RACKS

Printed circuit cards are assembled in orderly rows in a card rack. The card rack supports the lateral edges of the cards during the life of the product, provides guide rails for the initial insertion of the cards during assembly and alignment of the cards during substitution involved in servicing. A typical card rack, illustrated in Fig. 8.32, intended for commercial applications, is usually fabricated from sheet metal stampings and/or aluminum extrusions to form a rectangular opening much like a book shelf. The card racks are usually mounted in some type of an enclosure. A back panel, with the receptacle half of the rows of connectors, often forms the rear wall of the enclosure as shown in Fig. 8.33. Card guides are spaced at uniform intervals along the length of the card rack to facilitate the alignment of the PCB with the connector during insertion. The top and bottom surfaces or the card rack are usually open to permit air flow between the cards if direct impingement convection cooling is used. The structural rigidity of the card racks varies with the application of the system. For systems intended to be used in the normal office environment relatively thin gage sheet metal that is spot welded together to form a suitable rectangular structure is sufficient. In industrial applications, the top and bottom plates, containing the slots for the card guides are usually fabricated from aluminum extrusions. In military applications involving shock qualification testing, the card racks are usually machined from higher strength aluminum plate thick enough to withstand shock loading of 100 G's or more without undergoing plastic distortion.

Fig. 8.32 A typical sheet metal card rack intended for commercial applications.

Card racks are usually mounted into an enclosure. The assembly is rapid as the enclosure hardware accommodates the flange mounts on the sub rack, as illustrated in Fig. 8.33.

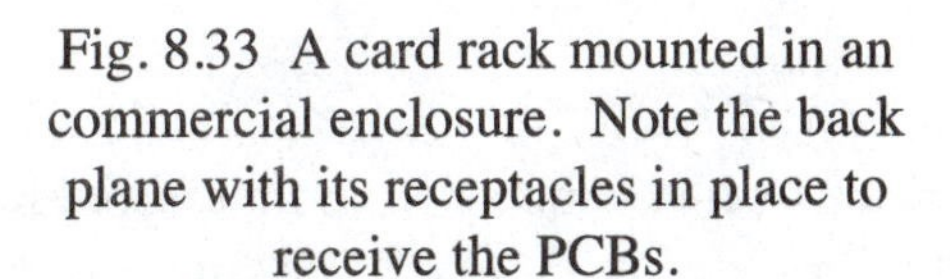

Fig. 8.33 A card rack mounted in an commercial enclosure. Note the back plane with its receptacles in place to receive the PCBs.

8.5.1 Card Guides and Retainers

Card guides are often divided into three groups depending upon the mechanical and thermal requirements imposed on the design. In the first group, the devices act only to guide the card into the connector and to loosely support the board. The board is maintained in its engaged position by a retainer bar attached to the sub rack across one edge of the PCB's. Low cost guides molded from nylon, shown in Fig. 8.34, are adequate in this application.

In the second group, the PCB is inserted into machined or extruded slots that are provided in the card rack. Card edge retainers reside in these slots and serve to guide the card into the receptacles located on the backplane and to clamp the card against one side of the slot. The use of these clamping type edge retainers is essential in vibration environments. Clamping one or more edges of the boards significantly increases the natural frequency of the circuit board and eliminates damage that may occur at lower resonant frequencies. Also, in some installations conduction cooling is employed and it is necessary to clamp the heat frame that is bonded to the PCB against a cold rail. Clamping the card is achieved by placing the edge of the PCB in a lever actuated cam device as shown in Fig. 8.35. This type of retaining guide has two basic parts. The first is a channel like housing that has a series of ramps stamped on its back side. The second part is a cam made of a strip of beryllium copper with a matching set of ramps. Rotating the lever pulls the cam up onto the channel ramps and wedges the PCB between the cam and the housing. Retaining forces of 30 to 45 lb (133 to 200 N) can be achieved with spring like retainers. The thermal contact resistance per unit length of the clamp varies from 4 to 17 (°C/W)/in. or 0.16 to 0.67 (°C/W)/mm depending on the detail design of the retainer.

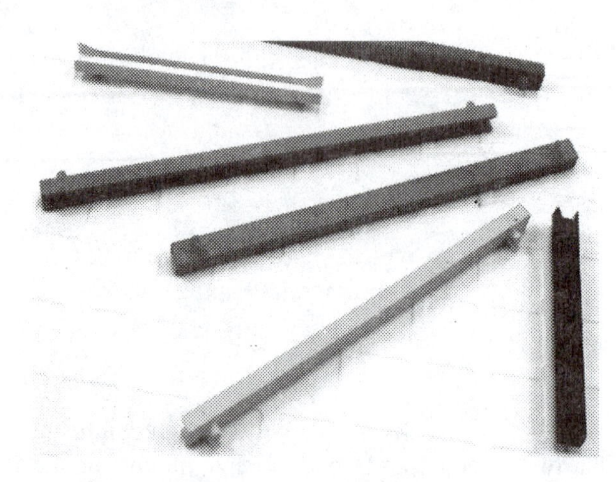

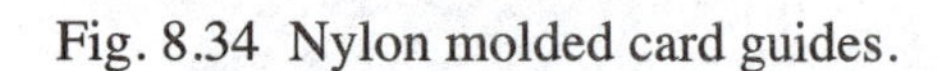

Fig. 8.34 Nylon molded card guides.

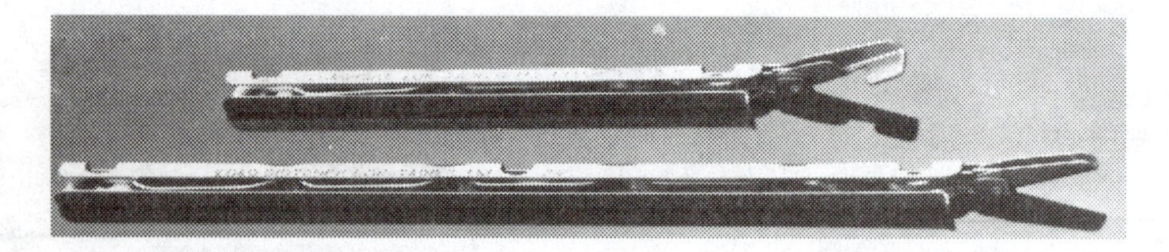

Fig. 8.35 Cam-spring retaining clamps.

The third type of retainer is for very heavy duty applications with clamping forces measured in hundreds of pounds. These devices use a wedge locking design illustrated in Fig. 8.38. The retainer is fabricated from aluminum bar stock with 45° angles cut on the end of each piece. A screw runs the length of the retainer. Tightening the screw causes the wedges to slide on one another and expand the width of the retainer. When the retainer and circuit board are both contained in a slot that is cut in a cold rail, the expansion of the retainer forces the PCB against the wall of the slot. When conduction cooling is employed, a heat frame bonded to the card is clamped in place with a large uniformly distributed retaining force. Thermal contact resistance per unit length achieved with wedge lock devices range from 2.7 to 8.5 (°C/W)/in. or 0.11 to 0.25 (°C/W)/mm depending on the detail design of the retainer.

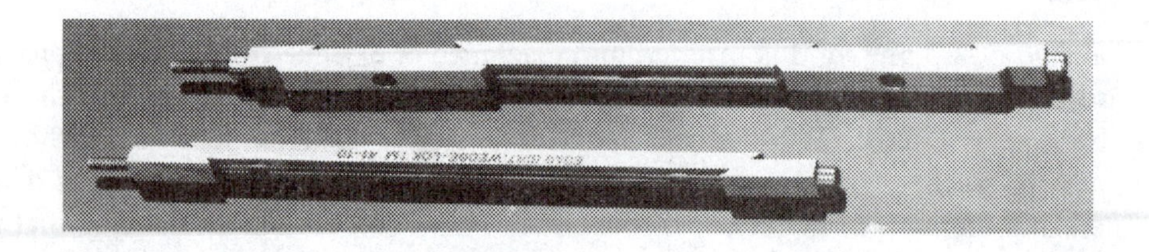

Fig. 8.36 Screw and wedge edge card retainers.

8.6 ELECTRONIC ENCLOSURES

Electronic enclosures serve to house the entire electronic assembly. The enclosures range in size from a case for a watch to large rugged military enclosures that weigh more than a ton when filled with cards and power supplies. While standards have been adapted for commercial instrument cases and cabinets, the range of product which must be enclosed is large and wide spread standardization is difficult. Another, factor that impedes standardization is the volume of the product produced. When large annual volume is involved, it is possible to design unique cabinets with special features that facilitate assembly and reduce both weight and volume. These specially designed enclosures are often lower in cost when compared to a commercially available standard cabinet. A final but critical argument for unique design of the enclosures is the appeal of the package. In some office products, the enclosure must serve as a piece of furniture and blend with the colors and function of the office. And while image is beyond the scope of this text, we should be well aware of its importance in marketing a product.

In this coverage of enclosures, we will first emphasis features common to the design of standard instrument cases and cabinets used in the office environment. As a second part, we will describe the common electronic equipment enclosure (CE^3) that was developed for ship board electronics employed on naval vessels.

8.6.1 Commercial Enclosures

Commercial enclosures, which include standardized instrument cases and cabinets, are intended for office and laboratory environments. These enclosures protect the circuits in a friendly environment. The environment is considered friendly if the temperature is maintained in a range comfortable for personnel, if the air is free of dirt and dust, the floors do not vibrate and the operating personnel are cooperative in that they respect ordinary locking devices.

A commercial instrument case, illustrated in Fig. 8.33, is intended to house an electronic system small enough to be placed on a table, desk or laboratory bench. A commercially available cabinet which houses larger systems, shown in Fig. 8.37 stands on the floor. Both enclosures are designed to be: (1) attractive, (2) to contain the PCBs and power supplies, (3) to permit easy access for assembly and service personnel, (4) to be structurally sound and (5) to accept standard card racks, panels and accessories such as slides, drawers and shelves. Many features available on standard cabinets are shown in Fig. 8.37.

Containment of the circuitry is important in keeping dust and dirt from accumulating on the circuit boards and back panel. It is also essential in protecting the operating personnel when high voltages and/or high currents that may exist within the enclosure. Closure is usually accomplished by using 16 gage steel sheets for the top, side and bottom panels. The back is often closed with a door and the front is closed with panels that support circuitry and operating controls. When high voltages or currents are a concern, the door is fitted with a lock to prevent entry by unauthorized personnel. In most cases it is necessary to allow air flow through the enclosure to cool the components. Cooling is enhanced by louvering the doors permitting air flow into the enclosure. A louvered top panel allows the discharge of this air flow by natural convection. When additional cooling air is required, fans are provided to enhance the flow rate.

Access during assembly and servicing is a necessary requirement in any enclosure. The enclosures are furnished with panels that can be removed leaving only the frame, which consists of the corner posts, headers and braces. In the stripped down condition, accessibility during initial assembly is insured. Accessibility during servicing is a different issue because the time and location of the repair usually does not permit the service person to disassemble the cabinet. Instead, the circuits are made available to the service person through a door at the rear of the enclosure, and by either drawers or gates that open to the front. Opening the drawer or gate brings the PCBs into view allowing the service person to use test probes and substitute cards as required for repair.

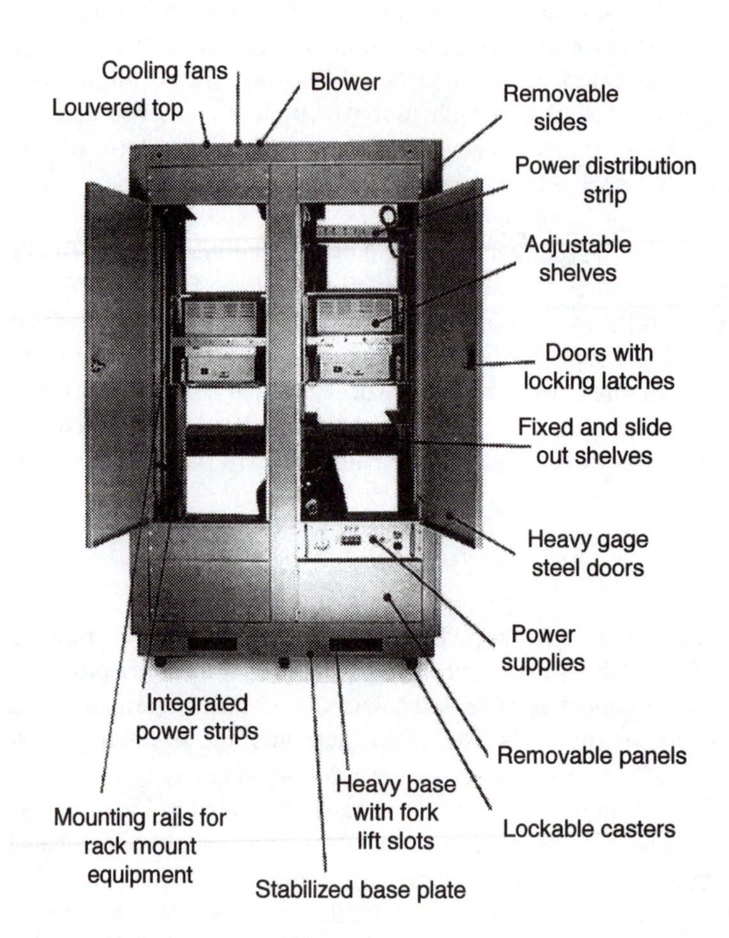

Fig. 8.37 Illustration of a large electronic enclosure showing many features available to the designer.

The structural strength and rigidity of an enclosure is determined by the corner posts and the interior headers and braces. Typically corner posts are fabricated from 14 gage steel and the headers and braces are fabricated from 16 gage steel. Additional support is provided by 12 gage panel mounting angles that are positioned along the vertical edges of the front and rear openings. Weight capacities are not usually specified by cabinet suppliers, but interior loading in excess of 1,000 lb (4485 N) is possible. Recall that additional support for the structure is provided during assembly when shelves and slides are bolted to the corner posts. While the commercial enclosures are sufficiently rugged for the office or even some factory environments, they will not perform well under most vibration and shock imposed on military electronics. The framing deforms plastically under the action of high shock loads which increase the static loading by a factor of 10 to 100 or more. In vibration, the light sheet metal panels and shelving resonate and fatigue failures often occur at the fasteners, the welded joints or the bracing.

The enclosures are fitted with vertical mounting angles along the edges of the front and rear openings. These angles are drilled and tapped with No. 10-32 threads at standard intervals. These tapped holes accommodate screws used to attach card racks, panels and control and display devices. As shown in Fig. 8.37, the enclosures also accept a wide range of accessories such as casters, leveling feet, writing surfaces, power strips, bus bars, shelves and drawers which facilitate assembly, servicing and/or operation of the electronic system.

The definition of attraction changes with time as we improve our understanding of features which appeals to the customer and those that improve the appearance of the product. Today, it is possible to provide cases which are painted with non-glare color coordinated finishes. Wood grain simulation in a variety of furniture woods is also available. Trim molding from anodized aluminum with vinyl inserts add attractive markings and contrast at select locations. Clearly, the gray enamel box with rub on labels is long gone.

For large volume products intended for the consumer, the enclosure is often styled to enhance market appeal. The computer packaging by Apple Inc. in the design of a recent Mac computer is presented in Fig. 8.38 as an example. The enclosure incorporates the CRT tube, the computer circuitry, hard drive and other drives into a stylish color coordinated package. This styling appeals to some customers; consequently, the design of the enclosure is used by Apple as a marketing tool. On the other hand, the author's Dell computer is packaged in a rectangular black and gray box with rounded corners. The CRT tube is not integrated with the circuitry or the various drives. The unique feature of Dell's enclosure is the ease of disassembly. Push a single button on the top of the box and the top and both sides are released. The box literally falls apart. This feature reduces both assembly and service time.

Fig. 8.38 The iMac and iPod enclosures illustrates the use of style in designing enclosures for sales appeal.

8.6.2 Military Enclosures

The design environment for most military electronics is not friendly and the enclosure must accommodate intense and enduring vibrations, high-level shock loading, many large range temperature cycles, fog, humidity, salt spray etc. Many of the enclosures for military applications are unique because they are designed to fit on a certain weapons platform with unusual space constraints. The annual volume of a particular product for the military is often small with production of only 50 to 100 units per year. With unique enclosure designs, exacting durability requirements and small production runs, the cost for designing and manufacturing military enclosures is very high.

In some cases, the weapons platforms are nearly the same and some commonalty in the enclosure with significant cost savings is possible. An example is the US Navy development of the common electronic equipment enclosure (CE^3) for deployment aboard ships and submarines. Aboard a ship electronic enclosures are bolted to the deck in an electronics room. The enclosures are provided with an ample supply of chilled water for conduction cooling. Many of the electronic systems used by the Navy are advanced signal processors that are high end products. They are large high-performance systems. In some integrated detection and control systems, the processors involve circuits containing thousands of circuit cards housed in several enclosures and connected together by massive signal busses.

Let's consider the description of the common electronic equipment enclosure CE^3 that was developed for the U. S. Navy. The enclosure is a free standing unit (i.e. deck bolting but no top supports) with the dimensions given in Fig. 8.39. The enclosure has nine vertical drawers each of which accommodates three rows of modules (circuit cards are often referred to as modules in Navy circles). The enclosure can be loaded with about 1,100 modules having a 0.3 in. pitch. The modules are conduction cooled and the drawers are fabricated with cold rails that remove heat from two opposite edges of the module heat frames. Individual power supplies are mounted on the front of each drawer to provide the dc voltages required. The enclosure is capable of 18 kW dissipation with 25 GPM of chilled water pumped through its drawers. The enclosure is built to withstand the severe vibration and shock requirements imposed by the Navy. The unit completely assembled with a full load of modules weighs slightly more than one ton.

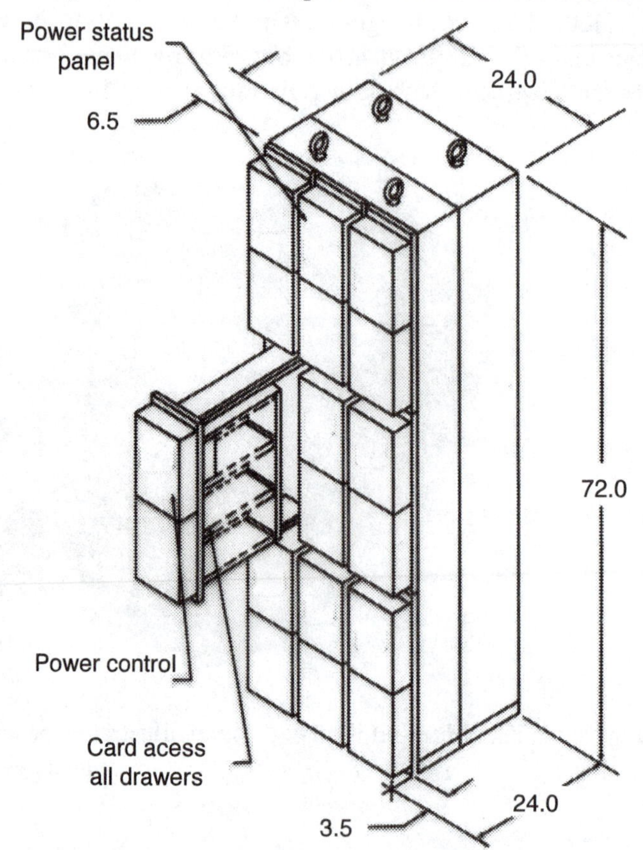

Fig. 8.39 The U. S. Navy's common electronic equipment enclosure CE^3.

The modules (circuit cards) that are used with the CE^3 were standardized by the Navy and designated as format D modules. A format D module with a 0.3 in. pitch is shown in Fig. 8.40. This module includes a multi-layer PCB with foot prints for surface mounted chips, a copper heat frame, a 100 pin connector and a test header. The 20 W of heat dissipated by the chips is conducted by the frame to its vertical edges. The card is clamped in the drawer slots with the spring type retainers described previously in Section 8.5.1.

With very dense packing of modules in each drawer, the power requirements are large—each drawer is capable of supporting up to 2 kW of power. Power supplies are mounted on the front face of each drawer. These supplies are connected to a controller that contains a relay for switching the power on or off for the entire drawer. The controller interfaces with a bus bar that extends along a back panel and distributes supply voltages at select locations along it and at a single point ground. The current distributed ranges from 300 to 500A depending on the particular current requirements of the circuits housed.

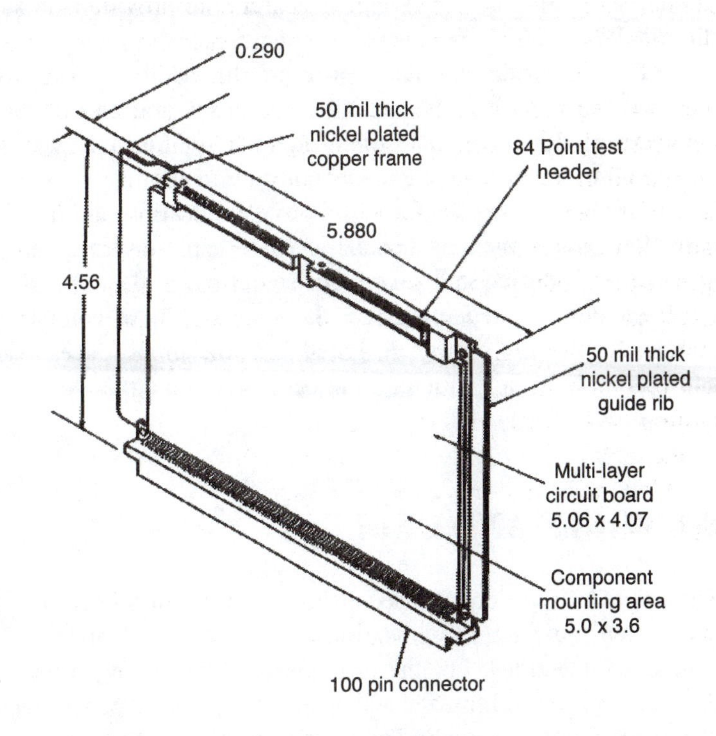

Fig. 8.40 Standard electronic module (SEM)
Format D, 0.300 in. (7.6 mm) pitch.

8.6.3 Automotive Enclosures

There are a number of different electronic systems in a typical automobile and even more systems in a luxury model. The electronic systems are employed for controlling the brakes (ABS) and many engine functions. Automatic controls for adaptive cruising and for dynamic stabilization are becoming more common. Development is underway on vision or radar based systems to make it more difficult for drivers to run off a highway.

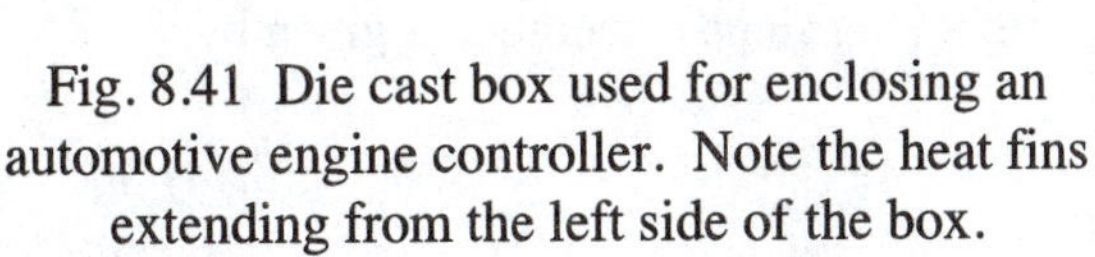

Fig. 8.41 Die cast box used for enclosing an automotive engine controller. Note the heat fins extending from the left side of the box.

Operational environments for automotive systems are much different than that encountered in office benign environments. Many systems are located close to the automotive function they control and in some cases directly mounted on-engine or on-transmission. For example, the engine controller, presented in Fig. 8.41 is positioned in the engine compartment and is subjected to relatively high and low temperatures and an abusive environment containing dust, dirt and oil vapors. Aluminum die cast boxes are commonly employed to house electronics in the engine compartment. The boxes are bolted to a relatively cool part of the engine or frame to provide firm support and a conduction cooling path. A die cast lid covers the box preventing entrance of dirt and oil vapors. In many instances, the electronics are potted with a flexible polymer to provide addition protection against the ingress of moisture and oil vapors. In many designs, heat fins are an integral part of the die casting to enhance heat transfer by convection. When the enclosure is equipped with heat fins, it is placed so that the fins are exposed to an air stream to increase to amount of convection cooling.

Other electronics for automotive applications are housed in the passenger compartment. In this case, the environment is not an issue, except for large temperature variations that occur when the automobile is sealed and parked for sometime in the sun. Enclosure is usually accomplished in small compartments located under the dashboard. These compartments are usually made from plastic or composites that are custom designed to fit the small spaces available for them. Protection from voltage and current is usually not an issue because the magnitude of both quantities is relative low in a typical application.

8.7 WIRES AND CABLING

The printed circuit board, used either as a PCB or a back panel, was introduced to circumvent the cost and lessen the circuit degradation due to ordinary wiring. Originally insulated wiring was used for making connections, but the efficiency of the wiring process was increased enormously and circuit degradation was minimized when PCBs and back panels were introduced. However, wiring and/or cables were not totally eliminated as it is necessary to connect units that are too far apart to effectively use either rigid or flexible printed circuits in a connection scheme. Examples of this type of wiring includes drawer to drawer connections in an enclosure, cabinet to cabinet connections in a large system, and the transport of signals from room to room, building to building, city to city, country to country and continent to continent. We will concentrate on the design of the more local wiring systems within an enclosure and within a system. Fiber optic cables are usually employed when the distance among the points to be connected becomes larger or for very high frequency signals.

8.7.1 Wire Conductors

The conductor used in almost all of the wiring is electrolytic tough pitch copper that is coated with tin to a thickness of about 50 μin. (1.25 μm). This coating inhibits oxidation, other corrosive effects and assists in soldering. If the wire does not undergo flexing or vibration in operation, solid conductors are used because they are lower in cost than stranded cable and can be handled and shaped more easily in assembly operations. The difficulty with solid conductors is their tendency to fail in fatigue due to cyclic bending. Recall, that the strain produced by bending a long slender member is given by:

$$\varepsilon = c/R \tag{8.14}$$

where $c = d/2$ is one half the wire diameter and R is the radius of curvature of the deformed wire.

The stress developed by bending a wire is:

$$\sigma = E\varepsilon = Ec/R \tag{8.15}$$

For long life, in excess of 10^6 cycles, the bending stresses are compared to the endurance strength S_e of copper wire. The endurance strength of non ferrous alloys is often taken as:

$$S_e = 0.35\ S_u \tag{8.16}$$

where S_u is the tensile strength of the material.

For copper the tensile strength depends strongly on the degree of cold working in fabricating the wire as indicated in Table 8.5. The tensile strength ranges from 36 ksi (248 MPa) for annealed wire to 97 ksi (669 MPa) for hard-cold-worked wire.

Specifying stranded wire is nearly imperative when it is necessary to flex the cable assembly for a large number of cycles. Great increases in the fatigue life of stranded wire occur for two reasons. First, the diameter of the strand is an order of magnitude smaller than the wire diameter. Because c in Eq (8.15) is related to the diameter of the strand, the stress imposed on the copper wire is reduced by the ratio of the conductor diameter to the strand diameter. The second reason is less obvious, but small nicks sometimes occur when removing the insulation from solid copper conductors. These nicks drastically reduce the fatigue life of the solid wire because they serve to concentrate the stresses. Nicks may also occur with stranded wire, but they only eliminate a strand or two of the many strands and result in small decreases in effective fatigue strength of the conductor. For a given size conductor, an increase in the number of strands increases the flexibility of the conductor. A listing of stranded conductors, designed for use in applications involving significant cable flexing, is presented in Table 8.6. Note, that the number of strands used increases from 7 for AWG 38 wire to 259 for AWG 20 wire. The diameter of the strand changes little with increasing wire size because only AWG 44 and 46 wires are used in fabricating stranded conductors.

Table 8.5
Properties of copper wire

State	Coating	Strength ksi (MPa)	Elongation %	Solder-ability	Oxidation Resistance
Annealed	None	36 (248)	15	Fair	Good
Annealed	Tin	36 (248)	15	Good	Good
Annealed	Silver	36 (248)	15	Good	Good
Annealed	Nickel	36 (248)	15	Poor	Good
Medium Hard	None	60 (414)	0.88	Fair	Poor
Hard	None	97 (669)	0.86	Fair	Poor

Table 8.6
Parameters describing highly flexible wire for fatigue applications

AWG	Strand No./Size	Insulation Wall Thickness mil (mm)	Maximum Voltage V	Nominal OD mil (mm)
38	7/46	5 (0.125)	200	15 (0.375)
36	7/44	5 (0.125)	200	16 (0.400)
34	10/44	5 (0.125)	200	18 (0.450)
32	18/44	7 (0.175)	300	23 (0.575)
30	28/44	10 (0.250)	600	31 (0.775)
29	51/46	10 (0.250)	600	32 (0.800)
28	41/44	10 (0.250)	600	35 (0.875)
27	48/44	10 (0.250)	600	37 (0.925)
26	66/44	10 (0.250)	600	42 (1.050)
24	105/44	15 (0.375)	1000	58 (1.450)
22	168/44	15 (0.375)	1000	64 (1.600)
20	259/44	10 (0.250)	600	61 (1.525)

The lay of the individual strands, which is the axial length of one turn of a single strand, affects both the flexibility and the cost of the wire. Wires with short lays exhibit high flexibility as the strands are approaching the shape of a helix found in a coil spring. Of course, the cost increases because more time is required to fabricate the conductor. In some applications, the strands are coiled around a high strength thread which serves as core and carries any tensile loads which might be imposed on the wire.

Finite fatigue life of cable assemblies is always difficult to predict and testing is essential to insure reliable performance when long cyclic life is involved. However, it is possible to adapt the Manson's fatigue life equation to this application. This empirical equation permits one to relate strain range $\Delta\varepsilon$ and cyclic life N by:

$$\Delta\varepsilon = 3.5\frac{S_u}{E}N^{-0.12} + \left(\frac{D}{N}\right)^{0.6} \tag{8.17}$$

where D is the ductility of the conductor material given by:

$$D = \log_e\left(\frac{1}{1-RA}\right) \tag{8.18}$$

and RA is the reduction in area measured in a simple tension test.

Combining the results of Eqs (8.14) with (8.17) permits the fatigue life of either solid or stranded wire to be estimated.

8.7.2 Wire Insulation

Conductors are coated with a dielectric that serves to insulate and protect it from corrosive attack from the environment. A large number of different polymers are used as insulating materials for special purpose wiring. We will cover only the more commonly utilized polymers including PVC, polyethylene, rubber, polyurethane and Teflon. Polyvinyl chloride (PVC) is one of the most common insulating materials because it is low in cost and easy to process. PVC has a high dielectric strength (500V/mil), as indicated in Table 8.7, and it is resistant to flame, water, oil and abrasion. Its primary disadvantage is a limited temperature range −20 to +80 °C. Recently, some of the PVC's have been irradiated that increases the degree of cross linking in the molecular chains and raises their temperature rating to +105 °C.

Polyethylene has a low dielectric constant, high dielectric strength, good resistance to solvents, low density and very good low temperature characteristics. Its primary disadvantages are its tendency to creep when subjected to stress and its poor flame retarding properties. Polyethylene is slightly higher than PVC in cost.

Natural and synthetic rubbers are often used as jackets for cables. They have excellent flexibility, very good resistance to abrasion and reasonable strength over a wide range of temperature. There are several different types of rubbers including natural, neoprene, hypalon, nitrile, butyl and silicone. The properties and costs vary widely and care should be exercised to select the most suitable of the large number of polymers and blends that are commercially available.

Polyurethane is an excellent insulation material with exceptional resistance to abrasion. It is frequently used as a thin coating insulation for magnet wire. Its main disadvantages are low operating temperature and poor flame retarding properties.

Teflon is an outstanding insulating material with superb properties in almost every category. However, it is so expensive that it is used only in demanding applications. The advantage of the wide operating temperature range of Teflon – 70 to + 250° C is evident.

Table 8.7
Properties of some insulation materials used on wire and cable

Material	Specific Gravity	Volume Resistivity Ω-cm	Breakdown Voltage V/mil	Abrasion Resistance	Dielectric Constant	Flame Retarding	Flexi-bility	Temp. Range (°C)
Rubber	0.93	10^{15}	150-500	Exc.	2.3-3.0	Poor	Exc.	–40 to 70
Silicone Rubber	0.97	10^{14}	100-600	Poor	3.2	Poor	Exc.	–60 to 200
Neoprene	1.25	10^{11}	150-600	Exc.	9.0	Good	Exc.	–30 to 90
Hypalon[1]	1.15	10^{14}	500	Exc.	7.0-10.0	Good	Good	–30 to 105
PVC Standard	1.3	10^{11}	500	Fair	7.0	Exc.	Good	–20 to 80
PVC Premium	1.3	10^{12}	500	Good	7.0	Exc.	Good	–55 to 105
PVC Irradiated	1.3	10^{12}	500	Good	5.0	Exc.	Good	–55 to 115
Polyethylene	0.95	10^{13}	600	Exc.	2.5	Poor	Fair	–60 to 80
Teflon[1] TFE & FEP	2.2	10^{13}	600	Exc.	2.1	Exc.	Fair	–70 to 250
Teflon[1] PFA	2.1	10^{18}	600	Exc.	2.1	Exc.	Fair	–70 to 250
Nylon	1.07	10^{14}	450	Exc.	4.0	Poor	Poor	–40 to 120
Polypropylene	0.91	10^{15}	650	Exc.	2.2	Poor	Poor	–40 to 105
Polyolefin Irradiated	1.3	10^{15}	600	Good	2.5	Exc.	Good	–50 to 125
Kynar[2]	1.76	2×10^{15}	250	Exc.	5.0-8.0	Exc.	Good	–40 to 150
Polyurethane	1.1	10^{11-14}	500	Exc.	5.0-8.0	Poor	Exc.	–50 to 80
Polysulfone	1.24	5×10^{15}	400	Good	3.1	Exc.	Good	–55 to 150
Kapton[1]	1.4	10^{13}	1.7 kV	Exc.	3.5	Exc.	Exc.	–40 to 200
Fluorosilicone	1.4	10^{14}	350	Exc.	7.0	Exc.	Exc.	–60 to 200
Tefzel[1]	1.70	10^{16}	400	Exc.	2.6	Exc.	Fair	–70 to 180
Halar[3]	1.68	10^{15}	490	Exc.	2.6	Exc.	Fair	–70 to 150

[1] Trademark of DuPont; [2] Trademark of Arkema; [3] Trademark of Solvay Solex

8.7.3 Wire to Cable

Wires of many different types are used in connecting components and circuit boards within an electronic enclosure. In this description, we will begin with the simplest form of insulated wire, magnet wire, and add features to the wire or the insulation needed to develop cable assemblies with different degrees of complexity. These wires and cables, illustrated in Fig. 8.42, show important characteristics of several different design methods followed in these upper level connections.

The simplest connector is magnet wire that consists of a solid copper conductor with a thin layer of varnish or urethane insulation. It is used primarily to wind coils on motors and inductors. However, it is often used as a soft (or white) wire on printed circuit boards when corrections to a circuit require a small degree of rewiring. The thin wires conform to the surface of the board and substitute for circuit traces without detracting seriously from the appearance of the PCB.

Fig. 8.42 Several examples of woven cable assemblies.

Single conductor hook-up wire is used to make point to point connections when the points are distributed over a relatively large volume of the enclosure. The hook up wire is insulated and it may be shielded. To facilitate assembly and to minimize the space required in the enclosure, the hook-up wires are tied together to form a wiring harness as illustrated in Fig 8.43.

Fig. 8.43 Wiring harness terminating in three round and rugged elbow connectors.

In many applications, the points to be connected are arranged at locations equally spaced along a line. Connections to a line of contacts are made most effectively with a flat ribbon cable. This cable, illustrated in Fig. 8.44, consists of a number of parallel conductors with the insulation serving to space the wires and to form a wide ribbon. The flat ribbon cable has many advantages when compared to connections made with discrete wires. It requires only about half of the weight and volume as a wiring harness. Also, the precise positioning of the conductors permits mass termination into connectors with a minimum of labor. Another advantage is the flexibility and handling characteristics. The flat ribbon can be rolled, or flat folded like an accordion allowing for compact storage in an enclosure.

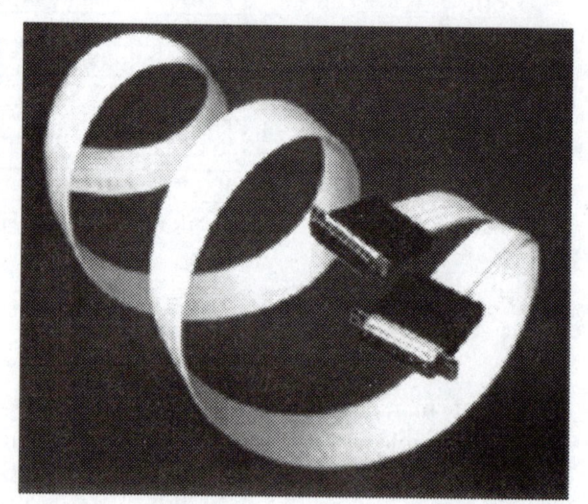

Fig. 8.44 Flat cable terminated in two type D connectors.

Woven wire cable, illustrated in Fig. 8.45, has many of the same characteristics as flat ribbon cable. It is fabricated from individually insulated round wires by weaving them with strong thread in a loom. The woven flat cable has the advantage of individual tailoring. That is the individual wires in a cable can be different depending on the requirements. The flexibility of the manufacturing process permits custom cable assemblies for specific packaging applications, like the fiber optics cable with different light emitters on each end as shown in Fig. 8.46.

Fig. 8.45 Woven wire cable assembly terminated with two pinned connectors.

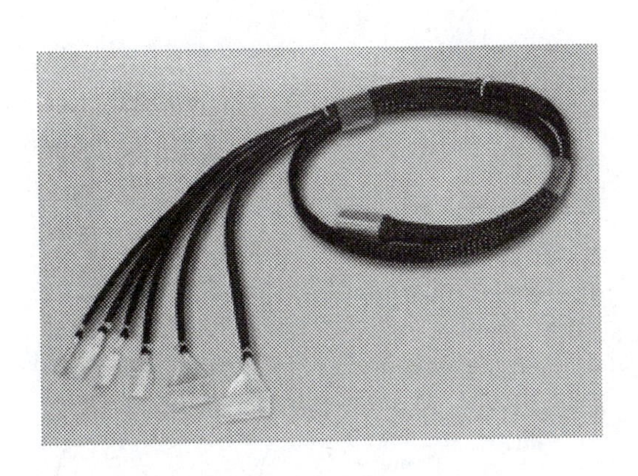

Fig. 8.46 An array of woven fiber optic light emitters.

Flat cables either ribbon or woven have electrical characteristics, distributed inductance and capacitance, which are important in circuit performance of the electronic system. These characteristics depend upon insulation material, conductor lay, and the proximity of the wires to a ground. Teflon insulation with its low dielectric constant reduces capacitance and increases the propagation velocity of the signal pulse. The presence of a nearby ground increases capacitance, decreases inductance and the net effect is to increase propagation delay. Twisting wire pairs, to control the characteristic impedance of the cable, results in an increase in the propagation delay and capacitance when compared to a flat lay ribbon cable. The electrical characteristics of six commonly employed flat cable assemblies are given in Table 8.8.

Table 8.8
Electrical characteristics of flat ribbon and woven cable fabricated with AWG 28 wire.

Cable Type	Insulation Material	Impedance Ω	Capacitance pF/ft (pF/m)	Inductance μH/ft (μH/m)	Propagation Delay ns/ft (ns/M)
Flat ribbon Parallel lay Ground plane	PVC	65	28 (91.9)	0.20 (0.66)	1.65 (5.41)
Flat ribbon Parallel lay	PVC	142	8.0 (28.2)	0.30 (0.98)	1.40 (4.59)
Woven Parallel lay	PVC	144	7.3 (24.0)	0.27 (0.89)	1.34 (4.40)
Woven Parallel lay	FEP	150	8.2 (20.3)	0.23 (0.75)	1.26 (4.13)
Woven Twisted pair	PVC	108	11 (38.1)	0.26 (0.85)	1.60 (5.25)
Woven Twisted pair	FEP	128	8.0 (28.2)	0.26 (0.85)	1.45 (4.76)

8.7.4 Wire Shielding

A shield is a conductor which is placed around a wire or group of wires to provide a barrier against external electromagnetic and electrostatic fields that could affect the signal being transmitted. The shield also serves to confine the transmitted signal to the central conductor reducing the fields generated in the region external to the shield. In coaxial cables, the shield acts as the return wire necessary to complete a circuit.

Two different shielding techniques used with individual wires to control the effects of electrostatic fields are presented in Fig. 8.47. Flat tape shields are the most commonly used. The tape is fabricated by bonding a thin foil of copper or aluminum to a thin film of polyester to form a laminated, high-strength conductive tape. The tape is wound in a spiral about the insulated central conductors as shown in Fig. 8.47a. A drain wire runs along the metal foil that is connected to ground to terminate the shield.

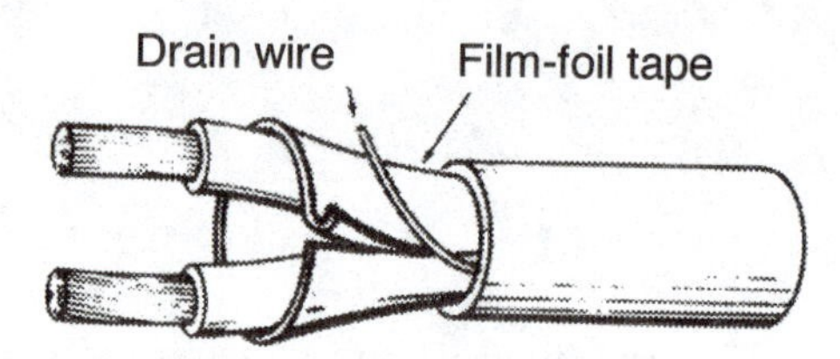

(a) Spiral tape with drain wire

(b) Braided shield

Fig. 8.42 Two types of shielded wire.

Braided shields, shown in Fig. 8.47b, are woven conducting envelops fabricated from small diameter (34 to 38 AWG) copper wire. These braided shields are used in most coaxial cables because they are effective as the return conductor and because they can be connected to the ground of an appropriate connector. The braid is an effective shield over a wide range of frequencies from 1 to 140 kHz. However, the openings in the braid can affect its performance. At the lower audio frequencies a 75% area coverage of the perimeter is sufficient. However, as the frequencies increase the perimeter area coverage by the shield is increased to 90-95% by using smaller diameter wires and a tighter weaving pattern.

Shielding against magnetic fields is much more difficult because the shielding materials must exhibit high permeability. Heavy layers of low carbon steel tape or Permalloy are effective but they add significantly to the cost of the wiring. Availability of a wide range of magnetically shielded wire is another issue. A much lower cost approach, applicable in some applications, is to twist two conductors tightly together to achieve a balanced pair. When magnetic field lines intersect in a twisted pair, they produce equal and opposite changes in the voltage that cancel.

8.8 FANS

In the design of an enclosure system, provisions must be made to dissipate the heat generated within the cabinet or case. For very low power systems, the heat is dissipated by natural convection and the primary concern is the placement of the vents that assist the natural flow of the buoyant heated air. It is evident that placement of vents on the bottom of the enclosure for air intake and at the top for exhaust are the most suitable locations. However, customer habits and preference often dictate the placement of vents at less desirable locations. In office environments, particularly with table top cases, the clearance allowed under the case is usually not sufficient for adequate air flow; consequently, the vents are placed along the bottom edge of the side walls and rear panel. Also, in the office environment top horizontal surfaces tends to be used as a tables that accommodates anything from coffee service to stacks of papers. Clearly, it is not advisable to place vents on this surface. To avoid air blockage and spillage of liquids, it is necessary to locate vents along the top of the side and rear walls to insure the required air flow.

For forced air cooling, fans are used to move larger volumes of air at higher velocities through the enclosure past PCBs and power supplies. With forced convection, it is possible to accommodate significantly larger amounts of heat. We will cover the theoretical aspects of forced convection cooling in Chapter 10. The purpose of this section is to show fan placement in an enclosure and discuss the problem of noise generation by the fans.

Conduction cooling is used for higher levels of heat dissipation in an enclosure. From an engineering view point, conduction cooling is extremely effective if the coolant, usually chilled water or a refrigerant, can be located very close to the source of heat. In this case, and the thermal path and resistance is reduced to a minimum. The use of conduction cooling is usually limited to high performance systems where the requirement for low junction temperatures with high power chips can only be achieved with highly effective cooling procedures. For these systems, it is necessary to design cold plates and/or cold rails to provide suitable heat sinks at select locations within the enclosure. We will show approaches for the design of these heat sinks in Section 8.8.2.

8.8.1 Fans and Fan Noise

Fans

Propeller fans and blowers are available in a wide ranges and sizes and capacities to increase air flow through the enclosure. Convenient fan assemblies, as shown in Fig. 8.48, are available as accessory items from suppliers of commercial cabinets. These fan assemblies enable the designer to specify fans mounted at the bottom, one or more of the sides, or top of the enclosures. It is relatively easy to select appropriate fans to deliver a specified quantity of air. It is more difficult to ensure that the fan noise will not become objectionable. The rotating fan blades produce an oscillating pressure which generates back ground noise. This noise can be measured as a sound pressure level given by:

$$L_p = 10 \log_{10}\left(\frac{p^2}{p_0^2}\right) \tag{8.19}$$

where L_p is the sound pressure level in Db relative to p_0, p is the RMS sound pressure in Pa, and p_0 is the reference RMS sound pressure = 2×10^{-5} Pa.

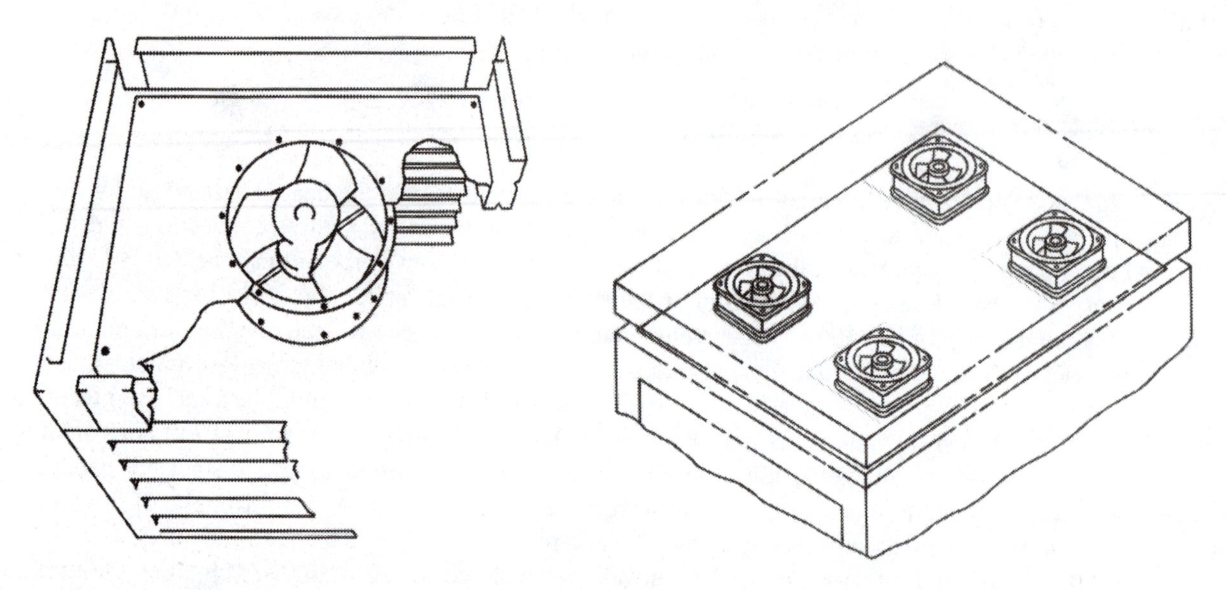

(a) Fan in cabinet base. (b) Pagoda top on an enclosure incorporating four fans.
Fig. 8.48 Fan assemblies are available as enclosure accessories.

There are two basic types of fans—axial and centrifugal blowers. The primary main difference between axial fans and centrifugal blowers is in their flow and pressure characteristics. Axial fans deliver air in a direction parallel to the axis of rotation of the fan blades. These fans produce high flow rates, but develop relatively low pressures. Blowers deliver air flow that is perpendicular to the axis of rotation of its blades at lower flow rates, but they develop higher air pressures.

There are several types of axial fans, which include the propeller, tube axial and vane axial. Propeller fans represent the simplest design as they consist of only a motor, propeller and support structure. They have the disadvantage that tip vortices are produced by the pressure differential across the airfoil section. A tube axial fan, shown in Fig. 8.49, is commonly used for cooling electronic systems. It is similar to a propeller fan, but it incorporates a shroud around the propeller to reduce the development of vortices. The vane axial fan uses vanes in the airflow trail behind the propeller to straighten the swirling flow created by the propeller. The motor in all three types of is mounted at the center of the assembly.

Fig. 8.49 A small axial fan used to cool electronic systems.

Centrifugal blowers are used less frequently in cooling electronic systems because their air flow rate is less than axial fans. However, they are commonly employed when large pressure drops are encountered in passing the air through an electronic assembly. In these cases, the higher pressures developed by the centrifugal blowers are required to insure adequate airflow to downstream components. An example of a centrifugal blower is presented in Fig. 8.50.

Fig. 8.50 A centrifugal blower generates higher pressure airflow than an axial fan.

Fan Noise

An excessive level of noise is detrimental because it interferes with speech, causes annoyance and interferes with completing important tasks. Recommended background noise levels that avoid speech interference have been developed for indoor applications. These results, known as the preferred noise criteria (PNC), are shown in Fig. 8.51. An examination of these curves indicates that the sound pressure level measured in each octave band should not exceed a specified Db level that varies with the frequency of the noise. Clearly, higher frequencies interfere with speech more than the lower frequencies. The particular PNC curve that defines an acceptable level of noise pollution depends on the function of the room in which the fans operate as indicated in Table 8.9.

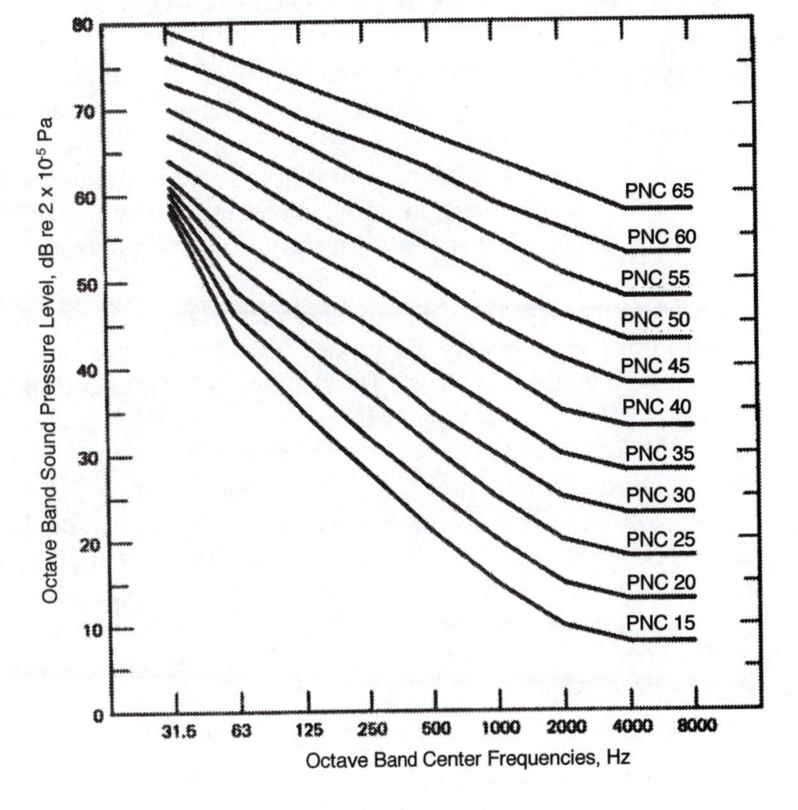

Fig. 8.51 Recommended background noise levels at different octave bands for indoor rooms based on PNC curves.

Table 8.9
Preferred A-weighted sound pressure levels for indoor rooms with different functions

Function of room	A-weighted Sound level (dB A)
Concert halls, opera houses, recital halls	21 – 30
Churches, large auditoriums, large drama theaters	less than 30
Broadcast, television and recording studios	less than 34
Small auditoriums, small theaters, small churches, conference rooms, large meeting rooms	less than 42
Bedrooms, hospitals, residences, apartments, hotels, motels	34 – 47
Private or semi-private offices, small conference rooms, classrooms, libraries	38 – 47
Living rooms, conversation areas	38 – 47
Large offices, reception areas, retail shops and stores, cafeterias, restaurants	42 – 52
Lobbies, laboratory work areas, engineering rooms, secretarial areas	47 – 56
Light maintenance shops, office and computer rooms, kitchens, laundries, etc	52-61
Shops, garages, factory floor, control rooms	56-66

From this table it is clear that electronic systems placed in large offices should follow a PNC 45 curve, while equipment placed in small semi-private offices should follow a PNC 40 curve. Annoyance and task interference are effects that depend upon many parameters, which affect the attitude of individuals. Because of this fact, it is difficult to specify an appropriate PNC curve. Addition information is given in Reference 13 pertaining to influence of noise on personnel behavior and productivity.

8.9 COLD PLATES AND COLD RAILS

Cold plates and cold rails are often fabricated by drilling an interconnecting series of holes in an aluminum plate. Another method of fabrication involves pressing copper or stainless steel tubes into a channeled aluminum extrusion, as shown in Fig. 8.52. These tube type cold plates are cost-effective and offer a good method for removing heat from devices with low to medium power densities. With a cold plate, heat is conducted from the component (a power supply or a large module) through a heat conducting frame to the surface of the plate, and then through the plate to the cooling fluid flowing in the tubing.

Another type of cold plate, illustrated in Fig. 8.53, is the flat tube cold plate. In this design, the coolant flows in a circular tube on one side, through a rectangular channel (flat tube) and out another circular tube on the other side. The flat tube cold plates are compact and offer extremely low thermal resistance. They contain internal fins that extend into the fluid flow to increase performance. They offer excellent thermal uniformity as coolant flows below the entire surface of the flatten tubes. They are employed for cooling small, high power components

The merit of a cold plate is judged by its size and its thermal resistance. It is evident that the larger cold plates can dissipate more heat than the smaller plates if the flow rates of the fluids through both are adequate. The thermal resistance of a cold plate should be as low as possible. The thermal resistance is determined by the materials employed, the plate thickness and the construction details. High conductivity materials such as copper, aluminum or copper alloys are usually employed. Plate thickness is minimized consistent with providing the cold plate adequate structural strength and rigidity. The higher performance cold plates provide for coolant flow across the entire plate. They also are fabricated with flat copper tubes with fins etched on their interior surfaces. Typical normalized thermal resistance for three different types of cold plates is presented in Fig. 8.54.

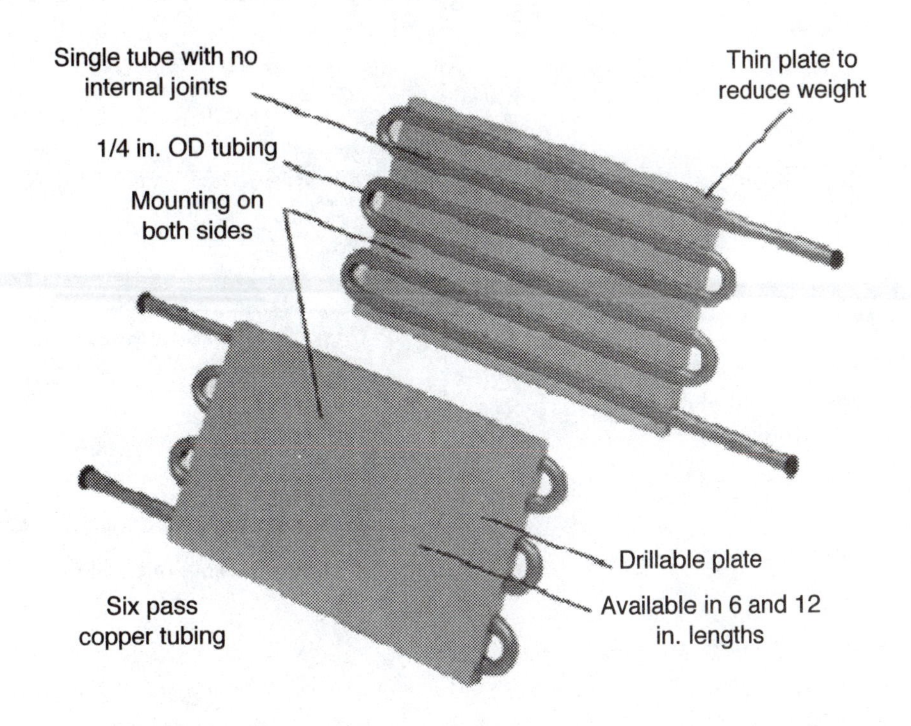

Fig. 8.52 Example of a cold plate fabricated by pressing copper or stainless steel tubing into extruded aluminum plates.

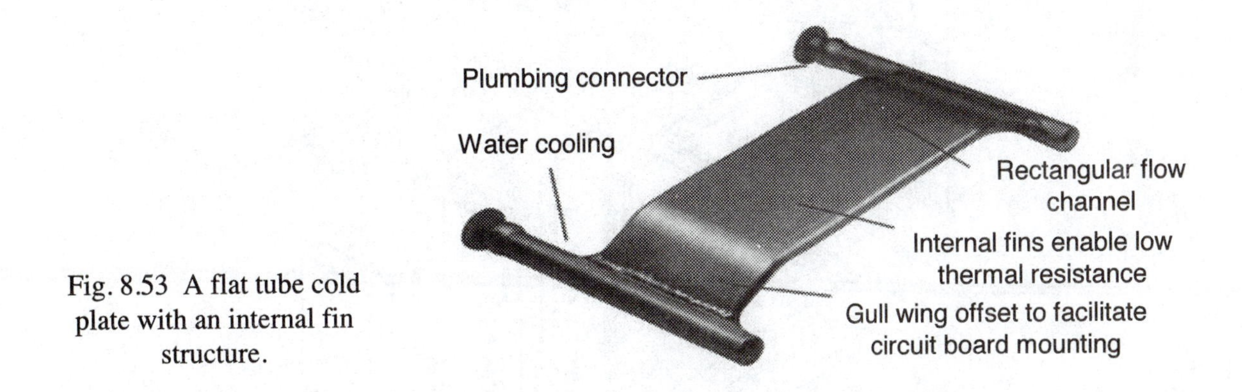

Fig. 8.53 A flat tube cold plate with an internal fin structure.

Cold rails are usually incorporated on both sides of a drawer or a gate for cooling circuit cards with attached heat frames. Slots are cut into the cold rail to accommodate the edges of the heat frames and card retainers such as wedge locks. Heat is conducted across the heat frame-slot interface though the rail to the fluid flowing in the internal channels. Water is the most common fluid used, although in some high-performance systems a refrigerant may be employed to obtain lower cold rail temperatures. In systems where direct impingement of air on the components is not permitted and water or refrigerants cannot be employed, cold air is used as the cooling fluid in cold rails or other heat exchangers. An example of a water cooled cold plate and four cold rails employed in the drawer of the CE^3 enclosure is presented in Fig. 8.55.

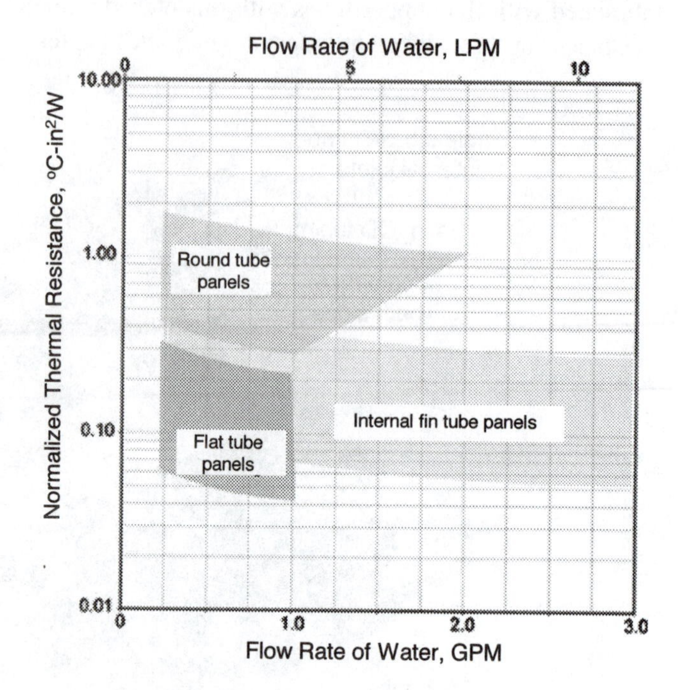

Fig. 8.54 Normalized thermal resistance for three different types of cold plates as a function of flow rate.

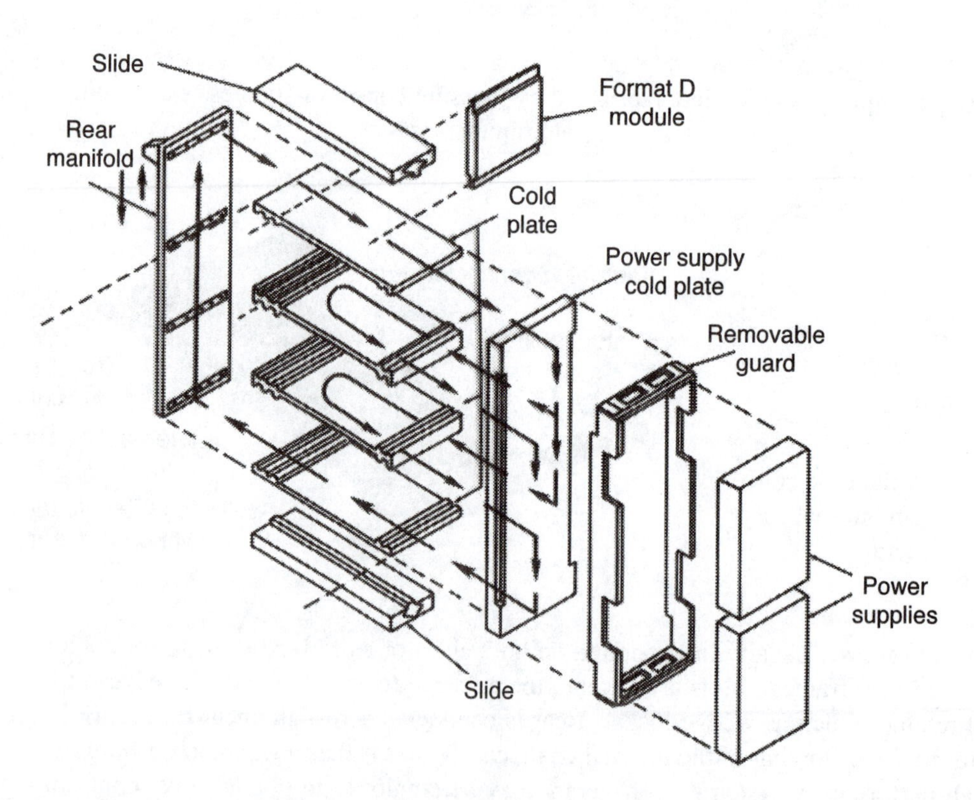

Fig. 8.55 An exploded view showing design details of a drawer for the CE[3]. The drawer provides a cold plate for attachment of an external power supply and four cold rails for retention of the format D modules.

REFERENCES

1. Pecht, M. (ed.), Handbook of Electronic Package Design, Marcel Dekker, New York, NY, 1991.
2. Blackwell, G. R., The Electronic Packaging Handbook, CRC Press, November, 1999.
3. Witaker, J. C., The Electronics Handbook, CRC Press, December, 1998.
4. Walsh, R. A., Electromechanical Design Handbook, McGraw-Hill, 2000.
5. Lau, J. H., et al, Electronics Manufacturing, McGraw-Hill, 2002.
6. Chapman, S., Fundamentals of Microsystems Packaging, McGraw-Hill, 2001.
7. A. J. Bilotta, Connections in Electronic Assemblies, Marcel Dekker Inc. New York, 1985.
8. J. H. Whitley, The Mechanics of Pressure Connections, AMP Inc., Dec. 1964.
9. R. H. Zimmerman, Engineering Considerations of Gold Electrodeposits in Connector Applications, AMP Inc., May 1973.
10. A. H. Graham and S. W. Updegraff, "Properties of Palladium Nickel Alloy and Pure Palladium for Connector Applications," Proceedings 1984 National Electronic Packaging and Production Conference.
11. T. Colomina, "Selecting Circuit Card Guides and Retainers," Machine Design, June 9, 1988.
12. S. S. Manson, "Fatigue: A Complex Subject--Some Simple Approximations," Experimental Mechanics, vol. 5, No. 7, 1965, pp 193-228.
13. P. F. Cunniff, Environmental Noise Pollution, John Wiley, New York, 1977, pp. 101-114.
14. L. L. Beranek, W. E. Blazier and J. J. Figawer, "Preferred Noise Criteria (PNC) Curves and Their Application to Rooms," Journal Acoustical Society America, vol. 50, 1971, pp. 1223-1228.

EXERCISES

8.1 List ten hardware items involved in third level packaging. Can you list 20 or more items?

8.2 Consider a system housed in a single standard enclosure which contains 60 circuit boards. Each circuit card is 150 mm by 200 mm. The cards are aligned on 25 mm centers. Air flow, generated by top mounted fans, is used to cool the cards and the power supplies. Sketch the design of the enclosure so that replacement of a malfunctioning board can be made within five minutes after arrival of the service person. List all of the factors which you have incorporated in your design to facilitate rapid access.

8.3 For the enclosure discussed in Exercise 8.2, design the duct work required to insure adequate air flow over the circuit cards.

8.4 List the number of ways an accident might occur that could inflict injury to either an operator or a service person dealing with an electronic system.

8.5 In Figs. 8.4 through Fig. 8.7, we have shown connections from the circuit card to cables, back panels and chassis. Why do we encounter so many different types of connectors?

8.6 Determine the force that can be exerted on a pin with a circular cross section 0.60 mm in diameter and 20 mm long if it is fabricated from beryllium copper. Consider the pin prior to and after entry into a corresponding socket in the receptacle.

8.7 Prepare a graph showing buckling force on a pin with a circular cross section if d = 0.40 mm to 2.00 mm and length L = 8.0, 12.5 and 18.0 mm. Suppose the pins are fabricated from brass.

8.8 Describe the difference between thermoplastics and thermosetting polymers. What are some of the differences in the properties of these two classes of polymers?

8.9 Describe the injection molding process and indicate why it is so effective in producing connector inserts.

8.10 Go to the library or Internet and find a complete description of the thermoplastic polyphenyleneoxide. Write a short (one or two page description) suitable for briefing your supervisor on advantages and disadvantages of this polymer.

8.11 Go to the library or Internet and find a complete description of the thermoplastic polyethylene terephthalte (PET) polyester. Write a short (one or two page description) suitable for briefing your supervisor on advantages and disadvantages of this polymer

8.12 Describe the important mechanical and electrical properties you would expect in a material suitable for fabricating pin, blades and contacts. Write a specification which could be used by a purchasing department in procuring these materials for manufacturing division.

8.13 If a connector pin has a resistance of 50 mΩ and is conducting 20 amps, determine the power dissipated at the connection. What would happen to this connector? How long would it take for the situation to develop?

8.14 For a connector pin that is conducting 5 A, determine the connection resistance requirement if the power loss is to be limited to: (a) 0.1 W, (b) 0.2 W, (c) 0.5 W and (d) 1 W.

8.15 Using the results presented in Fig. 8.10, determine κ in Eq (8.4) for copper, tin and nickel. Comment on the significance of these results in the selection of a material for a contact.

8.16 Describe the effect of the pores in a plated coating on the development of a film resistance in a connector contact. What procedures can be followed in reducing the effect of the pores in plating? What is the most cost effective solution?

8.17 You plan to gold plate a very local region on the blade and contact where the two parts touch (selective plating). If the blade and contact each are 25 mil (0.625 mm) wide by 100 mil (2.5 mm) long and the plating thickness is 0.5 μm, determine the cost of the gold required to plate 1,000 pairs of parts. Take the price of gold at $600/ troy oz. Note that there are 12 troy ounces in a pound.

8.18 What are the advantages of using a palladium nickel alloy in place of hard gold as a plating material? What is hard gold?

8.19 Tin is used to plate contacts in low cost applications. Is tin always effective as a plating material? Indicate the application where tin is often the cause of result in connection failures.

8.20 Why is the power through a connector an important factor to consider in the design of the pins and/or contacts? Describe a dry circuit condition for a connector. Should a connector be used in place of a switch for power on power off applications? Why not?

8.21 Why is lubrication effective in extending the life of tin plated contacts? Will lubrication extend the life of contacts plated with more precious metals? Explain your answer.

8.22 Describe the difference between passive and active lubrication for connectors.

8.23 Prepare a family of curves showing the normalized insertion force F_i/N_i as a function of the taper angle θ with the coefficient of friction $\mu = 0.1, 0.2$ and 0.3 as a parameter.

8.24 Verify Eqs (8.5) to (8.9).

8.25 Derive an equation for the bending stress developed in the contact as a function of the interference Δh.

8.26 Determine the insertion force for a blade being inserted into the contact shown in Fig. 8.18. The interference Δh = 10 mil (0.25 mm), the contact is fabricated from beryllium copper, the taper angle is 20°, the contact length L = 600 mil (15 mm) and the cross section of the contact arm is 30 by 30 mil (0.75 by 0.75 mm). Let the friction be a variable ranging from μ = 0.10 to 0.3 in steps of 0.05.

8.27 For the single pin connection described in Exercise 8.26, find the force to insert and extract the connector during the wiping phase of engagement.

8.28 How many pin of the type described in Exercise 8.26 could be placed in a multi-pin connector before it would be necessary to provide mechanical assistance for insertion and withdraw of the plug portion of the connector. State the assumptions involved in your answer.

8.29 Sketch the design of a simple pair of screws that could be used to both engage and disengage the two halves of a connector requiring a 100 lb (448 N) insertion force. Indicate the minimum size screw which would be required. What size screw would you specify? Why?

8.30 Determine the interference necessary to develop a normal force of 2.0 N during the wiping phase for the contact described in Exercise 8.26. Find the maximum bending stress developed in the contact by this interference. Comment on the magnitude of this stress as it relates either to the selection of the material or the size of the contact.

8.31 Briefly describe a back panel and distinguish it from a circuit board.

8.32 What are the advantages of employing a very dense back panel?

8.33 Compare wire wrap panels with back panels citing similarities and indicating differences. Cite the advantages and disadvantages of each.

8.34 What are the important differences between a pin plate and a wire wrap panel?

8.35 When can cable connected circuit boards be employed instead of back panels.

8.36 Can you propose another method for connecting together circuit boards not covered in Section 8.3. Prepare an engineering sketch showing the key features of your design. Hint: Compliant connections.

8.37 Describe five types of non-rechargeable batteries and indicate their important characteristics. Give examples of products which use each type of battery.

8.38 Describe three types of rechargeable batteries and indicate their important characteristics. Give examples of products which use each type of battery.

8.39 A large signal processor requires 12 kW of power to supply its various components. Estimate the size and weight of the power supplies. Justify your result and comment on the significance of the size and weight.

8.40 Disconnect the power cord and remove the cover from your PC. After locating its power supply, prepare a sketch showing the method used to connect the power and ground leads to it.

8.41 A power supply services 8 back panels each with a capacity of 300 A. The back panels are arranged to form a 2 wide by 4 high vertical panel that is housed in a standard 19 in. wide electronic enclosure 70 in. high. Prepare a sketch showing the locations of the back panels. Next, design a bus bar that will distribute the power from the supply to each back panel. Prepare a drawing of the buss bar. Be sure to include details pertaining to the terminations of the bus bar at the power supply and at the back panels.

8.42 Perform an analysis for the buss bar designed in Exercise 8.41 showing the voltage drop from the power supply to the back panels. Also determine the power losses for this bus bar design.

8.43 Prepare a sketch of a design for a card rack intended to be used in an office environment. The rack is intended to support 12 circuit cards each 200 by 400 mm in size.

8.44 Prepare a three page description of the card rack designed in Exercise 8.43. Justify your design decisions in this engineering brief.

8.45 Describe the various types of card guides and retainers used in card racks. Why does one encounter such a wide range of selection for something as simple as a card guide?

8.46 Prepare a sketch of a retainer-guide which will accommodate PCB's 60 mil (1.5 mm) thick. The retainer-guide should be a single piece part and clamp the edge of the card with a force of 2 lb/in. The goal of the design should be simplicity and ease of manufacturing.

8.47 Prepare a design analysis and description intended for a critical engineering review of your project (Exercise 8.46).

8.48 A heat frame 150 mm long and 75 mm wide is clamped on both of its edges into slots in a pair of cold rails using wedge lock retainers. Estimate the temperature difference from the center of the heat frame to the cold rail interface if 20W is transferred.

8.49 Obtain a copy of a catalog from a local electronics supply house and find the entry for an instrument case to accept a standard 5 by 19 in. (125 by 475 mm) panel. Determine the cost of the case and its weight. Determine the price per lb and note if the case is made of steel or aluminum. Is this price realistic? Why?

8.50 Prepare an engineering brief which supports the cost of the instrument case described in Exercise. 8.49.

8.51 Prepare an engineering brief which gives arguments indicating that the cost of the instrument case described in Exercise 8.49 is too high.

8.52 After disconnecting the power cord, carefully remove the cover from your PC and examine its construction. Prepare a list of parts of those components that you would classify as part of the enclosure. Compare your list with others in the class and find the enclosure that exhibits the simplest design. Does this enclosure have the fewest number of parts? Can you make at least one suggestion which would reduce the cost of a one of the components found in the enclosure? Can you make another suggestion which would modify a part to permit lower assembly cost? Can you make still another suggestion for a modification that would permit more rapid removal of the cover by service personnel? Do you believe the team designing the enclosure for your PC considered all of these design issues?

8.53 Prepare a sketch showing the layout of the cabinet to hold 90 circuit cards. Include racks in your layout and take the dimensions of the PCB's as 0.6 by 9 by 14 in. (15 by 225 by 350 mm). The average power to be dissipated is 20W/card. Size the cabinet selecting one of the standard sizes from a electronic supply house catalog. Remember to leave space for the power supplies and the cables.

8.54 What material would you employ in fabricating the CE^3? Give reasons for this selection.

8.55 Estimate the temperature differential ΔT at the cold rail if a 0.3 format D module dissipating 20W is clamped in place with Birtcher spring type retainers. What would the temperature differential be if wedge locks retainers were used in place of the spring type retainers?

8.56 Why is it good practice to use a relay when switching large currents? What is a large current?

8.57 Determine the stress produced by bending a copper wire over a mandrel of variable radius. Show your results in a graph of stress as a function of the mandrel radius R for wire sizes of AWG 20, 24, 28 and 32.

8.58 Superimpose, on the graph prepared for Exercise 8.57, a series of lines showing the fatigue strength S_e for copper wire. Consider that the wire may have tensile strengths which vary between 40 and 65 ksi (276 and 448 (MPa) depending on the degree of cold working during manufacturing the wire. Interpret the results shown on this new graph.

8.59 Determine the stress in AWG 20 stranded wire (259/44) if it is flexed with a radius of 4 in. Compare this to the stress developed in AWG 20 wire with a solid conductor. Estimate the fatigue life for both wires.

8.60 A series of wires are terminated in a connector with soldered joints. The solder is not to be stressed by applications of load in service. The cable is attached to a drawer that is opened periodically by a service person and the cable may be stretched in this process. Design a strain relief device that will prevent the loads imposed on the cable from being transmitted to the solder connections.

8.61 Prepare a graph, using Manson's equation, which shows $\Delta\varepsilon$ as a function of cycles to failure N_f for copper wire.

8.62 For AWG 36 wire, used as a strand in a larger size conductor, write an equation showing fatigue life as a function of the radius of curvature.

8.63 Repeat Exercise 8.62 but use AWG 32 wire for the strand in the larger gage wire.

8.64 Describe the common polymeric materials used in insulating wires. What are the important properties of the polymers when used in wiring applications?

8.65 Teflon is an excellent insulating material for wiring applications. Why is its use limited to relatively few applications?

8.66 Consider a washing machine found in an ordinary Laundromat. Lay out a sketch showing a wiring harness for this machine. Will the harness be two dimensional (i.e. all the wires lie in a plane) or will it be three dimensional with wires in two or more planes. If the harness is two dimensional, indicate a procedure for manufacturing the harness. If the harness is three dimensional, modify the manufacturing procedure required to produce the cable assembly.

8.67 Compare woven and ribbon flat cable assemblies used in connecting sub assemblies in electronic enclosure. Cite advantages and disadvantages for each approach.

8.68 Why is it necessary to shield wires in certain applications? Can you give two or three examples of problems you have experienced due to inadequate shielding?

8.69 A small instrument case 20 × 22 × 8 in. (500 × 550 × 200 mm) in size is used to house circuits cooled by natural convection. Sketch a design of the case showing the vents you plan to use for air flow. Describe in a short engineering brief the key features of your design.

8.70 What is the pressure of the fluctuation required to produce a sound pressure level of 60 Db? Prepare a graph of sound pressure level as a function of pressure p over the range of 40 to 140 Db.

8.71 A field engineer provides you with a the following measurements for the fan system you have recently designed:

Sound pressure Level (dB)	**Octave band Center frequency (Hz)**
65	63
60	125
55	250
52	500
48	1000
45	2000
44	4000
42	8000

Prepare a graph of these results and superimpose on this graph the appropriate PNC curve that is satisfied. Prepare an engineering brief describing the types of rooms in which the fan system can be used without interfering with speech.

PART 3

ANALYSIS

METHODS

CHAPTER 9

THERMAL ANALYSIS METHODS–CONDUCTION

9.1 INTRODUCTION

Thermal analysis of electronic systems involves predicting the junction temperatures of the integrated circuits used in both analog and digital circuits. It also involves predicting the temperature of other circuit components like resistors, capacitors and transformers. The analysis is extremely important as the system reliability and system availability are strongly dependent upon component temperatures and relatively small temperature changes can have a marked affect on both component and system reliability.

Heat is generated by the flow of current through the circuit components because of the I^2R losses. This heat is transferred within the enclosure by conduction, convection and to a lesser degree by radiation. Finally the heat is dissipated to the environment (a heat sink) by convection and radiation. It is important to reduce the temperature differential required to transfer the heat from the components through the enclosure to the environment. Lower temperature differentials enhance reliability and system availability. However, the task of maintaining low temperature differences is growing more difficult as the circuit density increases with each new generation of chips. The power to be dissipated has been increasing with each new generation while the volume of the electronic products incorporating these chips has been decreasing.

There are several different methods used in designing cooling systems for dissipating the heat generated to the environment during system operation. Natural convection, forced air convection, conduction to heat exchangers, radiation and boiling are all considered in thermal management schemes. The particular design, which evolves to control the component temperatures, depends to a large extent on the price and performance characteristics of the end product. For price driven products, low cost cooling methods with natural and forced convection are the most common design approach. For high performance products, conduction from the back of the chip to cold plates is often used to efficiently transfer large quantities of heat with small temperature differences ΔT. Intermediate systems often use forced convection with specially designed duct systems and heat exchangers to reduce the temperature gradients required to transfer the heat load.

This treatment is divided into two parts, to facilitate the presentation of the material in a manner easily interpreted by the reader. Because conduction is involved in every thermal design or analysis, it is treated in this chapter. Designs involving heat transfer by convection, radiation and boiling are covered later in Chapter 10. The presentation emphasizes a relatively simple analytical technique because this procedure best illustrates the key design approaches in developing an effective cooling system for electronic products. A more complete but complex approach is possible, if specialized software employing finite difference methods is used in determining more precise estimates of junction temperatures. A detailed treatment of these software programs is beyond the scope of this textbook.

The importance of control of the component temperatures on component and system reliability cannot be over emphasized. The failure rate λ that controls component reliability is a function of temperature and relatively small increases in temperature result in a significant increase in failure rate and a subsequent loss in system availability.

9.2 STEADY STATE HEAT TRANSFER BY CONDUCTION

Heat is generated in electronic devices primarily by I^2R losses, and it must be dissipated to avoid excessive temperatures that may impair the reliability of the system. The amount of heat generated depends on the size and complexity of the system and can vary from a few mW in a wrist-watch to 10 kW or more in a large server or signal processor. The design for dissipating the heat generated will depend upon the product, the environment, the accessibility of a coolant, the reliability, the system availability requirements and the costs involved in providing an adequate cooling system. There is no single easy solution for dissipating heat generated during the operation of an electronic system. Moreover, as the circuit densities increase with the scale of integration, the problem of reducing the thermal penalty ΔT, incurred in dissipating the heat becomes more difficult.

Heat is transferred from the component (the source) to the environment (the sink) by four basic mechanisms which include conduction, convection, radiation and boiling. Of these four mechanisms conduction is the most important, because it is always involved to some extent in transferring heat from the circuits on the chip to at least one of the external surfaces of the chip carrier where a different heat transfer process may be involved.

9.2.1 Heat Transfer Through a Plane Body

Fourier's law for the rate of heat transfer by conduction is given by:

$$q = -kA\frac{dT}{dx} \tag{9.1}$$

where q is the **rate of heat transfer** in watts (W); A is the area normal to the path of heat flow in (m^2); $\frac{dT}{dx}$ is the temperature gradient (°C/m) and k is the coefficient of thermal conductivity (W/m-°C).

Equation 9.1 describes the rate of transfer of heat in one dimension (along x) as illustrated in Fig. 9.1. The temperature gradient $\frac{dT}{dx}$ along the x-axis may be written as:

$$\frac{dT}{dx} = \frac{T_1 - T_2}{0 - L} = \frac{T_2 - T_1}{L} \tag{9.2}$$

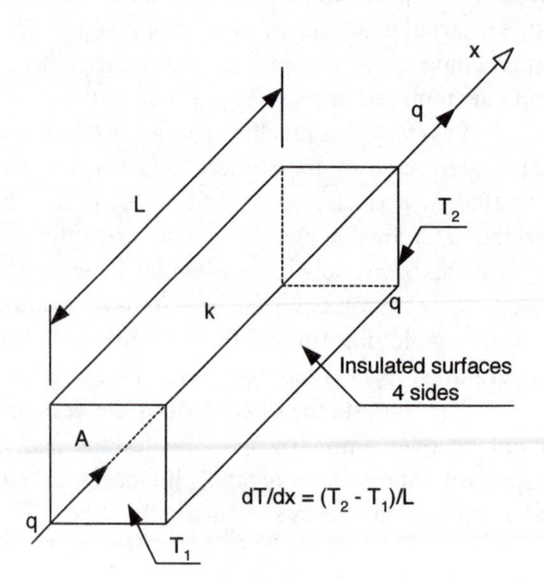

Fig. 9.1 Illustration of parameters affecting one-dimensional heat transfer along the x-axis.

The temperature gradient $\frac{dT}{dx}$ is a negative quantity because $T_1 > T_2$ and the heat flow is from the hot side of the insulated bar to its cold side in the positive x direction. Substituting Eq. (9.2) into Eq. (9.1) gives:

$$q = \frac{kA}{L}(T_1 - T_2) = K\Delta T \qquad (9.3)$$

The term kA/L is the effective thermal conductance K of a one-dimensional conductor. The thermal conductance K is the reciprocal of the thermal resistance R_T given by:

$$R_T = \frac{1}{K} = \frac{L}{kA} \qquad (9.4)$$

where R_T is expressed in terms of °C/W.

It is possible to draw an analogy between heat and current flow by comparing Eq. (9.3) with Ohm's law Eq. (3.3) and equating corresponding terms:

$$I = \frac{V}{R} \qquad q = \frac{\Delta T}{R_T} \qquad (9.5)$$

where I the current is the analog of q the rate of heat transfer.
V the voltage is the analog of $\Delta T = (T_1 - T_2)$.
R the electrical resistance is the analog of the thermal resistance $R_T = L/kA$.

The coefficient of thermal conductivity k is a material property. It is extremely important because of the very wide range of values that occur in materials commonly employed in packaging electronic systems. This fact is illustrated in Tables 9.1 to 9.4 where the coefficients of thermal conductivity for metals, ceramic, polymers and their composites, air and water are listed. It is significant that coefficient of thermal conductivity k ranges over five orders of magnitude from air at its low to silver at its high. To illustrate the importance of k, consider a conducting strip 2 mm thick, 10 mm wide and 100 mm long that conducts heat at the rate of 1 W. If the strip is fabricated from pure copper with k = 400 W/m-°C and its four sides are insulated, then the temperature difference necessary to achieve the specified rate of heat transfer is given by Eq. (9.5) as:

$$\Delta T = T_1 - T_2 = qR_T = (1 \times 0.1)/(400 \times 2 \times 10 \times 10^{-6}) = 12.5\ °C \qquad (a)$$

However, if the strip is fabricated from alumina with k = 40 W/m-°C, then the ΔT increases significantly because the thermal conductivity is ten times lower.

$$\Delta T = 12.5\ (400/40) = 125\ °C \qquad (b)$$

This simple example shows the marked influence of the coefficient of thermal conductivity on the temperature difference (a thermal penalty) necessary to transfer heat at a specified rate. The importance of the choice of materials in the design of packaging for electronic components and circuits cannot be over emphasized.

Table 9.1
Thermal conductivity k for several metals and their alloys[1]

Material	Thermal Conductivity k W/m-°C	Material	Thermal Conductivity k W/m-°C
Metals		Gold	296
Alloy 42	15.7	Au (98%) Si (2%)	50
Aluminum		Iron	
Pure	216	Cast	55
356 T6	150	Pure	75
2024 T4	121	Wrought	59
5052	144	Kovar	17.5
6061 T6	156	Lead	33
7075 T6	121	Magnesium	157
Beryllium	164	Molybdenum	134
Beryllium Copper	83	Monel	35
Brass		Silicon	153
Red	110	Silver	429
Yellow	94	Solder Spheres	50.6
Cu (70%) Zn (30%)	100	Steel	
Copper		SAE 1010	58
Alloy MF 202	160	SAE 1020	55
Alloy CDA 194	263	SAE 1045	45
Cu (90%) W (10%)	190	Stainless	16
Foil	390	Tin	62
Pure	400	Titanium	15.6
Drawn Wire	287	Zinc	102

Table 9.2
Thermal conductivity and temperature coefficient of expansion for glass and ceramic materials

Material	Thermal Conductivity k W/m-°C	Temperature coefficient of expansion $\times 10^{-6}$/°C
$Al_2 O_3$	40	8.0
Al N	180	4.4
$B_4 C$	25	4.3
BeO	210	7.4
SiC	50	4.3
$Si_3 N_4$	17	3.0
WC		5.2
Diamond	2,000	0.5
E glass fiber	0.32	
Fused silica	1.4	
Glass		
Sealing	0.6	6.6
Soft	0.98	
Pyrex	1.01	
Silver filled	270	8
Mica	0.75	

[1] Values cited for thermal conductivity k varies from one reference source to another. Slight differences in alloy content and measurement methods account for these variations.

Table 9.3
Thermal conductivity for polymers and polymer composites

Material	Thermal Conductivity k W/m-°C	Material	Thermal Conductivity k W/m-°C
Encapsulant	0.52	Nylon 6	0.25
Epoxy	0.21	Phenolic paper	0.28
Epoxy Ag filled	1.60	Plexiglas	0.19
FR-4 laminate	0.35	Polyethylene HD	0.50
Glass-PTFE	0.26	Polyimide	0.20
Glass polyimide	0.35	PVC	0.19
Kevlar	0.25	Rubber-silicone	0.19
Molding compound	0.80	Teflon	0.25
Mylar	0.19	Underfill	0.50

Table 9.4
Thermal conductivity for air and water

Material	Thermal Conductivity k W/m-°C
Air	0.0257
Water	0.658

9.2.2 General Equations Governing Conduction

The previous section described a very important but specialized case of conduction dealing with steady state, one-dimensional heat flow without heat generation within the body of the conductor. The more general case of heat transfer is transient, with temperatures changing relative to time, with heat flow in the three directions x, y and z and with internal heat generated at a rate of q_i throughout the volume of the conductor. A heat balance for a volume element of this conductor leads to the following classical differential equation that governs the most general heat transfer solution:

$$\frac{\partial^2 T}{\partial x^2}+\frac{\partial^2 T}{\partial y^2}+\frac{\partial^2 T}{\partial z^2}+\frac{q_i}{k}=\frac{1}{\alpha}\frac{\partial T}{\partial t} \tag{9.6}$$

where q_i is the heating rate per unit volume (W/m^3); t is time (s).
α is the thermal diffusivity (m^2/s) that is given by:

$$\alpha = k/c\rho \tag{9.7}$$

where ρ is the density of the material (kg/m^3) and c is the specific heat of the material (J/kg-°C).

In the absence of internal heating, with $q_i = 0$, Eq. (9.6) reduces to:

$$\nabla^2 T = (1/\alpha)\,\partial T/\partial t \tag{9.8}$$

where $\nabla^2 = \left[\frac{\partial^2}{\partial x^2}+\frac{\partial^2}{\partial y^2}+\frac{\partial^2}{\partial z^2}\right]$ is Laplace's operator.

Under steady state conditions, the temperatures have stabilized with respect to time $\frac{\partial T}{\partial t} = 0$, and Eq. (9.6) reduces to Poisson's equation:

$$\nabla^2 T = -q_i/k \qquad (9.9)$$

Finally, without internal heat generation in the conductor and steady state temperature conditions, Eq. (9.6) simplifies to Laplace's classic equation:

$$\nabla^2 T = 0 \qquad (9.10)$$

In most packaging applications, the geometry and boundary conditions for components such as chip carriers, circuit boards and heat sinks are complex. The complexities limit the availability of exact closed form solutions to the general equations. Because these exact solutions are difficult, simplified models have been developed that approximate the electronic assembly and permit reasonable estimates of the temperature of critical components. These simplified models also indicate design approaches to reduce ΔT for a specified heating rate. Another method of analysis utilizes specialized software to obtain numerical solutions to problems with complex geometries. These codes, based either on finite elements or finite differences, are written to provide solutions for specific problems arising in electronic packaging. Capabilities of two of the many available programs are shown in footnotes[2]; [3].

The approach followed in this text is to emphasize the development of relatively simple models of complex electronic assemblies. The primary purpose of the analysis is to show **design features** and **material selections** that markedly affect temperatures and to give a first approximation of the temperature of the components.

9.2.3 Heat Transfer Through a Hollow Cylinder

Consider the hollow cylinder shown in Fig. 9.2 and assume that steady state temperatures have been achieved and that no internal heat is generated in the body of the cylinder ($q_i = 0$). In this case, the heat transfer is one-dimensional, in the radial direction from the inside to the outside wall. Modifying Eq. (9.1) to account for the radial heat flow by letting $r = x$ and the area $A = 2\pi rL$ gives:

$$q = -2\pi kLr\left(\frac{dT}{dr}\right) \qquad (a)$$

Rearranging Eq. (a) yields:

$$dT = -\left(\frac{q}{2\pi kL}\right)\left(\frac{dr}{r}\right) \qquad (9.11)$$

Equation (9.11) can be integrated between the limits of r_1, r_2 and T_1, T_2 to obtain:

---

[2] FLOTHERM is powerful 3D simulation software program for thermal design of electronic components and systems. It enables designers to create virtual models of electronic equipment, perform thermal analysis and test design modifications in the early stages of the design process before building physical prototypes. The program uses advanced computational fluid dynamics techniques to predict airflow, temperature and heat transfer in components, boards and complete systems.

[3] CoolitPCB CFD software is designed specifically for predicting airflow and heat transfer in board designs. It permits advanced thermal analysis on a desk top computer. With this program, designers can determine air flow requirements and temperatures of critical components.

$$\Delta T = T_1 - T_2 = \frac{q}{2\pi kL} \log_e\left(\frac{r_2}{r_1}\right) \quad (9.12)$$

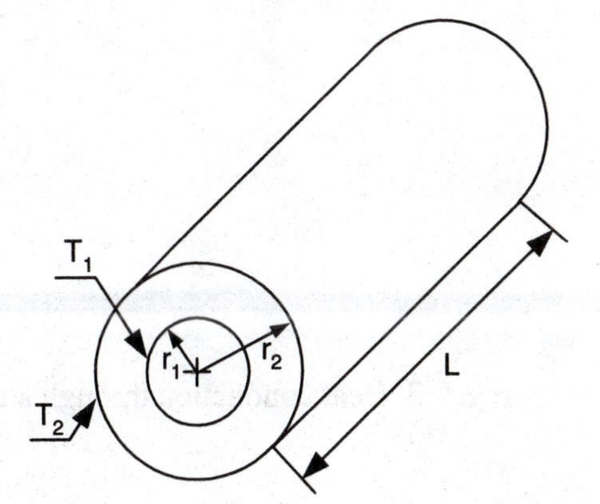

Fig. 9.2 Conduction through the wall of a hollow cylinder.

For the hollow cylinder, the thermal resistance is obtained from Eq. (9.4) and (9.12) as:

$$R_T = \frac{1}{2\pi kL} \log_e\left(\frac{r_2}{r_1}\right) \quad (9.13)$$

The results from Eqs (9.12) and (9.13) are useful in determining the temperature difference ΔT required to conduct heat radially outward from cylindrical packages containing electronic circuits that act as heat sources.

9.2.4 Heat Transfer Through Layered Structures

Consider a composite made from four layers of different materials as shown in Fig. 9.3. Heat is transferred into the left side of the first layer where the surface temperature of the wall is T_i. The heat is transmitted in the x direction through the four layers with a ΔT developed so as to maintain a constant rate of heat transfer through the composite assembly. This problem is classified as one-dimensional steady state conduction with $q_i = 0$. Applying Eq. (9.3) to each of the layers gives:

$$T_i - T_{1-2} = \frac{qL_1}{k_1 A} \quad \text{(a)}$$

$$T_{1-2} - T_{2-3} = \frac{qL_2}{k_2 A} \quad \text{(b)}$$

$$T_{2-3} - T_{3-4} = \frac{qL_3}{k_3 A} \quad \text{(c)}$$

$$T_{3-4} - T_0 = \frac{qL_4}{k_4 A} \quad \text{(d)}$$

Adding these four equations yields:

$$\Delta T = T_i - T_0 = q\sum_{i=1}^{4}\frac{L_i}{k_i A} = q\sum_{i=1}^{4} R_{Ti} \qquad (9.14)$$

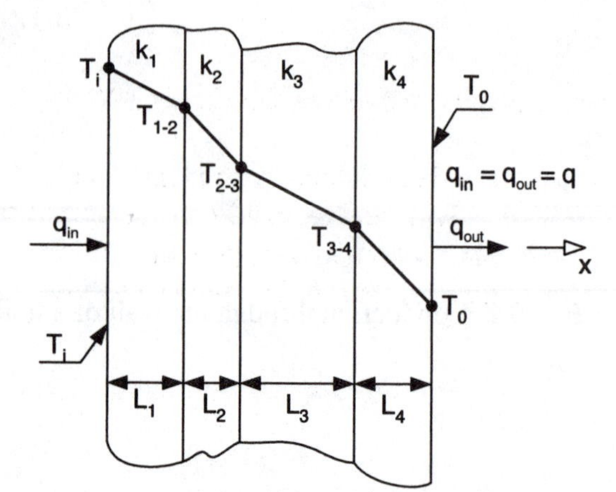

Fig. 9.3 Heat conduction through a composite slab.

Examination of Eq. (9.14) shows that ΔT across a layered composite depends on the rate of heat transfer and the sum of the thermal resistance of each of the four layers. The electrical analogy corresponding to this result is presented in Fig. 9.4. The use of this approach to analyze heat conduction through layered components will be described in more detail in Section 9.4.

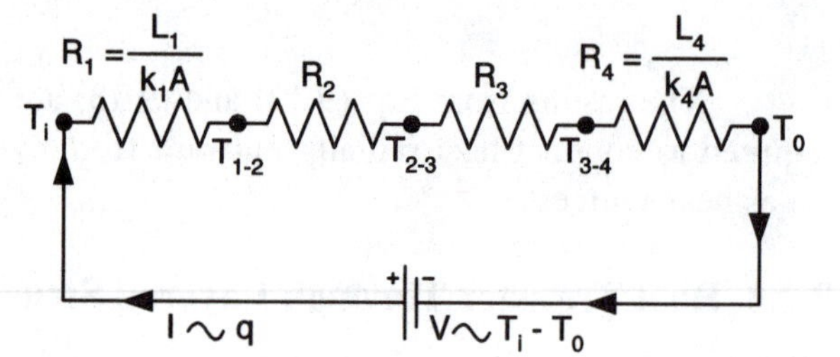

Fig. 9.4 Electrical analogy for heat conduction through a four-layer composite slab.

9.3 OVERALL COEFFICIENT OF HEAT TRANSFER

The examples illustrated in Figs. 9.1, 9.2 and 9.3 indicated well defined temperatures at the input and output surfaces of the heat source and sink. This procedure is correct but not very realistic as surface temperatures are not usually known. Instead, the temperature of the fluid adjacent to these surfaces is usually the known quantity. These fluid temperatures are different from the surface temperatures due to the development of a boundary layer at the surface. A more realistic representation of the temperature distribution that accounts for the effect of the boundary layer between the fluid and the conducting surface is presented in Fig. 9.5.

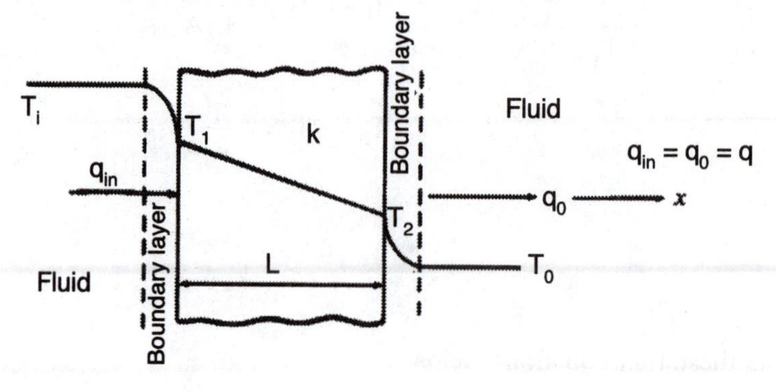

Fig. 9.5 Heat transfer by convection and then conduction.

Heat transfer occurring between the fluid and the solid surfaces is governed by the convection equation that is written as:

$$q = hA(T_i - T_1) \tag{9.15}$$

where h is the convection coefficient (W/m^2-°C).

The determination of h is involved and will be deferred until Chapter 10, where heat transfer by convection will be treated in considerable detail. At this stage in the development, consider h as a known quantity and use Eqs (9.3) and (9.15) to write the relations for the temperature differences across the boundary layers and the solid layer shown in Fig. 9.5 to obtain:

$$T_i - T_1 = \frac{q}{h_1 A} \qquad \text{(a)}$$

$$T_1 - T_2 = \frac{qL}{kA} \qquad \text{(b)}$$

$$T_2 - T_0 = \frac{q}{h_0 A} \qquad \text{(c)}$$

Adding Eqs. (a), (b) and (c) gives:

$$\Delta T = T_i - T_0 = \frac{q}{A}\left[\frac{1}{h_i} + \frac{L}{k} + \frac{1}{h_0}\right] \tag{9.16}$$

This result shows that the coefficients h_i, h_0 and k affect the rate of heat transfer for a specified ΔT. The coefficients can be combined by writing Eq. (9.3) in a slightly modified form as:

$$q = UA\Delta T \tag{9.17}$$

where U is the combined coefficient for the rate of heat transfer (W/m^2-°C):

$$U = \frac{1}{\left[\frac{1}{h_i} + \frac{L}{k} + \frac{1}{h_0}\right]} \tag{9.18}$$

The thermal resistance R_T for the electrical analog corresponding to this case is determined from Eqs. (9.16), (9.17) and (9.18) as:

$$\Delta T = qR_T \tag{9.4 bis}$$

where

$$R_T = \frac{1}{A}\left[\frac{1}{h_i} + \frac{L}{k} + \frac{1}{h_0}\right] \tag{9.19}$$

9.4 CONTACT RESISTANCE

The previous discussions of heat conduction through composite slabs assumed that the layers were bonded together giving perfect thermal contact at the interfaces. The result of perfect thermal contact at the junction of two materials is to reduce the temperature difference across the interface to zero as indicated in Fig. 9.6a. However, when the two layers are not bonded together but are brought into simple contact, as illustrated in Fig. 9.6b, the fit is far from perfect. The surface roughness on both surfaces produces voids which inhibit the flow of heat across the interface and result in an addition thermal contact resistance R_c. The thermal contact resistance must be considered in the determination of R_T for a thermal path that includes imperfect bonding at the interface.

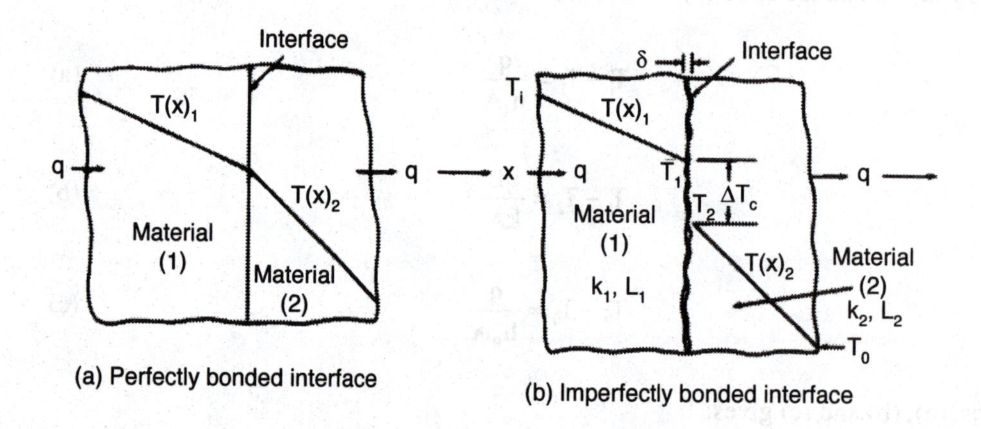

Fig. 9.6 Examples of perfectly bonded and imperfectly bonded interfaces showing temperature gradients across interfaces when heat is transferred by conduction.

To show the effect of an imperfect bond consider the heat flow through the composite body shown in Fig. 9.6b. Application of Eqs (9.3) and (9.15) gives:

$$T_i - T_1 = \frac{qL_1}{k_1 A} \qquad \text{(a)}$$

$$T_1 - T_2 = \frac{q}{h_c A} \qquad \text{(b)}$$

$$T_2 - T_0 = \frac{qL_2}{k_2 A} \qquad \text{(c)}$$

Adding these three relations gives:

$$\Delta T = T_i - T_0 = \frac{q}{A}\left[\frac{L_1}{k_1} + \frac{1}{h_c} + \frac{L_2}{k_2}\right] \qquad (9.20)$$

where h_c is the contact coefficient (W/m^2-°C).

The area A across the interface is not continuous and must be divided into the area A_c where contact occurs and the area A_v where voids exist. Thus:

$$A = A_c + A_v \tag{9.21}$$

Heat flow across the interface gap δ is due to conduction across the contact area A_c and the void area A_v. Assume the interface gap is δ is due to the sum of the surface roughness of the two contacting surfaces that can be written as:

$$\delta = \delta_1 + \delta_2 \tag{9.22}$$

where δ_1 and δ_2 are the surface roughness of materials 1 and 2.

Considering parallel heat transfer paths through A_c and A_v yields:

$$q = \frac{\Delta T_c}{\delta_1/(k_1 A_c) + \delta_2/(k_2 A_c)} + \frac{\Delta T_c}{\delta/(k_v A_v)} = h_c A \Delta T_c \tag{9.23}$$

where k_v is the thermal conductivity of the fluid filling the voids. Solving Eq. (9.23) for h_c gives:

$$h_c = \frac{(A_c/A)}{[(\delta_1/k_1) + (\delta_2/k_2)]} + \frac{(A_v/A)}{(\delta/k_v)} \tag{9.24}$$

The first term in Eq. (9.24) is due to conduction through the contact area and the second term is due to conduction through the fluid contained in the voids. While this equation is important in conceptual understanding of conduction across an interface, it is not usually employed to compute h_c. It is not possible to determine the quantities δ, δ_1, δ_2, A_c and A_v with sufficient accuracy to adequately describe the surfaces in contact.

A more pragmatic approach utilizes an empirical equation such as the one advanced by Chang-Lin Tien[4]:

$$h_c = 0.55m\left(\frac{k}{\sigma}\right)\left(\frac{p}{H}\right)^{0.85} \tag{9.25}$$

where $k = (2k_1k_2)/(k_1 + k_2)$ and $\sigma^2 = \sigma_1^2 + \sigma_2^2$

σ is the RMS value of the surface roughness (μm); m is the RMS value of the slope of the contacting asperities; p is the contact pressure; H is the hardness of the softer of the two materials in contact, which is usually taken as three times the yield strength.

Experimental results showing h_c as a function of contact pressure are presented in Fig. 9.7 for aluminum to aluminum and steel to steel interfaces. The effect of surface finish is extremely important. Improving the surface finish from 120 to 10 μin. RMS, results in a three fold increase in h_c at p = 0.69 MPa.

---

[4] Additional empirical relations for the contact coefficient proposed by several investigators are described in Reference [4].

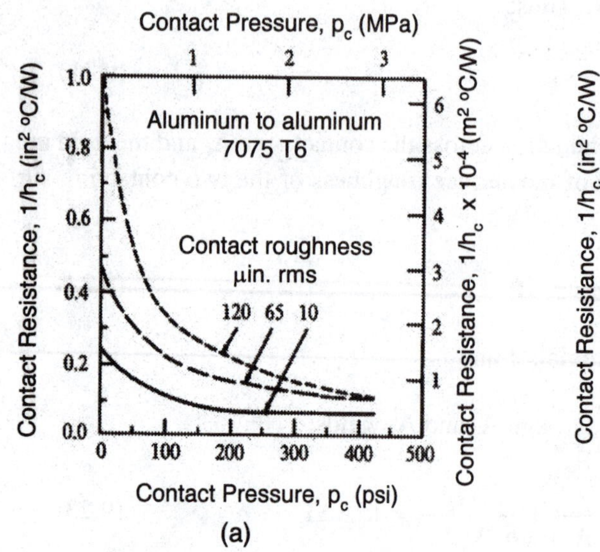

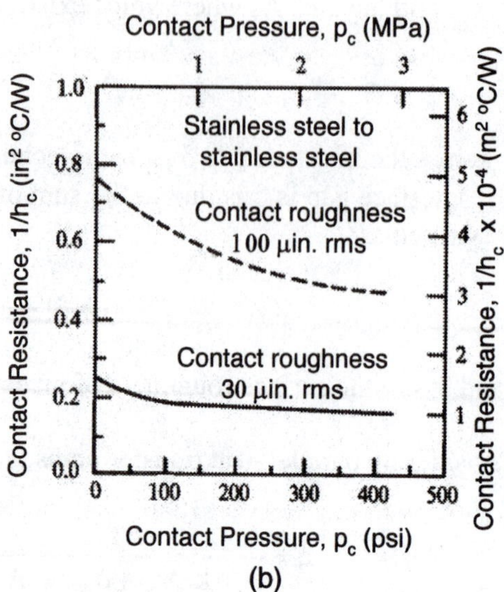

Fig. 9.7 Contact resistance per unit area as a function of contact pressure.

Several techniques can be used to reduce the thermal resistance across an interface. The first is to plate a coating of a soft material such as tin or zinc on the contact area that decreases the surface hardness H. The plating process also concentrates the coating in the surface's valleys reducing the roughness σ and its slope m. Both of these factors improve the contact coefficient h_c. The second technique is to use a thin layer of thermally conducting paste to fill the voids at the interface. The value of k for silicone grease[5] is 0.208 W/m-°C that is much higher than that of air (0.0257 W/m-°C). Reductions in thermal resistance by a factor of at least two to three can be achieved by filling the voids with paste. A third method involves increasing the interface pressure p by using wedge lock type clamping.

The effects of interfaces that are not thermally bonded and intended for space applications are even more significant because hard vacuum is encountered in space. The air in the voids is lost and the value of h_c decreases significantly. In these applications, the use of non-volatile pastes to fill the voids is even more important as indicated in Table 9.5.

Table 9.5
Contact coefficient h_c for 6 × 250 × 500 mm aluminum plates bolted at their corners

Interface material	h_c (W/m²-°C)	
	Air	Vacuum
0.05 mm aluminum foil	1700	567
0.5 mm rubber sheet	1133	227
0.12 mm beryllium copper foil	1133	227
No insertion material	1133	113

[5] Other filled pastes have been developed with thermal conductivity k as high as 3.8 W/m-°C.

9.5 CONDUCTION FROM DISCRETE HEAT SOURCES

The heat source in many electronic devices is often small in area when compared to the area of the channel used to conduct the heat from the source to the sink. The exact solution for problems of this type involves a three dimensional analysis with a significant increase in complexity. Approximate solutions are possible, by retaining the one-dimensional form of the analysis, if an additional thermal resistance R_{Ts} is added to R_T to account for the effect of the relatively small area of the discrete heat source. Following this approach the temperature difference may be written as:

$$\Delta T = q(R_{Ts} + R_T) \qquad (9.26)$$

where R_{Ts} is the thermal resistance due to the constriction.
R_T is the resistance of the conducting body.

To illustrate this approach, consider a heat source placed on the surface of a half space as illustrated in Fig. 9.8. For the circular heat spot, shown in Fig. 9.8a, with a uniform distribution of q, the thermal resistance R_{Ts} is given by:

$$R_{Ts} = \frac{16}{3\pi^2 Dk} = \frac{0.54}{Dk} \qquad (9.27)$$

For the square heat source shown in Fig. 9.8b, thermal resistance R_{Ts} is given by:

$$R_{Ts} == \frac{0.55}{Dk} \qquad (9.28)$$

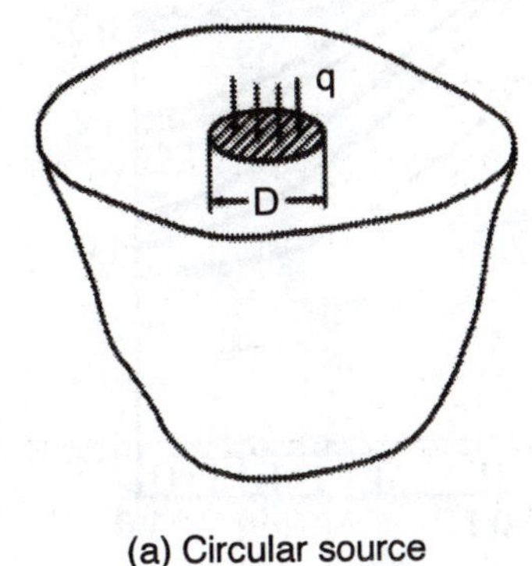

(a) Circular source

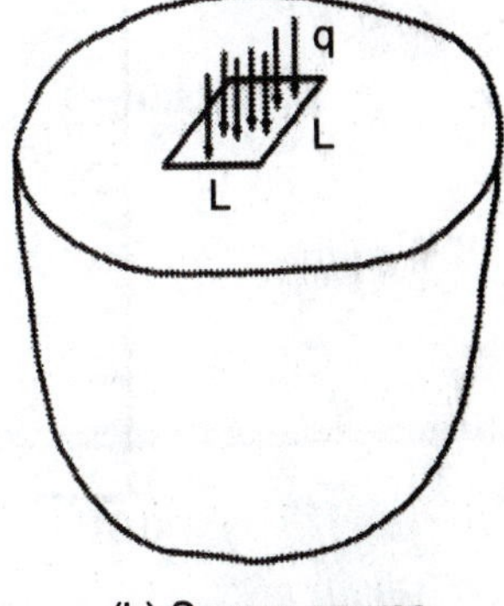

(b) Square source

Fig. 9.8 Discrete small area heat sources on the surface of a half-space.

If the conducting body is finite, the determination of the thermal resistance R_{Ts} is more difficult to estimate as the details of the geometry of the body must be taken into account. For a small circular source with a radius a on a cylindrical conductor with a radius b, shown in Fig. 9.9, the thermal resistance due to the constrictive effect is given by:

$$R_{Ts} = \left[\frac{1}{2\sqrt{\pi}\,ak}\right]\left[1-\left(\frac{a}{b}\right)\right]^{3/2} \qquad (9.29)$$

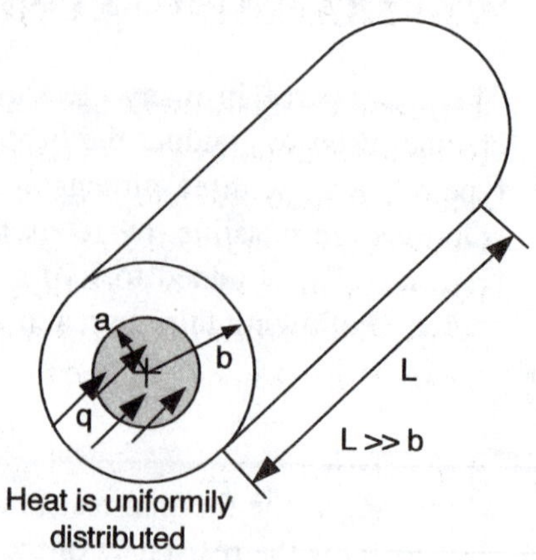

Fig. 9.9 Symmetrical uniform heat input over a small circular area on one face of a circular cylinder.

The thermal resistance for a small square source of heat on a square plate with finite dimensions has been determined by Ellison. His results are shown in the graph presented in Fig. 9.10.

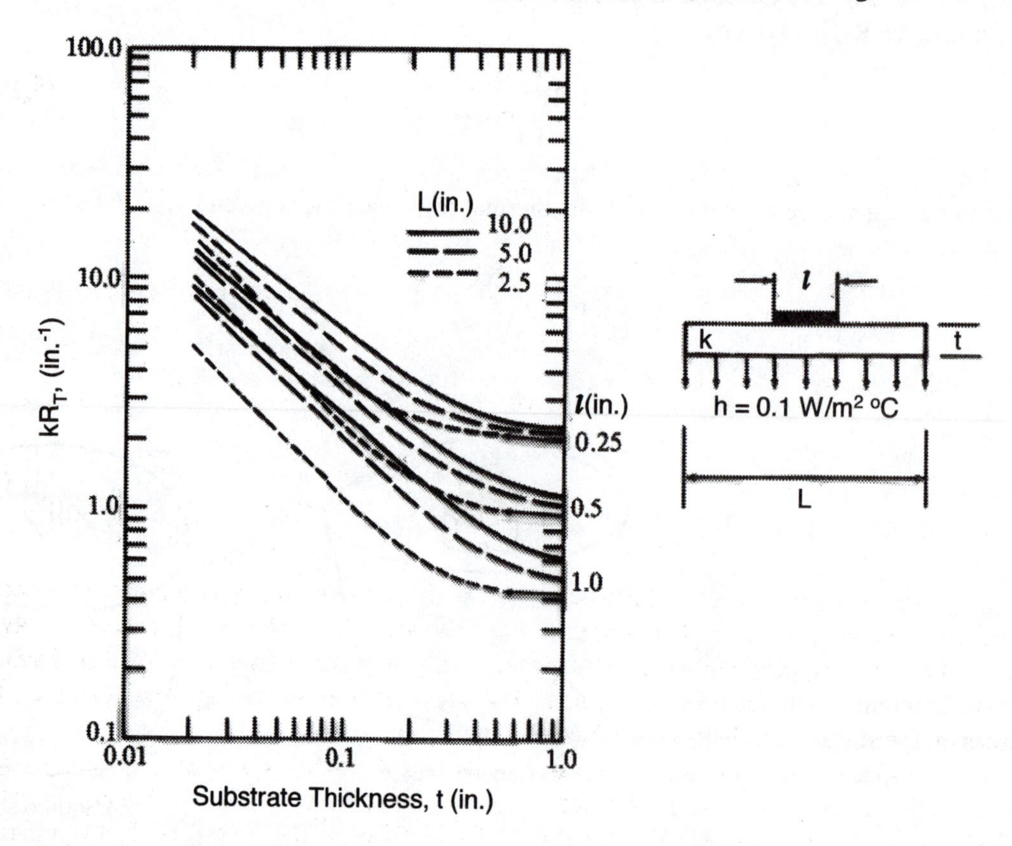

Fig. 9.10 Thermal resistance for a square heat source with dimensions $l \times l$ on a square conductor $L \times L$ in size with a finite thickness .

9.6 TRANSIENT CONDUCTION

Transient thermal conditions occur during the start up and shut down of an electronic system. Transient behavior is important as it controls the time required for warm-up and thermal equilibrium of the system. After warm-up, the temperature reaches a maximum and steady state conditions prevail. The maximum temperature controls the reliability of the various electronic devices as discussed in Section 9.1. Also, power-on and power-off cycling is very important in determining the reliability of connectors and solder joints. These joints may fail due to thermal strains induced by differential expansions and contractions between the chip carrier and the circuit board that occur during warm-up and cool-down.

9.6.1 Initial Warm Up Time

To illustrate the concept of warm-up time in a conduction cooled system, consider a device that generates heat at a rate q_g mounted on a totally insulated assembly with a weight W. The temperature of the device and the component are at the ambient temperature T_0 at time t = 0 when the system powered is applied. Under these conditions, Eq. (9.6) leads to:

$$q_i = \frac{k}{\alpha}\frac{dT}{dt} \tag{9.30}$$

where $q_i = q_g/\boldsymbol{V}$; $W = \rho \boldsymbol{V}$ and $\boldsymbol{V}$ is the volume of the assembly.

From Eqs. (9.7) and (9.30), we obtain:

$$q_g = (Wc)\frac{dT}{dt} \tag{9.31}$$

Integrating Eq. (9.31) gives:

$$T - T_0 = \left(\frac{q_g}{Wc}\right)t \tag{9.32}$$

Equation (9.32) predicts a linear increase in temperature with time until failure occurs due to excessively high temperature. This situation is obviously not satisfactory and occurs only when the system is insulated and without a means to transfer heat from the component to the environment. This result demonstrates, in a very simple manner, the importance of providing a suitable path for heat to transfer from the system to the environment.

If a suitable path for heat to transfer to the environment is introduced into the system, the maximum temperature is bounded. We note that this path of heat transfer exhibits a thermal resistance R_T; hence, it is necessary to modify Eq. (9.31) to account for the heat flow to the environment as indicated below:

$$q_g - \frac{(T - T_0)}{R_T} = (Wc)\frac{dT}{dt} \tag{9.33}$$

Rearranging this equation to show the differential equation in standard form yields:

$$\frac{dT}{dt} + \frac{T}{(WcR_T)} = \frac{1}{(Wc)}\left[q_g + \frac{T_0}{R_T}\right] \tag{9.34}$$

The homogeneous solution to Eq. (9.34) is given by:

$$T_c = Ce^{-(t/WcR_T)} \tag{9.35}$$

The particular solution providing the steady state value of the temperature T is given by:

$$T_p = q_g R_T + T_0 \tag{9.36}$$

Combining Eqs. (9.35) and (9.36) yields the general solution as:

$$T = T_c + T_p = Ce^{-(t/WcR_T)} + q_g R_T + T_0 \tag{9.37}$$

The integration constant C is determined to satisfy the initial condition that $T = T_0$ at $t = 0$, which leads to:

$$C = -q_g R_T \tag{a}$$

Substituting $C = -q_g R_T$ into Eq. (9.37) gives the relation for the temperature T as a function of time t.

$$T = q_g R_T\left[1 - e^{-(t/WcR_T)}\right] + T_0 \tag{9.38}$$

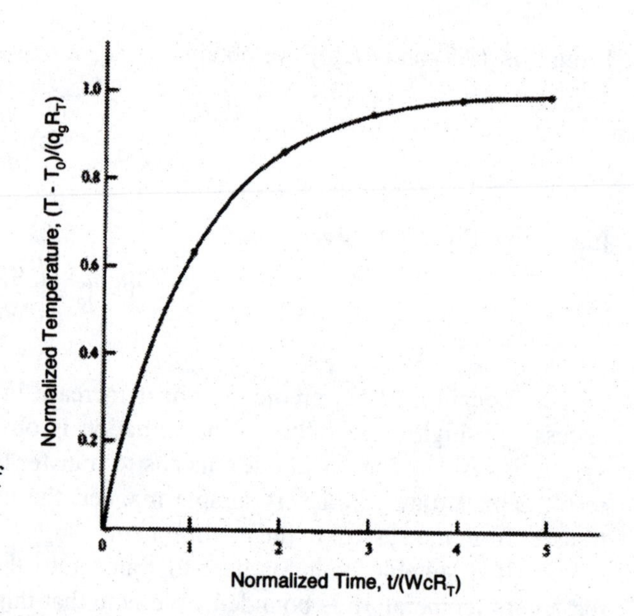

Fig. 9.11 Normalized temperatures as a function of normalized time after power is applied.

Characteristic results for Eq. (9.38), illustrated in Fig. 9.11, show the temperature of the system as a function of time after the device power is applied. The temperature increases exponentially with time and reaches a steady state value T_p, given by Eq. (9.36), after a relatively long period of time. It is customary to take the warm-up time equivalent to three time constants (WcR_T is the time constant) when the temperature has achieved 95 per cent of its steady state value. The warm-up time t_w to achieve 95% of steady state temperature is approximated by:

$$t_w = 3WcR_T \tag{9.39}$$

9.6.2 Cool Down Time

To determine the cool-down time after power disconnected, let $q_g = 0$ in Eq. (9.13) to obtain:

$$\frac{dT}{dt} + \frac{T}{(WcR_T)} = \frac{T_0}{(WcR_T)} \tag{9.40}$$

The homogeneous solution to Eq. (9.40) is the same as that given by Eq. (9.35) and the particular solution changes to $T_p = T_0$. The general solution describing the time temperature relation during cool down is then given by:

$$T = Ce^{-(t/WcR_T)} + T_0 \tag{9.41}$$

The integration constant C is determined from the initial conditions for cool down, which are $T = T_s$ at $t = 0$, where T_s is the steady state temperature given by Eq. (9.36). Substituting this initial condition into Eq. (9.41) gives $C = T_s - T_0$. It is now possible to recast Eq. (9.41) as:

$$\frac{(T - T_0)}{(T_s - T_0)} = e^{-(t/WcR_T)} \tag{9.42}$$

This result indicates that the temperature of the system will decrease from T_s when powered down with an exponential decay to nearly the ambient temperature T_0, in a time equivalent to three time constants. It is clear that warm-up and cool-down times are identical. The time to achieve steady state conditions dependent upon the weight of the assembly, the effective specific heat of the materials employed and the thermal resistance from the source to the sink.

9.7 HEAT TRANSFER IN CHIP CARRIERS

DIPs

An approximate solution for the temperature difference ΔT between the chip and the chip carrier is possible by using the simple methods of conduction described in Sections 9.2. To illustrate this method of solution, consider a chip dissipating 0.5 W packaged in a dual in line package (DIP), as illustrated in Fig. 9.12.

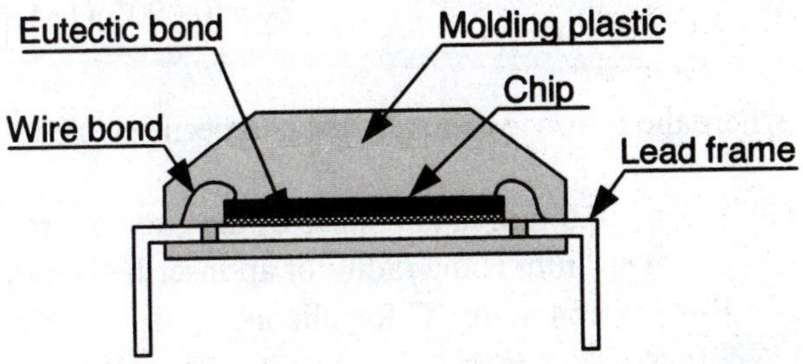

Fig. 9.12 Conduction from a chip packaged in a DIP.

The first step is to model the package with a resistance network representing the individual thermal constraints associated with each design feature involved in the heat flow path. The resistor network representing the DIP package is presented in Fig. 9.13. In constructing this resistor network, assumptions are made concerning the predominant heat paths and the heat transfer mechanisms involved. In this case, it was assumed that the heat generated by the device flows down through the chip, eutectic bond and lead frame. Then the heat passes through the plastic insulation between the lead frame to the leads where it is conducted outward along the leads to the outside of the package. The heat is dissipated by convection to air flowing along the surface of the PCB. In this example, heat transfer from the leads to the air will not be considered, because our treatment of convection as a heat transfer mechanism will be deferred until Chapter 10. Instead, the temperature of the leads will be taken as a constant T_L at the point where they exit the DIP body. It is also assumed that the heat transferred through the top and the base of a plastic molded DIP is negligible. Heat transfer through the top and the base is not significant because of the relatively low thermal conductivity of the plastic molding compound from which the case is fabricated. Also the presence of a layer of stagnant air between the base of the DIP and the printed circuit board limits heat transfer through the DIP's base.

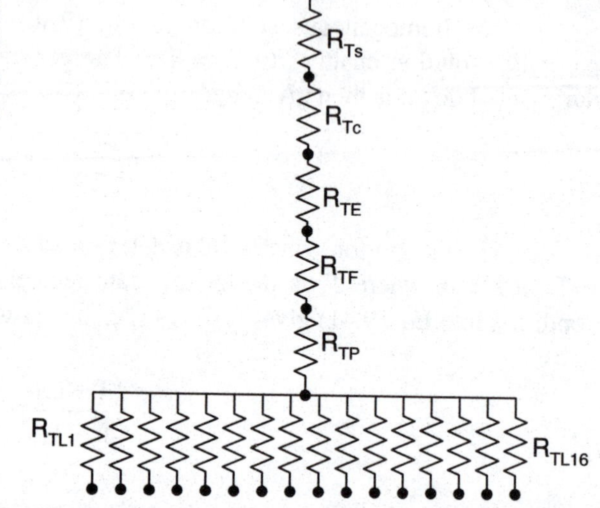

Fig. 9.13 Resistance network representing an analog of heat conduction from a DIP type chip carrier.

The second step in the solution for the temperature difference ΔT is to determine the individual resistances in the network by using the geometry of each design feature and the thermal properties of each material in the heat path. Let's begin at the top of the resistance network with the first resistor R_{Ts}. This thermal resistance, due to the spreading of the heat from a small device located on the surface of a larger chip, is estimated from Eq. (9.29) as:

$$R_{Ts} = \left[\frac{1}{2\sqrt{\pi}\, ak}\right]\left[1-\left(\frac{a}{b}\right)\right]^{3/2}$$

$$R_{Ts} = \frac{1}{\left[2\sqrt{\pi}(0.0005)(154)\right]}\left[1-\frac{0.5}{2}\right]^{3/2} = 2.38\ ^{\circ}\mathrm{C/W}$$

where the following quantities have been selected to describe the geometry of the conductor:

a = 0.5 mm is the radius of the circular area of the device.
b = 2 mm is the radius of an inscribed circle on a 4 × 4 mm chip.
k = 154 W/m-°C for silicon.

The resistance of the chip, eutectic bond and the lead frame are determined from $R_T = \dfrac{L}{kA}$. Hence, for the silicon chip, R_{Tc} is given by:

$$R_{Tc} = \frac{0.508 \times 10^{-3}}{\left[154 \times 16 \times 10^{-6}\right]} = 0.21\ ^\circ C / W$$

where L = 0.508 mm is the chip thickness and A = 16 mm^2 is the chip area.

For the eutectic bond R_{TE} is given by:

$$R_{TE} = \frac{0.050 \times 10^{-3}}{\left[296 \times 16 \times 10^{-6}\right]} = 0.01\ ^\circ C / W$$

where L = 0.05 mm is the bond thickness.
k = 296 W/m °C is the coefficient of thermal conductivity for AuGe solder.

For the lead frame R_{TL} is given by:

$$R_{TL} = \frac{0.25 \times 10^{-3}}{\left[263 \times 16 \times 10^{-6}\right]} = 0.06\ ^\circ C / W$$

where L = 0.25 mm is the frame thickness.
k = 263 W/m-°C is the coefficient for Alloy CDA 194.

The thermal resistance from the base of the lead frame to the leads can only be approximated because of the complexity of the geometry as indicated in Fig. 9.14. Several simplifications are made in determining the thermal resistance R_{TP}:

$$R_{TP} = \frac{0.2 \times 10^{-3}}{\left[0.8 \times 4 \times 10^{-6}\right]} = 62.5\ ^\circ C / W$$

where L = 0.2 mm the average space between the base and the leads.
k = 0.8 W/m-°C for the plastic insulation filling this space.
A = 4 mm^2 for all 16 leads each 0.25 mm thick by 1 mm wide.

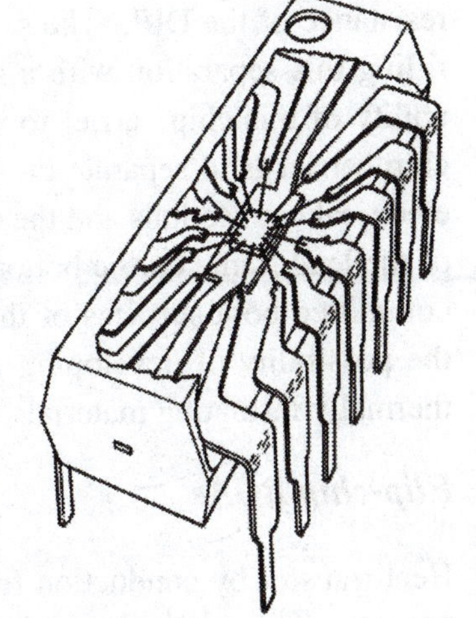

Fig. 9.14 Internal geometry of a DIP showing spacing between the lead frame and the leads.

This thermal resistance (62.5 °C/W) is extremely high and while the calculation is approximate it does indicate the difficulty in conducting a significant quantity of heat in a molded plastic chip carrier. The final thermal resistance to be determined is for the individual leads from the heat frame to the exterior of the DIP:

$$R_{TL} = \frac{6\times 10^{-3}}{\left[263\times 0.25\times 10^{-6}\right]} = 91.3\ ^\circ C/W$$

where L = 6 mm is the average lead length

The 16 individual leads are equivalent to 16 resistances in parallel, which can be combined to give and effective resistance R_{TLe} as:

$$R_{TLe} = R_{TL}/N = 91.3/16 = 5.70\ ^\circ C/W$$

Table 9.6
Listing of estimated resistances for various components in the conduction path of a plastic DIP

Component	Symbol	Resistance °C/W
Spreading	R_{Ts}	2.38
Chip	R_{Tc}	0.21
Eutectic bond	R_{TE}	0.01
Lead Frame	R_{TF}	0.06
Plastic	R_{TP}	62.5
Leads (effective)	R_{TLe}	5.70
Total	R_T	70.86

The results of summing the resistances in the network to obtain the total resistance R_T = 70.86 °C/W is shown in Table 9.6. For a chip dissipating 0.5 W housed in this plastic DIP type chip carrier, the ΔT between the junction and the case is determined from Eq. (9.4) as:

$$\Delta T = qR_T = 0.5 \times 70.86 = 35.4\ ^\circ C$$

These results show the dominance of the frame to lead separation resistance R_T in the thermal resistance of the DIP. The separation is necessary to electrically isolate the individual leads; however, filling this separation with a material having very low thermal conductivity significantly degrades the ability of this chip carrier to house higher powered chips. A much better approach to the design of a chip carrier is to separate the electrical and thermal requirements by using one side of the chip for the electrical connections and the other side for a heat transfer path. In the design of the DIP, the placement of the lead frame on the bottom side of the chip and the bonding pads on the top of the chip effectively committed both surfaces of the chip to electrical connection requirements. This placement eliminated the possibility of developing a low resistance path for heat flow by surrounding the heat source with thermally insulating materials[6].

Flip-chip BGAs

Heat transfer by conduction from a DIP type chip carrier made from molded plastic is not an efficient process. The relatively thick molded plastic over the top of the chip prevents heat conduction in that direction. The air pocket between the DIP and the PCB acts as an effective insulator. The only heat path available is through the lead frame to the leads. Consequently thermal penalties measured in terms of thermal resistance R_T are relatively high. Newer packaging methods provide a much better approach

[6] Flip-chip packaging reserves the back side of the chip for heat transfer. First level chip carriers utilizing flip-chip bonding with much better heat conduction characteristics will be demonstrated in Section 9.10.

for conducting heat from the chip to the exterior surfaces of the package. To illustrate this approach consider the ceramic BGA, which houses a flip-chip as shown in Fig. 9.15.

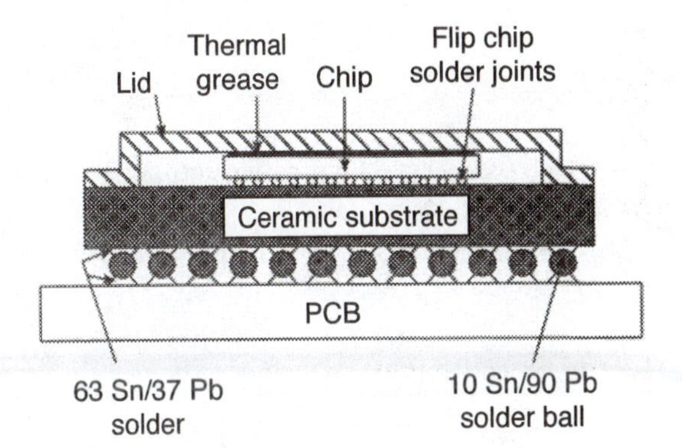

Fig. 9.15 Ceramic BGA carrier with flip-chip connections.

The first step is to model the BGA chip carrier with a resistance network that represents the individual thermal constraints associated with each design feature involved in the heat flow path. The resistor network representing the BGA package is presented in Fig. 9.16. In constructing the resistor network, assumptions are made concerning the predominant heat paths and the heat transfer mechanisms involved. In this case, it was assumed that the heat generated by the device flows up through the chip, the thermal grease and the lid where it is dissipated by convection to air flowing parallel to the surface of the circuit board. For BGAs with high power output, a heat exchanger is often bonded to the lid to enhance heat transfer by convection. In this example, convection from the lid or a heat exchanger will not be considered as our treatment of convection as a heat transfer mechanism will be deferred until Chapter 10. Instead, the temperature of the lid will be taken as a constant T_L. It was also assumed that the heat transferred through the sides and the base of the BGA was negligible. Heat transfer through the sides is not significant because of the low thermal conductivity of the air (or vacuum) that surrounds the chip within the housing. The heat conduction through the flip-chip solder balls, ceramic base, the BGA solder balls and the circuit board may be significant in some designs. It is possible to enhance heat flow from the base of a BGA with thermal vias and by using a cold plate upon which the circuit board is mounted. However, in this example, we will assume that heat is transferred by air flow that interacts with a lid mounted heat exchanger.

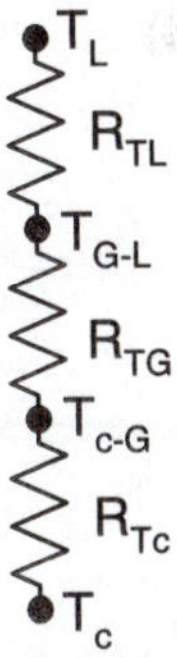

T_L
R_{TL}
$T_{G\text{-}L}$
R_{TG}
$T_{c\text{-}G}$
R_{Tc}
T_c

Fig. 9.16 Resistance network representing an analog of heat conduction from a flip-chip housed in a ceramic BGA.

The second step is to determine the individual resistances in the network by using the geometry of each design feature and the thermal properties of each material involved. Let's begin at the bottom of the resistance network with the first resistor R_{Tc}. Note in this example we have assumed that the heat is generated over most of the chip surface that is usually consistent with chip design. Consequently the thermal resistance due to the spreading of the heat from the circuits on the surface the chip is negligible.

The resistance of the chip, eutectic bond and the lid frame are determined from $R_T = \dfrac{L}{kA}$ as listed below:

For the silicon chip R_{Tc} is given by:

$$R_{Tc} = \frac{0.50 \times 10^{-3}}{\left[154 \times 100 \times 10^{-6}\right]} = 0.0325\ ^\circ C / W$$

where L = 0.50 mm is the chip thickness and A = 100 mm^2 is the area of a 10 × 10 mm chip.

For the thermal grease R_{TG} is given by:

$$R_{TG} = \frac{0.10 \times 10^{-3}}{\left[2.5 \times 100 \times 10^{-6}\right]} = 0.4\ ^\circ C / W$$

where L = 0.10 mm is the thickness o the thermal grease.
k = 2.5 W/m-°C is the coefficient of thermal conductivity for the filled silicone thermal grease.

For the lid R_{TL} is given by:

$$R_{TL} = \frac{0.60 \times 10^{-3}}{\left[160 \times 100 \times 10^{-6}\right]} = 0.0375\ ^\circ C / W$$

where L = 0.60 mm is the lid thickness.
A = 100 mm^2 is the effective area of the lid above the 10 × 10 mm size chip.
k = 160 W/m-°C is the coefficient for copper alloy MF 202.

Table 9.7
Listing of estimated resistances for various components in the conduction path of a plastic DIP

Component	Symbol	Resistance °C/W
Chip	R_{Tc}	0.0325
Thermal Grease	R_{TG}	0.40
Lid	R_{TL}	0.0375
Total	R_T	0.470

The sum of the three series resistors in the network, R_T = 0.470 °C/W, is shown in Table 9.7. For a chip dissipating 50 W housed in this ceramic BGA with flip-chip connections, the ΔT between the circuit junctions and the package lid is determined from Eq. (9.4) as:

$$\Delta T = qR_T = 50 \times 0.470 = 23.5\ ^\circ C$$

These results show that it is possible to dissipate large amounts of heat (50 W) without large thermal penalties when the back of the chip is available to conduct heat in an efficient manner. The small separation between the lid and the back of the chip is filled with thermal grease that is compounded to give it a relatively high thermal conductivity. The lid is fabricated from a copper alloy with a high thermal conductivity. The conduction path is short and filled with materials with relatively high thermal conductivities. The circuit side of the chip is bonded to the BGA with flip-chip connections that allow for high I/O count because its small diameter pads are placed on a small pitch.

9.7.1 Commercial Data for Thermal Resistance

Most manufacturers of IC's package their chips in appropriate carriers and specify the heat transfer characteristics of these packages in terms of a thermal resistance termed $\theta_{j\text{-}a}$. A typical curve for $\theta_{j\text{-}a}$ as a function of air velocity over a DIP mounted on a PCB is shown in Fig. 9.17. The term $\theta_{j\text{-}a}$ represents the total thermal resistance that includes both conduction and convection. The term is referenced to the temperature difference between the junction temperature T_j of the electronic device and the ambient temperature T_a of the air flowing over the chip carrier. This temperature difference is given by:

$$T_j - T_a = q\,\theta_{j\text{-}a} \tag{9.43}$$

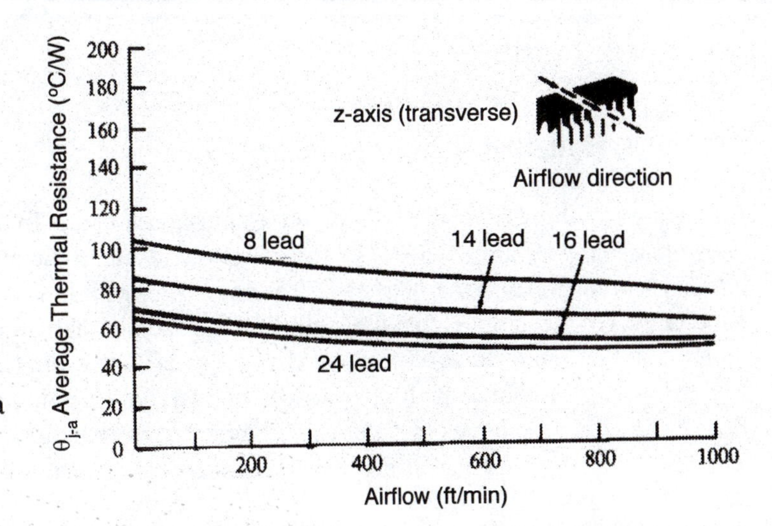

Fig. 9.17 Thermal resistance $\theta_{j\text{-}a}$ for a plastic DIP as a function of the velocity of the air flow.

The form of Eq. (9.43) is similar to Eq. (9.4). However, the thermal resistance $\theta_{j\text{-}a}$ incorporates two different thermal resistances. The first is due to conduction from the chip to the exterior surfaces of the case as described previously in this section. The first thermal resistance included in $\theta_{j\text{-}a}$ involves the detailed design of the chip carrier. The second part of the thermal resistance is due to convection and the formation of a boundary layer between the case and its leads and the air. Thus, we may write:

$$\theta_{j\text{-}a} = \theta_{j\text{-}c} + \theta_{c\text{-}a} \tag{9.44}$$

where $\theta_{j\text{-}c}$ is the thermal resistance of the junction to the case and $\theta_{c\text{-}a}$ is the resistance of the case/leads to the ambient air. The term $\theta_{j\text{-}c}$ is equivalent to the thermal resistance R_T that was determined in the example presented previously. The term $\theta_{c\text{-}a}$ will be treated later in Chapter 10.

In some products cooling accomplished by passing air directly over the chip carriers is not possible. For example, with some military and industrial systems direct air flow over the devices is prohibited because of the high concentration of dust, dirt or oil vapors in the environment. In these instances, it is necessary to transfer the heat by conduction from the package, through the circuit board, along a heat frame to an air or water cooled sink. Contact with the circuit board enhances the heat transfer from the chip carrier to the circuit board. The DIP, designed to stand off the circuit board, may require thermal enhancement. The enhancement, illustrated in Fig. 9.18, which reduces the thermal resistance from the package to the circuit board, may be used with higher powered DIPs. Another method for improving the thermal path is to fill any air gap with thermal grease. Thermal grease is used at the two interfaces to reduce the contact resistance between the heat dissipator and the DIP and PCB in Fig. 9.18.

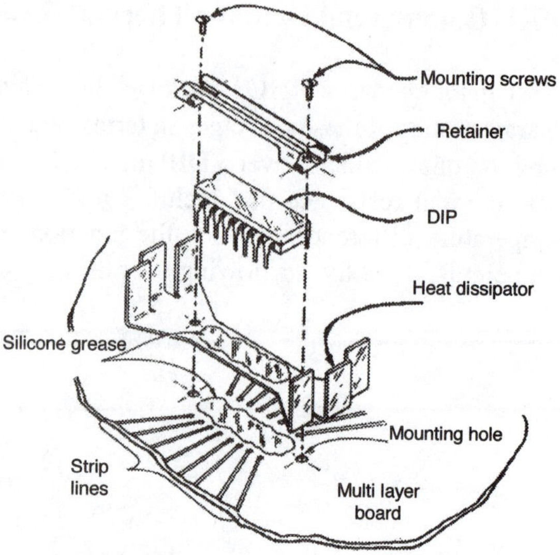

Fig. 9.18 Hardware for enhancing heat conduction from a DIP by utilizing conduction through the base and convection from a heat dissipator.

9.7.2 Description of Heat Transfer in BGAs

Conduction of heat from the circuits on a chip housed in a BGA is usually modeled with two series resistors as shown in Fig. 9.19. The resistance $\theta_{j\text{-}c}$ is due to the material in the thermal path between the circuits on the chip and the outside of the case. The resistance $\theta_{j\text{-}b}$ is due to the material in the thermal path between the circuits on the chip and the solder ball attachments to the circuit board. The temperature on the outside of the case is denoted as T_c and the temperature of the circuit board is T_b in Fig. 9.19. Numerical results for $\theta_{j\text{-}c}$ depend on material selection and construction details for the BGAs. For BGAs with wire bonds and a housing fabricated from molded plastic $\theta_{j\text{-}c}$ is typically 10 to 12 °C/W. The resistance $\theta_{j\text{-}b}$ depends strongly on the placement of thermal vias under the chip. These vias must be attached to the PCB with solder balls to be effective conductors of heat. With properly designed thermal vias the $\theta_{j\text{-}b}$ can be reduced to 8 to 12 °C/W. Other BGAs, designed without plastic encapsulation of the flip-chips, utilize thermal grease to contact the chip carrier lid with the back of the chip . In these thermally improved designs cases, $\theta_{j\text{-}c}$ can be reduced to less than 0.1 °C/W.

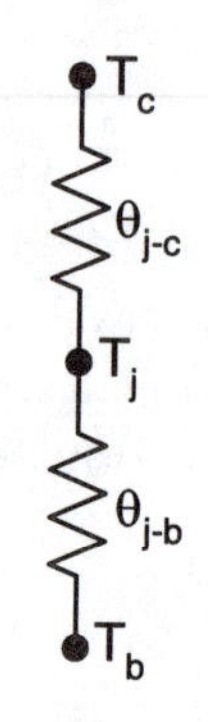

Fig. 9.19 Two resistor model for BGAs depicting heat conduction to the case and the PCB.

For BGAs with direct air impingement, the thermal resistance due to convection must be added to the model shown in Fig. 9.19. This thermal resistance $\theta_{c\text{-}a}$ depends on the method used to dissipate the heat into the air. Efficient heat exchangers often are adhesively bonded to the lid of a BGA to reduce $\theta_{c\text{-}a}$ significantly.

9.8 CONDUCTION IN CIRCUIT CARDS

In many conduction cooled systems, the heat flows from the chip through the chip carrier to the circuit board, where it must be conducted to the next element in a heat dissipation system. Circuit boards are manufactured from electrically insulating materials such as glass-epoxy or glass-ceramic and are not good conductors of heat. The thermal penalties involved in including the circuit card in the path of heat flow are considerable as shown in the following subsection.

9.8.1 Heat Conduction in the Plane of a Circuit Board

To show the poor performance of a conducting system containing a section of circuit board, consider the example illustrated in Fig. 9.20. In this example, we attempt to conduct heat in the plane of the circuit board. The thermal resistance due to heat conduction in the plane of the circuit board is given by:

$$R_T = \frac{L}{kA} = \frac{100\times10^{-3}}{\left(0.35\times2\times10\times10^{-6}\right)} = 14{,}286 \quad {}^{\circ}C/W$$

where k = 0.35 W/(m-°C) for FR-4 laminate.

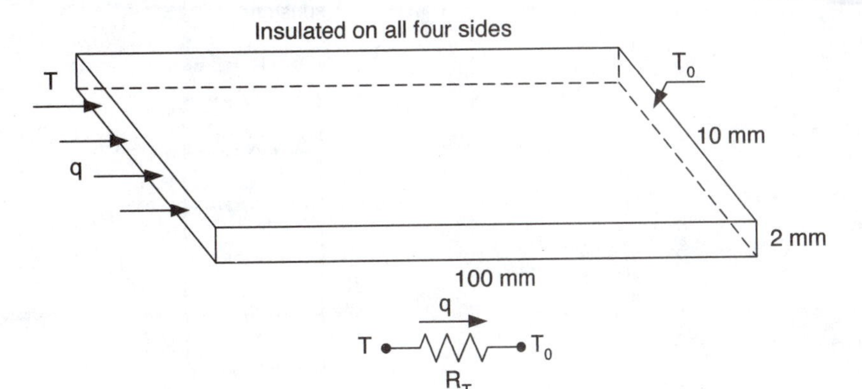

Fig. 9.20 Heat conduction along the length of a circuit board.

This result (14,286 °C/W) shows an extremely high value of the thermal resistance. It is clear that it is not possible to conduct heat in the plane of the PCB for any significant distance without incurring an excessively high thermal penalty.

The design of a multi-layer circuit board (MLB) incorporates several ground and power planes embedded through the thickness as illustrated in Fig. 9.21. To determine if these copper planes improve the thermal resistance, consider the same conducting geometry as shown in Fig. 9.20 except for the inclusion of four ground and power planes each fabricated from 2 oz. copper cladding. The resistance network which provides the thermal model for this conductor is given in Fig. 9.21b. The first resistance network contains 9 resistors, 4 for the copper planes and 5 for the intermediate glass-epoxy layers. This network can be simplified by neglecting the small amount of heat that flows through the glass-epoxy layers. The simplified resistance network consists of four parallel resistances R_{cu} each of which is given by:

$$R_{cu} = \left[\frac{100\times10^{-3}}{(390)(0.0027)(25.4)(10\times10^{-6})}\right]\left[\frac{1}{0.717}\right] = 521.5 \quad {}^{\circ}C/W$$

Note, that the term [1/0.717] in the equation above is a correction to account for the perforations in the ground and power planes, illustrated in Fig. 9.21c. These perforations are 60 mils in diameter on 100 mil centers and require removal of 28.3% cent of the copper. The perforations serve to electrically isolate the power planes from the plated-through-holes.

Using the rule for parallel resistance, gives the total thermal resistance for the MLB containing four copper planes as:

$$R_T = 521.5/4 = 130.4 \text{ °C/W}$$

This result shows that the 2 oz/ft^2 copper planes markedly improved heat conduction along the plane of the circuit board, but the thermal resistance is still much too high for planar conduction in MLBs to be seriously considered in the design of a heat dissipation system.

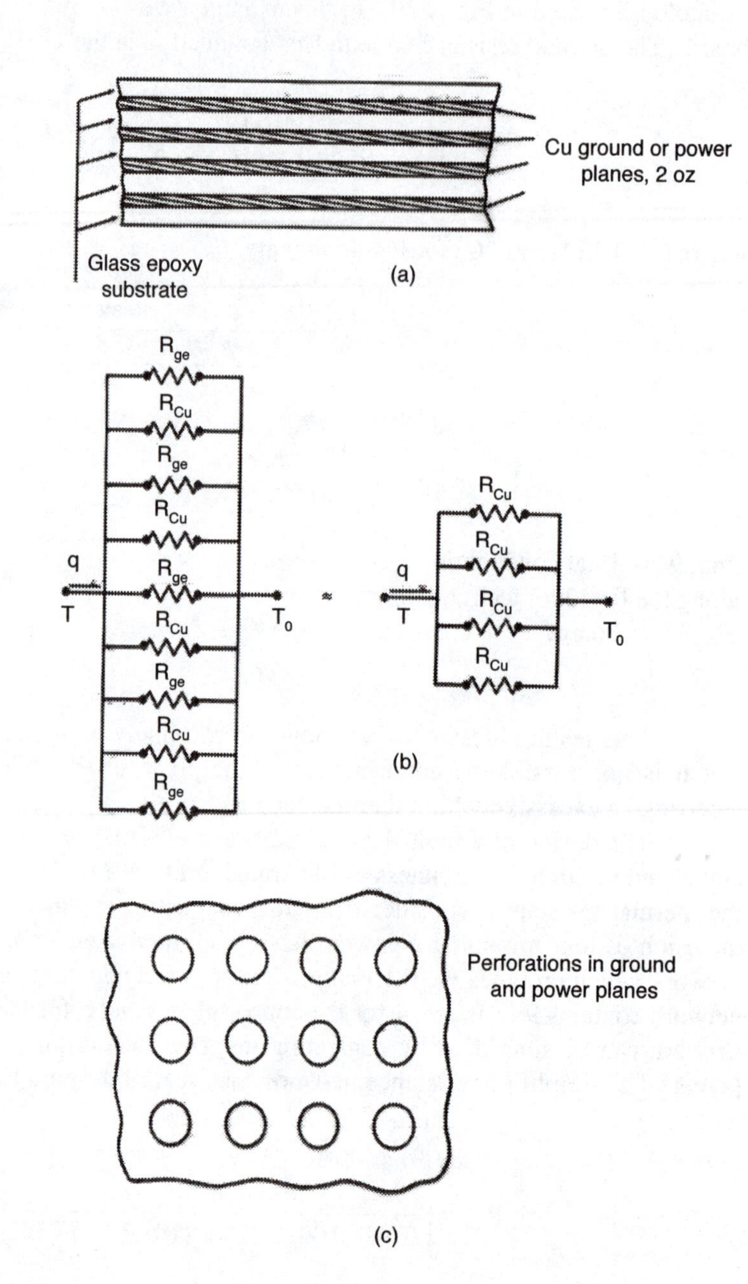

Fig. 9.21 Geometry and thermal modeling of a multi-layer-board.

9.8.2 Conduction Through the Thickness of a Circuit Board

Consider a small 25 by 25 mm segment of circuit board shown in Fig. 9.22, where the heat q is transmitted through the 1.0 mm thickness of the board. In this case the thermal resistance is given by:

$$R_T = \frac{L}{kA} = \frac{1.0\times 10^{-3}}{\left(0.35\times 25\times 25\times 10^{-6}\right)} = 4.57 \quad °C/W$$

This solution indicates a much lower thermal resistance than the results described previously. It is clear that if a circuit board must be in the path of heat flow it should be orientated so that the flow is through the thickness. Also, the thermal resistance of the board in the through thickness direction can be reduced substantially by incorporating thermal vias into the design.

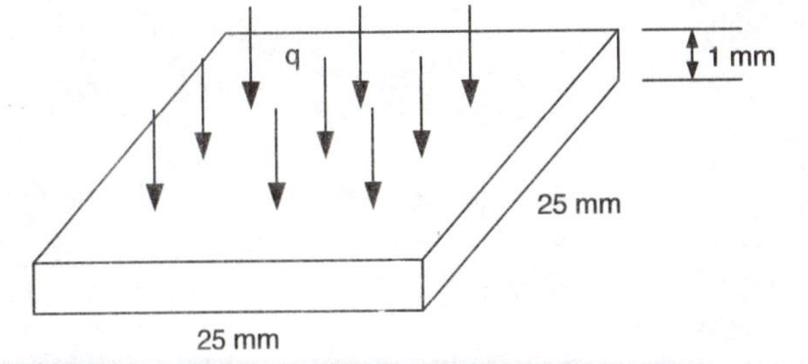

Fig. 9.22 Heat transmission through the thickness of a segment of FR-4 laminate.

Thermal vias are typically plated-through-holes (PTH's) that have been filled with solder. Suppose the 81 PTHs are 1 mm in diameter and are on 2.5 mm centers over the entire area of the board segment shown in Fig. 9.23.

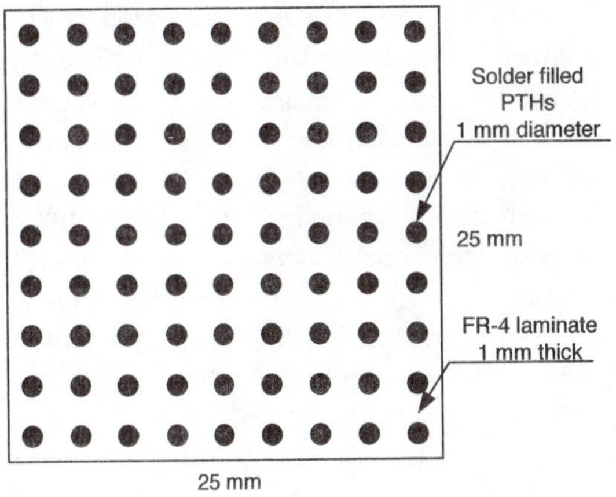

Fig. 9.23 Thermal vias to enhance through thickness heat transfer.

The fraction of the area covered by these solder filled vias in Fig. 9.23 is:

$$\frac{A}{A_v} = \frac{\pi(1)^2(81)}{(4\times 25\times 25)} = 0.1018$$

The heat is conducted in parallel paths through the vias and the circuit board. The resistance due to the vias is:

$$R_{Tv} = \frac{1.0\times 10^{-3}}{\left(50\times 25\times 25\times 10^{-6}\times 0.1018\right)} = 0.314 \quad {}^{\circ}C/W$$

where k = 50 W/m-°C for the solder filled copper clad via.

The resistance of the circuit board surrounding the vias is given by:

$$R_{Tb} = 4.57\left[\frac{1}{(1-0.1018)}\right] = 5.09 \quad {}^{\circ}C/W$$

Using the parallel resistance rule to find the effective thermal resistance yields:

$$R_{Te} = \frac{R_{Tv} R_{Tb}}{(R_{Tv} + R_{Tb})} \tag{9.45}$$

$$R_{Te} = \frac{(0.314)(5.09)}{(0.314 + 5.09)} = 0.296 \ ^{\circ}\mathrm{C/W}$$

The effect of the thermal vias is dramatic. The through thickness thermal resistance of the glass-epoxy laminate has been reduced by a factor of 4.57/0.296 = 15.5 and the thermal penalty imposed on the heat dissipation system by the circuit board has been reduced to less than 0.3 °C/W.

9.8.3 Conduction Cooling with a Heat Frame

In the design of a product where it is not possible to pass air directly over the chip carriers to dissipate heat, a conduction method illustrated in Fig. 9.24 is employed. In this system, the heat is transmitted from the chip carriers to the circuit board and then to a heat frame. The heat frame provides a low resistance path to a heat sink. We will consider a typical circuit card consisting of a multi-layer board bonded to a heat frame with the dimensions given in Fig. 9.25. The board is divided at the center line taking into account symmetry of the construction and the heat input. The right side of the board is divided into 8 segments with a heat input q_1 to q_8 imposed on each segment. Because the PCB is a multi-layered board, it is assumed that sufficient copper ground and power planes exist to spread the heat input uniformly over each area segment. With these assumptions, the resistance network, shown in Fig. 9.26, represents a thermal model that can be used to predict temperatures at different locations on the board. The network contains resistances R_a, R_f, R_{f1} and R_c, each of which may be determined by considering the detailed features of the electronic module.

Consider first the resistance R_a that represents the combined resistance to the heat flow through the thickness of the board, through the bonding adhesive and to the center of the heat frame. We compute R_a from the well known relation for series resistors as:

$$R_a = R_b + R_{ep} + R_{cu\downarrow} \tag{a}$$

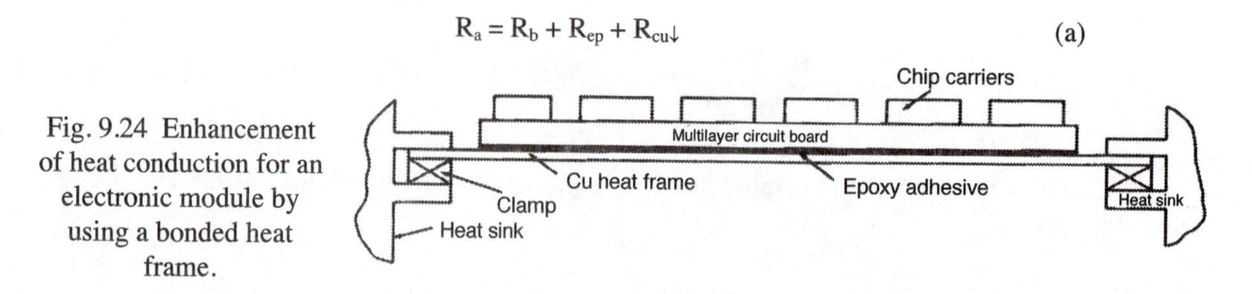

Fig. 9.24 Enhancement of heat conduction for an electronic module by using a bonded heat frame.

To determine R_b, note that the board thickness is 2 mm and that the area of a segment is 750 mm^2. Because thermal vias cover the entire area of the board, as illustrated in Fig. 9.23, we may use the results for the effective thermal resistance given by Eq. (9.45) to write:

$$R_e = \frac{(0.296)(2 \times 25 \times 25)}{(750)} = 0.4933 \ ^{\circ}\mathrm{C/W} \tag{b}$$

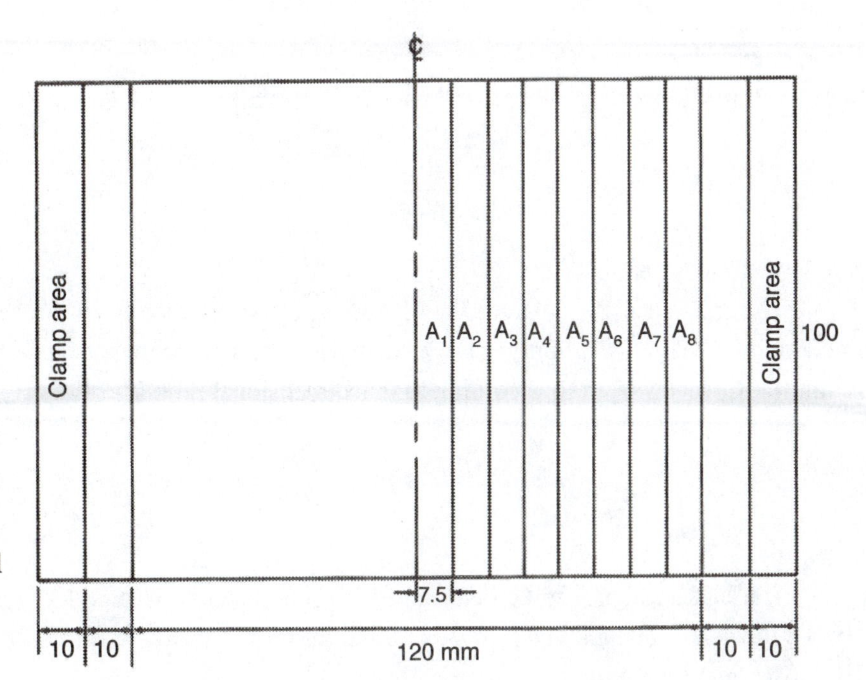

Fig. 9.25 Front view of an electronic module showing the right side of the circuit card divided into eight equal size areas A_1 to A_8.

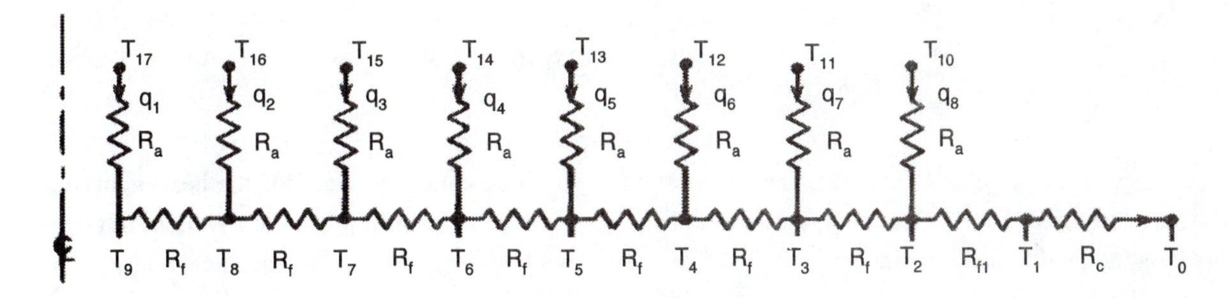

Fig 9.26 Resistance network representing a thermal model of the module shown in Fig. 9.25.

For the thermal resistance for a copper filled epoxy adhesive R_{ep}, we take its thickness L = 0.125 mm and k = 1.60 W/m-°C and then write:

$$R_{ep} = \frac{0.125\times 10^{-3}}{\left(1.6\times 750\times 25\times 10^{-6}\right)} = 0.1042 \text{ °C/W} \qquad \text{(c)}$$

To determine $R_{cu}\perp$, we take half the frame thickness as L = 0.5 mm and let k = 263 W/m-°C for alloy CDA 194 and then write:

$$R_{cu\downarrow} = \frac{0.5\times 10^{-3}}{\left(263\times 750\times 10^{-6}\right)} = 0.0025 \text{ °C/W} \qquad \text{(d)}$$

Adding $R_b + R_{ep} + R_{cu\downarrow}$ gives:

$$R_a = 0.600 \text{ °C/W} \qquad \text{(e)}$$

Next, we determine the resistance R_f due to the heat frame as:

$$R_f = \frac{7.5 \times 10^{-3}}{\left(263 \times 1.0 \times 100 \times 10^{-6}\right)} = 0.2852 \ \ ^\circ C/W \quad \text{(f)}$$

The resistance R_{f1} is larger than R_f because the distance from the center of A_8 to the center of the clamp area (L = 18.75 mm) is larger than the center to center distance between the segment areas (7.5 mm). R_{f1} scales from R_f according to this dimensional ratio as:

$$R_{f1} = \frac{18.75}{(75)} 0.2852 = 0.7129 \ \ ^\circ C/W \quad \text{(g)}$$

The last resistance to be determined is the contact resistance R_c, which is due to the interface between the heat frame and the heat sink. Taking $h_c = 10^4$ W/m^2-°C and noting that the clamp area is 10 × 100 = 10^3 mm^2, we write the equation for R_c as:

$$R_c = \frac{1}{(h_c A)} = \frac{1}{\left(10^4 \times 10^3 \times 10^{-6}\right)} = 0.10 \ \ ^\circ C/W \quad \text{(h)}$$

It is now possible to compute the temperature at any location on the board if the heat inputs q_1 to q_8 are specified. In this example, take each of the q's to be equal to 5 W. This assignment implies that the total heat to be dissipated over the module is 80 W with 40 W conducted to each heat sink.

To determine the temperature T_1 note from Eq. (9.4) that:

$$T_1 = R_c \, \Sigma q_i + T_0 = 40 \times 0.1 + 20 = 24.0 \ ^\circ C \quad \text{(i)}$$

where the heat sink temperature T_0 was taken as 20 °C.

The temperatures at several other locations are determined in a similar manner as:

$$T_2 = R_{f1} \, \Sigma q_i + T_1 = 40 \times 0.7129 + 24.0 = 28.52 + 24.0 = 52.52^\circ C \quad \text{(j)}$$

$$T_3 = R_f \sum_1^7 q_i + T_2 = 35 \times 0.2852 + 52.52 = 9.98 + 52.52 = 62.50 \ ^\circ C \quad \text{(k)}$$

$$T_4 = R_f \sum_1^6 q_i + T_3 = 30 \times 0.2852 + 62.50 = 8.56 + 62.50 = 71.06 \ ^\circ C \quad \text{(l)}$$

$$T_{10} = R_a q_8 + T_2 = 5 \times 0.60 + 52.52 = 3.00 + 52.52 = 55.52 \ ^\circ C \quad \text{(m)}$$

Table 9.8
Listing of temperatures in °C at locations defined in Fig 9.26

$T_0 = 20$	
$T_1 = 40R_c + T_0 = 24.0$	
$T_2 = 40R_{f1} + T_1 = 52.52$	$T_{10} = q_8\, R_a + T_2 = 55.52$
$T_3 = 35R_f + T_2 = 62.50$	$T_{11} = q_7\, R_a + T_3 = 65.50$
$T_4 = 30R_f + T_3 = 71.06$	$T_{12} = q_6\, R_a + T_4 = 74.06$
$T_5 = 25R_f + T_4 = 78.19$	$T_{13} = q_5\, R_a + T_5 = 81.19$
$T_6 = 20R_f + T_5 = 83.89$	$T_{14} = q_4\, R_a + T_6 = 86.89$
$T_7 = 15R_f + T_6 = 88.17$	$T_{15} = q_3\, R_a + T_7 = 91.17$
$T_8 = 10R_f + T_7 = 91.02$	$T_{16} = q_2\, R_a + T_8 = 94.02$
$T_9 = 5R_f + T_8 = 92.45$	$T_{17} = q_1\, R_a + T_9 = 95.45$

A complete listing of the temperatures determined at 17 locations on the electronic module is presented in Table 9.8. Note, that the maximum temperature is in the center of the board with $T_{max} = T_{17} = 95.45$ °C. The temperature difference ΔT necessary to transfer the heat from this central location to the heat sink was 75.45 °C. The temperature at the edge of the circuit card was much lower $T_{10} = 55.52$ °C with $\Delta T = 35.52$ °C.

This example illustrates that a circuit board can be incorporated into a heat dissipation system while maintaining reasonable ΔT's to conduct a heat flux of 6,670 W/m^2. The detrimental effect of the poor conductivity of the glass-epoxy laminate was mitigated by reducing the length of travel through thickness of the laminate, by incorporating an array of thermal vias on 2.5 mm centers and by employing a heat frame to conduct the heat from all locations on the board to the heat sink.

9.9 COLD PLATES AND COLD RAILS

Reference to Fig. 9.24 shows that the heat frame is clamped to a heat sink, called a cold rail, which is maintained at a temperature $T_0 = 20$ °C. The cold rail temperature is maintained at T_0 by passing chilled water through channels bored in the rail or through tubing welded to a plate as illustrated in Fig. 9.27.

The heat transferred through the cold rail can be approximated by:

$$q = kS\,(T_2 - T_1) \qquad (9.46)$$

where S is a shape factor dependent upon the geometry of the cross section of the cold rail. For the shape illustrated in Fig. 9.27, the shape factor S is given by:

$$S = \frac{2\pi L}{\log_e\left[\left(\frac{e}{\pi r}\right)\sinh\left(\frac{2\pi H}{e}\right)\right]} \qquad (9.47)$$

where L is the length of the fluid channel. The other symbols in Eq. (9.47) are defined in Fig. 9.27.

The temperature T_1 of the wall of the hole in the cold rail is slightly higher than the water temperature due to the ΔT across the fluid's boundary layer. The wall temperature depends upon the convection coefficient associated with the flow conditions. As the coolant flows through the cold rail, heat is transferred through the rail and across the boundary layer to the fluid. The temperature increase in the fluid is given by:

$$\Delta T = T_0 - T_i = \frac{q}{(dm/dt)c_p} \tag{9.48}$$

where dm/dt is the mass flow rate (kg/s), c_p is the specific heat (J/kg-°C), and the subscripts i and o refer to the inlet and outlet fluid temperatures. For water as a coolant at an average temperature of 27 °C, c_p = 4177 J/(kg-°C) or 9.78 W-min/(lbm-°C). Also, the mass flow rate (dm/dt) in units of lbm/min is given by:

$$dm/dt = 8.31\ G \tag{9.49}$$

where G is the volume flow rate in gallons per min. Using these conversion factors in Eq. (9.48) gives a relation which can be used to determine ΔT in commonly employed units of W for q and GPM for the flow rate:

$$\Delta T = 0.0123\ q/G \tag{9.50}$$

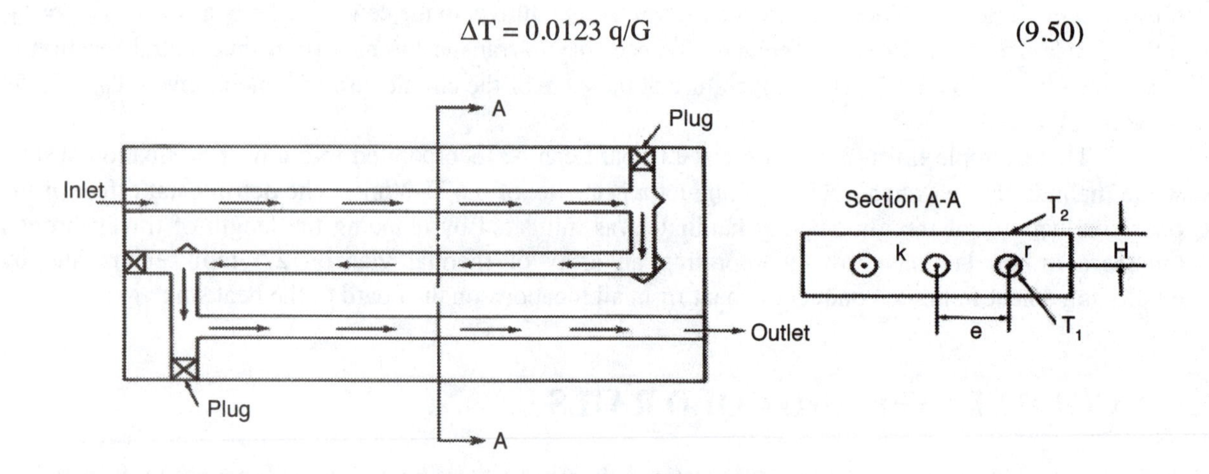

Fig. 9.27 Construction details of a cold rail.

Cold rails are very effective heat sinks because they offer low thermal resistance when properly designed and operate at relatively low flow rates. For example a heat rate of 1kW can be dissipated with a flow rate of 2 GPM with an increase in the water temperature from inlet to outlet of only about 6 °C. For very high reliability and high performance products, refrigerants can be employed as the coolant. Refrigerated coolant reduces the temperature of the heat sink to about −30 °C and lowers the junction temperatures by about 50 °C. Refrigerated cold rails were used to great advantage in the heat management system employed in the Cray I computer in the early 70's and they are still used today for large high-performance processors requiring dissipation of very large quantities of heat.

9.10 ADVANCED COOLING METHODS

For dense logic chips, the heat flux is usually quite high with power dissipation on high-frequency, high-performance chips exceeding 200 W. It is extremely difficult to maintain low junction temperatures in these high powered devices. The more common thermal management techniques involving traditional chip carriers and circuit boards introduce large thermal resistance that result in high junction temperatures. To successfully accommodate high powered ICs requires the use of flip-chip technology and new approaches to thermal management. The new approaches will be based on two fundamental premises. First, one side of the chip is reserved for electrical connections to accommodate the signal I/O and the power requirements. It is implied that the other side of the chip is reserved for heat dissipation. Second, the new thermal management methods bring the coolant to the chip and largely eliminate the thermal resistance between the sink (i.e. the coolant) and the chip.

Today flip-chip technology is implemented by most chip manufacturers together with BGAs CSPs to provide the dense electrical connections that are required in these high-performance ICs. There are several approaches for exploiting the back of the chip to remove large quantities of heat with small thermal penalties. We will describe three of these techniques in the following sub-section.

9.10.1 Advanced Cooling Concepts

Three advanced cooling concepts are illustrated in Fig. 9.28 to 9.30. All of these concepts are based on flip-chip bonding to a multi-layer substrate that forms the base of a BGA. The first concept, presented in Fig. 9.28, involves a heat spreader bonded to the back of a silicon chip with a thermal adhesive. The heat spreader in turn in coupled to a lid with a high conductivity compliant thermal compound. The lid encloses the module. A heat sink or cold plate is attached to the lid to provide the final step in transferring the heat to the environment.

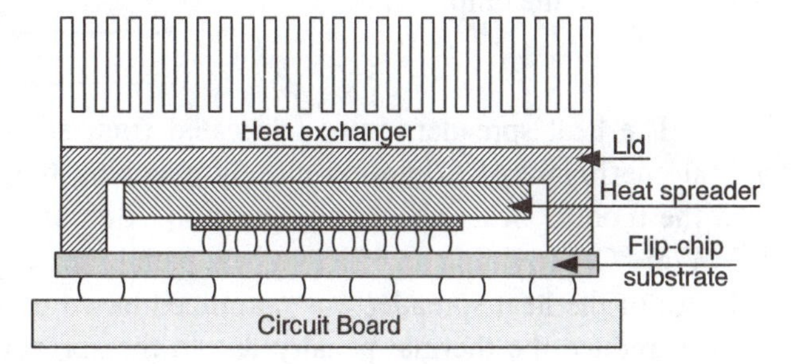

Fig. 9.28 Cooling concept using a heat spreader, lid and heat exchanger to dissipate heat with low thermal penalty.

The second advanced cooling concept is illustrated in Fig 9.29. In this design a micro-channel high efficiency heat exchanger, which will be described in more detail in Section 9.7.3, is bonded directly to the back of a silicon chip with a thermal adhesive. Fluid flow through the micro-channels transports the heat to the environment. The thermal penalty is extremely low because of the short conduction path and the high efficiency of micro-channel heat exchangers.

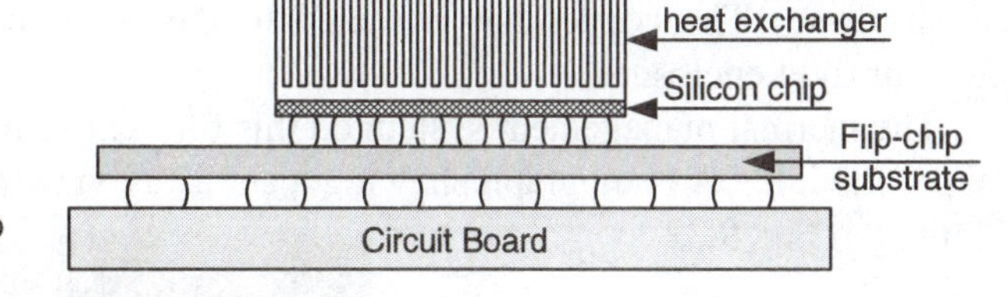

Fig. 9.29 Cooling concept using a micro-channel heat exchanger bonded directly to the back of a silicon chip.

The third advanced cooling concept is illustrated in Fig. 9.30. In this design a liquid is sprayed directly onto a thin lid that is bonded to the back of a chip with a thermally enhanced adhesive. The liquid is collected, cooled outside the system and recycled. This is a very effective system, but it has the disadvantage of possible leaks that may occur in service.

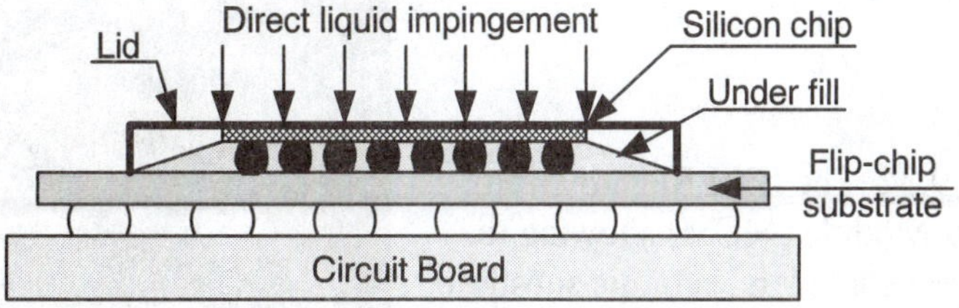

Direct liquid impingement
Lid
Silicon chip
Under fill
Flip-chip
substrate
Circuit Board

Fig. 9.30 Cooling concept using direct liquid impingement onto a thin lid bonded to the back of a chip.

9.10.2 The IBM Multi-Chip Module

In recent years, IBM has phased out its Thermal Conduction Module (TCM) and replaced it with the Regatta Multi-Chip-Module (MCM). The Regatta MCM utilizes four high powered chips each with two processors and an integrate memory cache. The design produces a chip power distribution that is non uniform with local regions on the chip exceeding 100 W/cm^2. Heat spreaders were employed in the design of the heat management system to address the problems caused by these localized hot spots. A schematic drawing of the design of this the IBM MCM is presented in Fig. 9.31.

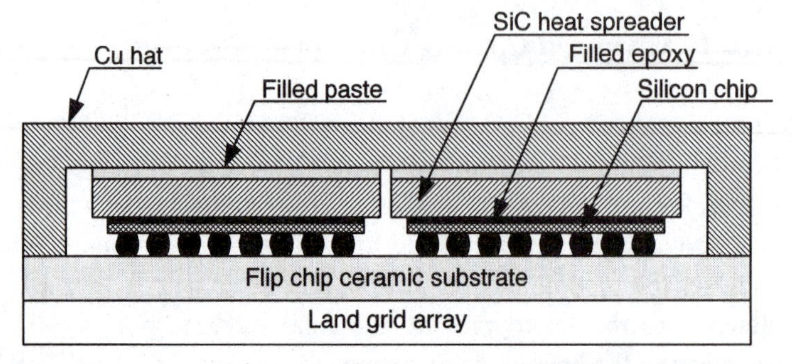

Fig. 9.31 Cross sectional view of a high performance MCM utilizing heat spreaders to mitigate local hot spots on the chip.

The heat spreaders were fabricated from silicon carbide because this material exhibits a low thermal coefficient of expansion closely matching the TCE of the chip thus avoiding thermal stresses when the module heats up during operation. Silicon carbide also exhibits a high thermal conductivity (275 W/m-°C) providing an efficient heat path to the copper hat that serves as a lid for the module. The thickness of the heat spreader was optimized based on computations using three dimensional computer codes to reduce the thermal penalty due to the spreader. The SiC heat spreaders were bonded to the backs of the chips using a filled conductive epoxy with a thermal conductivity of k = 1.23 W/m-°C. The thermal resistance due to the adhesive was minimized by using a very thin layer of adhesive.

Because of chip tilt and variations in stacking height of the solder joints, chip, adhesive layer and heat spreader, it was necessary to provide a small gap between the spreaders and the copper hat. This gap was filled with a thermal compliant layer exhibiting a thermal conductivity of 3.8 W/m-°C. The outer shell of the module is a thick copper hat that cools and protects the module. The top outer surface of the hat is mechanically and thermally connected to an impingement heat sink that receives ducted air flow. The copper hat is sealed to the ceramic substrate with an adhesive to provide a complete air tight enclosure.

The thermal management system on this four chip MCM dissipates 156 W per chip for a total of 624 W per module. A photograph showing a cut away view of the internal configuration of this module is presented in Fig. 9.32.

Fig. 9.32 The IBM Regatta Multi-Chip-Module (MCM) showing the SiC heat spreaders, ceramic substrate and Cu hat.

9.10.3 Pease and Tuckerman Experiments

David Tuckerman in an outstanding Ph D thesis (1984) showed the feasibility of an entirely new concept for cooling chips with extremely low thermal resistance. This concept, illustrated in Fig. 9.33, utilizes micro channels etched into a silicon substrate. With narrow channels about 50μm wide, the flow of the coolant through the channels is laminar and the rate of heat transfer is excellent. Fluids such as water, mercury, refrigerants and liquid nitrogen all show merit as coolants for these high performance heat exchangers.

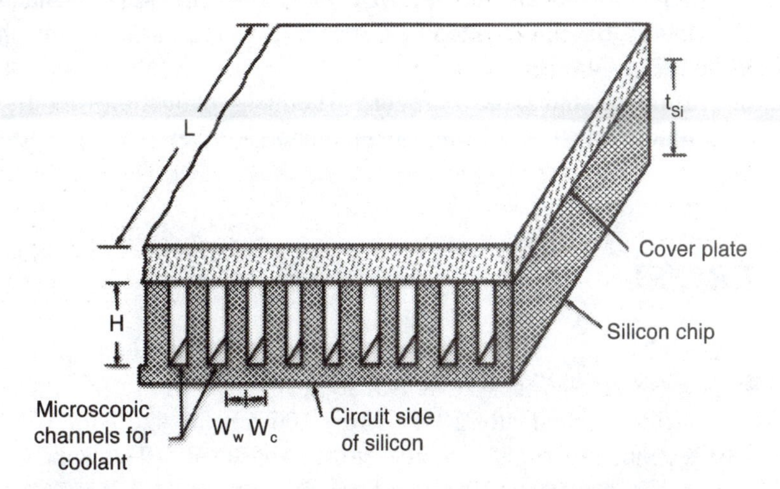

Fig. 9.33 Micro channel heat exchanger fabricated from silicon for cooling an IC.

Typical interfacial thermal resistances of 0.02 to 0.08 °C/W were measured in a series of carefully controlled experiments incorporating the micro channeling in the back of the chip and in a substrate in bonded to the back of a test chip. These results indicate that thermal loading of 10^7 W/m^2 can be removed from wafer devices at normal operating temperatures (100° C).

REFERENCES

1. G. N. Ellison, Thermal Computations for Electronic Equipment, Van Nostrand Reinhold Co., New York, NY. 1984, pp. 183-368.
2. J. B. Fourier, Theorie Analytique de la Chaleur, Paris, 1822, (trans.: Dover, New York, 1955).
3. F. M. White, Heat Transfer, Addison-Wesley Publishing Co., Reading Ma., 1984.
4. A. D. Krauss and A. Bar-Cohen, Thermal Analysis and Control of Electronic Equipment, Hemisphere Publishing Corp. Washington, DC, 1983.
5. C. L. Tien, "A Correlation for Thermal Contact Conductance of Nominally Flat Surfaces in a Vacuum," National Bureau of Standards, Special Publication 302, September 1968.
6. D. B. Tuckerman, Heat Transfer Microstructures for Integrated Circuits, Ph. D. Dissertation, Stanford University, Feb. 1984, University Microfilms International, Ann Arbor, MI.
7. J. L. Sloan, Design and Packaging of Electronic Equipment, Van Nostrand Reinhold Company, New York, NY. 1985.
8. A. Bejan and A. D. Kraus, Heat Transfer Handbook, John Wiley & Sons, New York, NY, 2003.
9. G. Comini, S. del Giudice and C. Nonino, Finite Element Analysis in Heat Transfer, Taylor Francis, UK, 1994.
10. L. Zhang, K. E. Goodson and T. W. Kenny, Silicon Microchannel Heat Sinks, Springer, New York, NY 2004.
11. G. R. Blackwell, The Electronic Packaging Handbook, CRC press, Boca Raton, FL, 1999.

12. K. J. Kim, Air Cooling Technology for Electronic Equipment, CRC Press, Boca Raton, FL, 1996.
13. R. Remsburg, Thermal Design of Electronic Equipment, CRC Press, Boca Raton, FL, 2000.
14. J. E. Sergent and A. Krum, Thermal Management Handbook, McGraw Hill, New York, NY, 1998.
15. J. U. Knickerbocker et al, "An Advanced Multi-Chip-Module (MCM) for High Performance UNIX Servers," IBM Journal of Research and Development, Vol. 46, No. 6, pp. 779-804, 2002.
16. P Singh et al, "A Power Packaging and Cooling Overview of the IBM eServer z900," IBM Journal of Research and Development, Vol. 46, No. 6, pp. 711-738, 2002.
17. J. U. Knickerbocker et al, "Development of Next Generation System-on-Package (SOP) Technology Based on Silicon Carriers," IBM Journal of Research and Development, Vol. 49, No. 4/5, pp. 725-744, 2005.
18. A. J. Blodgett and D. R. Barbour, "Thermal Conduction Module: A High Performance Package," IBM Journal of Research and Development, Vol. 26, No.1, pp. 30-36, 1982,.

EXERCISES

9.1 Describe why fault location methods are important in system availability in large digital processors. Define system availability.

9.2 Consider a conductor 10 by 12 by 100 mm in size with 10W to be conducted in the 100 mm direction. Consider its sides to be insulated. The conductor is fabricated from aluminum 6061T6, Kovar, a fiberglass composite and alumina. The cold end of the conductor is maintained at 20 °C. Find the temperature of the hot end. Prepare a summary of your results and draw conclusions from this simple exercise.

9.3 Determine the thermal resistance for the four conductors described in Exercise 9.2.

9.4 Examine the thermal coefficient of conductivity given in Table 9.1 to Table 9.4 and select from this list those materials that are good electrical insulators. Tabulate k for these materials and draw a conclusion on the merits of the design of a thermal path that must also provide for insulating circuits.

9.5 An alumina cylinder contains a thyratron that dissipates 800W. The cylinder has an inside and outside diameter of 50 mm and 64 mm respectively and a length of 110 mm. Determine the thermal resistance of the cylinder and the temperature difference ΔT. Discuss the validity of the cylinder as a thermal model for this tube.

9.6 A transmission tube for a small radio station is rated at 1,200 W. If its components are mounted in an alumina tube with inside and outside diameters of 80 and 100 mm, respectively, determine the thermal resistance of the cylinder and the temperature difference ΔT. The tube is 160 mm in length. Discuss the validity of the cylinder as a thermal model for this tube.

9.7 Consider a composite plate with an area $A = 0.04\ m^2$ made of five layers of material defined in Table Ex. 9.7. If the rate of heat transfer is 100 W, determine the temperature at each interface. Consider $T_0 = 20$ °C and assume that the interfaces are in perfect thermal contact.

Table Ex. 9.7

Thickness L (mm)	Material
$L_1 = 10$	Aluminum 5052
$L_2 = 0.2$	Epoxy
$L_3 = 6$	FR- 4 Laminate
$L_4 = 4$	Epoxy Ag Filled
$L_5 = 7$	Silicon Carbide

9.8 Repeat Exercise 9.7 but change the rate of heat transfer from 100 W to 340 W and increase the area of the plate to 0.075 m^2.

9.9 Determine the heat transferred across the plate defined in Fig. 9.5 if its dimensions are 8 mm × 0.4 m × 0.6 m. The plate is made from aluminum 2024-T4. The convection coefficient h is 6 W/m^2-°C on one side of the plate and 3 W/m^2-°C on the other side. The temperature difference available is 20 °C. Examine the three components of the thermal resistance and discuss the factor limiting the amount of heat being transferred across the plate. What design changes would you suggest to increase the amount of heat transferred?

9.10 Determine the contact resistance at the interface between two aluminum 2024-T4 plates. Each plate is 6 mm thick, the surface finish is 125 μin. RMS, the contact area is 900 mm^2 and the contact pressure is 0.69 MPa. If the designer changed materials from aluminum to stainless steel, determine the new contact resistance.

9.11 If you were to increase the pressure on the contact area described in Exercise 9.10 from 0.69 to 2.07 MPa, what would be the new contact resistance? If the contact pressure is reduced to 0.14 MPa, determine the reduction in the contact resistance. Prepare an engineering brief describing the effect of contact pressure on contact resistance for aluminum/aluminum and stainless steel/stainless steel interfaces.

9.12 A small heat source with an area A is to be placed on a large conductor that is sufficiently large to be treated as a half space. The heat source is distributed over a circle with a diameter D or a small square with a side L. Determine the difference in the thermal resistance due to constriction between these two options. Plot your results on Fig. 9.10 and comment on the position of your points relative to the curves for a body with finite dimensions.

9.13 Determine the thermal resistance due to the constrictive effect if a circular hot spot 3 mm in diameter is centered on a Si_3N_4 cylinder having a diameter of 9 mm and a length of 30 mm.

9.14 The thermal resistance of a cooling system is 2 °C/W. The system employs conduction and dissipates 80W through an aluminum assembly with a mass of 0.5 kg. Determine the RC constant for this system and prepare a graph of the temperature as a function of time if the ambient temperature is 22 °C.

9.15 Write Eq. (9.38) in a non dimensional form and discuss the significance of each side of the new equation. Compare this result with Eq. (9.42) and comment.

9.16 Suppose a 16 pin DIP is fabricated from an alumina shell, and the chip is mounted on a ceramic substrate inside the cavity formed by this shell. The chip carrier is sealed with a lid fabricated from Kovar. The case is filled with a thermally conductive and electrically insulating fluid with k = 2.0 W/m-°C. Prepare a thermal model represented by the resistive network, and determine the thermal resistance due to conduction R_T for this chip carrier. Use the dimensions of a 16 pin dip to estimate the cavity size required to house the chip.

9.17 A silicon chip 8 by 12 mm by 0.5 mm thick is soldered to a ceramic circuit card that is bonded to a cold plate as shown in Fig. 9Ex17. The surface temperature of the cold plate is maintained at 20 °C. Determine the temperature of the junctions on a chip as a function of the heat dissipated. Consider the power range from 10 to 100 W per chip.

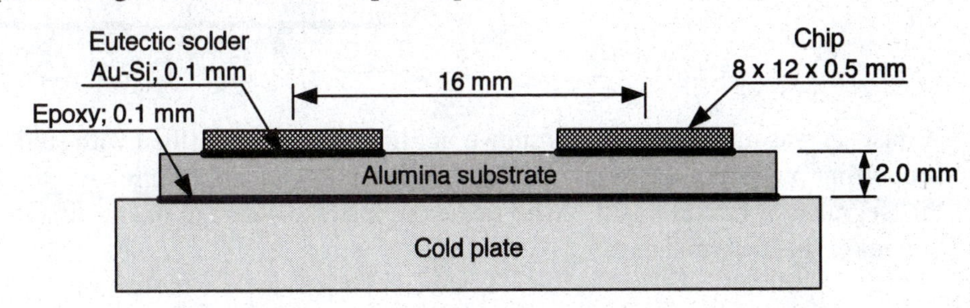

Fig. 9Ex17

9.18 Discuss heat dissipation in a DIP type chip carrier. Would it improve the thermal resistance if alumina was used in place of the plastic for the carrier material? Quantify this answer.

9.19 Sketch the essential features of a design modification for a DIP which improves the thermal resistance by a factor of 5. The basic outline of the DIP and the pin spacing cannot be changed in the new design. Write an engineering brief that describes your approach and includes an analysis of the new design.

9.20 A 24 lead plastic molded DIP that is rated at 0.6W is cooled by direct exposure to an air stream. If the flow velocity is 600 fpm with the air at 28 °C, find the junction temperature. Comment on the result.

9.21 For the BGA chip carrier defined in Fig. 9.15, determine its thermal resistance if a square hot spot 1.2 by 1.2 mm occurs on a 15 by 15 mm chip. The silicon chip thickness is 0.8 mm. The lid enclosing the BGA is fabricated from copper that is 0.8 mm thick. The lid has a clearance of 0.05 mm from the back of the chip, which is filled with thermal grease. Assume that the heat is discharged to the environment with an air cooled heat exchanger bonded to the lid of the BGA.

9.22 Write an engineering brief describing the advantages and disadvantages of using a flip-chip BGA.

9.23 Write an engineering brief describing the advantages and disadvantages of using a wire bonded BGA.

9.24 Determine the thermal resistance of a thermal conductor 20 by 1.5 mm in cross section by 150 mm long. The following circuit board materials are to be considered: FR-4, XXXP, Teflon-glass, Polyimide-glass, alumina, silicon and silicon carbide. Prepare a table for your results and write a paragraph with the conclusions you have drawn from this exercise.

9.25 A long unit strip of multi layer board, shown in Fig. 9.Ex25, is made from 5 planes of 2 oz copper and 4 layers of glass epoxy. A source of heat uniformly distributed over an area 1 mm by unity is dissipated through the thickness of the board. The bottom plane of copper is held at a constant temperature T_s = 20 °C. Find the temperature in the x-z plane for coordinates ranging from 0.5 to 6 mm. Are the copper planes useful in distributing the heat in the x direction. Prepare a sketch of the isothermal lines which demonstrates this fact.

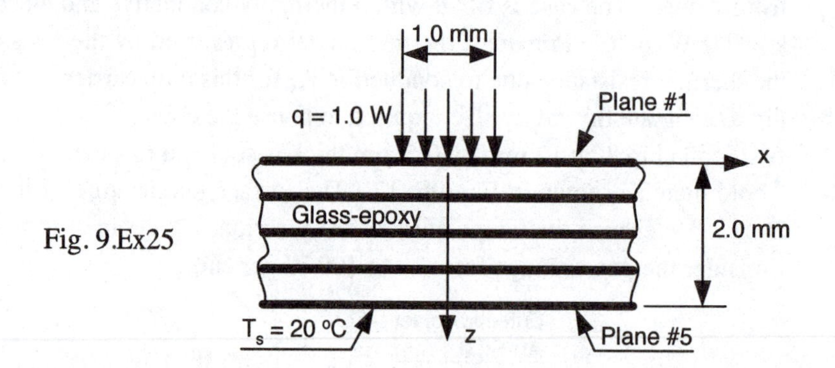

Fig. 9.Ex25

9.26 A mistake was made and the vias shown in Fig. 9.23 were not filled with solder. Only those vias supporting pins from the chip carriers were filled (14%). Determine the thermal resistance for this segment of circuit board. Who made the mistake—design or manufacturing? How would you make this determination?

9.27 Explain the purpose of a heat frame. Why would you use a heat frame in air cooled systems? Would you use a heat frame in space applications? Why?

9.28 Weight is a significant design factor in space applications. What material would you use for a heat frame design if you were told it cost $8,000 to put a pound into high space orbit? Prepare an engineering brief showing the procedure used to select the material and include a cost analysis to support your position.

9.29 If the heat input to each segment of the thermal model shown in Fig. 9.26 is: q_1 = 3.0 W, q_2 = 3.0W, q_3 = 3.0W, q_4 = 4.3W, q_5 = 7.0W, q_6 = 6.5W, q_7 = 6.8W, and q_8 = 7.2W, determine the temperatures T_1 to T_{17}.

9.30 The heat input to the board and heat frame given in Exercise 9.29 is not uniform. Those devices with the highest rate of dissipation were placed near the heat sink. Comment on the advisability of the heat distribution in shown in the Exercise 9.29. Be certain to include reliability in your discussion.

9.31 A cold plate is made by drilling 12 mm diameter holes on 60 mm in. centers in an aluminum plate 16 mm in. thick and 600 mm long. Determine the shape factor S for this plate. What is the thermal resistance of the cold rail?

9.32 If the cold plate in Exercise. 9.31 is 200 mm wide determine the amount of heat which can be transferred as a function of ΔT.

9.33 Water flowing at 1.5 GPM is used with a cold plate for cooling a wide variety of power supplies. Prepare a graph showing the rate of heat transfer with this cold plate as a function of ΔT for this cold plate.

9.34 Examine Eqs (9.46) and (9.48) and explain the difference between T_1, T_2, T_0 and T_i. Draw a sketch of the cold plate and locate the position of these four temperatures. Are you entirely satisfied that these four temperatures completely describe the temperature distribution in the cold plate? Explain?

9.35 Using the advanced cooling concept defined in Fig. 9.28, design a first level chip carrier for a 12 by 15 mm size chip that is rated at 180 W. The chip has a local hot spot 2 by 1.5 mm in size where the power flux is 120 W/cm^2. The chip is 0.8 mm thick and is flip-chip bonded to a multilayer ceramic substrate. You are to select materials and dimensions for the heat spreader, the lid and the thermal adhesives and pastes. Prepare a drawing of the assembly and then perform an analysis to estimate the highest junction temperature on the chip.

9.36 Write an engineering brief describing the advanced cooling concept illustrated in Fig. 9.29. Include in your discussion the advantages of the concept. Also include the problems that may be encountered.

9.37 Write an engineering brief describing the advanced cooling concept illustrated in Fig. 9.30. Include in your discussion the advantages of the concept. Also include the problems that may be encountered.

9.38 Prepare a term paper based on "system on a package". You may wish to use Reference [17] as a starting point.

9.39 Design a microchannel heat exchanger that is to be fabricated by etching a wafer of silicon. The chips are to be bonded to one side of the wafer heat exchanger with Au-Si eutectic solder. Show an arrangement to accommodate the manifolds with attention to detail to reduce the pressure required to drive a fluid through the micro channels. Select the fluid and prepare an analysis that shows design trade offs on size, heat capacity, flow rates, pressure drop and thermal resistance. You may wish to refer to Reference [6] before designing the micro-channel heat exchanger..

CHAPTER 10

THERMAL ANALYSIS METHODS: RADIATION AND CONVECTION

10.1 INTRODUCTION TO RADIATION HEAT TRANSFER

In addition to conduction, heat transfer by radiation and convection are important methods for dissipating heat from electronic systems to the environment. Indeed, in space applications, radiation is the final mechanism used to dissipate heat through the vacuum of outer space. Radiation is the electromagnetic transmission of energy from one opaque surface to another. The rate of heat transfer is dependent upon the fourth power of the absolute temperature T_a^4 of the transmitting and the receiving bodies. This fourth power relationship makes radiation very important in high temperature applications like combustion and incandescent lighting; however, in electronics where the design goal is to minimize junction temperatures, radiation is not a major factor in heat dissipation schemes. Only in space, where it is the only mechanism available, is radiation a dominate factor. In electronic products used here on earth, radiation is usually a minor factor that is often ignored because its contribution to heat transfer rates is small. Thermal analysis including conduction and convection are usually sufficient to accurately predict the heat transfer rates and the temperatures of critical components.

The rate of heat transmitted from a radiating surface depends upon the fourth power of its absolute temperature T_a^4 and the surface's emissivity. Emissivity in turn is a function of the wave length of the radiation, the material from which the radiating surface is fabricated and its surface characteristics such as roughness and coating color. Emittance is a numerical factor used to characterize the emissivity of a surface. Emittance is defined as the ratio of the radiant energy emitted by a body to that emitted by a perfect radiator, which is known as a black body. Values for the emittance range from about 0.05 for polished metals to 0.95 surfaces coated with black lacquer.

When the radiation energy arrives at the receiving body, it may be transmitted, absorbed, or reflected. The partitioning of the energy among these three categories depends on the temperature, the material from which the receiving body is fabricated, its surface geometry, surface characteristics and the wavelength of the radiation. Metal and plastic used in packaging electronics are opaque and tend to absorb most of the incoming radiation. The absorption occurs near the surface as the energy in the radiation wave only propagates into the packaging a short distance. To prevent absorption from the sun and the moon during space travel, highly polished metallic surfaces are employed on shields to increase the amount of heat that is reflected back into space.

In the design of a heat management system using radiant cooling, it is essential to accommodate the radiation energy arriving at the electronic package as well as the energy transmitted to some receiver (a radiant heat sink).

We will describe both radiation and convection as heat transfer mechanisms in this chapter. Then show methods utilizing the basic heat transfer relations that are applied to the cooling of electronic systems packaged in enclosures.

10.2 LAWS GOVERNING HEAT TRANSFER BY RADIATION

Heat transfer by radiation is very important in high temperature applications where the radiant flux density[1] $E_\lambda{}^b$ is relatively high. Plank's radiation law, shown graphically in Fig. 10.1a, indicates the sharp drop in $E_\lambda{}^b$ with the lower operating temperatures that are typical for electronic devices and enclosures. Also evident, is the shift of the peak flux density to a higher wave length λ as the temperature decreases. These features are shown more clearly in Fig. 10.1b where the data over the temperature range from 25 to 100 °C is presented. Integrating the relation for $E_\lambda{}^b$ over λ from 0 to ∞ leads to the Stefan-Boltzmann's relation given by:

$$E_\lambda{}^b = \sigma\, T_a{}^4 \tag{10.1}$$

where $\sigma = 5.670 \times 10^{-8}$ W/(m^2-K^4) is the Stefan-Boltzmann's constant.

$E_\lambda{}^b$ is the total energy radiated from a black body (W/m^2).

$T_a = T + 273$ the absolute temperature (°K) of the surface temperature T (°C).

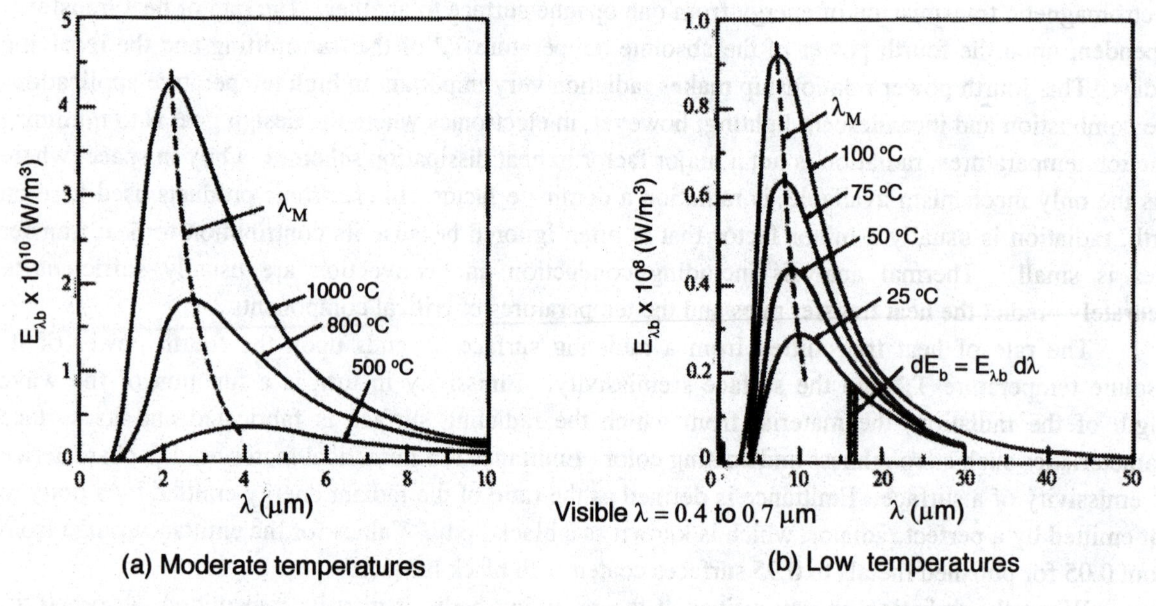

Fig. 10.1 Distribution of emissive power from a black body as a function of radiation wave length λ.

From Eqs (9.15) and (10.1) it is evident that the equivalent radiation heat transfer coefficient h_r is given by:

$$h_r = \frac{q^b/A}{\Delta T} = \frac{E_\lambda{}^b}{\Delta T} \tag{10.2}$$

The value of h_r depends strongly on the surface temperature of a body, because $E_\lambda{}^b$ increases as the fourth power of the absolute temperature of its surface.

---

[1] The units for flux density used in this discussion are (W/m^2).

10.2.1 Shape Factors and Energy Exchange

Heat transfer by radiation is more complex than indicated by Eqs. (10.1) and (10.2) because there is radiation exchange between two bodies and the geometries of both of these bodies affects this exchange of heat. The exchange concept is illustrated in Fig. 10.2 where surfaces S_1 and S_2 are shown and heat is exchanged between incremental areas ΔA_1 and ΔA_2. The incremental heat transferred from body 1 to body 2 is given by:

$$\Delta q_{1-2} = E_1^b \Delta A_1 \left(\frac{\Delta A_2}{\pi r^2} \cos\theta_1 \cos\theta_2 \right) \tag{a}$$

and from body 2 to body 1 the incremental heat transferred is given by:

$$\Delta q_{2-1} = E_1^b \Delta A_2 \left(\frac{\Delta A_1}{\pi r^2} \cos\theta_2 \cos\theta_1 \right) \tag{b}$$

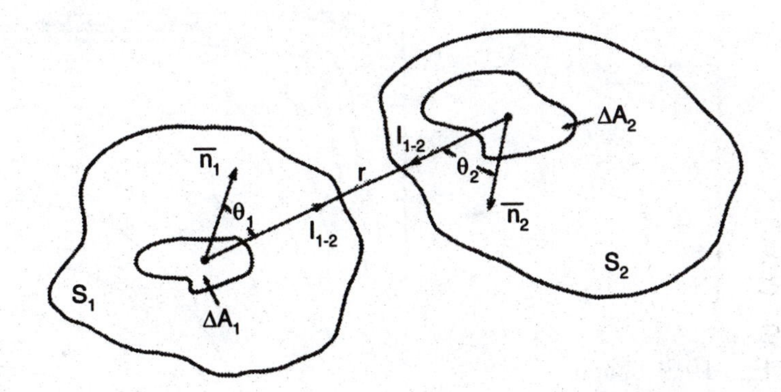

Fig. 10.2 Radiant energy exchange between two bodies.

The net exchange of heat between the two incremental areas is:

$$\Delta q = \Delta q_{1-2} - \Delta q_{2-1} \tag{c}$$

Substituting Eqs. (a) and (b) into (c) gives:

$$\Delta q = \left(E_1^b - E_2^b \right) \left(\frac{\Delta A_1 \Delta A_2}{\pi r^2} \cos\theta_1 \cos\theta_2 \right) \tag{10.3}$$

Equation (10.3) is integrated over the areas of the surfaces S_1 and S_2 to obtain:

$$q^b = (E_1{}^b - E_2{}^b) A_1 \boldsymbol{F}_{1-2} \tag{10.4}$$

where q^b is the black body heat that is transferred; $\boldsymbol{F}_{1-2}$ is a shape factor and A_1 is the area of the surface S_1.

The shape factor indicates that portion of the total radiation emitted from surface 1 that is intercepted by surface 2. Because the product of the shape factor and the area exhibits reciprocity, we may write:

$$A_1 \boldsymbol{F}_{1-2} = A_2 \boldsymbol{F}_{2-1} \qquad (10.5)$$

Combining Eqs. (10.1) and (10.4) gives:

$$q^b = \sigma \boldsymbol{F}_{1-2} A_1 (T_{a1}^4 - T_{a2}^4) \qquad (10.6)$$

The shape factors for common geometries have been determined by integration and are represented in Figs. 10.3 to 10.5. These results can be used with Eq. (10.6) to evaluate the black body rate of heat transferred q^b by radiation between two bodies. For more complex applications involving multiple bodies, the reader is referred to Reference [2], where methods are described to obtain shape factors for multiple body radiation and for more complex geometries.

Equation (10.6) is valid for black surfaces where either reflection does not occur or it is sufficiently small to be neglected without introducing significant error. When the surfaces are "gray", reflection takes place and it is usually necessary to account for this effect to improve the accuracy of the thermal analysis.

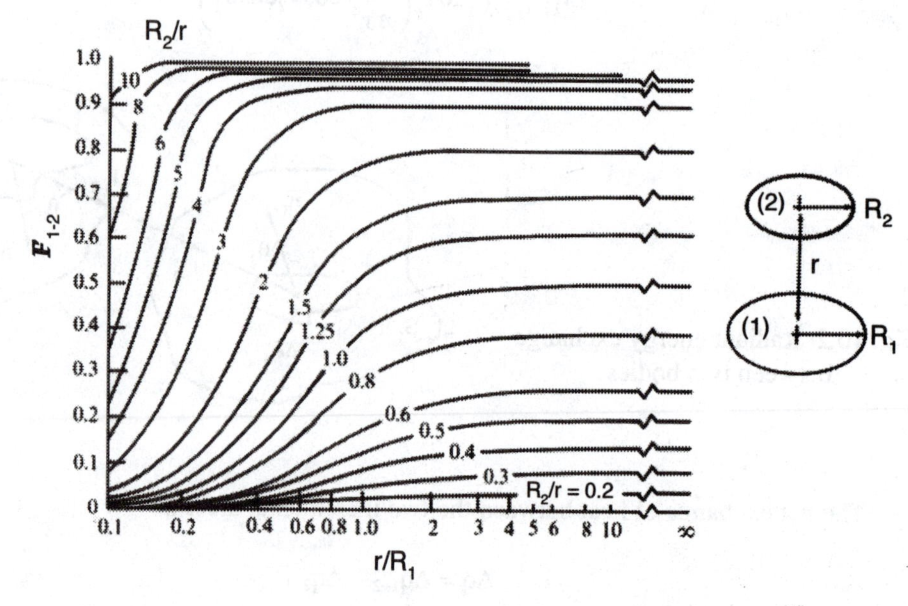

Fig. 10.3 Shape factor for parallel coaxial disks. From reference [1].

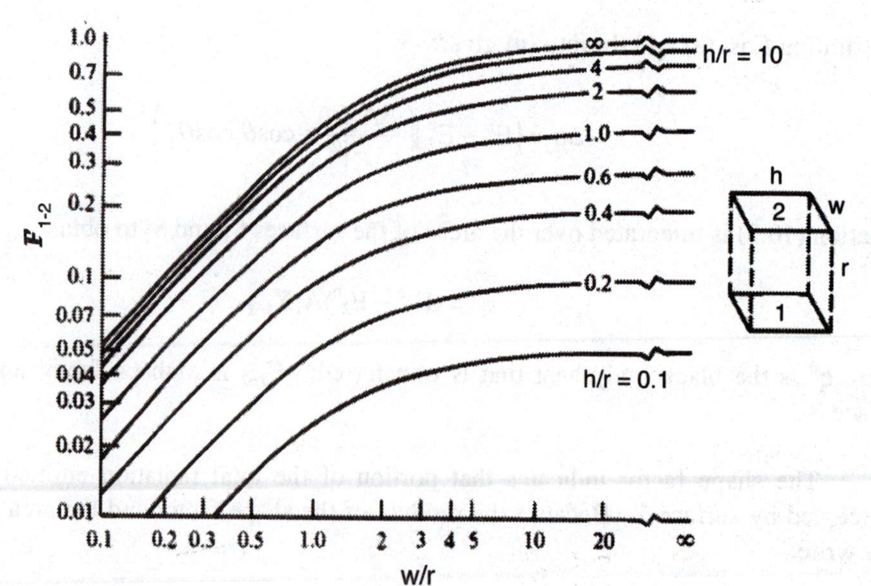

Fig. 10.4 Shape factor for equal size opposing rectangles. From reference [1].

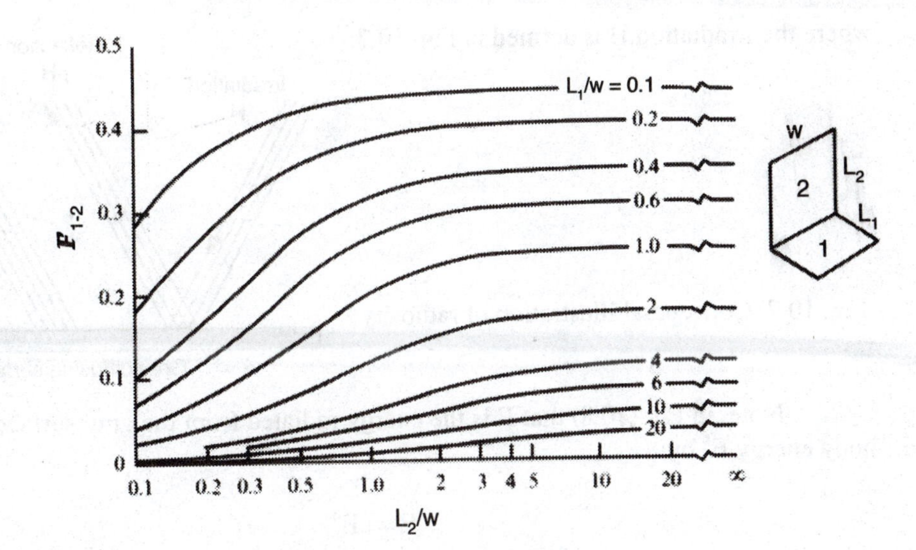

Fig. 10.5 Shape factor for perpendicular rectangles. From reference [1].

10.2.2 Influence of Surface Emissivity on Radiation Heat Transfer

In reality surfaces are never totally black and it is necessary to modify Eq. (10.6) to account for surface emittance and reflection. Noting from Fig. 10.6 that irradiation impinging on a surface can be absorbed, transmitted or reflected leads to three coefficients which sum to equal 1 as indicated by:

$$\alpha + \rho + \tau = 1 \qquad (10.7)$$

where α is the ratio of the irradiation absorbed; ρ is the ratio reflected and τ is the ratio transmitted.

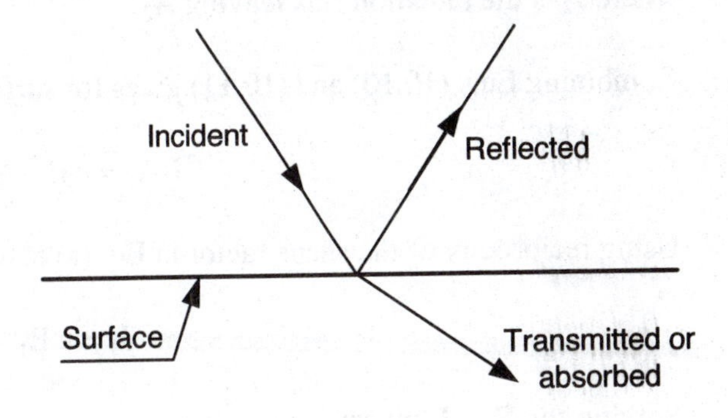

Fig. 10.6 Partitioning of the incident radiation impinging on a surface.

For the typical gray body with an opaque and diffusing surface it is clear that:

$$\tau = 0 \qquad \alpha = \varepsilon \qquad \text{and} \qquad \rho = 1 - \varepsilon \qquad (a)$$

where ε is the emissivity.

Consider a gray body with a diffusing surface receiving energy and emitting its own energy as indicated in Fig. 10.7. The total energy propagating away from this surface is termed the radiosity J, which is given by:

$$J = E + \rho H \tag{10.8}$$

where the irradiation H is defined in Fig. 10.7.

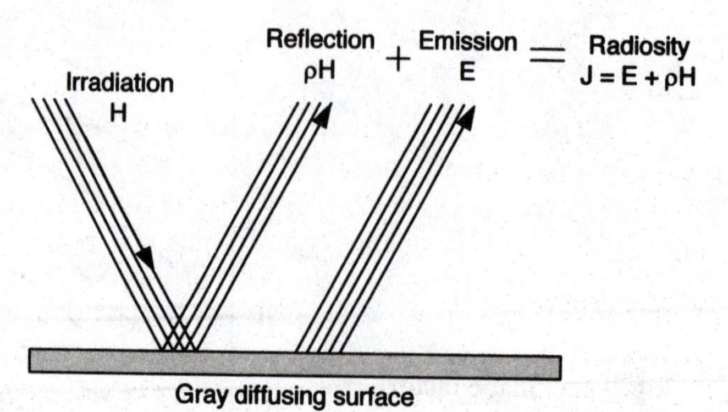

Fig. 10.7 Conceptual illustration of radiosity.

Note, in Eq. (10.8) that E is the energy radiated from the gray surface that is related to the black body energy E^b by:

$$E = \varepsilon E^b \tag{10.9}$$

Combining these two equations yields:

$$J = \varepsilon E^b + \rho H \tag{10.10}$$

Next, observe that the irradiation H_1 falling on the surface with an area A_1 is due to only one other surface with an area A_2. Then it follows that:

$$A_1 H_1 = \boldsymbol{F}_{2\text{-}1}\, A_2\, J_2 \tag{10.11}$$

where J_2 is the radiation flux leaving A_2.

Combining Eqs. (10.10) and (10.11) gives for surface A_1:

$$J_1 A_1 = \varepsilon\, E_1^{\,b}\, A_1 + \rho_1\, \boldsymbol{F}_{2\text{-}1}\, A_2\, J_2 \tag{10.12a}$$

Using reciprocity of the shape factor in Eq. (a) leads to:

$$J_1 = \varepsilon\, E_1^{\,b} + \rho_1\, \boldsymbol{F}_{1\text{-}2}\, J_2 \tag{10.12b}$$

Solving for $\boldsymbol{F}_{1\text{-}2}\, J_2$ gives:

$$\boldsymbol{F}_{1\text{-}2}\, J_2 = (J_1 - \varepsilon E_1^{\,b})/\rho_1 \tag{10.13}$$

The net heat transferred from surface S_1 is given by:

$$q_1 = J_1\, A_1 - H_1 A_1 = J_1\, A_1 - \boldsymbol{F}_{2\text{-}1}\, J_2\, A_2 \tag{10.14}$$

Noting reciprocity of the shape factor permits simplification of Eq. (10.14) to:

$$q_1 = A_1\, (J_1 - \boldsymbol{F}_{1\text{-}2}\, J_2) \tag{10.15}$$

Substituting Eq. (10.13) into Eq. (10.15) results in:

$$q_1 = \frac{\varepsilon_1 A_1}{1-\varepsilon_1}\left(E_1^b - J_1\right) \tag{10.16}$$

These equations may be used to determine the heat transferred by radiation; however, it is usually easier to use an electrical analogy developed by Oppenheim [3] that follows the concepts developed previously in Chapter 9. We again consider the simple case with only two surfaces, as shown in Fig. 10.8a. The analogy indicates that q is analogous to the current and that the radiation fluxes E_1 and E_2 are analogous to voltages. In the two surface problems, there are three resistances to be considered, namely R_1, R_{sf} and R_2 as indicated in Fig. 10.8b. Note that the intermediate voltages between the resistors are analogous to the energy densities J_1 and J_2.

The resistance R_1 is determined by noting from the electrical analogy that:

$$q_1 = \frac{\left(E_1^b - J_1\right)}{R_1} \tag{10.17}$$

Direct comparison of Eqs. (10.16) and (10.17) shows that:

$$R_1 = \frac{1-\varepsilon_1}{\varepsilon_1 A_1} \tag{10.18}$$

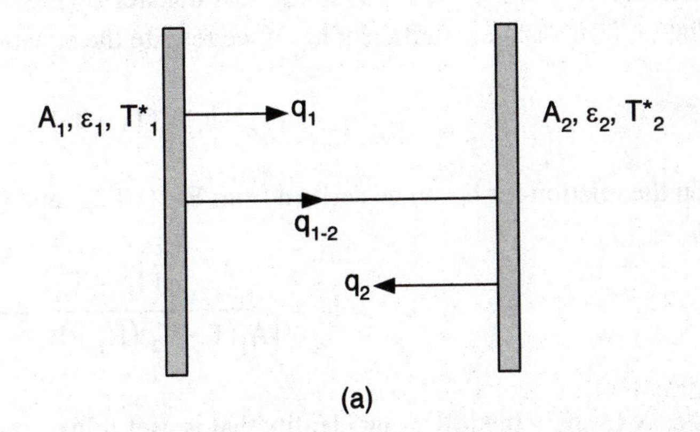

Fig. 10.8 (a) Radiation heat transfer between two perpendicular surfaces. (b) Electrical network for the analogy.

Because R_1 is the resistance due to emissivity ε_1 associated with surface S_1, it is evident that the resistance for surface S_2 may be written as:

$$R_2 = \frac{1-\varepsilon_2}{\varepsilon_2 A_2} \tag{10.19}$$

Finally, R_{sf} is related to q_1 by:

$$q_1 = \frac{(J_1 - J_2)}{R_{sf}}$$

where $$R_{sf} = 1/(A_1 \mathcal{F}_{1\text{-}2}) \quad (10.20)$$

The rate of heat transfer by radiation between the two gray surfaces is calculated from:

$$q_{1-2} = \frac{E_1^b - E_2^b}{R_1 + R_{sf} + R_2} \quad (10.21)$$

From Eq. (10.1) it is clear that:

$$E_1^b - E_2^b = \sigma(T_{a1}^4 - T_{a2}^4) \quad (10.22)$$

The electrical analogy provides a simple and direct approach for calculating the radiation heat transferred by relating it to the black body surface energies while minimizing the algebra involved. Values of the emissivity ε which are employed in Eqs (10.18) and (10.19) are given in Table 10.1.

10.2.3 The Radiation Heat Transfer Coefficient

It is interesting to consider the rate of heat transfer by radiation between two gray bodies in terms of a radiation heat transfer coefficient h_r. If we rewrite the equation for q_{1-2} as:

$$q_{1-2} = h_r A_1(T_1 - T_2) \quad (10.23)$$

Then the relation for h_r can be derived from Eq. (10.21) and (10.23) as:

$$h_r = \frac{\sigma(T_{a1}^4 - T_{a2}^4)}{[A_1(T_1 - T_2)(R_1 + R_{sf} + R_2)]} \quad (10.24)$$

It is easy to show the following identity that is useful in developing Eq. (10.24):

$$a^4 - b^4 = (a - b)(a^3 + a^2b + ab^2 + b^3) \quad (a)$$

Substituting Eq. (a) into Eq. (10.24) gives:

$$h_r = \frac{\sigma(T_{a1}^3 + T_{a1}^2 T_{a2} + T_{a1} T_{a2}^2 + T_{a2}^3)}{[A_1(R_1 + R_{sf} + R_2)]} \quad (10.25)$$

Table 10.1
Emissivity of different materials[2] at T = 100 °C

Material	Condition	Emissivity, ε
Aluminum	Commercial Sheet	0.09
Aluminum	Polished	0.095
Aluminum	Roughened	0.18
Brass	Polished	0.059
Carbon	Rough Plate	0.77
Carbon (Graphite)	Rough Plate	0.76
Chromium	Polished	0.075
Copper	Polished	0.052
Copper-nickel	Polished	0.059
Iron	Dark Gray Surface	0.31
Iron	Rough Polish	0.27
Lamp Black	Rough Deposit	0.84
Molybdenum	Polished	0.071
Nickel	Polished	0.072
Nickel-Silver	Polished	0.135
Paint-Aluminum	Clean	0.29
Paint-White	Clean	0.79
Paint-Cream	Clean	0.77
Paint-Black	Clean	0.84
Paint-Bronze	Clean	0.51
Silver	Polished	0.052
Stainless Steel	Polished	0.074
Steel	Polished	0.066
Tin	Polished	0.069
Tin	Commercial Coating	0.084
Tungsten	Polished Coating	0.066
Zinc	Commercial Coating	0.21
Fuzed Quartz	1.99 mm Thick	0.775
Convex D Glass	3.40 mm Thick	0.83
Nonex Glass	1.57 mm Thick	0.835

Next, if substituting Eqs. (10.18), (10.19) and (10.20) into Eq. (10.25) we obtain:

$$h_r = \frac{\sigma\left(T_{a1}^3 + T_{a1}^2 T_{a2} + T_{a1} T_{a2}^2 + T_{a2}^3\right)}{\left[(1-\varepsilon_1)/\varepsilon_1 + \left(1/\boldsymbol{F}_{1-2}\right) + (A_1/A_2)(1-\varepsilon_2)/\varepsilon_2\right]} \tag{10.26}$$

Note, also that the thermal resistance R_r associated with the radiation heat transfer rate is given by:

$$R_r = \frac{1}{h_r A_1} \tag{10.27}$$

The application of the radiation heat transfer coefficient h_r is illustrated in the exercises following this chapter.

---

[2] Values for the emissivity vary considerably from source to source. This variation is probably due to the difficulties in characterizing the surface finish that markedly affects the value of emissivity ε.

10.3 INTRODUCTION TO CONVECTION HEAT TRANSFER

Convection is extremely important as a means of heat dissipation in electronic systems because air cooling is used in most low to medium performance systems. With convection, the heat is transferred from a solid interface to a fluid and then the fluid is moved from the system to the atmosphere that serves as a heat sink. When possible, free or natural convection is used where the buoyancy of the air serves to transport the air and the heat away from the system. For higher heat transfer rates, buoyancy does not provide sufficient air velocity or flow rate and forced convection is employed. In these cases, fans drive the air at higher velocities increasing the volume flow rate of the air and the heat transfer coefficient.

Convection cooling and fluid mechanics are closely coupled. To understand the mechanism of heat transfer from a solid to a fluid requires an appreciation of the influence of boundary layers, their thickness and the type of flow involved. Thick boundary layers impede the transmission of heat across the solid-fluid interface. Moreover the boundary layer is dependent upon whether the flow is laminar or turbulent. And the transition from laminar to turbulent flow is not a precise event that can be predicted with high accuracy.

Analysis methods in fluid mechanics differ considerable from the methods used in solid mechanics. In fluid mechanics dimensionless numbers due to Reynolds, Rayleigh, Prandtl, Grashof and Nusselt are used to characterize type of flow, transition, buoyancy, heat transfer coefficients, etc. The relations developed using these dimensionless numbers all contain functions that are empirical. Results from experiments conducted over many years are used to generate the empirical relations used to predict heat transfer coefficients required for the analysis of heat transferred by convection.

10.4 CONVECTION HEAT TRANSFER

In the preceding section, we considered heat transfer by radiation and noted that a fluid was not necessary as a transport media for heat to be dissipated by radiation. In Chapter 9, we considered conduction where heat transfer occurred in a solid or in a layered structure, which consisted of a combination of solids in thermal contact. In this section, we introduce convection as a means for heat transfer. With convection, heat is transferred from a solid to a fluid where a boundary layer forms at the interface. Because the boundary layer impedes the amount of heat transferred, it is an important parameter that must be incorporated in determining the convective heat transfer coefficient.

To illustrate the importance of convection, consider the boundary layer shown conceptually in Fig. 10.9 where a wall with an area A is maintained at a constant temperature T_w. The fluid (air) in the boundary layer is moving and heat is transferred by this motion across the layer to the external sink which is at a temperature T_0.

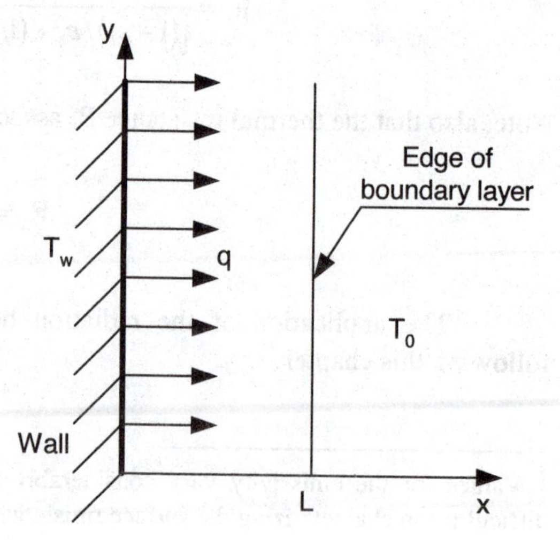

Fig. 10.9 Heat transfer across the boundary layer that separates a wall from an external heat sink.

The heat transferred q_{cv} across the boundary layer is given by:

$$q_{cv} = h\,A(T_w - T_0) \qquad (10.28$$

where h is the convective heat transfer coefficient.

Consider again the same boundary layer but without fluid motion. In this case; convection cannot occur without fluid motion; consequently, heat is transferred across the layer entirely by conduction through the fluid. From Eq. (9.3) it is evident that:

$$q_{cd} = \frac{kA(T_w - T_0)}{L} \qquad (10.29)$$

The ratio q_{cv}/q_{cd} gives the **Nusselt** number Nu defined by:

$$Nu_L = \frac{q_{cv}}{q_{cd}} = \frac{hL}{k} \qquad (10.30)$$

where L is a characteristic length parameter

Low values for the Nusselt number, about one, indicates a small contribution due to convection in the heat transfer process because of low velocity laminar flow in the boundary layer. On the other hand, if the Nusselt number is large, about 100, then convection is dominant and the boundary layer exhibits turbulent flow.

Convective heat transfer is divided into two different categories, namely free or natural convection where buoyancy produces the movement of the fluid and forced convection where fans or blowers drive the fluid. In electronic systems, the fluid is usually air which serves as the transport media to carry heat to the environment. Air can be blown over the electronic devices directly to dissipate the heat from the chip carriers, or air can be passed over plate or duct type heat exchangers to remove the heat from one or more subsystems. In either case, it is necessary to determine the convective heat transfer coefficient, h, associated with dissipating heat from the circuit boards or heat exchangers.

A mathematical theory for boundary layers and convection across these layers has been developed and is presented in standard textbooks on Heat Transfer [4]. However, the theory is very complex and the few meaningful solutions that exist are not in a form that can be easily employed. To obtain more useful solutions, experiments were conducted and empirical relations were developed in terms of dimensionless parameters. The dimensionless parameters used in describing the convective heat transfer coefficient include the Nusselt, Prandtl, Grashof and Reynolds numbers.

Another factor which markedly affects the rate of heat transfer is the type of flow. Is the flow laminar with a velocity profile which varies smoothly with distance across the boundary layer, or is the flow turbulent with strong mixing occurring in the boundary layer? In performing an analysis, we classify the type of flow as laminar or turbulent and then use the most suitable empirical relation to obtain a solution that is sufficiently accurate for engineering design. This approach permits analysis of air cooled electronic systems. It should be recognized that convective heat transfer with a much wider range of fluids (inert gasses, oils, molten salts or liquid metals) is a much more difficult topic; however, we avoid this added complexity by restricting the cooling fluid to air in this treatment.

10.5 NATURAL CONVECTION

An example of free convection is presented in Fig. 10.10, where the boundary layer for a heated vertical plate is shown. Note, the free stream velocity U is zero and that the velocity u of the flow in the boundary layer is achieved by the local increase in the temperature of the air. The heated air has a lower density ρ and buoyant forces develop producing convection velocities u in the boundary layer. The convection velocity, u, increases with the vertical position parameter y according to:

$$u_{ave} = \sqrt{\left(1 - \frac{\rho_w}{\rho_0}\right) gy} \qquad (10.31)$$

where ρ is the density of air; g is the gravitational constant and subscripts w and o refer to the wall and ambient locations.

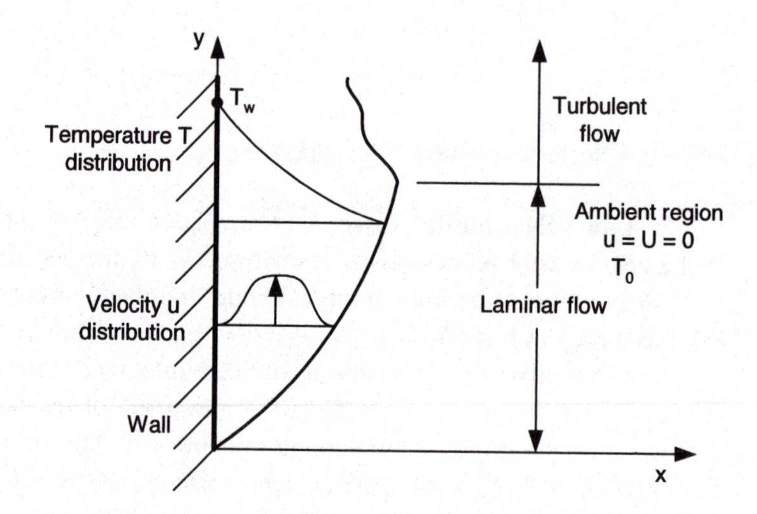

Fig. 10.10 Boundary layer, velocity and temperature distribution for natural convection associated with a heated vertical plate.

If the wall is sufficiently high, the velocities become large and turbulent flow occurs. Clearly as the boundary layer develops, its thickness Δ and its mass flux both increase with y. The added mass of fluid is due to entrainment from the ambient region into the boundary layer.

The local Nusselt number that is a function of the position y associated with free convection depends on the Grashof number, the Prandtl number and the shape of the surface under consideration. The local Grashof number Gr_y is defined as:

$$Gr_y = \left(\frac{g\beta}{\nu^2}\right)(T_w - T_0)y^3 \qquad (10.32)$$

where β is the coefficient of thermal expansion and ν is the kinematic viscosity.

The term $g\beta/\nu^2$ depends only on temperature for any fluid. For air, which is our primary concern, this term together with other useful properties is given in Table 10.2 and Table 10.3.

Table 10.2
Properties of dry air at atmospheric pressure (U S Customary Units)

T		ρ	c_p	μ	ν	k	Pr	β	g/ν^2	$g\beta/\nu^2$
°C	°F	g/in.^{3}	J/(g-°C)	10^{-4}g/(in.-s)	in.^{2}/s	10^{-4}W/(in.-°C)		10^{-3}/(°C)	10^6 /in.^{3}	10^3/(in.^{3}-°C)
−18	0	0.0227	1.000	4.195	0.0187	5.803	0.73	3.916	1.120	4.38
0	32	0.0212	1.003	4.403	0.0209	6.168	0.72	3.661	0.899	3.29
19	50	0.0204	1.004	4.519	0.0220	6.374	0.72	3.528	0.812	2.90
38	100	0.0186	1.004	4.856	0.0259	6.945	0.72	3.216	0.569	1.83
66	150	0.0171	1.008	5.150	0.0301	7.473	0.72	2.952	0.446	1.36
93	200	0.0158	1.008	5.442	0.0344	8.000	0.72	2.729	0.324	0.885
121	250	0.0147	1.012	5.780	0.0392	8.440	0.71	2.538	0.260	0.674
149	300	0.0137	1.012	6.085	0.0441	8.968	0.71	2.370	0.195	0.462
177	350	0.0129	1.016	6.305	0.0491	9.495	0.70	2.214	0.161	0.366
204	400	0.0121	1.024	6.614	0.0544	9.979	0.69	2.094	0.128	0.269

Table 10.3
Properties of dry air at atmospheric pressure (SI units)

T °C	ρ kg/m^3	c_p kJ/(kg-°C)	ν m^2/s	k W/(m-°C)	β 10^{-3}/°C	Pr
−150	2.793	1.026	3.08	0.0116	8.21	0.760
−100	1.980	1.009	5.95	0.0160	5.82	0.740
−50	1.534	1.005	9.55	0.0204	4.51	0.725
0	1.293	1.005	13.30	0.0243	3.67	0.715
20	1.205	1.005	15.11	0.0257	3.43	0.713
40	1.127	1.005	16.97	0.0271	3.20	0.711
60	1.067	1.009	18.90	0.0285	3.00	0.709
80	1.000	1.009	20.94	0.0299	2.83	0.708
100	0.946	1.009	23.06	0.0314	2.68	0.703
120	0.898	1.013	25.23	0.0328	2.55	0.700
140	0.854	1.013	27.55	0.0343	2.43	0.695
160	0.815	1.017	29.85	0.0358	2.32	0.690
180	0.779	1.022	32.29	0.0372	2.21	0.690
200	0.746	1.026	34.63	0.0386	2.11	0.685
250	0.675	1.034	41.17	0.0421	1.91	0.680
300	0.616	1.047	47.85	0.0454	1.75	0.680
350	0.566	1.055	55.05	0.0485	1.61	0.680
400	0.524	1.068	62.53	0.0515	1.49	0.680

Next, recall the definition the **Prandtl** number as:

$$Pr = \mu c_p / k \tag{10.33}$$

Substituting Eq. (10.7) into Eq. (10.33) gives:

$$Pr = \nu/\alpha \tag{10.34}$$

Reference to Table 10.2 shows that the Prandtl number is nearly constant at Pr = 0.72 over the temperature range normally encountered in air cooling electronic components. Recognizing this simplification, permits Pr to be treated as a constant (0.72) in all subsequent developments for both free and forced convection.

10.5.1 Free Convection on a Vertical Plate with Laminar Flow

If the buoyant flow along the vertical plate is laminar, experimental results lead to an expression for the local Nusselt number as:

$$Nu_y = f(Pr)Gr_y^{1/4} \tag{10.35}$$

This equation is valid in the laminar flow region $10^4 \leq Gr_y \leq 10^9$. The functional relation for $f(Pr)$ varies to some degree depending on the reference used. Here we will use the results of Ostrach [5] and LeFevre [6] to obtain:

$$f(Pr) = \frac{0.75\,Pr^{1/2}}{\left[2.435 + 4.884\,Pr^{1/2} + 4.953\,Pr\right]^{1/4}} \tag{10.36}$$

If we set Pr = 0.72, then $f(Pr) = 0.357$ and Eq. (10.35) may be written as:

$$Nu_y = 0.357\ Gr^{1/4} \tag{10.37}$$

The local value of the convection heat transfer coefficient h_y is given by:

$$h_y = \frac{kNu_y}{y} = \frac{0.357kGr^{1/4}}{y} \tag{10.38}$$

Substituting Eq. (10.32) into Eq. (10.38), we obtain:

$$h_y = 0.357k\left(\frac{g\beta}{\nu^2}\right)^{1/4}\left(\frac{T_w - T_0}{y}\right)^{1/4} \tag{10.39}$$

where k and $g\beta/\nu^2$ are determined for $T_{ave} = (T_w + T_0)/2$

We note from the form of Eq. (10.39) that the convective heat transfer coefficient decreases with increasing y as shown in Fig. 10.11. This reduction in h_y is due to the increase in the thickness of the boundary layer with y that causes an increase in the thermal resistance. The average convection coefficient h_y over the height H of the vertical plate is given by:

$$h_{ave} = \frac{1}{H}\int_0^H h_y dy \tag{10.40}$$

Integrating Eq. (10.40) gives:

$$h_{ave} = \frac{4}{3}\left[h_y\right]_{y=H} \tag{10.41}$$

This result indicates that the average heat transfer coefficient is larger by a factor of 4/3 than the value of h_y at y = H.

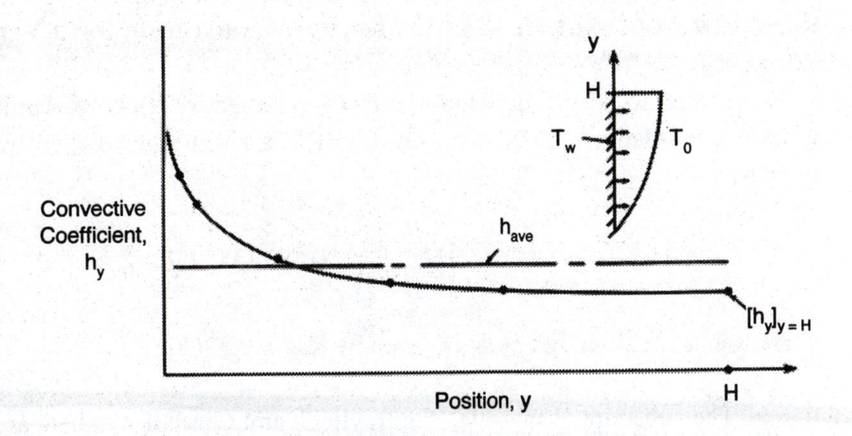

Fig. 10.11 Variation in h_y with position y on a vertical plate with laminar flow due to free convection.

The results of Eq. (10.39) are for a constant temperature T_w at the wall. However, if the wall temperature varies while the heat flux transferred remains constant with position y, the relation for the local Nusselt number changes to:

$$Nu_y = f_1(Pr)Gr^{*}{}_y{}^{1/5} \qquad (10.42)$$

where $f_1(0.72) = 0.487$ and $Gr_y^* = \frac{q}{A}\left(\frac{g\beta y^4}{k\nu^2}\right)$

The results from Eq. (10.42) indicate slightly higher values of Nu_y and h_y than the results for situation with a constant wall temperature given in Eq. (10.37). Exercises are provided at the end of this chapter to show a distribution of wall temperature for this case and to indicate the difference in h between the constant temperature and the constant heat flux case.

10.5.2 Free Convection on a Vertical Plate with Turbulent Flow

When the length of the vertical plate is extended, a transition from laminar to turbulent flow occurs that initiates when $Gr = 10^7$. The turbulence becomes fully developed at $Gr = 10^9$. The **Rayleigh** number Ra that is given by:

$$Ra = Gr\ Pr \qquad (10.43)$$

is often used to indicate the onset of turbulence in free convective flow. The conversion from laminar to turbulent flow depends upon local disturbances and it is common to account for these local effects by considering transition at a value of $Ra = 10^9$. For values of $Ra \geq 10^9$ where the flow is considered turbulent, the local Nusselt number is given by:

$$Nu_y = f_2(Pr)Gr^{2/5} \qquad (10.44)$$

where $f_2(0.72) = 0.0221$

Substituting Eqs. (10.30) and (10.32) into Eq. (10.44) indicates that h_y is proportional to $y^{1/5}$; hence, integration with respect to y gives:

$$Nu_{ave} = (5/6)[Nu_y]_{y=H} \qquad (10.45)$$

Exercise 10.24 is provided to show the difference between heat transfer under laminar and turbulent flow conditions.

10.5.3 An Approximate Relation for Free Convection for a Vertical Plate

It is possible to avoid determining if the convective flow is laminar or turbulent by using an approximate relationship for the average Nusselt number developed by Churchill and Chu [7] that is given by:

$$\left[Nu_H\right]_{ave}^{1/2} = 0.825 + 0.325 Ra_H^{1/6} \tag{10.46}$$

where Ra_H is the Rayleigh number given by $Ra_H = 0.72 Gr_H$.

The subscript H indicates the Rayleigh and Grashof numbers are evaluated at y = H. Equation (10.46) is for a constant temperature wall and is valid for $0.1 \le Ra_H \le 10^{12}$. Some accuracy is lost in employing this approximate relation but it should give predictions accurate to ± 30% that is within the variation in the data reported for convection coefficients.

10.5.4 Free Convection Heat Transfer with Other Shapes

We noted in previous sections of this chapter that the shape of the body was important in determining the coefficient for heat transfer by radiation. The shape of the body also influences the convection heat transfer coefficient. In this sub section, we will present a summary of the empirical results for the convection heat transfer coefficient for shapes of general interest in cooling electronic systems. In all of the relations listed below, we have used air as the cooling fluid and set Pr = 0.72.

Vertical Cylinder

Consider first a vertical cylinder of diameter D and height H. The average Nusselt number is given [8] as:

$$\left[Nu_H\right]_{ave}^{cyl} = \left[Nu_H\right]_{ave}^{plate}\left[1 + \zeta^{0.9}\right] \tag{10.47}$$

where $\zeta = (H/D) Gr_H^{-1/4}$ and the average Nusselt number for the plate is determined from Eq. (10.46).

Horizontal Cylinder

For a horizontal cylinder of diameter D [9], gives Nu_{ave} as:

$$\left[Nu_D\right]_{ave}^{1/2} = 0.60 + 0.322 Ra_D^{1/6} \tag{10.48}$$

This relation is valid for $10^{-5} \le Ra_D \le 10^{12}$.

Horizontal Plate

For a horizontal plate with a characteristic length L, [10] gives the following empirical relations:

With a hot upper surface and a cool lower surface:

$$[Nu_L]_{ave} = 0.54Ra_L^{1/4} \qquad \text{for } 10^4 \leq Ra_L \leq 10^7 \tag{10.49}$$

$$[Nu_L]_{ave} = 0.15Ra_L^{1/4} \qquad \text{for } 10^7 \leq Ra_L \leq 10^{11} \tag{10.50}$$

With a cool upper surface and a hot lower surface:

$$[Nu_L]_{ave} = 0.27Ra_L^{1/4} \qquad \text{for } 10^5 \leq Ra_L \leq 10^{11} \tag{10.51}$$

The characteristic length L is taken as the average of the length and width of a rectangular plate. It is equivalent to 0.9D for a circular plate. For a plate with an irregular shape, L is determined by dividing the plate area by its perimeter.

Equally Spaced Vertical Fins

Finally consider an array of equally spaced fins (vertical plates) as illustrated in Fig. 10.12a. Elenbaas [11] developed an empirical solution to this problem which is given by:

$$[Nu_W]_{ave} = \left(\frac{\psi}{24}\right)\left(1 - e^{-35/\psi}\right)^{3/4} \tag{10.52a}$$

where $\psi = \left(\frac{W}{H}\right)Gr_W Pr$; H is the height of each fin and W is the distance between fins.

Results for Nu_W as a function of ψ are presented in Fig. 10.12b. The heat transfer coefficient is given by:

$$h = kNu_W/W \tag{10.52b}$$

The total heat transferred is $q = hA\Delta T$.

The value of q/q_{max}, presented as a function of ψ in Fig. 10.12c, shows a maximum at $\psi = 46$ indicating an optimum spacing W for the fins. The heat transferred increases initially by decreasing the spacing between fins and adding more fins for a specified length L of heat exchanger. However, a point is reached where the fins are so dense that they inhibit flow and the capacity for heat transfer begins to decrease. The optimum spacing W for the fins is obtained from:

$$W = 63.9H/Gr_W \tag{10.53}$$

with $Pr = 0.72$. At this optimum value for W, $[Nu_W] = 1.19$.

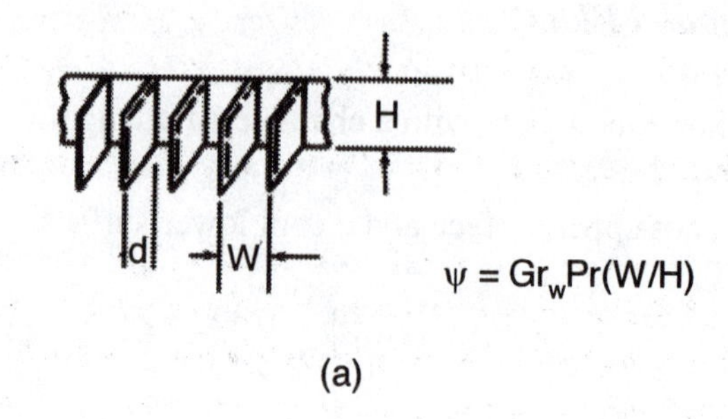

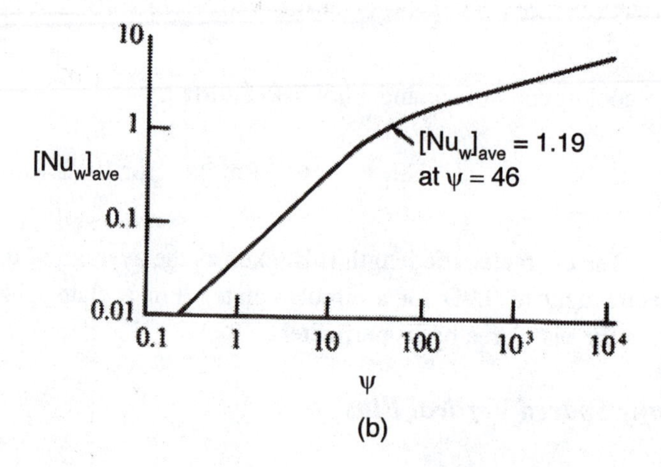

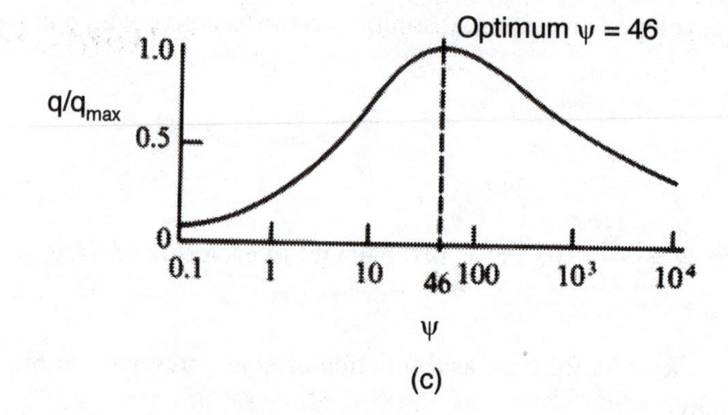

Fig. 10.12 $[Nu_W]_{ave}$ and q/q_{max} as a function of ψ showing the optimum spacing of the fins.
(a) Geometry of a fin type heat exchanger.
(b) $[Nu_W]_{ave}$ as a function of ψ.
(c) q/q_{max} as a function of ψ.

10.6 FORCED AIR CONVECTION COEFFICIENTS

While heat dissipation from electronic systems by natural convection has the advantage of simplicity and economy, the amount of heat which can be transferred at a specified ΔT is not always sufficient to accommodate the heat generated by the system. In these cases, the convective heat transfer coefficient is enhanced by increasing the velocity of the air flow over the heat exchangers or in some cases over the chip carriers directly. The increased air velocity is achieved by the use of one or more fans placed in the electronic enclosure with ducts formed to carry the proper mass flow of air to the various subsystems housed in the enclosure.

The approach used in determining the heat transferred by forced convection is to calculate the Nusselt number and then the convection coefficient h. Again the complexities of the problem require approximate solutions based on empirical equations developed from experimental results. In this treatment, we will cover two different geometries—flat plates and rectangular ducts—because they are the configurations commonly employed in design of air cooled electronic systems.

10.6.1 Laminar Flow Over a Flat Horizontal Plate

Consider a flat plate inserted in a flow field as shown in Fig. 10.13. The flow field is external to the left side of the plate with a velocity U in the x direction and a temperature T_0. The heat is transferred from the plate to the flow field through the boundary layers which exhibit thickness $\delta(x)$ and $\Delta(x)$ in the laminar and turbulent regions, respectively. The local velocity u is a function of position y in the velocity boundary layer and is represented by the following polynomial:

$$u/U = 2\eta - 2\eta^3 + \eta^4 \qquad (10.54)$$

where $\eta = y/\delta(x)$

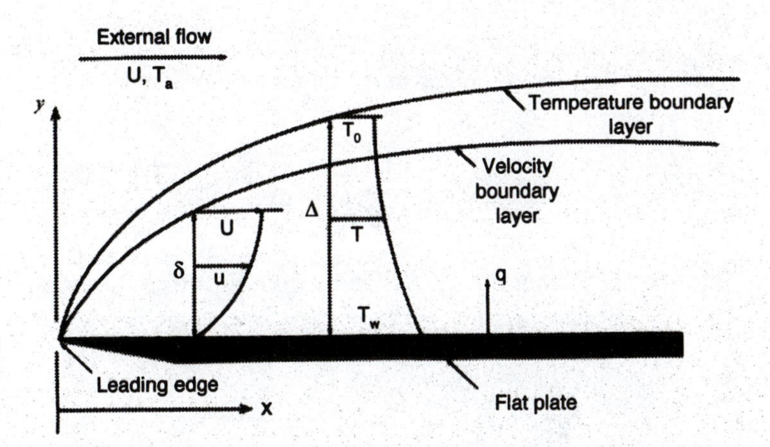

Fig. 10.13 Boundary layers for velocity u and temperature T shown with distribution profiles for both parameters.

By considering the momentum integral equation for a flat plate boundary layer, we may write:

$$\tau_W = \frac{d}{dx}\int_0^\infty \rho u(U - u)dy = \mu\left(\frac{\partial u}{\partial y}\right)_{y=0} \qquad (10.55)$$

where τ_W is the shear stress at the surface of the plate.

Substituting Eq. (10.54) into Eq. (10.55) leads to:

$$\delta = 5.84\ x/Re_x^{1/2} \qquad (10.56)$$

$$\tau_W = 2\mu U/\delta \qquad (10.57)$$

where
$$Re_x = \rho Ux/\mu = Ux/\nu \qquad (10.58)$$

Inspection of these results shows that the thickness δ of the boundary layer increases with $x^{1/2}$. However, experiments indicate that the constant 5.84 is too high. In practice the constant in Eq. (10.56) is replaced with 5.0 to give improved accuracy. Hence:

$$\delta = 5.0\ x/Re_x^{1/2} \qquad (10.59)$$

Noting that the friction factor *f* for the plate is:

$$f = 2\tau_W / \rho U^2 \tag{10.60}$$

Then combining Eq. (10.57) and Eq. (10.60) gives:

$$f = 0.685 / Re_x^{1/2} \tag{10.61}$$

The heat transferred from a flat horizontal plate is based on a polynomial approximation of the temperature profile given by:

$$\frac{T - T_W}{T_0 - T_W} = 2\zeta - 2\zeta^3 + \zeta^4 \tag{10.62}$$

where $\zeta = y/\Delta$

Note that the energy integral equation for a boundary layer is given by:

$$\frac{q_W}{A_x} = \frac{d}{dx}\int \rho\, c_p u (T - T_0) dy = \left[-k\frac{\partial T}{\partial y}\right]_{y=0} \tag{10.63}$$

Substituting T from Eq. (10.62) and u from Eq. (10.54) into Eq. (10.63) leads to:

$$\frac{q_W}{A_x} = \frac{2k(T_W - T_0)}{\Delta} \tag{10.64}$$

The local convection coefficient h_x is given by:

$$h_x = \frac{q_W}{(T_W - T_0)} = \frac{2k}{\Delta} \tag{10.65}$$

This equation is not used because of the presence of the unknown dimension of the boundary layer Δ in the denominator. Instead, h_x is determined from the local Nusselt number which was derived by Pohlhausen [12] as:

$$Nu_x = 0.332\, Re_x^{1/2} f(Pr) \tag{10.66}$$

where $f(Pr) = Pr^{1/3}$. For air with $Pr = 0.72$, Eq. (10.66) reduces to:

$$Nu_x = 0.298\, Re_x^{1/2} \tag{10.67}$$

If we convert Eq. (10.67) into a dimensional form and solve for h_x, we obtain:

$$h_x = 0.298k\left(\frac{\rho U}{\mu x}\right)^{1/2} \tag{10.68}$$

It is evident from this relation that the convection coefficient increases with velocity as $U^{1/2}$ and decreases with distance x along the length of the plate as $1/x^{1/2}$.

The average value of the convection coefficient for a plate of length L is obtained from:

$$h_{ave} = \frac{1}{L}\int_0^L h_x dx = \frac{f(T)}{L}\int_0^L x^{-1/2} dx \tag{10.69}$$

Integrating Eq. (10.69) gives:

$$h_{ave} = \frac{2f(T)}{\sqrt{L}} \tag{10.70}$$

Next, evaluate h at the position x = L from Eq. (10.68) to obtain:

$$h_L = \frac{f(T)}{\sqrt{L}} \tag{10.71}$$

It is clear from Eqs. (10.70) and (10.71) that:

$$h_{ave} = 2h_L \tag{10.72}$$

The procedure is to calculate $Re_L = UL/\nu$ and to substitute this value in Eq. (10.67) to obtain Nu_L. Then $h_L = kNu_L/L$ is substituted into Eq. (10.72) to determine h_{ave}. Finally, the heat transferred from the plate to the flow field is given by:

$$q = h_{ave}A(T_W - T_0) \tag{10.73}$$

The properties of the air μ, ν, k and ρ are functions of temperature which changes with y across the boundary layer. This complication is avoided by using the average temperature

$$T_{ave} = (T_W + T_0)/2 \tag{10.74}$$

across the boundary layer in determining μ, ν, k and ρ from Tables 10.2 or 10.3.

10.6.2 Turbulent Flow over a Flat Horizontal Plate

If the velocity U is sufficiently high, the boundary layer is more involved than that shown in Fig. 10.13. At the higher velocities, the flow in the boundary layer undergoes a transition from laminar to turbulent flow. The form of the boundary layer, as it develops from the leading edge of the plate, includes laminar, transition and turbulent regions as presented in Fig. 10.14. The position of the transition x_c is a function of the Reynolds number and the surface roughness of the plate. Usually the transition position for a flat plate is determined from the critical Reynolds number that is usually taken as:

$$Re_{xc} = 500{,}000 \tag{10.75}$$

where the free field velocity U is used in calculating Re.

Using Eq. (10.58) together with Eq. (10.75), gives the location of the transition region as:

$$x_c = \left(\frac{5\nu}{U}\right) \times 10^5 \tag{10.76}$$

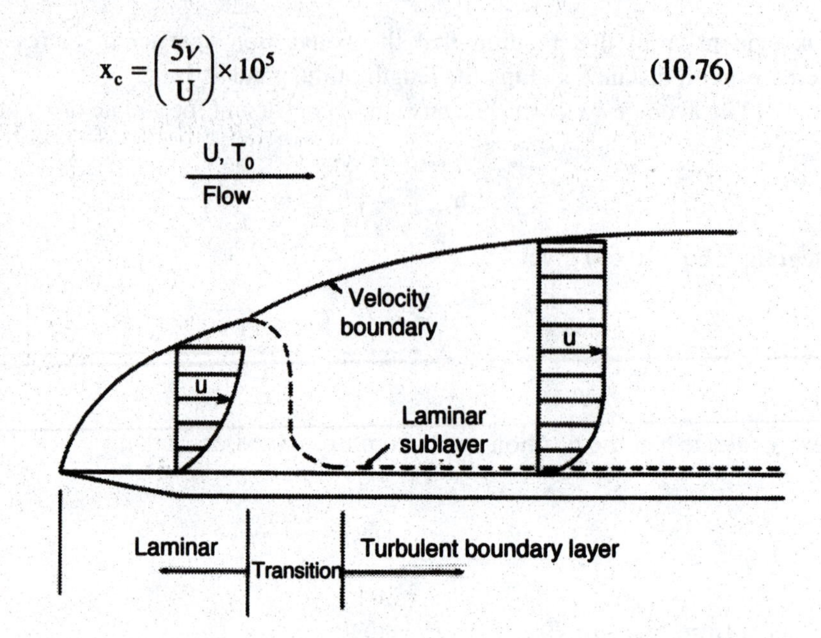

Fig. 10.14 Development of laminar and turbulent boundary layers on a horizontal flat plate.

For flat plates with $L > x_c$, the flow is turbulent and different relations exist for the friction factor and the Nusselt number than those obtained for laminar flow. The friction factor is:

$$f_x = 0.0592\ Re_x^{-1/5}$$
$$f_{ave} = 0.074\ Re_L^{-1/5} \tag{10.77}$$

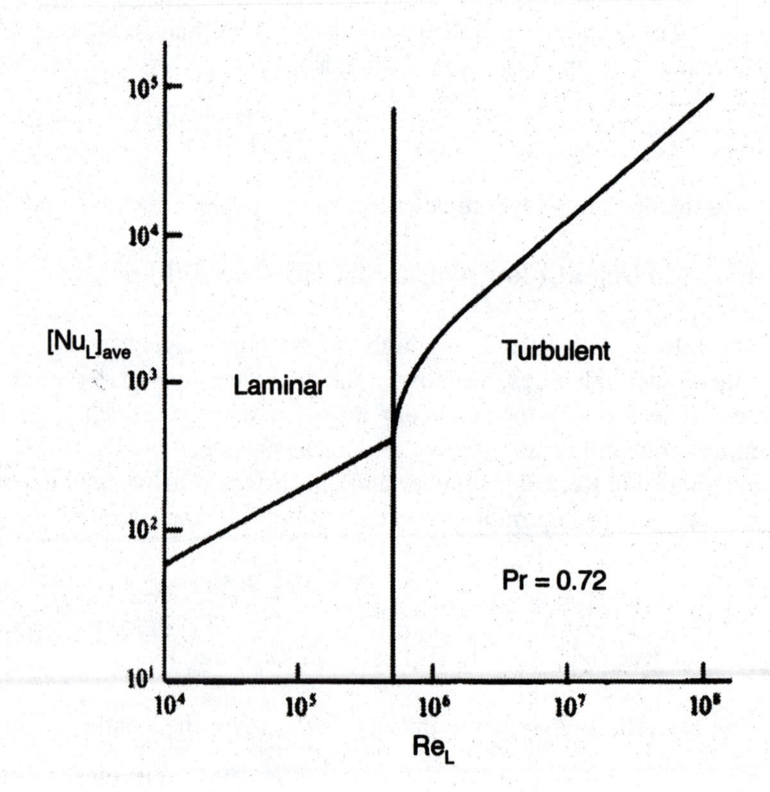

Fig. 10.15 Change in the average Nusselt number with Reynolds number for a flat plate with Re_c = 500,000.

If the Reynolds number is $5 \times 10^5 < Re_x < 10^7$, the Nusselt number is given by:

$$Nu_x = 0.0296\, Re_x^{4/5}\, Pr^{1/3} \qquad (10.78)$$

Noting Pr = 0.72 for air enables Eq. (10.78) to be written as:

$$Nu_x = 0.0265\, Re_x^{4/5} \qquad (10.79)$$

Finally, the average value for the Nusselt number over a plate of length L is:

$$[Nu_L]_{ave} = 0.0332\, Re_L^{4/5} \qquad (10.80)$$

A comparison of $[Nu]_{ave}$ for turbulent and laminar flow is presented in Fig. 10.15. It is clear that $[Nu_L]_{ave}$ increases markedly with Reynolds number and that velocities sufficiently high to produce turbulence markedly increases the convection heat transfer coefficient $[h_L]_{ave}$.

10.6.3 Combined Laminar and Turbulent Flow over a Flat Horizontal Plate

We have treated laminar and turbulent flow as two separate cases and introduced turbulence when $Re_L >$ 500,000. When the flow velocity produces a Reynolds number slightly larger than 500,000, both laminar and turbulent flow occur, and the heat transferred in the laminar region of the plate must be considered. The convection coefficient averaged over the length of the plate is the sum of:

$$h_{ave} = \frac{k}{L}\left[\int_0^{x_c} \frac{Nu_x^L}{x}\,dx + \int_{x_c}^{L} \frac{Nu_x^T}{x}\,dx\right] \qquad (10.81)$$

where the superscripts L and T refer to laminar and turbulent flow conditions. It is possible to write Eq. (10.81) as:

$$[Nu_L]_{ave} = \left(\frac{h_{ave}L}{k}\right) = \left[Nu_{x_c}^L\right]_{ave} + \left[Nu_L^T\right]_{ave} - \left[Nu_{x_c}^T\right]_{ave} \qquad (10.82)$$

Substituting Eqs (10.67), (10.72) and (10.80) into Eq. (10.82) gives:

$$\left(\frac{h_{ave}L}{k}\right) = 0.596 Re_{x_c}^{1/2} + 0.332\left(Re_L^{4/5} - Re_{x_c}^{4/5}\right) \qquad (10.83)$$

Noting that $Re_{xc} = 5 \times 10^5$ in Eq. (10.83) gives:

$$\left(\frac{h_{ave}L}{k}\right) = 0.332 Re_L^{4/5} - 781.7 \qquad (10.84)$$

These results clearly indicate that higher heat transfer coefficients can be achieved by increasing Reynolds number since h is proportional to $U^{4/5}$. However, higher velocities require more power from the fans because the shear stresses (drag) produced on the surface of the plate τ_W is proportional to $U^{9/5}$. More importantly, the noise level is increased with higher velocity fans and this noise must not become objectionable when compared to the background sound levels associated with the operating environment.

10.6.4 Convection from a Flat Plate with Uniform Heat Flux

The previous developments of heat transfer from the flat plate to an external flow field considered the plate at a constant temperature. However, in some applications the heat flux to the plate is held constant and the plate temperature increases with x. The relations describing the heat transfer coefficients are obtained in a similar manner and only the final results will be summarized here:

For Laminar Flow:

$$Nu_x = 0.406\ Re_x^{1/2} \tag{10.85a}$$

For Turbulent Flow:

$$Nu_x = 0.0276\ Re_x^{4/5} \tag{10.85b}$$

The temperature of the plate varies with position x according to:

$$T_W = T_0 + (q^*x)/(kNu_x) \tag{10.86}$$

where $q^* = q/A$ is the heat flux transferred by the plate.

10.7 FORCED CONVECTION WITH FLOW IN DUCTS

In most design applications, it is necessary to transport the air to electronic subsystems that are arranged in an orderly manner within an enclosure. The air passes through a heat exchanger for the subsystem and under goes an increase in temperature and is then transported to another subsystem or exhausted into the atmosphere. The passages used in transporting the air within the enclosures are called ducts. In some cases, portions of the duct walls also serve as heat exchangers. Components are mounted on the duct walls and heat is conducted through the wall to be dissipated into the air stream. Analysis of the head (pressure) required to drive the air and the heat transferred in a duct is affected by the flow field. It is necessary to first consider the duct entrance where the flow is developing and then to classify the flow as laminar or turbulent. The equations used in the analysis depend upon the characteristics of the flow field.

Let's first examine the entrance of a duct and observe that the requirements for continuous temperature distribution and zero velocity at the wall produce boundary layers similar to those described in Section 10.6. As we proceed along the duct, the boundary layers increase in thickness and at some location x_{fd} the layer boundaries meet, as illustrated in Fig. 10.16. For $x > x_{fd}$, the flow field is fully developed and the duct is filled with boundary layer flow. In the fully developed region, the velocity profile is independent of position x along the length of the duct. The shape of the temperature profile is independent of x, but the magnitude of the temperature usually varies due to the heat transferred from the duct's walls into the air stream.

The entrance region $x < x_{fd}$, exhibits velocity and temperature profiles which change with location x. The heat transfer coefficient and the friction factor in the entrance regions are high because the boundary layer is thin. In general, $Pr \neq 1$ and the values of x_{fd} are different for velocity and temperature. However, for air in the temperature range of interest, $Pr = 0.72$ that we consider close enough to 1 to permit the assumption that the fully developed conditions for temperature and velocity occur at the same position, which is given by:

$$x_{fd} = 0.04\ Re_D D \tag{10.87}$$

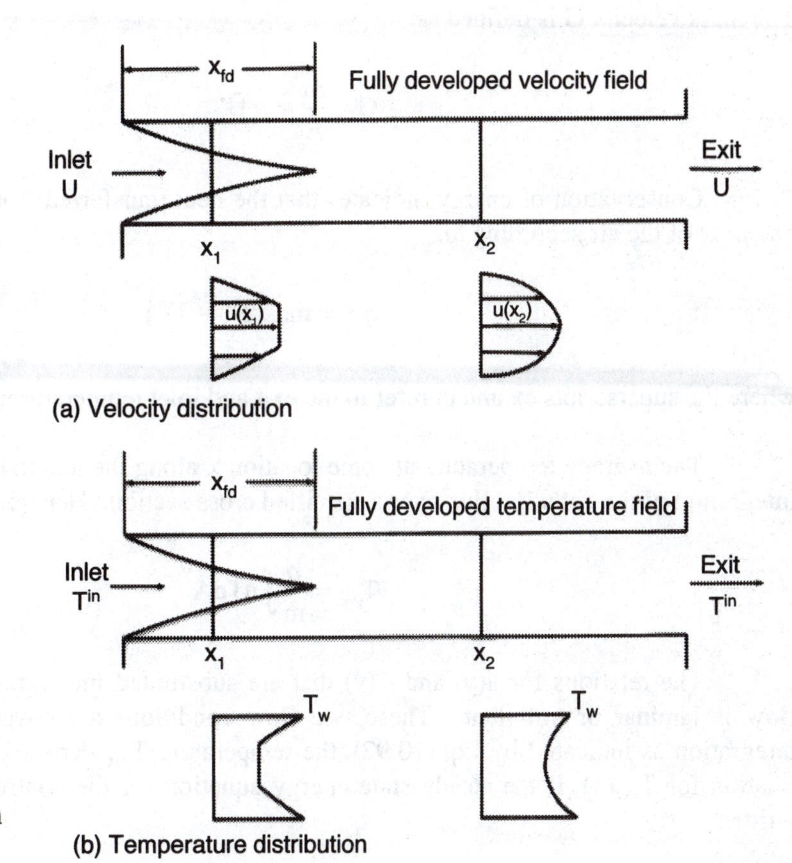

Fig. 10.16 Temperature and velocity distributions for air flow in a duct with Pr = 1.0.

10.7.1 Average Velocity and Temperature

We observed in Fig. 10.16 that both the temperature and the velocity varied with distance from the walls in regions with developing flow and in fully developed flow. These profiles affect the head requirements and the heat transferred, but we normally do not seek information on the detail of the flow field. Instead we work with average values of the velocity and temperature and find suitable relations for the unknowns Nu and *f* in terms the Reynolds's number Re_D, the position x and duct shape.

For steady flow in a straight duct with a cross sectional area A, the mass flow rate $\dot{m}$ is constant and is given by:

$$\dot{m} = \rho U A = G A \tag{10.88}$$

where ρ the density is considered as a constant with respect to position x along the length of the duct.

The velocity U is an average over the cross section defined by:

$$U = \frac{Q}{A} = \frac{1}{A}\int u\,dA \tag{10.89}$$

where Q is the volume flow rate through the duct.

The mass velocity G is defined as:

$$G = \frac{\dot{m}}{A} = \rho U \qquad (10.90)$$

Conservation of energy indicates that the heat transferred from the duct walls q_W increases the enthalpy of the air according to:

$$q_W = \dot{m}c_p\left(T^{ex} - T^{in}\right) \qquad (10.91)$$

where the superscripts ex and in refer to the exit and inlet temperature, respectively.

The average temperature at some location x along the length of the duct is determined from the integration of the enthalpy flux over a specified cross section. Hence:

$$T_{ave} = \frac{\rho}{\dot{m}}\int uTdA \qquad (10.92)$$

The relations for u(y) and T(y) that are substituted into Eq. (10.92) depend upon whether the flow is laminar or turbulent. These two flow conditions are covered later in Section 10.8. After integration as indicated by Eq. (10.92), the temperature T_{ave} depends only on x. We can develop the relation for $T_{ave}(x)$, if the steady state energy equation for the control volume shown in Fig. 10.17 is written:

$$dq_W = h\left(T_W - T_{ave}\right)pdx = \dot{m}c_p dT_{ave} \qquad (10.93)$$

where p is the perimeter of the duct.

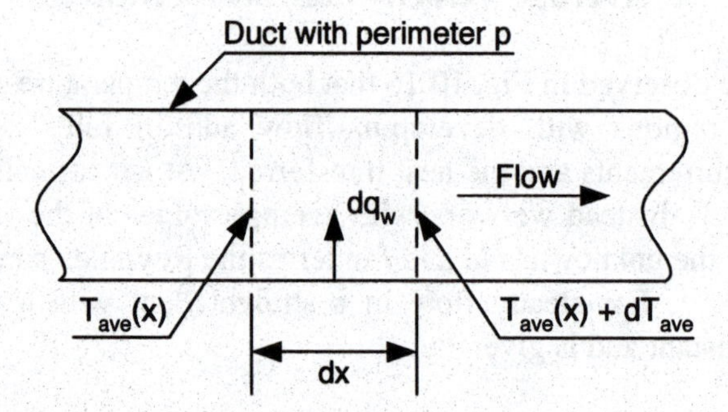

Fig. 10.17 Control volume for heat exchange between the duct walls and the air stream.

If we take the case of a constant wall temperature T_W, the integration of Eq. (10.93) leads to:

$$\int_{T^{in}}^{T_{ave}} \frac{dT}{T - T_W} = -\left(\frac{p}{c_p\dot{m}}\right)\int_0^x h_x dx \qquad (10.94)$$

Equation (10.94) reduces to:

$$T_{ave} = T_W + \left(T^{in} - T_W\right)e^{-\beta(x)} \qquad (10.95)$$

where $$\beta(x) = \left(\frac{p}{c_p \dot{m}}\right)\int_0^x h_x dx \qquad (10.96)$$

The distribution of T_{ave} with x is presented in Fig 10.18. In the fully developed region, h is a constant with respect to x and Eq. (10.96) indicates that β increases as a linear function of x. For this reason T_{ave} approaches T_W as an exponential for large values of x.

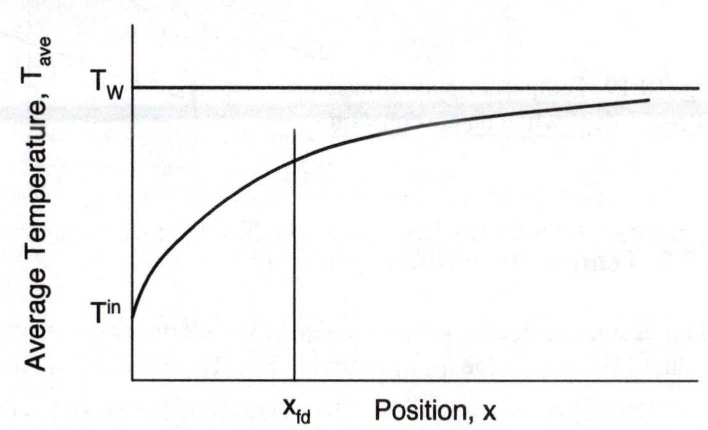

Fig. 10.18 Temperature T_{ave} as a function of position x for steady state air flow in a duct with T_W constant.

For a second case, consider that the heat flux q_{W*} is a constant with x along the wall of the duct. Noting that:

$$dq_W = q_{W*}\, p dx \qquad (10.97)$$

Integration of Eq. (10.93) with respect to x leads to:

$$\left(\frac{p q_{W*}}{c_p \dot{m}}\right)\int_0^x dx = \int_{T^{in}}^{T_{ave}} dT \qquad (10.98)$$

After integrating Eq. (10.98), we obtain:

$$T_{ave} = T^{in} + \left(\frac{q_{W*} p}{c_p \dot{m}}\right) x \qquad (10.99)$$

This result indicates that the average temperature of the air increases linearly with position as it flows along the duct as shown in Fig. 10.19. With q_{W*} constant, the wall temperature of the duct must increase as a function of position. The relation for T(x) is determined from Eq. (10.93) as:

$$T_W = \frac{q_{W*}}{h_x} + T_{ave} \qquad (10.100)$$

The distribution of the wall temperature is illustrated in Fig. 10.19. In the region of fully developed flow, the difference between T_{ave} and T_W is constant for $x_{fd} < x < L$.

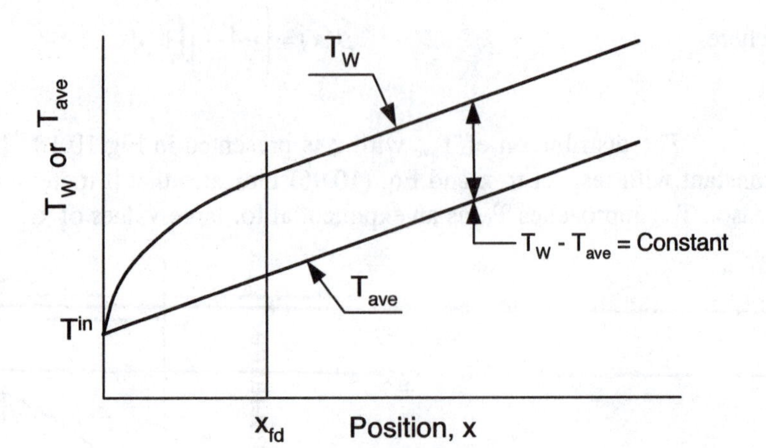

Fig. 10.19 Temperature distribution along a duct when q_{W*} is constant.

10.7.2 Temperature Difference

When designing ducts, we are interested in determining the total heat transferred over the length L of the duct. We determine q_W by applying Eq. (10.28) which is recast as:

$$q_W = h_x\, pL(T_W - T_{ave})_x \tag{10.101}$$

to show the relation with subscripts previously introduced in this section. For the case of a duct with a constant temperature wall, the application of Eq. (10.101) is difficult because T_{ave} varies with x as shown in Fig. 10.18. To simplify the approach, (10.101) is rewritten in terms of average values for h_x and $(T_W - T_{ave})_{ave}$ as:

$$q_W = h_{ave}\, pL(T_W - T_{ave})_{ave} \tag{10.102}$$

In Eq. (10.102), we are considering two averages. The first is across the cross section to obtain T_{ave} and the second is over the length L of the duct to obtain $(T_W - T_{ave})_{ave}$. The total heat transferred to the air is given by:

$$q_a = \dot{m}c_p\left(T^{ex} - T^{in}\right) = \dot{m}c_p\left[\left(T^{ex} - T_W\right) - \left(T^{in} - T_W\right)\right] \tag{10.103}$$

When x = L, Eq. (10.98) gives:

$$T_{ave} = T^{ex} = T_W + \left(T^{in} - T_W\right)e^{-\beta} \tag{10.104}$$

where

$$\beta = \left(\frac{pLh_{ave}}{c_p\dot{m}}\right) \tag{10.105}$$

From Eqs (10.104) and (10.105), it is evident that:

$$\left(\frac{pLh_{ave}}{c_p\dot{m}}\right) = \log_e \frac{T^{in} - T_w}{T^{ex} - T_W} \tag{10.106}$$

Now, if we set $q_a = q_W$ and use Eq. (10.106), it is clear that:

$$(T_W - T_{ave})_{ave} = \frac{\left[\left(T^{ex} - T_W\right) - \left(T^{in} - T_W\right)\right]}{\log_e\left[\left(T^{in} - T_W\right)/\left(T^{ex} - T_W\right)\right]} \quad (10.107)$$

This expression is often called the log-mean temperature difference. If the inlet and exit temperatures do not differ by more than 50%, Eq. (10.107) can be replaced without introducing significant error by the arithmetic average:

$$(T_W - T_{ave})_{ave} = \frac{1}{2}\left[\left(T_W - T^{ex}\right) + \left(T_W - T^{in}\right)\right] \quad (10.108)$$

The temperatures T^{in}, T^{ex} and T_W are either known or easily measured and are used with Eqs. (10.108) and (10.102) to determine the heat transferred as the air passes through the duct.

10.8 HEAT TRANSFER AND FRICTION COEFFICIENTS FOR AIR FLOW IN DUCTS

The coefficients of interest for duct flow are determined from the Nusselt number and the Reynolds number with both of these numbers referenced to the diameter D of the duct by:

$$Nu_D = hD/k \quad (10.109)$$

$$Re_D = UD/\nu \quad (10.110)$$

For ducts with cross sections which are not circular, D is the hydraulic diameter given by:

$$D = 4A/p \quad (10.111)$$

First, let's consider the coefficients associated with fully developed flow and then summarize the design equations for laminar and then turbulent flow.

For fully developed laminar flow in circular ducts:

$$f = 16/Re_D \quad (10.112)$$

$$Nu^q_D = 4.36 \quad \text{with constant heat flux} \quad (10.113a)$$

$$Nu^T_D = 3.66 \quad \text{with constant wall temperature} \quad (10.113b)$$

$$Re_D < 2{,}100 \quad \text{for laminar duct flow} \quad (10.114)$$

For fully developed laminar flow in rectangular ducts [13]:

Equations (10.112) to (10.114) can be used if the flow is fully developed in rectangular ducts; however the constant in each of these relations must be changed to reflect the rectangular cross section and its width to height ratio. The tabular data listed below provides the necessary constants as a function of w/h.

w/h	Nu^q_D	Nu^T_D	$f Re_D$
1	3.61	2.98	14.23
2	4.12	3.39	15.55
3	4.79	3.96	17.09
4	5.33	4.44	18.23
6	6.05	5.14	19.70
8	6.49	5.60	20.58
∞	8.24	7.54	24.00

The symbols w and h refer to the width and height of the duct's rectangular cross section.

For fully developed turbulent flow [14]:

$$(4f)^{-1/2} = -1.8\log_{10}\left[\left(\frac{\varepsilon}{3.7D}\right)^{1.11} + \frac{6.9}{Re_D}\right] \qquad (10.115)$$

$$Nu_D = 0.023\, Re_D^{0.8} \quad \text{and} \quad Nu^q_D = Nu^T_D \qquad (10.116)$$

$$Re_D > 10{,}000 \quad \text{for turbulent duct flow} \qquad (10.117)$$

where ε is the surface roughness

Consider next the entrance region of the duct where the flow is developing. The boundary layers are thin in this region and the friction factor f and the convection coefficient are elevated. The theory and the experiments are also much more complex and the analysis of the developing flow region is beyond the scope of this text. However, we will indicate some empirical results that may be useful in the analysis of air flow in circular ducts.

For laminar developing flow in circular duct:

The friction coefficient is shown as a function of a normalized position parameter in Fig. 10.20. The average Nusselt numbers for a constant temperature wall and for a uniform heat flux are presented in Fig. 10.21. These graphs are both based on Pr = 0.7, which is a close approximation to Pr for air cooled electronic systems. The graphical results can be used to adjust results for the effects of the entrance. Recall that the extent of the developing region is given by x_{fd} and is determined from Eq. (10.87).

For developing turbulent flow:

$$\frac{[Nu_x]_{ave}}{[Nu]_{fd}} = 1 + \frac{2D}{x} \qquad (10.118)$$

where $[Nu]_{fd}$ is given in Eq. (10.116)

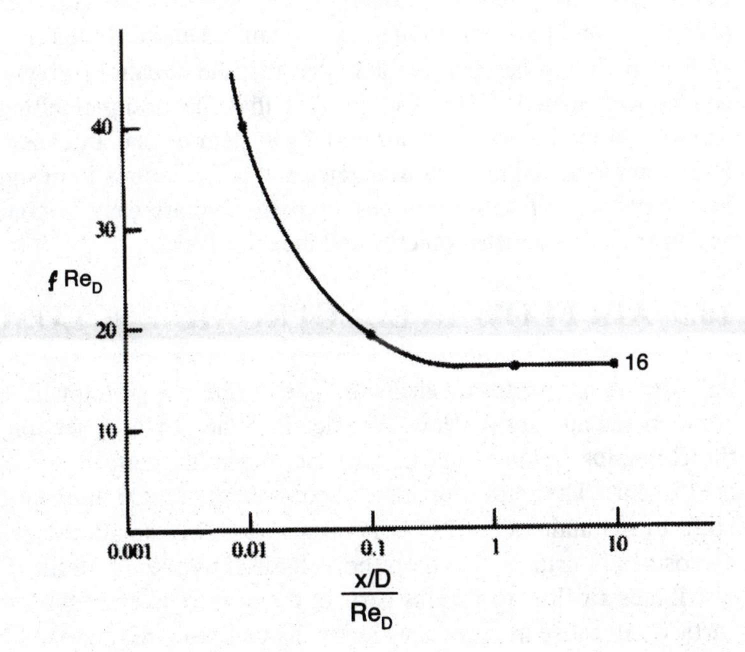

Fig. 10.20 Friction factor for developing laminar flow in a circular duct.

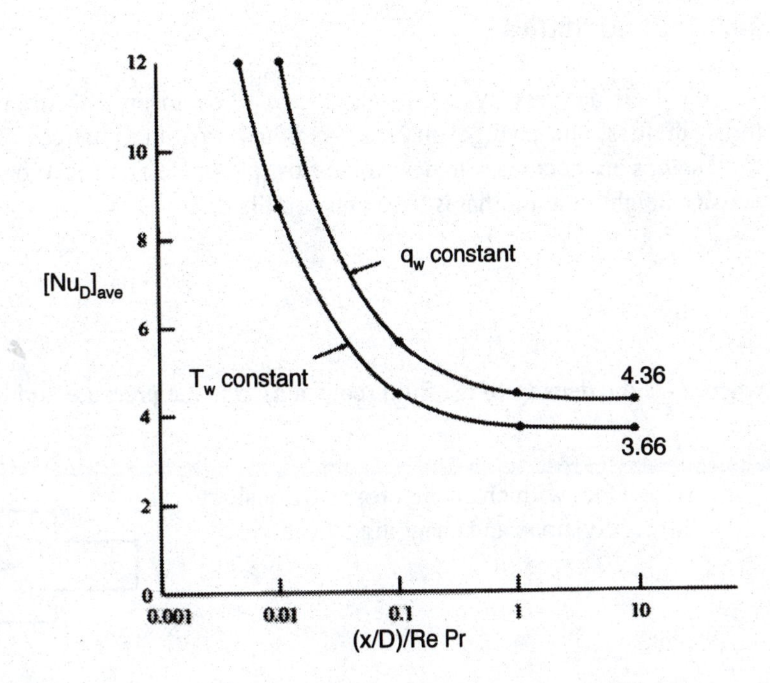

Fig. 10.21 Average Nusselt number for developing laminar flow in a circular duct.

10.8.1 Summary

We have presented results in this section for both free and forced convection that can be used to design cooling systems for electronic equipment. In both cases, we showed that the convection coefficient was related to the Nusselt number. For free convection, empirical equations were given that showed the Nusselt number was related to the Grashof number and the shape of the heat exchanger. The influence of the Prandtl number was eliminated by setting Pr = 0.72 for air, which is a reasonable approximation. For forced convection, the Nusselt number was related to the Reynolds number and the shape of the heat exchanger, and the distinction between laminar and turbulent flow becomes more important.

It should be clear to the reader that the equations given here are approximate. Errors of ± 30% can be anticipated. This fact implies that the designer either provides for comfortable margins of capacity in the layout of the air cooling system or that an experimental model be constructed and tested to determine actual heat transfer rates and temperatures in designing cooling systems with more modest heat capacities. Fortunately these experiments are easy to conduct and the results of the approximate analyzes can be verified quickly and inexpensively.

10.9 AIR FLOW IN ELECTRONIC ENCLOSURES

Air flow in an electronic enclosure is extremely important in reducing the ΔT associated with cooling components and subsystems. We noted in the previous section that the velocity U strongly influenced the Reynolds number and in turn the Reynolds number affects Nusselt number and the convection coefficient. The mass flow rate $\dot{m}$ controls the temperature of the cooling air and must be sufficiently large to maintain $\Delta T = T^{ex} - T^{in}$ at a reasonably small value. The analysis of air flow in electronic enclosures is usually based on the volume flow rate Q, defined in Eq. (10.89). For a duct system that distributes air flow to various parts of the electronic enclosure, we determine Q at each heat exchanging surface. Because the pressures in the duct system are relatively low, we treat air as an incompressible to simplify the analysis.

10.9.1 Fluid Statics

A typical air delivery system in an electronic enclosure, illustrated in Fig 10.22, includes a duct that turns, divides, and changes in cross sectional area and changes in elevation. Some basic relations in fluid statics are necessary to determine flow parameters in this type of duct. First, consider the pressure-density-height relation that is based on equilibrium:

$$z_2 - z_1 = h = \frac{(p_1 - p)}{\omega} \tag{10.119}$$

where ω is the density of the fluid (constant); p is the pressure and h and z are defined in Fig 10.23.

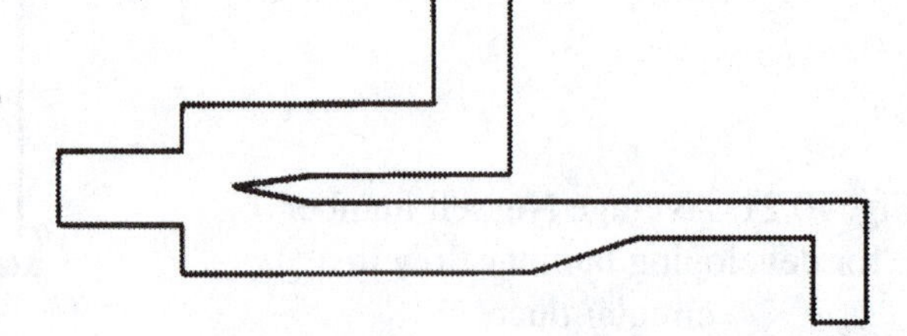

Fig. 10.22 Duct with changing cross sectional area, turns, divisions and changing elevation.

Continuity of the mass flow through a duct with changing cross sectional areas leads to:

$$A_1U_1 = A_2U_2 \tag{10.120}$$

The volume flow rate is:

$$Q = UA \tag{10.89}$$

Conservation of energy leads to Bernoulli's equation:

$$\frac{p_1}{\omega}+\frac{U_1^2}{2g}+z_1=\frac{p_2}{\omega}+\frac{U_2^2}{2g}+z_2 \quad (10.121)$$

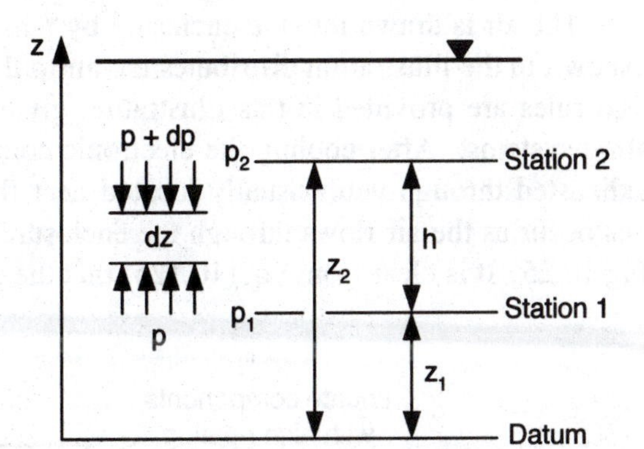

Fig. 10.23 Definition of pressures at station z_1 and z_2.

This relation implies that energy is not added or lost as the fluid flows between station 1 and station 2. In practice energy is added by inserting a fan in the system and energy is lost due to wall friction and due to changes in the duct configuration. To account for these gains from fans h_f and friction losses h_L, we modify Eq. (10.121) when appropriate by:

$$\frac{p_1}{\omega}+\frac{U_1^2}{2g}+z_1+h_f=\frac{p_2}{\omega}+\frac{U_2^2}{2g}+z_2+h_L \quad (10.122)$$

Each term in this relation represents a vertical linear distance, called a head, h, as shown in Fig. 10.23. We rewrite Eq. (10.122) in terms of these heads as:

$$h_{s1}+h_{d1}+h_{P1}+h_f=h_{s2}+h_{d2}+h_{P2}+h_L \quad (10.123)$$

where h_s is the static or pressure head.
h_d is the dynamic or velocity head.
h_P is the potential head.
h_f is the head due to a fan between station 1 and 2.
h_L is the total head loss between station 1 and 2.

As an example of the head loss due to friction, recall from your study of fluid mechanics that:

$$\Delta p=4\left(\frac{L}{D}\right)\left(\frac{\rho U^2}{2}\right)f \quad (10.124)$$

where ρ is the mass density equal to ω/g; D is the hydraulic diameter of the duct and f is its friction factor.

This relation reduces to:

$$h_L=\frac{\Delta p}{\omega}=4\left(\frac{L}{D}\right)\left(\frac{U^2}{2g}\right)f \quad (10.125)$$

We will use these relations later in an analysis of forced air flow in enclosures.

10.9.2 Enclosure Impedance and Fan Characteristics

A typical electronic enclosure with a system for heat dissipation by forced air cooling is presented in Fig. 10.24. The air is drawn into the enclosure by fans mounted at the bottom of the rear panel. Duct work not shown in the illustration distributes the air to the appropriate locations within the enclosure. Several design rules are provided in this illustration giving practical details regarding the layout of forced air cooling systems. After cooling the electronic components distributed throughout the enclosure, the air is exhausted through vents usually located near the top of the side panels into the atmosphere. Head losses occur as the air flows through the enclosure that depends on the volume flow rate Q as indicated in Fig 10.25. It is clear from Eq. (10.125) that the head losses increase in proportion to Q^2.

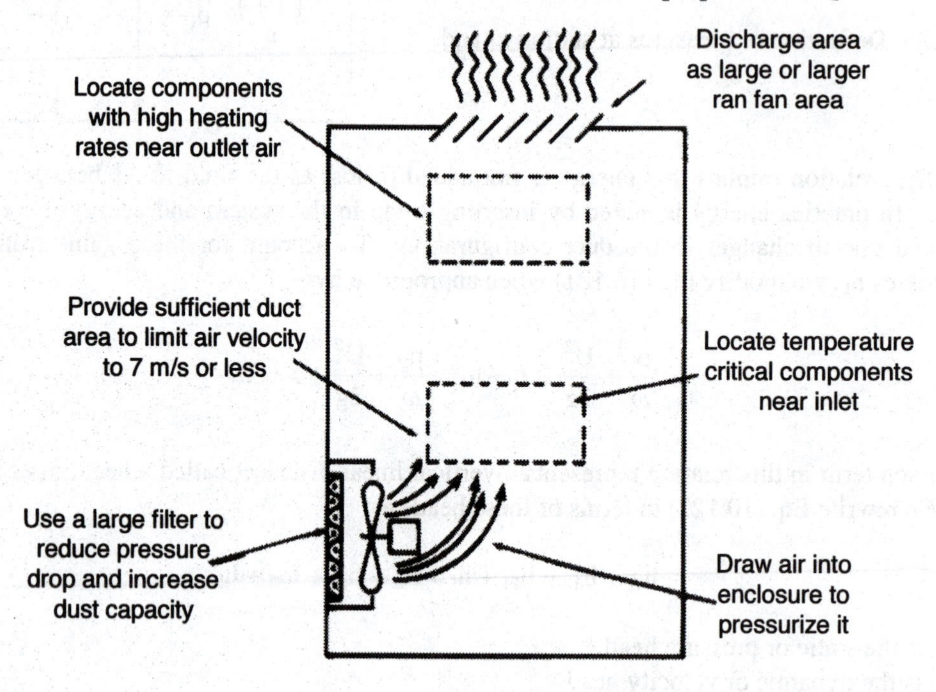

Fig. 10.24 Typical electronic enclosure with forced air cooling. Some design rules for cooling system layout are included.

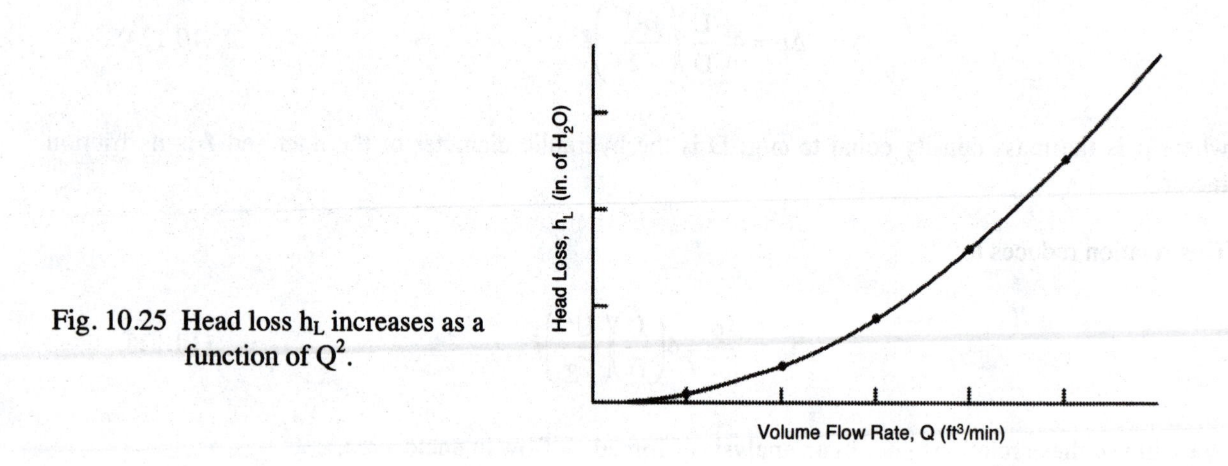

Fig. 10.25 Head loss h_L increases as a function of Q^2.

To provide the pressure necessary to offset the head losses, fans are employed. The fan manufacturers provide performance curves for fans that give the static head h_s developed, the power required and the efficiency as a function of volume flow rate Q. A typical set of performance curves is shown in Fig. 10.26 for an axial propeller type of fan. The system volume flow rate Q is determined by superimposing the h_L – Q curve for the enclosure with the h_s – Q curve for the fan. The intersection of the two curves, as illustrated in Fig 10.27, defines the operating point with the corresponding volume flow rate Q_0. Fans are selected so that Q_0 is within the recommended range of operation for the specified fan that results in efficiencies between 30 and 40%.

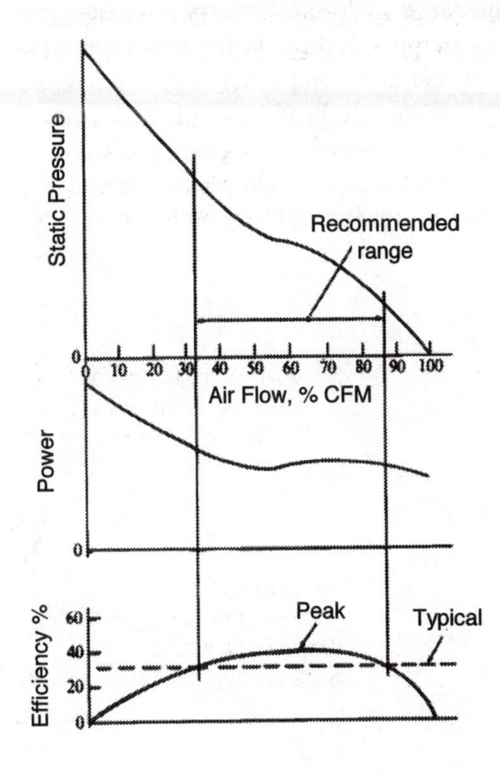

Fig. 10.26 Performance characteristics of an axial propeller type fan.

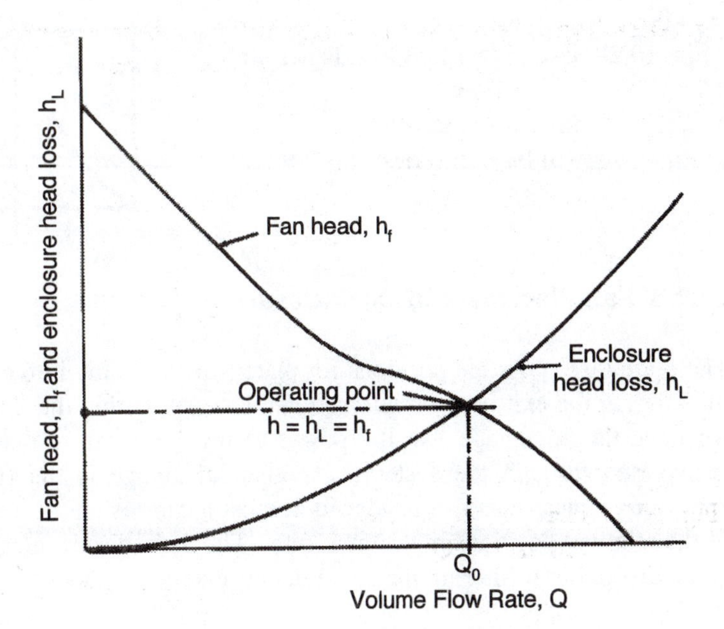

Fig. 10.27 Intersection of the fan head and enclosure loss curves determines the operating point for a forced air cooling system.

In some instances, sufficient flow cannot be achieved with a single fan and it is necessary to consider adding a second fan to the system to enhance the flow rate. Two possibilities exist for the placement of the second fan. It can be placed in parallel (side by side arrangement) or it can be placed in series where first fan provides the input for the second fan. The correct arrangement for placement depends upon the relative head loss of the enclosure. If the head loss is low as illustrated in Fig 10.28a, then the parallel arrangement provides a significant increase in air flow with a modest increase in head loss. However, if the system impedance is high a parallel arrangement is not effective and a series arrangement is necessary. Reference to Fig 10.28b shows that flow enhancement is possible in high impedance enclosures, but the increase, even with the preferred series fan arrangement, is small and is accomplished with a relatively large increase in head loss. In some cases it may be necessary to redesign the ducts to decrease the system impedance to obtain sufficient flow rate.

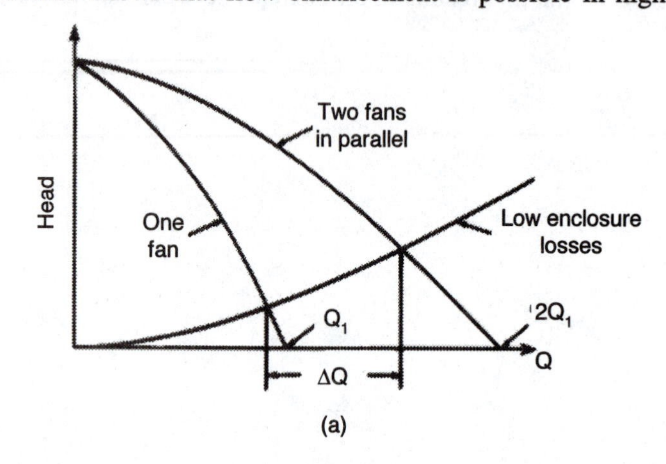

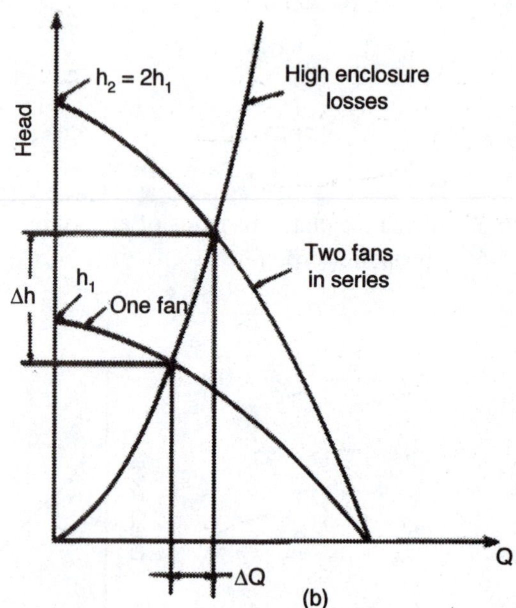

Fig. 10.28 Use of two fans to enhance air flow.
(a) Fans in parallel
(b) Fans in series.

10.9.3 Fan Placement in an Enclosure

There are three possible positions for placing a single fan in the duct system of an enclosure, namely at the inlet, at the exit or at an intermediate location within the ducts. When the fan is placed at the exit, we have the advantage that the power to drive the fan is dissipated into the atmosphere and is not transported through the system. Another advantage is the flexibility of the air distribution in the enclosure. Inlet vents can be placed at most locations on the vertical panels to achieve cooling at each of the required positions in the enclosure. The main disadvantage with fans mounted at the exhaust side is the difficulty in filtering the air. Inlet air enters the enclosure from many inlet vents and from cracks or other openings not intended for air flow. In many environments, unfiltered air carries dirt and dust particles that collect on circuit components and in time may cause malfunctions.

When the fan is placed at the inlet the enclosure is pressurized and the air can be filtered as it is drawn into the enclosure. This placement enables the electronic systems to be operated in a cleaner environment and enhances life and reduces maintenance. Another advantage is that the fan handles cooler and denser air increasing its cooling capacity. The primary disadvantage is that the power to operate the fan is converted to heat that is transported through the enclosure. Of course, this heat load adds to the heat which must be dissipated from the electronics and increases the size of the cooling system requirements.

The intermediate fan has the disadvantages of both the exhaust and inlet fan positions. Its only advantage is the safety provided by the inaccessibility afforded by its location within the enclosure. To insure the safe operation of fans positioned at inlet and exit locations, grills are employed which prevent the accidental insertion of objects or fingers into the fan blades.

The operating point of the system may be affected by the placement location. Consider first the placement of the fan at the inlet position as shown in Fig 10.29a. Applying Eq. (10.123) across the fan from station 1 to 2, we obtain:

$$h_f = h_{s2} + h_{d2} \tag{10.126}$$

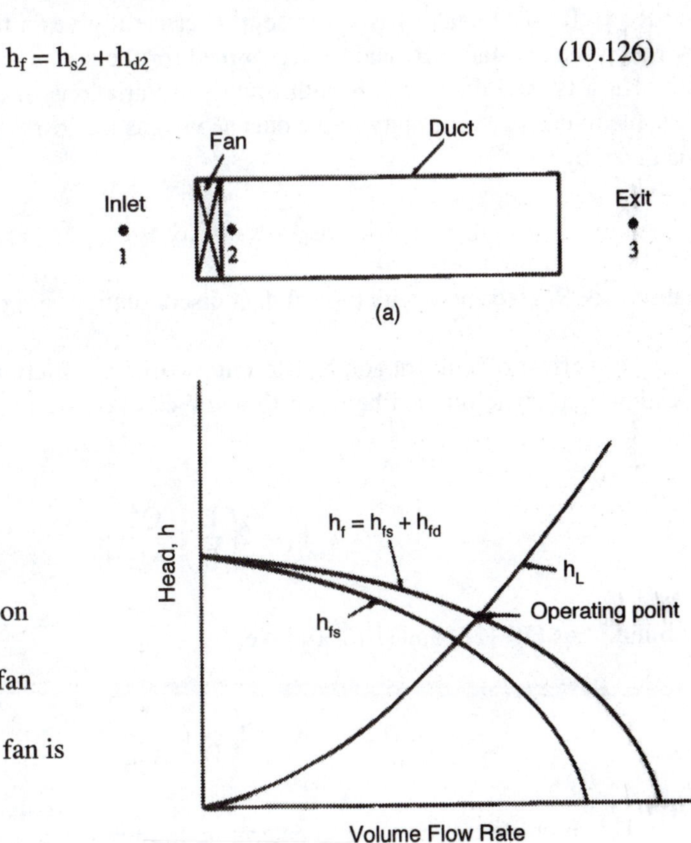

Fig. 10.29 Influence of fan position on volume flow rate.
(a) Station locations 1, 2 and 3 for a fan placed at the inlet.
(b) Change in operating point when the fan is placed at the inlet.

Clearly, the head produced by the fan is divided into two parts. The static head as given in the rating curve of Fig. 10.27 and the dynamic head which depends upon the volume flow rate Q. Next, write Eq. (10.123) again between station 2 and 3 to show:

$$h_L = h_f = h_{s2} + h_{d2} \tag{10.127}$$

This result indicates that a fan positioned at the inlet, produces a head loss h_L equal to the sum of both the static and dynamic head developed by the fan. This fact causes the operating point of the system to change as shown in Fig 10.29b with an increase in the flow rate Q_0. A similar analysis, for the fan placed at the exit, shows that the duct losses are equal to the static head developed by the fan and the operating point shown in Fig 10.27 is correct.

10.9.4 An Electrical Analog for Head Loss Determination

We noted in Eq. (10.125) that the head loss due to flow in a duct is proportional to Q^2 or U^2. This fact leads to:

$$h_L = R_A Q^2 \tag{10.128}$$

where R_A is the constant of proportionality.

Examination of Eq. (10.128) shows that it is analogous to Ohms law V = IR with $h_L \Rightarrow V$, $Q \Rightarrow I$ and $R_A \Rightarrow R$. The analogy is very useful because it gives a relation for the head loss for any volume flow rate if the resistance R_A can be determined for the duct system.

In a typical duct system with turns and variations in cross sectional area, we encounter head losses due to each discontinuity in the duct as well as the losses due to wall friction. The total resistance R_A is given by:

$$R_A = \Sigma R_i \tag{10.129}$$

Relations for R_i associated with typical duct discontinuities are presented in Fig. 10.30.

The effect of wall friction is determined from the friction factor f which was covered previously in Section 10.8. The relation between R_i and f can be determined from Eqs. (10.89) and (10.125) that gives:

$$h_L = 2\left(\frac{L}{D}\right)\left[\frac{Q^2}{A^2 g}\right] f \tag{10.130}$$

Combining Eqs (10.128) and (10.130) gives:

$$R_i = 2\left(\frac{L}{D}\right)\left[\frac{1}{A^2 g}\right] f \tag{10.131}$$

This relation can be used to determine the resistance to air flow due to wall friction along the length L of the duct.

The resistance due to circuit cards with electronic components exposed to the air stream is difficult to characterize because the pattern of components is so variable. A first approximation to the resistance R_i is:

$$R_i = \frac{2L \times 10^{-3}}{A^2} \tag{10.132}$$

where L is the length of the circuit card and A is the cross sectional area of the channel over the card.

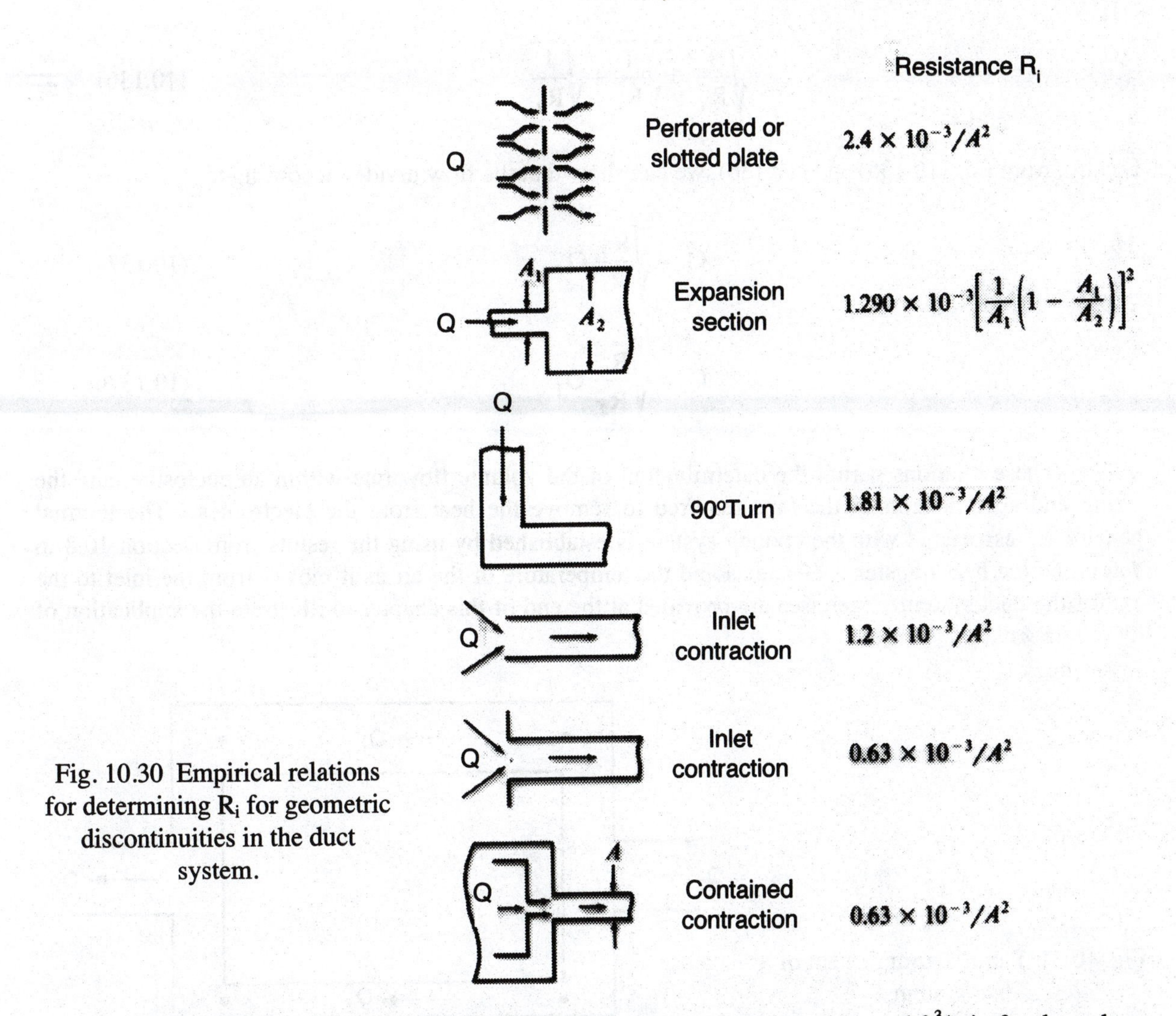

Fig. 10.30 Empirical relations for determining R_i for geometric discontinuities in the duct system.

Dimensions used in these relations for R_i are inches for the duct sizes and ft³/min for the volume flow rate. While these are mixed units, they are consistent with common practice in determining the head loss in inches of water.

In the analysis of head loss in a duct, it is clear that the combined resistance R_T of segments of the duct that are arranged in a series combination is given by:

$$R_T = R_1 + R_2 + \ldots.. + R_n \tag{10.133}$$

On the other hand if the ducts are in a parallel arrangement as shown in Fig (10.31), then the head loss over each element is the same:

$$h_L = h_{L1} = h_{L2} \tag{10.134}$$

However, the volume flow rate Q_T divides into Q_1 and Q_2 with:

$$Q_T = Q_1 + Q_2 \tag{10.135}$$

Combining Eqs (10.128), (10.134) and (10.135) leads to:

$$\sqrt{\frac{1}{R_T}} = \sqrt{\frac{1}{R_1}} + \sqrt{\frac{1}{R_2}} \tag{10.136}$$

Finally, from Eqs (10.135) and (10.136), we can show that the flow divides according to:

$$Q_1 = \sqrt{\frac{R_T}{R_1}}\, Q_T \tag{10.137a}$$

$$Q_2 = \sqrt{\frac{R_T}{R_2}}\, Q_T \tag{10.137b}$$

These relations permit the determination of the volume flow rate within an enclosure, and the sizing and specification of the fans required to remove the heat from the electronics. The thermal penalty ΔT associated with the cooling system is established by using the results from Section 10.8 to determine the heat transfer coefficients and the temperature of the air as it moves from the inlet to the exit of the duct system. Exercises are provided at the end of this chapter to illustrate the application of these relations.

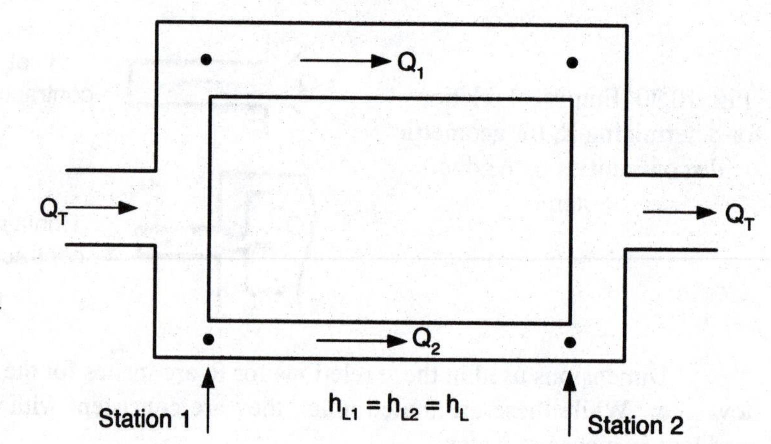

Fig. 10.31 Parallel arrangement of a duct system.

10.10 INTRODUCTION TO HEAT TRANSFER BY EVAPORATION, BOILING AND CONDENSATION

The strategy for dissipating the heat generated by most electronic systems today utilizes a combination of conduction and convection heat transfer processes. Heat is transferred by conduction through solids and their interfaces to some type of a heat exchanger. At that point the heat is transferred by convection to a stream of water (in a cold plate) or a stream of air flowing through the heat exchanger. Both of these mechanisms have relatively low transfer rates unless the physical dimensions associated with the heat transfer path are extremely small as is the case when micro-channel heat exchangers are etched into the back side of a silicon chip. Much higher heat transfer rates are possible if we consider changing the phase of the fluid used in to cool the system. Changing phase of a fluid from a liquid to a vapor (gas) at a constant pressure and temperature requires the addition of heat—called the latent heat of vaporization. The term latent is used because the temperature remains constant during the transfer of heat as the phase changes. The temperature of a fluid that is heated slowly at a constant pressure, presented in Fig. 10.32, shows the effect of the latent heat of vaporization during a phase change from liquid to vapor.

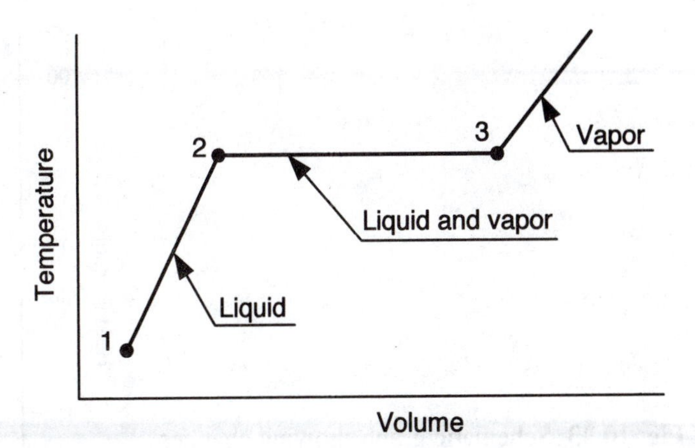

Fig. 10.32 Temperature of a fluid subjected to slow heating at constant pressure.

The addition of heat to the fluid begins at point 1 causing an increase in temperature until the phase change from liquid to vapor begins. Further additions of heat from point 2 to point 3 supplies the latent heat of vaporization at constant temperature. At point 3 all of the liquid is converted to vapor and additional heat causes the temperature of the vapor to increase. Clearly the stability of the temperature during the phase change is an interesting phenomenon that might be employed in a heat management strategy if the problems of handling liquids, vapors and the combination of the two could be handled.

10.10.1 Evaporation, Boiling and Condensation

Evaporation and boiling both refer to changing phase of a fluid from a liquid to a vapor. Evaporation occurs at a liquid-gas interface such as air over water. As water is converted to water vapor, the latent heat of vaporization is drawn from the air thereby cooling it. The swamp coolers used in the dry parts of the world are examples of evaporation used for low cost air conditioning. On the other hand, boiling refers to changing the phase of a fluid from a liquid to a vapor at a solid liquid-interface. The interaction of the liquid with the solid interface is complex; consequently, the boiling process is an involved phenomenon comprised of four stages.

1. Natural convection occurs as heat is transferred across the solid–fluid interface increasing the liquid temperature at this interface.
2. Nucleate boiling occurs when small bubbles begin to form at nucleation sites on the solid surface. More bubbles form as the interface temperature increases; they increase in size, break free from the solid surface, rise and stir the liquid. Marked increases (a factor of 100 or more) in the heat flux that can be transfer occur in this stage of boiling.
3. Partial film boiling occurs when the bubbles present of the surface of the solid merge forming a vapor layer between the solid surface and the liquid over a portion of the interface surface. The heat flux that can be transmitted across this film decreases due to the presence of the vapor layer even with increasing temperatures.
4. Film boiling occurs when a layer of vapor covers the entire surface of the solid and heat is transferred from the solid, through the vapor layer and into the liquid. The heat flux transmitted during this stage increases, but significant temperature increases are necessary to achieve the higher heat flux.

A schematic graph, presented in Fig. 10.33, shows the variation in heat flux with temperature, defines the four stages of boiling and shows the trends during each of these four stages of boiling.

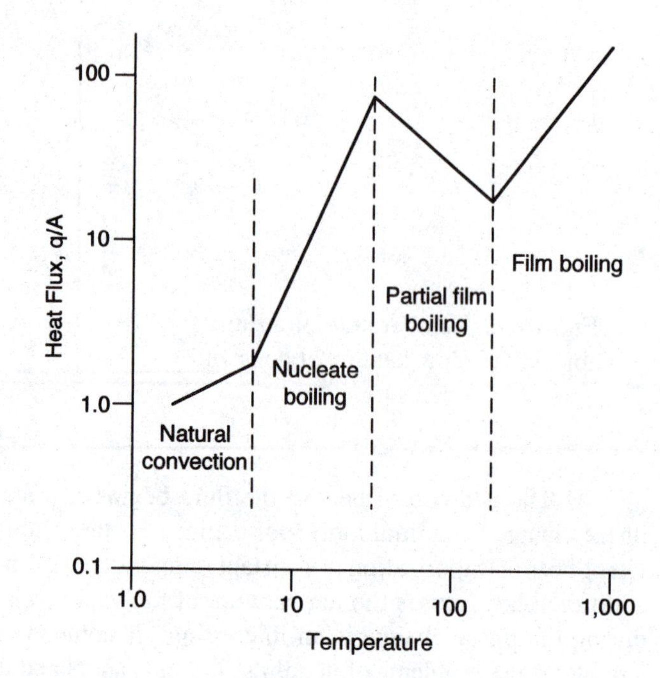

Fig. 10.33 Schematic graph showing changes in the heat flux with increasing temperature during the four stages of boiling.

10.10.2 Immersion Cooling

For the past six decades, the development of faster and denser microelectronic devices has resulted in increasing heat fluxes that must be dissipated with low temperature differences (gradients). Significant advances have been made in the application of air cooling techniques to handle these high heat fluxes. Although air cooling continues to be the most widely used method for heat management, it is recognized that much higher heat fluxes can be accommodated by using liquid cooling.

Liquid cooling for microelectronics may be categorized as either indirect or direct. With indirect liquid cooling, the fluid does not contact the microelectronic chips or the circuit board upon which the chips are mounted. Instead an effective conduction channel is provided from the heat sources to a liquid cooled cold-plate attached to the circuit board as described in Section 8.9. Because the liquid does not contact the electronics directly, water is usually used as the coolant.

Immersion cooling is often called direct liquid cooling because no physical barriers separate the microelectronic chips or the surface of the circuit board from the liquid coolant. This approach for cooling offers the opportunity to remove heat from the chips without additional thermal conduction resistance. Immersion cooling enables a high heat transfer coefficient reducing the temperature increase of the chip surface above the liquid coolant temperature. The value of the heat transfer coefficient is affected by both the coolant and the mode of convective heat transfer (natural convection, forced convection or boiling) as shown in Fig. 10.34. Water is the most effective coolant and the boiling mode offers the highest heat transfer coefficient. Direct liquid immersion cooling also more uniform chip temperatures than is possible with air cooling.

The liquid used for direct immersion cooling cannot be made on the basis of heat transfer characteristics alone, because the coolant must be chemically compatibility with the chips and other materials exposed to the liquid. Several coolants provide adequate cooling, but only a few are chemically compatible. For example, water is has very desirable heat transfer characteristics, but is unsuitable for direct immersion cooling because of it is a conductor. Fluorocarbon liquids (FC-72, FC-86, FC-77, etc.) are considered to be the most suitable liquids for direct immersion cooling. These coolants are clear, colorless per-fluorinated liquids with a relatively high density and low viscosity. They also exhibit a high dielectric strength and a high volume resistivity. The boiling points for the commercially available "Fluorinert" liquids manufactured by the 3M Company, range from 30 to 253

°C. These liquids should not be confused with the Freon coolants that are chlorofluorocarbons (CFCs). Some of the Freon, such as R-113, exhibit similar cooling characteristics; however, concern over their environmental effect on the ozone layer preclude their use.

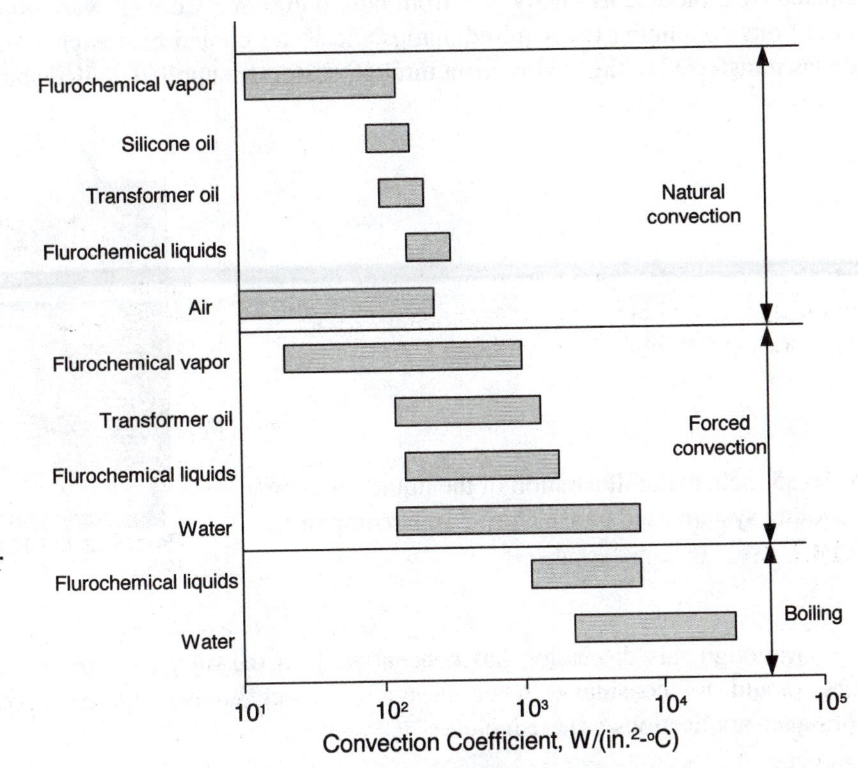

Fig. 10.34 Relative magnitude of heat transfer coefficients for different coolants and types of convection.

From Fig. 10.34, it is clear that boiling is the most effective method for heat transfer with significantly higher convection coefficients than those for either natural or forced convection. The boiling heat transfer rate q_B may be expressed as:

$$q_B = hA\,(T_w - T_{BP})^3 \qquad (10.138)$$

where h is a constant depending on the fluid-surface combination
A is the heat transfer surface area
T_w is the temperature of the heated surface, and T_{BP} is the boiling point of the liquid.

The typical critical heat fluxes q/A encountered in saturation temperature boiling of fluorocarbon liquids ranges from about 10 to 15 × 10^4 W/m^2, depending upon the characteristics of the solid surface where boiling nucleates. The allowable critical heat flux may be extended by sub cooling the liquid below its saturation temperature. Experiments results indicate that it is possible to increase the critical heat flux to 25 × 10^4 W/m^2 by lowering the liquid temperature to − 25 °C. Even higher critical heat fluxes are possible with flow boiling where the liquid impinges on the surface at some velocity. Heat fluxes from 25 to 30 × 10^4 W/m^2 have been measured for liquid flowing at 0.5 to 2.5 m/s over the solid surface.

While there has been considerable interest in direct immersion liquid cooling for many years, there have been only a few applications of the method. These applications involve the highest circuit densities and the highest generation of heat found on large mainframe and supercomputers.. For example, a large scale forced convection (not boiling) fluorocarbon cooling system was provided by the CRAY-2 supercomputer. A schematic drawing of the cooling arrangement, presented in Fig. 10.35, shows stacks of modules that were cooled by a forced flow of FC-77 in parallel across each module. Each module consisted of 8 printed circuit boards that served to carry arrays of single chip carriers. A total flow rate of 70 GPM was used to cool 14 stacks containing 24 modules each. The power dissipated by a module assembly was from 600 to 700 W. Coolant was supplied to the system by two parallel units containing the required pumps and water-cooled heat exchangers. The total system heat load was transferred to the environment through customer supplied chilled water.

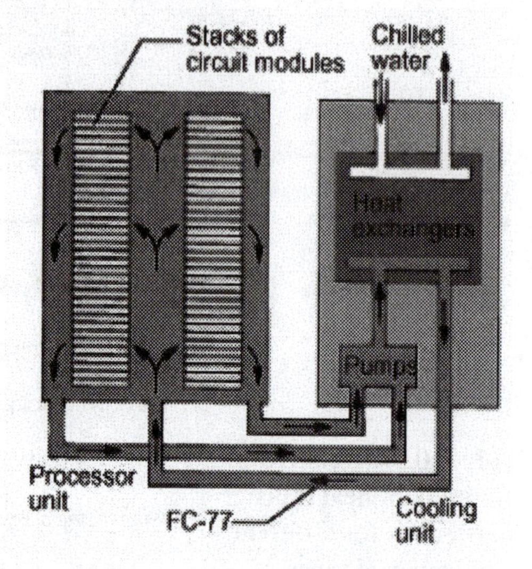

Fig. 10.35 Schematic illustration of the liquid immersion cooling system used on the Cray-2 supercomputer.

Although this discussion has concentrated on the merits of immersion cooling, several other factors should be considered when designing direct liquid immersion cooling system for high-performance applications. These include:

1. Fluorocarbon liquids are expensive; hence, they should only be used in closed systems.
2. Care must be taken to ensure that the seal materials are compatible with the liquid.
3. If boiling occurs, the design must incorporate a means to condense the resulting vapors.
4. Flow systems require a pump and care must be exercised to prevent cavitations due to the relatively high vapor pressure of low boiling point fluorocarbons.
5. In forced circulating liquid systems, it essential to add a particulate and a chemical filter to ensure the long-term purity of the coolant.

The advantages of direct immersion cooling are apparent. Thermal management systems can be designed accommodating higher heat flux. However, materials compatibility, added hardware with condensers, pumps and filters add to the complexity. Sealing liquids over a long term service life is always an issue. Considering these difficulties, it is understandable that direct immersion cooling has been employed on few applications involving very expensive, high-performance, high-end processors and supercomputers.

10.10.3 Heat Pipes

Heat pipes offer a relatively simple approach for using evaporation of a liquid to increase the effectiveness of a heat management system. While a heat pipe is classified as a closed system it does not require a pump or any moving hardware because the fluid is returned to the evaporation chamber by capillary action. Basically a heat pipe is a sealed tube containing an evaporation chamber on one end and a condensing chamber on its other end. The inner perimeter of the tube supports a wick that serves to return the liquid from the condenser chamber to the evaporation chamber by capillary action. The tube is of sufficient length to place the condenser end outside the electronic enclosure so the heat can be dissipated to the environment. A section view of a heat pipe, presented in Fig. 10.36, illustrates the design concepts for transporting vapor and fluid without moving hardware. Note the ends of the pipe serve as open chambers for evaporation and condensation.

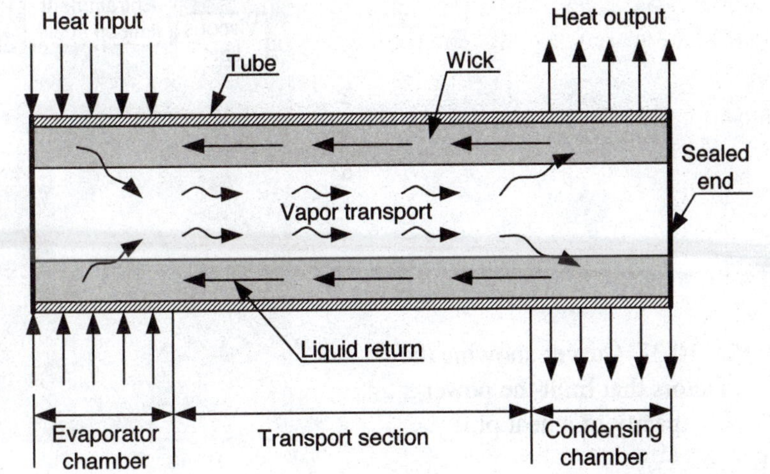

Fig. 10.36 Cross sectional view of a heat pipe showing the vapor and liquid transport mechanisms.

The fluids used in a heat pipe depend on the temperature range over which it is to operate. In electronic cooling applications where it is desirable to maintain junction temperatures below about 125 to 150°C, copper tube pipes with water are typically used. Copper tube pipes with methanol are employed if the heat pipe is to operate below 0°C. Because the containment tube is sealed, it is not necessary to replace the fluids over heat pipe's service life. When heat is applied at the evaporator end of the pipe, the liquid vaporizes the pressure in the pipe increases and the vapor is forced to the condenser end of the pipe. A heat exchanger is attached to the condenser end to maintain its temperature below the condensation temperature of the fluid. As the vapor condenses, it gives up its latent heat of vaporization, and the liquid is absorbed in the annual wick. Capillary action draws the fluid from the condenser end back to the evaporation chamber.

The wick is the critical component in the design because it must function as a capillary pump. Porosity and continuity of the internal passages through which the liquid flows are the parameters that control the effectiveness of the wick. Porous ceramics, powdered metals and woven stainless steel mesh are used in some units. In other designs, micro-channels are formed on the inside wall of the tube that are sufficiently narrow and deep to support the required capillary action.

Heat pipes function effectively as long as the fluid remains pure. However, if the tube, wick and ends were contaminated before or during assembly when the tube is sealed under a partial vacuum, impurities transfer to the fluid and the performance of the heat pipe is degraded.

Heat Pipe Design

There are only two major design considerations if heat pipes are fabricated from copper tubing and supplied with water—the power capacity of the heat pipe and its effective thermal resistance. Properly designed heat pipes can dissipate power levels ranging from a few watts or several kilowatts. A heat pipe can transfer much higher power for a specified ΔT than a solid copper cylinder of the same diameter; however, if it is driven beyond its capacity the performance is degraded significantly.

The maximum power capacity is governed by five factors that limit performance, which include viscous, sonic, capillary pumping, entrainment or flooding, and boiling. Curves corresponding to the five limiting factors, presented in Fig. 10.37, show that capillary action is the limiting factor for most heat pipe designs. The capacity of this specific heat pipe ranges from about 30 W at an operating temperature of 10 °C to about 100 W at 140 °C.

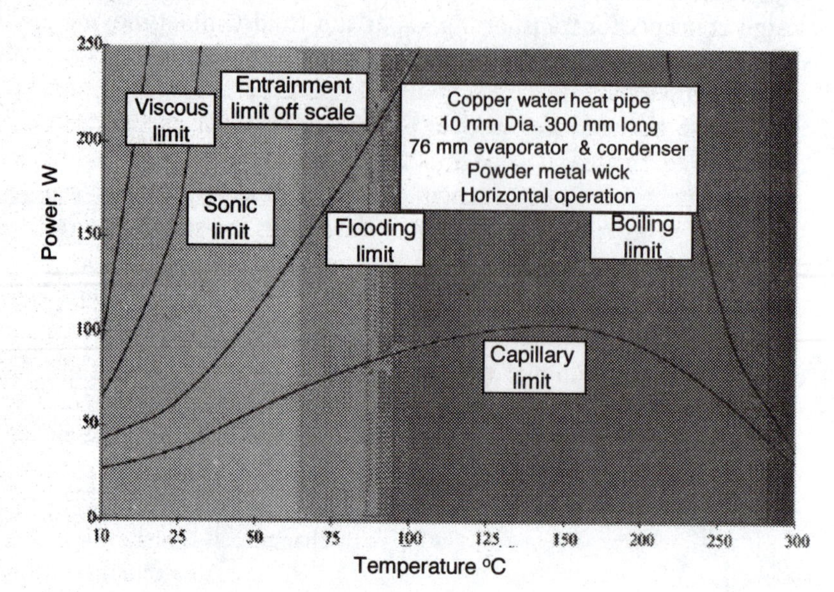

Fig. 10.37 Curves showing the factors that limit the power capacity of a heat pipe.

Capillary pumping capacity, usually the limiting factor in a heat pipe performance, is dependent on the orientation of the pipe during operation and the type of wick structure. The two most important properties of a wick are the pore radius and the permeability. The pore radius determines the pumping pressure the wick develops, and the permeability establishes the frictional losses of the fluid as it flows through the wick. There are several types of wick structures available including: grooves, screen, fibers, and sintered powder metal in order of decreasing permeability and decreasing pore radius. Grooved wicks have a large pore radius and a high permeability that enables low friction losses, but their pumping pressure is low. Grooved wicks can transfer high amounts of power in a horizontal or gravity aided position, but they do not perform well against gravity. The powder metal wicks, on the opposite end of the performance rankings, have small pore radii and relatively low permeability. Powder metal wicks are limited by pressure loses when oriented in the horizontal position, but they can transfer large loads against gravity.

The power capacity of a typical heat pipe is shown as a function of orientation angle for both screen and powder metal wicks in Fig. 10.38. The superior performance of the screen wick is notable for most orientation angles. Also note the loss of capacity as the orientation of the heat pipe changes from horizontal to vertical.

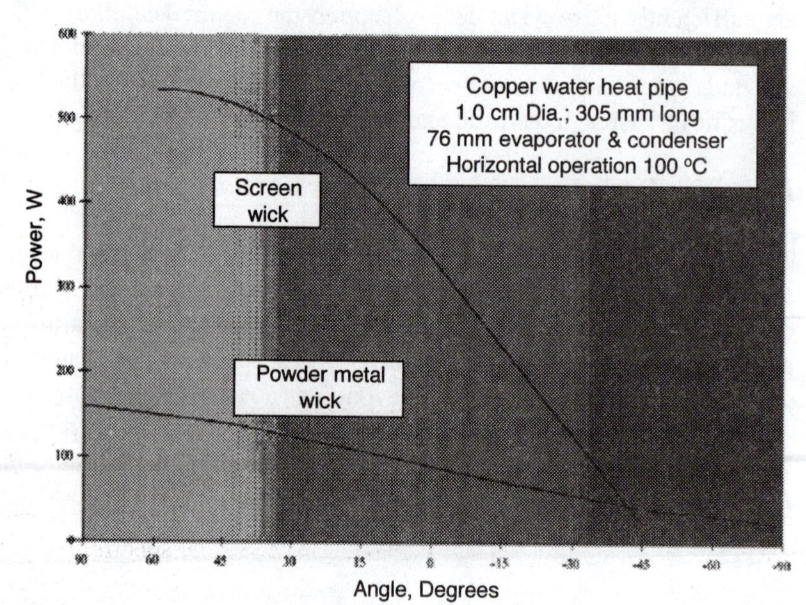

Fig. 10.38 Influence of orientation angle on power capacity of a specific heat pipe with screen and powder metal wicks.

Effective Thermal Resistance of Heat Pipes

The other important consideration in heat pipe design is its effective thermal resistance or the magnitude of the temperature difference required to transfer a specified amount of power. Because a heat pipe is a two-phase heat transfer device, a constant value of its thermal resistance cannot be determined. The effective thermal resistance is not constant but a function of a large number of variables, such as heat pipe geometry, evaporator length, condenser length, wick structure and working fluid.

The thermal resistance of a heat pipe is the sum of the resistances due to conduction through the wall, conduction through the wick, evaporation or boiling, axial vapor flow, condensation and conduction losses back through the condenser section wick and wall. The detailed thermal analysis of heat pipes is involved; however, there are a few simple rules that are useful for estimating the effective thermal resistance. For a copper - water heat pipe with a powder metal wick, the area dependent thermal resistance at the evaporator and condenser is estimated at 20 (°C-mm^2)/W and at 2.0 (°C-mm^2)/W for axial resistance. The evaporator and condenser area dependent resistances are based on the outer surface area of the heat pipe and the area dependent axial thermal resistance is based on the cross-sectional area of the vapor chamber.

To show an example of this approximation, consider a copper- water heat pipe with a 12.7 mm diameter, 305 mm long with a 1.0 mm diameter vapor chamber diameter. Let's assume the heat pipe is dissipating 75 watts with a 50 mm evaporator and a 50 mm condenser length. The evaporator and condenser heat flux (q/A) is given by:

$$\left(\frac{q}{A_{EV}}\right)_{EV} = \left(\frac{q}{A_{CON}}\right)_{CON} = \frac{q}{\pi DL} = \frac{75}{\pi(50)(12.7)} = 37.6\times10^{-3}\ \text{W/mm}^2 \tag{a}$$

The axial heat flux is given by:

$$\left(\frac{q}{A_V}\right)_V = \frac{4q}{\pi D_V^2} = \frac{4(75)}{\pi(10)^2} = 0.955\ \text{W/mm}^2 \tag{b}$$

From Eqs. (9.5) and (9.14), we write the expression for the temperature difference required to transfer this power as:

$$\Delta T = \left(\frac{q}{A_{EV}}\right)_{EV} R^*_{EV} + \left(\frac{q}{A_V}\right)_V R^*_V + \left(\frac{q}{A_{CON}}\right)_{CON} R^*_{CON} \tag{c}$$

where R^*_{EV}, R^*_V and R^*_{CON} are area dependent thermal resistances for the three sections of heat pipe.

Substituting numerical values into Eq. (c) yields:

$$\Delta T = (0.0376)(20) + (0.955)(2) + (0.0376)(20) = 3.41\ ^\circ\text{C} \tag{d}$$

These results indicate that the temperature differences required to transfer significant power are relatively small and that the heat pipes are a viable approach for designing an effective heat management system. However, the values cited for the area dependent thermal resistances are approximate and should only be used as estimates. Heat pipe manufacturers can provide more accurate values for the area dependent thermal resistances for specific designs.

Heat Pipe Electronic Cooling Applications:

One of the highest volume applications for heat pipes is cooling the Pentium processors in notebook computers. Weight, space and battery power are limited in notebook computers and all other portable electronic products. Air cooling by natural convection is usually not adequate and fans associated with forced air cooling consume too much power and make too much noise. Heat pipes offer a high efficiency, passive, light-weight and compact heat management system for cooling high power processors in portable electronic equipment. Heat pipes 3 to 4 mm in diameter are sufficient to transfer the high flux heat from a Pentium processor. The heat pipe spreads the heat load over a relatively large area heat sink, which is designed so that the heat flux is sufficiently low that it can be dissipated through the notebook case to ambient air. A photograph of a heat pipe design for notebook computers is presented in Fig. 10.39. Note that the shape of the condenser is a relatively large flat square to reduce the heat flux before dissipating the heat to the environment.

Fig. 10.39 Heat pipe designs for a notebook computers.

REFERENCES

1. D. C. Hamilton and W. R. Morgan, "Radiant Interchange Configuration Factors", NACA Technical Note 2836, 1952.
2. Frank M. White, Heat Transfer, Addison Wesley, Reading, MA, 1984.
3. A. K. Oppenheim, "Radiation Analysis by the Network Method," ASME Transactions, vol. 78, pp. 725-735, 1956.
4. Y. Jaluria, Natural Convection Heat and Mass Transfer, Pergamon Press, New York, 1980.
5. S. Ostrach, "An Analysis of Laminar Free Convection Flow and Heat Transfer about a Flat Plate Parallel to the Direction of the Generating Body Force," NACA Report 1111, 1953.
6. E. J. LeFevre, "Laminar Free Convection from a Vertical Plane Surface," Proc. 9th International Congress of Applied Mechanics vol. 4, p. 168, 1956.
7. S. W. Churchill and H. H. S. Chu, "Correlating Equations for Laminar and Turbulent Free Convection from a Vertical Plate," Journal Heat Transfer, vol. 18, pp. 1323-1329, 1975.
8. W. J. Minkowycz and E. M. Sparrow, "Local Non-similar Solutions for Natural Convection on a Vertical Cylinder," Journal Heat Transfer, Vol. 96, pp. 178-183, 1974.
9. S. W. Churchill and H. H. S. Chu, "Correlating Equations for Laminar and Turbulent Free Convection from a Horizontal Cylinder," International Journal Heat and Mass Transfer, vol. 18, pp. 1049-1053, 1975.

10. R. J. Goldstein, E. M. Sparrow and D. C. Jones, "Natural Convection Mass Transfer Adjacent to Horizontal Plates," International Journal Heat and Mass Transfer, vol. 16, pp. 1025-1035, 1973.
11. W. Elenbaas, "Heat Dissipation of Parallel Plates by Free Convection," Physica, vol. 9, no. 1, pp. 1-28, 1942.
12. E. Pohlhausen, "Der Warmeeaustausch Zwischen Fasten Korpern und Flussigkeiten mit Kliener Reibung und Kliener Warmeleitung," Z. Angew Mathematics and Mechanics, vol. 1, pp. 115-121, 1921.
13. R. K. Shah and A. L. London, Laminar Flow Forced Convection in Ducts, Academic Press, New York, 1979.
14. A. D. Krauss and A. Bar-Cohen, Thermal Analysis and Control of Electronic Equipment, Hemisphere Publishing Corp. Washington, DC, 1983.
15. A. Bejan and A. D. Kraus, Heat Transfer Handbook, John Wiley & Sons, New York, NY, 2003.
16. G. R. Blackwell, The Electronic Packaging Handbook, CRC press, Boca Raton, FL, 1999.
17. K. J. Kim, Air Cooling Technology for Electronic Equipment, CRC Press, Boca Raton, FL, 1996.
18. R. Remsburg, Thermal Design of Electronic Equipment, CRC Press, Boca Raton, FL, 2000.
19. J. E. Sergent and A. Krum, Thermal Management Handbook, McGraw Hill, New York, NY, 1998.
20. P Singh et al, "A Power Packaging and Cooling Overview of the IBM eServer z900," IBM Journal of Research and Development, Vol. 46, No. 6, pp. 711-738, 2002.
21. G. P. Peterson, An Introduction to Heat Pipes Modeling, Testing, and Applications, John Wiley and Sons, Inc., 1994.
22. D. S. Steinberg, Cooling Techniques for Electronic Equipment, 2nd Edition, John Wiley 1ne Sons, New York, NY, 1991.

EXERCISES

10.1 A flat plate 100 by 150 mm in size serves as a radiation heat exchanger for an electronic assembly. The heat dissipated is q^b = 12W, if the plate is considered to be a black body. The heat is radiated to another flat, parallel, and black plate of the same size positioned 160 mm away.

(a) Determine the shape factor $\boldsymbol{F}_{1-2}$.

(b) If the temperature of the receiving plate is maintained at −100 °C, find the temperature of the transmitting plate.

10.2 Consider the heat exchanger described in Exercise 10.1. If a temperature limit T_1 = 80 °C is imposed on the heat exchanger, find the maximum q^b that can be dissipated.

10.3 A flat plate 100 by 200 mm radiates heat to another parallel flat plate of equal size positioned 400 mm away. Both plates are considered as black bodies. Construct a family of curves showing q^b as a function of T_1 for $0 < T_1 < 150$ °C if:

(a) $T_2 = -100$ °C

(b) $T_2 = -50$ °C

(c) $T_2 = 0$ °C

(d) $T_2 = 50$ °C

10.4 A long rectangular duct is designed with 100 mm wide and 50 mm high sides. The top side is maintained at 100 °C and the other three sides are held at 0 °C. Treat the duct as a black body and determine the heat transferred from the top surface per meter of duct to:

(a) One vertical side.

(b) The bottom.

(c) Both sides and the bottom.

10.5 A pair of circular disks 150 mm in diameter are concentric and parallel. If one disk is maintained at 50 °C and the other at –50 °C, determine their separation distance if q^b is: (a) 1.0W; (b) 5.0W; (c) 10 W; (d) 20 W.

10.6 A 1 × 2 in. angle section that is 10 in. long is used as a radiation heat exchanger. The 2 in. side, which is horizontal, is the heat sink that is maintained at 20 °C. Determine the temperature of the vertical side if the black body heat transferred is: (a) 0.5 W; (b) 1.0 W; (c) 2.0 W; (d) 5.0 W.

10.7 Repeat Exercise 10.1 if the two plates have gray surfaces with an emissivity $\varepsilon = 0.4$.

10.8 Repeat Exercise 10.2 using gray surfaces with emissivity $\varepsilon = 0.5$.

10.9 Repeat Exercise 10.3 but consider that the plates are fabricated from polished copper.

10.10 Repeat Exercise 10.4 with the duct constructed from commercial sheet aluminum.

10.11 Repeat Exercise 10.5 with emissivity $\varepsilon_1 = 0.8$ for one disk and $\varepsilon_2 = 0.4$ for the other.

10.12 Repeat Exercise 10.6 with $\varepsilon = 0.9$ for the vertical side and $\varepsilon = 0.3$ for the horizontal side.

10.13 Determine the radiation heat transfer coefficient for the heat exchanger described in Exercise 10.1 if the gray surfaces both have an emissivity of 0.4. Take the heat transfer as q = 5.0 W and the temperature of the receiving plate as T = 10 °C.

10.14 Determine the radiation heat transfer coefficient for the heat exchanger described in Exercise 10.3 if the emissivity of the plates is 0.5 and 0.3. Consider temperatures T_1 and T_2 in °C as listed in the Table below:

Exercise	T_1 °C	T_2 °C
10.14a	50	20
10.14b	100	20
10.14c	150	20
10.14d	200	20

10.15 Using the results from Exercise 10.13 determine the thermal resistance R_r due to the radiation heat exchanger. Compare this value of R_r with the thermal resistance offered by a copper bar 1 × 1 × 25 mm conducting heat with the same temperature differences.

10.16 A vertical plate 100 mm wide and 150 mm high is maintained at a temperature of 80 °C. The temperature of the ambient air on one side of the plate is 20 °C. Consider the other side of the plate insulated. Determine:

(a) The convection coefficient as a function of the position y and show a graph of h(y).
(b) The average convection coefficient for the plate.
(c) The heat transferred from the plate to the ambient air.

10.17 A vertical plate 25 mm high and 200 mm wide is held at a constant temperature of 90 °C. Both sides of the plate are exposed to ambient air at 15 °C. Determine:

(a) The convection coefficient as a function of the position y and show a graph of h(y).
(b) The average convection coefficient for the plate.
(c) The heat transferred from the plate to the ambient air.

10.18 Repeat Exercise 10.17 with the plate turned through 90 degrees so that the 200 mm edge is vertical. Describe why there is a difference in the results.

10.19 Find the Grashof number as a function of position y for the plate described in Exercise 10.16. Determine the average and the maximum value of Gr. Is the flow over the plate laminar or turbulent?

10.20 Consider a plate of height H with $T_W = 100$ °C and $T_0 = 20$ °C. Find the position y_t where the flow undergoes a transition from laminar to turbulent flow assuming $y_t < H$.

10.21 Determine the heat flux q/A for a single sided vertical plate heat exchanger with its height H equivalent to the position y_t where the laminar flow undergoes a transition to turbulent flow.

10.22 A vertical plate 100 mm wide and 150 mm high transfers heat from one side. The plate has a constant heat flux q/A with a wall temperature which varies with position y. Determine the value of $(T_W - T_0)$ as a function of the heat flux q/A. Truncate your set of solutions when the flow becomes turbulent near the top of the plate.

10.23 Compare the results of Exercise 10.22 and Exercise 10.1 and show the differences between constant temperature and constant flux heat transfer for free convection.

10.24 Consider a vertical plate heat exchanger which is sufficiently high to have both laminar and turbulent flow regions. For the case where $(T_W - T_0) = 100$ °C:

(a) Construct a graph showing Nu(y).
(b) Determine the average value of Nu in the laminar flow region.
(c) Determine the average value of Nu in the turbulent flow region.
(d) Establish the height of the plate so that the portions of the plate subject to laminar and turbulent flow are equal and then find the relative amount of heat transfer from each part.

10.25 Repeat Exercise 10.24 but use instead Eq. 10.44 to determine the average value of Nu_H.

10.26 A cylindrical electronic component 20 mm in diameter and 60 mm long is mounted with its axis in the vertical direction. If the component dissipates 5 W to the ambient air that is at a temperature of 25 °C, find the surface temperature of the component.

10.27 Repeat Exercise 10.27 but consider the cylindrical component to be oriented in the horizontal direction.

10.28 A flat plate which serves as the top panel for an electronic enclosure also acts as a heat exchanger. The panel is 0.3 m by 0.5 m in size. If its surface temperature is limited to 60 °C, find the capacity of the heat exchanger. Why would we limit the temperature of the top panel of an enclosure to the value specified here?

10.29 Repeat Exercise 10.28 but consider the flat plate to be the bottom panel of an electronic enclosure. The surface temperate is limited to 75 °C when the plate is below the enclosure.

10.30 A fin type heat exchanger, illustrated in Fig 10.11a, is constructed with 20 fins having W = 5 mm, H = 40 mm and d = 15 mm. Determine the average value of Nu for the heat exchanger and the heat capacity as a function of $(T_W - T_0)$.

10.31 Are the fins used in the heat exchanger of Exercise 10.30 optimized? If not, determine the spacing W necessary to optimize the configuration. With this new spacing, find the new capacity of the 20 fin exchanger. What is its new length? If we increased the number of fins holding the original length the same, determine the new capacity. Use $(T_W - T_0) = 60$ °C in this analysis.

10.32 Consider a flat plate 0.6 m long placed in an air stream with a temperature $T_0 = 20$ °C. Prepare a graph showing the Reynolds number as a function of position x along the plate. The graph should include separate curves for free stream velocities of 1, 2, 5, 10 and 20 m/s. Identify on each curve the position of the transition from laminar to turbulent flow.

10.33 For a plate with single sided heat transfer in an air stream with a velocity U = 5 m/s, find the average value of h. The temperatures are $T_W = 90$ °C and $T_0 = 25$ °C. The plate is 150 mm long and 100 mm wide. Find also the heat transferred from the plate to the air stream.

10.34 Repeat Exercise 10.33 with U = 1 m/s and U = 10 m/s. Comment on the effect of air stream velocity on the capacity of the plate as a heat exchanger.

10.35 Repeat Exercise 10.33 but consider the plate to have lengths of 50, 100 and 200 mm. Prepare a graph showing the heat flux q/A as a function of the length of the plates.

10.36 For the plate in Exercise 10.33 find the maximum velocity U before turbulent flow occurs. Is there anything wrong with permitting turbulent flow to occur in a flat plate heat exchanger for use in cooling electronic components?

10.37 A flat plate, one sided heat exchanger 150 mm wide by 300 mm long is in an air stream with a velocity of 25 m/s. If the air stream is at T_0 = 300 °K and the plate is at T_W = 400 °K determine:
  (a) The location of the transition region.
  (b) The average Nu over the laminar region.
  (c) The average Nu over the turbulent region.
  (d) The average Nu over the entire plate.
  (e) The heat transferred from the plate to the air stream.

10.38 Repeat Exercise 10.37 but consider the plate to dissipate a uniform heat flux q/A = 2×10^{-3} W/mm^2. In addition to finding solutions to parts (a) to (e), determine the temperature of the plate T_W as a function of the position x.

10.39 For a rectangular duct with dimensions of 50 by 100 mm, determine the distance that is required for fully developed flow in the duct. The temperature of the air is 20 °C and its velocity is 10 m/s.

10.40 For the duct described in Exercise 10.39 determine Q the volume flow rate, $\dot{m}$ the mass flow rate and G the mass velocity. Express the results in both US customary and SI units.

10.41 Consider a duct with a 20 by 90 mm rectangular cross section. The two 90 mm walls are maintained at a temperature T_W = 75 °C, and the two 20 mm walls are insulated. Determine the temperature of the air as it moves from the inlet to the exit of the duct. The velocity of the air stream U = 10 m/s and the inlet temperature is 20 °C. The duct is 0.6 m long. Construct a graph showing ΔT as a function of position x.

10.42 Repeat Exercise 10.41 but consider a constant heat flux q/A = 1,600 W/m^2 on the two 90 mm duct walls rather than a constant temperature.

10.43 Determine $(T_W - T_{ave})_{ave}$ from Eqs (10.104) and (10.105) with T^{in} = 20 °C and T_W = 75 °C. Let T^{ex} vary in 10 degree increments from 30 to 70 °C and show the differences between the results from these two equations as a function of T^{ex}.

10.44 For the duct described in Exercise 10.41, find the heat transferred to the air stream. Also determine the friction factor *f*.

10.45 For the duct described in Exercise 10.42, find the heat transferred to the air stream.

10.46 Repeat Exercise 10.41 with U = 4 m/s. Explain the difference in the results when the velocity of the air steam is lowered.

10.47 Repeat Exercise 10.41 with U = 25 m/s. Explain the difference in the results when the velocity of the air stream is increased.

10.48 Find the head in inches of air corresponding to a pressure of 1 psi. Find the head in inches of H_2O for a pressure of 1 psi.

10.49 Find the head in inches of H_2O corresponding to an air stream velocity of 10 m/s. Determine the head if the velocity is change to 100 ft/s.

10.50 For the duct system illustrated in Fig. E10.50, show the equivalent resistance net work for the electrical analogy pertaining to head loss.

10.51 Determine the value of the resistances R_i for the duct system illustrated in Fig. E10.50. Note that adjustable baffles on the exhausts permit control of the volume flow rate so that Q_1 and Q_2 are 30 and 20 ft^3/min respectively. Assume an average temperature of 50 °C of the air moving through the ducts.

10.52 Determine the head loss for the duct system in Fig. E10.50 for the volume flow rates given in Exercise 10.51.

10.53 Prepare a graph showing the impedance of the duct system in Fig. E10.50 for total volume flow rates which range from 10 to 200 ft^3/min. Maintain the ratio Q_1/Q_2 = 1.5.

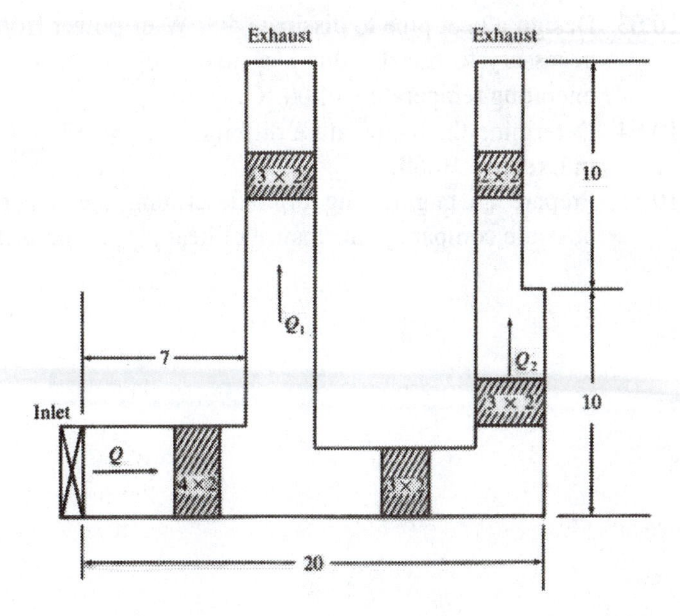

Fig. E10.50

10.54 Show the resistance network for the electrical analogy to determine head loss in the duct shown in Fig. E10.54.

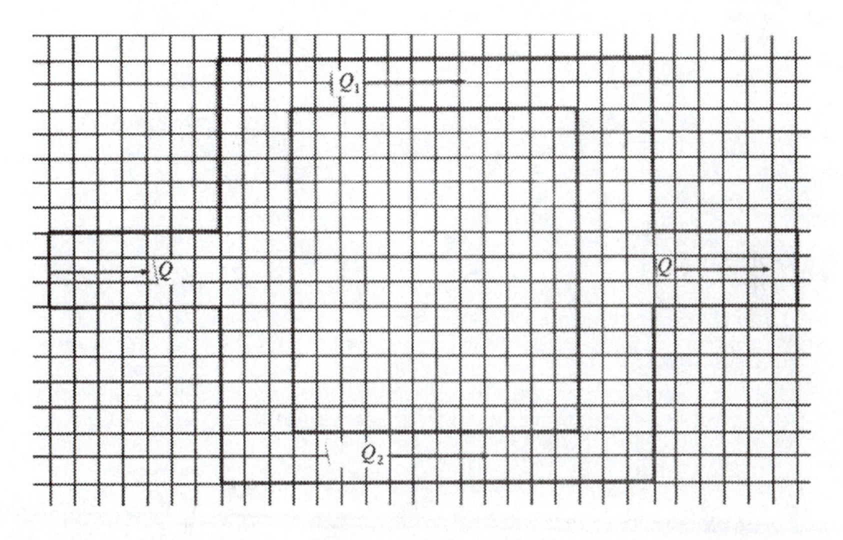

Fig. E10.54

10.55 Determine the resistances R_i for the network established in Exercise 10.54. Combine these resistances to obtain the equivalent resistance for the entire system.

10.56 Prepare a graph of the system impedance as the flow rates vary from 5 to 50 ft^3/min showing data points at increments of 5 ft^3/min.

10.57 Write an engineering brief explaining why heat transfer using phase transitions of fluids from liquid to vapor (gas) are more effective than heat transfer by conduction or convection.

10.58 Describe the four stages of boiling.

10.59 Describe the difference between fluorinert compounds and Freon. Why is this distinction important?

10.60 Discuss the engineering problems encountered when implementing a heat management system that uses direct liquid immersion.

10.61 Sketch a cross sectional view of a heat pipe intended for cooling electronic equipment.

10.62 What are the factors that control the capacity of a heat pipe.

10.63 Design a heat pipe to dissipate 400 W of power from a module that is 50 × 75 mm in size. It is necessary to transfer this heat load a distance of 400 mm to a suitable heat sink. Assume an operating temperature of 60 °C.

10.64 Determine the temperature difference ΔT required to transfer the heat in the heat pipe described in Exercise 10.63.

10.65 Prepare an engineering brief describing the materials used for wicks in heat pipes. When possible compare performance of heat pipes with different wicking materials.

CHAPTER 11

STRESS AND FAILURE ANALYSIS OF MECHANICAL COMPONENTS

11.1 INTRODUCTION

In the design analysis, we compute the forces acting on a structural member, the amount it deformed under load, and the stresses developed. Next, we compare the stresses acting on this member with its strength and determine its safety factor. The magnitude of the safety factor is evaluated to determine if this member is safe to be employed in a product. This chapter describes methods that will enable you to size mechanical components used in electronic equipment providing enclosures, support structures, fasteners and printed circuit card with adequate safety margins.

11.2 STRESS, STRAIN AND DEFORMATION IN AXIAL MEMBERS

The usual approach when performing a design analysis of a structure is disassemble it and consider its members individually. The structural members fall into several different categories including axial rods, beams, torsion bars, frames, plates and shells. These structural members are modeled with free body diagrams and the forces acting them are determined using equilibrium relations or a combination of equilibrium relations together with deformation equations. After the forces acting on each member are determined the stresses, strains and deformations are computed using relatively simple closed form equations that are based on equilibrium and on the assumption that plane sections in a body remain plane after it deforms.

In this section, we will address uniaxial structural members that are long and thin. The members are called bars if the direction of the forces coincide with its axis, beams if the forces are transverse to its axis and a torsion rod if the forces produce a torque that acts about the center line of the rod. We will introduce closed form equations enabling you to determine the stress, strain and deformation for these three structural elements in this section.

In Section 11.3, we will extend this approach to two-dimensional structural elements and present similar relations that enable you to determine the stress, strain and deformation in plates and shells. Analytical methods for addressing more complex two- and three-dimensional structures are covered in Chapter 12 where numerical methods are introduced.

11.2.1 Bars in Tension and Compression

When we apply an axial tension force on a long thin member it stretches by a small amount δ. We define δ, the **stretch** or **deformation** of the structural member, as:

$$\delta = L_f - L_o \tag{11.1}$$

where L_f is the length of the member under load and L_o is the original length.

The deformation δ of the long thin member is determined from:

$$\delta = \frac{PL}{AE} \tag{11.2}$$

where P is the axial force, L is the length of the member A is its cross sectional area and E is the **modulus of elasticity** of the material from which the axial member is fabricated.

Consider the simple tension test where a uniaxial member is subjected to a monotonically increasing tensile load. If we measure the applied load P and the deformation δ of the member as it is stretched in a universal tensile machine, we can construct the graph shown in Fig. 11.1a.

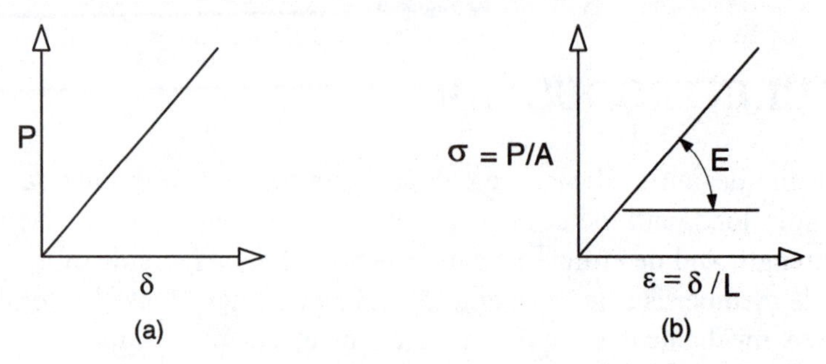

Fig. 11.1 (a) Graph of load versus deformation δ; (b) Stress versus strain.

Note a linear relationship exists between P and δ as indicated by Eq. (11.2). This P-δ curve is a straight line until the member begins to fail by **yielding**. The linear portion of the P-δ relation is the **elastic region** of the load-deformation response of the tension bar. Equation (11.2) is valid only in this elastic region.

If the axes in Fig. 11.1 are modified by dividing the force P by the area A and the deformation δ by the length L, as indicated below, it is possible to determine the stress and strain to which the tension bar is subjected.

$$\varepsilon = \delta/L \tag{11.3}$$

$$\sigma = P/A \tag{11.4}$$

where ε is the strain and σ is the stress in the member, respectively.

These relations show that the **normal strain,** ε, is the change in length divided by the original length; strain is a dimensionless quantity. The normal stress σ is the internal force P divided by the area A, over which it acts. The stress is expressed in terms of Pa (Pascal) in SI units or psi (lb/in^2) in U. S. Customary units.

We have shown a graph of stress versus strain in Fig. 11.1b. As expected there is a linear relation between stress and strain until the stress is sufficient to cause the tension bar to yield. The **slope** of the σ - ε line is the **modulus of elasticity** E of the material from which the bar is fabricated. The **modulus of elasticity** is a **material property** that is independent of the shape of the body.

It is evident from the linear response in the stress-strain diagram that:

$$\sigma = E\varepsilon \tag{11.5}$$

This stress-strain relation is known as **Hooke's law**. A word of caution—Hooke's law is valid only for uniaxial states of stress that arise in long thin structural members. We will introduce another form of the stress-strain relation to accommodate two- and three-dimensional stress fields later in this chapter.

Next combine Eqs. (11.3), (11.4) and (11.5) in the manner shown below:

$$\sigma = \frac{P}{A} = E\varepsilon = \frac{E\delta}{L} \qquad \text{(a)}$$

and solve this expression for δ to derive Eq. (11.2):

$$\delta = \frac{PL}{AE} \qquad (11.2)$$

Strength and Failure of Tension Bars

Suppose we take a tension bar, grip it in a universal-testing machine and pull until it fails. The behavior of the bar under increasing load depends on the material from which the bar was fabricated. Nearly all materials exhibit a **linear elastic** response like that shown in Fig. 11.1 at lower stress levels; however, at higher stresses significantly different behavior is observed for different materials. We have classified these behaviors as:

- **Brittle** with abrupt failure and small, elastic deformations.
- Yielding with **strain hardening** and plastic deformation prior to rupture.
- Yielding with **strain softening** and plastic deformation prior to rupture.

These three behaviors are illustrated graphically in Figs. 11.2 and 11.3.

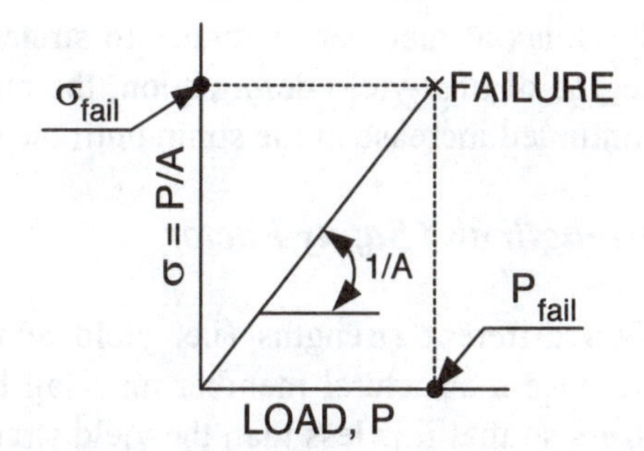

Fig. 11.2 Linear-elastic response until brittle failure.

The stress-load diagram in Fig. 11.2 indicates the linear elastic response of a brittle material. The stress increases linearly with both axial load and strain. The slope of the stress-load line is the reciprocal of the area (1/A), and the slope of the stress-strain line is the modulus of elasticity E as indicated in Fig. 11.1. When the load is increased to a critical value P_{fail}, the uniaxial member breaks. The failure occurs at a specific value of stress, σ_{fail}. We define this failure stress as the **ultimate tensile strength**, S_u of the tension bar's material.

$$S_u = \sigma_{fail} = \frac{P_{fail}}{A} \qquad (11.6)$$

The two stress-strain diagrams in Fig. 11.3 illustrate material behavior when yielding occurs. The diagram in Fig. 11.3a is typical of a material that yields and then strain hardens. When the axial load on the tension bar increases, the stress increases linearly until the material begins to yield at σ_{yield}. At that point, the uniaxial member continues to stretch, but the load and stress remain essentially constant. After some degree of post-yield stretch, the material stiffens and the load and stress begin to increase. The stress increases until the wire ruptures at σ_{fail}.

Two different strengths are determined for the bar's material using data from this stress-strain diagram. The **yield strength** S_y given by:

$$S_y = \sigma_{yield} = \frac{P_{yield}}{A} \tag{11.7}$$

and the **ultimate tensile strength,** S_u given by Eq. (11.8).

$$S_u = \sigma_{fail} = \frac{P_{fail}}{A} \tag{11.8}$$

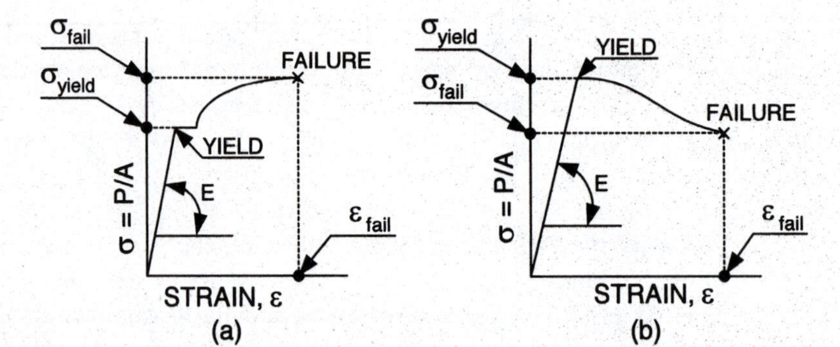

Fig. 11.3 Material yielding behavior with (a) strain hardening; (b) strain softening.

The stress –strain diagram in Fig. 11.3b is typical of a material that yields and then strain softens. As the load increases, the stress increases linearly until the material begins to yield at σ_{yield}. At that point, the uniaxial member continues to stretch, but the load and stress remain nearly constant. After some degree of post-yield deformation, the material softens and the load and stress begin to decrease with a continued increase in the strain until the tension bar ruptures at σ_{fail}.

Strength and Safety Factor

Two different strengths (i.e. yield and ultimate tensile) have been defined for the bar's material. Because a structural member may fail by excessive deformation, we may choose to limit the applied stress so that it is less than the yield strength S_y. On the other hand, we can tolerate plastic deformation in one or more members of in some structures without compromising the structure's function. In these cases, we can tolerate stresses exceeding the yield strength, S_y but they must be less than the ultimate strength, S_u.

In designing a structural member to carry a specified load, the member is sized so that the **design stress,** σ_{design}, is less than the strength based on either the yield or another failure criteria. It would not be prudent to permit the design stress to equal the strength of the member. To size structural members so they are safe, a **factor of safety, SF** is employed in the analysis. The safety factor is defined as the ratio of a strength divided by the design stress. This definition leads to the two relations given below:

$$\mathbf{SF_u} = S_u / \sigma_{design} \tag{11.9}$$

$$\mathbf{SF_y} = S_y / \sigma_{design} \tag{11.10}$$

where S_u and S_y are the ultimate strength and yield strength of the material, respectively.
σ_{design} is the design stress given by P_{design}/A.
P_{design} is the load specified for the structural member.

The value for the factor of safety used in designing a structure depends upon its application. Usually, the factor of safety varies from two to four depending on the uncertainties in the anticipated loading of the structure. However, excessively large safety factors must be balanced by practical considerations such as costs, weight, aesthetics, functionality and ease of assembly. The factor of safety specified should reflect these concerns.

11.2.2 Beams in Bending

A beam is similar to a rod because both have the same geometry—long and slender. However, the beam is loaded differently. A rod is subjected to tensile or compressive forces applied in the axial direction. A beam is subjected to transverse forces applied in a perpendicular direction to the longitudinal axis of the long, thin member. The difference in the loading is depicted in Fig. 11.4.

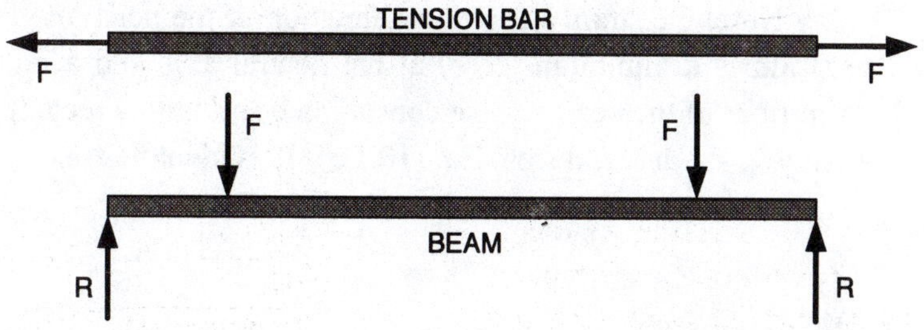

Fig. 11.4 Both the rod and beam are fabricated from long, thin members, but the direction of the applied loads is different.

Next consider the beam to be subjected to forces that produce **pure** bending. A beam is in pure bending if it is subjected to a constant bending moment over its length. A straight beam subject to a constant moment deforms into a circular arc with a radius of curvature ρ. This discussion of stresses, strains and deformations in beams is limited to structural members with cross sections that are symmetric with respect to the vertical or y axis.

Equilibrium concepts are employed to show that equal and opposite internal forces P act on the surface exposed by the section cut across the beam. The forces produce an internal couple, $M_i = (P)(d)$ that counteracts the externally applied bending moment M. The internal forces P are due to normal stresses σ_x, which act over the surface of the exposed cross sectional area as shown in Fig. 11.5.

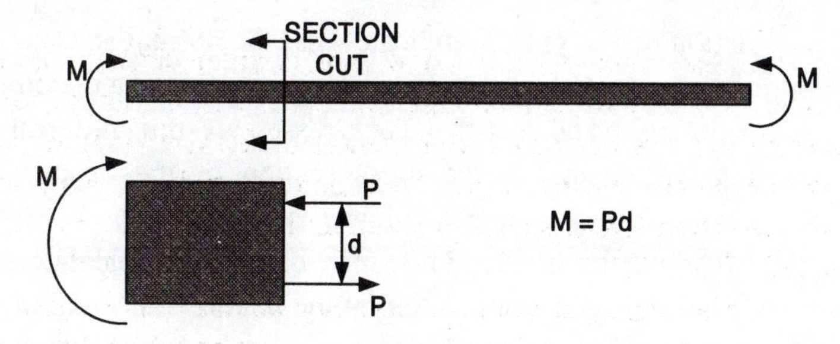

Fig. 11.5 The left-hand portion of the beam showing a couple (P)(d) acting on the area exposed by a section cut.

Clearly, the upper most force in Fig. 11.5 is compressive since it presses against the surface exposed by the section cut. Similarly, the lower force is tensile. The distributions of normal stresses presented in Fig. 11.6 produce equal and opposite internal forces P separated by some distance d.

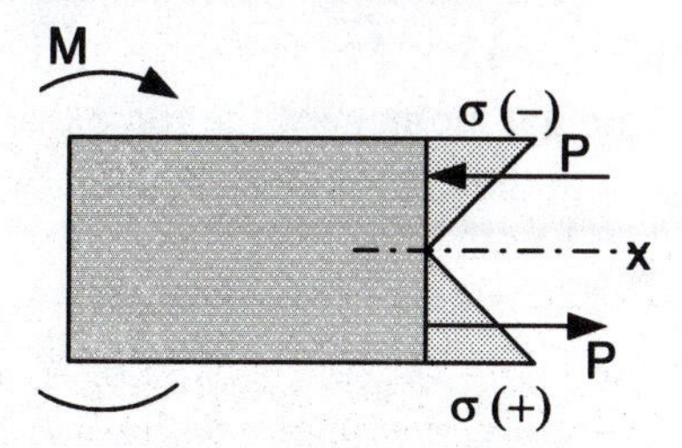

Fig. 11.6 Distributions of stress σ_x acting on an internal surface producing the couple $M = P\ d$.

Stresses and Strains in Beams

When a beam is subjected to a pure bending moment, it bends into a circular arc. Parallel vertical lines scribed on the sides of the beam remain straight after the beam undergoes the bending deformation.

From the geometry of the bending deformations, it is easy to show that the strain ε produced on the beam is given by:

$$\varepsilon = -y/\rho \qquad (11.11)$$

where y is the distance from the neutral axis to the point of strain.

Note, the strain ε is a linear function of the position y, which is measured from the neutral axis. The strain is a minimum (zero) at the neutral axis and a maximum when y locates either the top or bottom fiber of the beam. If we consider a beam with a rectangular cross section and define its height as h, then $y_{max} = -h/2$ and from Eq. (10.1), it is evident that:

$$\varepsilon_{max} = \frac{h}{2\rho} \qquad (11.12)$$

The sign of the maximum strain is positive (tensile) for fibers along the bottom surface of the beam if the curvature is positive.

It is easy to convert the strains in the beam to stresses by using Hooke's Law—Eq. (11.5). The longitudinal fibers in the beam are subjected to a uniaxial state of stress with fibers on the convex side of the neutral axis in tension and those on the concave side in compression. By combining Eqs. (11.11) and (11.5), we obtain:

$$\sigma = E\varepsilon = -yE/\rho \qquad (11.13)$$

Examination of Eq. (11.13) indicates that the normal stress σ is linearly distributed with respect to y; a minimum of zero at the neutral axis where y = 0, and a maximum at the outer fibers in the beam where y is a maximum. The fact that the stresses are not uniformly distributed has significant implications regarding the design of the cross section of the beam. We will discuss this important design consideration in much more detail later in this section.

The results in Eq. (11.3) may be used to calculate the stress on a beam if the material from which it is fabricated, the location of the neutral axis and the radius of curvature are known. However, in most cases, the applied forces are known—not the radius of curvature. The neutral axis passes through the centroid of the area of the transverse section of the beam. For common cross section shapes such as circular, rectangular, I, H, wide-flange, etc., the location of the centroid is at their intersection of the two axes of symmetry.

The bending stresses σ are distributed linearly so as to produce an internal couple equal to the applied moment M, as shown in Fig. 11.6. We will use this fact in the following derivation. The incremental moment of the force due to the stress σ acting over an element dA, shown in Fig. 11.7, with respect to the neutral axis is given by:

$$dM = -\sigma\, dA(y) = \left(\frac{Ey}{\rho}\right)(dA)(y) \qquad (a)$$

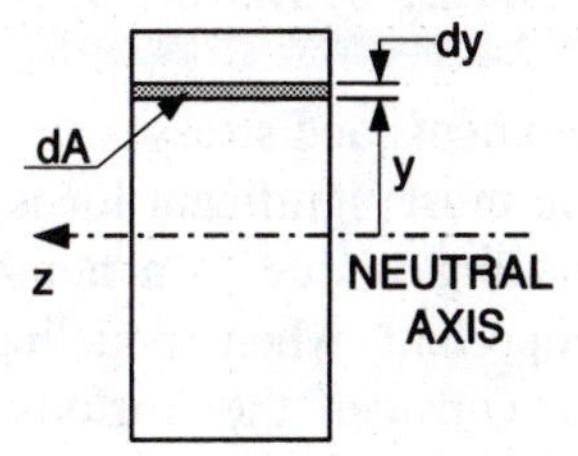

Fig. 11.7 Cross sectional area of a rectangular beam with elemental area dA.

By integrating Eq. (a) and recalling that the second moment[1] of the area A relative to the neutral axis is given by $I_z = \int y^2\, dA$, we can derive the relation between the radius of curvature δ and the moment M as:

$$M_i = (EI_z/\rho) = M \tag{11.14}$$

$$\kappa = 1/\rho = M/(EI_z) \tag{11.15}$$

where $\kappa = 1/\rho$ is the curvature of the beam and M is the moment.

From Eq. (11.15), it is evident that the curvature κ is directly proportional to the moment M and inversely proportional with the flexural rigidity (EI_z) of the beam. Finally, by substituting Eq. (11.14) into Eq. (11.13) the relationship between the applied moment and the internal stress distribution can be obtained:

$$\sigma = -\frac{My}{I_z} \tag{11.16}$$

In Eq. (11.16), M is positive when the beam is deflected so that the convex side is on the bottom and y is positive in the upward direction. This relation is considered to be one of the most important of the many equations employed by engineers when designing structures that contain beams.

Stresses in Beams with Rectangular and Circular Cross Sections

Consider a beam with a rectangular cross section with a width b and a height h. From Eq. (11.16) it is evident that the maximum bending stresses acting on the beam can be determined from:

$$\sigma_{max} = \frac{6M}{bh^2} \tag{11.17}$$

For a beam with a circular cross section with a diameter d, the maximum bending stresses acting on the beam can be determined from:

$$\sigma_{max} = \frac{32M}{\pi d^3} \tag{11.18}$$

If the stress distribution shown in Fig. 11.6 is examined, it is apparent that the beam with a rectangular cross section is not an efficient structural element. Only the top and the bottom fibers of the beam are stressed to the maximum. The interior region of the beam is under stressed and near the neutral axis the stresses approach zero. The beam with a rectangular cross section is extremely inefficient because the average magnitude of stress over the section is only $\sigma_{max}/2$. To improve the effectiveness of beams, the under-stressed, central-region of the cross section is removed to produce I and flanged beams.

---

[1] The second moment of the area is often called the moment of inertia.

Bending of Beams with Transverse Forces

Moments and stresses are induced in beams when they are subjected to transverse forces. In structures, the most significant loads are usually due to gravitational forces although in machine components the loading is due to a number of different effects and gravitational forces are often negligible. It is important, when modeling a beam, to consider the various types of loading and supports for the beam. Typical examples of different loads and supports for the beams are illustrated in Fig. 11.8.

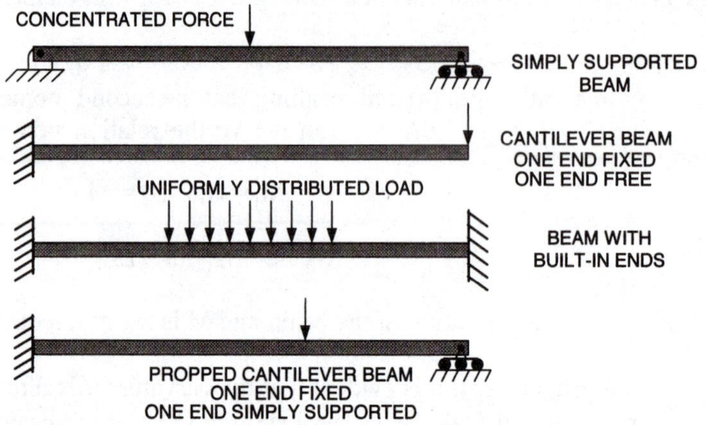

CONCENTRATED FORCE
SIMPLY SUPPORTED BEAM
CANTILEVER BEAM
ONE END FIXED
ONE END FREE
UNIFORMLY DISTRIBUTED LOAD
BEAM WITH BUILT-IN ENDS
PROPPED CANTILEVER BEAM
ONE END FIXED
ONE END SIMPLY SUPPORTED

Fig. 11.8 Various types of transverse loading and supports for beams.

The transverse forces acting on a beam produce internal shear forces and bending moments that vary as a function of position over the length of the beam. To ascertain the distribution of the shear forces and bending moments over the length of the beam, equilibrium relations are used to construct shear and bending diagrams, as illustrated in Fig. 11.9.

a
F
A
L
V = Fc/L
V
SHEAR DIAGRAM
x
V = - Fa/L
M_{MAX} = Fac/L
M
MOMENT DIAGRAM
x

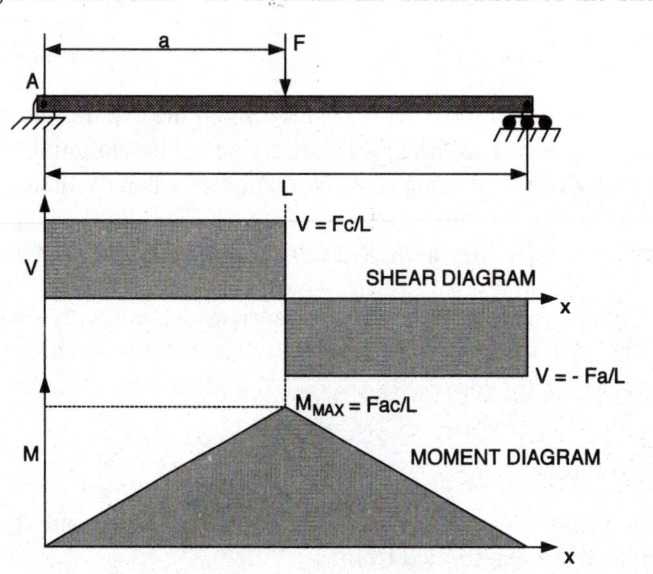

Fig. 11.9 Shear and bending moment diagrams show the distribution of the shear force V and the bending moment M as a function of position x.

A shear stress τ also develops in the beam due to the shear force, V. For long thin beams, the shear stress τ is small when compared to the normal stress σ and is usually ignored without compromising the safety of the structure.

Beam Deflections

When a beam deflects under the action of a bending moment, the neutral axis becomes curved. Locally it is a circular arc with a radius of curvature ρ that depends on the bending moment M(x). As the local curvature varies with the bending moment, the neutral axis generates an elastic curve over the length of the beam. By writing an equation for the elastic curve (the deformed shape of the neutral axis), the deflection of the beam becomes be apparent.

Let's rewrite Eq. (11.14) as:

$$\frac{1}{\rho} = \frac{M}{EI_z} \quad 1/\rho = (M/EI_z) \tag{11.19}$$

A well-known relation from calculus gives the curvature κ of any curve as a function of the derivatives of y as:

$$\kappa = \frac{1}{\rho} = \frac{\dfrac{d^2y}{dx^2}}{\left[1+\left(\dfrac{dy}{dx}\right)^2\right]^{3/2}} \tag{11.20}$$

Well-designed beams are stiff and consequently the deflections are very small compared to the length of the beam. Because of this fact, the slope of the elastic curve (dy/dx) is small when compared to 1 and $(dy/dx)^2$ is extremely small compared to 1. Hence, Eq. (11.20) can be approximated with:

$$\kappa = \frac{1}{\rho} = \frac{d^2y}{dx^2} \tag{11.21}$$

Substituting Eq. (11.19) into (11.21) yields:

$$\frac{d^2y}{dx^2} = \frac{M(x)}{EI_z} \tag{11.22}$$

Next note that shear force $V(x) = \dfrac{dM(x)}{dx}$, and substitute this relation into Eq. (11.22), to obtain:

$$\frac{d^3y}{dx^3} = \frac{V(x)}{EI_z} \tag{11.23}$$

Similarly recall that the load $q(x) = -\dfrac{dV(x)}{dx}$ and substitute this relation into Eq. (11.23) to obtain:

$$\frac{d^4y}{dx^4} = \frac{-q(x)}{EI_z} \tag{11.24}$$

Clearly the elastic curve y(x) can be determined by integrating any one of the three equations given above. Equation (11.22) is usually employed because only two integrations are necessary to obtain the results for the elastic curve y(x) for the beam.

Equations for Deflection of Simply Supported Beams and for Cantilever Beams

The integration method for determining beam deflections is tedious and it can often be avoided for beams with common supports and common loading. Solutions for the slope and deflection of the elastic curve have been developed and are presented in Tables 11.1 for simply supported beams and in Table 11.2 for cantilever beams.

11.2.3 Torsion Bars or Shafts

Circular shafts are commonly employed to transmit power from a motor to some appliance. For example, two shafts transmit power from the gearbox to the drive wheels on an automobile. The relation between the power P transmitted and the torque T applied to the shaft as:

$$P = T\omega \tag{11.25}$$

where ω is the angular velocity of the shaft, P is the power and T is the torque.

We are particularly interested in the torsion load (torque) produced when a shaft transmits power. The torque produces stresses and angular deformations (twisting) in the shaft. If either of these two quantities becomes excessive, the shaft will fail in service.

As a motor rotates it acts on a shaft producing a torque T. This action is illustrated in Fig. 11.10 where two circular disks are connected by a circular shaft. A couple with forces F_1 act on the larger diameter disk producing a torque $T_1 = F_1\, d_1$. Another couple with forces F_2 act on the smaller diameter disk producing a torque $T_2 = F_2\, d_2$. If the shaft is either stationary or rotating at a constant angular velocity, it is in equilibrium and the torques T_1 and T_2 are equal in magnitude and opposite in sign. The shaft transmits the torque from one disk to the other.

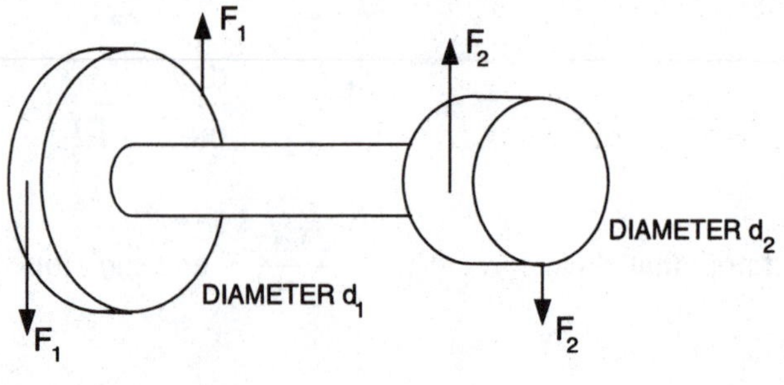

Fig. 11.10 A circular shaft transmitting torque $T_1 = F_1\, d_1 = T_2 = F_2\, d_2$.

Deformation of a Circular Shaft Due to Torsion

Consider a circular shaft of length L with one end built-in (fixed) and the other end free. A twisting moment (torque) is applied to its free end as illustrated in Fig. 11.11. A straight line A-B is scribed on the outside surface of the shaft prior to the application of the torque T. Upon the application of the torque, the shaft deforms, the scribe line rotates through an angle γ and point B moves to a new location given by point C. The length of the arc B-C is given by:

$$BC = \gamma\, L \tag{a}$$

The angle ϕ is known as the angle of twist for the shaft. Let's examine the deformation of the circular shaft more closely by considering a square scribed on its surface at some arbitrary position along line A-B as shown on Fig. 11.12.

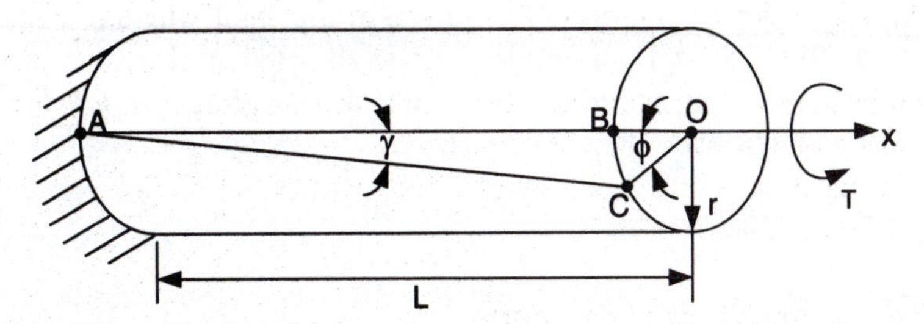

Fig. 11.11 Deformation of a circular shaft due to the application of a torque T at its free end.

When the shaft deforms under the action of the torque, the straight line A-B rotates, remains straight, and assumes the orientation of line A-C. The two sides of the inscribed square that are perpendicular to the axis of the shaft do not rotate. However, both of the sides parallel to the axis of the shaft rotate through the angle γϕ. Clearly, the small square is distorted as indicated in Fig. 11.12. The included angles of the square have changed 90° ± γ. Therefore, it is clear that a shearing strain γ occurs on the external surface of the circular shaft.

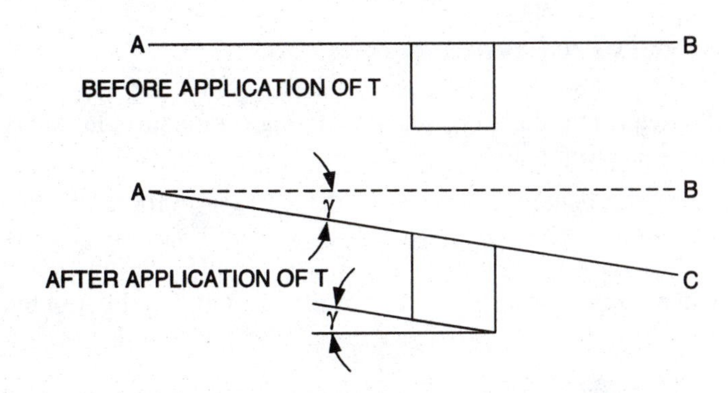

Fig. 11.12 Deformation of a small square scribed on the surface of the shaft before and after application of the torque.

To determine a relation for the magnitude of the shearing strain γ, consider the free end of a circular shaft subjected to an applied torque T as indicated in Fig. 11.13. The line OB rotates through an angle ϕ sweeping through an arc BC. The length of the arc B-C is given by:

$$BC = r\phi \qquad \text{(b)}$$

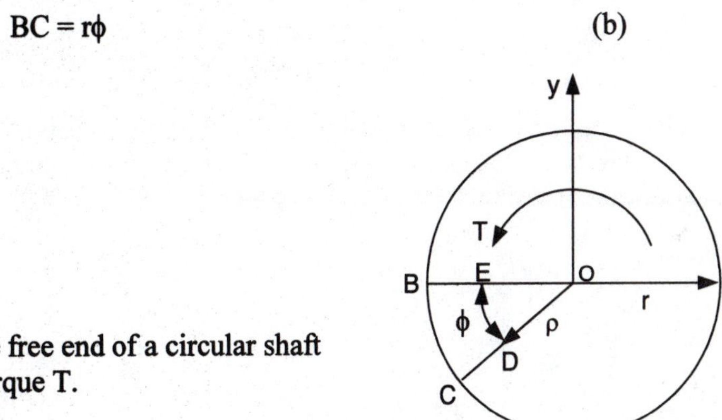

Fig. 11.13 Rotation of line OB on the free end of a circular shaft subjected to a torque T.

Substituting Eq. (a) into Eq. (b) yields:

$$\gamma_{\rho=r} = \frac{r\phi}{L} \qquad (11.26)$$

The shear strain γ on the surface of the shaft where ρ = r is given by Eq. (11.26). The shear strain at some point D on the interior of the shaft is less because the arc length D-E is decreased. Following the same procedure, we write the shearing strain γ for an arbitrary point defined by the position radius ρ as:

$$\gamma_\rho = \frac{\rho\phi}{L} \tag{11.27}$$

Examination of Eq. (11.27) shows that the shearing strain varies from zero at the centerline of the shaft to a maximum value at the surface that is given by:

$$\gamma_{max} = \frac{r\phi}{L}\gamma \tag{11.28}$$

Because the applied torque is constant along the shaft's length, the strain is also constant along its length.

Stresses Produced by Torsion

To determine the shear stresses acting on the circular shaft, the relation between shear stress and shear strain is written as.

$$\tau = G\gamma \tag{11.29}$$

where G is the shear modulus of the shaft material given by:

$$G = \frac{E}{2(1+\nu)} \tag{11.30}$$

Substituting Eq. (11.27) into Eq. (11.29) yields:

$$\tau_\rho = \frac{G\phi\rho}{L} \tag{11.31}$$

The shear stress τ increases linearly with the position radius ρ from zero at the centerline to a maximum value at the surface of the shaft. This linear distribution of shear stress is illustrated in Fig. 11.14a.

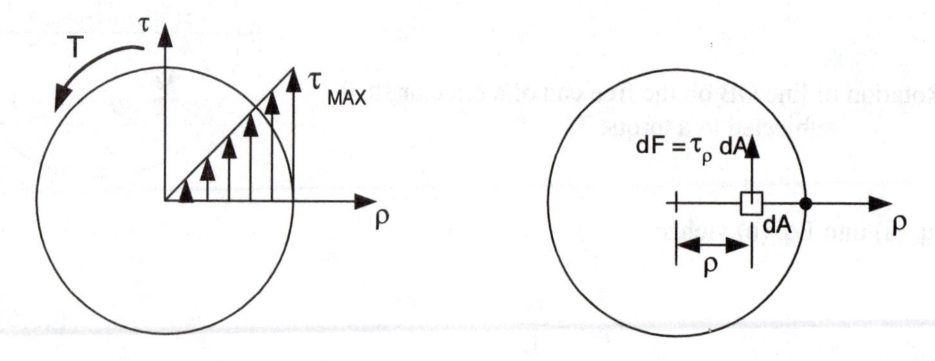

Fig. 11.14a Linear distribution of shear stress τ.

Fig. 11.14b The incremental force dF = τdA.

Table 11.1
Equation for the slope and deflection of simply supported beams

SIMPLY SUPPORTED BEAMS LENGTH, L	SLOPE	DEFLECTION EQUATIONS
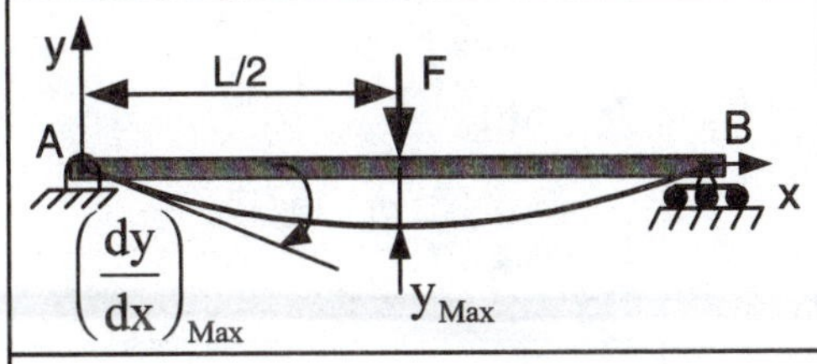 y, x, A, B, L/2, F, $\left(\frac{dy}{dx}\right)_{Max}$, y_{Max}	$\left(\frac{dy}{dx}\right)_{Max} = \frac{-FL^2}{16EI_z}$	$y = \frac{-Fx}{48EI_z}(3L^2 - 4x^2)$ $0 \le x \le L/2$ $y_{Max} = \frac{-FL^3}{48EI_z}$
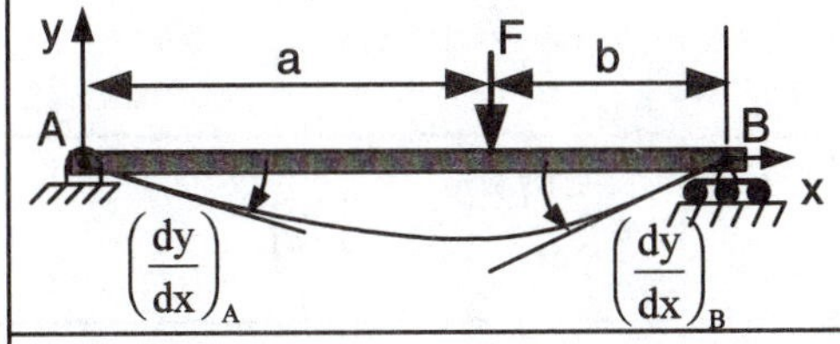 y, x, A, B, a, F, b, $\left(\frac{dy}{dx}\right)_A$, $\left(\frac{dy}{dx}\right)_B$	$\left(\frac{dy}{dx}\right)_A = \frac{-Fab(L+b)}{6EI_zL}$ $\left(\frac{dy}{dx}\right)_B = \frac{Fab(L+a)}{6EI_zL}$	$y = \frac{-Fbx}{6EI_zL}(L^2 - b^2 - x^2)$ $0 \le x \le a$ $y_{x=a} = \frac{-Fba}{6EI_zL}(L^2 - b^2 - a^2)$
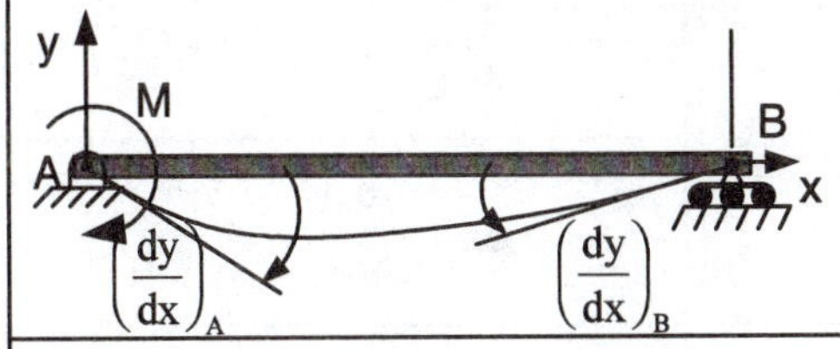 y, x, M, A, B, $\left(\frac{dy}{dx}\right)_A$, $\left(\frac{dy}{dx}\right)_B$	$\left(\frac{dy}{dx}\right)_A = -\frac{ML}{3EI_z}$ $\left(\frac{dy}{dx}\right)_B = \frac{ML}{6EI_z}$	$y = \frac{-Mx}{6EI_zL}(x^2 - 3Lx + 2L^2)$ $y_{Max} = \frac{-ML^2}{\sqrt{243}EI_z}$
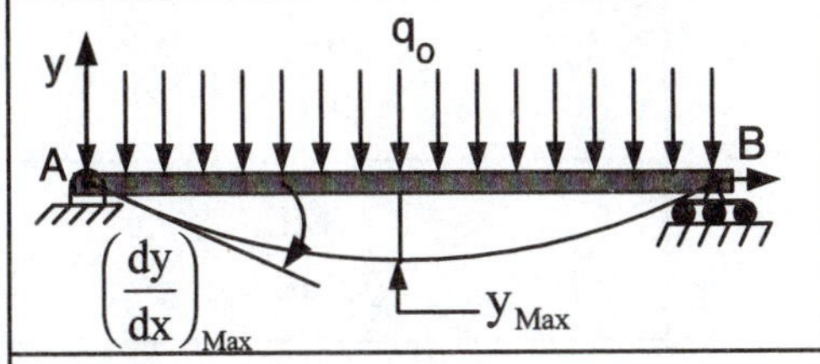 y, q_o, A, B, $\left(\frac{dy}{dx}\right)_{Max}$, y_{Max}	$\left(\frac{dy}{dx}\right)_{Max} = -\frac{q_0L^3}{24EI_z}$	$y = \frac{-q_0x}{24EI_z}(x^3 - 2Lx^2 + L^3)$ $y_{Max} = \frac{-5q_0L^4}{384EI_z}$
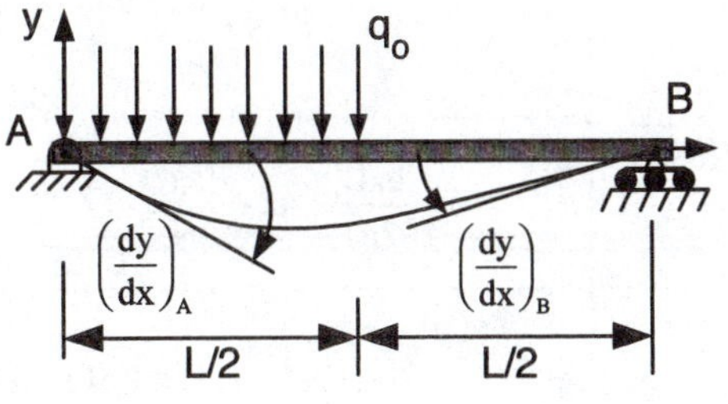 y, q_o, A, B, $\left(\frac{dy}{dx}\right)_A$, $\left(\frac{dy}{dx}\right)_B$, L/2, L/2	$\left(\frac{dy}{dx}\right)_A = -\frac{3q_0L^3}{128EI_z}$ $\left(\frac{dy}{dx}\right)_B = \frac{7q_0L^3}{384EI_z}$	$y = \frac{-q_0x}{384EI_z}(16x^3 - 24Lx^2 + 9L^3)$ $0 \le x \le L/2$ $y = \frac{-q_0L}{384EI_z}(8x^3 - 24Lx^2 + 17L^2x - L^3)$ $L/2 \le x \le L$ $y_{Max} = \frac{-6.563 \times 10^{-3} q_0L^4}{EI_z}$ at $x = 0.4598L$
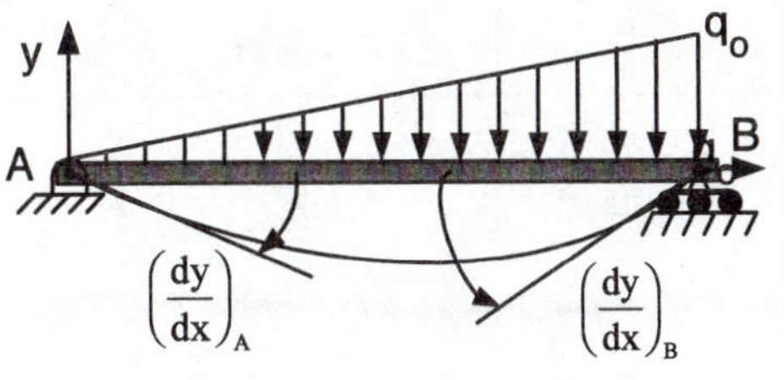 y, q_o, A, B, $\left(\frac{dy}{dx}\right)_A$, $\left(\frac{dy}{dx}\right)_B$	$\left(\frac{dy}{dx}\right)_A = \frac{-7q_0L^3}{360EI_z}$ $\left(\frac{dy}{dx}\right)_B = \frac{q_0L^3}{45EI_z}$	$y = \frac{-q_0x}{360EI_zL}(3x^4 - 10L^2x^2 + 7L^4)$ $y_{Max} = \frac{-6.52 \times 10^{-3} q_0L^4}{EI_z}$ at $x = 0.5193$

Table 11.2
Equations for the slope and deflection of cantilever beams

CANTILEVER BEAMS LENGTH, L	SLOPE	DEFLECTION EQUATIONS
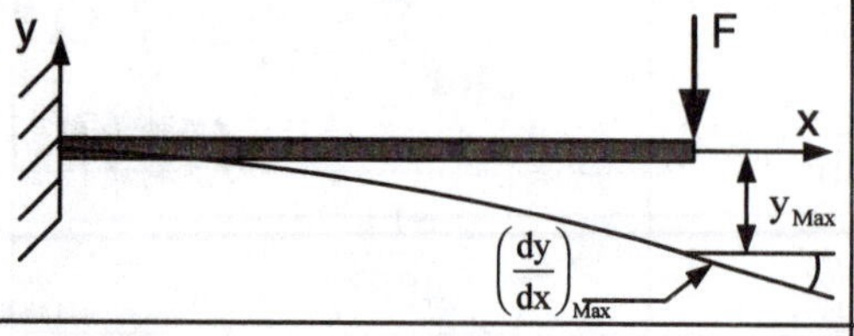	$\left(\frac{dy}{dx}\right)_{Max}=\frac{-FL^2}{2EI_z}$	$y=\frac{-Fx^2}{6EI_z}(3L-x)$ $y_{Max}=\frac{-FL^3}{3EI_z}$
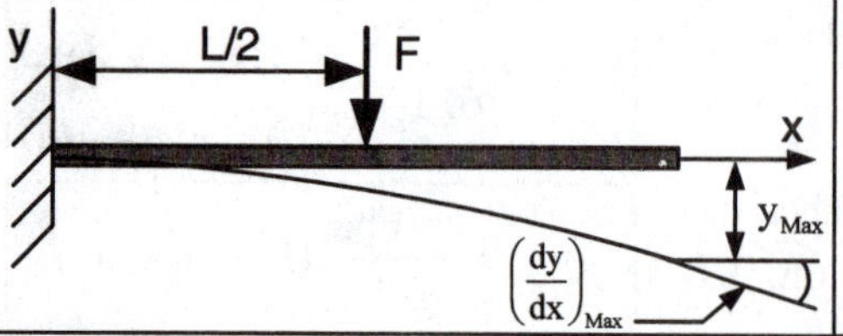	$\left(\frac{dy}{dx}\right)_{Max}=\frac{-FL^2}{8EI_z}$	$y=\frac{-Fx^2}{6EI_z}\left(\frac{3}{2}L-x\right)$ $0\le x\le L/2$ $y=\frac{-FL^2}{24EI_z}\left(3x-\frac{1}{2}L\right)$ $L/2\le x\le L$
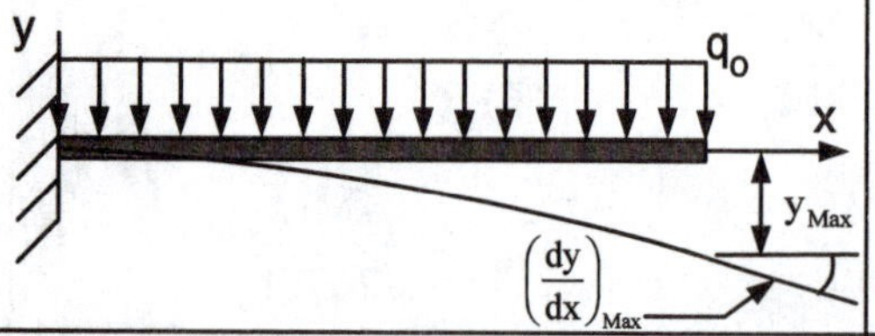	$\left(\frac{dy}{dx}\right)_{Max}=\frac{-q_0L^3}{6EI_z}$	$y=\frac{-q_0x^2}{24EI_z}\left(x^2-4Lx+6L^2\right)$ $y_{Max}=\frac{-q_0L^4}{8EI_z}$
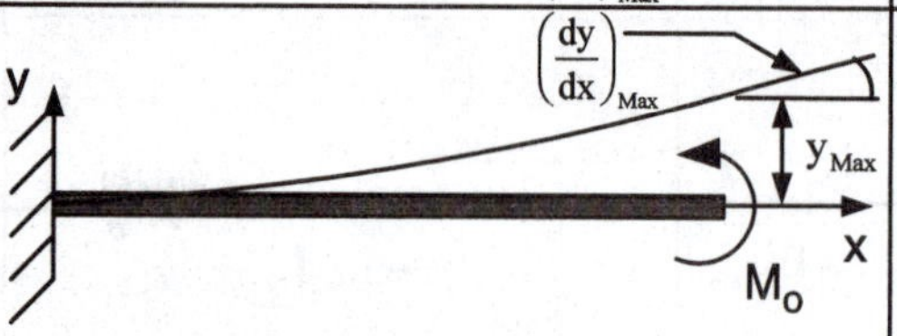	$\left(\frac{dy}{dx}\right)_{Max}=\frac{ML}{EI_z}$	$y=\frac{Mx^2}{2EI_z}$ $y_{Max}=\frac{ML^2}{2EI_z}$
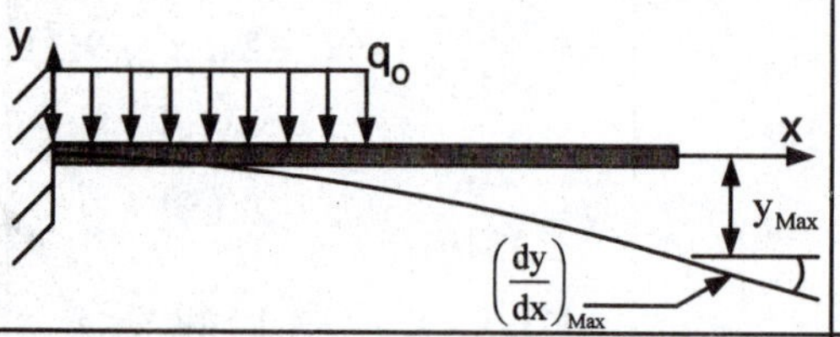	$\left(\frac{dy}{dx}\right)_{Max}=\frac{-q_0L^3}{48EI_z}$	$y=\frac{-q_0x^2}{24EI_z}\left(x^2-2Lx+\frac{3}{2}L^2\right)$ $0\le x\le L/2$ $y=\frac{-q_0L^3}{192EI_z}\left(4x-\frac{L}{2}\right)$ $L/2\le x\le L$
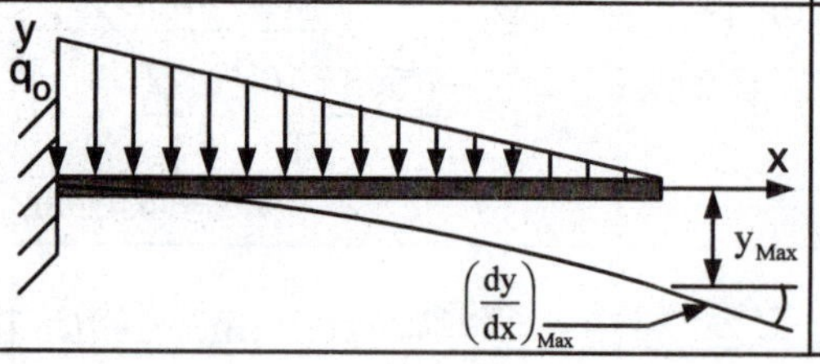	$\left(\frac{dy}{dx}\right)_{Max}=\frac{-q_0L^3}{24EI_z}$	$y=\frac{-q_0x^2}{120EI_zL}\left(10L^3-10L^2x+5Lx^2-x^3\right)$ $y_{Max}=\frac{-q_0L^4}{30EI_z}$

The shear stress produces an equal and opposite moment to the applied torque at the far end of the shaft. From the equilibrium equations for the moments it is possible to show that:

$$T = (G\phi \int \rho^2 \, dA)/L \qquad (11.32)$$

The term $\int \rho^2 \, dA$ is the polar moment of inertia J of a circular area about its center. Hence, we can rewrite Eq. (11.32) as:

$$T = G\phi J/L \qquad (11.33)$$

Finally, we note from Eq. (11.31) that $(G\phi/L) = \tau_\rho/\rho$. Accordingly, we can write:

$$\tau_\rho = T\rho/J \qquad (11.34)$$

Equation (11.34) is employed to determine the shear stresses in circular shafts when the applied torque is known. Clearly, the maximum shear stress occurs when $\rho = r$. In this instance:

$$\tau_{Max} = Tr/J = Td/(2J) \qquad (11.35)$$

Because $J = \pi r^4/2 = \pi d^4/32$ for solid circular shafts, Eq. (11.35) is written as:

$$\tau_{Max} = 2\,T/(\pi r^3) = 16\,T/(\pi d^3) \qquad (11.36)$$

In many engineering applications weight is an important consideration because it often detrimentally affects performance and usually increases cost. It is possible to reduce the weight of torsion members by using tubes instead of solid rounds. The weight penalty, associated with solid round bars, is due to the reduced stresses in the bar's central region that are much smaller than the maximum shear stress occurring on its outer diameter. With a tube, the central region of the shaft is removed and the resulting member is more uniformly stressed and structurally more efficient.

Equation (11.34) derived for a solid circular shaft is also valid for a hollow circular shaft. The only difference that arises is in the expression for the polar moment of inertia J. For a solid shaft $J = \pi d^4/32$; however, for a hollow shaft the polar moment of inertia is given by:

$$J = (\pi/32)(d_o^4 - d_i^4) \qquad (11.37)$$

where d_i and d_o represent the inside and outside diameters, respectively.

Angle of Twist (Deformation)

Finally, substituting Eq. (11.35) into Eq. (b) gives:

$$\phi = (TL)/(GJ) \qquad (11.38)$$

Noting that $J = \pi d^4/32$, we may rewrite Eq. (11.35a) as:

$$\phi = (32\,TL)/(\pi d^4\,G) \qquad (11.39)$$

where G is the shear modulus defined in Eq. (11.9a).

If any of the parameters (T, G or J) are not constant along the length of the shaft, then Eq. (11.38) must be modified to account for the variation.

$$\phi_{Total} = \sum_{i=1}^{n} \frac{T_i L_i}{G_i J_i} \tag{11.40}$$

In this instance, the shaft must be divided into several sections and Eq. (11.38) used to determine ϕ for each section. The total angle of twist, ϕ_{Total} is then determined by summing ϕ for each section.

11.3 STRESS CONCENTRATIONS

In our discussion of stresses in tension bars subjected to axial loading, we assumed that the stresses were uniformly distributed over the cross sectional area of the bar. For a tension bar with a uniform cross section along its entire length, this is a valid assumption except near the ends of the bar where the external forces are applied. However, if a discontinuity such as a hole is drilled into the bar, stresses concentrate at the discontinuity. As a consequence, the using Eq. (11.4) to compute the stresses seriously underestimates their value. We must account for the effect of the structural discontinuities by determining a suitable stress concentration factor.

Stress concentrations occur at discontinuities in tension bars, beams in bending and torsion members. We will briefly discuss stress concentration factors in tension members. The reader is referred to an excellent book by R. E. Peterson for additional information on stress concentrations arising in tension bars and other structural elements [1].

1.3.1 Stress Concentration Factors Due to Circular Holes in Tension Bars

Consider the bar with a centrally located circular hole subjected to an axial tension force as shown in Fig. 11.15. The stress distribution in a section removed three or more diameters from the hole is uniform with a magnitude given by Eq. (11.4) as $\sigma_o = P/A = P/(bw)$. However, on the section through the center of the hole the stress distribution shows significant variation. The stresses increase sharply adjacent to the discontinuity (the hole) and concentrate at this location. The maximum value of the normal stress occurs adjacent to the hole as indicated in Fig 11.16

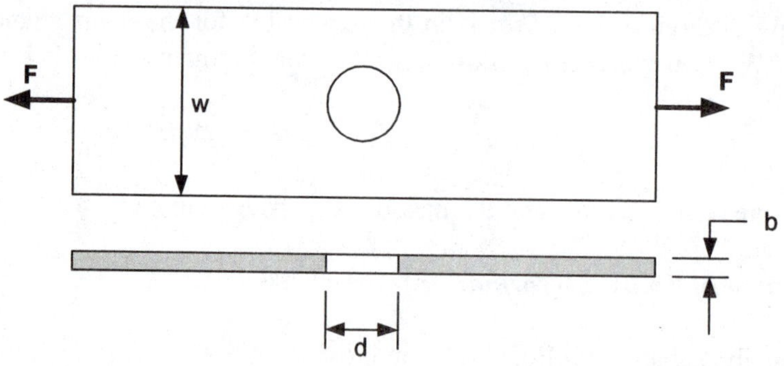

Fig. 11.15 A centrally located circular hole in an axially loaded bar.

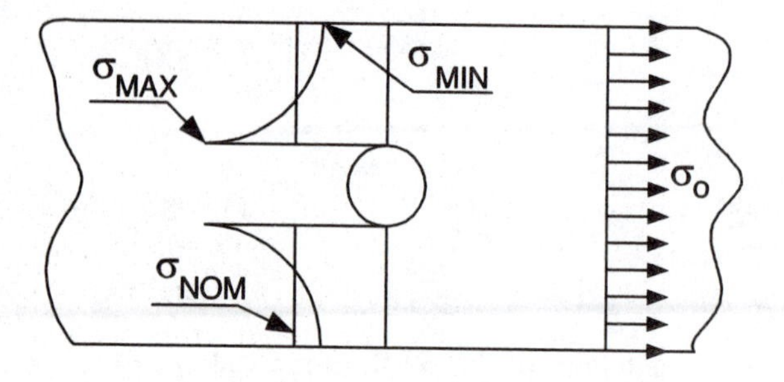

Fig. 11.16 Distribution of stress across a section through the center of the hole shows the concentration of stresses adjacent to the hole's boundary.

To determine the maximum stress σ_{MAX} adjacent to the hole, the maximum stresses are expressed in terms of a stress concentration factor by employing:

$$\sigma_{MAX} = K\,\sigma_{NOM} \tag{11.41}$$

where K is the stress concentration factor and σ_{NOM} is the nominal stress.

The **nominal stress** is the average stress across the net section containing the hole, and is given by:

$$\sigma_{NOM} = \frac{P}{A_{NOM}} = \frac{P}{(w-d)b} \tag{11.42}$$

where b is the thickness of the bar, w is the bar width and d is the hole diameter.

The uniform stress σ_o and the nominal stress σ_{NOM} are related by:

$$\sigma_{NOM} = \left(\frac{w}{w-d}\right)\sigma_0 \tag{11.43}$$

The nominal stress is always greater than the uniform stress because the factor $w/(w-d) > 1$.

The **stress concentration factor** K for a uniform thickness bar with a central circular hole subjected to axial loading is a function of the geometry depending on the ratio of d/w as shown in Fig. 11.17.

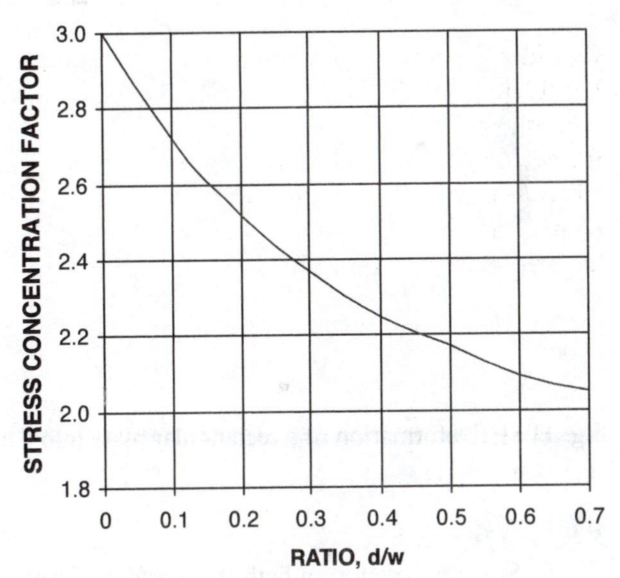

Fig. 11.17 Stress concentration factor for a central circular hole in an axially loaded bar.

11.4 BENDING OF THIN RECTANGULAR PLATES

Rectangular plates and beams behave in a similar manner when subjected to a pure bending moment. They deflect into a circular arc with tension stress generated on the lower surface and compression stress on the upper surface. The difference in the analytical approach between plate and beam theory is due to the marked difference in the cross section of these two structural elements. Beam cross sections are narrow and deep whereas a plate cross section is wide and shallow, as shown in Fig. 11.18.

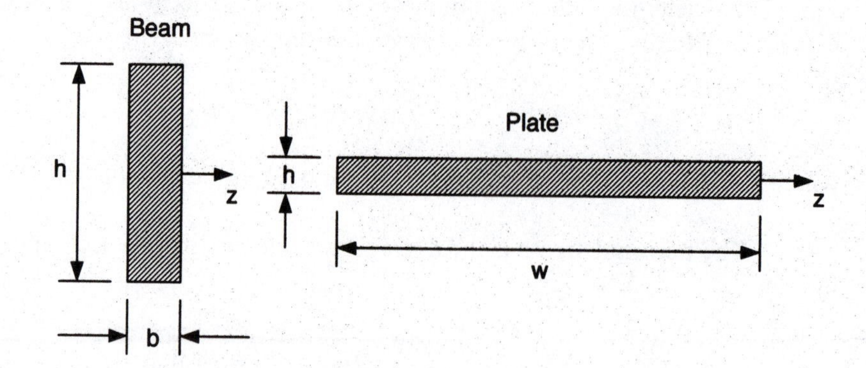

Fig. 11.18 Cross sections from a beam and a plate showing an important distinction.

The longitudinal stresses developed in a beam are uniaxial because the section is too narrow to support normal stresses in the z direction. However, the plate section is wide and normal stresses in the z direction occur when the plate is subjected to bending moments. The state of deformation differs as well. In a beam strains develop in the z direction with $\varepsilon_z = -\nu\varepsilon_x$. The width of the plate constrains its lateral deformation and a state of plane strain exists in a plate with $\varepsilon_z = 0$.

To develop the relations describing the stresses and deformation in a thin rectangular plate, consider it to be deformed into a circular arc with a radius of curvature of ρ as shown in Fig. 11.19. A strip of unit width is sectioned from the plate. Because plane sections remain plane before and after bending, we can express the strains as:

$$\varepsilon_x = \frac{-y}{\rho} \quad \text{and} \quad \varepsilon_z = 0 \tag{11.44}$$

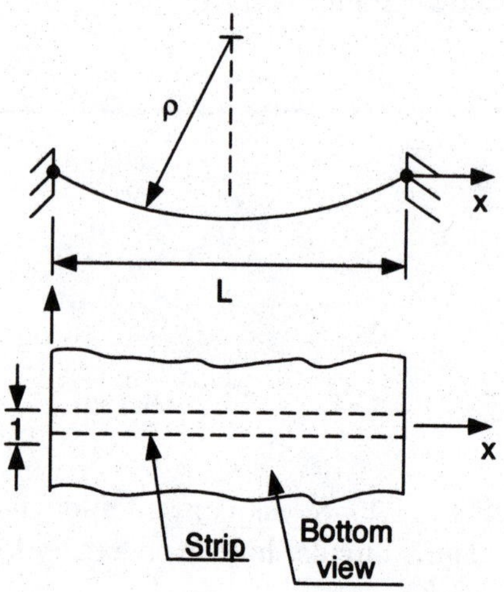

Fig. 11.19 Deformation of a rectangular plate into a circular arc.

Stresses develop in both the x and z directions within the plate; hence, the biaxial stress strain relations are required.

$$\varepsilon_x = \frac{1}{E}(\sigma_x - \nu\sigma_z)$$
$$\varepsilon_z = \frac{1}{E}(\sigma_z - \nu\sigma_x) \tag{11.45}$$

$$\sigma_x = \frac{E}{(1-\nu^2)}(\varepsilon_x + \nu\varepsilon_z)$$
$$\sigma_z = \frac{E}{(1-\nu^2)}(\varepsilon_z + \nu\varepsilon_x) \qquad (11.46)$$

Substituting Eq. (11.44) into Eq. (11.46) yields:

$$\sigma_x = \frac{E\varepsilon_x}{(1-\nu^2)} = -\frac{Ey}{(1-\nu^2)\rho}$$
$$\sigma_z = \frac{\nu E\varepsilon_x}{(1-\nu^2)} = -\frac{\nu Ey}{(1-\nu^2)\rho} = \nu\sigma_x \qquad (11.47)$$

From equilibrium equations, it is clear that the bending moment required to impose this deformation is given by:

$$M = \frac{Eh^3 12}{(1-\nu^2)\rho} \qquad (11.48)$$

This relation is rewritten as:

$$\frac{1}{\rho} = \frac{M}{D} \qquad (11.49)$$

where D is the flexural rigidity of the plate given by:

$$D = \frac{Eh^3 12}{(1-\nu^2)} \qquad (11.50)$$

For relatively small deflections the plate curvature (1/ρ) can be approximated by $\frac{d^2w}{dx^2}$ and Eq. (11.49) can be written as:

$$D\frac{d^2w}{dx^2} = -M \qquad (11.51)$$

where w is the deflection of the flat plate normal to its surface due to transverse forces.

Solutions for the plate displacement w in Eq. (11.48), which lead to results for stresses and displacements, are usually obtained by numerical methods that incorporate specified ratios of plate dimensions, the loading and boundary conditions. Results of these numerical solutions for flat rectangular plates with straight edges subjected to uniform pressure over the plate's entire surface are presented in Table 11.3.

Solutions for Flat Rectangular Plates in Bending

Ten solutions for flat plates in bending are presented in Table 11.3. All of the solutions pertain to a uniform pressure p applied over the entire surface of the plate. Results are given for the maximum stress σ_{max}, the maximum deflection δ_{max} and the maximum reaction force per unit length R_{max} that occurs at the supports. The results are presented in equations shown below:

$$\sigma_{max} = \frac{\beta p b^2}{h^2} \quad (11.52)$$

$$\delta_{max} = \frac{\alpha p b^4}{E h^3} \quad (11.53)$$

$$R = \gamma p b \quad (11.54)$$

where β, α and γ are dimensionless constants that depend on the aspect ratio a/b of the plate. Note a is the dimension of the long side of the plate and b is its short side. The thickness of the plate is h and E is its modulus of elasticity.

11.5 STRESSES IN THIN WALLED PRESSURE VESSELS

Thin walled[2] pressure vessels (shells) are widely employed as containers for liquids and gasses. They are usually subjected to internal pressure, which generates membrane stresses in their walls. Bending stresses do not occur unless the shell is designed with discontinuities such as nozzles or skirts. Because the state of stress is primarily membrane, the pressure vessels are very efficient structures with stresses that are uniformly distributed through the thickness of the wall. Pressure vessels are fabricated in various sizes and shapes. While the most common shape is cylindrical, they may be designed with spherical, conical or toroidal shapes.

11.5.1 Spherical Pressure Vessels

Let's consider a portion of a thin spherical shell to illustrate the concept of membrane stresses. The portion of a spherical shell with a radius r and wall thickness t subjected to an internal pressure p is illustrated in Fig. 11.20. Writing the equilibrium equation $\Sigma F_y = 0$ yields:

$$p\pi r_x^2 - (N \sin \phi)(2\pi r_x) = 0 \quad (11.55)$$

where N is the membrane force per unit length and r_x is the radius defined in Fig. 11.20

---

[2] A pressure vessel is considered to be thin walled if the ratio of $r/t > 10$. The membrane stresses in thin walled pressure vessels are uniformly distributed through the thickness of the wall.

Table 11.3
Flat plate solutions

Support Conditions — **Specific values for β, α and γ**

Case 1: Simply supported all edges

a/b	1.0	1.2	1.4	1.6	1.8	2.0	3.0	4.0
β	0.2874	0.3762	0.4530	0.5172	0.5688	0.6102	0.7134	0.7410
α	0.0444	0.0616	0.0770	0.0906	0.1017	0.1110	0.1335	0.1400
γ	0.420	0.455	0.478	0.491	0.499	0.503	0.505	0.502

Case 2: Three edges simply supported, one free

a/b	0.50	0.667	1.0	1.5	2.0	4.0
β	0.36	0.45	0.67	0.77	0.79	0.80
α	0.080	0.106	0.140	0.160	0.165	0.167

Case 3: Three edges simply supported, short edge clamped

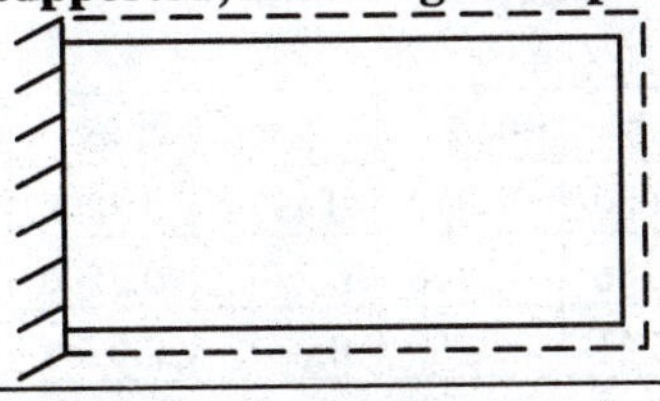

a/b	1.0	1.5	2.0	2.5	3.0	3.5	4.0
β	0.50	0.67	0.73	0.74	0.75	0.75	0.75
α	0.030	0.071	0.101	0.122	0.132	0.137	0.139

Case 4: Three edges simply supported, long edge clamped

a/b	1.0	1.5	2.0	2.5	3.0	3.5	4.0
β	0.50	0.66	0.73	0.74	0.74	0.75	0.75
α	0.030	0.046	0.054	0.056	0.057	0.058	0.058

Case 5: Two long edges simply supported, two short edges clamped

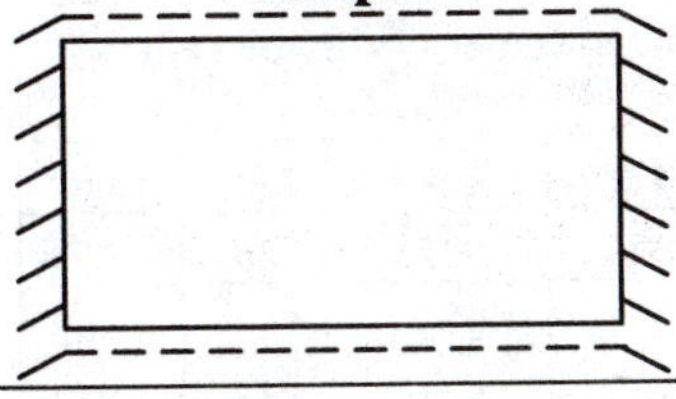

a/b	1.0	1.2	1.4	1.6	1.8	2.0	∞
β	0.4182	0.5208	0.5988	0.6540	0.6912	0.7146	0.750
α	0.0210	0.0349	0.0502	0.0658	0.0800	0.0922	

Table 11.3 (Continued)
Flat plate solutions

Support Conditions	Specific values for β, α and γ

Case 6: Two edges simply supported, two edges clamped

a/b	1.0	1.2	1.4	1.6	1.8	2.0	∞
β	0.4182	0.4626	0.4860	0.4968	0.4971	0.4973	0.500
α	0.0210	0.0243	0.0262	0.0273	0.0280	0.0283	0.0285

Case 7: One edge clamped, two edges simply supported, one free

β_1 and β_2 refer to clamped and free edges, respectively.
γ_1 and γ_2 refer to clamped and free edges, respectively.

a/b	0.25	0.50	0.750	1.0	1.5	2.0	3.0
β_1	0.044	0.176	0.380	0.665	1.282	1.804	2.450
β_2	0.048	0.190	0.386	0.565	0.730	0.688	0.434
γ_1	0.183	0.368	0.541	0.701	0.919	1.018	0.1055
γ_2	0.131	0.295	0.526	0.832	1.491	1.979	2.401

Case 8: All edges clamped

β_1 gives σ_{max} at center of long edge
β_2 gives σ_{max} at center of plate

a/b	1.0	1.2	1.4	1.6	1.8	2.0	∞
β_1	0.3078	0.3834	0.4356	0.4680	0.4872	0.4974	0.5000
β_2	0.1386	0.1794	0.2094	0.2286	0.2406	0.2472	0.2500
α	0.0138	0.0188	0.0226	0.0251	0.0267	0.0277	0.0284

Case 9: Three edges clamped, one edge simply supported

β_1 gives σ_{max} at center of plate: β_2 gives σ_{max} at clamped edge
γ_1 and γ_2 gives R_{max} at center of plate and simply supported edge.

a/b	0.25	0.5	0.75	1.0	1.5	2.0	3.0
β_1	0.020	0.081	0.173	0.307	0.539	0.657	0.718
β_2	0.031	0.121	0.242	0.343	0.417	0.396	0.318
γ_1	0.115	0.230	0.343	0.453	0.584	0.622	0.625
γ_2	0.125	0.256	0.382	0.471	0.547	0.549	0.530

Case 10: Three edges clamped, one edge free

β_1 gives σ_{max} at center of plate: β_2 gives σ_{max} at center of clamped edge
γ_1 and γ_2 gives R_{max} at center of plate and at center of clamped edge.

a/b	0.25	0.5	0.75	1.0	1.5	2.0	3.0
β_1	0.020	0.081	0.173	0.321	0.727	1.226	2.105
β_2	0.031	0.126	0.286	0.511	1.073	1.568	1.962
γ_1	0.114	0.230	0.341	0.457	0.673	0.845	1.012
γ_2	0.125	0.248	0.371	0.510	0.859	1.212	1.627

Solving Eq. (11.55) for N and noting that $r_x/\sin\phi = r$ gives:

$$N = pr_x/(2\sin\phi) = pr/2 \quad (11.56)$$

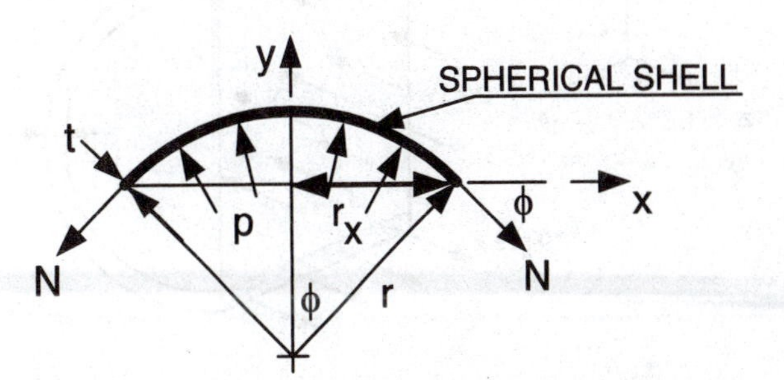

Fig. 11.20 A portion of a spherical shell subjected to an internal pressure p.

It is evident from Eq. (11.56) that the value of N is independent of the angle ϕ; hence, the membrane force per unit length is independent of position. The membrane stress σ_m for a spherical pressure vessel is obtained from Eq. (11.56) as:

$$\sigma_m = \frac{N}{t} = \frac{pr}{2t}\sigma \quad (11.57)$$

Because the spherical pressure vessel is symmetric, it is evident that the membrane stresses in the meridian σ_{my} and tangential σ_{mx} directions are the same, and the stress state is biaxial with:

$$\sigma_1 = \sigma_2 = \sigma_{my} = \sigma_{mx} = \frac{pr}{2t} \quad (11.58)$$

These are also principal stresses because their axes of symmetry coincide with principal planes.

11.5.2 Cylindrical Pressure Vessels

Cylindrical pressure vessels are the most common type employed because the cylindrical shape is easy to fabricate. Thin metal plates are formed into rounds and the longitudinal (axial) seam is welded shut to produce a thin walled tube. End caps are welded on this tube to produce a thin walled vessel capable of maintaining internal pressures.

To determine the membrane stresses σ_h (hoop or circumferential direction) and σ_a (axial or longitudinal direction) in a thin walled cylindrical pressure vessel consider a cylinder with a radius r and a wall thickness t as illustrated in Fig. 11.21. Next remove a semi-circular segment of the cylinder as a FBD. The forces due to pressure p and the membrane forces acting on this FBD are shown in Fig. 11.22.

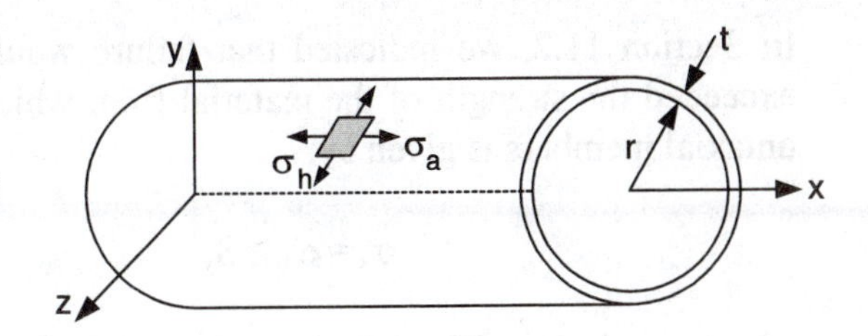

Fig. 11.21 A cylindrical pressure vessel with a radius r and a wall thickness of t.

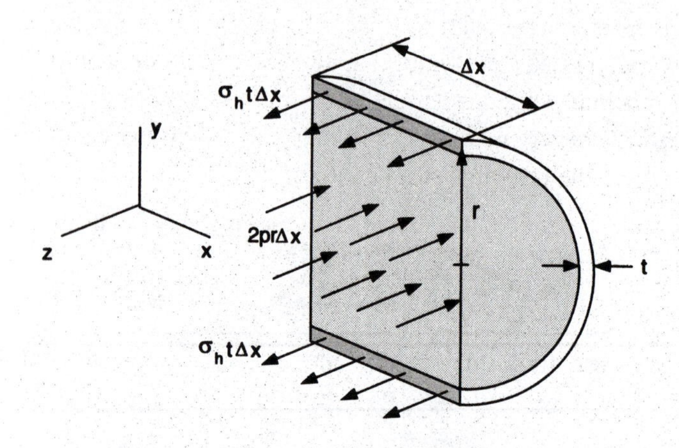

Fig. 11.22 A free body diagram of a semi-circular segment from the thin walled cylindrical pressure vessel.

To solve for the hoop stress σ_h, write $\Sigma F_z = 0$:

$$2(\sigma_h t\Delta x) - 2pr\Delta x = 0 \tag{a}$$

$$\sigma_h = \frac{pr}{t} \tag{11.59}$$

The axial stresses are determined by considering a segment of the cylinder formed by a section cut perpendicular to the axis of the cylinder as shown in Fig. 11.23. This free body diagram (FBD) shows the forces due to pressure and the stresses in the axial direction. Writing $\Sigma F_x = 0$ yields:

$$\sigma_a(2\pi rt) - p(\pi r^2) = 0 \tag{b}$$

Fig. 11.23 Free body diagram of a cylindrical pressure vessel showing the forces in the axial direction.

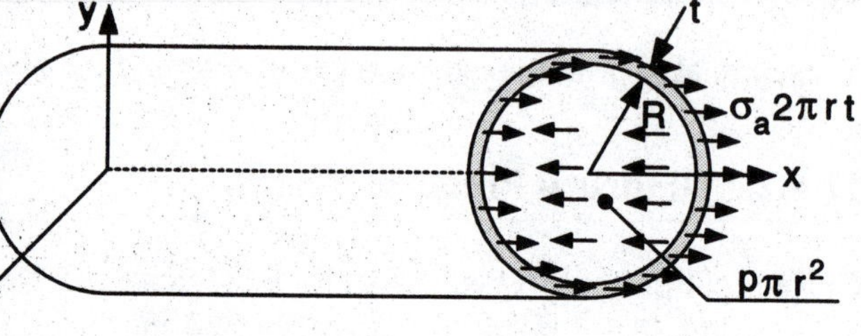

Solving for σ_a gives:

$$\sigma_a = \frac{pr}{2t} \tag{11.60}$$

Comparing Eq. (11.59) and Eq. (11.60) indicates that the hoop stress is twice as large as the axial stress in a thin walled circular pressure vessel.

11.6 FAILURE THEORIES

In Section 11.2, we indicated that failure would occur when the stress σ imposed on a tension bar exceeded the strength of the material from which it was fabricated. The equation for failure for these uniaxial members is given by:

$$\sigma_x = \sigma_1 \geq S_y \qquad\qquad \sigma_x = \sigma_1 \geq S_u \tag{11.61}$$

where the choice of S_y or S_u depends on the admissible mode of failure.

For uniaxial members the stress state is also uniaxial with $\sigma_y = \sigma_z = 0$ and $\sigma_2 = \sigma_3 = 0$; hence, the conditions for structural failure given in Eq. (11.61) are valid. However, for more complex structural components such as plates, shells or machine components, the state of stress is two- or three-dimensional and more involved theories of failure are required. We will consider a two-dimensional state of stress—plane stress— in this section where the principal stresses σ_1, σ_2 and σ_3 are ordered so that:

$$\sigma_1 \geq \sigma_2 \qquad \sigma_3 = 0 \tag{11.62}$$

Several different theories of failure are used today, but in the interest of brevity, we will describe only the three most commonly used.

1. The maximum principal stress theory.
2. The maximum shear stress theory.
3. The maximum distortional energy theory.

We will also assume that the failure occurs upon yielding and that the yield strength in tension is the same as the yield strength in compression.

11.6.1 The Maximum Principal Stress Theory

The maximum principal stress theory states that a structural component fails by yielding when the maximum principal stress σ exceeds the yield strength of the material from which it is fabricated.

$$|\sigma_1| \geq S_y \qquad |\sigma_2| \geq S_y \tag{11.63}$$

If we consider all possible combinations of σ_1 and σ_2, it is possible to construct the yielding diagram presented in Fig. 11.24.

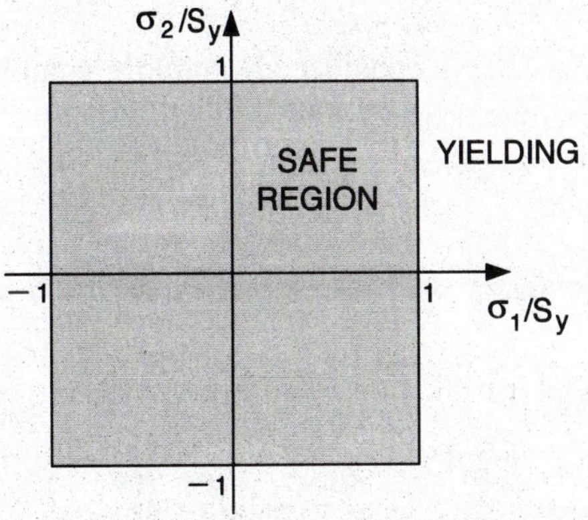

Fig. 11.24 Failure diagram for the maximum principal stress theory for yielding.

If we plot a point with coordinates σ_1/S_y and σ_2/S_y and it falls within the square in Fig. 11.24, the structural component will not yield. However, if the point falls on the boundary or outside the boundary of the box, yielding will occur. This theory is employed in the analysis of uniaxial structural members where the influence of the second principal stress σ_2 is small. Unfortunately, failure of ductile materials subjected to biaxial states of stress occurs due to shear on planes of maximum shear stress. In these cases, the maximum principal stress theory is not conservative. Yielding by shear occurs at stress states lower than predicted by the maximum principal stress theory.

11.6.2 The Maximum Shear Stress Theory

The maximum shear stress theory for yielding, sometimes called the Tresca theory, is often employed to predict the onset of yielding in machine components fabricated from ductile materials. These ductile materials fail on maximum shear planes that are oriented at 45° to the principal planes. This theory indicates that yielding will occur when τ_{Max} equals the maximum shear stress achieved at yield in a simple tensile test of the material.

Consider first the tension test where:

$$\sigma_1 = S_y \qquad \sigma_2 = \sigma_3 = 0 \qquad \tau_{Max} = (\sigma_1 - \sigma_2)/2 = S_y/2 \tag{11.64}$$

Next, consider the maximum shear stress in the structural element. Two possibilities must be considered:

When σ_1 and σ_2 are of opposite signs τ_{Max} is given by:

$$\tau_{Max} = (\sigma_1 - \sigma_2)/2 \tag{11.65}$$

Substituting Eq. (11.64) into Eq. (11.65) gives the equation governing yielding as:

$$|\sigma_1 - \sigma_2| \geq S_y \tag{11.66}$$

When σ_1 and σ_2 are the same sign, τ_{Max} is given by:

$$\tau_{Max} = (\sigma_1 - \sigma_3)/2 = \sigma_1/2 \tag{11.67}$$

Substituting Eq. (11.64) into Eq. (11.67) gives the equation governing yielding as:

$$|\sigma_1| \geq S_y \tag{11.68}$$

If we consider all possible combinations of σ_1 and σ_2, it is possible to construct the yielding diagram presented in Fig. 11.25.

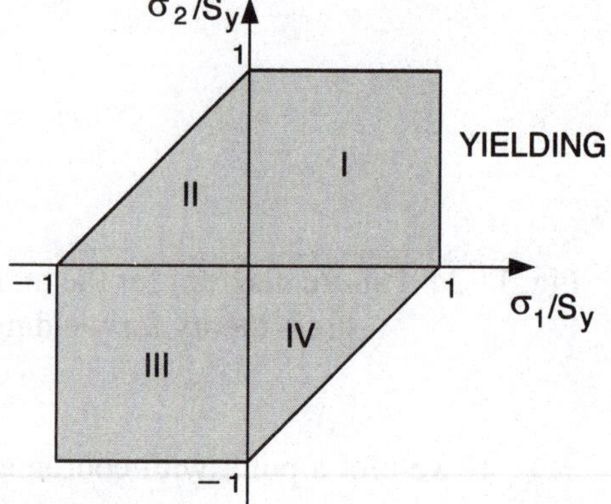

Fig. 11.25 Failure diagram for the Maximum shear stress theory of yielding.

When the principal stresses σ_1 and σ_2 are of like sign, the stress state corresponds to quadrant I or III in Fig. 11.25, and the maximum shear theory for yielding corresponds to the maximum principal stress theory. When the signs of σ_1 and σ_2 are opposite, the stress state corresponds to quadrant II and IV. In this instance, there is a marked difference between the two theories, and the maximum shear stress theory is much more conservative.

11.6.3 Maximum Distortion Energy Theory

The maximum distortion energy theory, often called the von Mises theory, predicts yielding in a structural element when the distortion energy per unit volume equal the distortion energy per unit volume achieved at yielding of the same material in a tensile test.

For a two-dimensional stress state with $\sigma_3 = 0$, the distortion energy per unit volume U_d is given by:

$$U_d = (1/6G)[\sigma_1^2 - \sigma_1\sigma_2 + \sigma_2^2] \tag{11.69}$$

In a tensile test with $\sigma_1 = S_y$ and $\sigma_2 = \sigma_3 = 0$, the distortion energy per unit volume is:

$$U_d = (1/6G)S_y^2 \tag{11.70}$$

By substituting Eq. (11.70) into Eq. (11.69), we obtain the von Mises equation for yielding.

$$(\sigma_1/S_y)^2 - (\sigma_1/S_y)(\sigma_2/S_y) + (\sigma_2/S_y)^2 = 1 \tag{11.71}$$

If we consider all possible combinations of σ_1 and σ_2, it is possible to construct the yielding diagram presented in Fig. 11.26. The boundary of the failure diagram is an ellipse that is slightly less conservative than the maximum principal stress theory for yielding in quadrants I and III. In quadrants II and IV, where the principal stresses are of opposite sign, the von Mises theory is less conservative than the maximum shear theory but more conservative that the maximum principal stress theory. The distortional energy theory is widely accepted as the most accurate theory for predicting yielding in ductile materials commonly used to fabricate structural elements and machine components.

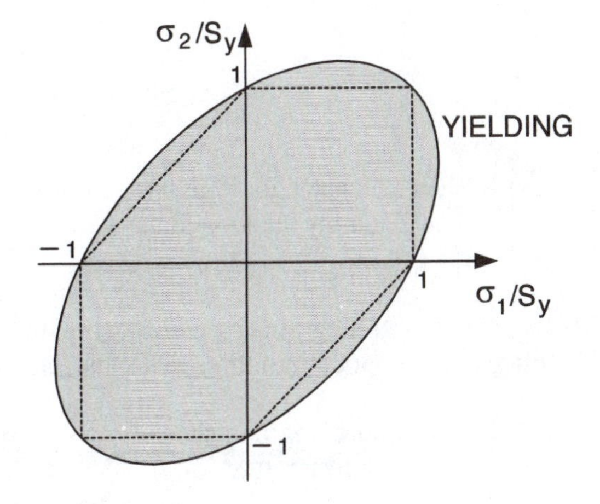

Fig 11.26 Failure diagram for the maximum distortion energy theory for yielding.

11.7 FATIGUE FAILURE

Many structures are loaded only once. For example, the structural members in an existing high rise apartment building were loaded by the dead weight of all of the concrete and steel as the building was constructed. The people, furnishings and equipment going into the building added some weight, but this additional load was small when compared to the weight of the basic structure. For structures and machine components loaded only once, either the yield strength or the ultimate tensile strength is adequate to predict the safety of the structure.

However, many structures and machine components are subjected to repeated loading. Consider the engine of your automobile. If you are driving along a highway with the tachometer registering 3,000 RPM, you are subjecting the crankshaft in your engine to 3,000 cycles of stress for each minute of operation or nearly 1.5 million cycles in an eight hour trip.

There are two detrimental effects due to the cyclic loading. First, the design strength of the material from which the structure is fabricated is lowered. We design to a fatigue strength S_f that is a function of the cyclic stresses and the number of cycles of load imposed onto the structure.

Second, fatigue failures in **ductile** materials are of a **brittle** nature (i.e., structural failure and collapse occur catastrophically). The failure mechanism in fatigue is markedly different from that observed in yielding or rupture. In fatigue, microscopic cracks are initiated due to accumulated irreversible slip in a very thin layer of material adjacent to the surface. These cracks grow larger and extend into the material until reaching critical size. At this point, the cracks become unstable and extend at very high speed across the structural member producing sudden and catastrophic collapse. Examination of a surface of a fatigue failure shows the brittle nature of the phenomena. Although the material may be classified as ductile based on the results of a tensile test, there is no visible sign of coarse slip or necking associated with failure in fatigue.

11.7.1 High Cycle Fatigue and the S_f – N Diagram

To accommodate for the degrading effects of cyclic loading, we compare the applied stresses with the **fatigue strength, S_f** of the material from which the structural element is fabricated. A typical example of the fatigue strength for low carbon steel is presented in the S_f-N diagram of Fig. 11.27.

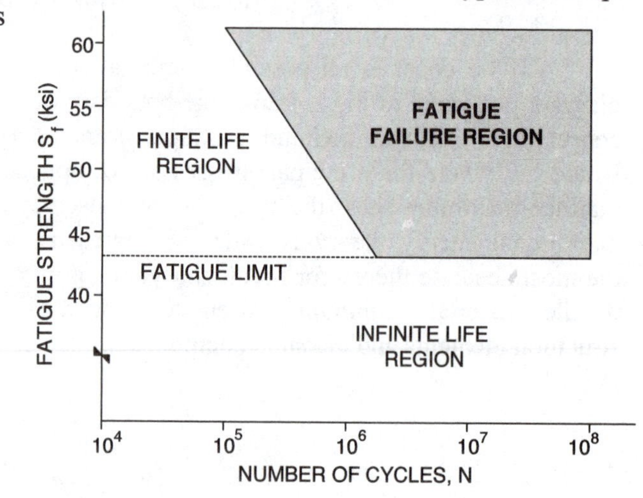

Fig. 11.27 S_f – N curve for high strength low alloy steel.

The S_f - N diagram is a graphical representation of both safe and critical states of cyclic stresses. The diagram is divided into three different regions:

1. A failure region (shaded) where many cycles of high stresses produce a fatigue crack leading to fracture.
2. Safe regions where component stresses lower than the fatigue (endurance) limit ensure infinite cyclic life.
3. A finite-life region where a specified number of cycles can be endured at a specified stress level that is greater than the fatigue limit.

For infinite life, the fatigue strength is often called the endurance limit S_e where:

$$S_e = S_f \qquad \text{for } N > 10^6 \text{ cycles} \qquad (11.72)$$

For finite life, with the number of cycles less than 10^6, the fatigue strength S_f is larger than S_e. The value of S_f used in a design analysis is determined from the S_f–N diagram for the material used to fabricate the structural elements.

In comparing the cyclic stresses with the strength as defined in Fig. 11.27, we use the alternating portion of the applied stresses. The alternating stress, σ_a, and the mean stress, σ_m, for different types of cyclic loading are defined in the stress-time diagrams presented in Fig. 11.28. We compute the mean stress for cyclic loading by identifying the maximum and minimum stress in a given cycle of applied loading. The mean stress is the average of the maximum and minimum stress. The relation used to determine the mean cyclic stress is given by:

$$\sigma_m = (\sigma_{max} + \sigma_{min})/2 \qquad (11.73)$$

The alternating stress is determined from the difference in the maximum and minimum stress during a typical load cycle. The relation used to determine the alternating cyclic stress is given by:

$$\sigma_a = (\sigma_{max} - \sigma_{min})/2 \qquad (11.74)$$

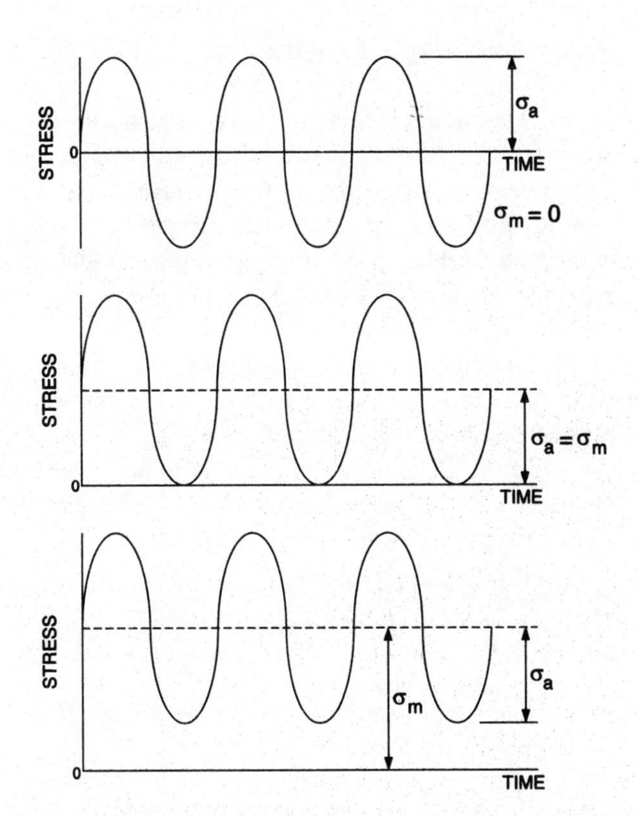

Fig. 11.28 Different types of cyclic stresses imposed on a structure.

11.7.2 The Goodman Diagram

In some instances, the cyclic loading on a structural member produces a combination of alternating and mean stresses. When mean stresses are superimposed on the alternating stresses, the effect is to decrease the fatigue strength of the material. Goodman developed an empirical method to determine a modified fatigue strength that accounts for the detrimental effects of combined alternating and mean stresses. The modified fatigue strength S_a is presented as a function of the cyclic mean stress in Fig. 11.29. Examination of Fig. 11.29 reveals that the modified fatigue strength S_a is equal to the endurance limit when $\sigma_m = 0$. The decrease in the modified fatigue strength is linear with the increase in σ_m, and the strength becomes zero when $\sigma_m = S_u$.

We may also characterize the modified fatigue strength S_a in equation format as:

$$S_a = S_e\,[1 - (\sigma_m/S_u)] \qquad (11.75)$$

where S_e is the endurance limit when $\sigma_m = 0$.
S_u is the ultimate tensile strength.
σ_m is the cyclic mean stress.

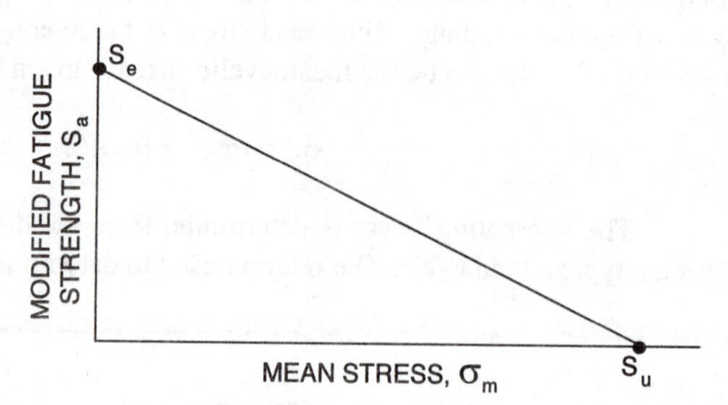

Fig. 11.29 Modified fatigue strength S_a decreases as the cyclic mean stresses increase.

11.7.3 Low-Cycle-Fatigue

Low-cycle-fatigue refers to failures that occur after a relatively short cyclic life, which ranges from about 10^2 to 5×10^4 cycles of stress. The stresses imposed on electronic components are due to thermal cycles between two different temperatures or to repeated mechanical loading at constant temperature. In both situations, the stresses exceed the yield strength of the material involved and plastic strains are generated. The linear relation between stress and strain (Hooke's Law) is not valid. For this reason, the relations developed to predict life in the low-cycle-fatigue regime relate life N_f to plastic strain ε_p or total strain range $\Delta\varepsilon$.

A typical stress-strain diagram for a metallic material cycled at stress levels that exceed the yield strength is shown in Fig. 11.30. A hysteresis loop is formed that becomes stable (asymptotic) after a relatively small number of cycles. The stress is a non-linear function of the strain, and the maximum strain is the sum of both elastic and plastic strains.

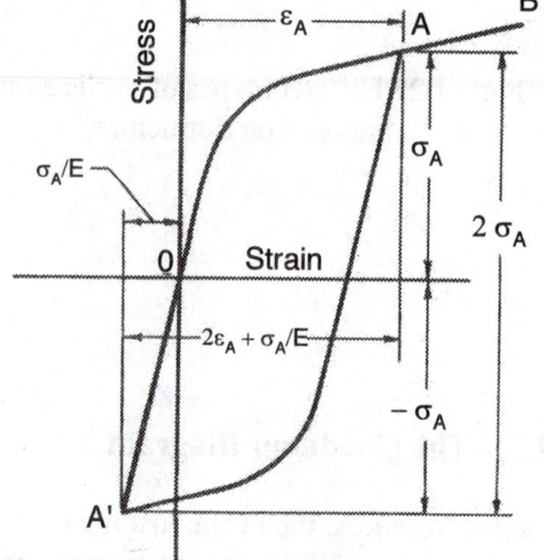

Fig. 11.30 Cyclic stress-strain diagram for a metallic material stressed beyond its yield strength.

To determine the relation between cyclic strain and the number of cycles to failure Manson [9] employed the asymptotic cyclic stress-strain diagram to characterize the elastic and plastic response of the material. A typical graph showing the stress range $\Delta\sigma$ as a function of the strain range $\Delta\varepsilon$, obtained from the asymptotic cyclic stress-strain diagram, is presented in Fig. 11.31.
Part of the strain range is elastic ($\Delta\varepsilon_e = \Delta\sigma/E$) and part is plastic. The plastic strain $\Delta\varepsilon_p$ is determined from:

$$\Delta\varepsilon_p = \Delta\varepsilon - \frac{\Delta\sigma}{E} \tag{11.76}$$

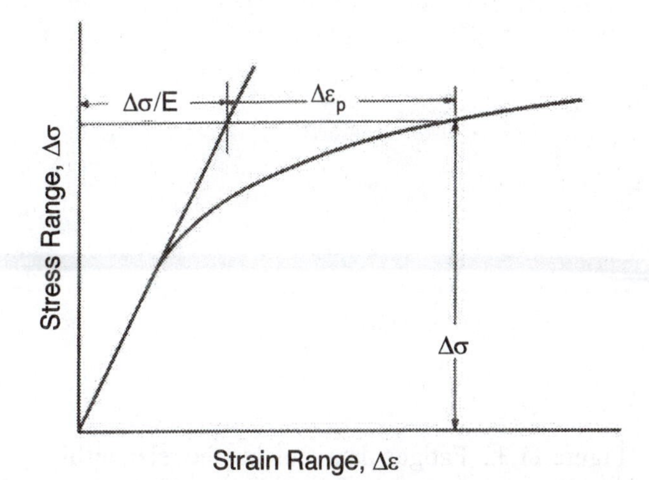

Fig. 11.31 Representation of cyclic stress-strain response of a material.

Many low-cycle fatigue studies conducted by Manson and Coffin [9 to 12] in the mid 1950s showed that the cyclic life N of steel and other materials was related to the plastic strain by a power law of the form given below:

$$\Delta\varepsilon_p = MN_f^z \tag{11.77}$$

where M and z are material constants.

The constants M and z are measured by conducting low-cycle fatigue experiments to generate a log-log graph of plastic strain as a function of cycles to failure. An example of the results of low-cycle fatigue experiment for 4130 steel is presented in Fig. 11.32 with fatigue life ranging from 10 to 5×10^4 cycles. It is evident that the correlation between the experimental data and the power law relation between cyclic life and plastic strain is excellent. The slope of the straight line on the log-log graph enables the determination of the material's exponent constant z. The intercept permits the determination of the material's ductility constant M.

Another approach for predicting the fatigue life of components is to use the total strain range $\Delta\varepsilon$ instead of the plastic strain range. Using the total strain range as a parameter, we recast Eq. (11.77) as:

$$\Delta\varepsilon = \Delta\varepsilon_p + \frac{\Delta\sigma}{E} = MN_f^z + \frac{G}{E}N_f^\gamma \tag{11.78}$$

where G and γ are material constants that account for the influence of elastic strain range on fatigue life.

Experimental data shows that both the elastic and plastic parts of the total strain generate straight lines when plotted against fatigue life with log-log scales. Moreover, examination of Fig. 11.33 indicates that the sum of $\Delta\varepsilon_p$ and $\Delta\varepsilon_e$ is not a straight line on the log-log graph. In the shorter life region, the elastic strain range is nearly negligible when compared to the plastic strain range and $\Delta\varepsilon$ is closely approximated by $\Delta\varepsilon_p$. However, in the longer life region, the plastic strain range is negligible relative to the elastic strain range. The cross over point for the elastic and plastic strain range curves on this graph is in the neighborhood of 10^4 cycles.

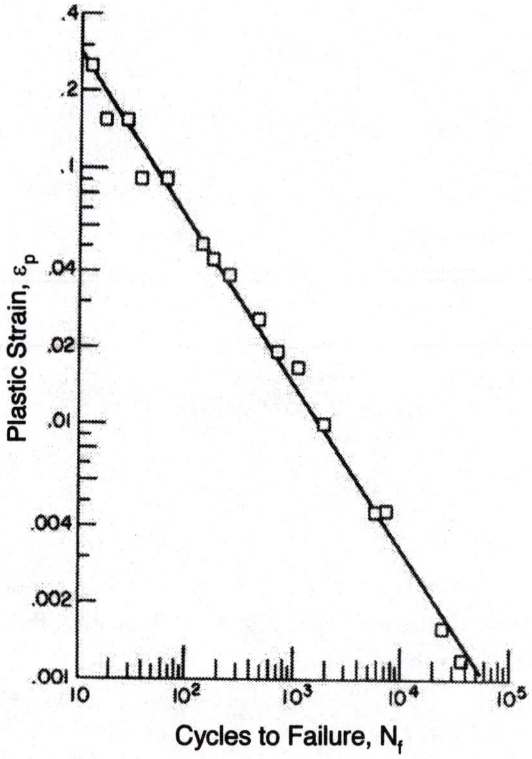

Figure 11.32 Fatigue data showing the relationship between plastic strain ε_p and cyclic life N_f for 4130 steel.

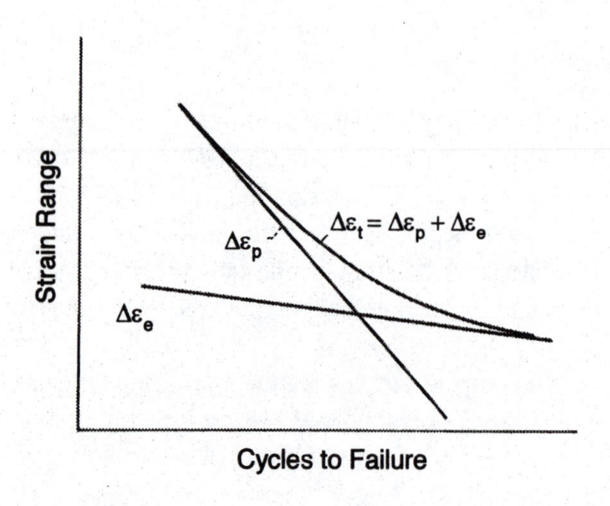

Fig. 11.33 Elastic, plastic and total strain ranges as a function of cycles to failure.

The material constants M, z, G and γ in Eq. (11.78) are determined from strain cycling experiments. The procedure is to conduct experiments at two strain range levels ($\Delta\varepsilon_1$ and $\Delta\varepsilon_2$) and measure the corresponding stress range ($\Delta\sigma_1$ and $\Delta\sigma_2$) and cycles to failure (N_1 and N_2). The material constants in Eq. (11.78) in terms of these measurements are given by:

$$\gamma = \frac{\log_e(\Delta\sigma_1) - \log_e(\Delta\sigma_2)}{\log_e N_1 - \log_e N_2}$$

$$G = \Delta\sigma_1 N_1^{-\gamma} \quad \text{or} \quad G = \Delta\sigma_2 N_2^{-\gamma}$$

$$z = \frac{\log_e\left[\Delta\varepsilon_1 - (\Delta\sigma_1/E)\right] - \log_e\left[\Delta\varepsilon_2 - (\Delta\sigma_2/E)\right]}{\log_e N_1 - \log_e N_2}$$

$$M = \left(\Delta\varepsilon_1 - \frac{\Delta\sigma_1}{E}\right)N_1^z - \left(\Delta\varepsilon_2 - \frac{\Delta\sigma_2}{E}\right)N_2^z \tag{11.79}$$

In some cases, it is not feasible to conduct fatigue experiments to determine the required material constants. In these situations, Manson proposed using a low-cycle life predicting relation based on the total strain range $\Delta\varepsilon$ and commonly known material properties including ultimate tensile strength, modulus of elasticity, and reduction in area. His approximation of fatigue life is given by:

$$\Delta\varepsilon = 3.5\left(\frac{S_u}{E}\right)N_f^{-0.12} + D^{0.6}N_f^{-0.6} \tag{11.80}$$

where D is the ductility of the material that is given by:

$$D = \log_e\left[\frac{1}{(1-RA)}\right] \tag{11.81}$$

where RA is the reduction in area measured in a standard tension test.

Low-cycle Fatigue of Solder Joints

The widespread usage of leadless chip carriers, ball grid arrays and chip scale packages has dramatically increased the number of devices that can be mounted on a printed circuit board and decreased the overall cost of an assembly. However, the ability of the solder joints to withstand the thermally imposed strains over the life of the product has decreased. These thermal strains are produced by the mismatch in the temperature coefficient of expansion between the chip carrier and the printed circuit board. When the temperature cycles, in operation or storage, shearing strains at the solder joints are generated. Because the solder has a much lower modulus of elasticity than either the PCB or the chip carrier, nearly all of the cyclic shearing strains occur in the solder. Unfortunately, fatigue failure of solder joints may occur after a relatively few number of thermal cycles.

Low-cycle fatigue failure of solder joints has been recognized for many decades. The results of lap shear fatigue experiments shown in Fig. 11.34 show the marked decrease in life of a solder joint with increasing shearing strain range $\Delta\gamma$. The influence of temperature and frequency of the cyclic strains imposed on the solder joint is also evident. These and other experiments demonstrated that solder joints failed in low-cycle fatigue and that the work of Manson, Coffin and others could be adapted to model behavior of solder under cyclic thermal strains.

Models for the Low-Cycle-Fatigue of Solder Joints

Many models have been proposed to predict the low-cycle-life of solder joints under the action of thermal strain cycling. These models are usually based on the plastic strain range $\Delta\varepsilon_p$ because the relatively low number of cycles (10^2 to 10^4) of interest where the plastic strain dominates the material behavior. Moreover, the plastic strain range is easy to measure because it is the width of the hysteresis loop in tensile tests under strain range control. The six models are described below follow the development of the predictive equations first proposed by Coffin and Manson.

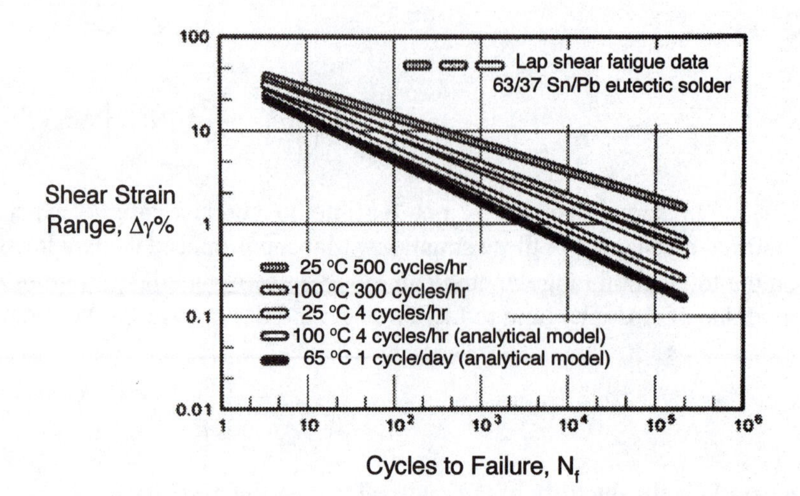

Fig. 11.34 Fatigue of lap shear eutectic tin-lead solder joints at different temperatures and cyclic frequency.

Coffin–Manson Model (Plastic Strain Range–Fatigue Life)

The relationship between the plastic strain range $\Delta\varepsilon_p$ and the number of cycles to failure N_f follows the Coffin–Manson model expressed in equation form below:

$$\Delta\varepsilon_p N_f^{\alpha} = M \tag{11.82}$$

where $\Delta\varepsilon_p$ α and M are material constants.

The adequacy of this relation between $\Delta\varepsilon_p$ and the fatigue life N_f has been investigated for 96.5Sn-3.5Ag solder by conducting strain controlled fatigue tests with different frequencies and temperatures. The test results showed that the exponent constant α was independent of cyclic frequency, but that ductility coefficient M increased with increasing frequency and decreasing temperature.

To improve the Coffin-Manson model a frequency multiplier was incorporated in Eq. (11.82) to better account for the influence of frequency on fatigue life. The frequency-modified Coffin–Manson model is expressed in equation form below:

$$(N_f \nu^{k-1})^{\alpha} \Delta\varepsilon_p = M \tag{11.83}$$

where ν and k are the frequency and frequency exponent, respectively.

Results from strain controlled fatigue tests at different frequencies showed that the data for $\Delta\varepsilon_p$ and N_f can be fit to a single curve; however, the data obtained at different temperatures indicate a dependence on this parameter. It appears that only isothermal low-cycle-fatigue behavior of Sn-Ag eutectic solder can be described by the frequency-modified Coffin–Manson relationship.

Smith–Watson–Topper Model

The Smith–Watson–Topper model assumes that the fatigue life is dependent on the product ($\sigma_{max}\,\Delta\varepsilon_t$). The model incorporates this product to modify the Coffin Manson relation by recasting Eq. (11.82) as:

$$(\sigma_{max}\,\Delta\varepsilon_t)N_f^{m} = D \tag{11.84}$$

where σ_{max} is the maximum stress, $\Delta\varepsilon_t$ is the total strain range, m and D are the fatigue exponent and ductility coefficient, respectively.

Results from strain controlled fatigue tests at different frequencies showed that the fatigue exponents for different temperatures and frequencies are similar, but the ductility coefficient increases with decreasing temperature and increasing frequency. It appears that the Smith–Watson–Topper Model cannot account for the either effects of temperature or frequency on low-cycle-fatigue behavior of Sn-Ag eutectic solder.

Morrow Energy Model (Plastic Strain Energy Density–Life)

The plastic strain energy density, the distortion energy associated with a change in shape of a volume element, is related to ductile failure. It is measured by determining the area inside a stable stress-strain hysteresis loop obtained in a strain controlled uniaxial fatigue test. With this model, the fatigue life is described in terms of the plastic strain energy density W_p by writing:

$$W_p N_f^q = C \tag{11.85}$$

where q and C are the fatigue exponent and coefficient, respectively. Results from strain controlled fatigue tests at different frequencies showed that fatigue exponents for different temperatures and frequencies are similar, but the fatigue coefficient increases with decreasing temperature and increasing frequency. It is clear that the Morrow energy model cannot account for either the effects of temperature or frequency on low-cycle-fatigue behavior for Sn-Ag eutectic solder.

To account for the effects of frequency, a frequency term was incorporated into the Morrow energy model and Eq. (11.84) was recast as:

$$(N_f v^{k-1})^{\alpha} W_p = C \tag{11.86}$$

where v and k are the frequency and frequency exponent, respectively.

Results from strain controlled fatigue tests at different frequencies and temperatures are shown in Fig. 11.35. The test data corresponding to W_p and N_f are concentrated within a narrow band indicating that the model accommodates the effects of both temperature and frequency.
The final predictive relation incorporates a flow stress-modified and frequency-modified adaptation of the Morrow energy model where both temperature-dependent and frequency-dependent material parameters are introduced into the plastic strain energy density model as indicated below:

$$(N_f v^{k-1})^q \frac{W_p}{2\sigma_f} = C \tag{11.87}$$

where σ_f is the flow stress taken as the average value between the yield stress the maximum stress on the hysteresis loop and q, k and C are material constants.

Because the flow stress is strongly dependent on temperature and strain rate, this modification includes this parameter to better depict the effect of temperature and frequency on low-cycle fatigue behavior of a visco-plastic material like solder. Experimental results for W_p and the fatigue life N_f at different temperatures and frequencies are located within a narrow band with less scatter than that shown in Fig. 11.35. It appears that the flow stress-modified, frequency-modified Morrow energy model accommodates the effect of temperature and frequency on low-cycle-fatigue behavior better than the other models described above. This model provides relatively good predictions for the fatigue life N_f of Sn-Ag eutectic solder in the temperature range of 20–120°C and frequency range of 10^{-3} to 10^{-1} Hz.

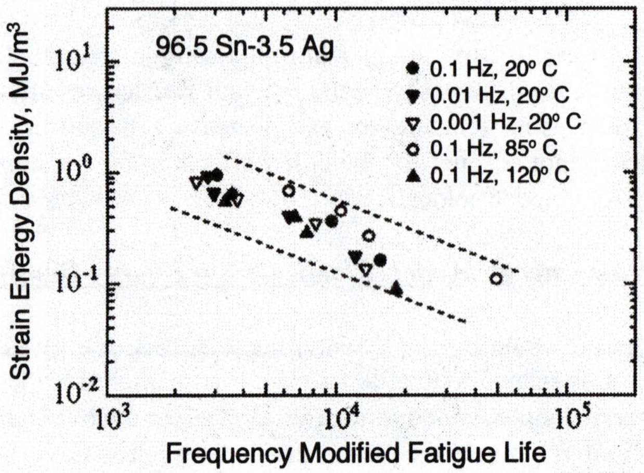

Fig. 11.35 Experimental results showing the fit of data with the frequency-modified Morrow energy model for 96.5Sn-3.5Ag eutectic solder.

11.8 FASTENERS

Threaded fasteners, bolts and screws, are used in large numbers in the assembly of electronic equipment. When the unit must be disassembled for repair or service, screws are often used; however, if the fastening is to be permanent, rivets are usually preferred[3]. The standards followed in specifying a fastener and defining the thread forms in common usage are reviewed. Screws are so common that they are often treated like a commodity. As a commodity they do not have any distinguishing characteristics and can be supplied by any firm producing screws to the specification standards. The advantage of being able to specify screws as commodities is in reducing fastener cost to a minimum.

There are two standards in common usage, namely the Unified and the SI thread. The basic form of the thread for the Unified system, presented in Fig. 11.36, has a 60° included angle between adjacent threads. The crest and the root are usually rounded although they are shown as flats in drawings of thread forms. The SI threads are similar in that they have the 60° include angle and the same size flats for the crest and the roots. However the pitch, the distance between adjacent threads, is specified in millimeters for SI threads. Because the pitch for the Unified and SI threads differs, screws with these two different thread forms are not interchangeable.

Listings of the standard dimensions for the most common sizes of the Unified and SI threads are shown in Tables 11.4 and 11.5 respectively. We note in both tables that coarse and fine threads are available. The coarse threads are used in the lower modulus materials like aluminum or in plastics. The fine thread is used only in steel. The stress area A_t is determined by using the average of the pitch and minor diameters to compute the effective cross sectional area at the threads. The bolt area A_b, not shown in the tables, is determined using the major diameter of the bolt.

The standards allow an engineer to specify a threaded fastener with efficiency. For example, 8-32 UNC is sufficient to identify a unified coarse thread, number 8 screw with 32 threads/in. Of course, it is also necessary to indicate the length of the screw, and its thread and its type of head. With the SI fasteners, the threads are designated by diameter and pitch—M3 × 0.5. This designation specifies a metric screw 3 mm in diameter with threads on a 0.5 mm pitch.

[3] For cases made from molded plastic snaps are often used to fasten panels. These plastic snaps reduce assembly costs, but often pose a challenge when equipment must be disassembled for repair or routine maintenance.

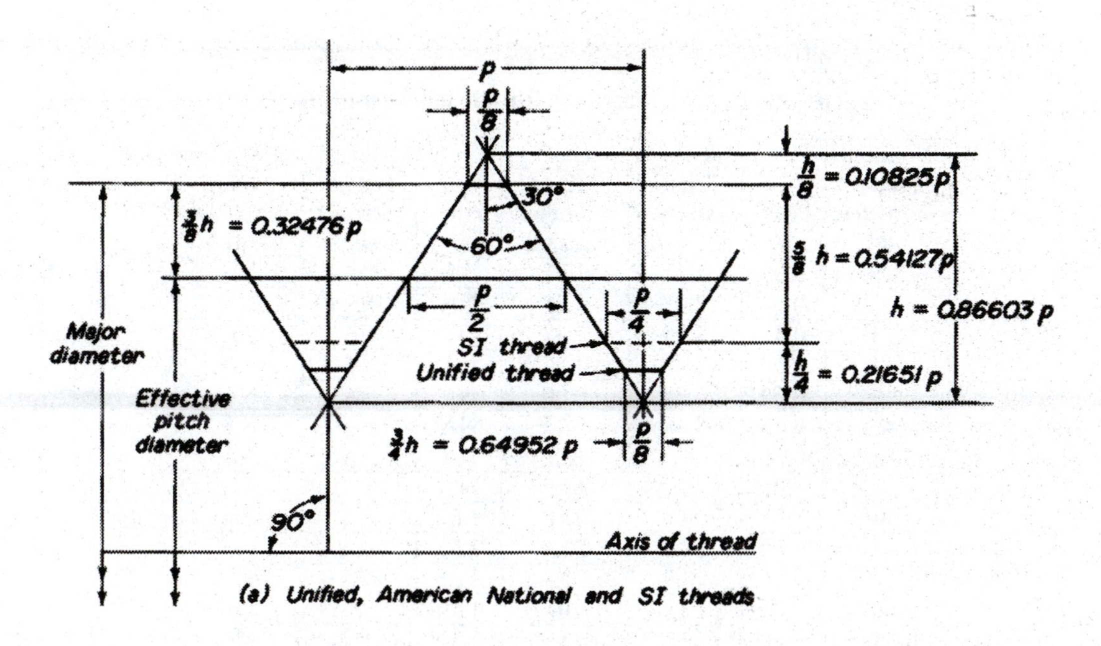

Fig. 11.36 Unified American National and SI thread profiles.

Table 11.4
Dimensions of Unified National threads

Size	Major diameter d (in.)	Coarse-thread Series			Fine-thread Series		
		Threads per inch	Pitch diameter (in.)	Stress area A_t (in.^{2})	Threads per inch	Pitch diameter (in.)	Stress area A_t (in.^{2})
0	0.0600	NA	NA	NA	80	0.0519	0.0018
1	0.0730	64	0.0629	0.0026	72	0.0640	0.0028
2	0.0860	56	0.0744	0.0037	64	0.0759	0.0039
3	0.0990	48	0.0855	0.0049	56	0.0874	0.0052
4	0.1120	40	0.0958	0.0060	48	0.0985	0.0066
5	0.1250	40	0.1088	0.0080	44	0.1102	0.0083
6	0.1380	32	0.1177	0.0091	40	0.1218	0.0102
8	0.1640	32	0.1437	0.0140	36	0.1460	0.0147
10	0.1900	24	0.1629	0.0175	32	0.1697	0.0200
12	0.2160	24	0.1889	0.0242	28	0.1928	0.258
1/4	0.2500	20	0.2175	0.0318	28	0.2268	0.0364
5/16	0.3125	18	0.2764	0.0524	24	0.2854	0.0580
3/8	0.3750	16	0.3344	0.0775	24	0.3479	0.0878
7/16	0.4375	14	0.3911	0.1036	20	0.4050	0.1187
1/2	0.5000	13	0.4500	0.1419	20	0.4675	0.1599
9/16	0.5625	12	0.5084	0.1820	18	0.5264	0.2030
5/8	0.6250	11	0.5660	0.2260	18	0.5889	0.2560
3/4	0.7500	10	0.6850	0.3340	16	0.7094	0.3730
7/8	0.8750	9	0.8028	0.4620	14	0.8286	0.5090
1	1.0000	8	0.9188	0.6060	12	0.9459	0.6630

Table 11.5
Dimensions of SI threads (all dimension in millimeters)

Nominal major diameter d	Coarse-thread series			Fine-thread series		
	Pitch p	Tensile stress area, A_t	Minor diameter area, A_r	Pitch p	Tensile stress area, A_t	Minor diameter area, A_r
1.6	0.35	1.27	1.07	NA	NA	NA
2	0.04	2.07	1.79	NA	NA	NA
2.5	0.45	3.39	2.98	NA	NA	NA
3	0.50	5.03	4.47	NA	NA	NA
3.5	0.60	6.78	6.00	NA	NA	NA
4	0.70	8.78	7.75	NA	NA	NA
5	0.80	14.2	12.7	NA	NA	NA
6	1.00	20.1	17.9	NA	NA	NA
8	1.25	36.6	32.8	1.00	39.2	36.0
10	1.50	58.0	52.3	1.25	61.2	56.3
12	1.75	84.3	76.3	1.25	92.1	86.0
14	2.00	115	104	1.50	125	116
16	2.00	157	144	1.50	167	157
20	2.50	245	225	1.50	272	259
24	3.00	353	324	2.00	384	365
30	3.50	561	519	2.00	621	596
36	4.00	817	759	2.00	915	884
42	4.50	1,120	1,050	2.00	1,260	1,230
48	5.00	1,470	1,380	2.00	1,670	1,630
56	5.50	2,030	1,910	2.00	2,300	2,250
64	6.00	2,680	2,520	2.00	3,030	2,980
72	6.00	3,460	3,280	2.00	3,860	3,800
80	6.00	4,340	4,140	1.50	4,850	4,800
90	6.00	5,590	5,360	2.00	6,100	6,020
100	6.00	6,990	6,740	2.00	7,560	7,470

11.8.1 Strength of Fasteners

Threaded fasteners are produced by automatic machines that cold form the head and roll the threads. The screw materials are usually low to medium carbon steels and elevated strengths are achieved by heat treating. The strength is designated by the SAE grade number as indicated in Table 11.6, with numbers ranging from 1 for the softer lower strength screws to 8 for the harder and higher strength products. Some of the grades have line markings as indicated in Table 11.5. These lines, embossed on the hex heads, permit the assessment of the strength of the material either before or after installation of the bolt.

The proof strength S_p of the bolt material is expressed in terms of the proof load P_p as:

$$S_p = P_p/A_t \qquad (11.88)$$

The proof load is the maximum force that can be applied to a screw before it undergoes a permanent deformation. In most instances, the proof strength is between 80 and 90 per cent of the yield strength of the material from which the bolt is manufactured.

Table 11.6
Strength and grade markings for bolts and screws

SAE grade number	Proof strength S_p (ksi)	Yield strength S_y (ksi)	Tensile strength S_u (ksi)	Elongation (%)	Hardness Rockwell	Nominal diameter (in.)	Grade Marking
1	33	36	60	18	B70/B100	¼ to 1½	None
2	55	57	74	18	B80/B100	¼ to ¾	None
4	65	100	115	10	C22/C32	¼ to 1½	None
5	85	92	120	14	C25/C34	¼ to 1	
5.1	85	----	120	14	C25/C40	#6 to 5/8	
5.2	85	92	120	14	C26/C36	¼ to 1	
7	105	115	133	12	C28/C34	¼ to 1½	
8	120	130	150	12	C33/C39	¼ to 1½	
8.1	120	130	150	10	C32/C38	¼ to 1½	None
8.2	120	130	150	10	C35/C42	¼ to 1	

11.8.2 Bolt Preload

In assembly, the bolts are torqued to tighten the fastener and to draw together the mating surfaces making the joint. In effect, the bolts clamp the joint tightly together. The magnitude of clamping pressure depends upon the preload applied through the fastener, which in turn is controlled by the torque applied in tightening the assembly. The relationship between the preload F_i and the torque T is given by:

$$T = 0.20F_i d \quad (11.89)$$

where d is the major diameter of the bolt.

The amount of preload applied to the fastener depends on the type of loading imposed on the joint. For static loading, a high preload is recommended with:

$$0.6P_p \leq F_i \leq 0.9P_p \quad (11.90)$$

With fatigue loading, the preload is usually lower and is dependent upon the proportioning of the externally applied load between the bolt and the joint. Load proportioning will be discussed in the next section.

11.9 FASTENED JOINTS LOADED IN TENSION

The joint formed when two or more mating surfaces are fastened together can be loaded in either tension or shear. The method of analysis of the fastener system differs significantly with the type of loading. Let's consider the joint shown in Fig. 11.37, with a tensile load P applied so as to separate the mating surfaces. When the load is applied, both the bolt and the joint respond by deflecting a distance δ. The surface of the joint remains in contact and $\delta_b = \delta_j$. This equality leads directly to:

$$P_b/k_b = P_j/k_j \qquad (11.91)$$

where P_b is the proportion of the load P that is carried by the bolt. P_j is the proportion carried by the joint, which is obtained by reducing the interface pressure. The spring rates k_b and k_j are given by:

$$k_b = \frac{A_b E_b}{L_b} \quad \text{and} \quad k_j = \frac{A_j E_j}{L_j} \qquad (11.92)$$

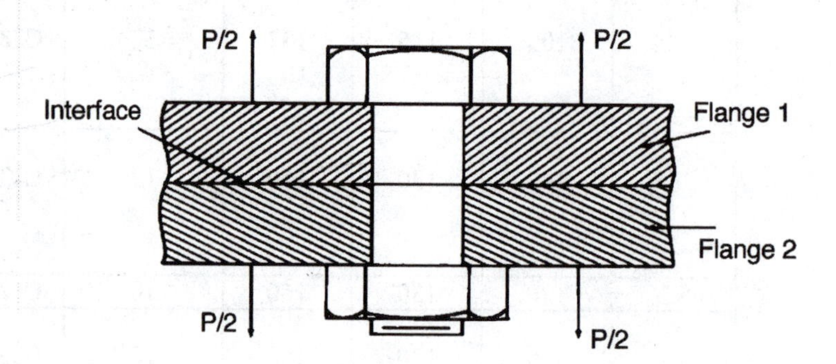

Fig. 11.37 A bolted joint subjected to a tensile load P that tends to separate the interface.

The area of the joint depends on the thickness L_j of the two edges being clamped and may be approximated by:

$$A_j = \frac{\pi}{4}\left(d_j^2 - d_b^2\right) \qquad (11.93)$$

where

$$d_j = (3d_b + L_j)/2 \qquad (11.94)$$

Noting that $P = P_b + P_j$ and using Eq. (11.115) gives:

$$P_b = \frac{k_b}{k_b + k_j} P = CP \qquad (11.95a)$$

$$P_j = \frac{k_j}{k_b + k_j} P = (1 - C)P \qquad (11.95b)$$

where $C = \dfrac{k_b}{k_b + k_j}$

Examination of Eqs (11.95) shows that the applied loaded is partitioned in accordance with the relative stiffness of the two components. The stiffer component carries the larger share of the load. Next, we super impose the preload to obtain:

$$F_b = P_b + F_i \quad \text{and} \quad F_j = P_j - F_i \tag{11.96}$$

The force F_j in the joint must always remain negative to insure a clamping pressure that maintains the mating surfaces in contact. If P_j becomes greater than the bolt preload F_i the joint will open and the entire load P is carried by the bolt. If this situation occurs, the fastening system almost always fails. The bolt force F_b produces stresses that can vary with the external load P causing cyclic loading of the bolt. The effect of these cyclic stresses on the fatigue behavior of the bolt will be considered in the next section.

11.9.1 Stress and Fatigue Analysis

The stresses produced in the bolt by the externally applied load P may be determined from:

$$\sigma_b = F_b/A_t \tag{11.97}$$

If we consider P to vary with time as shown in Fig. 11.38a, then the forces F_b and the stresses σ_b will vary with time in the manner illustrated in Fig. 11.38b. Two different effects of the fluctuating stresses are of concern. First, the maximum value of σ_b must be less than the proof stress of the screw to avoid static failure by permanently deforming the screw. The safety factor for static failure by a single cycle of overload is:

$$\mathbf{S} = S_p/(\sigma_b)_{max} \tag{11.98}$$

Second, the stress σ_b must be limited to avoid fatigue failure. The fatigue analysis involves determining the mean and alternating stresses imposed on the bolt, which are defined as:

$$\sigma_m = \frac{\sigma_{max} + \sigma_{min}}{2} \quad \text{and} \quad \sigma_a = \frac{\sigma_{max} - \sigma_{min}}{2} \tag{11.99}$$

where the subscripts m and a refer to mean and alternating components of stress as defined in Fig. 11.37b.

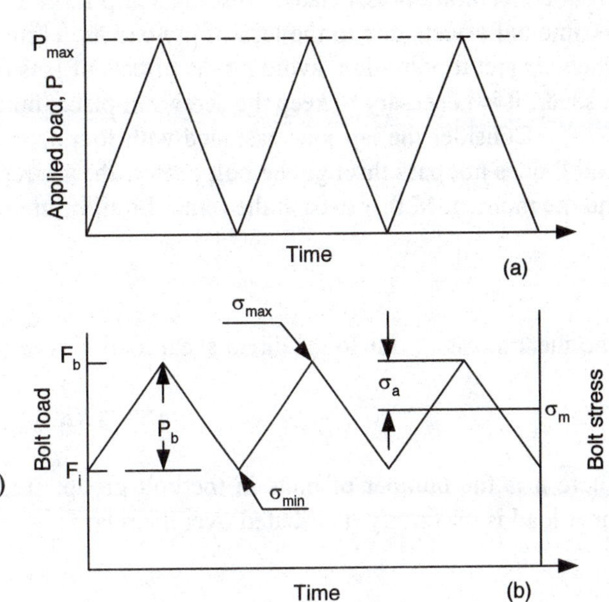

Fig. 11.38 (a) Load applied to a bolted joint. (b) Load ands stress acting on the bolt.

Combining Eqs (11.96),(11.95a) and (11.97) with Eq. (11.99) gives:

$$\sigma_m = \frac{CP + 2F_i}{2A_t} \quad \text{and} \quad \sigma_a = \frac{CP}{2A_t} \tag{11.100}$$

The modified Goodman theory accounts for the effect of both the mean and the alternating stresses acting simultaneously on the fatigue strength. The equation for the modified Goodman line is:

$$\frac{\sigma_u}{S_u} + \frac{\sigma_a}{S_e} = \frac{1}{\boldsymbol{S}} \tag{11.101}$$

Let's use the approximate relation between the endurance strength S_e and the tensile strength S_u and set $S_e = ½S_u$. Then substitute Eq. (11.100) into Eq. (11.101) to obtain:

$$\boldsymbol{S} = \frac{2A_tS_u}{3CP + 2F_t} \tag{11.102}$$

Inspection of this relation shows that the safety factor ***S*** for fatigue is increased by selecting larger screws (higher A_t) made from higher strength materials. The effect of increasing the preload F_i is to decrease the safety factor. This result does not imply that very low preload should be employed. Remember that F_i must be greater that the maximum value of P_j to keep the joint from separating.

The safety factor employed should be relatively high to account for the effects of the stress concentration at the root of the threads. Because this stress concentration is about 3 for rolled threads, safety factors of 4 to 5 are commonly employed in designing joint assemblies subjected to fluctuating tensile loads.

11.10 FASTENED JOINTS LOADED IN SHEAR

Design of joints that place screws in shear rather than tension are preferred for two reasons. First, the stress concentration associated with the sharp radii at the roots of the threads is avoided. Second, the detrimental effects due to the loss of preload are eliminated. Because bolted assemblies tend to loosen when subject to vibration, avoiding the effects of loss of preload is important.. When the joint is loaded in shear, it is necessary to keep the screws in place, but preload is not required for load partitioning.

Consider the lap joint fastened with four screws, illustrated in Fig. 11.39. Because the applied load P does not pass through the bolt center, shear stresses are produced by both the direct shear force V and the moment M that exist at the joint. Equilibrium relations give:

$$V = P \quad \text{and} \quad M = PL \tag{11.103}$$

The shear stress τ^* due to the direct shear load V is:

$$\tau^* = V/(nA_b) \tag{11.104}$$

where n is the number of bolts in the bolt group. In writing Eq. (11.104), we have assumed that the shear load is uniformly distributed over the n bolts.

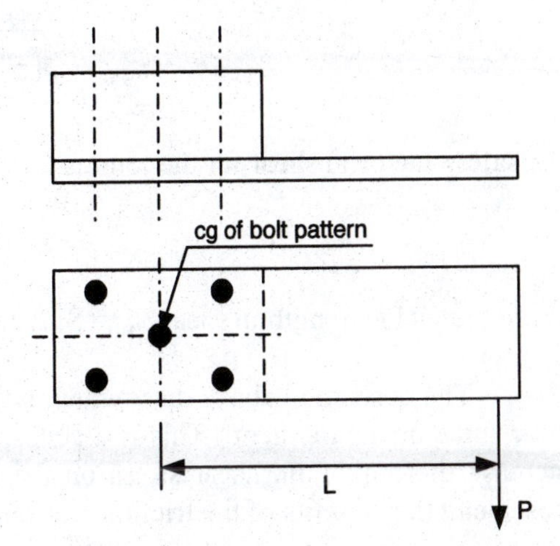

Fig. 11.39 A bolted joint loaded in shear.

The moment M is resisted by developing the forces F_1, F_2, F_n, as shown in Fig. 11.40. The moment force relation is:

$$M = F_1 r_1 + F_2 r_2 + ... + F_n r_n \qquad (11.105)$$

Next, it is assumed that the force F is proportional to the radial position of the individual bolt from the cg of the bolt group. This assumption leads to:

$$F = cr \qquad (11.106)$$

where c is the constant of proportionality.

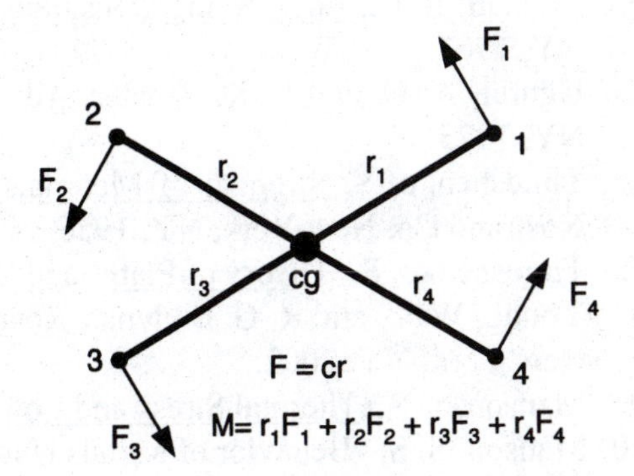

Fig. 11.40 Shear forces F_1, ... F_4, developed due to moment M.

Combining Eqs (11.105) and (11.106) gives the force F_n as:

$$F_n = \frac{M r_n}{\left(r_1^2 + r_2^2 + ... + r_n^2\right)} \qquad (11.107)$$

The stress τ^{**} due to the moment M is then:

$$\tau_n^{**} = F_n / A_{bn} \qquad (11.108)$$

Note, that F_n and τ_n are maximums for the bolt furthest removed from the center of gravity of the bolt group. It is this bolt that determine if failure in shear will occur. We combine the contributions of the two shear stresses by using vector addition to obtain:

$$\tau_{max} = \sqrt{\left(\tau_x^{**}\right)^2 + \left(\tau_x^{*} + \tau_y^{**}\right)^2} \tag{11.109}$$

The safety factor in shear for the joint is :

$$\boldsymbol{S} = \tau_{max}/S_{ys} \tag{11.110}$$

where the yield strength in shear $S_{ys} = S_y/2$ according to the Tresca theory of yielding.

The derivation above discounted the effect of friction forces in providing resistance to either the shear force or the moment. This is a conservative approach that is recommended. Friction forces will decrease the maximum shear stress on the bolts; however, when exposed to vibrations bolts tend to loosen and the benefits of the friction resistance is lost.

REFERENCES

1. Peterson, R. E., Stress Concentration Design Factors, 2nd Edition, Wiley & Sons, New York, NY 1990.
2. Dally, J. W. and R. J. Bonenberger, Design Analysis of Structural Elements, 4th Edition, College House Enterprises, Knoxville, TN, 2004.
3. Riley, W. F., Sturges, L. D. and D. H. Morris, Static and Mechanics of Materials, John Wiley & Sons, New York, NY, 1995.
4. Juvinall, R. C., Engineering Considerations of Stress Strain and Strength, McGraw Hill, New York, NY, 1967.
5. Ugural, A. C. and S. K. Fenster, Advanced Strength and Applied Elasticity, Elsevier, New York, NY, 1975.
6. Timoshenko, S. Strength of Materials, Part II Advanced Theory and Problems, 3rd Edition, Van Nostrand Co. New York, NY, 1956.
7. Timoshenko, S., Theory of Plates and Shells, McGraw Hill, New York, NY, 1940.
8. Young, W. C. and R. G. Budynas, Roark's Formulas for Stress and Strain, 7th Edition, McGraw Hill, New York, NY, 2002.
9. Manson, S. S., Thermal Stress and Low-Cycle Fatigue, McGraw Hill, New York, NY, 1966.
10. Manson, S. S. "Behavior of Metals Under Conditions of Thermal Stress," Heat Transfer Symposium University of Michigan Press, pp. 9-75, 1953.
11. Coffin, L. F., "A Study of Cyclic Thermal Stresses in a Ductile Material", Transactions of the ASME, Vol. 76, pp. 931-950, 1954.
12. Coffin, L. F., "Design Aspects of High Temperature Fatigue with Particular Reference to Thermal Stresses," Transactions of the ASME, Vol. 78, pp. 527-532, 1956.
13. Kanchanomai, C. and Y. Mutoh, "Low-Cycle Fatigue Prediction Model for PB-Free Solder 96.5 Sn-3.5 Ag," Journal of Electronic Materials, Vol. 23, No. 4, 2004.
14. Smith, R. N. P. Watson and T. H. Topper, "Journal of Materials, Vol. 5, p. 767, 1970.
15. Morrow, J. D. ASTM STP 378, ASTM, Philadelphia, PA , p. 45, 1965.

EXERCISES

11.1 A uniaxial rod fabricated from an aluminum alloy is subjected to an axial force of 400 N. Design the rod's diameter and select an alloy of aluminum that will provide a safety factor 2.75 against failure by yielding.

11.2 Determine the axial deformation of a long slender rod subjected to an axial force of 400 N. The is fabricated from aluminum with a diameter of 1.0 mm and length of 600 mm.

11.3 Write Hooke's law for uniaxial, biaxial and triaxial states of stress. Write the equations for the stresses in terms of strains.

11.4 Sketch a stress-strain diagram and identify the modulus of elasticity, the yield strength and the ultimate tensile strength on the diagram.

11.5 A copper wiring trace on a flexible circuit is formed around a mandrel with a radius of 14 mm. Determine the strain and stress in the trace and a function of the thickness of the copper for cladding with weights of 0.5, 1.0 and 2.0 oz/ft ^{2}.

11.6 A beam, subjected to a pure moment of 200 N-m, spans an opening 800 mm wide. Select the beam material and its cross section to minimize the weight of the beam while maintaining a safety factor of 2.5.

11.7 A small shaft transmits 40 W of power to drive hard disk at 7,000 RPM. Determine the torque applied to this shaft.

11.8 If the shaft in Exercise 11.7 is 2.0 mm in diameter, determine the shear stresses produced in the shaft when operating at full power.

11.9 If the shaft in Exercises 11.7 and 11.8 is 8.0 mm long between the motor coils and the hard disk, determine its angle of twist when the disk is operating at full power. Assume the shaft is fabricated from steel.

11.10 You are assigned the task of designing a small leaf spring as part of a mechanism. This mechanism will accommodate a cantilever, as shown in Fig. Ex11.10. The spring rate is specified as 16 N/mm. The beam's length is constrained to 14 mm. Design the beam's cross section and select the beam material to best meet the spring rate specification.

P

L

Fig. Ex11.10

11.11 A refrigeration unit for a heat management system for a large high performance computer is to be placed on the roof of the building that houses it. It is to be supported by two parallel beams with a span of 8.5 m as shown in Fig. Ex11.11. The weight of the refrigeration unit is 20 kN that is divided equally between the two beams. Select the material for the beam and its cross section to provide a support structure with a safety factor of 3.5.

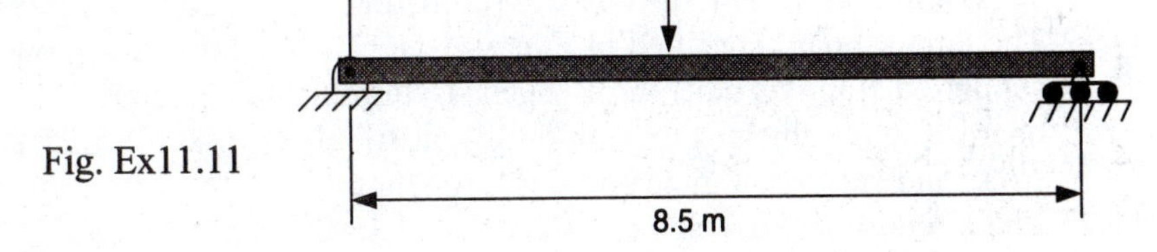

Fig. Ex11.11

11.12 Extend Exercise 11.11 to include the determination of the maximum deflection of one of the beams due to the loading of the refrigeration unit.

11.13 The beam shown in Fig. Ex11.13 is one of several beams used to support the floor in a small electronics assembly plant. Select the material for the beam and its cross section to provide a floor support system with a safety factor of 3.5.

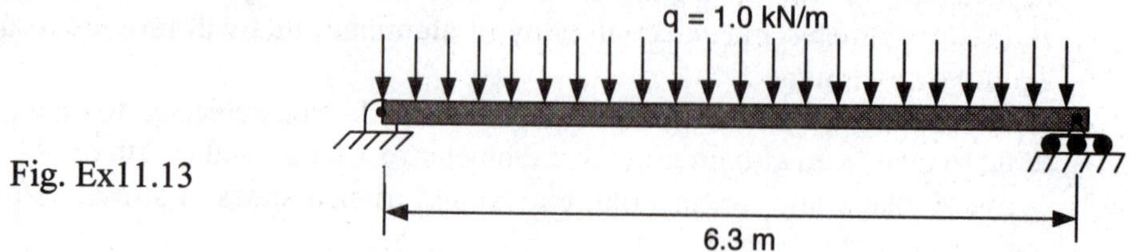

Fig. Ex11.13

11.14 Extend Exercise 11.13 to include the determination of the maximum deflection of one of the beams due to the uniformly distributed floor loading.

11.15 A rectangular flat, fabricated from aluminum 2024-T4 with yield strength of 296 MPa, is used to support a heavy truck mounted transformer. The bar is 25 mm wide, 200 mm long and 4 mm thick. In supporting the weight of the transformer the bar is subjected to a static stress of 140 MPa. During repair of the equipment a technician drills a 12.7 mm hole in the center of the bar to route a new cable past the transformer assembly. Determine the effect of the presence of the hole and the maximum stress in the critical section of the tension bar. Discuss the possible consequences of the presence of the hole.

11.16 A circuit board, 150 mm wide by 250 mm long by 1.5 mm thick, is fabricated from glass-reinforced epoxy determine its flexural rigidity. The modulus of elasticity for the glass reinforced epoxy is 20 GPa and its Poisson's ratio is 0.25.

11.17 A circuit board, 100 mm wide by 200 mm long by 1.0 mm thick, is fabricated from glass-reinforced epoxy determine its flexural rigidity. The modulus of elasticity for the glass reinforced epoxy is 18 GPa and its Poisson's ratio is 0.27.

11.18 A circuit board, 200 mm wide by 300 mm long by 1.5 mm thick, is fabricated from glass-reinforced epoxy determine its flexural rigidity. The modulus of elasticity for the glass reinforced epoxy is 23 GPa and its Poisson's ratio is 0.28.

11.19 The loading on the circuit board in Exercise 11.16 during a vibration test is simulated by summing the weight of the board and all of its components and dividing that weight by the area of the board to obtain an effective uniform pressure. If the components mounted on this PCB have a mass of 400 grams and the glass reinforced epoxy has a specific gravity of 2.0, determine the effective uniform pressure acting on it.

11.20 During the vibration test the circuit board described in Exercise 11.19 is subjected to a resonance dwell with a force transmission coefficient of 24. Determine the effective uniform pressure when the PCB is resonating.

11.21 The circuit board specified in Exercise 11.16 is clamped along one of its long edges and simply supported along its other three edges. Determine the maximum stress and deflection of the circuit board if it is subjected to a uniformly distributed pressure of 30 kPa. Determine the maximum stress and deflection if the board is resonating with a force transmission coefficient of 22.

11.22 The circuit board specified in Exercise 11.17 is clamped along one of its long edges and simply supported along its other three edges. Determine the maximum stress and deflection of the circuit board if it is subjected to a uniformly distributed pressure of 25 kPa. Determine the maximum stress and deflection if the board is resonating with a force transmission coefficient of 18.

11.23 The circuit board specified in Exercise 11.18 is clamped along one of its long edges and simply supported along its other three edges. Determine the maximum stress and deflection of the circuit board if it is subjected to a uniformly distributed pressure of 40 kPa. Determine the maximum stress and deflection if the board is resonating with a force transmission coefficient of 26.

11.24 The circuit board specified in Exercise 11.16 is clamped along one of its long edges and simply supported along two edges with the remaining long edge free. Determine the maximum stress and deflection of the circuit board if it is subjected to a uniformly distributed pressure of 35 kPa.

Determine the maximum stress and deflection if the board is resonating with a force transmission coefficient of 20.

11.25 The circuit board specified in Exercise 11.17 is clamped along one of its long edges and simply supported along two edges with the remaining long edge free. Determine the maximum stress and deflection of the circuit board if it is subjected to a uniformly distributed pressure of 25 kPa. Determine the maximum stress and deflection if the board is resonating with a force transmission coefficient of 18.

11.26 The circuit board specified in Exercise 11.18 is clamped along one of its long edges and simply supported along two edges with the remaining long edge free. Determine the maximum stress and deflection of the circuit board if it is subjected to a uniformly distributed pressure of 40 kPa. Determine the maximum stress and deflection if the board is resonating with a force transmission coefficient of 26.

11.27 Determine the membrane stress in spherical pressure vessel if it is filled with hydrogen and pressurized to 5 MPa. The r/t ratio for this spherical vessel is 20. Select a material to provide a suitable safety factor. Explain your rationale for selecting a numerical value for a "suitable" safety factor.

11.28 Determine the membrane stress in cylindrical pressure vessel if it is filled with a toxic chemical and pressurized to 6 MPa. The r/t ratio for the spherical vessel is 14. Select a material to provide a suitable safety factor. Explain your rationale for selecting a numerical value for a "suitable" safety factor.

11.29 Describe the failure theory most commonly used to predict the onset of yielding in machine component.

11.30 Describe the failure theory most commonly used to predict fracture of a machine component fabricated from a brittle material.

11.31 Prepare a sketch showing the difference between the theory of failure due to maximum shear stresses and maximum distortion energy.

11.32 A rule for approximating the fatigue strength is $S_f = S_u/2$. Using this rule, determine the maximum stress that can be imposed on a machine component with a dynamic load that is twice the static load as illustrated in Fig. Ex11.32. The machine component is fabricated from steel with $S_u = 340$ MPa. Neglect the influence of the cyclic mean stresses on the fatigue behavior.

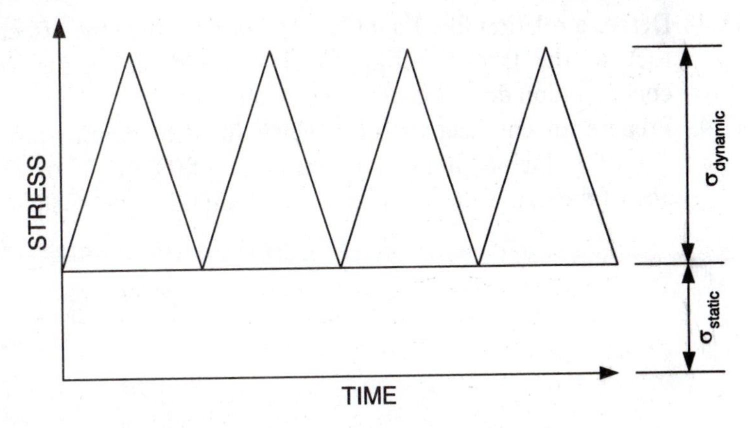

Fig. Ex11.32

11.33 Reconsider Exercise 11.32 taking into account the effect of the cyclic mean stress.

11.34 Prepare a sketch of a hysteresis loop on a stress-strain diagram similar to the one shown in Fig. 11.30. Assume stress and strain values for point A in the diagram of 320 MPa and 0.03, respectively. From this stress-strain diagram determine the elastic strain range $\Delta\varepsilon$ and the plastic strain range $\Delta\varepsilon_p$.

11.35 Using the data presented in Fig. 11.32, determine the material constants M and z used in Eq. (11.77).

11.36 A test laboratory has conducted stain controlled fatigue tests at two strain range levels $\Delta\varepsilon_1$ 0.05 and $\Delta\varepsilon_2 = 0.09$ and measured the corresponding stress range $\Delta\sigma_1 = 300$ MPa and $\Delta\sigma_2 = 340$ MPa and cycles to failure $N_1 = 2.5 \times 10^4$ and $N_2 = 1.2 \times 10^3$. Determine the material constants used in Eq. (11.78) for this material.

11.37 An aluminum alloy 2024-T4 is specified with a tensile strength of 448 MPa, an elastic modulus of 72 GPa and a reduction of area of 40% in a tensile test to rupture. Write a predictive equation for the low cycle fatigue life N_f for this alloy.

11.38 Write an engineering brief explaining why it is more difficult to model the low-cyclic fatigue behavior of solder than that of alloys with much higher melting points.

11.39 Why is flow stress important in formulating a model showing the behavior of solder subjected to strains generated by thermal cycling.

11.40 Explain why rivets are usually preferred as the fastener when it is not necessary to provide for the opportunity for disassembly.

11.41 Take out one of the screws holding the cover of one of your electronic products and completely describe the screw in terms of the terminology presented in Section 11.8.

11.42 Prepare an engineering brief giving arguments for using a grade 5 bolt rather than a grade 1. Can you support this position even if the loads do not require a high strength fastener?

11.43 Sketch a curve showing the cost of a fastener as a function of the size. Include in this estimate the labor associated with tapping a hole for the screw and the assembly operation. Would you ever use a 1-72 UNF? Cite the application and explain why.

11.44 Specify the torque to be applied to a 1/4-20 UNC, grade 5 bolt if it is used to clamp a joint subjected to a static load.

11.45 Determine the spring rates k_b and k_j for a joint designed with a 10-32 screw that is clamping two steel flanges each 3 mm thick.

11.46 If the joint of Exercise 11.45 is subjected to a load which varies from 0 to 2,000 N in a vibratory application, find the maximum, minimum, mean and alternating stresses acting on the screw. If a preload of $F_i = P_p/2$ is applied to the screw determine the safety factor for static loading and for fatigue loading.

11.47 Verify Eq. (11.102).

11.48 Derive a relation like Eq. (11.102), but introduce the effect of the stress concentration, due to the root of the thread, acting to elevate the alternating stress. Also, assume that the stress concentration does not affect the mean stress.

11.49 Prepare an engineering brief which justifies elimination of the friction force in deriving Eq. (11.107). Include in this brief the effect of 0.0001 in. of vibration wear under the head of the bolt and of the nut. Consider that the total joint thickness is 0.4 in.

CHAPTER 12

THERMO-MECHANICAL ANALYSIS

12.1 INTRODUCTION

A typical printed circuit board is fabricated with many different components which in turn are fabricated from many different materials. These different materials are attached to one another by adhesive bonding or by soldering in the production of the chip carriers or in the assembly of the components on the printed circuit board. Because each of these materials exhibits different properties—modulus of elasticity, Poisson's ratio, temperature coefficient of expansion, etc.—differential strains are induced in each component when the temperature changes. The temperature changes can be due to power on/off or they can be due to changes in the ambient conditions. In either circumstance the temperature differential can be large and the thermal strains result in thermal loads, thermal deformations (warping) and thermal stresses. Thermal cycles due to repeated power on and power off as well as those imposed in qualification testing impose cyclic stresses on the components and on the solder joints. Consequently, fatigue failures are a serious concern in almost all system designs.

There are several topics of concern to engineers designing a first level package or a printed circuit board. These include failure of the chip, failure of the solder joints when using flip chip technology, delamination of the barrier layer metallization, thin film stresses, warpage of the printed circuit board, failure of plated through holes, and failure of the solder joints between the chip carrier and the PCB. Closed form analysis methods are available, but they are at best approximate, because of the non-linearity in material properties over the temperature ranges imposed on the components. More accurate analysis involves finite element methods using software which enables the engineer to account for the non-linear behavior of the materials involved. In some cases, experimental methods are employed to measure the displacements, or strains induced by the temperature changes. Finally, tests are often conducted with thermal cycling of a relatively large number of components that are mounted to simulate a board design. These tests are conducted with a sufficient number of thermal cycles to fail as significant portion of the population. The results are plotted on Weibull graph paper enabling relatively accurate predictions of system reliability.

12.2 THE DNP RELATION

Let's consider a simple two component structure illustrated in Fig. 12.1a. In this case, we have a chip carrier soldered to a printed circuit board. We assume it to be in a zero stress state at a temperature of T_0. If the ambient temperature increase to T where $T > T_0$, both the chip carrier and the PCB expand by an amount equal to $\alpha L \Delta T$. However, the temperature coefficient of expansion α is different for the chip carrier and for the PWB. For a PCB fabricated from a typical composite(glass-reinforced-epoxy) the temperature coefficient of expansion $\alpha_{PCB} = 20 \times 10^{-6}$/°C. The temperature coefficient of expansion (TCE) of the chip carrier depends on the material from which it is fabricated. The TCE can be much lower than that of the glass-reinforced epoxy if it is made from 92% alumina (ceramic), where $\alpha_{CC} = 6.8 \times 10^{-6}$/°C. The illustrations presented in Fig. 12.1 assume that $\alpha_{PCB} > \alpha_{CC}$.

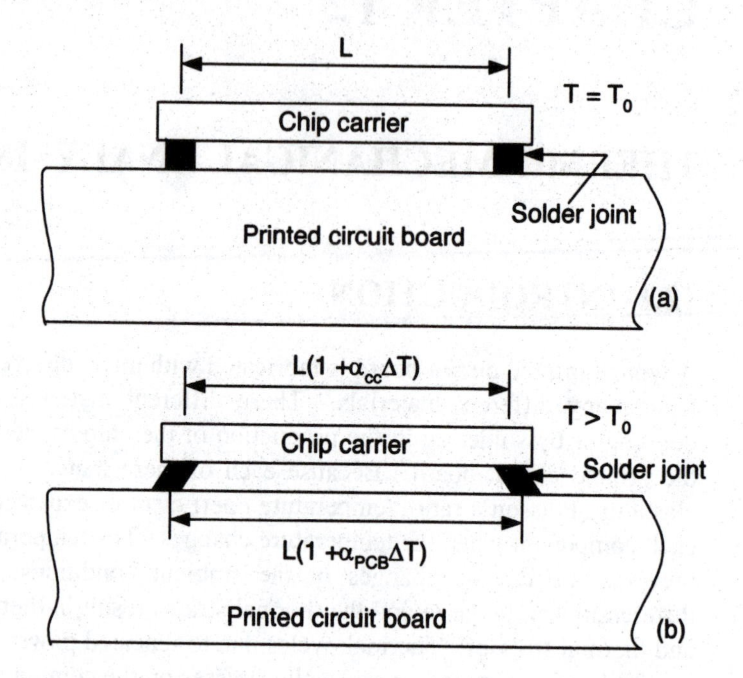

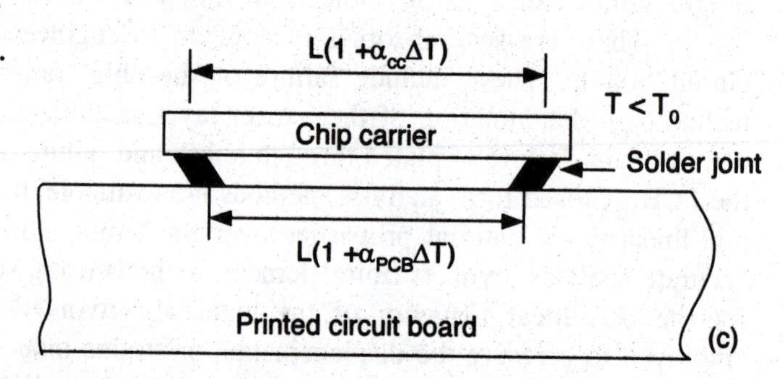

Fig. 12.1 Three illustrations depicting solder joint deformation due to temperature changes.
(a) Neutral state with zero stresses.
(b) Shear strains induced in solder joint due to a temperature increase.
(c) Shear strains induced in solder joint due to a temperature decrease.

When the temperature increases both the chip carrier and the PCB expand as presented in Fig. 12.1b. However, the PCB expands more than the chip carrier because its TCE is larger. This differential expansion causes the solder joints to deform (shear) as indicated in Fig. 12.1b. When the temperature decreases the PCB contracts more than the chip carrier, the differential contraction causes the solder joints to undergo a shear strain in the opposite direction as shown in Fig. 12.1c.

An estimate of the shear strain produced in the solder joint by the difference in the thermal expansion between the chip carrier and the PCB may be made by referring to Fig. 12.2. The expansion of the printed circuit board measured from the neutral point on the chip carrier centerline is given by:

$$\Delta S_{PCB} = (L/2)[\alpha_{PCB}\, \Delta T] \tag{12.1}$$

The expansion of the chip carrier measured from the neutral point is given by:

$$\Delta S_{CC} = (L/2)[\alpha_{CC}\, \Delta T] \tag{12.2}$$

Subtracting Eq. (12.2) from Eq. (12.1) yields the differential expansion d as:

$$d = (L/2)[\alpha_{PCB} - \alpha_{CC}]\Delta T \tag{12.3}$$

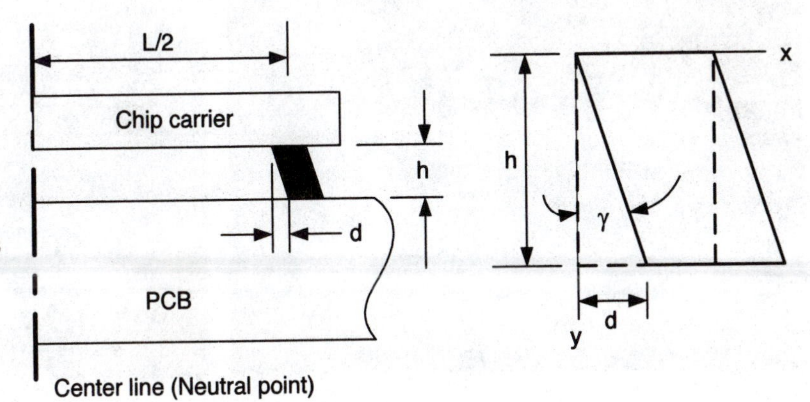

Fig. 12.2 Geometric parameters defining the shear strain in a solder joint.

The differential expansion (or contraction) produces a shear strain in the solder joints. Recalling that the shear strain is the change in the right angle between the x and y axes, as shown in Fig. 12.2, we can write:

$$\tan\gamma \approx \gamma = \frac{d}{h} = \frac{L(\alpha_{PCB} - \alpha_{CC})\Delta T}{2h} \tag{12.4}$$

where h is the height of the solder joint

(L/2) is the distance from the solder joint to the neutral point (DNP)

It is evident from Eq. (12.4) that the shear strain γ decreases as the height of the solder joint increases. It is also apparent that increases in the DNP increases the shear strain. This fact is the reason why fatigue failures usually occur at solder joints located at the corners of leadless chip carriers when they are subjected to a large number of thermal cycles.

Depending upon the temperature extremes, the shear strain γ is composed of an elastic part and a plastic part. At the higher temperatures the shear strain is mostly plastic and at the lower temperatures it is mostly elastic. The temperature cycle depends on the environment to which the equipment is subjected as well as the maximum temperature reached when the chip is operating. Commercial products are frequently cycled from 0 to 100 °C in test programs to verify reliability. Military products are often exposed to much more severe environments and temperature cycling from −55 °C to + 150 °C is common.

While testing a large number of components in cyclic thermal fatigue tests is recommended to establish a Weibull relation upon which to base reliability predictions, it is possible in some cases to approximate cyclic thermal fatigue life using Eq. (12.4) together with the Coffin Manson relation written as:

$$N_f = 1.29(\Delta\gamma_p)^{-1.96} \tag{12.5}$$

where $\Delta\gamma_p$ is the plastic strain range due to the temperature changes in thermal cycling. The constants 1.29 and 1.96 in Eq. (12.5) are dependent upon the material (solder) and will change with the composition of the alloy employed.

An example of a failure of a solder joint failure of ceramic chip resistor soldered to a FR-4 printed circuit board is illustrated in Fig. 12.3.

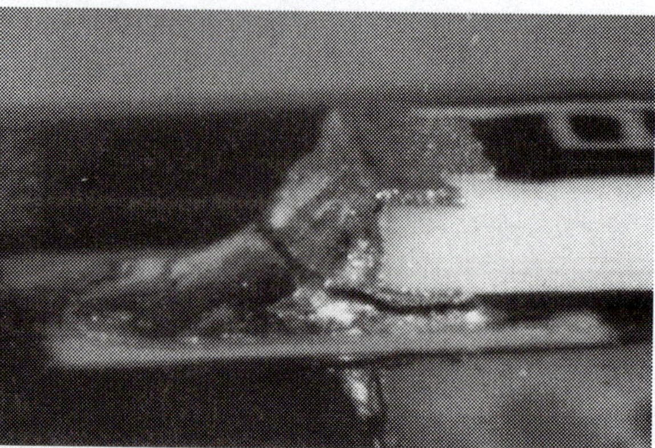

Fig. 12.3 Example of a solder joint failure in a ceramic resistor mounted on a FR-4 printed circuit board.

Let's determine the approximate number of cycles to produce a failure of this type by using Eq. (12.4) and Eq. (12.5). Assume that the ceramic resistor is 0.500 in. long (DNP = 0.250 in.) and made of 92% alumina with α_{CC} = of 6.8×10^{-6}, and is attached with solder joints with a height h = 0.004 in. Also assume that the printed circuit board is fabricated from FR-4 with $\alpha_{PCB} = 20 \times 10{-}6$ and that the board is cycled in temperature from 0 to 100 °C. We will assume that all of the shearing strain is plastic to provide a conservative estimate of the life. Substituting these values into Eq. (12.4) gives:

$$\gamma = \frac{DNP(\alpha_{PCB} - \alpha_{CC})\Delta T}{h} = \frac{0.250(20 - 6.8)10^{-6}(100)}{(0.004))} \qquad (a)$$

$$\gamma = 0.0825$$

Substituting Eq. (a) into Eq. (12.5) yields:

$$N_f = 1.29(\Delta\gamma_p)^{-1.96} = 1.29(0.0825)^{-1.96} = 172 \text{ cycles} \qquad (b)$$

This result of 172 cycles is probably conservative because part of the shear strain is elastic and would not contribute to the failure mechanism to the extent that slip due to plastic strain causes fatigue damage. More exact analysis is possible using finite elements to solve for the strain distribution in the solder joint as a function of the temperature. Most of the finite element codes in use today enable the engineer to consider material properties that vary with temperature. Hence the finite element solutions accommodate the creep and plastic strains that occur in the solder joints due to repeated temperature cycles. An example of a finite element solution for the solder joint shown in Fig. 12.3 is presented in Fig. 12.4.

The results of this finite element analysis indicate a very high concentration of strain at the reentrant corner under the resistor where the solder joint begins. A crack forms at this location and slowly propagates with each temperature cycle through the solder at the interface between the two solder pads on the PCB and the chip carrier. The crack then angles upward and extends into the fillet made at the ends of the resistor.

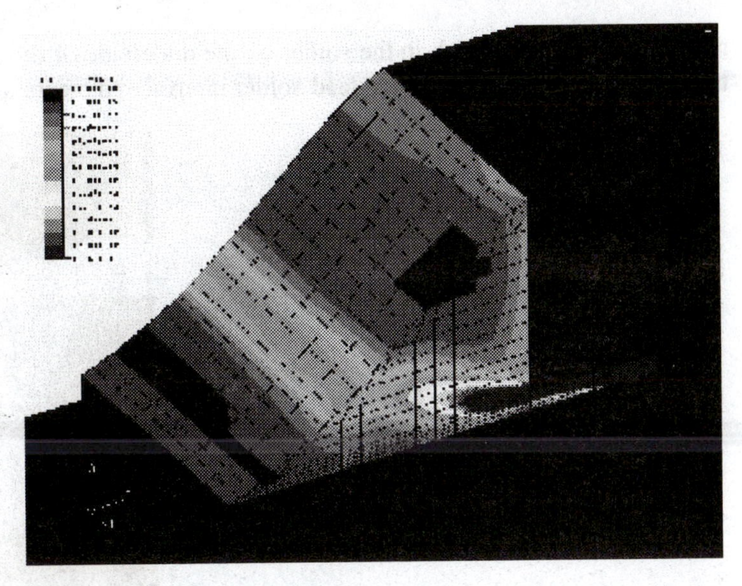

Fig. 12.4 Finite element solution for the strains at a solder joint due to temperature differentials.

12.2.1 Failure of Leaded Chip Carriers due to Thermal Cycling

When leaded chip carriers are employed in the design of a product, the thermal strains imposed on the solder joints are reduced by a significant amount. The reason for the reduction in strain is the compliance of the assembly. The differential expansion of the PCB and the chip carrier is accommodated by both the flexure of the leads and the shearing of the solder joint. An illustration of the effect of the compliance of the lead in accommodating the differential expansion is presented in Fig. 12.5. In this figure s_d is the differential displacement that is comprised of s_s and s_L the displacement of the solder due to shear and the flexure of the J lead, respectively. Clearly the flexure of the J type lead reduces the amount of differential displacement that must be accommodated by the solder joint and increases its fatigue life under the action of cyclic thermal shearing strains.

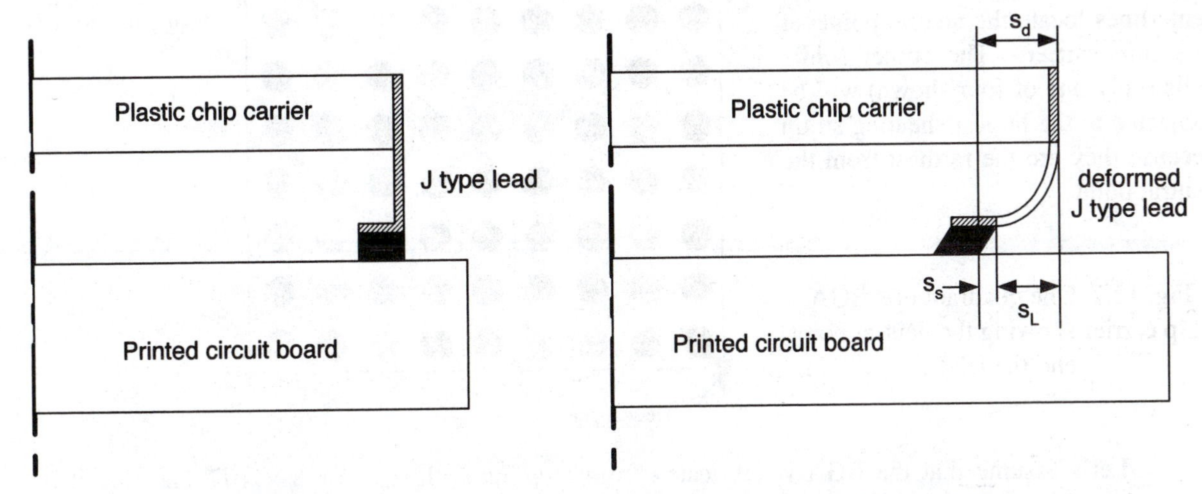

Fig. 12.5 Bending of the lead wire aids in accommodating the differential thermal displacements.

Leaded chip carriers of all types can accommodate larger differential displacements between the PCB and the chip carrier than leadless components. However, if the temperature difference ΔT is sufficiently large and/or the cycle count is very high fatigue failures of the solder joints will occur. Such a failure for a gull wing chip carrier is depicted in Fig. 12.6. In this case the fatigue crack initiated at a relatively sharp notch in the solder on the underside of the lead. Upon reaching the lead, the crack turned and propagated along the lead-solder interface to create an open circuit.

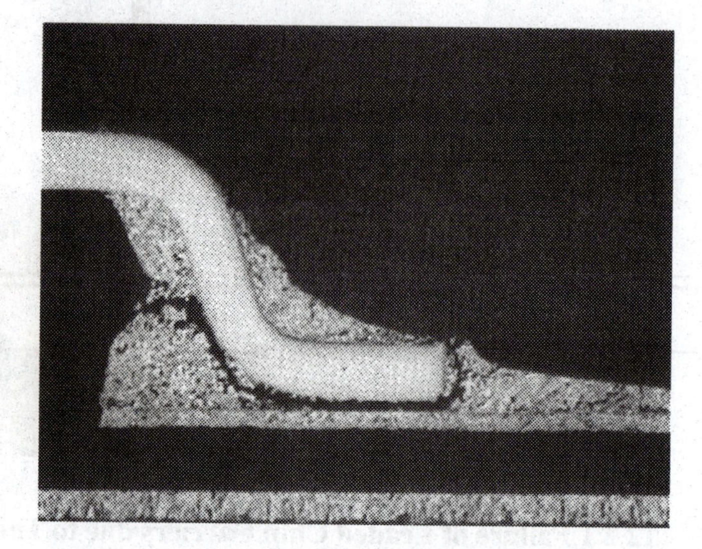

Fig. 12.6 Failure of a solder joint on a chip carrier with gull-wing leads.

12.2.2 Failure of BGA Chip Carriers due to Thermal Cycling

The DNP expression given in Eq. (12.4) can be applied to BGA type chip carriers with two limiting qualifications. First, that the diagonal of the chip carrier is used to locate the solder ball that is the farthest from the neutral point. Second that underfill has not been employed to mitigate the thermal strains due to temperature cycling. To illustrate the location of the solder ball that is subjected to the largest shear strains due to thermal cycling reference the drawing of one quadrant of a BGA chip carrier presented in Fig. 12.7. The two centerlines locate the neutral point for this chip carrier. The corner solder balls (only one of four shown) will be subjected to the largest shearing strain because they are the farthest from the neutral point.

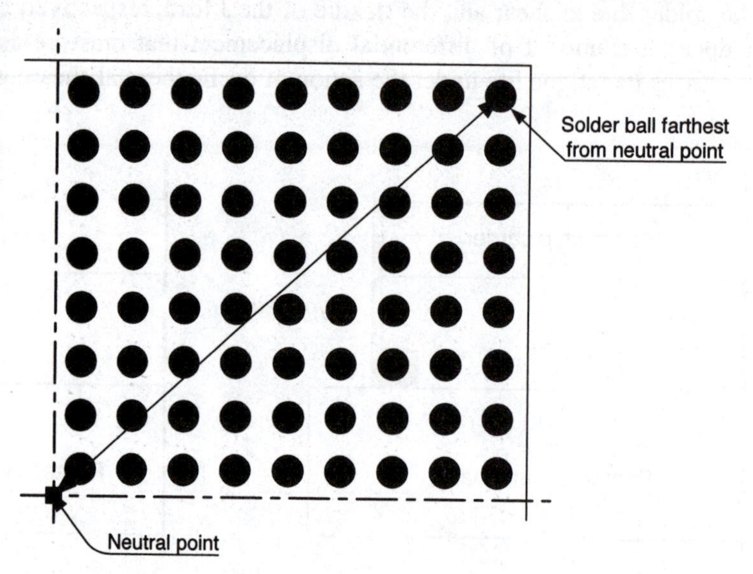

Fig. 12.7 One quadrant of a BGA chip carrier showing the neutral point and the DNP.

Let's assume that the BGA is fabricated from alumina with $\alpha_{CC} = 6.8 \times 10^{-6}$ and that it is soldered to a PCB fabricated from FR-4 with $\alpha_{PCB} = 20 \times 10^{-6}$. The 25 mil solder balls flatten during reflow to provide a space between the chip carrier and the PCB equal to 18 mil. The DNP = 750 mil. If the temperature is cycled from 0 to 100 °C, determine the cyclic thermal fatigue life.

Begin by substituting the numerical values provided above into Eq. (12.4) to give:

$$\gamma = \frac{\text{DNP}(\alpha_{\text{PCB}} - \alpha_{\text{CC}})\Delta\text{T}}{\text{h}} = \frac{0.750(20 - 6.8)10^{-6}(100)}{(0.018))} \qquad \text{(a)}$$

$$\gamma = 0.055$$

We will assume that all of the shearing strain is plastic to provide a conservative estimate of the life. Substituting Eq. (a) into Eq. (12.5) yields:

$$N_f = 1.29(\Delta\gamma_p)^{-1.96} = 1.29(0.055)^{-1.96} = 380 \text{ cycles} \qquad \text{(b)}$$

This result indicates that failures of the solder balls at or near the corner of the ceramic BGA chip carrier is relatively short. In this case, underfill should be specified to reduce the magnitude of the shearing strains due to the ΔT of 100 °C.

Another possible solution is to employ a BGA with a plastic package. In this case the $\alpha = 24 \times 10^{-6}$, and all other parameters are identical to those cited above. Begin the analysis by substituting the numerical values provided above into Eq. (12.4) to give:

$$\gamma = \frac{\text{DNP}(\alpha_{\text{PCB}} - \alpha_{\text{CC}})\Delta\text{T}}{\text{h}} = \frac{0.750(20 - 24)10^{-6}(100)}{(0.018))} \qquad \text{(c)}$$

$$\gamma = -0.01667$$

We will assume that all of the shearing strain is plastic to provide a conservative estimate of the life. We also ignore the minus sign because failure will occur as readily with negative shear strain as with positive shear strain. Substituting the results from Eq. (c) into Eq. (12.5) yields:

$$N_f = 1.29(\Delta\gamma_p)^{-1.96} = 1.29(0.01667)^{-1.96} = 3{,}941 \text{ cycles} \qquad \text{(d)}$$

This is a conservative estimate that indicates a fatigue life exceeding 10 years providing one temperature cycle is imposed each day.

12.2.3 Discussion of the DNP Relation

The DNP solution provides a reasonably accurate prediction of shearing strain induced by temperature changes for leadless resistors, capacitors and chip carriers. It is also a good predictor for flip chip solder connections and for ceramic BGAs providing underfill is not employed. If underfill is used, Eq. (12.4) is no longer valid. Plastic BGAs soldered to glass reinforced plastic PCBs do not usually undergo large thermally induced shearing strains because their TCEs are closely match as indicated by the above example. The non-linearity of the plastic molding compound at elevated temperatures is not considered in the derivation of Eq. (12.4); therefore the use of this relation is questionable for PBGAs.

Surface mounted leaded chip carriers and components mounted with through-hole solder joints usually do not fail due to thermal cycling. The lead geometry introduces compliance into the system and lead flexure accommodates much of the differential displacement thereby relieving the solder joint. If Eq. (12.4) is employed the deflection of the leads must be subtracted from the differential displacement d before this relation can be used to predict solder joint shear strain..

12.2.4 Failure of Plated-Through-Holes

Plated-through-holes may also fail when subjected to large temperature changes. The failure is due to tensile stresses imposed on the copper barrel of the plated-through-hole that occur when the printed circuit board is heated. The highest stresses usually occur in assembly operations when the PCB is heated to relatively high temperatures. The temperature coefficient of expansion of the PCB in the out of plane direction (along the z axis) is much higher than the in-plane TCE. The glass fibers in the laminate constrain in-plane (x and y) expansions causing the out-of-plane expansion to markedly increase with temperature. The TCE is non linear with temperature increasing from about 40×10^{-6} at 20 °C to 100×10^{-6} at 140 °C to 200×10^{-6} at 185 °C.

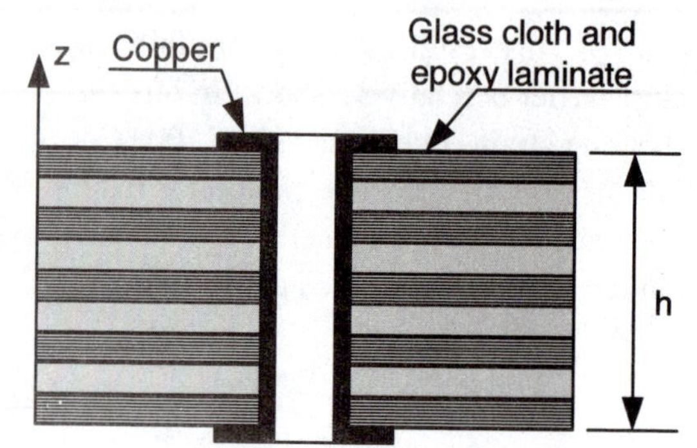

Fig. 12.8 Cross sectional view of a plated-through-hole in a multilayered board.

When a plated-through-hole assembly (see Fig. 12.8) is heated the copper undergoes a free expansion Δh_{Cu} given by:

$$\Delta h_{Cu} = h\alpha_{Cu}\,\Delta T \tag{a}$$

The glass-fiber-epoxy laminate also undergoes a free expansion Δh_{PCB} that is given by:

$$\Delta h_{PCB} = h\alpha_{PCB}\,\Delta T \tag{b}$$

The differential expansion d of these two materials is given by:

$$d = \Delta h_{PCB} - \Delta h_{Cu} = h(\alpha_{PCB} - \alpha_{Cu})\Delta T \tag{12.6}$$

The strain on the copper plated through hole produced by this differential expansion is:

$$\varepsilon = d/h = (\alpha_{PCB} - \alpha_{Cu})\Delta T \tag{12.7}$$

and the stress is given by Hooke's law as:

$$\sigma = E\varepsilon = E\,(\alpha_{PCB} - \alpha_{Cu})\Delta T \tag{12.8}$$

To demonstrate the magnitude of the stresses imposed on a plated-through-hole consider an assembly similar to the one shown in Fig. 12.8 that is subjected to a temperature cycle from 20 to 185 °C. The TCE of copper is $\alpha_{Cu} = 17 \times 10^{-6}$ and the average TCE for the PCB over this temperature range is taken as $(\alpha_z)_{PCB} = 140 \times 10^{-6}$. The assembly thickness is 0.060 in. We determine the strain from Eq. (12.7) as:

$$\varepsilon = d/h = (140 - 17)(165) \times 10^{-6} = 0.02030 \tag{c}$$

and the stress is given by Eq. (12.8) as:

$$\sigma = E\varepsilon = 17.5 \times 10^6\ (0.0203) = 355{,}000 \text{ psi} \tag{d}$$

The result shown in Eq. (d) clearly exceeds the yield strength of plated copper. The integrity of the copper plated through hole depends on the ability of the copper to undergo plastic flow at the elevated temperatures encountered in assembly processes. In some cases, discontinuities exist in the plating and the strains are localized causing separation of the copper barrel. If the plated through hole is used as a signal or power via the separation will produce an open and cause a failure of the system.

12.3 CHIP STRESSES IN PLASTIC PACKAGES

Stresses are produced on the chips in plastic chip carriers during the manufacturing process. In this process the chip is bonded to a lead frame as shown in Fig. 12.9. Electrical connections are made by welding gold or aluminum bonding wires between pads about the perimeter of the chip to the leads which extend out of the package. The lead frame and chip are then placed in a die and the molding compound is injected into the die cavity to form the package. The injection process is performed at elevated temperature—180 °C for 120 seconds at a pressure of 900 psi. The molding compound is post cured at 175 °C for four hours to complete the polymerization of the epoxy. The molding compound becomes rigid as it is cooled to room temperature (20 °C).

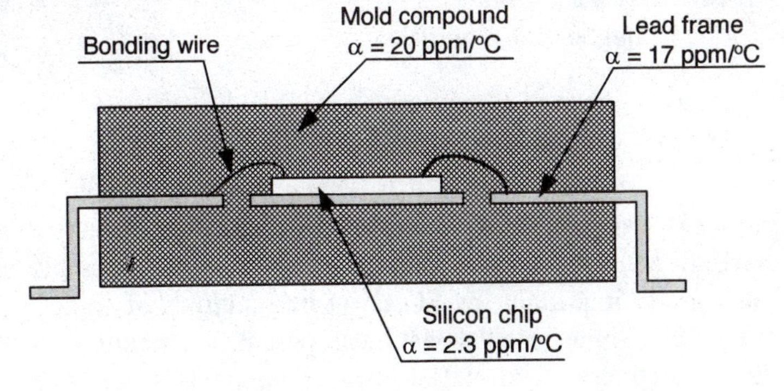

Fig. 12.9 Cross sectional view of a leaded chip carrier formed with molding compound.

Upon cooling to room temperature the differences in the TCE of the molding compound and the silicon chip cause stresses on the chip and on the molding compound. Cracks formed on the die can produce electrical failure and cracks in the molding compound permit entry of moisture, which eventually leads to corrosion and electrical failure. Examples of cracks through the molding compound are illustrated in Fig. 12.10.

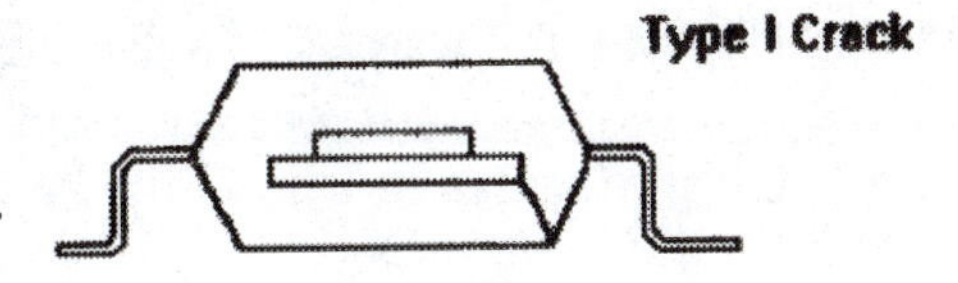

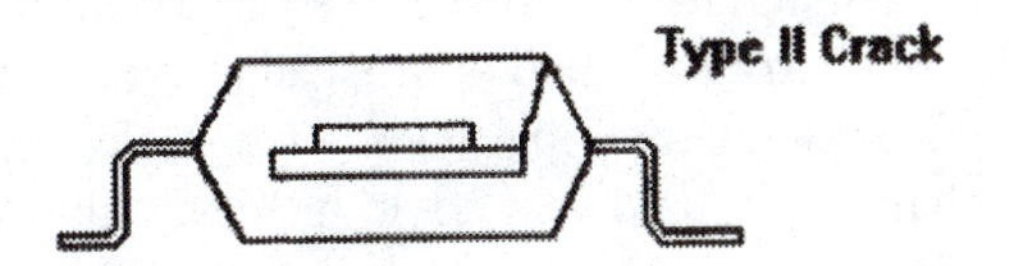

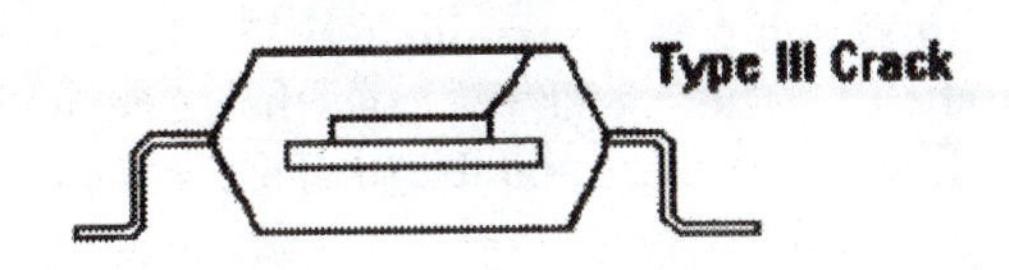

Fig. 12.10 Three types of cracks formed in the molding compound of plastic encapsulated packages.

Efforts to mitigate cracking due to thermal stresses and polymerization shrinkage are usually focused on the formulation of the molding compound. The molding compounds are epoxy based with a resin that exhibits a TCE of 60×10^{-6}/°C. Silica particles with a TCE of 0.5×10^{-6}/°C are blended with the epoxy with sufficient volume to reduce the effective TCE of the mixture to approximately 20×10^{-6}. Additional silica cannot be added to further reduce the TCE of the molding compound because the decreased viscosity would tend to wipe of the delicate bonding wires shown in Fig. 12.9. An example of the silica particles in an epoxy matrix is shown in Fig. 12.11. Note that the particles are of irregular size and shape.

Fig. 12.11 The molding compound is a composite material containing silica particles as well as other particles that toughen the material.

A second problem that occurs due to differential expansions of the materials within a chip carrier is warpage. It is essential that the chip carrier be reasonably flat on its bottom surface so that the leads all contact the solder pads on the PCB in the assembly process. Planarity of the bottom surface is particularly important for BGAs where hundreds of solder balls must all contact the planar surface of the PCB. Some small deviation is possible, depending on the thickness of the solder paste, but the combined tolerance for out-of-plane components is only a few mils.

One approach for improving the planarity of the chip carrier is to introduce silicone into the molding compound. With silicone additions to the base epoxy the flexural modulus of the molding compound is reduced from 1.3 GPa to 0.9 GPa at 260 °C. The polymer shrinkage is also reduced from 0.15% to 0.10%. The warpage is reduced by a factor of 5.

12.3.1 Chip Stresses in Flip Chip Packaging

If a flip chip is mounted on a glass reinforced plastic substrate without underfill, failure will result due to temperature induced stresses. To mitigate these types of failures the space between the chip and the chip carrier is filled with an adhesive. This adhesive is called an underfill and is composed of an epoxy filled with silica particles. The underfill markedly reduces the shearing strain imposed on the solder balls, but in doing so it increases the strains transmitted to the chip. Die cracking occurs more frequently when the chips are tested at very low temperatures (−55 °C). Two types of failure are observed. The half moon failure that occurs at a solder ball location where localized forces are applied to the chip is shown in Fig. 12.12. The second type of failure is edge cracking as illustrated in Fig. 12.13. The cracks are exacerbated by two factors: the modulus of elasticity of the underfill polymer at low temperature and the fillet angle the underfill makes with the edge of the chip.

The modulus of elasticity of the unfilled epoxy in its glassy state is about 2 to 4 GPa. The addition of 65 to 70% of silica particles increases the modulus to about 9 to 15 GPa. The correct modulus of the underfill material is difficult to determine. A high underfill modulus is necessary to reduce the shear stresses on the solder bumps and eventual cracking by temperature cycling. However, if the modulus is too high cracking of the chip may occur when the packages are exposed to very low temperatures (–55 °C).

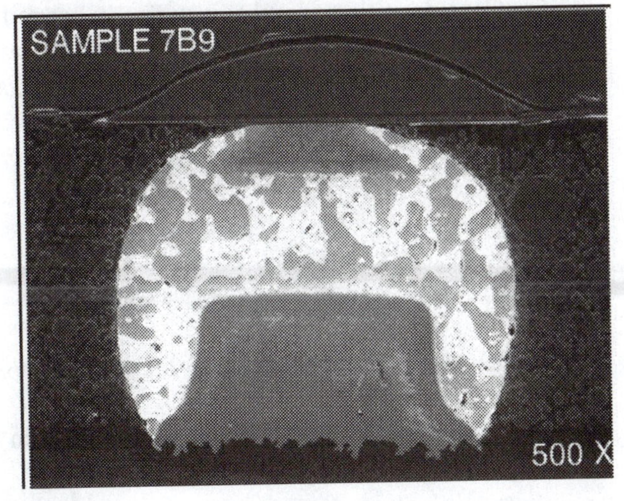

Fig. 12.12 Half moon cracking of the die above a solder ball due to stresses transmitted by the underfill.

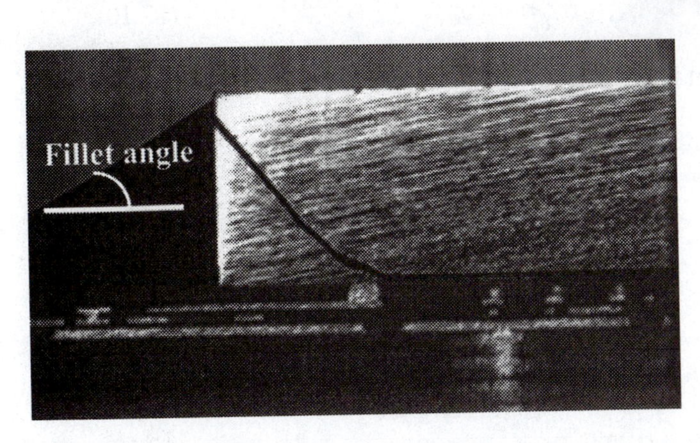

Fig. 12.13 Edge cracking of die due to stresses transmitted by the underfill.

To reduce edge cracking and delamination, it is important that the underfill edge coverage be sufficient. In addition voids and stress concentrations in the fillet region must be avoided. Reducing the elastic modulus of the underfill is often accomplished by the addition of silicone or rubber modifiers. However, this approach may increase the TCE of the underfill, reduce its ability to flow and decrease its glass transition temperature. Clearly, improved underfill adhesives would reduce the failure rate of dies incorporating flip chip technology.

12.4 THIN FILM STRESSES ON RIGID SUBSTRATES

Thin metallic and or oxide films are often placed on a silicon wafer to produce circuit lines, pads for electrical connections or insulating layers. The temperature coefficient of expansion of these metals and oxides often differ from that of silicon. Consequently thermal stresses are created by changes in temperature from that used in placing the thin films. Depending on the TCE of the two materials either biaxial tensile or compressive stresses are introduced in the thin films when the wafer undergoes a temperature change. Biaxial compressive stresses can cause failure when the thin films buckle and delaminate from the substrate. Biaxial tensile stresses can cause failure by cracking the brittle oxide coatings. Usually the thin metallic films are sufficiently ductile to yield thus relieving the tensile stresses without failing.

It is relatively easy to determine the magnitude of the stresses generated by temperature changes (ΔT). Consider a thin film that is perfectly bonded to a rigid substrate. Assume the x-y plane coincides with the plane of the substrate and the z axis is normal to this plane. Also assume a state of plane stress where $\sigma_{zz} = 0$. Next, write the equations for the strains in term of the stresses:

$$\varepsilon_{xx} = (1/E)[\sigma_{xx} - \nu\sigma_{yy}]$$

$$\varepsilon_{yy} = (1/E)[\sigma_{yy} - \nu\sigma_{xx}] \qquad (12.9)$$

$$\varepsilon_{xx} = (-\nu/E)[\sigma_{xx} + \sigma_{yy}]$$

where E is the modulus of elasticity and ν is Poisson's ratio.

The in-plane stresses and strains in the thin film are equal; hence we may write

$$\varepsilon_{xx} = \varepsilon_{yy} = \varepsilon_{\text{in-plane}} \quad \text{and} \quad \sigma_{xx} = \sigma_{yy} = \sigma_{\text{in-plane}} \qquad (12.10)$$

From Eq, (12.9) and Eq. (12.10), it is evident that:

$$\varepsilon_{xx} = \varepsilon_{\text{in-plane}} = \frac{1-\nu}{E}\sigma_{\text{in-plane}} \qquad (12.11)$$

and

$$\sigma_{xx} = \sigma_{\text{in-plane}} = \frac{E}{1-\nu}\varepsilon_{\text{in-plane}} \qquad (12.12)$$

The thermal strains in the substrate $[\varepsilon_s]_{\Delta T}$ are given by:

$$[\varepsilon_s]_{\Delta T} = \alpha_s\,\Delta T \qquad (12.13)$$

where α_s is the temperature coefficient of the substrate material.

Because the thin film is bonded to the substrate, it undergoes the same thermal strain. Hence:

$$[\varepsilon_{TF}]_{\text{Bonded}} = [\varepsilon_s]_{\Delta T} = \alpha_s\,\Delta T \qquad (12.14)$$

The strain $[\varepsilon_{TF}]_{\Delta T}$ due to the free expansion of the thin film material under the same ΔT is given by:

$$[\varepsilon_{TF}]_{\Delta T} = \alpha_{TF}\,\Delta T \qquad (12.15)$$

where α_{TF} is the temperature coefficient of the thin film material.

The strain due to the mismatch in the temperature coefficients of expansion is given by:

$$[\varepsilon_{TF}]_{\Delta T} = (\alpha_{TF} - \alpha_s)\Delta T \qquad (12.16)$$

Substituting Eq. (12.16) into Eq. (12.12) yields the in-plane thin film stresses $\sigma_{\text{in-plane}}$ due to ΔT as:

$$\sigma_{\text{in-plane}} = \frac{E}{1-\nu}(\alpha_{TF} - \alpha_s)\Delta T \tag{12.17}$$

The out-of-plane strain ε_z due to the in-plane stresses is given by Eq. (12.9) and Eq. (12.17) as:

$$\varepsilon_z = -\frac{2\nu}{E}\sigma_{\text{in-plane}} = -\frac{2\nu}{1-\nu}(\alpha_{TF} - \alpha_s)\Delta T \tag{12.18}$$

Finally we add the strain due to the free expansion in the z direction of the thin film resulting from ΔT to Eq. (12.18) to obtain the total strain as:

$$\varepsilon_z = -\left[\alpha_{TF} + \frac{2\nu}{1-\nu}(\alpha_{TF} - \alpha_s)\right]\Delta T \tag{12.19}$$

The analysis of stresses in thin films is accurate as long as the temperature extremes are not excessive. However, at higher temperatures the elastic constant of metallic materials diminishes and the materials flow (yield) at a lower stress level. On the other hand at very low temperatures, the thin film materials are elastic and the equations derived above predict the stresses and strains accurately.

12.5 STRESSES IN BI-MATERIAL LAMINATES DUE TO ΔT

In the previous section, we consider two different materials bonded together: however one material was so thin compared to the other material that the in-plane thermal strains in the thin film were dictated by the behavior of the much thicker substrate. In this section, we again consider two different materials that are bonded together; however they are both comparable in size. Because their TCEs are different the bi-material composite will bend when it undergoes a temperature change ΔT.

Consider a beam of width w and depth h fabricated from materials A and B with a cross section as depicted in Fig 12.14.

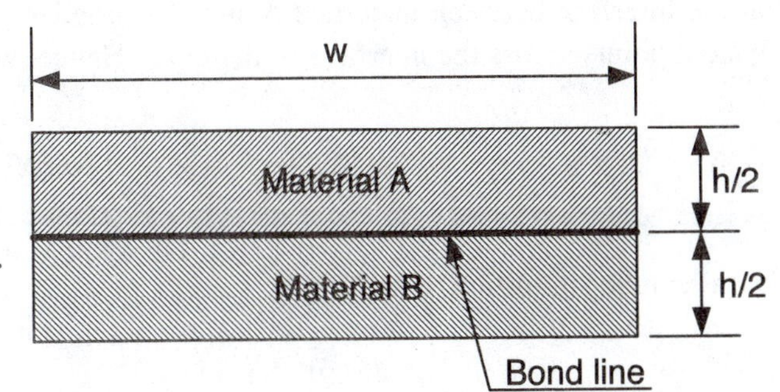

Fig. 12.14 Cross section of a bi-material beam.

When the beam undergoes a temperature change ΔT, it bends and a self equilibrated stress state is induced. We examine this self equilibrated state of stress with a section cut from the length of the beam as illustrated in Fig. 12.15. We have shown the moments and the forces on the section cut for the case where material B has a higher TCE than material A. Consequently when ΔT is positive, material B will be subjected to a bending moment M_B and a compression force P_B. On the other hand, the positive ΔT causes material A to be subjected to a bending moment M_A and a tensile force P_A. The self equilibrated stresses are depicted by the plus and minus triangular regions shown along the centerline of the section cut.

To begin the analysis, consider equilibrium of internal forces over the right side of the section cut and write:

$$\sum F_x = 0; \qquad P_A - P_B = 0; \qquad P_A = P_B = P \tag{12.20}$$

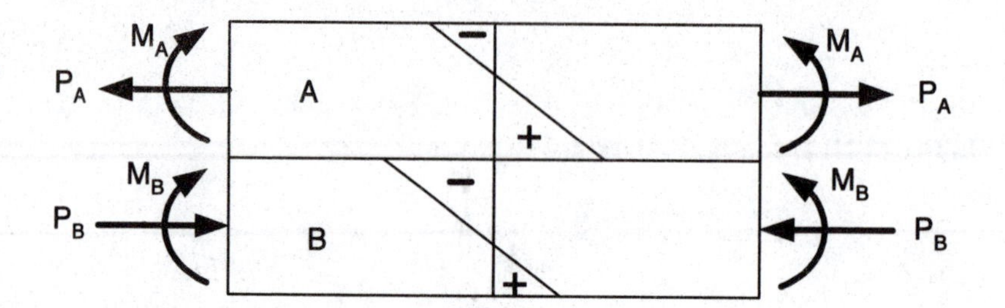

Fig. 12.15 Section cut from the length of the bi-material beam showing the self equilibrated stress state induced by a temperature change ΔT.

The moment acting on the right side of the section cut can be written as:

$$M_A + M_B = Ph/2 \tag{12.21}$$

Recall the well known relation between moments and radius of curvature ρ for beams and write:

$$M_A = \frac{E_A I_A}{\rho} \qquad \text{and} \qquad M_B = \frac{E_B I_B}{\rho} \tag{12.22}$$

Substituting Eq. (12.22) into Eq. (12.21) yields:

$$\frac{Ph}{2} = \frac{E_A I_A}{\rho} + \frac{E_B I_B}{\rho} \tag{12.23}$$

At this stage of the analysis it is clear that we have two unknowns—P and ρ. However, we note that the interface between materials A and B is bonded. This fact implies that the displacement of two adjacent points across the interface is identical. Hence, we may write:

$$\alpha_A \Delta T + \frac{2P}{E_A hb} + \frac{h}{4\rho} = \alpha_B \Delta T - \frac{2P}{E_B hb} - \frac{h}{4\rho} \tag{a}$$

Eq. (a) can be simplified to read:

$$\frac{2P}{hb}\left(\frac{1}{E_A} + \frac{1}{E_B}\right) = (\alpha_B - \alpha_A)\Delta T - \frac{h}{2\rho} \tag{12.24}$$

Substituting Eq. (12.23) into Eq. (12.24) yields:

$$\frac{4}{bh^2\rho}(E_A I_A + E_B I_B)\left(\frac{1}{E_A} + \frac{1}{E_B}\right) = (\alpha_B - \alpha_A)\Delta T - \frac{h}{2\rho} \tag{12.25}$$

Because we have assumed that the thickness of both materials A and B is the same, we may write:

$$I_A = I_B = I = (bh^3)/96 \qquad \text{(b)}$$

Substituting Eq. (b) into Eq. (12.25) and simplifying yields:

$$\frac{h}{24\rho}\frac{(E_A + E_B)^2}{E_A E_B} = (\alpha_B - \alpha_A)\Delta T - \frac{h}{2\rho} \qquad \text{(c)}$$

Solve Eq. (c) for the curvature $1/\rho$ to obtain:

$$\frac{1}{\rho} = \frac{(\alpha_B - \alpha_A)\Delta T}{h\left[\frac{1}{2} + \frac{1}{24}\frac{(E_A + E_B)^2}{E_A E_B}\right]} \qquad (12.26)$$

Substituting Eq. (12.26) into Eq. (12.23) gives the expression for the axial loading P produced by the temperature change ΔT as:

$$P = \frac{bh}{2}(E_A + E_B)\frac{(\alpha_B - \alpha_A)\Delta T}{\left[12 + \frac{(E_A + E_B)^2}{E_A E_B}\right]} \qquad (12.27)$$

The moments M_A and M_B are given by substituting Eq. (12.26) into Eq. (12.22) to obtain:

$$M_A = \frac{bh^2 E_A}{4}\frac{(\alpha_B - \alpha_A)\Delta T}{\left[12 + \frac{(E_A + E_B)^2}{E_A E_B}\right]} \qquad (12.28a)$$

$$M_B = \frac{bh^2 E_B}{4}\frac{(\alpha_B - \alpha_A)\Delta T}{\left[12 + \frac{(E_A + E_B)^2}{E_A E_B}\right]} \qquad (12.28b)$$

The maximum stress in this case occurs in material A which has a lower TCE than material B and is subjected to an axial tensile force P. The maximum stress is given by:

$$[\sigma_{max}]_A = \frac{2P}{bh} + \frac{hE_A}{4\rho} \qquad (12.29)$$

where $(1/\rho)$ and P are given by Eq. (12.26) and Eq. (12.27), respectively.

The distortion of the bi-material beams is probably more important than the stresses because of the planarity requirements for both the chip carriers and the printed circuit boards in the production process. To determine the out of plane deflection of a bi-material beam we use the radius of curvature as expressed in Eq. (12.26). The out of plane deflection δ defined in Fig. 12.16 is given by:

$$\delta = \rho(1 - \text{Cos}\,\theta) \tag{12.30}$$

where the angle θ is determined from the length of the beam using the expression given below:

$$\theta = \sin^{-1}(L/2\rho) \tag{12.31}$$

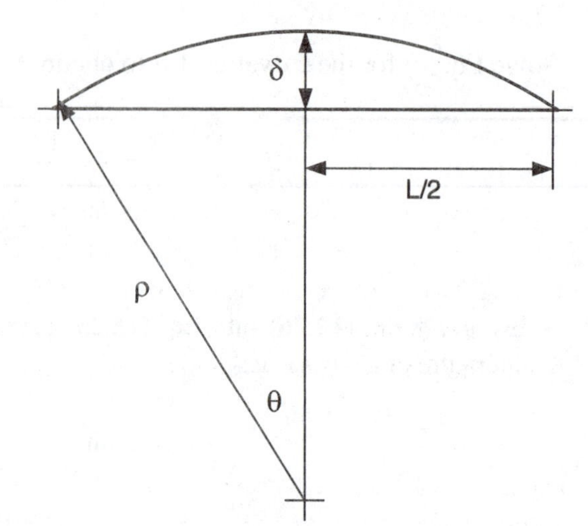

Fig. 12.16 Maximum out of plane defection due to temperature induced bending of a bi-material beam.

12.6 WARPAGE OF CHIP CARRIERS AND PCBS

The warpage problem affects both the chip carrier and the printed circuit boards. The leads or solder balls from surface mounted chip carriers must lie in a plane if they are to make contact simultaneously with the solder pads on a PCB. The planarity problem is exacerbated as the chip carriers become larger and as the compliance of the solder balls on a BGA approaches zero. A similar problem exists for the printed circuit boards, particularly multi-layered boards. Their surface must be flat over the area occupied by a chip carrier. Flatness of the chip carrier and the PCB are two different issues; therefore we will treat them separately.

12.6.1 Warpage in Chip Carriers

The flatness is not an issue with ceramic chip carriers because the ceramic carrier is formed (sintered) and checked for flatness before a chip is soldered into its cavity. Moreover, the temperature coefficient of expansion of the ceramic and the silicon chip are closely matched. The relatively small moments induced by temperature changes are accommodated with only minute deformation because of the very high elastic modulus of the ceramic.

The flatness of plastic chip carriers is a significant issue for both wire bonded and flip chip packages. In both types of chip carriers, a molding compound is injected into a die that contains the chip and the lead frame or interposer. The transfer molding process involves injecting the molding compound into the die at a temperature of about 180 °C. The encapsulated chip is subjected to post mold curing at 175 °C for several hours to complete the polymerization of the molding compound. When the encapsulated chip cools the differential contractions of the molding compound, chip and lead frame or interposer may cause warpage of the chip carrier.

Methods to limit the warpage usually involve materials used in formulating the molding compound. Studies have shown that the warpage correlates well with the flexural modulus and the glass transition temperature of the molding compound. The warpage decreases as glass transition temperature of the molding compound T_g increases from 150 to 180 °C. This improvement is due to the fact that the $(TCE)_1$ of plastic is lower below its T_g than the $(TCE)_2$ above the T_g. Additions of modifiers such as silicone or rubber to the epoxy-silica molding compound decreases it flexural modulus. The lower flexural modulus decreased the warpage by a significant amount.

12.6.2 Warpage in Printed Circuit Boards

Normal board applications require a warpage of less than 1% of any linear dimension (0.01 inch/inch or 0.01 mm/mm), assuming that the overall board dimensions still fit within the board clearance and width specifications. Maximum board warpage cannot exceed 2 mm (0.08 in.). Board warpage of 0.5% or less is required to maximize throughput for some device types.

Warpage in printed circuit boards is usually due to the lack of symmetry in the lay-up of the various layers involved in the design of the laminate. Consider first a single layer board that is comprised of several layers of glass cloth impregnated with epoxy resin. Thin layers of copper foil are bonded to its top and bottom surfaces. This board is placed in a platen press between two flat plates where it is squeezed and heated. The board is cured at elevated temperature until the polymerization process is complete. When removed from the platen press there is no tendency for the laminate to warp because its lay-up is symmetric with equal thickness of copper foil on the top and bottom sides of the board. Its flatness has been insured by containment between two flat plates during the entire time that the epoxy resin has been held at elevated temperatures.

When the laminate is processed to produce a single layer (double sided) circuit board, much of the copper is etched away. However, warpage is still minimal because most of the copper is etched from both sides of the board. Another factor limiting the warpage is the relatively close match of the TCE of copper (17×16^{-6}/°C) with the TCE of glass reinforce plastic (20×10^{-6}/°C).

Multi-layered Circuit Boards

The design of multi-layer boards is more involved because they may contain many layers of cured laminates with copper on both sides in addition to layers of B staged pre-preg that are used to space and insulate the cured laminates. A typical multi-layered circuit board is illustrated in Fig. 12.16. This illustration shows a multi-layered circuit board with five cured laminates that provide a total of 10 wiring planes that can be used for signal, power or ground. The stack is held together with four layers of B-staged prepreg. The stack is placed in a platen press and heated under pressure until the pre-preg is cured. The stack is maintained in the press under pressure as the multilayer board is cooled to room temperature. Upon removal from the platen press, the multi-layer board remains flat at room temperature. Due to symmetry of the stack about its center line, the moments induced by differential contraction of the glass, epoxy and copper cancel out and there is no tendency to warp during cool down. One may argue that the copper pattern remaining on each of the wiring planes differs from plane to plane and layer to layer destroying perfect symmetry. Usually the remaining copper on each signal plane is small and its thickness is only 1.4 mil or less; consequently its effect usually can be neglected. Even the power plane wiring which entails thicker copper and wider traces does not usually affect the flatness of the stack upon cool down in any significant manner. In some cases, ground planes maintain a large fraction of the copper area and could possibly contribute to warpage. However, this effect is mitigated if two ground planes are used in the stack providing they are placed equidistant from its center line.

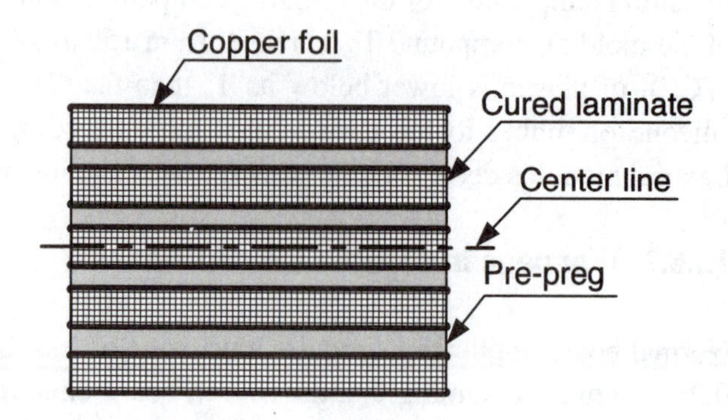

Fig. 12.17 Multi-layered circuit board showing stacking details.

Warpage Occurring in Assembly

In many cases printed circuit boards are warped in the assembly process. There are two basic reasons for the occurrence of warpage. First the placement of chip carriers and other components on the circuit board destroys its symmetry. Moreover, when ball grid arrays are used in a design, the compliance between the solder balls and the board is negligible. The expansion and contractions of the chip carrier are transmitted to the circuit board without significant mitigation causing warpage. Underfill often used to enhance the fatigue life of the solder balls couples the chip carrier even more closely to the printed circuit board. The best approach to solve reduce warpage due to chip carriers is to design the multilayer board with chips on both its sides. This practice restores, at least partially, the symmetry of the design and markedly reduces the amount of warpage.

The second basic reason for warpage is poor process control employed in the assembly operation. One of the key engineering considerations is the glass transition temperature of the glass cloth reinforced epoxy. For example, a traditional FR4 glass epoxy laminate exhibits a glass transition temperature between 115° and 135 °C. If during the assembly processes, the glass transition temperature is exceeded, the laminate loses rigidity and becomes pliable. When the temperature decreases below T_g the laminate becomes more rigid. However, if the laminate is permitted to deform when it is pliable, the deformations are locked in place as the laminate cools. It is clear that the board must be supported so that it does not deform when it is pliable.

Component placement may also cause warpage as unevenly distributed components will usually produce stresses in the PCB during cooling after soldering. Board mounted connectors and sockets can also restrict board expansion and contraction, causing warpage. The PCB shape can also affect the distribution of stresses in the laminate. Internal cut outs make soldering difficult and can produce an uneven board surface. All the sections of the board that are to be removed after the assembly process is complete should have continuous copper foil. This copper increases the stiffness of the board at the elevated temperatures which occur during soldering operations. The PCB manufacturer should perform tests to determine board's flatness both with continuous copper foil and after etching complex patterns. The number of glass layers in the laminate and pre-preg should be maximized as this provides better dimensional stability for the final assembly. Storage of the copper-clad laminate panels should be on flat and smooth shelves to avoid gravity induced distortions. When laminates are baked to cure resist or dry ink, it is essential that the laminates be supported and not permitted to sag.

In a typical multilayer lamination process, several multilayer boards are assembled in a stack. The lamination process consists of heating and cooling this "book" of boards under pressure. Poor control of lamination pressure and high temperature variations across the stack lock in stresses that are relieved during the next high temperature excursion. Relief of these stresses often produces warpage of the laminate. Reflow of solder coatings or solder leveling increases the temperature of the surface layers of the PCB above the glass transition temperature and may cause it to warp. The dimensional changes are not always noted at this stage of production because of the size of the panel being processed. The warpage becomes more evident when the circuit boards are cut to their final shape. The fast cooling during washing the flux from the board surface after leveling or reflow is also an issue. If cold water rather than hot water is used in the first washing operation, it is possible to introduce residual thermal stresses in the laminate.

Soldering subjects the PCB to temperatures well above the glass transition temperature of many laminates. Wave soldering is usually conducted between 235 and 255 °C, and vapor phase soldering is conducted at temperatures in the 215 to 225 °C range. If PCBs are not adequately supported during the soldering process, they will be left with a permanent warp or twist. PCBs expand as they pass through pre-heat and to increase their temperature to that required for the soldering operation. If the support mechanism for the PCBs in the soldering operation is incorrectly designed, it may clamp the boards without allowing for expansion. This type of support is more harmful than soldering without support.

In some cases it is necessary to bake laminates to prevent outgassing in the subsequent processing of the laminate. Baking can introduce distortion if high temperature or poor oven loading procedure is used. A test method often used to establish a suitable baking procedure requires that the board to be placed on a flat surface with its concave side down. The laminate is then baked and cooled with a prescribed temperature cycle prior to making measurements to determine the maximum out-of-flatness condition. If the planarity of the laminate is not within specification, the flatness of the oven shelf is checked, and the temperature cycle is modified.

12.7 FINITE ELEMENT SOLUTIONS

The finite element method (FEM) is widely used by the electronics industry in the development of new packaging, in the introduction of new materials and in the development of new assembly processes. The finite element methods have several advantages over the analytical techniques described in the previous sections. Most important is the ability to account for the non-linearity in material properties over the temperature range of importance to product reliability. Both two and three-dimensional analysis is possible, although three-dimensional analysis requires much more complex modeling. The output from a finite element analysis (FEA) includes all of the stress and strain components as well as the three displacements. The stresses, strains and displacements are displayed in color maps that are easy to interpret.

A large number of commercial programs are currently obtainable. A survey (1987) of some FEM software systems can be found in the Finite Element Handbook, Kardestuncer and Norrie (eds). Most of the commercial programs reviewed in this book are still current, and there is further information in the references cited. The large programs are usually very expensive ($100,000), but are often made available to academic institutions at a reduced rate. We will not attempt to compare these programs, but to show the general approach followed in conducting a FEA. Several examples will be described to show the type of results that can be achieved.

The general approach involves seven separate steps that include:

1. Defining the geometry of the body or bodies to be included in the analysis.
2. Preparing the mesh for each of the bodies.
3. Defining the location and the types of supports for the body or bodies (boundary conditions).
4. Defining the loading and the points of application of the loads.
5. Defining the temperature and the changes in temperature.
6. Defining the material properties over the temperature range specified in item 5.
7. Specifying the output from the program.

Most of the commercial finite element software interfaces with CAD software, which enables the engineer to import the geometry of the body or bodies to be analyzed. Automatic mesh generators reduce the time to construct the mesh because the location of the nodes is automatically incorporated into the program. The engineer specifies the density of the mesh in regions where high concentrations of stress are anticipated.

The supports for the body are located at nodes. They may be fixed where both displacements and rotations are prevented or they may be simple supports where displacements in one or more directions are fixed but rotations are permitted. When two bodies are in contact the node points coincide. If the bodies are bonded together the nodes from both bodies must undergo the same displacements.

The loads—distributed loads, concentrated loads and moments, are specified by providing their magnitude and direction. The loads must be applied at a node. In some cases displacements of a certain set of nodes are specified instead of loads. In other cases both loads and displacements are specified.

For thermal stress problems, the initial and final temperatures are specified. The thermal strains are produced by the differential expansion of the bodies included in the model. When the temperature range is relatively large it is necessary to specify the material properties as a function of temperature. This is particularly important in the analysis of thermal stresses in electronic packaging because the properties of solder and of the plastics employed vary significantly with temperature. Special provisions must be incorporated in the program to handle the complications introduced by material non-linearity with temperature.

Finally, the output from the program is specified. This output is usually the maximum (principal) stress and strain and the displacements. The output can be tabular or graphical. In most cases, the graphical output is preferred because it is in vivid color and easy to interpret.

12.7.1 Example Finite Element Solutions

Three finite element solutions will be described in the following subsections. These include: die stresses that are produced when a chip is encapsulated with molding compound; stresses in a solder joint formed with an inadequate amount of solder paste; and a full three-dimensional model of a module for automotive applications. These examples show the applicability of the finite element method to aid in the design of chip carriers and assembly processes.

Die Stresses Due to Encapsulation with Molding Compound

One quadrant of a quad flat pack was modeled with a finite element mesh shown in Fig. 12.18. The model included three materials namely copper, silicon and molding compound. The room temperature properties of these materials are presented in Table 12.1. The package was assumed to be stress free at a temperature of 155 °C, which corresponds to the glass transition temperature of the molding compound. The chip carrier was cooled from 155 °C to 25° C (room temperature). The stress distribution on the top surface of the die due to the contraction of the molding compound as it cools to room temperature is shown in Fig. 12.19. Note the stresses are compressive over the entire body of the die.

Table 12.1
Room temperature properties of materials in the quad flat pack.

Material	TCE, α /°C	Elastic Modulus, E (psi)	Poisson's Ratio, ν
Copper	17.0×10^{-6}	19.2×10^{6}	0.34
Silicon	2.6×10^{-6}	19.0×10^{6}	0.28
Mold compound	18.1×10^{-6}	1.5×10^{6}	0.30

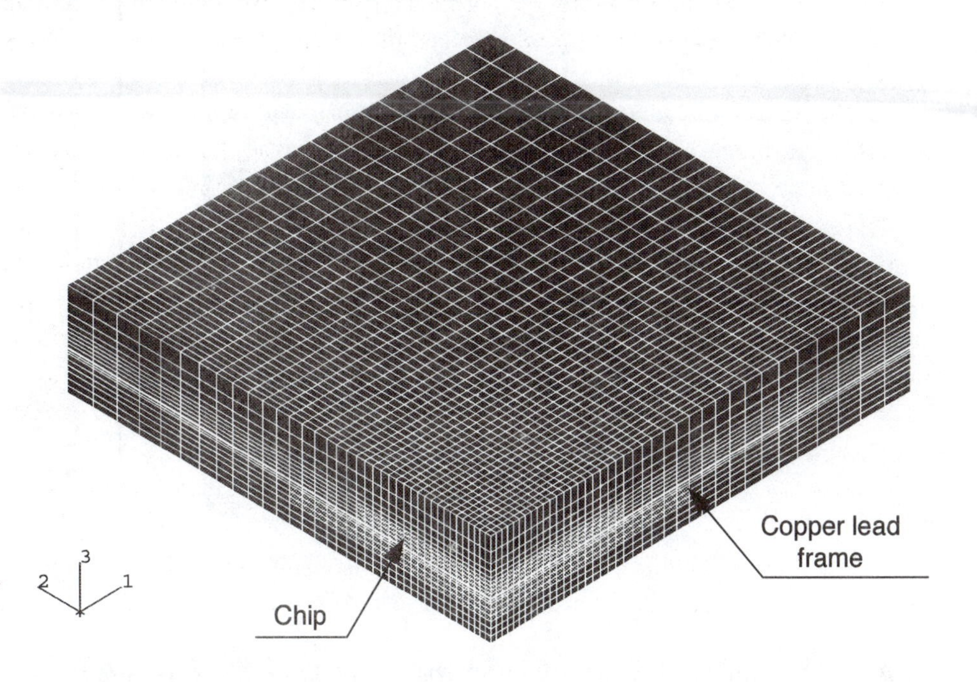

Fig. 12.18 Finite element model of a quadrant of a quad pack subjected to a temperature change.

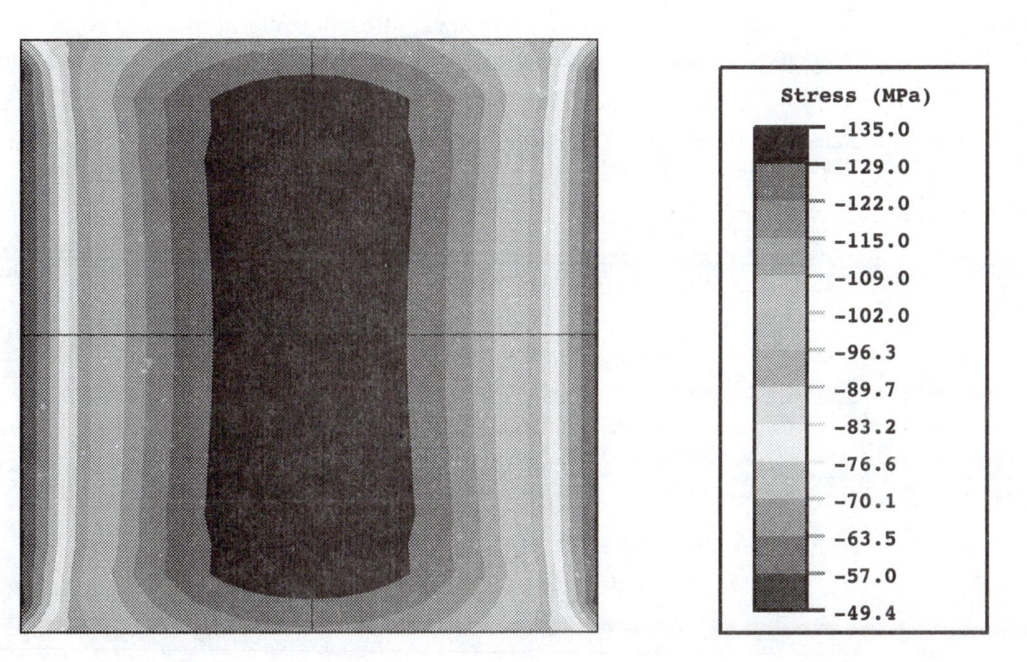

Fig 12.19 Stress distribution on the top surface of the die indicates compressive stresses due to the shrinkage of the molding compound as it cools to room temperature.

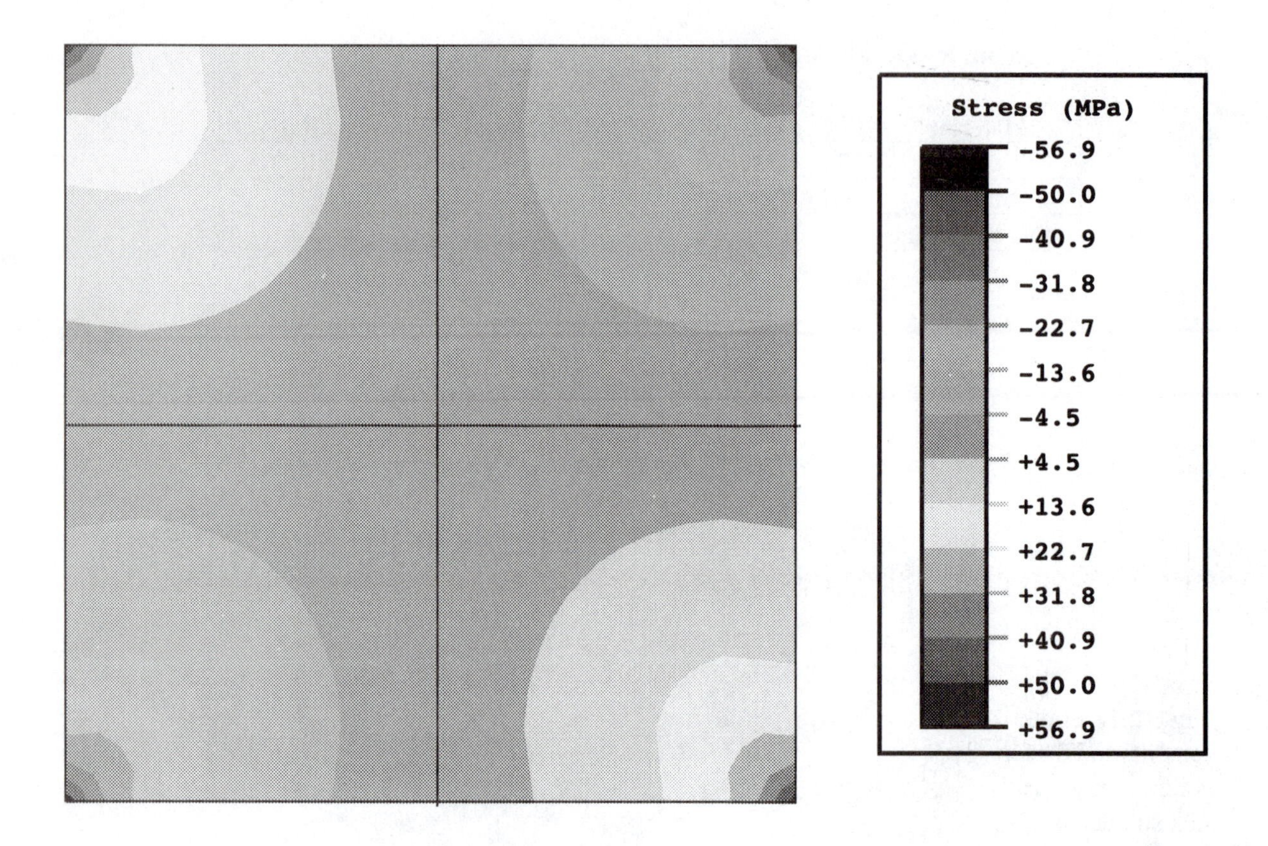

Fig. 12.20 Shear stresses distributed over the top surface of the die.

The shear stresses over the top surface of the die are illustrated in Fig. 12.20. Note the symmetry of the pattern about the diagonals. High shear stresses occur at the chip corners. These high stresses give an indication of possible delamination of the bond line between the chip and the encapsulation beginning at the corners.

Solder Joint Stresses in a Gull Wing Chip Carrier

The finite element model in this example, presented in Fig. 12.21, includes the chip carrier with its leads soldered to a printed circuit board containing vias and copper planes. The chip carrier includes the chip and the lead frame.

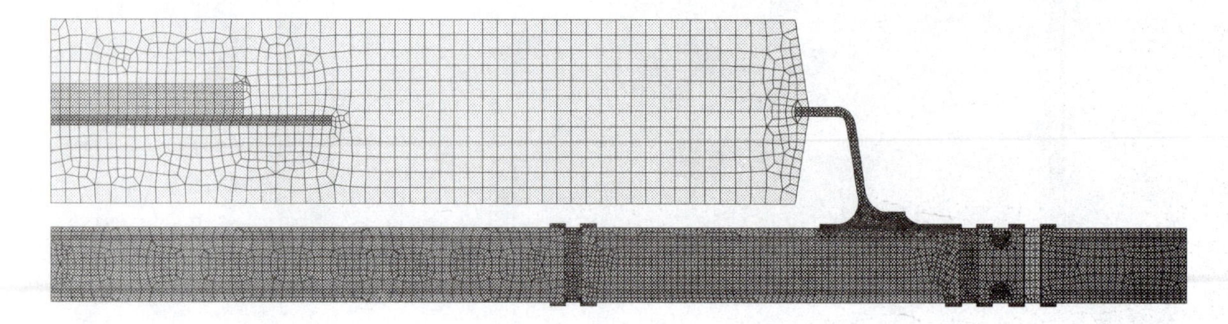

Fig12.21 Finite Element model includes the solder joint connecting the chip carrier to the PCB.

Note that the mesh size varies with the structural detail of the components with high mesh densities where either the geometry or the materials changes. The detail of the mesh for the lead where it enters the chip carrier and where it forms the connection with the solder joint and the PCB is presented in Fig. 12.22.

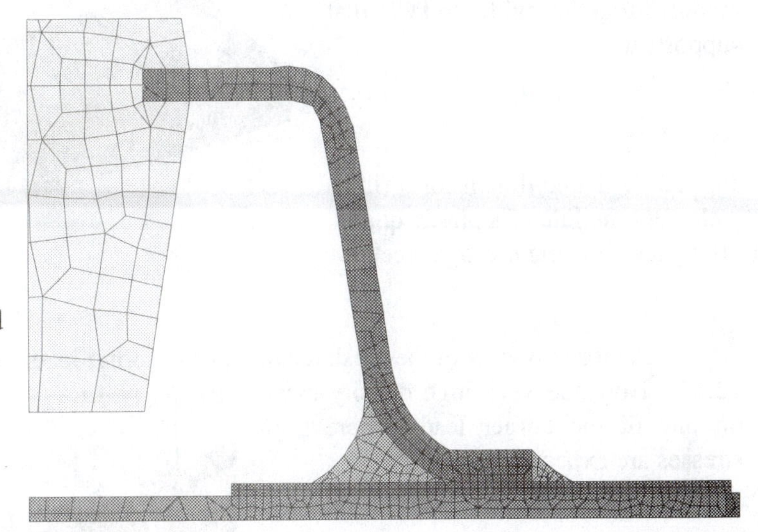

Fig. 12.22 Mesh detail of the lead and the solder joint on a gull wing chip carrier.

The results of the finite element analysis for the solder joint are presented in Fig. 12.23. Examination of this figure shows that the highest stress occurs at the toe of the lead where the lead ends and the solder begins. The second highest stress occurs near this discontinuity at the very tip of the solder joint. It should be noted that the amount of solder paste used in making this solder joint was inadequate. A layer of solder should be found between the lead and the pad on the PCB. This layer of solder is absent in this case. Another finite element analysis was conducted with an adequate amount of solder and the results showed a marked reduction in the stress levels with the maximum stresses occurring at the two lower tips of the solder joint.

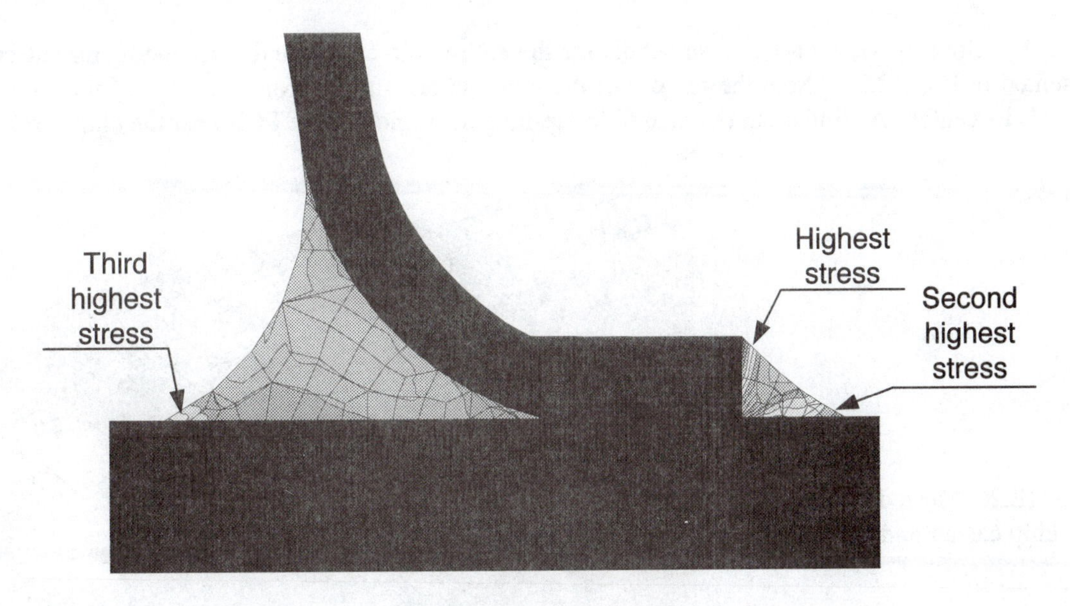

Fig. 12.23 Stress distribution in a solder joint formed with an inadequate amount of solder paste.

A Three-Dimensional Analysis of a Module Assembly

The third example pertains to the design of an assembly including a plastic quad flat pack to reduce the solder joint stresses. The design of the model for the finite element analysis is presented in Fig. 12.24. Note that the entire chip carrier is modeled together with the PCB that supports it.

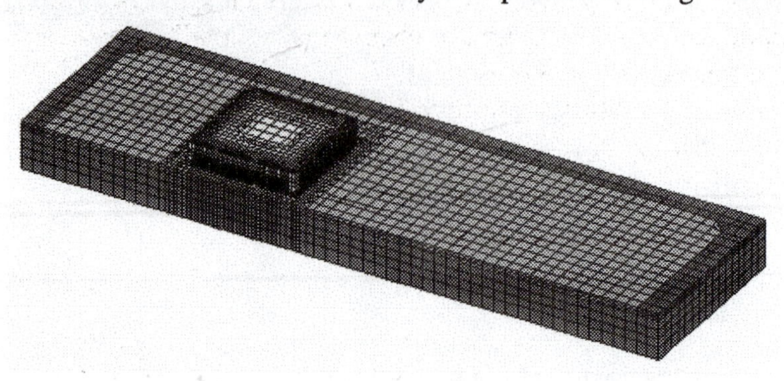

Fig. 12.24 Three-dimensional finite element module of a plastic quad flat pack showing mesh geometry.

A close up view of the mesh details for the corner leads and the solder joints is presented in Fig. 12.25. Note the very high density mesh on one of the corner leads where high stresses are expected to occur.

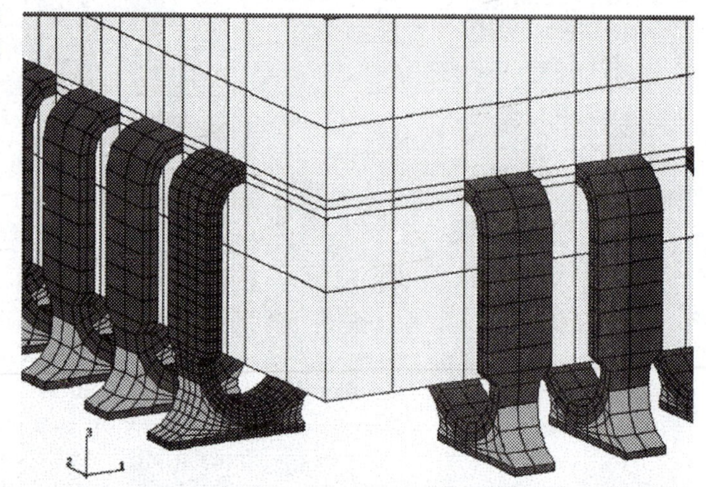

Fig. 12.25 Details of the mesh used in analyzing the leads and solder joints.

A close up view of the mesh details for the entire chip carrier and surrounding circuit board is presented in Fig. 12.26. Note the fan out of the mesh as one moves from the edge of the chip carrier towards its center. A similar fan out is used in meshing the region of the PCB near the chip carrier.

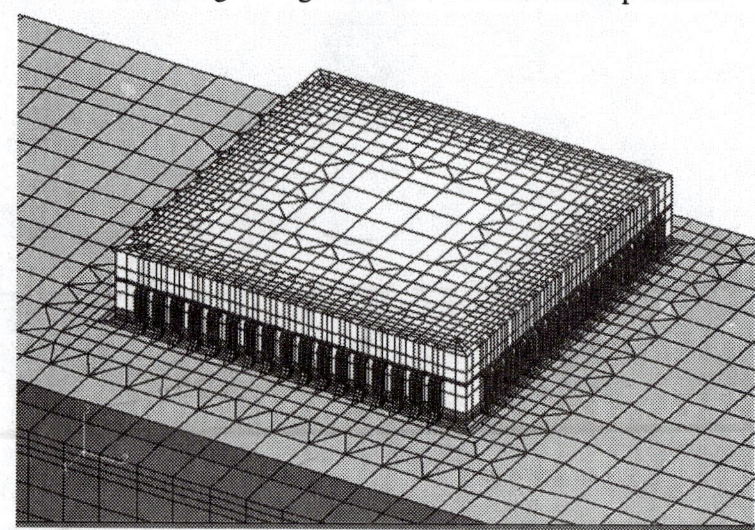

Fig. 12.26 Mesh details for the chip carrier and the PCB.

The finite element program was executed using temperature cycling between the limits anticipated in an automotive application. One of the outputs from the program was displacement of the nodes in the z (out-of-plane) direction, which show the warpage of the assembly. This warpage is illustrated in Fig. 12.27. The warpage is due to the unbalanced lay out of the circuit board. A single large package on one side of the board with a different effective TCE than that of the PCB introduces significant thermal stresses, which tend to warp the circuit board.

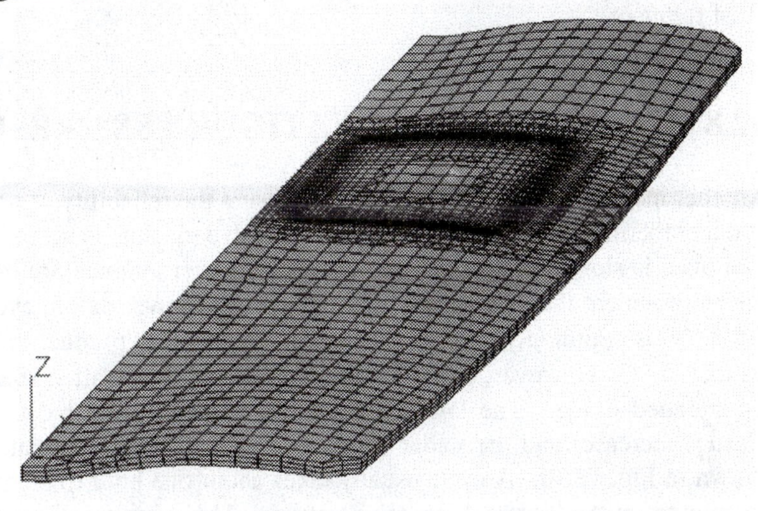

Fig. 12.27 Circuit board warpage due to temperature induced differential expansions.

The warpage of the printed circuit board markedly affects the stress distribution in the solder joints. The module is more rigid than the circuit board and as a consequence most of the displacement in the z direction, which causes the warpage, must be accommodated by deformation of the leads and solder in the joints. An illustration showing the z direction deformation of the leads and solder joints is presented in Fig. 12.28.

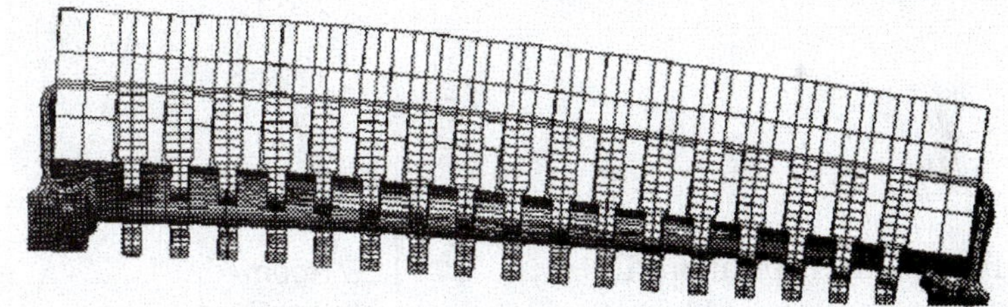

Fig. 12.28 Displacement of the solder joints across the width of the module due to thermally induced warpage.

The stresses in the corner solder joint where the thermal deformations were the maximum are presented in Fig. 12.29. Note the maximum stresses occur in the solder under the bottom point of the J lead at the interface with the solder and the solder pad and the interface between the lead and the solder.

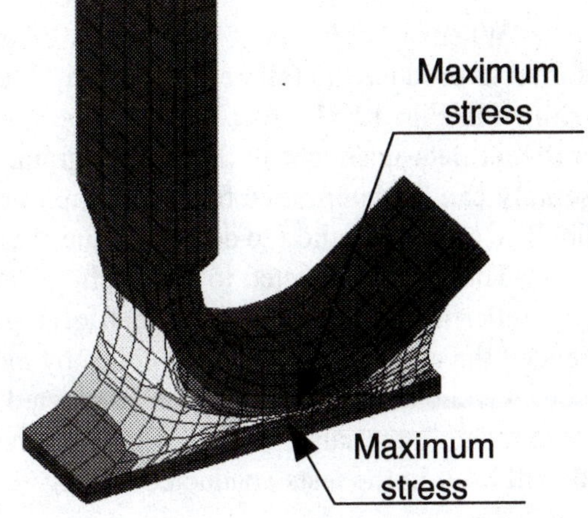

Fig. 12.29 Solder joint stresses due to warpage of the PCB due to temperature changes.

These three examples show that the finite element method can be employed effectively to study a number of different types of problems arising in the design of chip carriers, the layout of components, selection of materials, and in the optimization of production processes. Unfortunately space limitations do not permit the description of additional examples. Also printing in black and white precludes the use of color to show the rich detail in the stress distribution found in the solder joints where failures are the most frequent.

12.8 CYCLIC THERMAL FATIGUE TEST RESULTS

Another method that is frequently used to assure the reliability of a new chip carrier or an assembly is to conduct thermal fatigue tests. These tests involve placing an assembly in a temperature controlled oven. The oven is slowly cycled between upper and lower temperature limits. Temperature limits depend on the environment that the product will be exposed to over its life cycle. For commercial product – 40 to + 100 °C is commonly specified; however, for military product the temperature is often cycled from – 55 to 125 °C. However, different temperature ranges are often specified depending on the product and its intended usage. The lower temperature is usually the most critical because the modulus of the plastics increases and the solder becomes brittle at these temperatures. A typical temperature cycle is shown in Fig. 12.30. A cycle usually takes about one hour to complete, which includes dwell times of 15 minutes at the high and low temperatures. The components on the test articles are wired so that the resistance of each solder joint can be monitored. As a solder joint begins to fail its resistance increases as the fatigue damage progresses. A resistance of 10 Ω often is defined as a failure point for the solder joint.

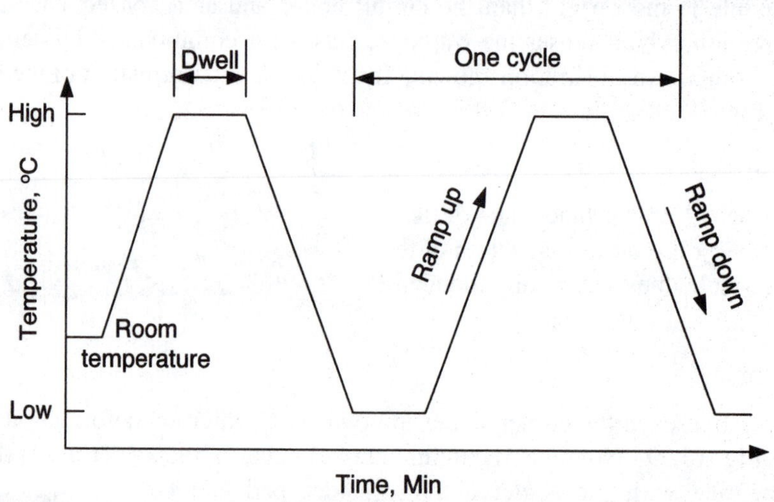

Fig. 12.30 Typical profile of thermal cycles used in testing PCB assemblies.

When a solder joint fails as its resistance exceeds a specified limit, the number of cycles required to produce the failure is recorded. The percent failures are then plotted on Weibull graph paper as shown in Fig. 12.31. A relatively large number of assemblies are included in the test if statistically significant data is sought in the test program. The Weibull constants used to predict reliability of the assembly can be determined from the graph presented in Fig. 12.31. The Weibull method to determine reliability and the method to determine the Weibull constants are covered later in Chapter 15.

The tests conducted to establish the data plotted in Fig. 12.31 were to measure the benefit of using underfill in mounting plastic ball grid arrays to glass reinforced laminates. Clearly, the underfill extended the cyclic life of the assemblies by more than 1,000 cycles.

Cyclic thermal tests can be conducted to verify the reliability of new chip carriers mounted on new laminate materials. They can be employed to test the effectiveness of new materials such as the underfill used in the tests conducted for Fig. 12.31. The thermal cyclic fatigue tests can also be used to establish data pertaining to the fatigue life of new solder alloys or to test the feasibility of introducing new production processes.

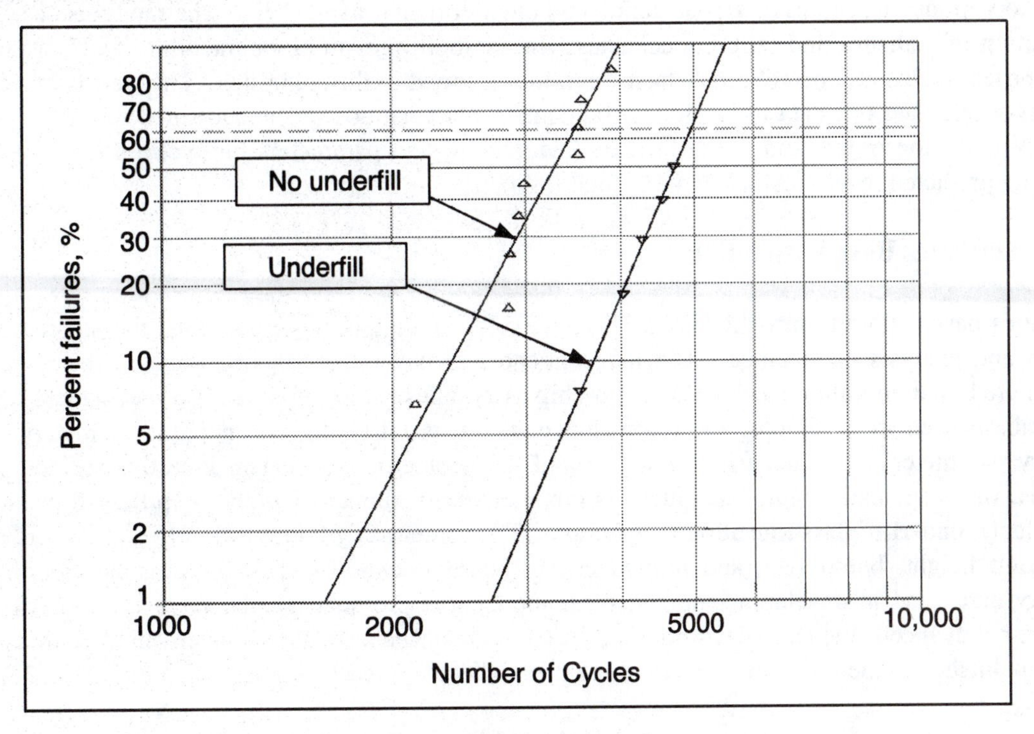

Fig. 12.31 The percent failures versus the number of cycles for temperature cycled from – 40 to 125 °C.

12.9 STATISTICAL PREDICTION OF THERMO-MECHANICAL RELIABILITY

Significant differences in the temperature coefficient of expansion of the silicon chip and organic-laminate substrate subjects the devices to several thermo-mechanical failure modes as described in previous sections of this chapter. Solder joint fatigue due to thermal cycling is one of the dominant failure mechanisms in flip chip packages. For this reason, there is a need for predictive tools and techniques in design for optimization and trade-off studies. Accelerated testing, described in the previous section, is a time consuming and resource-intensive process. However, modeling and simulation techniques are an attractive alternative for calculation of stresses, strains and life prediction. There have been many studies showing the effect of material and geometric parameters on the reliability of laminated PCBs. Several approaches are available today for predicting reliability of a design including non-linear finite element analysis and first-order closed form analysis. The finite element analysis, often requires expert users, are time-consuming and not immediately available when making design decisions. First-order closed form analysis provide timely results but offer limited accuracy.

The statistical approach is intended for circuit board designers who must select components and board configurations based on specified design requirements. In addition, the statistical model presented in this section provides guidance for the intelligent selection of component technologies and for modifying existing product designs to reduce risk when inserting of new packaging technologies into a product. The statistical model serves as an aid for assessing the reliability of a design to geometry, chip carrier design, material properties and board attributes. Using the statistical model enables the engineer to intelligently select components, materials and processes.

The approach used in this development enables higher-accuracy reliability predictions by perturbing known accelerated-test data-sets using factors which give the change in the reliability of a product to various design, material, packaging and environmental parameters. The model is based on a combination of statistics and failure mechanics. In addition, parameter interaction effects, which are often ignored in closed form solutions, have been incorporated in this approach. The statistics model is based on accelerated test data in harsh environments. The accuracy of the statistical model to predict reliability is demonstrated, and the influence of various design parameters on cyclic life is shown by comparing predicted results against experimental test data.

12.9.1 Statistics Based Modeling

A statistics-based closed-form model has been developed that is based on multivariate regression equation and analysis of variance. The independent design variables incorporated in the regression equation are based on failure mechanics of flip-chip assemblies subjected to thermo-mechanical stresses. The database used in developing the regression equation is fairly diverse in terms of materials and geometry parameters. The material properties and the geometric parameters investigated include die thickness, die size, ball count, ball pitch, bump metallurgy, underfill types (capillary-flow, reflow encapsulant), underfill glass transition temperature (T_g), solder alloy composition (SnAgCu, SnPbAg), solder joint height, bump size, and printed circuit board thickness. The flip-chip accelerated test reliability data used in developing the statistical model was provided by the researchers at the NSF Center for Advanced Vehicle Electronics (CAVE). This database was supplemented with various datasets published in the literature. The range of data collected in each case is shown in Table 12.2.

Table 12.2
Range of parameters included in the accelerated test data base

Parameter	Data Range
Die Size	3 to 12.6 mm
Ball Count	42 to 184
Ball Pitch	0.20 to 0.457 mm
Ball Diameter	0.04 to 0.195 mm
Ball Height	0.04 to 0.147 mm
Solder Composition	Sn63-Pb37, An96.5-Ag 3.5 Sn99.3-Cu0.7, Sn95.8-Ag3.5-Cu0.7
PCB Thickness	0.5 to 1.0 mm
T_{High} in Tests	100, 125, 150 °C
T_{Low} in Tests	−55, −40, 0 °C

Each data point in the database is based on the characteristic life of a set of flip-chip devices of a given configuration tested under harsh thermal cycling or thermal shock conditions. The influence of geometry and material parameters on reliability has been derived based on a linear multivariate regressions equation. The characteristic life of the various configurations of the flip-chip packages under different accelerated testing conditions has been used as the dependent variable. Independent variables include, different thermal conditions, design parameters and material attributes of the package as indicated in Table 12.2.

The linear regression relation representing the statistical model is given by:

$$N_{63.2\%} = a_1 + b_1x_1 + b_2x_2 + b_3x_3 + b_4x_4 + b_5x_5 + b_6x_6 + b_7x_7 + b_8x_8 \qquad (12.32)$$

where $N_{63.2\%}$ is the independent variable, which is a reliability parameter since it gives the number of cycles for 63.2% of the population of samples to fail.

$x_1, x_2, ... x_8$ are the independent variables (design parameters) defined by:

x_1 is the die size in mm.
x_2 is the ball count.
x_3 is the ball diameter in mm..
x_4 refers to the pad type
x_5 is the temperature differential ΔT in °C.
x_6 is the pitch of the solder balls in mm.
x_7.is the solder ball height in mm.
x_8 is the solder alloy composition

The term—a_1 from the equation of a line, it is the $N_{63.2\%}$ value where the best-fit line representing Eq. (12.32) intercepts the ordinate. It is the estimated value of the dependent variable if the independent variables all have a value of zero. The terms $b_1, b_2, ... b_8$ are the regression coefficients. These are numbers referenced to the independent variables in the statistical model that shows how much a unit change in the independent variable is estimated to change the value of the dependent variable.

The dependent (endogenous) variable $N_{63.2\%}$ on the left hand side of the regression equation represents the characteristic life of a three-parameter Weibull distribution for the flip-chip package when subjected to accelerated thermo-mechanical stresses. The independent quantities on the right hand side of the equation are the design parameters that influence the reliability of the package. The coefficient of each of these quantities is an indicator of the relative influence of that quantity on the characteristic life of the package. In this case, constants a_1 and b_i, for i = 1 to 8 are the coefficients in the linear regression equation. They indicate the relative sensitivity of the respective independent quantities on the reliability of the package. The dimensions given in Table 12.3 must be employed with the independent variables (e.g. x_1 is the die size in mm). Qualitative quantities such as pad type that influence the package reliability have been used as dummy variables to ascertain their influence. These qualitative quantities include:

1. x_4 refers to pad type which varies between 0 and 1, with the value 0 corresponding to a solder mask defined pad and the value of 1 corresponding to a non-solder mask defined pad.
2. x_8 refers to the solder alloy with values of 0, 1, 2, and 3 corresponding to the four alloys in the database including, Sn63Pb37, Sn95.8Ag3.5Cu0.7, Sn96.5Ag3.5, and Sn99.3Cu0.7, respectively.

Using the data from the accelerated test reliability data for flip chip packages provided by the researchers at the NSF Center for Advanced Vehicle Electronics (CAVE), the coefficients for the linear regression equation were determined. The results of a linear regression analysis are shown in Table 12.3. The first column in this table defines the independent design variable and its units if they are required in the analysis. The second column gives the regression coefficients associated with the eight different independent variables. A negative value of the regression coefficients indicates a decrease in cycles-to-failure with increase in value of the independent variable, where as a positive coefficient indicates an increase in the cycles-to-failure with an increase in the value of the independent variable.

The third column gives the standard error for each coefficient. The standard error is an estimate of the standard deviation of the coefficient. Essentially, it measures the variability in the estimate of the coefficient. Lower standard errors lead to more confidence in the estimate because the lower the standard error, the closer to the estimated regression coefficient is to the true regression coefficient.

Table 12.3
Multivariate Regression Models of BGA Thermal Fatigue Data.

Quantity	Regression Coefficient	Standard Error	t -Statistic	P-value
Constant	$a_1 = 11374$	1526	7.45	0
Die size (mm)	$b_1 = -456.18$	62.23	–7.32	0
Ball Count	$b_2 = 21.994$	4.413	4.98	0.001
Ball Diameter (mm)	$b_3 = 19873$	6553	3.03	0.016
Pad Type	$b_4 = 1781.2$	261.5	6.81	0
Delta T (°C)	$b_5 = -72.546$	6838	–10.61	0
Pitch (mm)	$b_6 = 11860$	2884	4.11	0.003
Ball Height (mm)	$b_7 = -16102$	4624	–3.48	0.008
Solder Alloy Composition	$b_8 = 124.17$	57.38	2.16	0.062

The fourth column gives the t-statistic, which is a statistical measure that yields the number of standard deviations the estimated regression coefficient is from zero. Assuming the data base is represented by a normal distribution, a t-statistic of 2 or more is generally accepted as statistical significance. Examination of Table 12.3 shows that all of the independent design variables are statistically significant. The final column gives the P-value, which is the probability that the estimated regression coefficient is equal to zero, assuming that the error in the estimate is normally distributed. Examination of Table 12.3 shows that the probability of the regression coefficient being equal to zero is nearly zero except for the solder alloy composition where the probability is only 6.2%.

This regression equation has a correlation coefficient R^2 equal to 99.4%. The correlation coefficient R^2 is a measure of how well the model explains the variability in the dependent variable. It is defined as explained sum of squares over total sum of squares. As such it measures the percent of the change in the dependent variable that is accounted for by the regression model. The value of 99.4% for the correlation coefficient R^2 indicates that a very large part of the variation in the characteristic life has been explained by the model. This conclusion was verified by an analysis of variance of the data set.

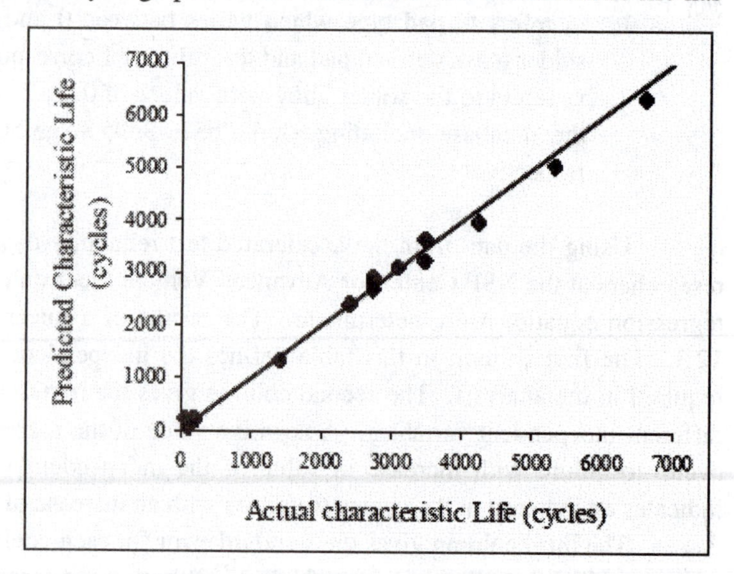

Fig. 12.32 Actual vs predicted characteristic life (cycles).

Characteristic life (63.2% failure of population) predicted by the regression equation has been plotted against the actual life from the experimental database to assess the goodness of fit of the regression equation. The straight line at 45° in Fig. 12.32 represents the perfect fit of the predicted values. After comparing various statistical models for the fit of the predicted values of $N_{63.2\%}$ with that of actual number of cycles for failure from the experimental data, the model with the best fit was selected. The graph presented in the Fig. 12.32 shows that the predicted cyclic life is remarkably close to the experimental results.

12.9.2 Validation of the Regression Model

The regression model (relation) presented in Eq. 12.32 has been validated against the experimental accelerated test failure data by comparing the predictions from the regression equation with experimental results for each independent variable. The regression coefficients that quantify the effect of design, material and environment parameters on thermal fatigue reliability have been used to compute cyclic life. The predictions from the regression model have been compared with the experimental data for each of the independent design parameters.

Die Size

The thermo-mechanical reliability of a flip-chip devices generally decreases with an increase in the die size. This effect has been demonstrated in flip-chip devices soldered to both FR4-substrate and BT-substrates. The regression equation has been used to evaluate effect of die size on cyclic life. Wafer-level devices with compliant elastomeric substrates (e.g. μBGA) may not exhibit this effect. In addition, the model predictions focus on failure of the solder joints. If the failure mode is different, such as underfill cracking, then the failure response will be different. The number of cycles for 63.2% failure has been plotted against the die size of various devices. The predicted values from the regression model follow the experimental values quite accurately as shown in Fig. 12.33. The trend identified for decreasing cyclic life for solder joints associated with larger die size is consistent with failure mechanics studies. The solder joints are subjected to higher strains due to the increased distance from the neutral point, to the solder joints located at the corner of the flip-chip.

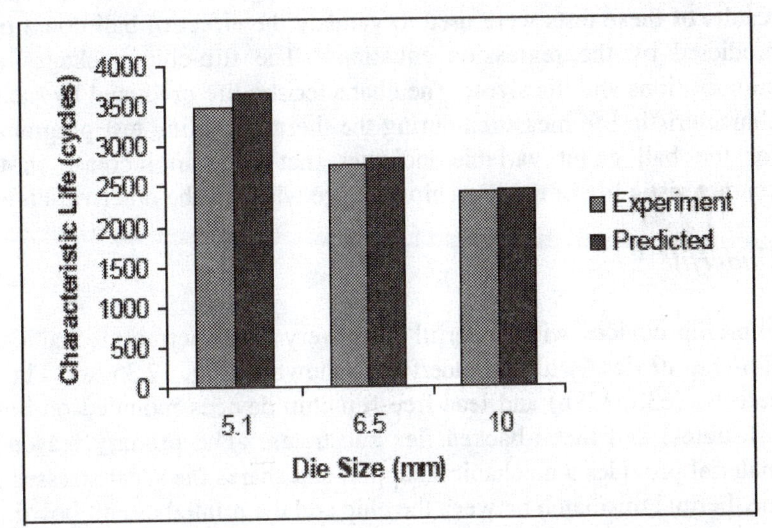

Fig. 12.33 Effect of die size on thermal fatigue reliability of encapsulated flip-chip with Sn37Pb63 solder joints.

Encapsulated flip-chip packages with die sizes of 5.1, 6.5 and 10 mm have been used in comparing the model predictions with the actual test data. All three packages had eutectic (Sn37Pb63) solder joints of different ball diameter, pitch and pad definition and were subjected to different thermal cycles of 0°C to 100°C, –40°C to 125°C, and –55°C to 125°C. Clearly the model has been tested for its ability to predict both single and coupled effects. A negative regression coefficient has been determined for the effect of die size. A negative regression coefficient indicates that the characteristic life of a flip-chip decreases when the die size increases with all the other independent variables remaining constant.

Ball Count

The effect of ball count on thermo-mechanical reliability of flip-chip packages is shown in Fig. 12.34. A trend of increasing reliability with an increase in the ball count is predicted by the statistical model and is supported by experimental data. This trend is also supported by failure mechanics theory, because the shear stress due to thermo-mechanical loading is distributed over a larger number of solder interconnects as the ball count increases. The subsequent reduction in the loading reduces the stress in the individual solder joints increasing their life. The predicted effect of ball count is only applicable in the case where the failure mode is cracking of the solder joints. The trend may be different for other failure modes such as underfill-delamination or trace-cracking.

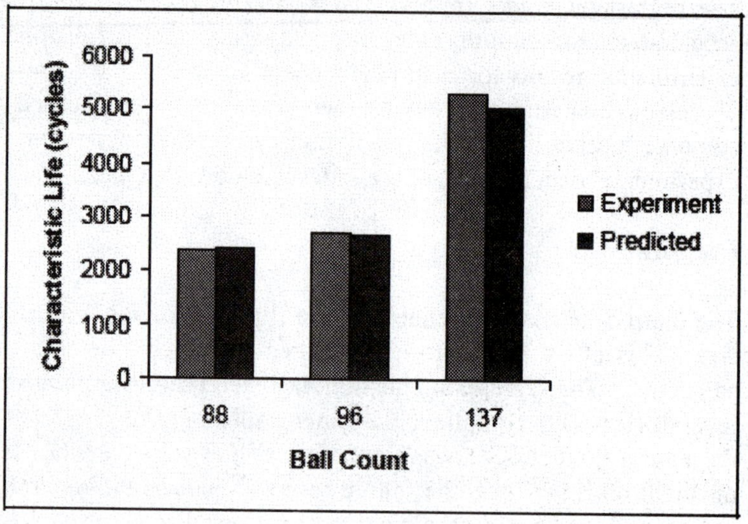

Fig. 12.34 Effect of ball count on thermal fatigue reliability of encapsulated flip-chip with leaded and lead-free solder joints.

Three different flip-chip packages with underfill and a ball count of 88, 96 and 137 were subjected to different thermal cycles including, – 40 to 125°C, – 55 to 125°C, and – 55 to 150°C. The results of these tests were used to validate the effect of ball count on the thermo-mechanical reliability predicted by the regression equation. The flip-chip packages also include different solder joint compositions and die sizes. The characteristic life predicted by the model compares very well with the characteristic life measured during the thermal cycling test program. A positive regression coefficient for the ball count variable indicates that with an increase in the total ball count increases the characteristic life of the flip-chip package when all the other variables are held constant.

Underfill

Flip-chip devices with underfill show very high thermo-mechanical reliability when compared to the flip-chip devices without underfill as shown in Fig. 12.35 and Fig. 12.36. This trend is true for both eutectic (63Sn37Pb) and lead-free flip-chip devices mounted on both rigid-organic (e.g. BT and FR-4 substrates) and metal-backed flex substrates. The primary reason for this trend is that the underfill material provides a mechanical support and shares the shear stresses generated in the solder joints due to the thermal mismatch between the chip and the printed circuit board.

However this trend is reversed for the case where the failure mode is trace cracking, because the application of underfill leads to much higher stresses on the Cu traces and can cause trace peeling. Underfill cracking and delamination are the other failure modes commonly associated with encapsulated flip-chip packages along with solder joint failure and Cu trace cracking. Figure 12.34 and Fig. 12.36 show the comparison of actual and predicted characteristic life for flip-chip packages with and without underfill. Data for these figures was obtained from 12.6mm flip-chip packages bumped with 99.3Sn0.7Cu and 95.5Sn4Ag0.5Cu solder joints. The devices were also subjected to different thermal cycles including, 0 to 100°C, – 40 to 125°C and – 55 to 150°C.

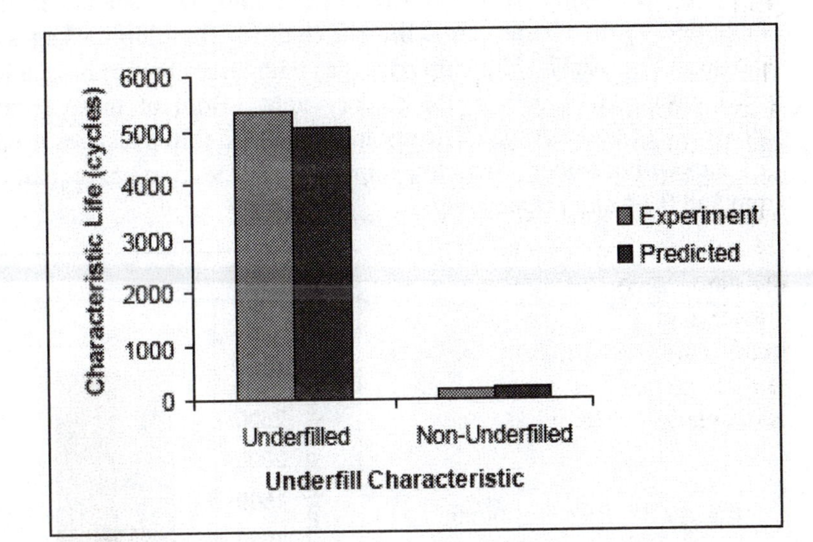

Fig. 12.35: Effect of underfill on thermal fatigue reliability of flip-chip devices with 99.3Sn0.7Cu solder joints.

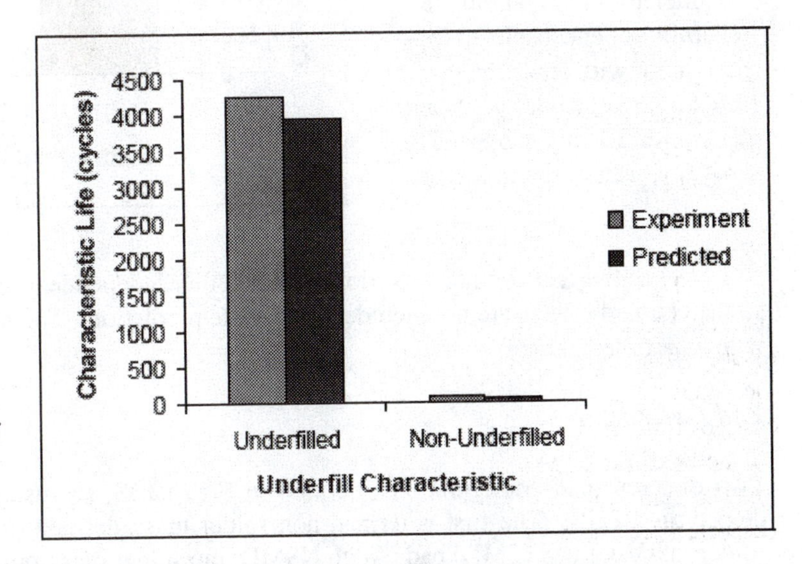

Fig. 12.36: Effect of underfill on thermal fatigue reliability of flip-chip with 95.5Sn4.0Ag0.5Cu solder joints.

The improvement in the characteristic life of flip-chip packages with underfill compared to those without underfill, by a factor of 30 to 40, is partially due to the coupled effects of die size and thermal cycling conditions. Hence it cannot be attributed to absence of underfill alone. The correlation demonstrates that the model is robust enough to accurately predict coupled effects. The added advantage of the approach demonstrated here is that the effects of a single design variable can be predicted for configurations that were never tested. The regression coefficient for underfill is positive, which indicates an increase in solder-joint reliability with presence of underfill[1]. The prediction for the characteristic life of the device from the regression equation closely corresponds to the failure data from the thermal cyclic test program.

---

[1] Underfill is not listed as an independent design variable in Eq. (12.32). However, its influence on reliability was studied in other statistical models that included underfill. The added life due to underfill is so large that it is assumed that underfill will be employed in all but exceptional cases. Hence, underfill is not included as a design parameter in Eq. (12.32).

Solder Joint Diameter

The thermo-mechanical reliability of the flip-chip devices is also influenced by the solder joint or bump diameter. Flip-chip packages with larger bump size usually result in higher reliability. This trend is supported by the characteristic life bar chart for flip-chip packages with two different ball diameters as shown in Fig. 12.37. Flip-chip packages with larger bumps have a lower stress concentration and longer crack propagation path in the solder joints. Both of these factors enhance the thermo-mechanical reliability of the package. The encapsulated flip-chip packages used for this validation had a die size of 5.1 mm with a solder joint alloy composition of 96.5Sn3.5Ag, ball count of 88, and ball diameter of 0.1 mm and 0.12 mm respectively.

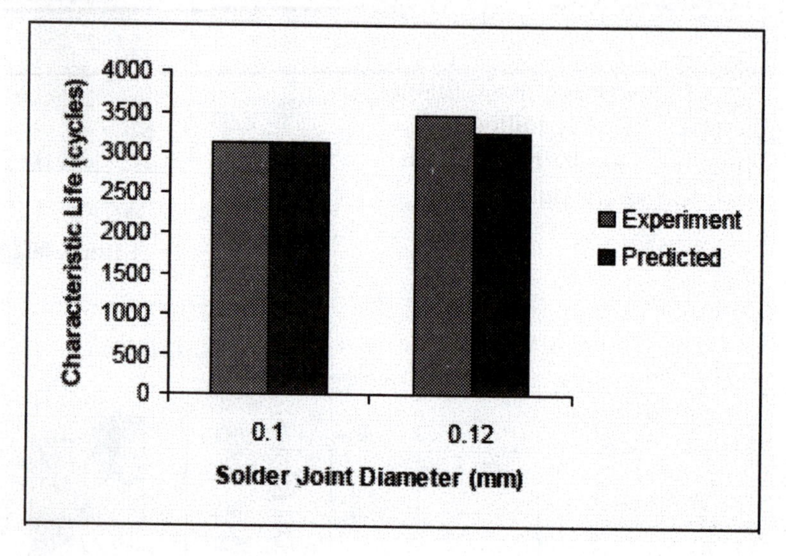

Fig. 12.37 Effect of bump diameter on thermal fatigue reliability of encapsulated flip-chip with lead-free (96.5Sn3.5Ag) solder joints and die size of 5.1 mm subjected to – 40°C to 125°C thermal cycle.

The thermal cycling conditions and all other independent variable are constant in both the cases, so the coupled effects are not included in this life prediction. The predictions show excellent correlation with experimental data.

Pad Definition

Two different pad configurations, defined in Fig. 12.38, are usually used for mounting the flip-chip device on a PCB. The first is termed non-solder mask defined (NSMD) pad and the second is called solder mask defined (SMD) pad. With NSMD, the solder mask opening is separated from the pad metal allowing a clearance of bare PCB substrate material between the pad and the solder mask, as shown in Fig. 12.38. With SMD, the solder mask opening overlaps the metal pads and the shape of exposed copper pad is defined by the mask, as shown in Fig. 12.38. With SMD there is greater probability of solder shorts because the solder ball may become larger than the intended joint if the solder flows over the solder mask. The NSMD pad layout is becoming the de-facto standard for layout of PCBs except for the case where excessive bending may occur.

Flip-chip devices mounted on PCB with NSMD pad configuration have been shown to have higher thermo-mechanical reliability. The solder mask in the SMD pad creates a sharp edge in the solder joint which leads to a stress concentration and early failure of the joint, resulting in poor reliability compared to solder interconnects with NSMD pad configuration. This is true only when the failure mode is solder joint cracking under thermo-mechanical stresses. On the other hand SMD pads provide better reliability in the cases where the failure mode is due pad peeling or cracking of the Cu trace. In these failure modes, the solder mask over the edges of the copper pad provides support for the pad and inhibits it from peeling. Trace peeling and cracking modes of failure occur with large out-of-plane deformation (bending) that is associated with shock and vibration. A comparison of the characteristic life of 12.6 mm x 7.46 mm flip-chip packages with underfill and different pad configurations is presented in Fig. 12.39.

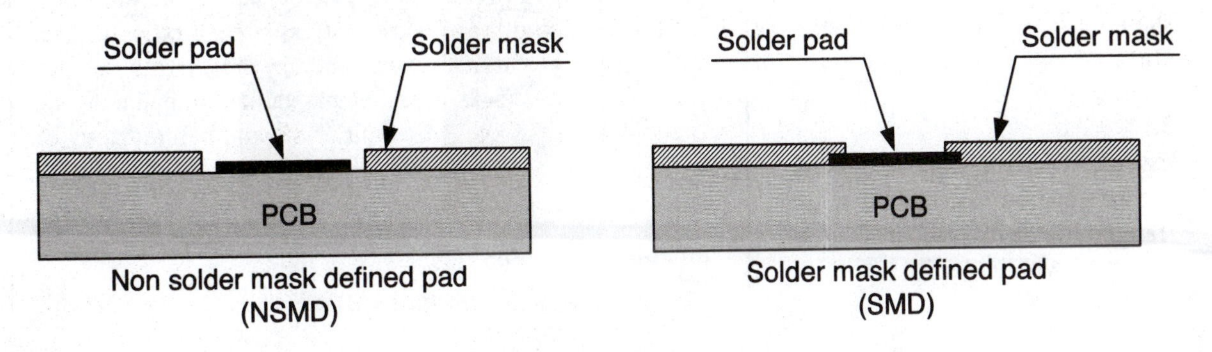

Fig. 12.38 Two types of pad layout for flip chip and BGA packages.

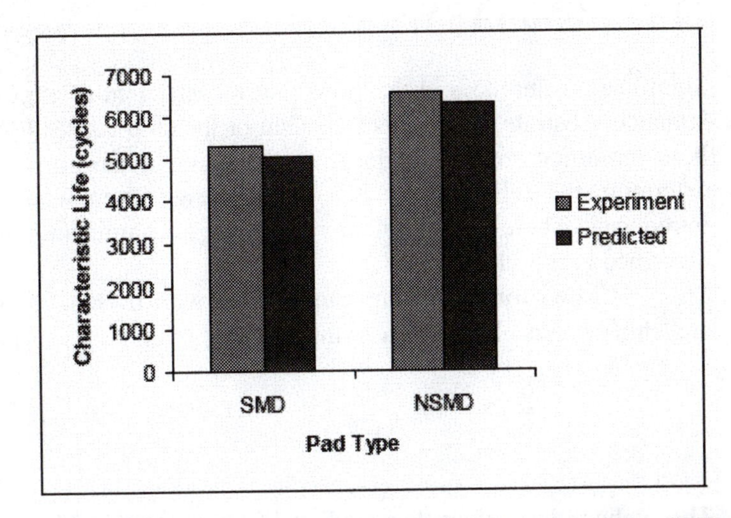

Fig. 12.39 Effect of pad configuration on thermal fatigue reliability of encapsulated flip-chip with die size of 12.6 mm subjected to 0°C to 100°C thermal cycle.

The type of pad definition has been modeled with dummy values with "0" indicating a SMD pad and "1" indicating a NSMD pad. This independent design variable has a positive regression coefficient indicating an increase in thermo-mechanical reliability for changing the pad definition from SMD to NSMD.

Magnitude of the Temperature Change

The thermo-mechanical life of the flip-chip devices, similar to other package types is a function of the environment to which it is subjected. The magnitude of the temperature range experienced during the accelerated test is an influential parameter as indicated by the results presented in Fig. 12.40.

The characteristic life of the flip-chip package decreases with an increase in the temperature range ΔT in the accelerated temperature cyclic tests. Temperature differential has a negative regression coefficient indicating that the thermo-mechanical reliability will decrease with increasing ΔT. The data presented in Fig. 12.40 includes coupled effects of other design variables such as, die size, ball diameter, ball count and cycle time. The values for characteristic life calculated from the regression equation correspond closely to the experimental results.

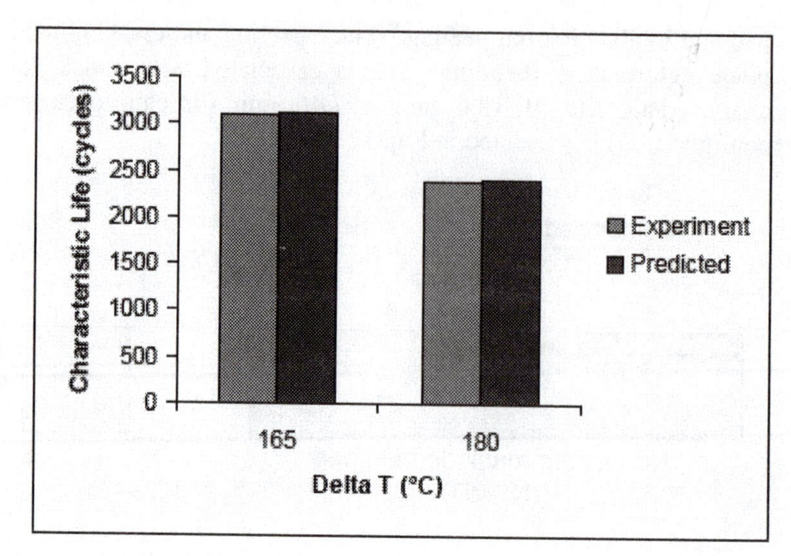

Fig. 12.40 Effect of the magnitude of the temperature cycle ΔT on the fatigue life of flip-chip packages with underfill.

12.10 SUMMARY

Significant differences in the temperature coefficient of expansion of the silicon chip and the organic-laminate substrate may subject the chip or the chip carrier thermo-mechanical failure by cyclic fatigue. It is important for the engineering designing a new product or modifying an existing product to determine the reliability of his or her design even when new components and or processes are introduced. There are several approaches for accomplishing the reliability studies that have been briefly described in this chapter.

Closed form solutions can be used with reasonable accuracy for leadless resistors, capacitors and chip carriers. In these cases the shearing strain γ in the solder joints is determined from:

$$\gamma = \frac{DNP(\alpha_{PCB} - \alpha_{CC})\Delta T}{h}$$

The cyclic fatigue life is then predicted from the Coffin-Manson relation given by:

$$N_f = 1.29(\Delta\gamma_p)^{-1.96}$$

The relations shown above are not valid when underfill is employed between a flip-chip and its substrate or between a chip carrier and a PCB. The underfill distributes the shearing strain more uniformly over the solder joints, which extends the fatigue life of the solder joint significantly. These relations cannot be employed with leaded chip carriers because the leads flex under load and accommodate part of the thermal induced displacements. Finite element methods are usually employed with leaded chip carriers to assess solder strains and to determine reliability.

These relations can be employed with BGAs providing that no underfill is employed in mounting the package to the PCB. However, when underfill is employed the life is extended by a significant amount. In these cases, it is necessary to employ a finite element analysis of a statistical model to predict reliability.

Thermal stresses generated in plated-through-holes can be determined from:

$$\sigma = E\varepsilon = E\,(\alpha_{PCB} - \alpha_{Cu})\Delta T \tag{12.8}$$

Because the difference between α_{PCB} and α_{Cu} is quite large at elevated temperatures, the thermal stresses are also very large usually exceeding yield. This fact implies that the copper plating must be smooth and continuous over the length of the plated-through-hole. Otherwise, strains can localize at an imperfection and failure results.

Plastic packages are formed by injection molding that is performed with a molding compound at relatively high temperature. When the molding compound is cured and cooled to room temperature thermal stresses are generated that tend to crack the plastic encapsulating the chip. These cracks can permit moisture to enter the package, which eventually leads to corrosion and failure. The problem is addressed by changing eth formulation of the molding compound.

Chips can also be damaged by cracking when packaging flip-chips. This damage is due to the underfill that is employed to distribute thermal stresses among the solder balls connecting the flip-chip to its organic substrate. The problem is remedied by using underfill with an improved formulation and by providing a void free fillet in the underfill that extends beyond the edges of the chip..

Thin metallic and or oxide films are often placed on a silicon wafer to produce circuit lines, pads for electrical connections or insulating layers. The temperature coefficient of expansion of these metals and oxides often differ from that of silicon. Consequently thermal stresses are created when the chips are employed at temperatures that differ from the temperatures involved in depositing these thin films. The in plane stresses that are produced by ΔT are given by:

$$\sigma_{\text{in-plane}} = \frac{E}{1-\nu}(\alpha_{TF} - \alpha_s)\Delta T$$

We again consider two different materials that are bonded together to form a bi-metallic strip. Because their TCEs are different the bi-material composite will bend when it undergoes a temperature change ΔT. For a beam of width w and depth h fabricated from materials A and B we showed the out-of-plane deflection of the beam is given by:

$$\delta = \rho(1 - \text{Cos}\ \theta)$$

where the angle θ is determined from the length of the beam using the expression given below:

$$\theta = \sin^{-1}(L/2\rho)$$

and ρ is given by:

$$\frac{1}{\rho} = \frac{(\alpha_B - \alpha_A)\Delta T}{h\left[\frac{1}{2} + \frac{1}{24}\frac{(E_A + E_B)^2}{E_A E_B}\right]}$$

The warpage problem affects both the chip carrier and the printed circuit boards. The leads or solder balls from surface mounted chip carriers must lie in a plane if they are to make contact simultaneously with the solder pads on a PCB. This problem is usually addressed by introducing materials in the molding compound that increased its T_g and decrease its flexural modulus. Component placement may also cause warpage of PCBs as unevenly distributed components will usually produce stresses in the PCB during cooling after soldering. Board mounted connectors and sockets can also restrict board expansion and contraction, causing warpage. Another reason for warpage is poor process control employed in the assembly operation.

The finite element method (FEM) is widely used by the electronics industry in the development of new packaging, in the introduction of new materials and in the development of new assembly processes. The finite element methods have several advantages over the analytical techniques described in the previous sections. Most important is the ability to account for the non-linearity in material properties over the temperature range of importance to product reliability. Both two and three-dimensional analysis is possible, although three-dimensional analysis requires much more complex modeling. The output from a finite element analysis (FEA) includes all of the stress and strain components as well as the three displacements. The stresses, strains and displacements are displayed in color maps that are easy to interpret.

The general approach involves seven separate steps that include:

1. Defining the geometry of the body or bodies to be included in the analysis.
2. Preparing the mesh for each of the bodies.
3. Defining the location and the types of supports for the body or bodies (boundary conditions).
4. Defining the loading and the points of application of the loads.
5. Defining the temperature and the changes in temperature.
6. Defining the material properties over the temperature range specified in item 5.
7. Specifying the output from the program.

Three finite element solutions have been described including: die stresses that are produced when a chip is encapsulated with molding compound; stresses in a solder joint formed with an inadequate amount of solder paste; and a full three-dimensional model of a module for automotive applications. These examples show the applicability of the finite element method to aid in the design of chip carriers and assembly processes.

Another method that is frequently used to assure the reliability of a new chip carrier or an assembly is to conduct thermal fatigue tests. These tests involve placing an assembly in a temperature controlled oven. The oven is slowly cycled between upper and lower temperature limits. Temperature limits depend on the environment that the product will be exposed to over its life cycle. For commercial product – 40 to + 100 °C is commonly specified; however, for military product the temperature is often cycled from – 55 to 125 °C.

A statistics-based closed-form model has been developed that is based on multivariate regression equation and analysis of variance. The independent design variables incorporated in the regression equation are based on failure mechanics of flip-chip assemblies subjected to thermo-mechanical stresses. The database used in developing this regression equation was diverse in terms of materials and geometry parameters. The material properties and the geometric parameters investigated include die thickness, die size, ball count, ball pitch, bump metallurgy, underfill types (capillary-flow, reflow encapsulant), underfill glass transition temperature (T_g), solder alloy composition (SnAgCu, SnPbAg), solder joint height, bump size, and printed circuit board thickness. The flip-chip accelerated test reliability data used in developing the statistical model was provided by the researchers at the NSF Center for Advanced Vehicle Electronics (CAVE). This database was supplemented with various datasets published in the literature.

The linear regression relation representing the statistical model is given by:

$$N_{63.2\%} = a_1 + b_1x_1 + b_2x_2 + b_3x_3 + b_4x_4 + b_5x_5 + b_6x_6 + b_7x_7 + b_8x_8$$

where $N_{63.2\%}$ is the independent variable, which is a reliability parameter since it gives the number of cycles for 63.2% of the population of samples to fail.

$x_1, x_2, \ldots\ldots x_8$ are the independent variables (design parameters) defined by:

x_1 is the die size in mm.
x_2 is the ball count.

x_3 is the ball diameter in mm..
x_4 refers to the pad type
x_5 is the temperature differential ΔT in °C.
x_6 is the pitch of the solder balls in mm.
x_7.is the solder ball height in mm.
x_8 is the solder alloy composition

This regression model has been validated against experimental test failure data by comparing the predictions from the regression equation with experimental results for each independent variable. The regression coefficients that quantify the effect of design, material and environment parameters on thermal fatigue reliability have been used to compute cyclic life. The predictions from the regression model compared closely with the experimental data for each of the independent design parameters.

REFERENCES

1. Lall, P., Singh, N., Strickland, M., Blanche, J. and Suhling, J.: "Decision-Support Models for Thermo-Mechanical Reliability of Lead Free Flip-Chip Electronics in Extreme Environments." IEEE 2005 Electronic Components and Technology Conference, pp. 127-136.
2. Lall, P., Hariharan, G., Shirgaokar, A., Suhling, J. Strickland, M. and Blanche, J.: "Thermo-Mechanical Reliability Based part Selection Models for Addressing Part Obsolescence in CBGA, CCGA, FLEXGBA and Flip-Chip Packages," Proceedings of IPACK2007, ASME InterPACK '07, July 8-12, 2007, Vancouver, British Columbia, Canada.
3. Adams, R. M., A. Glovatsky, T. Lindley, J. L. Evans and A. Mawer, "PBGA Reliability Study for Automotive Applications", Proceedings of the 1998 SAE International Congress and Exposition, Detroit, MI, pp. 11-19, February 23-26, 1998.
4. Bedinger, John M., "Microwave Flip Chip and BGA Technology", IEEE MTT-S International Microwave Symposium Digest, v 2, pp 713-716, 2000.
5. Clech, Jean-Paul, "Solder Reliability Solutions: A PC based design-for-reliability tool", Surface Mount International, San Jose, CA, Sept. 8-12, pp. 136-151, 1996.
6. Darveaux, R., "Effect of Simulation Methodology on Solder Joint Crack Growth Correlation," Proceedings of 50th ECTC, pp.1048-1058, May 2000.
7. Darveaux, R., How to use Finite Element Analysis to Predict Solder Joint Fatigue Life, Proceedings of the VIII International Congress on Experimental Mechanics, Nashville, Tennessee, June 10-13, pp. 41-42, 1996.
8. Darveaux, R., Banerji, K., Mawer, A., and Dody, G., "Reliability of Plastic Ball Grid Array Assembly, "Ball Grid Array Technology, J. Lau, ed., McGraw-Hill, Inc. New York, pp. 379-442, 1995.
9. Darveaux, R., and Banerji, K., "Constitutive Relations for Tin-Based Solder Joints," IEEE Trans-CPMT-A, Vol. 15, No. 6, pp. 1013-1024, 1992.
10. Engelmaier, W., "Functional Cycles and Surface Mounting Attachment Reliability", ISHM Technical Monograph Series, pp. 87-114, 1984.
11. Evans, J. L., R. Newberry, L. Bosley, S. G. McNeal, A. Mawer, R. W. Johnson and J. C. Suhling, "PBGA Reliability for Under-the-Hood automotive Applications", Proceedings of InterPACK '97, Kohala, HI, pp. 215-219, June 15-19, 1997.
12. Gustafsson, G., Guven, I., Kradinov, V., Madenci, E., "Finite Element Modeling of BGA Packages for Life Prediction", Proceedings of the 2000 Electronic Components and Technology Conference, Las Vegas, Nevada, pp. 1059-1063, May 21-24, 2000.

13. Hall, P. M., "Forces, moment, and displacements during thermal chamber cycling of leadless ceramic chip carriers soldered to printed boards", IEEE Transactions on CHMT, Vol. 7, No. 4, pp.314-327, December 1984.
14. Jung, E. Heinricht, K. Kloeser, J. Aschenbrenner, R. Reichl, H., Alternative Solders for Flip Chip Applications in the Automotive Environment, IEMT-Europe, Berlin, Germany, pp.82-91, 1998.
15. Knecht, S., and L. Fox, "Integrated matrix creep: application to accelerated testing and lifetime prediction", Chapter 16, Solder Joint Reliability: Theory and Applications, ed. J. H. Lau, Van Nostrand Reinhold, pp. 508-544, 1991.
16. Lall, P.; Islam, M. N. , Singh, N.; Suhling, J.C.; Darveaux, R., "Model for BGA and CSP Reliability in Automotive Under-hood Applications", IEEE Transactions on Components and Packaging Technologies, Vol. 27, No. 3, p 585-593, September 2004.
17. Lall, P., N. Islam, J. Suhling and R. Darveaux, "Model for BGA and CSP Reliability in Automotive Under-hood Applications", Electronic Components and Technology Conference, pp.189 –196, 2003.
18. Lau, J. H., "Flip chip technologies", McGraw Hill, New York, 1996.
19. Lindley, T. R., "BGA Solder Joint Reliability Study for Automotive Electronics", Proceedings of the 1995 International Conference on Multi-chip Modules, Denver, CO, pp. 126-133, April 19-21, 1995.
20. Liu X. S., S. Haque, and G.-Q. Lu, "Three-dimensional Flip-Chip on Flex Packaging for Power Electronics Applications," IEEE Transactions on Advanced Packaging, vol. 24, pp. 1-9, Feb. 2001.
21. Lu H., Bailey C., Cross M., "Reliability Analysis of Flip Chip Designs via Computer Simulation", Journal of Electronic Packaging, Transactions of the ASME, Vol. 122, No. 3, pp. 214-219, September 2000.
22. Mawer, A., N. Vo, Z. Johnson and W. Lindsey, "Board- Level Characterization of 1.0mm and 1.27mm Pitch PBGA for Automotive Under-Hood Applications", Proceedings of the 1999 Electronic Components and Technology Conference, San Diego, CA, pp. 118-124, June 1-4, 1999.
23. Pascariu G., Cronin P, Crowley D, "Next-generation Electronics Packaging Using Flip Chip Technology", Advanced Packaging, Nov. 2003
24. Popelar S. F., "Parametric study of flip chip reliability based on solder fatigue modeling: Part II - flip chip on organic", Proceedings of SPIE - The International Society for Optical Engineering, v 3582, pp 497-504, 1998.
25. Ray, S. K., Quinones, H., Iruvanti, S., Atwood, E., Walls, L., "Ceramic Column Grid Array (CCGA) Module for a High Performance Workstation Application", Proceedings - Electronic Components and Technology Conference, pp 319-324, 1997.
26. Sillanpaa, M., Okura, J. H., "Flip chip on board: assessment of reliability in cellular phone application", IEEE-CPMT Vol.27, Issue:3, pp. 461 – 467, Sept.2004.
27. Syed, A. R., "Thermal Fatigue Reliability Enhancement of Plastic Ball Grid Array (PBGA) Packages", Proceedings of the 1996 Electronic Components and Technology Conference, Orlando, FL, pp. 1211-1216, May 28-31, 1996.
28. Suhir, E., Calculated Thermally Induced Stresses in Adhesively Bonded and Soldered Assemblies, Proc. of the International Symposium on Microelectronics, Atlanta, Georgia, pp. 386-392, Oct. 1986.
29. Suhir, E., Thermal Stress Failures in Microelectronic Components – Review and Extension", Vol. 1, Chapter 5, in Advances in Thermal Modeling of Electronic Components and Systems, ed. A. Bar-Cohen and A. D. Kraus, Hemisphere Publishing Co., pp. 337-412, 1990.
30. Timoshenko, S. P., Analysis of Bi-Metal Thermostats, Journal of the Optical Society of America, Vol. 11, September 1925.

31. Van den Crommenacker, J., "The System-in-Package Approach", IEEE Communications Engineer, Vol 1, No. 3, pp. 24-25, June/July, 2003.
32. Yeh C., Zhou W. X., Wyatt K., "Parametric Finite Element Analysis of Flip Chip Reliability", International Journal of Microcircuits and Electronic Packaging, Vol. 19, No. 2, pp. 120-127, 1996.
33. Vandevelde, B., Christiaens F., Beyne, Eric., Roggen, J., Peeters, J., Allaert, K., Vandepitte, D. and Bergmans, J., "Thermo-mechanical Models for Leadless Solder Interconnections in Flip Chip Assemblies", IEEE Transactions on Components, Packaging and Manufacturing Technology-Part A, Vol.21, No. 1, pp.177-185, March 1998.
34. Wong, B., D. E. Helling and R. W. Clark, "A creep rupture model for two phase eutectic solders", IEEE CHMT Trans., Vol. II, No. 3, pp. 284-290, Sep. 1988.

EXERCISES

12.1 Estimate the cyclic life of a ceramic capacitor soldered to a printed circuit board fabricated from an organic laminate. The ceramic capacitor is 0.80 in. in length with metallic ends each 0.10 in. long. Its TCE is 6.8×10^{-6} /°C and its modulus of elasticity is 44×10^{6} psi.. The printed circuit board is fabricated from a glass epoxy laminate with a TCE of 20×10^{-6} and a modulus of elasticity of 1.5×10^{6} psi. The temperature is cycled from 10 °C to 110 °C once each day. Solder paste is applied to the copper pads to yield a solder joint thickness of 0.005 in. Comment on the expected life of the solder joints.

12.2 Estimate the cyclic life of a leadless ceramic chip carrier soldered to a printed circuit board fabricated from an organic laminate. The ceramic chip carrier is square with side lengths of 32 mm. Its TCE is 6.8×10^{-6} /°C and its modulus of elasticity is 44×10^{6} psi.. The printed circuit board is fabricated from a glass epoxy laminate with a TCE of 20×10^{-6} and a modulus of elasticity of 1.5×10^{6} psi. The temperature is cycled from 0 °C to 125 °C three times each day. Solder paste is applied to the copper pads to yield a solder joint thickness of 7 mils. Comment on the expected life of the solder joints.

12.3 Estimate the cyclic life of a plastic BGA soldered to a printed circuit board fabricated from an organic laminate. The BGA is square with sides 55 mm long. The centerline of the last row of solder balls in 5 mm from the edge of the chip carrier. The ball diameter is 0.10 mm. The TCE of the BGA chip carrier is 18.8×10^{-6} /°C. The printed circuit board is fabricated from a glass epoxy laminate with a TCE of 20×10^{-6} and a modulus of elasticity of 1.5×10^{6} psi. The temperature is cycled from 20 °C to 125 °C three times each day. Solder paste is applied to the copper pads to yield a solder joint height including the balls of 0.005 in. Comment on the expected life of the solder joints.

12.4 A copper plated-though-hole extends through a printed circuit board that is 4 mm thick. The PCB, which is fabricated from glass reinforced epoxy, exhibits and average out-of-plane TCE of $\alpha_z = 120 \times 10^{-6}$ over the temperature range of 20 to 200 °C. Determine the stresses in the copper barrel of the plated-through-hole.

12.5 prepare an engineering brief discussing the types of failure that can occur in the plastic encapsulating a silicon chip as it cools after being injection molded and cured at elevated temperature. For each type of failure describe the problem it creates in the operation of an electronic system.

12.6 Prepare an engineering brief describing the formulation of the compound used for encapsulating silicon chips in plastic by injection molding.

12.7 Prepare an engineering brief describing the problems encountered when underfill is used with flip chips to alleviate solder joint failures due to thermal cycling.

12.8 Describe the situations where underfill is not needed in packaging flip-chips.

12.9 Determine the in-plane stresses generated in a thin film of copper (10 μm thick) that is applied to a silicon wafer 10 mm thick. The temperature of the copper when it is deposited is 300 °C. The TCE for copper is 17 × 10^{-6} and for silicon 2.6 × 10^{-6}. As the copper cools to 20 °C will compressive or tensile stresses result? Does it matter? What mode of failure results when the thin film compressive stresses become excessive?

12.10 For the thin film described in Exercise 12.10 determine the out-of-plane strain e_z.

12.11 A piece of glass reinforced epoxy 0.6 mm thick, 50 mm wide and 150 mm long is bonded to a piece of aluminum of the same dimensions. The bonding operation occurred at room temperature 20 °C. If this bi-material strip is heated to 75 °C, determine its curvature. TCE for aluminum and the glass reinforced plastic are 23.2 × 10^{-6} and 20.0 × 10^{-6}, respectively. The modulus of elasticity for aluminum and the glass reinforced plastic are 10.1 × 10^6 and 1.5 × 10^6, respectively. Determine the radius of curvature due to the temperature change.

12.12 For the bi-material strip described in Exercise 12.11, determine the maximum stress and its location.

12.13 Write an engineering brief describing why warpage of chip carriers is a serious concern.

12.14 Write an engineering brief explaining why warpage of multi-layer boards is a serious concern.

12.15 What can be done to mitigate the problem of warpage in plastic chip carriers?

12.16 Why is warpage not a serious problem with ceramic chip carriers?

12.17 Cite the reasons for warpage in multilayered boards.

12.18 Describe the limitations of the DNP solutions described in this chapter.

12.19 Describe the advantages of using finite element methods for determining, stress, strain and displacements introduced by temperature changes in electronic packaging.

12.20 Cite the seven steps necessary for the engineer to perform in conducting a finite element analysis.

12.21 Describe precautions taken in preparing the mesh for a leaded chip carrier before conducting a finite element analysis.

12.22 Why is it more involved in conducting a three-dimensional finite element analysis that a two-dimensional analysis?

12.23 Describe the procedure for conducting cyclic thermal fatigue tests. How do you determine dwell time? How do you determine maximum and minimum temperatures? Is the rare of temperature increase and decrease important?

12.24 How is failure of solder joints determined in a cyclic thermal fatigue test?

12.25 How is the data from a cyclic thermal fatigue test processed at the conclusion of the test?

12.26 Using the multi variate linear regression equation given in Eq. (12.32), predict the characteristic life for the following flip chip packages.

Design Variable	Case 1	Case 2	Case 3	Case 4
x_1 is the die size in mm	6	8	10	12
x_2 is the ball count	48	84	124	180
x_3 is the ball diameter in mm	0.15	0.12	0.10	0.05
x_4 refers to the pad type	SMD	NSMD	NSMD	NSMD
x_5 is the temperature differential ΔT in °C	150	125	125	100
x_6 is the pitch of the solder balls in mm	0.4	0.3	0.3	0.2
x_7.is the solder ball height in mm	0.13	0.11	0.09	0.04
x_8 is the solder alloy composition	Sn96.5Ag3.5	Sn96.5Ag3.5	Sn96.5Ag3.5	Sn96.5Ag3.5

CHAPTER 13

ANALYSIS OF VIBRATION AND SHOCK OF ELECTRONIC SYSTEMS

13.1 INTRODUCTION

An electronic product may be subjected to shock and or vibration during manufacture, shipping and operational life. Products may be subjected to a combination of temperature cycling and random vibration as a screening environment to eliminate manufacturing flaws. Shipping from the assembly plant through the distribution chain to the customer may involve product transport in a shipping container, which may be loaded and off-loaded on several occasions, subjecting the product to shock, vibration, in conjunction with thermal stresses. Operational life may involve shock, vibration or accidental drop during normal usage.

The trend towards miniaturized fine pitch electronics, has increased the susceptibility of failure in shock, vibration and high-strain rate stresses. Fundamental understanding of the failure mechanisms and the analytical techniques is required to enable product reliability over the design life. To maintain a high reliability of the product in field usage and minimize warranty costs, the design defects and dominant failure mechanisms must be eliminated in the factory before the product is shipped. Manufacturing screening tests are often used in both high-reliability commercial and military systems to eliminate the most of these defects [1, 2]. Consumer products with short design-life often involve subjecting a product-sample to accelerated life test, but may not involve a 100-percent screen on the production volume. The screening tests are adapted to a specific product.

Vibration reliability is ascertained by subjecting the product to random vibration with specified power spectral density of the exposure varied with frequency as shown in Fig. 13.1. The product under test is positioned on a vibration table so that the table's axis of motion is perpendicular to the printed circuit boards. If the critical components in the product are oriented in more than one plane, the vibration test is repeated sequentially with the vibration applied along three orthogonal axes. The duration of the random vibration is may usually vary up to a few minutes for single axis or multiple axes of vibration. When it is necessary to repeat the vibration along the orthogonal axes, similar durations may be applied to each axis. The equipment under evaluation is usually operational during the test to identify incipient faults or system-failures. Malfunctions in the operation indicate that a latent defect has been exposed by random vibration. Subsequent inspection reveals the defect that can be corrected in the plant prior to shipping.

In addition to vibration, the accelerated tests and screens may also incorporate temperature cycling. Common ranges of test temperature in the cycling include –55 to +125 °C, –40 to 125°C, and 0 to 100°C. Again, the equipment is functional during the cycling, and malfunctions indicate failure due to a latent defect that can be identified, located and replaced. The number of temperature cycles used in the accelerated test or screen depends on the complexity of the product, intended design-life and the desired reliability in the end application. Temperature cycle screen often involves subjecting the product from one cycle to a few cycles in order to identify latent defects. One temperature-cycle is sufficient for screening simple products with less than 100 electronic components, but moderately complex equipment with as many as 500 parts employ three temperature cycles. As many as ten cycles are specified for very complex systems with 4,000 or more components. An example of the anticipated reduction in failures per unit as a function of the number of temperature cycles is presented in Fig. 13.2.

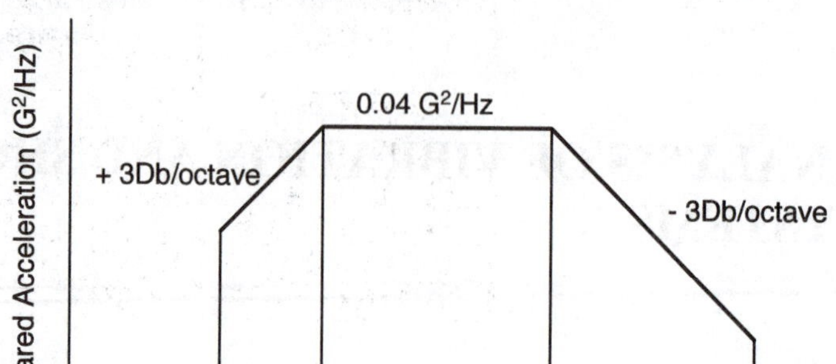
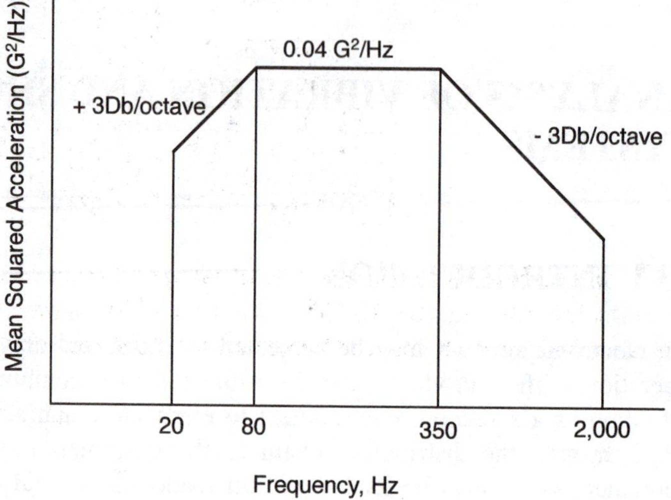

Fig. 13.1 Power spectral density for random vibration recommended as a manufacturing screening test [1].

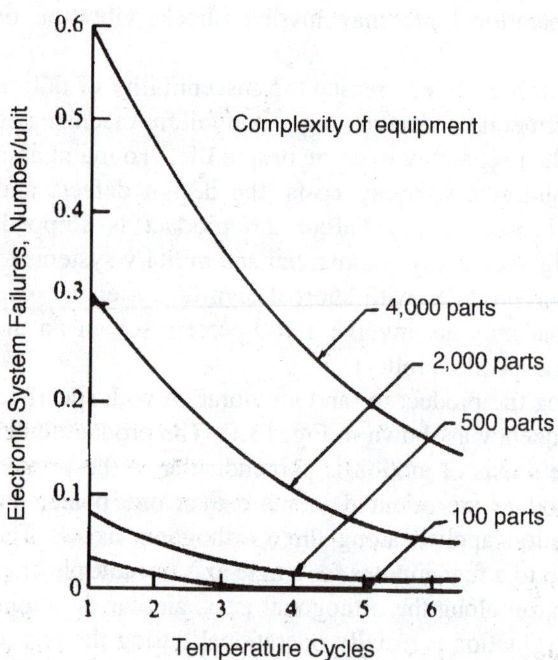

Fig. 13.2 Improvement in failure rate with the number of temperature cycles for electronic equipment with different degrees of complexity.

The shipping environment imposed on the product, as it moves from the factory through the distribution chain to the customer, differs from product to product and even from shipment to shipment. The differences are due to placement in the transport vehicle, vehicle operator, drops and/or tosses during handling, routing, road conditions, season, etc. The American Society for Testing Materials, ASTM, has characterized the shock and vibration exposures that may be encountered [3]. The shock environment, presented in Table 13.1, is due to handling when the product is transferred from one location to another. The drop height is related to the product weight with the lighter product likely to be dropped from a higher position. The assurance level is related to the number of units to be shipped and their value. Assurance level II is for average size shipments of moderately priced units.

Table 13.1
Shock environments

Shipping weight lb (kg)	Drop height [in. (mm)] Assurance level		
	I	II	III
0-20 (0-9.1)	24 (610)	15 (381)	9 (227)
20-40 (9.1-8.1)	21 (533)	13 (330)	8 (203)
40-60 (18.1-27.2)	18 (457)	12 (305)	7 (178)
60-80 (27.2-36.3)	15 (381)	10 (254)	6 (152)
80-100 (36.3-45.4)	12 (305)	9 (227)	5 (127)
100-200 (45.4-90.7)	10 (254)	7 (178)	4 (102)

The vibration environment is often specified in terms of a random vibration spectrum. The spectrum incorporated in ASTM standard D4169, shown in Fig. 13.3, is often used in screening programs that are considered representative for the random vibration environment produced by a truck [4].

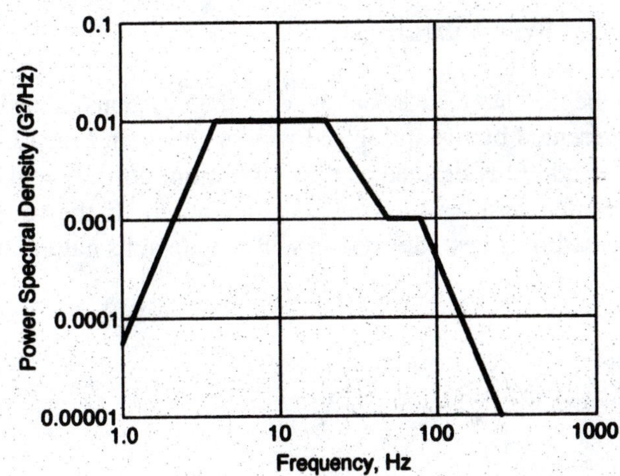

Fig. 13.3 Random vibration spectrum representing vibration produced by a truck.

The shock and vibration environment encountered in operation at the customer location clearly depends on the application. Consumer products such as cellular phones, mobile computing devices such as laptops, personal digital assistants, gaming devices, entertainment devices may be subjected to accidental drop during normal usage. The product may be subjected to excessive acceleration, and out-of-plane deformation, causing catastrophic failure during an accidental drop. Military electronics may be placed on a weapons platform such as a tank, missile, or armored vehicle. In such instances, the equipment must be designed initially to resist strong shocks and prolonged vibration at high G levels. Consumer products are often designed to survive a finite number of drops from ear-height of 5ft-6ft. Rugged electronic equipment is becoming more common as processors of all types are moving from the desktop to mobile workplaces. To facilitate the initial design, the military customer will define the shock and vibration environment in the procurement documentation. Examples of military specifications for shock and vibration are given in [5 - 8]. Another application with severe vibration and shock loading involves systems installed on machine tools and on vehicles. Indeed, electronics on automobiles for engine control include hundreds of millions of systems subject to vibration, shock and large temperature fluctuations.

The basic equations describing motion of an electronic system due to both shock and vibration are introduced in this chapter. Applications of these equations to the design of select electronic components are then described. For more detailed analysis, modeling printed circuit card assemblies with finite elements is introduced. Experimental methods to study failure and reliability for CSPs subjected to impact loading are described. Finally, exercises are provided to demonstrate the approach used in designing systems for shock and vibration environments.

13.2 VIBRATING SYSTEMS WITH ONE DEGREE OF FREEDOM

Mechanical systems respond to a dynamic disturbance imposed either by forces or displacements by vibrating. The vibratory response is often quite complex with several different time dependent motions occurring simultaneously. Significant insight into the analysis of complex systems can be gained by the study of such idealized systems. In some cases, the dynamic parameters of complex systems can be developed using lumped mass analysis for computational efficiency. In this case, the mechanical system can be treated as a single degree of freedom system without introducing significant error.

13.2.1 Free Vibrations

To begin the discussion of vibrating systems, consider a simple one degree of freedom system represented by the spring and weight shown in Fig. 13.4. After the weight W is attached to the spring, the origin O is defined as the equilibrium position and y, the time dependent position of the weight, is measured positive upward from this point. If the weight is deflected downward by a distance δ and released from rest, the system will vibrate at its natural frequency.

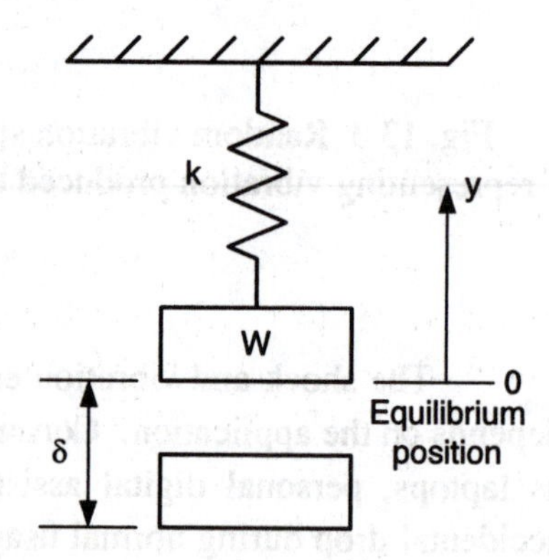

Fig. 13.4 Spring and weight assembly representing a single degree of freedom system without damping.

We analyze this vibration by writing the equation of motion of the weight as:

$$m\ddot{y} + ky = 0 \tag{13.1}$$

where $\ddot{y}$ is the second derivative of y with respect to time t and k is the spring rate.

Solution of this ordinary second order differential equation is given by:

$$y = A \sin \omega_n t + B \cos \omega_n t \tag{13.2}$$

where ω_n is the natural frequency of the system given by:

$$\omega_n = \sqrt{\frac{k}{m}} \qquad (13.3)$$

The coefficients A and B on Eq. (13.2) are constants of integration that are determined from the initial conditions: $t = 0, y = -\delta$ and $\dot{y} = 0$.

Substituting these initial conditions into Eq. (13.2) gives:

$$y = -\delta \cos \omega_n \qquad (13.4)$$

A graph of the initial response of the motion of the system, presented in Fig. 13.5, shows that the weight oscillates at constant amplitude δ with a period T. The relation between the period and the natural frequency f_n is given by:

$$T = 1/f_n \qquad (13.5a)$$

$$f_n = \frac{1}{2\pi}\omega_n = \frac{1}{2\pi}\sqrt{\frac{k}{m}} \qquad (13.5b)$$

Note that $mg = k\delta_s$, combining Eqs.(13.3) and (13.5b) leads to:

$$f_n = \frac{1}{2\pi}\sqrt{\frac{g}{\delta_s}} \qquad (13.6)$$

where g is the gravitational constant and δ_s is the deflection due to a static load.

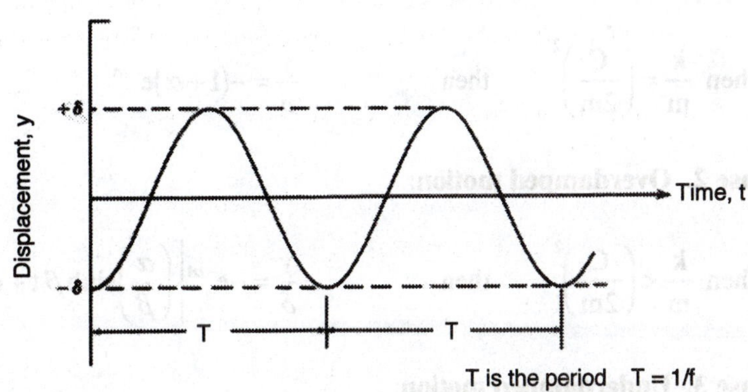

Fig. 13.5 Displacement y as a function of time for an undamped single degree of freedom system.

These results indicate that the vibratory motion expressed in terms of the displacement y will continue indefinitely. This is not realistic because some dampening is always present that will cause the amplitude of the oscillation to decay with time.

When viscous dampening is introduced in the vibrating system, a dash pot is added in parallel with the spring as shown in Fig. 13.6. The weight is again deflected downward by an amount δ and released from rest. The differential equation of motion for the weight is given by:

$$m\ddot{y} + C\dot{y} + ky = 0 \qquad (13.7)$$

where C is the dampening coefficient.

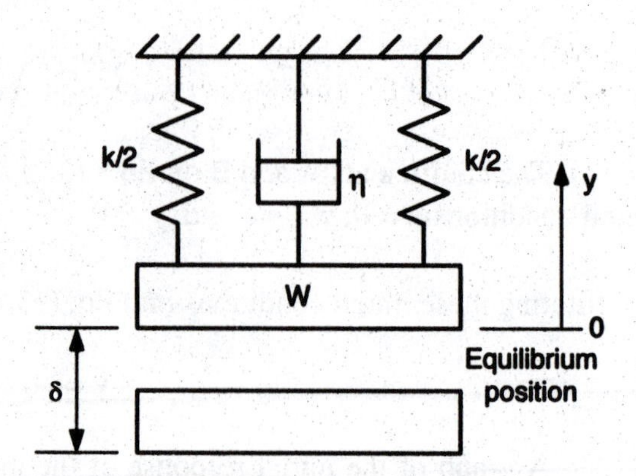

Fig. 13.6 Spring and mass assembly representing a single degree system with damping.

Substituting $y = e^{\lambda t}$ into Eq. (13.7) leads to a quadratic auxiliary equation in terms of λ. The roots of this equation are given by:

$$\lambda = -\frac{C}{2m} \pm \sqrt{\left(\frac{C}{2m}\right)^2 - \left(\frac{k}{m}\right)} \tag{13.8}$$

The solution for the displacement y depends on whether λ is real or a complex number. Three different solutions are described which characterize the type of dampened vibration that can occur, depending on the values of $\left(\frac{k}{m}\right)$ and $\left(\frac{\lambda}{2m}\right)^2$.

Case 1. Critical dampening:

When $\frac{k}{m} = \left(\frac{C}{2m}\right)^2$ then

$$\frac{y}{\delta} = -(1+\alpha)e^{-\alpha t} \tag{13.9}$$

Case 2. Overdamped motion:

When $\frac{k}{m} < \left(\frac{C}{2m}\right)^2$ then

$$\frac{y}{\delta} = -e^{-\alpha t}\left[\left(\frac{\alpha}{\beta}\right)\sinh\beta t + \cosh\beta t\right] \tag{13.10}$$

Case 3. Underdamped motion:

When $\frac{k}{m} > \left(\frac{C}{2m}\right)^2$ then

$$\frac{y}{\delta} = -e^{-\alpha t}\left[\left(\frac{\alpha}{\beta}\right)\sin\beta t + \cos\beta t\right] \tag{13.11}$$

where $\alpha = \frac{C}{2m}$ and $\beta^2 = \left(\frac{C}{2m}\right)^2 - \frac{k}{m}$

The response of the system to the initial displacement δ is presented in Fig. 13.7. It is clear, that the system recovers its initial equilibrium position, without oscillation for the critical and the overdamped cases. Vibration, with its oscillatory motion, occurs only in the underdamped case.

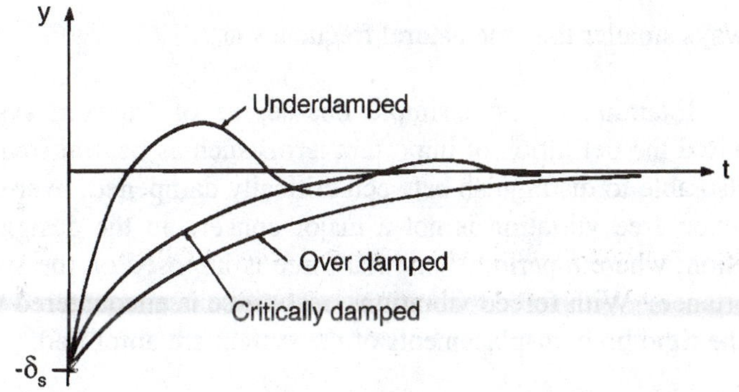

Fig. 13.7 Displacement y as a function of time for a single degree of freedom system with different amounts of damping.

Because the amount of dampening associated with mechanical systems is usually small, critical and overdamped motion are rare events and are not usually considered in a study of vibration of mechanical systems. We are concerned with the underdamped case where vibrations occur that can damage electronic components because of either excessive displacements or the large forces generated during the oscillatory motion.

Consider Eq. (13.11), which describes the motion for the underdamped case, and modify this relation by introducing the natural frequency and the damping ratio (ξ) that are defined by:

$$\xi = C/C_c \tag{13.12}$$

and

$$C_c = 2m\omega_n \tag{13.13}$$

Substituting Eqs.(13.3), (13.12) and (13.13) into Eq. (13.11) leads to:

$$y = e^{-\xi\omega_n t}\left[A\sin\omega_n\sqrt{1-\xi^2}\,t + B\cos\omega_n\sqrt{1-\xi^2}\,t\right] \tag{13.14}$$

The constants of integration A and B in this relation are determined by using the initial conditions:

$$y = -\delta \text{ and } \dot{y} = 0 \text{ at time } t = 0.$$

With $A = \dfrac{-\delta\xi}{\sqrt{1-\xi^2}}$ and $B = -\delta$, it is clear that Eq. (13.14) reduces to:

$$y = -\delta e^{-\xi\omega_n t}\left[\frac{\xi}{\sqrt{1-d^2}}\sin\omega_n\sqrt{1-\xi^2}\,t + \cos\omega_n\sqrt{1-\xi^2}\,t\right] \tag{13.15}$$

Examination of this relation indicates that the motion is expressed in terms of a product. The first term is the dampening multiplier $e^{-\xi\omega_n t}$ that reduces the amplitude of the oscillation exponentially with time. The second term, a cosine function gives the time variance of the oscillatory motion. Note, that the dampened natural frequency (ω_{nd}),

$$\omega_{nd} = \sqrt{1-\xi^2}\ \omega_n \tag{13.16}$$

is always smaller than the natural frequency ω_n.

Examination of a simple one-degree of freedom system with dampening in free vibration permitted the definition of important terms such as natural frequency, period and dampening ratio. We are also able to distinguish between critically dampened, over dampened and under dampened systems. However, free vibration is not a major concern in the design of electronic systems. Instead, forced vibration, where a periodic external force is imposed on the system for some time interval, is of major importance. With forced vibrations, resonance is encountered where the forces constraining the system and the rigid body displacements of the system are amplified.

13.2.2 Forced Vibration

To treat the problem of forced vibrations, consider a periodic external force applied to the weight W as illustrated in Fig. 13.8. An external force is applied as a cosine function, but the results would be essentially the same for a sine function. The equation of motion for the weight shown in Fig. 13.8 can be written as:

$$m\ddot{y} + C\dot{y} + ky = P\cos\omega t \tag{13.17}$$

where P is the amplitude of the external force.

Because the external force is applied for an extended interval of time, the system will achieve steady state motion. For this reason, we seek the particular solution to Eq. (13.17) that describes this steady state motion. The particular solution is written as:

$$y_P = A\cos\omega t + B\sin\omega t \tag{13.18}$$

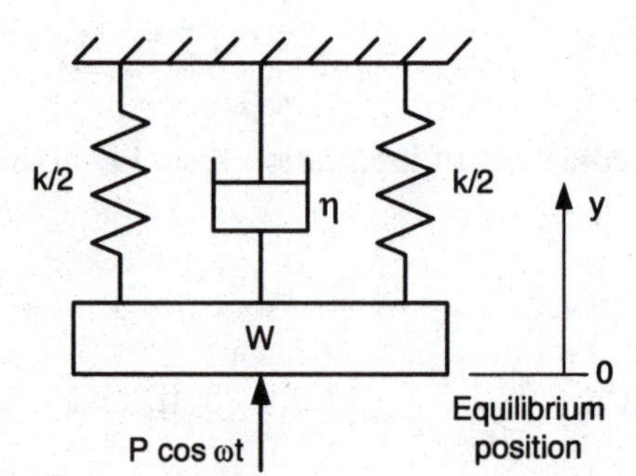

Fig. 13.8 Spring and weight assembly subjected to a periodic external force P.

The unknown coefficients A and B are determined by substituting Eq. (13.18) into Eq. (13.17) to obtain:

$$A = \frac{P\left[k - m\omega^2\right]}{\left[(C\omega)^2 + \left(k - m\omega^2\right)^2\right]} \tag{13.19a}$$

$$B = \frac{PC\omega}{\left[(C\omega)^2 + \left(k - m\omega^2\right)^2\right]} \tag{13.19b}$$

Next, substituting Eqs.(13.19) into Eq. (13.18) gives:

$$y = \frac{P\cos(\omega t - \phi)}{\sqrt{\left(k - m\omega^2\right)^2 + (C\omega)^2}} \qquad (13.20)$$

where ϕ is the phase angle given by:

$$\tan\phi = \frac{C\omega}{\left(k - m\omega^2\right)} \qquad (13.21)$$

The form of Eq. (13.20) is not convenient for analysis, so it is modified by introducing Eqs.(13.3), (13.12) and (13.13) to obtain:

$$y_p = A_\omega \cos(\omega t - \phi) \qquad (13.22)$$

The amplitude of the forced vibration A_ω is a function of the driving frequency ω given by:

$$A_\omega = \frac{P/k}{\sqrt{\left(1 - r^2\right)^2 + (2\xi r)^2}} \qquad (13.23)$$

where $r = \omega/\omega_n$.

We will defer the discussion of A_ω as it is an essential element in the amplitude transmission coefficient introduced in the next subsection.

13.2.3 Transmission Coefficients

When an electronic system is subjected to forced vibrations, we are concerned with two different aspects of the system response. The first is the amplitude of the vibration that is important because excessive amplitude can result in impact if an electronic sub-assembly strikes an adjacent structure. The second is the maximum force that is developed at the constraints. In resonance, this force may become large enough to cause plastic deformation or fracture of the support structure or the attachments of the electronic components. The transmission coefficients for both the amplitude and the forces involved will be determined as a function of the driving frequency ω and the dampening ratio d.

Consider the amplitude of the forced vibration A_ω given by Eq. (13.23) and normalize this quantity by:

$$\boldsymbol{A} = \frac{A_\omega}{\delta_s} \qquad (13.24)$$

Note, that $\boldsymbol{A}$ is the amplitude ratio and $\delta_s = P/k$ is the equivalent static deflection. Combining Eq. (13.23) and (13.24) yields the relation for the amplitude transmission coefficient $\boldsymbol{A}$ as:

$$\boldsymbol{A} = \frac{1}{\sqrt{\left(1 - r^2\right)^2 + (2\xi r)^2}} \qquad (13.25)$$

The amplitude ratio depends only on the frequency ratio r and the degree of dampening d. The ***A*** versus r relationship, shown in Fig. 13.9, indicates a resonance condition occurs when $r = \omega/\omega_n = 1$. At resonance $\omega = \omega_n$ and Eq. (13.25) reduces to:

$$\boldsymbol{A} = \frac{1}{2\xi} \tag{13.26}$$

It is evident that the response of the system can become very large as the dampening ratio ξ goes toward zero. Unfortunately, the dampening ratio in most electronic assemblies is small and values of ***A*** > 10 are common. With very low dampening, it is advisable to design the electronic assembly with a natural frequency that is higher than the driving frequency. To explore the influence of the frequency ratio for those cases where the dampening ratio is low, let $\xi = 0$ and note that Eq. (13.25) reduces to:

$$\boldsymbol{A} = \frac{1}{\left(1 - r^2\right)} \tag{13.27}$$

The results of Eq. (13.27), presented in Fig. 13.10, indicate that ***A*** < 2 can be achieved if r < 0.7. If it is not possible to avoid the resonance region (0.9 < r < 1.1), employing a vibration isolation system to protect the system from the detrimental influence of the excessive amplitudes may be required.

Now that the amplitude transmission coefficient ***A*** has been established, consider the constraining forces that are required to support the spring and dash pot shown in Fig. 13.8. The total force F transmitted to the support system is given by:

$$F = ky + C\dot{y} \tag{13.28}$$

Substituting Eqs. (13.22) and (13.23) into Eq. (13.28) yields:

$$F = \frac{(P/k)\sqrt{k^2 + (C\omega)^2}}{\sqrt{\left(1 - r^2\right)^2 + (2\xi r)^2}} \tag{13.29}$$

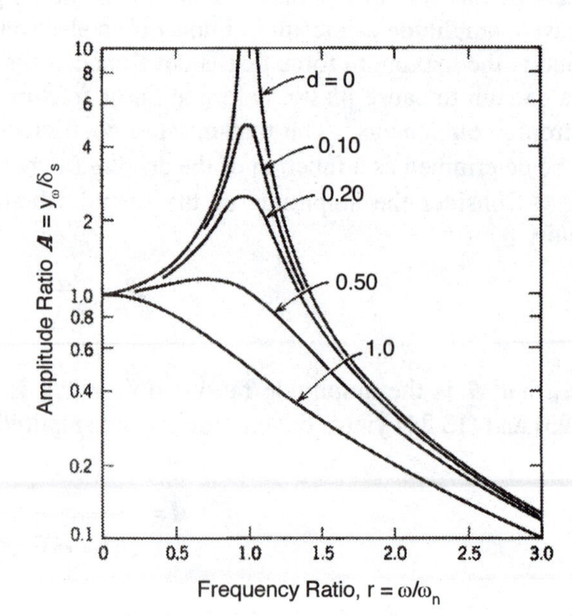

Fig. 13.9 Amplitude ratio ***A*** as a function of frequency ratio ω/ω_n for different degrees of damping.

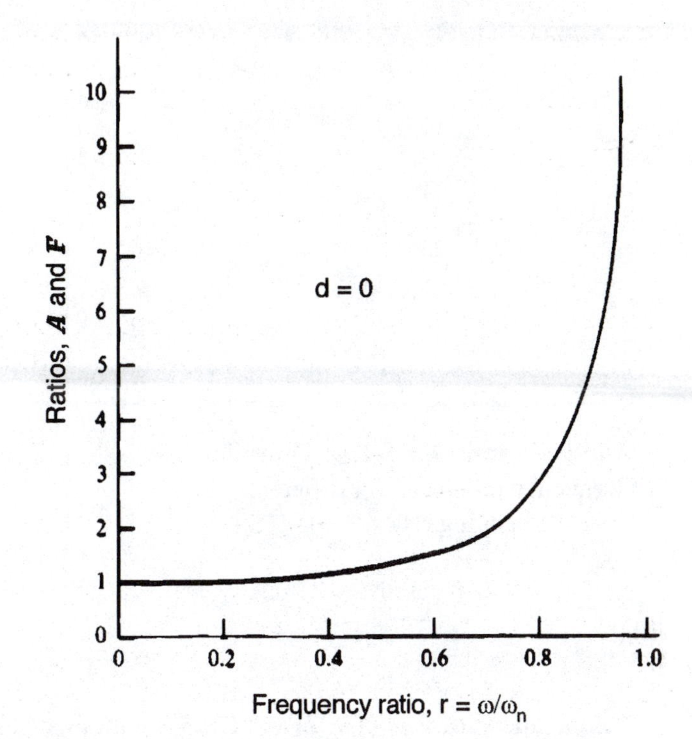

Fig. 13.10 Ratios $\boldsymbol{A}$ and $\boldsymbol{F}$ as a function of frequency for systems with no damping.

Next, define the force transmission coefficient $\boldsymbol{F}$ as:

$$\boldsymbol{F} = \frac{F}{P} \tag{13.30}$$

Then by substitution of Eq. (13.29) into Eq. (13.30), we obtain:

$$\boldsymbol{F} = \frac{\sqrt{1+(2\xi r)^2}}{\sqrt{(1-r^2)^2+(2\xi r)^2}} \tag{13.31}$$

A graph showing $\boldsymbol{F}$ as a function of the frequency ratio r presented in Fig. 13.11, indicates resonance at r = 1 with very high values of $\boldsymbol{F}$. The amplification of the forces depends strongly on the dampening ratio ξ. However, since the dampening ratio is small (i.e. less than 0.1) for most electronic assemblies extremely high forces will be generated if the system is exposed to frequencies near ω_n.

It is of interest to consider the resonance condition where r = 1. At resonance, Eq. (13.31) reduces to:

$$\boldsymbol{F} = \frac{\sqrt{1+(2\xi)^2}}{2\xi} \approx \frac{1}{2\xi} \tag{13.32}$$

With for small dampening $\xi \Rightarrow 0$, the force transmission coefficient can be approximated by:

$$\boldsymbol{F} = \frac{1}{(1-r^2)} \tag{13.33}$$

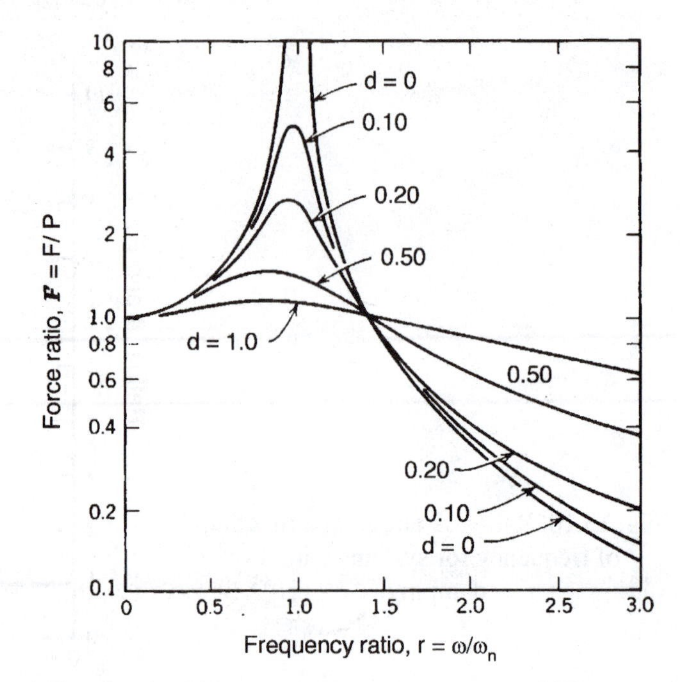

Fig. 13.11 Force ratio **F** = F/P as a function of frequency ratio ω/ω_n for different damping ratios.

Comparison of Eqs. (13.32) and (13.33) with Eqs. (13.26) and (13.27) shows that **A** = **F** either at resonance or if $\xi \Rightarrow 0$. A significant difference in **A** and **F** occurs only when the dampening ratio is large and the system is not at resonance.

Because dampening in electronic assemblies is usually very low, one can anticipate large amplification of the vibrating amplitude and large forces transmitted to the support structure if the system is driven at a frequency that approaches ω_n. Clearly, extended exposure of the system to resonance conditions must be avoided to prevent damage due to fatigue failure of solder joints, lead wires and fasteners. If it is not possible to design the system with a natural frequency that exceeds the frequency of the forcing function by a factor of two ($\omega_n > 2\omega$), then isolation of the electronic system from the disturbing force may be required.

13.3 ISOLATION OF SYSTEMS FROM EXCITING FORCES

The concept of vibration isolation is presented in Fig. 13.12 where an electronic system, with a mass m, is suspended with springs and a damper in an enclosure. One seeks to reduce the amplitude of vibration of the mass m and to minimize the forces imposed on the mass, the suspension and the enclosure. The motion of the electronic assembly is identified with the displacement coordinate x. The enclosure is subjected to a vibratory motion described in terms of the coordinate y. The suspension is entirely separate from the electronics. The independence of the spring and the dash pot enables the designer to select these components to isolate the mass m from the imposed vibrations. To analyze this isolator, the equation of motion for the mass m is written as:

$$m\ddot{x} = -k(x - y) - C(\dot{x} - \dot{y}) \qquad (13.34)$$

Next, the terms are rearranged to obtain:

$$m\ddot{x} + C\dot{x} + kx = ky + C\dot{y} \qquad (13.35)$$

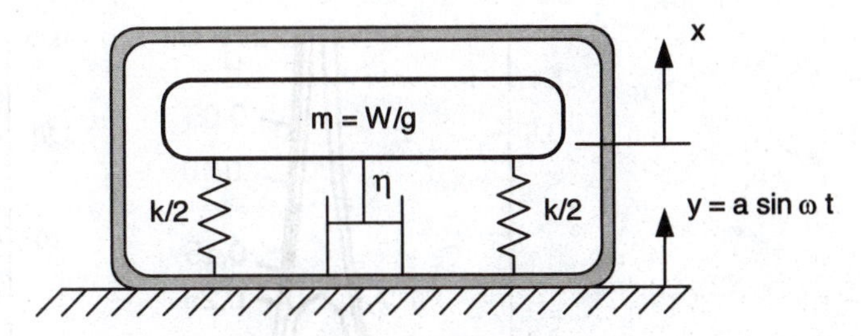

Fig. 13.12 Isolated system with springs and dampers separating the mass from an enclosure.

Now consider the input y(ω) to the housing to be periodic with:

$$y = y_0 e^{j\omega t} \tag{13.36}$$

The response of the mass will also be periodic of the form:

$$x = x_\omega e^{j(\omega t - \phi)} \tag{13.37}$$

where ϕ is the phase angle between the input and output motions.

Substituting Eqs.(13.36) and (13.37) into Eq. (13.35) leads to:

$$\frac{x_\omega}{y_0} = \frac{(k + j\omega C)e^{j\phi}}{\left[\left(k - m\omega^2\right) + j\omega C\right]} \tag{13.38}$$

Using methods from complex variables, it is easy to show that the absolute value of the amplitude ratio $|\boldsymbol{I}|$ is given by:

$$|\boldsymbol{I}| = \left|\frac{x_\omega}{y_0}\right| = \sqrt{\frac{\left[1 + (2\xi r)^2\right]}{\left[\left(1 - r^2\right)^2 + (2\xi r)^2\right]}} \tag{13.39}$$

where the phase angle ϕ is given by:

$$\tan\phi = \frac{2dr^3}{\left[\left(1 - r^2\right) + (2dr)^2\right]} \tag{13.40}$$

A graph of the isolation amplitude ratio $|\boldsymbol{I}|$ is presented in Fig. 13.13. Again the effects of a resonance condition are obvious with a large amplification of the input displacement. However, the isolation system as described by its suspension components (k and C) is designed to avoid a resonance condition and to reduce both the displacements and the forces imposed on the electronics enclosure. This isolation is accomplished by designing a suspension system with a frequency ratio $r > \sqrt{2}$ that provides an isolation ratio $|\boldsymbol{I}| < 1.0$ for systems regardless of the degree of damping.

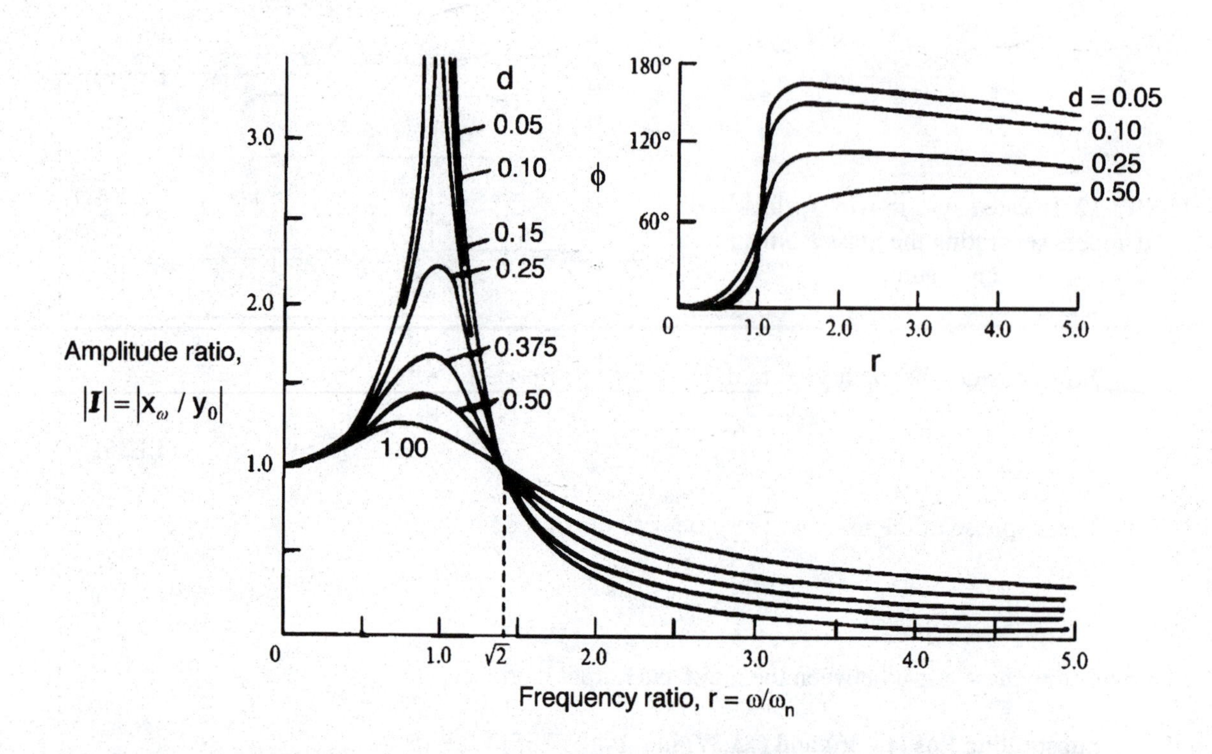

Fig. 13.13 Amplitude ratio and phase angle as a function of frequency for a vibration isolated system.

Effectiveness of isolators in reducing the amplitude of vibration is indicated by the transmissibility (amplitude ratio) of the system. A simplified transmissibility curve for an electronic system of weight W supported on an isolator with stiffness k and damping coefficient C that is subjected to a forcing function with a frequency ω is presented in Fig. 13.14. When the system is excited at its natural frequency, the system will be in resonance and the disturbance forces and the vibration displacement x will be amplified rather than reduced. Therefore, it is essential to select the proper isolation system so that its natural frequency does not coincide with any of the driving frequencies. Also the natural frequency of an isolation system should not coincide with any critical frequencies of the components in the electronic system. Referring to Fig. 13.14, it is evident that when the $\omega/\omega_n < \sqrt{2}$, the transmissibility is greater than one and the electronic system experiences amplification of the input. However, isolation begins when $\omega_d/\omega_n > \sqrt{2}$ with significant reductions in amplitude achieved if $\omega/\omega_n > 3$.

Most isolation systems are designed with some degree of damping. The damping factor $\xi = C/C_c$ for various materials is shown in Table 13.2. Damping is advantageous when the isolated system is operating at or near its natural frequency because damping reduces transmissibility. When electronic equipment is subjected to frequency sweeps and resonance frequencies are within the bounds of the sweep, high damping is important to reduce the buildup of forces that occur at resonance. The relationship between a highly damped and a lightly damped system is illustrated in Fig. 13.13. This graph shows that as damping is increased, isolation efficiency is reduced in the isolation region. While higher values of damping cause significant reduction of transmissibility at resonance, they result in only small increases of transmissibility in the isolation region.

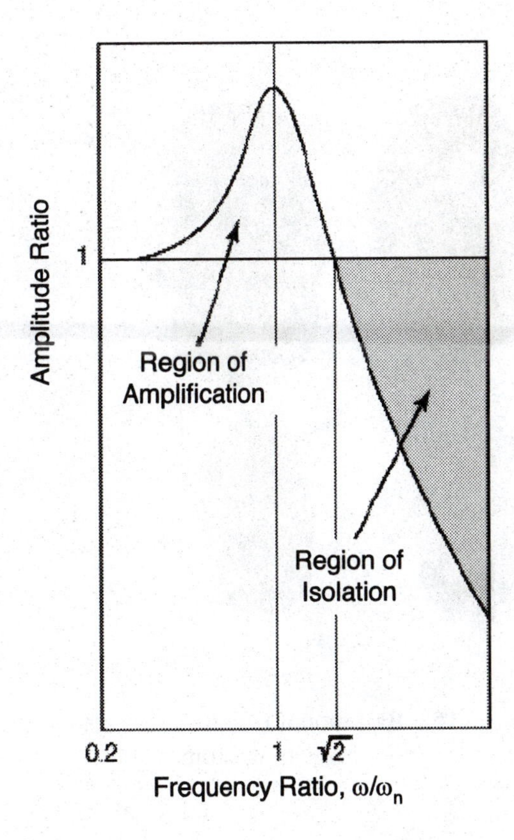

Fig. 13.14 Simplified version of Fig. 13.13 clearly showing regions of amplification and isolation.

Table 13.2
Damping factors for springs fabricated from various materials.

Material	Damping Factor $\xi = C/C_c$
Steel	0.005
Natural Rubber	0.05
Neoprene Rubber	0.05
Butyl Rubber	0.12
Silicone Rubber	0.15
Metal mesh	0.12
Cork and Felt	0.06

There are many spring dash pot assemblies called isolation mounts that are commercially available, as indicated in Fig. 13.15. An example of a molded elastomer mount, which is effective for many lower weight electronic applications, is presented in Fig. 13.15a. The natural frequency of these mounts at their rated loads is about 15 Hz, so that isolation of frequencies above 22 Hz will occur. Of course, the transmissibility of the isolation system improves at higher frequencies as indicated by the response curves shown in Figs. 13.13 and 13.14. Other commercial isolators are presented in Figs. 13.15b, c and d.

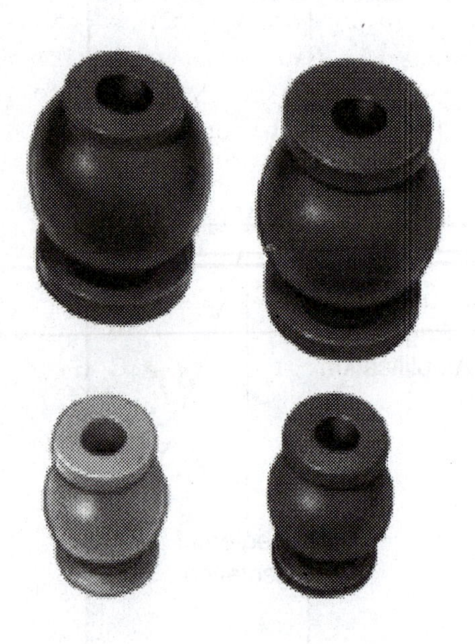

Fig. 13.15a Ball mount isolator fabricated from an elastomer.

Fig. 13.15b Cup mounted isolator with an elastomer spring element.

Fig. 13.15c Cylindrical stud mount isolator with spring and elastomeric motion limiters.

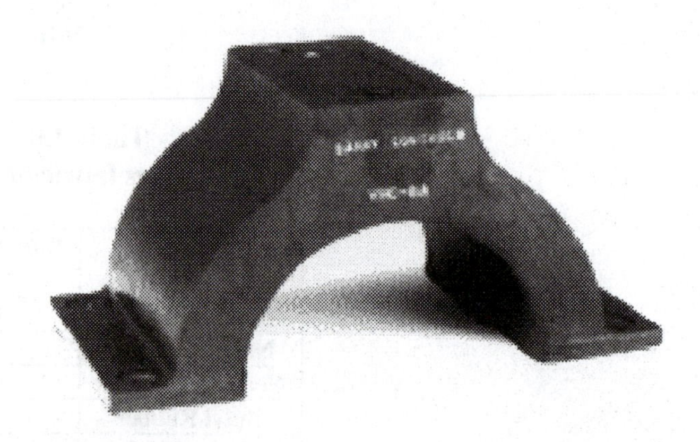

Fig. 13.15d High deflection isolator fabricated from an elastomer.

13.4 VIBRATION OF CIRCUIT BOARDS

Printed circuit boards, which have been described in Chapter 5, are also subject to vibration that can cause damage to the components mounted on the board or to the connectors used to connect it to the back panel. Because most PCB's are rectangular, they can be treated as rectangular plates. We are concerned with the out-of-plane vibration of the plate as indicated in Fig. 13.16. The natural frequency for a rectangular plate depends on the boundary conditions that include clamped, simply supported or free edges. Altogether, there are many possible combinations of these boundary conditions. Solutions to several of these boundary value problems exist in a form that can be readily used in design.

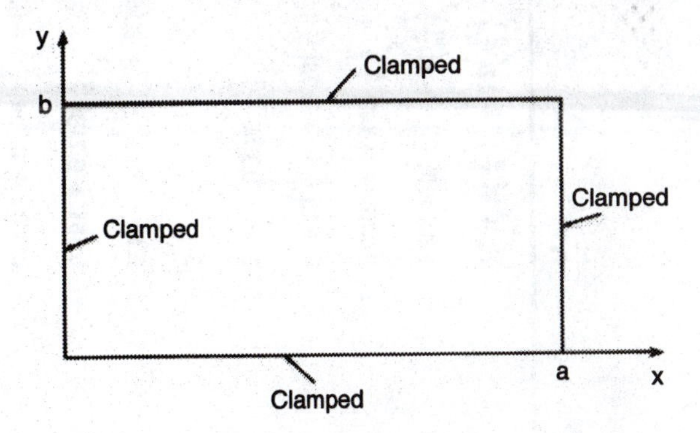

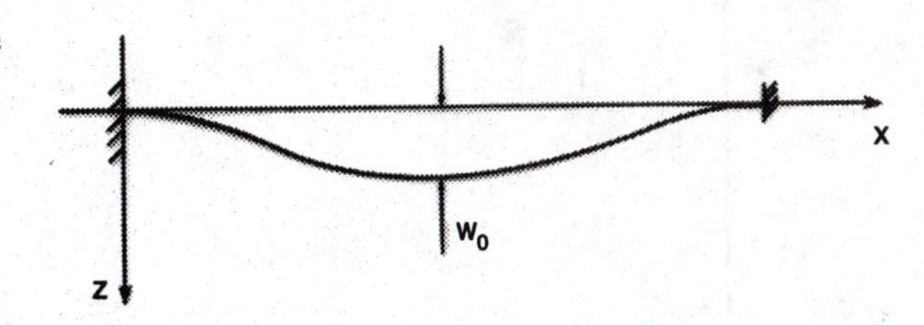

Fig. 13.16 Definition of plate dimensions and the coordinate system.

Leissa [9], in an excellent summary of available results for frequency and mode shapes, presents the solutions for the natural frequency ω_n of the rectangular plates in the following form:

$$\omega_n^2 = \frac{\pi^4 D c_1}{a^4 \rho c_2} \tag{13.41}$$

where c_1 and c_2 are given in Table 13.3; a and b are the plate dimensions; ρ is the mass density per unit area of the plate and D is its flexural rigidity defined by:

$$D = \frac{Eh^3}{12(1-\nu^2)} \tag{13.42}$$

where h is the thickness and ν is Poisson's ratio.

The out of plane deflection w is given in the following form:

$$w = w_\omega \, f[(x/a), (y/b)] \cos \omega t \tag{13.43}$$

where the deflection function f[(x/a), (y/b)] is listed in Table 13.3.

Table 13.3
Parameters c_1 and c_2 for rectangular plates with different boundary conditions used to compute the natural frequency ω_n from Eq. (13.41).

Boundary conditions	Deflection function or mode shape	c_2	c_1
1. 	$\left(\cos\frac{2\pi x}{a}-1\right)\left(\cos\frac{2\pi y}{b}-1\right)$	2.25	$12+8\left(\frac{a}{b}\right)^2+12\left(\frac{a}{b}\right)^4$
2. 	$\left(\cos\frac{3\pi x}{2a}-\cos\frac{\pi x}{2a}\right)\left(\cos\frac{2\pi y}{b}-1\right)$	1.50	$3.85+5\left(\frac{a}{b}\right)^2+8\left(\frac{a}{b}\right)^4$
3. 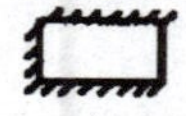	$\left(1-\cos\frac{\pi x}{2a}\right)\left(\cos\frac{2\pi y}{b}-1\right)$	0.340	$0.0468+0.340\left(\frac{a}{b}\right)^2+1.814\left(\frac{a}{b}\right)^4$
4. 	$\left(\cos\frac{2\pi x}{a}-1\right)\sin\frac{\pi y}{b}$	0.75	$4+2\left(\frac{a}{b}\right)^2+0.75\left(\frac{a}{b}\right)^4$
5. 	$\left(\cos\frac{2\pi x}{a}-1\right)\frac{y}{b}$	0.50	$2.67+0.304\left(\frac{a}{b}\right)^2$
6. 	$\cos\frac{2\pi x}{a}-1$	1.50	8
7. 	$\left(\cos\frac{3\pi x}{2a}-\cos\frac{\pi x}{2a}\right)\left(\cos\frac{3\pi y}{2b}-\cos\frac{\pi y}{2b}\right)$	1.00	$2.56+3.12\left(\frac{a}{b}\right)^2+2.56\left(\frac{a}{b}\right)^4$
8. 	$\left(\cos\frac{3\pi x}{2a}-\cos\frac{\pi x}{2a}\right)\left(1-\cos\frac{\pi y}{2b}\right)$	0.227	$0.581+0.213\left(\frac{a}{b}\right)^2+0.031\left(\frac{a}{b}\right)^4$
9. 	$\left(1-\cos\frac{\pi x}{2a}\right)\left(1-\cos\frac{\pi y}{2b}\right)$	0.0514	$0.0071+0.024\left(\frac{a}{b}\right)^2+0.0071\left(\frac{a}{b}\right)^4$
10.	$\left(\cos\frac{3\pi x}{2a}-\cos\frac{\pi x}{2a}\right)\sin\frac{\pi y}{b}$	0.50	$1.28+1.25\left(\frac{a}{b}\right)^2+0.50\left(\frac{a}{b}\right)^4$
11.	$\left(\cos\frac{3\pi x}{2a}-\cos\frac{\pi x}{2a}\right)\frac{y}{b}$	0.333	$0.853+0.190\left(\frac{a}{b}\right)^2$
12. 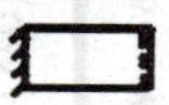	$\cos\frac{3\pi x}{2a}-\cos\frac{\pi x}{2a}$	1.00	2.56
13.	$\left(1-\cos\frac{\pi x}{2a}\right)\frac{\pi^2}{b^2}\sin\frac{\pi y}{b}$	0.1134	$0.0156+0.0852\left(\frac{a}{b}\right)^2+0.1134\left(\frac{a}{b}\right)^4$
14.	$\left(1-\cos\frac{\pi x}{2a}\right)\frac{y}{b}$	0.0756	$0.0104+0.0190\left(\frac{a}{b}\right)^2$
15.	$1-\cos\frac{\pi x}{2a}$	0.2268	0.0313
16. 	$\sin\frac{\pi x}{a}\sin\frac{\pi y}{b}$	0.25	$0.25+0.50\left(\frac{a}{b}\right)^2+0.25\left(\frac{a}{b}\right)^4$
17.	$\left(\sin\frac{\pi x}{a}\right)\frac{y}{b}$	0.1667	$0.1667+0.0760\left(\frac{a}{b}\right)^2$
18. 	$\sin\frac{\pi x}{a}$	0.50	0.50

From Eq. (13.43), it is evident that the deflection of the circuit board is a function of the size of the board given by a and b, the exciting frequency ω and the amplitude of the vibration w_ω that is controlled by the strength of the vibratory input.

Let's consider the simplest of the cases listed in Table 13.3 as an example. Case No. 18 of Table 13.3 shows a rectangular plate with two opposing edges simply supported and the other two opposing edges free. The deflection for this plate may be written as:

$$w = w_\omega \sin\left(\frac{\pi x}{a}\right)\cos\omega t \qquad (13.44)$$

The maximum amplitude will occur along the center line x = a/2 at resonance when $\omega = \omega_n$. With these conditions, Eq. (13.44) reduces to:

$$w_{max} = w_{\omega n} \cos \omega_n t \qquad (13.45)$$

Differentiating Eq. (13.45) twice with respect to time gives the maximum out of plane acceleration $\ddot{w}$ as:

$$\ddot{w}_{max} = -w_{\omega n}\, \omega_n^2 \cos \omega_n t \qquad (13.46)$$

Let us examine the amplitude of the acceleration and define a^*_{out} as:

$$a^*_{out} = \ddot{w}/g = -\, w_{\omega n}\, \omega_n^2/g \qquad (13.47)$$

In this form, the acceleration is given in terms of G's. Solving this relation for $w_{\omega n}$ gives the amplitude of the maximum displacement of the PCB as:

$$w_{\omega n} = -\frac{g a^*_{out}}{\omega_n^2} \qquad (13.48)$$

From Eq. (13.24) it is evident that:

$$a^*_{out} = \boldsymbol{A}\, a^*_{in} \qquad (13.49)$$

Substituting Eq. (13.49) into Eq. (13.48) yields:

$$w_{\omega n} = \frac{g \boldsymbol{A} a^*_{in}}{\omega_n^2} \qquad (13.50)$$

This relation permits the maximum displacement of the PCB to be determined for a given input acceleration a^*_{in} that is specified in a qualification test procedure. Because ω_n^2 is in the denominator, it is clear that the circuit boards should be designed with a high natural frequency to reduce the amplitude of the displacement.

One difficulty occurs in using Eq. (13.50)—the transmission coefficient ***A*** is often unknown. There are three different approaches for determining ***A***. The first is to construct a model of the PCB and to perform vibration sweeps on a shaker table. By measuring the ratio of the acceleration at the center of the board and at the shaker table, as illustrated in Fig. 13.17, the natural frequency of the board and the transmission coefficient ***A*** can be established. While vibration testing is the most accurate method of determining the displacement, natural frequency and the transmission coefficient it is expensive and time consuming.

Another approach due to Steinberg [10] is to estimate $\boldsymbol{A}$ at the resonance frequency as:

$$\boldsymbol{A} = [f_n]^{1/2} \tag{13.51}$$

In using this technique, we determine the natural frequency from Eq. (13.41), estimate $\boldsymbol{A}$ from Eq. (13.51), and then compute the maximum displacement for a given input a^*_{in} from Eq. (13.50).

The remaining technique for estimating $\boldsymbol{A}$ is based on the knowledge of the degree of dampening for a particular type of construction and the size of a PCB. If the degree of dampening can be estimated, the transmission coefficient $\boldsymbol{A}$ at resonance is determined from Eq. (13.26). Exercises are provided at the end of this chapter to demonstrate the application of these methods to vibration analysis of circuit boards.

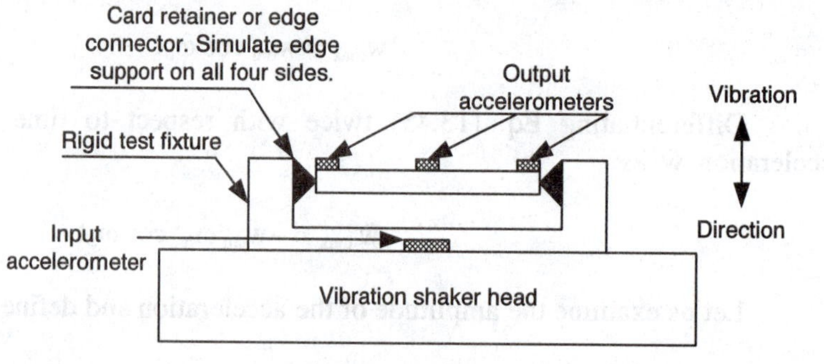

Fig. 13.17 Vibration testing arrangement for printed circuit cards.

13.4.1 Improving the Vibration Behavior of Circuit Cards

There are two characteristics of a PCB used to judge its merit to withstand a vibratory environment—the natural frequency and the degree of dampening both of which markedly affects the transmission coefficient. We seek to make the natural frequency as high as possible to limit the amplitude of the oscillation as shown by Eq. (13.50). Also, we attempt to increase the degree of dampening to reduce the transmission coefficient at resonance. Reference to Eq. (13.41) clearly indicates that the natural frequency decreases as the square of the size of the circuit board. Large circuit cards are advantageous because they can support many components. However, large circuit boards have relatively low natural frequencies. To illustrate this fact, consider a square card with all four edges simply supported. From the results listed in Case No. 16 of Table 13.3 and Eq. (13.41), it is clear that:

$$f_n = \left(\frac{\pi}{a^2}\right)\sqrt{\frac{D}{\rho}} \tag{13.52}$$

In this example, f_n, decreases with the square of the edge dimension a of the card. While large card areas have advantages for efficient packaging of a large number of components, the size must be limited if the system is to operate in a harsh vibratory environment.

The board thickness also affects ω_n. From Eq. (13.52) and (13.42), with $\nu = 1/4$ and $\rho = gh/\gamma$, we obtain:

$$f_n = \left(\frac{\pi}{a^2}\right)\sqrt{\frac{4Eg}{45\gamma}}\ h \tag{13.53}$$

where γ is the weight density of the board material.

Equation (13.53) shows that the natural frequency increases linearly with thickness of the circuit board. This relation also shows the effect of size with f_n decreasing with a^2 corresponding with the results given in Eq. (13.52).

Another approach, which markedly stiffens the card, is to bond it to a heat frame. This approach, often used in conductive cooling arrangements, provides a composite cross section with a flexural rigidity D that is much larger than an unsupported circuit card.

The edge supports of the circuit card are clearly important in determining the natural frequency. Reference to Table 13.3 indicates that clamped edges are preferred in enhancing ω_n. However, clamped edges are difficult to achieve in many installation as they require wedge lock type retainers (see Chapter 6) that are expensive. More common are inexpensive plastic edge guides that essentially provide simple support to two opposing sides of the card. Edge card connectors, frequently employed on one side of the circuit board, provide a simply supported edge condition. In many instances, one side of the card is free.

Ribs may be attached to circuit boards to increase their stiffness as shown in Fig. 13.18. The ribs are usually strips of aluminum or stainless steel that are adhesively bonded to the circuit board. Composite beam theory is used to determine the effective inertia I_e of the cross section. The flexural rigidity is then determined from:

$$D = EI_e/b \tag{13.54}$$

Dampening enhancement is another approach for mitigating vibration effects. Clearly, increasing the degree of dampening ξ decreases both $\boldsymbol{A}$ and $\boldsymbol{F}$; however, space is usually too limited to place mechanical dampening devices on the boards. Application of a thick viscous coating would improve the degree of dampening. Currently conformal coatings are applied to some circuit boards to protect the wiring traces on the surface from effects of humidity. These coatings are usually too thin and too elastic to absorb enough energy to provide significant dampening enhancement. A thicker more viscous coating that could also serve as a moisture barrier would be helpful in providing protection against vibration as well as humidity.

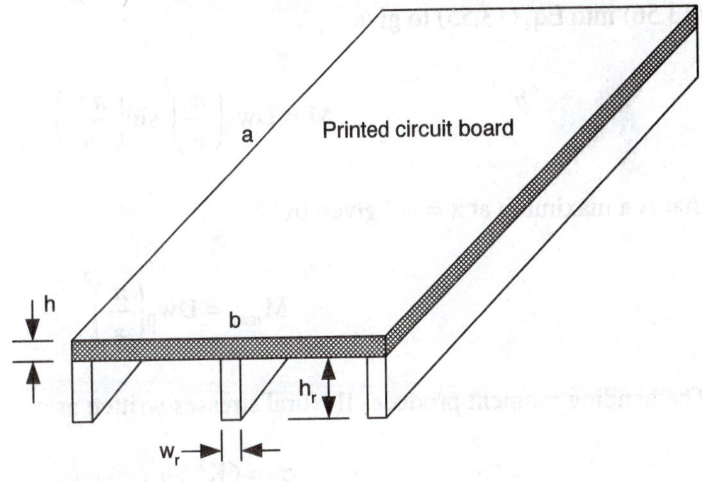

Fig. 13.18 Ribs bonded to a printed circuit board to increase its natural frequency.

13.4.2 Stress in Circuit Boards Due to Vibrations

When a circuit board is deflected during exposure to vibration, stresses are produced in the plane of the board that may produce failure by induced fatigue damage. The magnitude of the stresses produced depends upon the deflected shape of the board, which is given by w(x,y). The deflection function w(x,y) is controlled by the boundary conditions imposed in constraining the board as indicated in Table 13.3.

We will consider a simple example to first illustrate the method to follow in determining stresses produced by vibration and second to generalize on the relative importance of this topic. The example selected is a board with simple supports as shown in Fig. 13.19. This is an example illustrating severe vibration because of the board is completely free on two edges and free to rotate about its supported edges. This freedom leads to large deflections and relatively high stress at the center of the circuit board. For this simply supported PCB, the deflection function given in Table 13.3 for Case No. 18 as:

$$w = w_0 \sin(\pi x/a) \tag{13.55}$$

where w_0 is the displacement at $x = a/2$.

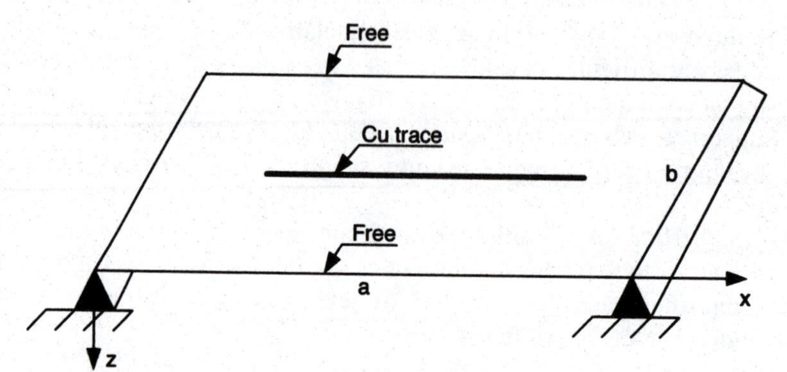

Fig. 13.19 Simply supported circuit board.

The moment per unit width of the board is related to w(x) by:

$$M = -D\frac{d^2w}{dx^2} \tag{13.56}$$

The distribution of the moment along the length of the board is obtained by substituting Eq. (13.56) into Eq. (13.55) to give:

$$M = Dw_0\left(\frac{\pi}{a}\right)^2 \sin\left(\frac{\pi x}{a}\right) \tag{13.57}$$

that is a maximum at $x = a/2$ given by:

$$M_{max} = Dw_0\left(\frac{\pi}{a}\right)^2 \tag{a}$$

The bending moment produces flexural stresses written as:

$$\sigma_0 = 6KM/h^2 \tag{13.58}$$

The symbol K is the stress concentration due to the geometric discontinuities on the board. Usually the plated through holes produce the largest value of $K = 3$. If we combine Eqs.(13.58) and (a) and let $K = 3$, the maximum stress can be written as:

$$\sigma_{max} = \frac{18Dw_0}{h^2}\left(\frac{\pi}{a}\right)^2 \tag{13.59}$$

This result is correct but it is not in a form that can be easily interpreted since w_0 is not known. From Eqs.(13.50) and (13.51), we can write:

$$w_0 = \frac{g a_{in}^*}{\sqrt{2\pi}\ \omega_n^{3/2}} \qquad (13.60)$$

Next, use Eq. (13.41) and results from Table 13.3 to show:

$$\omega_n = \left(\frac{\pi}{a}\right)^2 \sqrt{\frac{D}{\rho}} \qquad (b)$$

Substituting Eq. (13.60) and Eq. (b) into Eq. (13.59) yields:

$$\sigma_{max} = \frac{9\sqrt{2} D^{1/4} g a \rho^{3/4} a_{in}^*}{\pi^{3/2} h^2} \qquad (c)$$

Recall that $\rho = \gamma h/g$ and $D = 4Eh^3/45$ with $\nu = 1/4$. Substitute these equalities in Eq. (c) to obtain a result that can be interpreted in terms of basic dimensions and properties of the circuit board:

$$\sigma_{max} = \frac{1.248 E^{1/4} g^{1/4} a \gamma^{3/4} a_{in}^*}{h^{1/2}} \qquad (13.61)$$

Examination of Eq. (13.61) shows the stresses increase linearly with the input acceleration a^*_{in} and the dimension a. The maximum stress decreases with $\sqrt{h}$ and increase with E and density γ raised to powers of 1/4 and 3/4 respectively.

To examine the importance of these bending stresses on the performance of the circuit boards, consider the following characterization of a board in vibration:

$E = 3 \times 10^6$ psi; g = 386 in./s; a^*_{in} = 4 G's; a = 12 in.; h = 0.060 in.; γ = 0.065 lb/in^3.

Substitution of these values into Eq. (13.61) gives σ_{max} = 5,800 psi. Because the tensile strength of the common circuit board material, epoxy reinforced with fiberglass, is about 80 ksi, the safety factor **S** due to static deformation is 13.8. The S - N diagram for the glass fiber reinforced epoxy, shown in Fig. 13.20, indicates that the endurance strength S_e = 16 ksi. The safety factor for an infinite fatigue life with this 4 G exposure is:

$$\mathbf{S} = S_e/\sigma_{max} = 16/5.8 = 2.76 \qquad (d)$$

This example represents a severe vibratory input on a large and poorly supported circuit board. The resulting stresses were relatively low and the safety factors were more than adequate. We may generalize and conclude that epoxy-glass circuit boards can be designed as fatigue resistant. Indeed, structural failure of the boards is rarely observed. The most common form of damage that does occur in many cases is with the card connectors. The board displacements and rotations cause wear of the pin/contact combination and under more severe vibration pin breakage will occur.

This generalization should not be extended to circuit substrates fabricated from ceramics. The strength of ceramics, in either tensile or fatigue, is difficult to specify as a deterministic quantity and failure due to vibration induced deformation is difficult to predict. We are fortunate that ceramics used in chip carriers, hybrids and circuit boards are usually small in size. This small size together with their extremely high modulus gives components with very high ω_n that resist deformation.

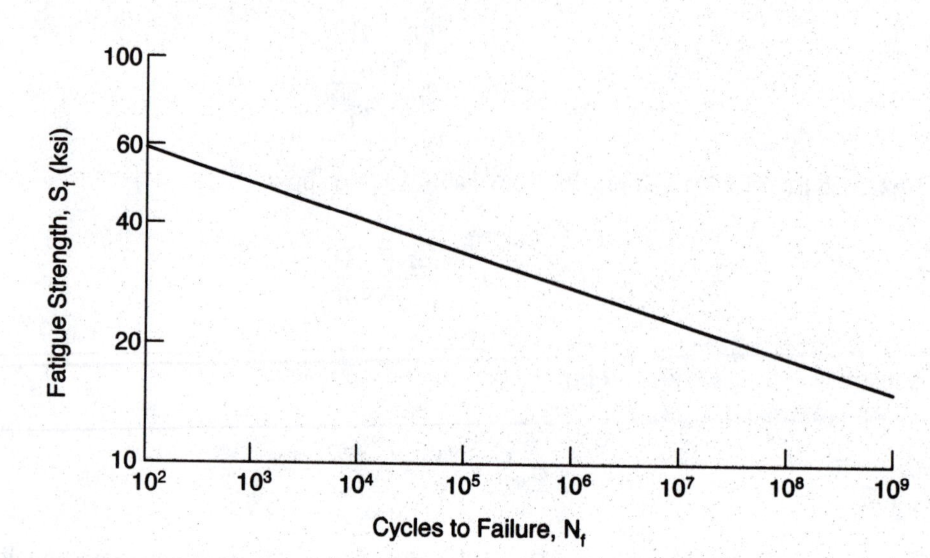

Fig. 13.20 S-N diagram for a typical glass-epoxy circuit board material.

13.4.3 Stresses in Copper Wiring Traces

In the previous discussion, we showed that the stresses induced in the PCB even under severe vibration conditions were relatively low in magnitude. However, this fact does not imply that the stresses in the copper traces that are bonded to the circuit board are also small. To explore this question, let's consider the same circuit board as shown in Fig. 13.19 with its x coordinate along the longitudinal axis of the board, assuming the board material is isotropic. The strain in the board is:

$$\varepsilon_x = \frac{1}{E}\left(\sigma_x - \nu\sigma_y\right) \tag{13.62}$$

Because the circuit board is modeled as a plate where $\varepsilon_y = 0$, we may write:

$$\sigma_y = \nu\sigma_x \tag{13.63}$$

Next, combine Eqs.(a) and (b) to give:

$$\varepsilon_x = \frac{1-\nu^2}{E}\sigma_x \tag{13.64}$$

The copper trace is bonded to the top surface of the circuit board, as illustrated in Fig. 13.19. We assume that the strain in the circuit board is transmitted to the copper trace without change in magnitude; hence:

$$(\varepsilon)_{cu} = (\varepsilon)_{ge} \tag{13.65}$$

where the subscripts cu and ge refer to copper and glass-epoxy.

If we take into account the uniaxial geometry of the copper trace, Hooke's law indicates:

$$\sigma_{cu} = E_{cu}(\varepsilon_x)_{cu} \quad (13.66)$$

Substituting Eqs.(13.64) and (13.65) into Eq. (13.66) gives:

$$\varepsilon_x = \frac{E_{cu}}{E_{ge}}\left(1-\nu_{ge}^2\right)\left(\sigma_x\right)_{ge} \quad (13.67)$$

This expression relates the stresses developed in the copper trace to those occurring in the circuit board. Let's return to the example used in Section 13.4.2 and note that $E_{cu} = 16 \times 10^6$ psi, $E_{ge} = 3 \times 10^6$ psi, $\nu_{ge} = 1/4$ and $(\sigma_x)_{ge} = 5{,}800$ psi. Substituting these values in Eq. (13.67) gives $\sigma_{cu} = (5)(5{,}800) = 29{,}000$ psi. Because the elastic modulus of copper is much higher than that of the glass reinforced epoxy, we develop much larger stresses (a factor of about five) in the copper trace. This stress is significant relative to tensile strength of the copper and indeed it is larger than its endurance strength S_e. Reference to the S – N diagram for cold drawn copper presented in Fig. 13.21 shows that the anticipated life for the copper trace is of the order of 10^4 cycles. If this circuit board were installed in a product exposed to 4 G's at the resonance frequency for more than a few minutes, fatigue failures of the circuit traces could be anticipated.

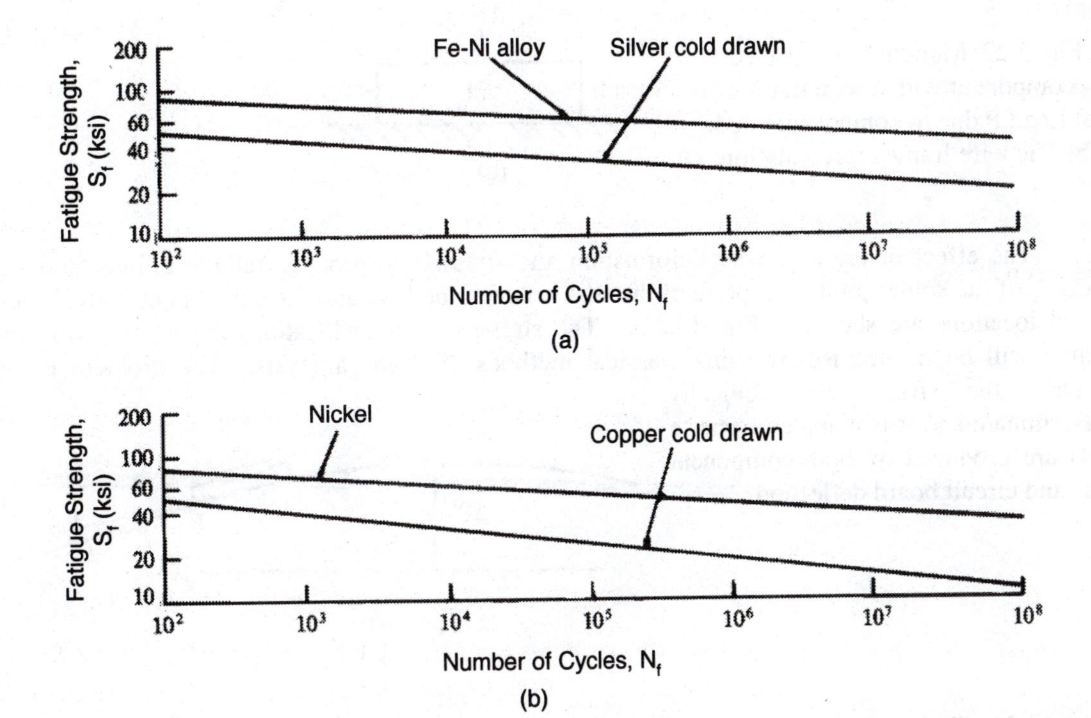

Fig. 13.21 Fatigue strength S_f as a function of the number of cycles for different types of metal wires subjected to reverse bending.

13.5 LEAD WIRE FAILURES ON VIBRATING PRINTED CIRCUIT BOARDS

When a printed circuit board vibrates, the components mounted on it are subjected to stress from two different effects. First, the mass of the component is subjected to an acceleration that produces a force P normal to the surface of the board as shown in Fig. 13.22a. The body of the component is kept in equilibrium with reactive forces developed in the lead wires. Second, the surface of the board flexes, which tends to bend the leads back and forth at their junction with the board. We will consider these two effects individually in determining component deflections and lead wire stresses.

Components with leads wires such as resistors, capacitors, DIPs or flat packs are modeled as frames. The body of the component is so rigid relative to the lead wire that we can assume that the deflection of the component is due entirely to the deformation of the lead wires. This assumption enables us to represent the component with a lead wire frame as shown in Fig. 13.22b. Because the leads are soldered to the boards and reinforced with a solder fillet, they are considered to be built-in when defining the boundary conditions for the frame. The built-in end condition requires that the leads remain locally perpendicular to the board and that they rotate with the board as it deflects during exposure to the vibratory input.

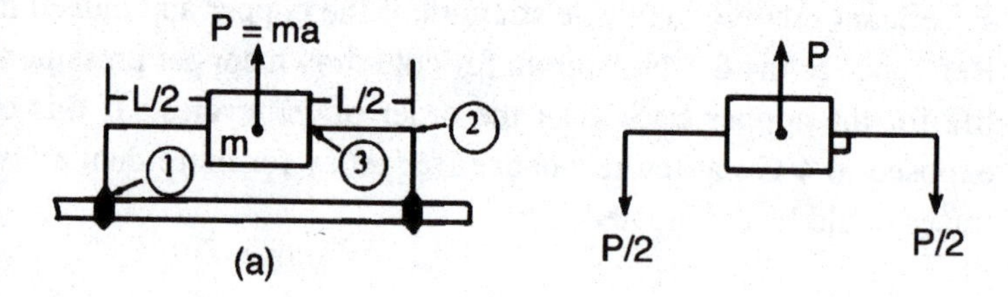

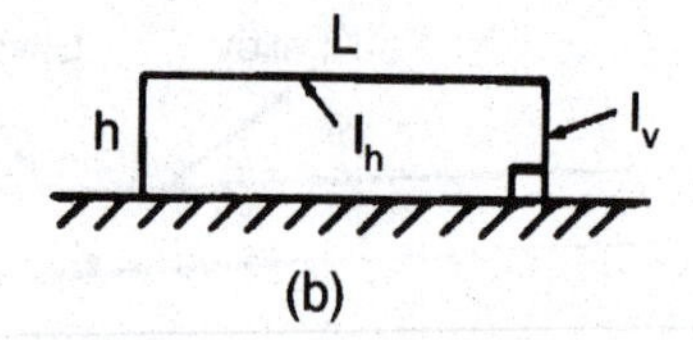

Fig13.22 Modeling of a leaded component with a wire frame. (a) Load P due to component mass. (b) The wire frame representation.

The effect of the lead wire deformation and stress is to produce fatigue failure of the leads, cracking of the solder joint, or rupture of the seal between the lead and the component body. The three critical locations are shown in Fig. 13.22a. The stresses at these locations due to the two types of loading will be determined by using classical methods of frame analysis. The problem is tedious because the frame is statically indeterminate and it is complex because loads are produced by both component mass and circuit board deflection.

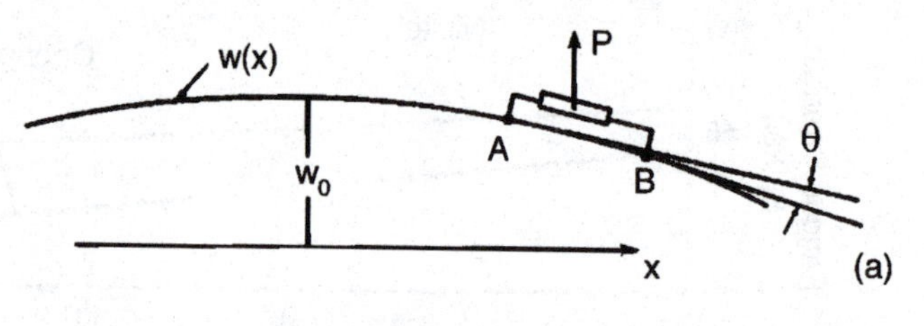

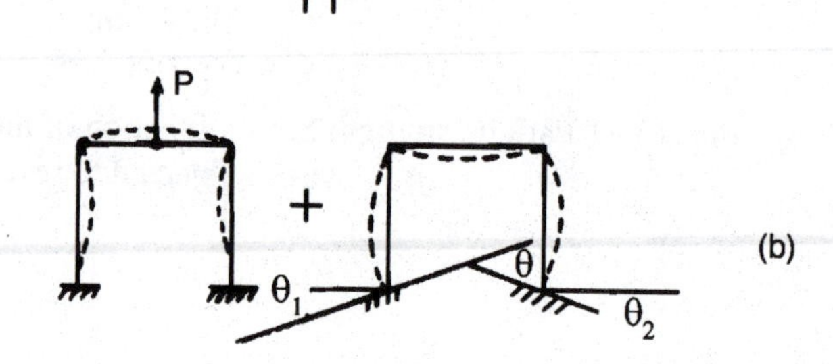

Fig. 13.23 Circuit board component interaction. (a) Circuit board deflection relative to the component. (b) Wire frame modeling with the load P and the rotation θ.

Let's divide the combined problem, depicted in Fig. 13.23a, into two simpler problems illustrated in Fig. 13.23b. Note the relative rotation of the leads θ is given by:

$$\theta = \left(\frac{\partial w}{\partial x}\right)_B - \left(\frac{\partial w}{\partial x}\right)_A \tag{13.68}$$

The worst case situation for lead rotation is when the component straddles the center line of the circuit board and the slopes at points A and B are equal in magnitude and opposite in sign. In this case, $\theta = 2\left(\frac{\partial w}{\partial x}\right)_B$ and the force P is given by:

$$P = \boldsymbol{F}\, a^*_{in} W \tag{13.69}$$

The force transmissibility coefficient $\boldsymbol{F}$ can be approximated by:

$$\boldsymbol{F} = 1.5\, \sqrt{f_n} \tag{13.70}$$

where f_n is the natural frequency of the component which is determined from:

$$f_n = \frac{1}{2\pi}\sqrt{\frac{192EIg}{WL^3}} \tag{13.71}$$

Substituting Eq. (13.70) into Eq. (13.69) gives:

$$P = 1.5\, \sqrt{f_n}\; a^*_{in}\, W \tag{13.72}$$

First, consider the wire frame with the central load P, and use statically indeterminate methods to find the unknown reactions. The free body diagrams, presented in Fig. 13.24, define these unknown reactions as Q and M at the built-in end. Castigliano's theorem can be employed to determine Q and M_1. The other moments and deflections follow from the well known equations from statics and materials of materials.

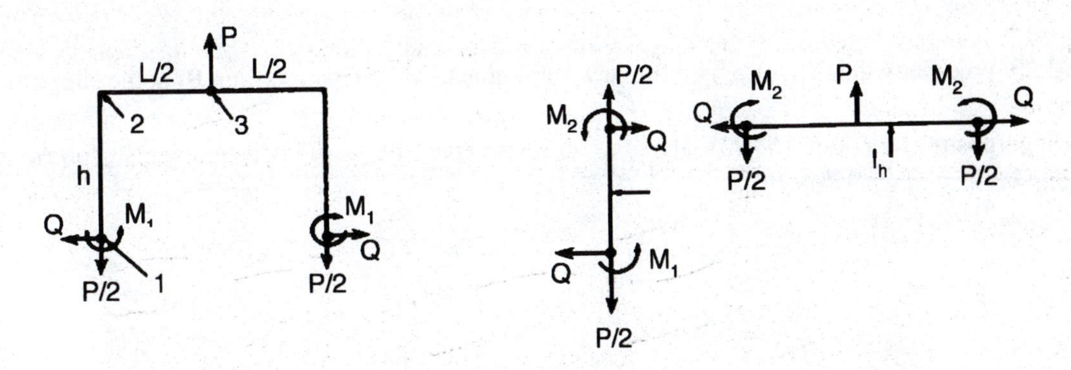

Fig. 13.24 Free-body diagram of a wire frame with the load P.

The results for the frame subjected to the load P described in Fig. 13.24 are:

$$Q^P = \frac{3PL}{8h(C+2)} \qquad M_1^P = \frac{PL}{8(C+2)} \qquad M_2^P = \frac{PL}{4(C+2)} \qquad M_3^P = \frac{PL(C+1)}{4(C+2)}$$

$$\delta_3^{P} = \frac{PL^3}{192EI_h}\left[\frac{C+1}{C+2}\right] \qquad C = \left(\frac{h}{L}\right)\left(\frac{I_h}{I_v}\right) \tag{13.73}$$

where M_3^P and δ_3^P are the moment and the deflection at the mid point of the horizontal member under the load P.

Consider the second part of the problem—the moments induced in the wire frame due to the rotation θ_1 imposed on the ends of the frame at the attachment points. The free body diagrams defining the unknown moments and forces are presented in Fig. 13.25. Using Castigliano's theorem gives the moment M_1 as:

$$M_1^{\theta} = \frac{2\theta_1(3+2C)EI_v}{h(2+C)} \qquad Q^{\theta} = \frac{M_1(1-C)}{h} + \frac{4EI_h\theta_1}{hL} \qquad M_2^{\theta} = \frac{4\theta_1 EI_h}{L} - CM_1^{\theta}$$

$$M_3^{\theta} = M_2^{\theta} \qquad \delta_3^{\theta} = \frac{\theta_1 L}{8C} \tag{13.74}$$

where M_3^{θ} and δ_3^{θ} are the moment and the deflection at the mid point of the horizontal member.

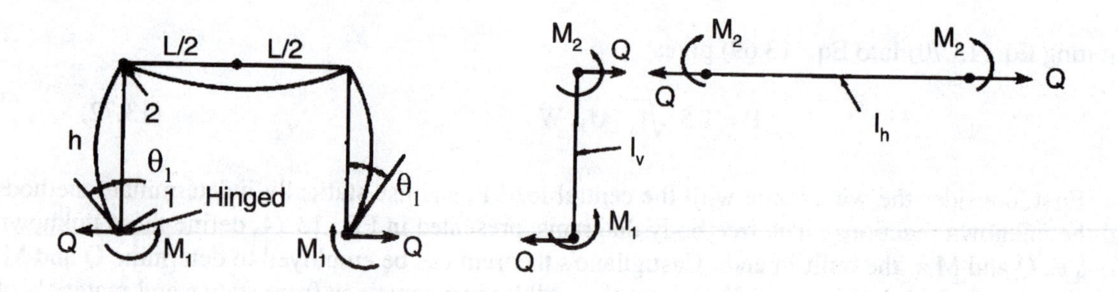

Fig. 13.25 Free-body diagram of a wire frame with moments M_1 giving rotations θ_1 at the supports.

Superposition of Eq. (13.73) and Eq. (13.74) give the combined moments, forces and deflections associated with P and θ_1 as:

$$M_1 = M_1^{P} + M_1^{\theta}$$

$$Q = Q^{P} + Q^{\theta} \tag{13.75}$$

$$\delta_3 = \delta_3^{P} + \delta_3^{\theta}$$

An exercise is given to illustrate the application of these equations. The stresses are determined from the moments by using the flexure formula (σ = Mc/I). These stresses are then compared to the fatigue strength S_f of the lead wire material for the number of cycles of imposed on the circuit board. If $S_f \geq \sigma$ the lead wire will not fail in fatigue.

13.5.1 Stresses in Solder Joints with Leaded Chip Carriers

The stresses in the solder joint are determined from the loading M_1 and P/2 that are imposed on the joint as defined by the geometry presented in Fig. 13.26a. The solder joint illustrated in this figure is for a double sided circuit board with plated-through-holes. The solder has formed a fillet above the board that has a critical location about one wire diameter d_w above the board's surface where failure usually occurs. Because the lead wire and solder have different moduli of elasticity, the solution for the stresses must account for the composite nature of the joint materials. The approach taken here is to assume that the moment M_1 is totally constrained by the stresses in the lead wire. This approach is justified because the elastic modulus of the solder is small relative to the modulus of the metals used for lead wires. This assumption permits us to write:

$$\sigma_w = \frac{M_1 c}{I_w} = \frac{32 M_1}{\pi d_w^3} \tag{13.76}$$

The lead wire strain is given by:

$$\varepsilon_w = \frac{\sigma_w}{E_w} = \frac{32 M_1}{E_w \pi d_w^3} \tag{13.77}$$

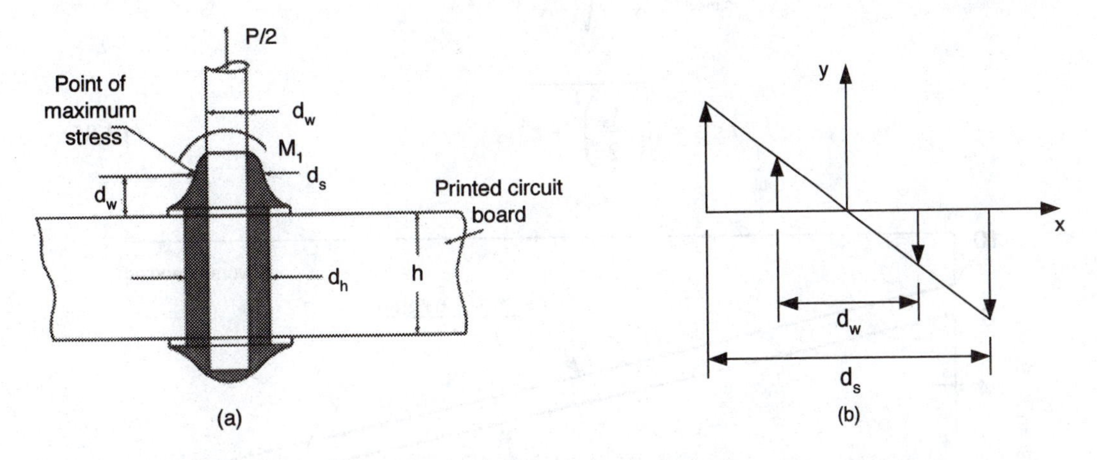

Fig. 13.26 (a) A solder joint on a double sided printed circuit board.
(b) Strain distribution at the critical location.

In bending, plane sections remain plane, which leads to the observation that the strain across the composite joint is linearly distributed as shown in Fig. 13.26b. From similar triangles, we can write the relation for the maximum strain in the solder as:

$$\varepsilon_w = (d_s/d_w)\varepsilon_w \tag{13.78}$$

and the stress as:

$$\sigma_s = E_s \sigma_s = \left(\frac{d_s}{d_w}\right)\left(\frac{E_s}{E_w}\right)\frac{32 M_1}{\pi d_w^3} \tag{13.79}$$

The failure initiation site in a solder joint is usually at a position where $d_s = 3d_w/2$. This fact leads to a simplification of Eq. (13.79) to:

$$\sigma_s = \left(\frac{E_s}{E_w}\right)\frac{48M_1}{\pi d_w^3} \tag{13.80}$$

In addition to the bending stresses, shear stresses τ_s due to the axial force are developed. The average shear stress are determined from:

$$\tau_s = F/A_s \tag{13.81}$$

Note that F = P/2 and the shear area A_s is given by:

$$A_s = \pi d_h(h + 2d_w) \tag{13.82}$$

where d_h is the diameter of the plated through hole and h is the board thickness. Combining Eqs.(13.81) and (13.82) gives:

$$\tau_s = \frac{P}{2\pi h d_h (h + 2d_w)} \tag{13.83}$$

The bending stresses and shear stresses combine to give maximum principal stress and shear stress as:

$$\sigma_{max} = \frac{\sigma_s}{2} + \sqrt{\left(\frac{\sigma_s}{2}\right)^2 + \tau_s^2} \tag{13.84}$$

$$\tau_{max} = \sqrt{\left(\frac{\sigma_s}{2}\right)^2 + \tau_s^2} \tag{13.85}$$

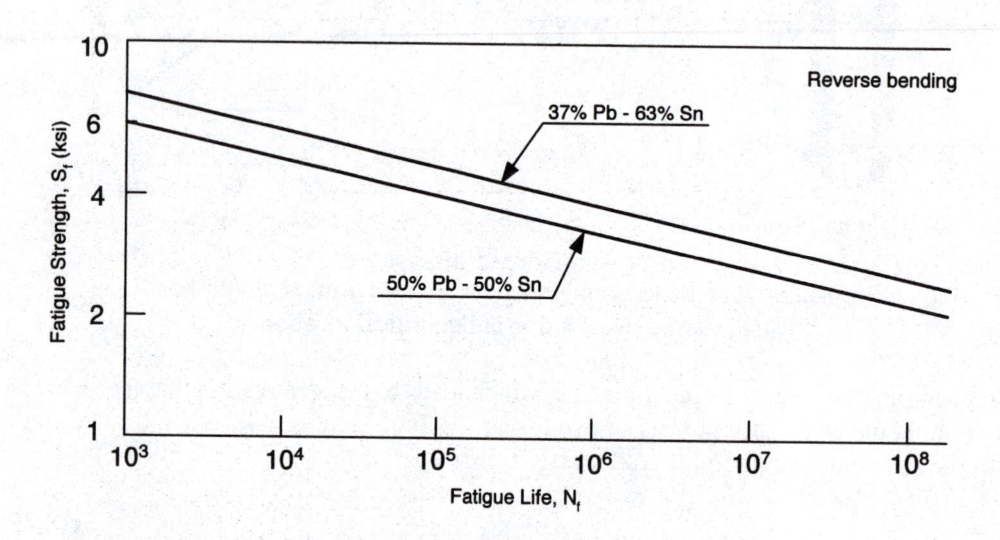

Fig. 13.27 S-N diagram showing fatigue strength as a function of life for lead-tin solders.

The maximum stress is equal to the alternating stress in reverse bending that occurs in vibratory exposure of a printed circuit board. The S-N diagrams for some common solders are presented in Fig. 13.27. The fatigue strength of the newer tin-free solder alloys containing silver is superior to the tin-lead solders. The safety factor for the solder joint is given by:

$$\boldsymbol{S} = S_f/\sigma_{max} \tag{13.86}$$

Because the solder joint may contain voids, inclusions and geometric imperfections a safety factor of at least 3 is advisable to account for the variability encountered in dealing with a very large number of solder joints exposed to a vibratory environment.

13.5.2 Stresses in Solder Joints on a Ball Grid Array

The attachment of a ball grid array (BGA) to a printed circuit board is through a large number of controlled-collapse solder joints deployed at a fine-pitch over most of the package footprint. The force acting on a BGA due either to shock or vibration is not very large because the mass of the chip carrier is small for most chip-scale ball-grid arrays. This force is given by:

$$P = \boldsymbol{F}\, a^*_{in} W \tag{13.69}$$

where W is the weight of the BGA, a^*_{in} is the input acceleration due to vibration and $\boldsymbol{F}$ is the force transmissibility factor.

The normal stress generated in the solder joints is given by:

$$\sigma = P/A_s \tag{13.87}$$

where A is the total area of the solder joints, which is given by:

$$A_s = n\pi d_s^2/4 \tag{13.88}$$

Substituting Eq. (13.69) and (13.88) into Eq. (13.87) yields:

$$\sigma = \frac{4\boldsymbol{F} a^*_{in} W}{n\pi d_s^2} \tag{13.89}$$

Let's consider an example BGA to ascertain the magnitude of the normal stress σ due to a vibration with a^*_{in} = 1000 G and a force transmissibility factor $\boldsymbol{F}$ = 18. The weight of the BGA is 3 g, its solder ball pitch is 1.00 mm with a nominal pad diameter of 0.45 mm. A total of 240 pads are arranged in a symmetric pattern over the base of the BGA.

First convert the weight of the package from grams to Newton by using $W = (3)(9.81 \times 10^{-3}) = 0.0294$ N. Next substituting quantities from the problem statement into Eq. (13.89) yields:

$$\sigma = \frac{4(18)(1000)(0.0294)}{(240)\pi(0.45)^2} = 13.87\frac{N}{mm^2} = 13.87 MPa \tag{a}$$

A comparison of the results presented in Eq. (a) with the yield strength (19 MPa) of a tin-silver eutectic solder indicates that the solder joints are highly stressed during this vibration event. Additional failure modes may include, board pad cratering, copper trace breakage, and chip fracture.

Thermal Stresses in the Solder Joints of BGAs

The example in the paragraph above indicated that the large number of solder joints on a ball grid array mitigated relatively large forces transmitted to the chip carrier by vibration. It is easy to show that this statement is true for shock induced forces as well. However, the stresses in some of the solder joints due to temperature fluctuations are of serious concern. The problem develops if there is a marked difference in the temperature coefficient of expansion between the chip carrier and the circuit board.

The circuit board assembly is subjected to large thermally induced strains due to a mismatch of the coefficients of expansion of the BGA and the circuit board materials. Thermal cycling may be encountered in operation due to power on power off, or cycling may be imposed as a manufacturing screening test or in qualification testing. During these thermal cycles, the solder joints are exposed to a thermally induced strain of relatively large magnitude. The effect of this strain is to induce fatigue failures of the solder joints. The failures begin with the solder joints located at the corners of the chip carrier and then move inward to more centrally located solder joints. The number of thermal cycles required to initiate fatigue failure depends on the temperature extremes involved in the thermal cycling, the rate of temperature change, the mismatch in the coefficients of thermal expansion and the size of the chip carrier. Usually this life is so short for the larger ceramic BGAs that their service is limited to relatively few thermal cycles if they are mounted on glass-epoxy circuit boards. To extend the life in this situation, the space between the ceramic chip carrier and the glass-fiber-epoxy circuit board is filled with a strong and rigid adhesive. The adhesive, called an underfill, couples the two dissimilar materials together and reduces the thermal expansion of the glass-fiber-epoxy material and increases the expansion of the ceramic BGA.

In some applications, the BGAs are fabricated from BT/FR5 cores with an epoxy over-mold. With this construction technique, the temperature induced stresses are mitigated because the temperature coefficient of expansion of the chip carrier and the circuit board are in close correspondence. However, wire bonding is usually employed in these applications, limiting the number of I/O that the BGA can accommodate.

13.6 SHOCK ISOLATION

Shock and vibration are usually considered together because the isolation system for both are modeled with the spring and dashpot supported mass, illustrated in Fig. 13.12. However, the input to the isolation system differs significantly. In studying vibration, we consider a steady state periodic input such as a sinusoid. In analyzing shock, we consider a transient input which results in the application of a velocity (usually time dependent) to the structure supporting the isolated electronic system.

Shock is defined as a transient condition where a single impulse of energy is transferred to a system in a short period of time (milliseconds) with large acceleration. The reduction of shock is obtained with isolators that essentially store the shock energy temporarily and then release this energy over a longer period of time as the system vibrates. The energy storage is accomplished by extending or compressing a spring contained in an isolator. The effectiveness of a shock isolator is measured not by transmissibility (as with vibration isolators) but instead by the magnitude of the force transmitted to the electronic system and the deflection imposed on the spring in the isolation assembly.

In isolating electronic systems, shock is characterized by motion of the support structure where a shock isolator is employed to reduce the severity of the shock experienced by the equipment mounted on the supporting surface. A simple example of an impact or transient condition is an object of a weight W being dropped through a vertical distance onto a floor. The maximum force that would be transmitted into the floor depends upon the deflection of the package containing the object and the deflection of the floor. The smaller the sum of these two deflections, the larger the impact force between the package and the floor.

In order to determine the force transmitted through the spring, which is the critical force transmitted to the isolated electronic assembly, we employ an energy balance. First, the kinetic energy of the object the instant before it contacts the floor is determined. This kinetic energy $\mathbf{E}_k$ is equal to the potential energy $\mathbf{E}_p$ of the weight which is given by:

$$\mathbf{E}_k = \frac{1}{2} m v^2 = \mathbf{E}_p = Wh \tag{13.90}$$

Solving Eq. (13.90) yields the impact velocity v as:

$$v = \sqrt{2gh} \tag{13.91}$$

The floor and the package absorb this kinetic energy in bringing the object to rest after impact. The impact energy is absorbed by the springs in the isolation system. The energy stored in the springs is given by:

$$\mathbf{E}_s = ½\, k y^2 \tag{13.92}$$

where y is the coordinate referenced in Fig. 13.12 and k is the total spring rate for springs used in the design of the isolator system. Note that the dash pot is ignored because damping in the design of an isolation system for shock loading is much less important than the design of its springs.

For an isolator with a linear spring, Eq. (13.90) and Eq. (13.92) are equated to establish the dynamic displacement y as:

$$y = \sqrt{\frac{2Wh}{k}} = \frac{v}{\omega_n} \tag{13.93}$$

The force transmitted after isolation is determined from Eq. (13.93) as:

$$F = ky = \sqrt{2Whk} = \frac{kv}{\omega_n} \tag{13.94}$$

The acceleration imposed on the electronic system is given by Newton's second law as:

$$a = \frac{F}{m} = \omega_n v \tag{13.95a}$$

The acceleration is usually expressed in terms of Gs; hence, Eq. (13.95a) is modified to read as:

$$G = \frac{a}{g} = \frac{F}{gm} = \frac{\omega_n v}{g} \tag{13.95b}$$

In designing an isolator system for shock damping is of minor importance when compared to the spring rate for the spring assembly. The spring rate controls the natural frequency of the system as well as the dynamic displacement according to Eq. (13.93). The equations listed above were derived for an isolated system dropped from some height unto a floor. However, the equations are employed for many different types of shock loading. In each case, the velocity v of the support structure is determined together with the natural frequency ω_n of the isolated system. These values are then employed with Eq. (13.93) to Eq. (13.95) to determine the dynamic displacements, the transmitted forces and the acceleration imposed on the electronic assembly.

Three different input pulses for shock loading are represented in Fig. 13.28 that include force, acceleration and velocity. The equation for the velocity v for each case is given together with the profile for the pulse.

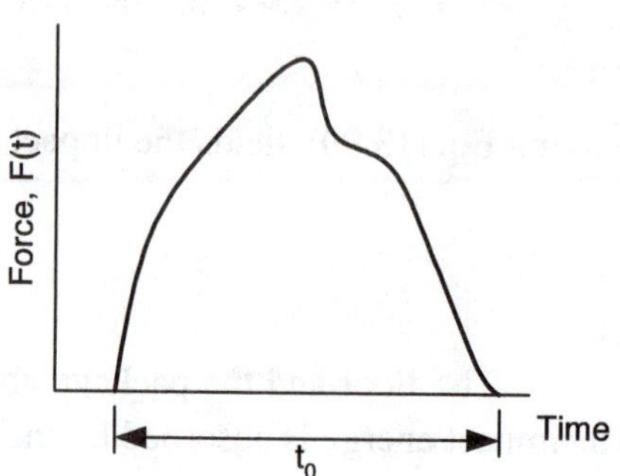

Fig. 13.28a Force pulse F(t).

$$v = \frac{1}{m}\int_0^{t_0} F(t)dt$$

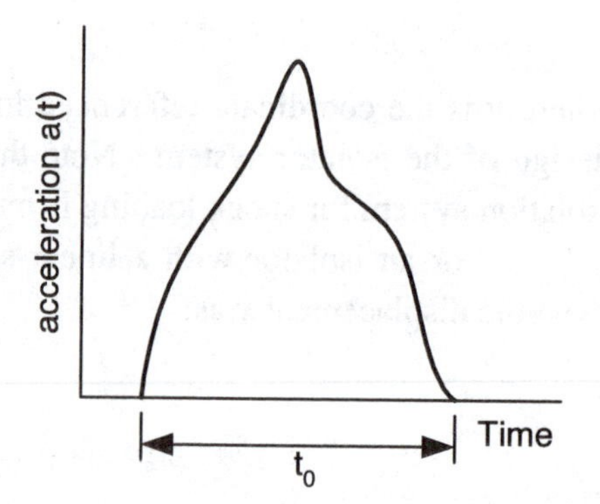

Fig. 13.28b Acceleration pulse a(t).

$$v = \int_0^{t_0} a(t)dt$$

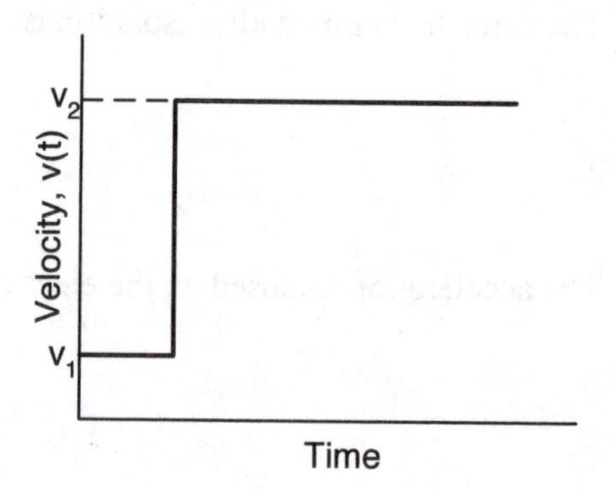

Fig. 13.28c Velocity pulse.

$$v = v_2 - v_1$$

There are many types of shock pulses encountered in service; consequently, several types of shock tests have been developed for testing electronic equipment. The shock test specified is usually associated with the environment that the system will be exposed to during its service life. Electronic systems installed in aircraft are normally tested on a free-fall shock machine that generates either a half-sine or terminal peak sawtooth form, as shown in Fig. 13.29. A typical test generates an acceleration pulse with an 11 ms half-sine waveform and a peak acceleration a_0 = 15 G's. For systems on missiles at locations where large shock pulses occur, a 6 ms sawtooth at 100 G's is often specified. If the electronic system is to operate on a Navy vessel, the normal test will be a 3,500 lb hammer blow to a simulated deck as specified in MIL-S-901. This impact produces a shock velocity of approximately 120 in./s.

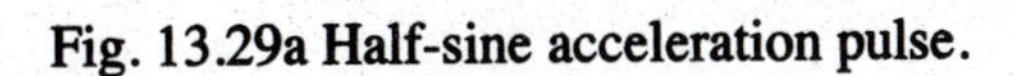

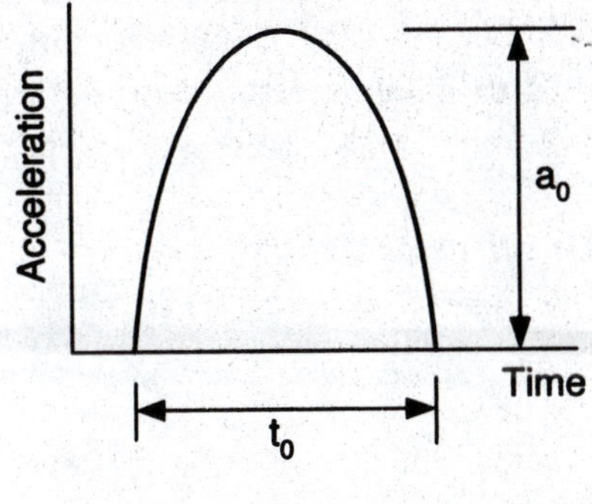

Fig. 13.29a Half-sine acceleration pulse.

$$v = \frac{2g}{\pi} a_0 t_0$$

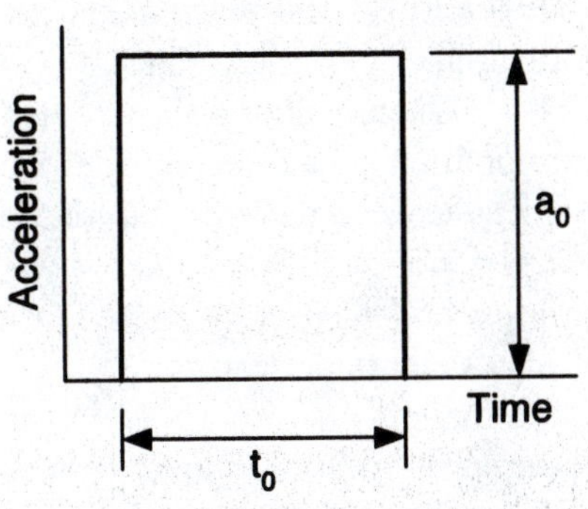

Fig. 13.29b Rectangular acceleration pulse.

$$v = g\, a_0\, t_0$$

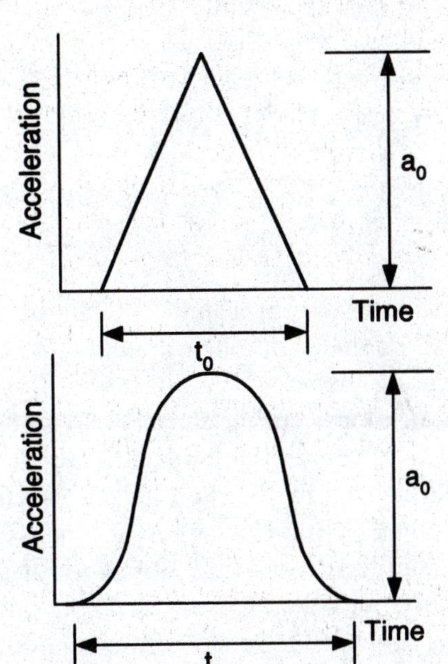

Fig. 13.29c Triangular acceleration pulse

$$v = \frac{g}{2} a_0 t_0$$

Fig. 13.29d Versed-sine acceleration pulse

$$v = \frac{g}{2} a_0 t_0$$

Fig. 13.29 Idealized forms of shock pulses for various types of shock waveforms.

A shipping container is normally tested by dropping it on a concrete floor, or by suspending it in a pendulum mechanism and swinging it against a concrete abutment. Other tests for shipment containers are edge and corner drops from various drop heights. All of these tests attempt to simulate the shock pulse or pulses that will be encountered in shipping the system or in its operation over its complete life cycle. These shock inputs are usually specified in the contractual requirements or in a work requirement. The isolation of shock inputs is considerably different from that of a vibratory input. The shock isolator is characterized as an energy storage device where the input energy, usually with a very steep wave front, is initially absorbed by the spring element in the isolator system. This energy is dissipated slowly as the mass (the electronic package) vibrates at the natural frequency of the spring-mass system. The usual procedure for designing a shock isolation system employs equations in Fig. 13.29 to determine the velocity, and Eq. (13.90) to Eq. (13.92) to determine the transmitted forces, accelerations ands rattle space requirements.

Nonlinear Isolators:

The preceding discussion of vibration and shock isolation assumes that the spring element in the isolator may be characterized with a linear force-deflection curve. This assumption is valid for most design analyzes. In the isolation of steady-state vibration, the displacement amplitude is usually small, and nonlinearity of the elastomeric spring elements can usually be neglected. In a few situations, the deflection of the spring elements is large and non-linear effects should be included in the analysis.

When isolating shock pulses, nonlinearity tends to be more important because large deflections are common. The degree of isolation is often affected by the ability, or lack thereof, of the spring element to accommodate the required deflection. In many applications of shock isolation, sufficient space is not available to allow for full travel of a linear isolator. Therefore, a nonlinear isolator is necessary. There are two types of non-linear spring elements that are available to alleviate the problem of insufficient "rattle space". The first non-linear spring becomes stiffer as deflection increases, which limits the amount of motion, but increases the G level imparted on the equipment. The second non-linear spring is stiff at small deflection, but becomes much softer when it buckles at higher deflections. This non-linear spring permits the isolator to store more energy in the same amount of deflection.

13.7 MODELING SHOCK WITH FINITE ELEMENTS

Solder joint failure in electronic devices subject to shock and impact due to dropping is an important concern for the telecommunications industry. The trend towards miniaturization and increased functional density has resulted in decreasing the I/O pitch of the chip carriers. The reduced pitch increases the probability of failure of the solder joints on the chip carrier under shock and vibration environments. Solder joint failure occurs due to a combination of PCB bending and direct shock loads during impact. Optimization of the design of the chip carrier is necessary to minimize the effects of shock loads during impact on its solder interconnections.

Currently, product testing depends on experimental methods that are employed. Test results are influenced by the drop height, product orientation during drop and variations in product design. The Joint Electron Device Engineering Council (JEDEC) drop-test, JESD22-B11 is the standard test used to determine board-level reliability of components. This test involves subjecting the board to a 1,500-g pulse with the PCB oriented horizontally. However, when a product is actually dropped, the PCB may be subjected to several different drop orientations. While the experimental approach in board-level drop testing provides useful data, it is not feasible to test all the board configurations during product development. Today, board-level and product drop-tests are employed to better understand shock-performance of fine-pitch interconnections. However, short development time, cost, and time-to-market constraints limit the number of configurations that can be fabricated and tested during a development cycle.

One of the challenges in modeling shock response of electronic products, is the scale differences between the dimensions of the individual layers, such as solder interconnects, copper pad, chip-interconnects, and the dimensions of the electronic assembly. These scale differences make the computational effort required to attain the fine mesh necessary to model solder joints while capturing the system-level dynamic behavior very difficult.

Various modeling approaches have been pursued to reduce the computational time required for simulation. These include:

1. Equivalent layer models to represent the solder joints and simulate their behavior under drop impact.
2. Solid-to-solid sub-modeling techniques to analyze ball grid array (BGA) reliability for drop impact using half the PCB board.
3. Shell-to-solid sub-modeling using a beam-shell-based quarter symmetry model to reduce the computational time.
4. Symmetry of load and boundary conditions is used to attain computational efficiency and decrease the model size.

The assumption of symmetry in the computational models yields symmetric deformation modes. The explicit time-integration is suitable for the solving wave propagation equations necessary to model drop impact event. The simulation time is determined by the size of the time step which is directly proportional to the length of the smallest element in the model.

Board-level impact simulations using the smeared property approach have been conducted by Lall et al [11] and validated with experimental data. Lall, et. al. [11] has shown that transient dynamic responses of board assemblies can be predicted fairly accurately while achieving computational efficiency. Conventional shell with Timoshenko-beam element models, continuum shell with Timoshenko-beam element models, and global-local explicit sub-models can be used to simulate the impact phenomenon, without assumptions of symmetry, and to predict the location and mode of failure of the critical solder interconnection using suitable failure theories.

The modeling approach described below enables prediction of both symmetric and anti-symmetric deformation modes, which may dominate during an actual drop event. This method's computational efficiency and accuracy has been verified with data obtained from actual drop-tests.

13.7.1 Finite Element Models

The method for determining the transient dynamic response of a printed circuit board under drop impact is described in this section. The method involves a finite element model with step-by-step direct integration in time for both explicit and implicit formulations. The governing differential equation of motion for a dynamic system can be expressed as:

$$[M]\{\ddot{D}\}_n + [C]\{\dot{D}\}_n + \{R^{int}\}_n = \{R^{ext}\}_n \tag{13.96}$$

For a linear problem, $\{R^{int}\}_n = [K]\{D\}_n$, where [M], [C] and [K] are the mass, damping, and stiffness matrices, respectively, and $\{D\}_n$ is the nodal displacement vector at each time step. Methods of explicit direct integration calculate the dynamic response at the time step (n + 1) from the equation of motion, the central difference formulation and known conditions at one or more preceding time steps as shown below:

$$\left[\frac{1}{\Delta t^2}M + \frac{1}{2\Delta t}C\right]\{D\}_{n+1} = \{R^{ext}\}_n - \{R^{int}\}_n + \frac{2}{\Delta t^2}[M]\{D\}_n - \left[\frac{1}{\Delta t^2}M - \frac{1}{2\Delta t}C\right]\{D\}_{n-1} \tag{13.97}$$

Equation (13.97) has been combined with Eq. (13.96) at time step n. In the implicit algorithm, the dynamic response at time step (n + 1) has been calculated from known conditions at present time-step, in addition to one or more preceding time-steps. Using Newmark's relations and the average acceleration method, the equation of motion can be written as:

$$\begin{aligned}\left[K^{eff}\right]\{D\}_{n+1} = \left\{R^{ext}\right\}_{n+1} + [M]\left\{\frac{1}{\beta\Delta t^2}\{D\}_n + \frac{1}{\beta\Delta t}\{\dot{D}\}_n + \left(\frac{1}{2\beta}-1\right)\{\ddot{D}\}_n\right\} \\ +[C]\left\{\frac{\gamma}{\beta\Delta t}\{D\}_n + \left(\frac{\gamma}{\beta}-1\right)\{\dot{D}\}_n + \Delta t\left(\frac{\gamma}{2\beta}-1\right)\{\ddot{D}\}_n\right\}\end{aligned} \tag{13.98}$$

where

$$\left[K^{eff}\right] = \frac{1}{\beta\Delta t^2}[M] + \frac{\gamma}{\beta\Delta t}[C] + [K] \tag{13.99}$$

and γ and β are numerical factors that control the characteristics of the algorithm such as accuracy, numerical stability, and amount of algorithmic damping. All the terms on the right-hand side of Eq. (13.97) are known and are calculated at earlier time steps; however, this statement is not true for Eq. (13.98). The mass matrix, [M] has been diagonalized, using a lumped approach, which improves computational efficiency, because each time step is executed very quickly without solving simultaneous equations. Use of the lumped mass approach increases the allowable step time, but this approach is limited to the explicit formulation. For the implicit formulation, the effective stiffness matrix $[K^{eff}]$ is not a diagonal matrix, even when the mass and damping matrices are diagonalized, because it contains the stiffness matrix [K]. The diagonal mass matrix in the implicit formulation provides very little computational economy. The implicit method is usually more accurate when [M] is the consistent mass matrix; however this approach requires increasing the computational time and storage requirements.

Element size in the explicit model is limited due the conditional stability of the explicit time-integration. The time step is limited to avoid instability and error accumulation in the time integration process. This limiting condition increases the number of time steps required to cover the time duration of an analysis. However, explicit time-integration is well suited to wave propagation that occurs in an impact event, because the dynamic response of the board decays within a few multiples of its longest period. Most implicit formulations are unconditionally stable, which means that the process is stable regardless of the size of the time step. This approach involves fewer time steps as compared to the explicit method. However, with the high deformation rates involved in impact, the implicit formulation with a large time step often permits too large of an increase in strain during a single time-step, causing divergence in large deformation analysis. The explicit formulation is better suited to accommodate material and geometric nonlinearity without any global matrix manipulation.

In a finite element analysis, the printed circuit board assembly is modeled as an orthotropic material, which consists of various layers such as the copper pad, solder interconnections, and solder mask, with multiple scale differences in their dimensions. Analysis of the stress and strain variations in the solder interconnections, while at the same time capturing the dynamic response of the assembly, requires a very finely meshed model. Unfortunately, a very fine mesh significantly increases computational time. To overcome these difficulties and achieve computational efficiency with reasonable accuracy, a sub-modeling technique is employed. This technique involves a global solution with a coarse mesh followed by an analysis of the local and critical part of the model that is defined with a refined mesh.

13.7.2 Explicit Models and Element Formulations

Reduced integration elements are usually employed in an analysis because they require fewer integration points to form the element stiffness matrices, thus reducing the computational-time for simulation of transient dynamic events. First-order elements perform better, when large strains or very high-strain gradients are expected as is often the case with impact events. Higher order elements have higher frequencies than lower order elements and tend to produce noise when stress waves move across a mesh element.

Two types of shell elements are available in Abaqus including conventional shell elements and continuum shell elements. The conventional shell elements discretize the surface by defining the element's planar dimensions, its surface normal and its initial curvature. Surface thickness is defined through section properties. The conventional shell element is a reduced integration element, which accounts for large strains and large rotations. Continuum shell elements resemble three-dimensional solid elements and discretize the entire three-dimensional body. The continuum shell elements are formulated such that their kinematic and constitutive behavior is similar to conventional shell elements. The continuum shell element (SC8R), has three-translational degrees of freedom at each node, and the element accounts for finite membrane strains and arbitrarily large rotations. Shell elements often are used to model printed circuit boards because their thickness dimension is significantly smaller than their planar dimensions.

Interconnect modeling can be accomplished with two element types—the three-dimensional, linear, Timoshenko beam element (B31), and the eight-node hexahedral reduced integration elements. Three-dimensional beam elements have six degrees of freedom at each node including, three translational and three rotational degrees of freedom. The rotational degrees of freedom have been constrained to model interconnect behavior. The B31 element allows shear deformation, because its cross section is not constrained to remain normal to the beam axis. Shear deformation is important for first-level interconnects, because it is recognized that shearing deformation is the predominant mode of failure. It is assumed in a finite element analysis (FEA) the radius of curvature of the beam is large compared to the thickness of its cross section, and that the beam cannot fold into a tight hinge.

Lall et al [11] have investigated three explicit models including:

1. Smeared property models
2. Timoshenko-beam element interconnect models with continuum and conventional shell element
3. Sub-models with combination of Timoshenko-beam elements and reduced integration hexahedral element corner interconnects

For each different type of element used for the PCB, the various component layers such as the substrate, die-attach, silicon die, and mold-compound have been modeled with Abaqus C3D8R elements.

The solder interconnections were modeled with two-node beam elements (B31). Smeared properties were derived for all the individual components based on volumetric averaging. The simulated weight of the model for the PCB and all the components closely approximated the actual weights. The concrete floor was modeled using rigid R3D4 elements. Node to surface contact was made between a reference node on the rigid floor and the impacting surface of the test assembly. The drop orientation has been varied from a 0° JEDEC drop to 90° free drop.

Explicit sub-modeling involved using a local model in addition to the global model. The local model was finely meshed and included all the individual layers of the chip scale package (CSP) and the corresponding portion of the PCB. The four corner solder interconnections were created using solid elements, and the remaining solder joints were modeled with beam elements. The shell-to-solid sub-modeling technique was employed to transfer the time history response of the global model to the local model. Displacement degrees of freedom from the global model are transferred to the local model and applied as boundary conditions. The corresponding initial velocities for the respective drop orientation were assigned to all the components of the sub-model.

13.7.3 Typical Results of a Finite Element Analysis

In Lall's [11] investigation, the printed circuit assemblies were subjected to a controlled drop from various heights onto a concrete floor. The transient dynamic motion was captured using a high-speed video system at 50,000 fps. Image tracking software was used to quantitatively measure displacements during the drop event. The continuity of the circuits of the components on the PCB was simultaneously monitored using a high-speed data acquisition system to detect component failure.

The field quantities were compared from the explicit finite-element models and experimental drop test data. The correlation of the board's relative displacement 2.4 ms after impact, from high-speed image analysis with the model predictions from smeared, continuum-shell with Timoshenko-beam, and conventional-shell with Timoshenko-beam models is shown in Fig. 13.30. The peak strain results exhibit error in the range of 10 to 30%. All three modeling approaches including smeared properties, conventional-shell with beam elements, and continuum-shell with beam elements exhibit similar results. The computational efficiency of various models in the free drop is presented in Table 13.4.

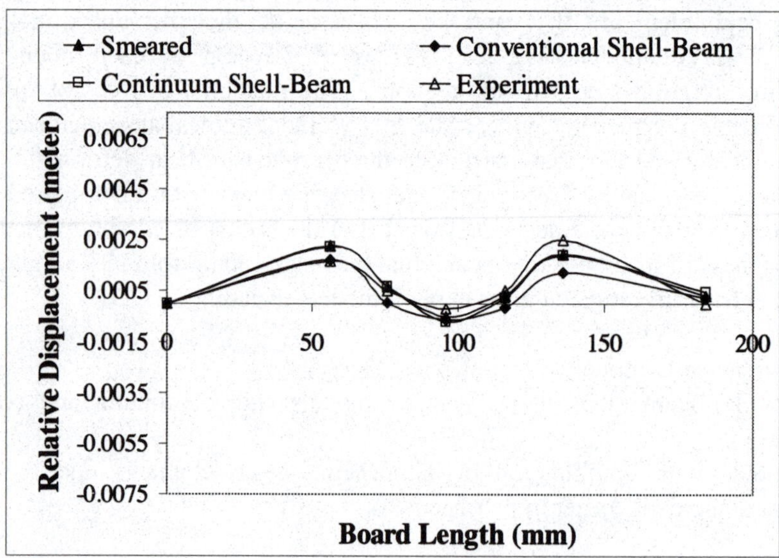

Fig. 13.30 Comparison of displacement along the length of the circuit board assembly among FEA and experimental results.

Table 13.4
Computational efficiency for the three explicit models (free drop)

Model	Number of Elements	Number of Nodes	Computational Time, (s)
Smeared	77,825	82,135	59,400
Conventional Shell-Beam	120,105	133,576	219,600
Continuum Shell-Beam	218,206	264,124	568,800

Solder Interconnect Strain Variation with Time

A graph showing the results of the strain as a function of time for the corner solder joints of a ball grid array is presented in Fig. 13.31. These results were obtained with the Timoshenko-beam element with conventional-shell model an impact event corresponding to a 0 JEDEC drop. Similar results for the solder balls at the corner of the fourth row from the outside are shown in Fig. 13.32.

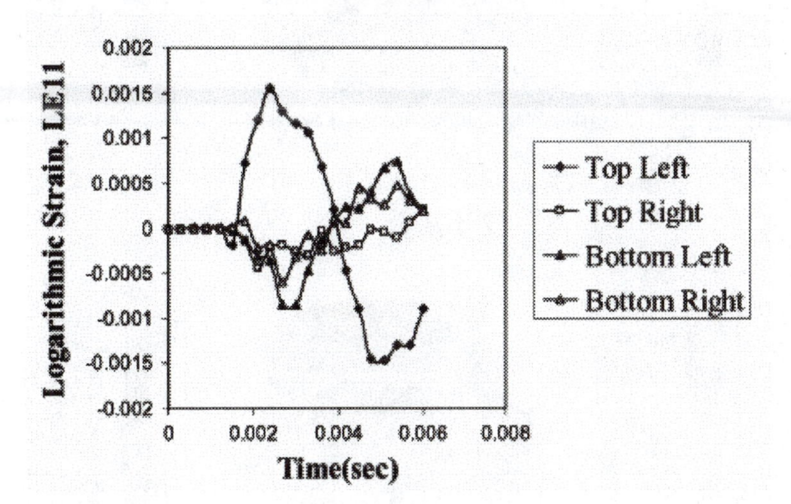

Fig. 13.31 Strain as a function of time for the outside corner solder joints on a chip scale package during a 0° JEDEC drop test.

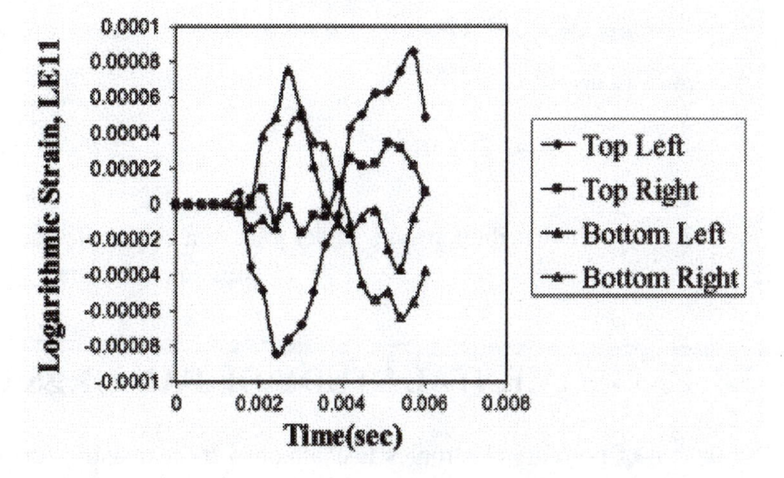

Fig. 13.32 Strain as a function of time for the corner solder joints located four rows from the outside on a chip scale package during a 0° JEDEC drop test.

A comparison of transient strain histories in Figs. 13.31 and 13.32 reveals that a large portion of the strain is carried by the outside row of the solder joints. For this reason, the explicit sub-model includes reduced-integration hexahedral elements for the corner solder interconnect. The hexahedral element mesh for the solder balls give the distribution of the logarithmic strain, LE23, over the solder interconnects, as shown in Fig. 13.33. The results indicate that the maximum strains occur at the solder joint to package interface and the solder joint to printed circuit board interface, indicating a high probability of failure at both of these interfaces. Failure analysis of the drop test samples reveals that the observed failure modes correlate well with the FEA predictions.

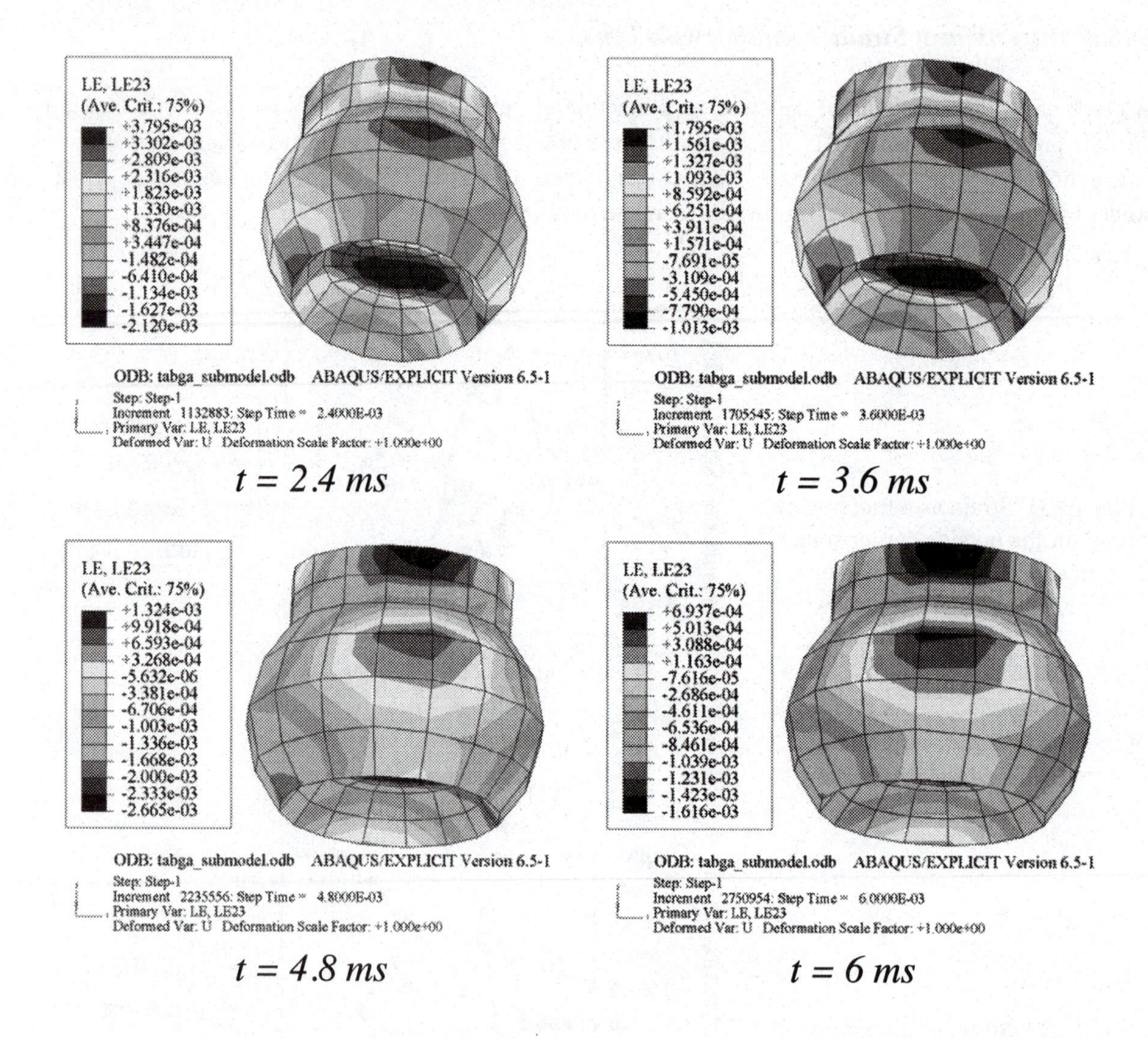

Fig. 13.33 Strain distributions in a solder joint of a chip scale package at four different times during a 0° JEDEC impact event.

13.8 EXPERIMENTAL STUDY OF FAILURES UNDER SHOCK LOADS

The design of portable electronics is challenged by increasing functional density, short product cycles, and concerns about cost. Increasing functional density required in portable electronic products has lead to the rapid growth in the use of chip scale packages (CSP). The expected life of a portable product is typically short compared to other electronic products; however, portable products typically must survive several drops over their life span. The input/output (I/O) pitch of CSPs and the small pads and solder joints, make survival of impact due to several accidental drops a difficult requirement.

There are two approaches to improving reliability of a product subjected to multiple drop impact. The first is the mechanical design of the product to minimize the shock loads and bending of the printed circuit board (PCB) that occurs due to impact. The second approach is to use underfill to mechanically reinforce the CSP solder joints. Underfilling adds cost and cycle time to the manufacturing process; hence, the development of a robust mechanical design, capable of resisting multiple drops is usually the preferred approach. Electronics inside portable products may be subjected to G loading that may range for a few hundred Gs to thousands of G's when dropped from about a five feet. Experiments are often employed to verify a design approach.

Test methods to insure drop reliability can be broadly classified into constrained and unconstrained or free drop. Examples of constrained drop include the JEDEC test method. The JEDEC standard is a component-level test, which is often used to evaluate and compare the drop performance of surface mount electronic components for handheld electronic product applications. The primary intent of the JEDEC test specification is to standardize the test specimen and methodology to provide a reproducible assessment of the drop performance of surface mount electronics. However, the correlation between drop-performance in the test and that at the product-level leaves much to be desired. Product-level failures are often influenced by housing design and drop-orientation, which may not always be perpendicular to the board surface. In this section the results of the transient dynamic behavior of the board assembly during a vertical-drop will be described.

13.8.1 Test Specimens and Testing Methods

Two test boards were been used by Lall et al [12] to study the reliability of BGAs and CSPs when subjected to impact loading. The first group of boards included 8-mm flex-substrate CSPs, 0.5-mm pitch, 132 I/O. Each board had 10 CSP locations on one side of the board only. The second group of boards included 15-mm CSPs, 16-mm C2BGAs and 27-mm BGAs. Two versions of the boards with 8-mm CSPs were tested with a tin-lead-silver solder, 62Sn36Pb2Ag on one version and lead-free solder balls 95.5Sn4.0Ag0.5Cu on the other version. The package attributes for both test boards are shown in Table 13.5. Both test boards had six-layers based on standard PCB technology with no build-up or HDI layers. The first group of boards was 74.93 × 183.9 mm by 1.067 mm thick and the second group was 101.6 × 88.9 mm by 1.067 mm thick.

The test boards were subjected to a controlled drop with heights varied from 36 to 72 in. (0.91 m to 1.83 m). Component locations on the test boards were instrumented with strain gages. Strain measurements were acquired during the impact event with a high-speed data acquisition system capable of 5×10^6 samples/s. The impact-event was also monitored with a high-speed video camera operating at 40,000 fps. Targets were mounted on the edge of the board to facilitate measurement of relative displacement of the board during impact. The test boards were dropped in their vertical orientation with a weight attached to their top edge. Image tracking software was used to measure displacements during impact the drop event. In addition to displacement determinations, velocity of the board prior to impact was measured.

Table 13.5
Description of components tested.

Body Size	I/O Count	Ball Pitch (mm)	Thermal Balls	Die Size (mm)	Substrate Thickness (mm)	Substrate Pad Dia. (mm)	Substrate Pad Type	Ball Dia. (mm)
Test Board A								
8 mm 62Sn36Pb2Ag	132	0.5	None	4.0×4.0	0.1	0.28	Thru Flex	0.3
8 mm 95.5Sn4Ag0.5Cu	132	0.5	None	4.0×4.0	0.1	0.28	Thru Flex	0.3
Test Board B								
27 mm BGA	388	1.0	36 (6×6)	10×10	0.56	0.5	SMD	0.60
16 mm C2BGA	240	0.8	None	5.0×5.0	0.56	0.45	SMD	0.53
15 mm BGA	193	0.8	25 (5×5)	8.6×8.6	0.56	0.5	SMD	0.60

13.8.2 Controlling the Specimens During the Drop Test

Repeatability of board orientation at impact is critical to measuring a meaningful response. Small variations in the board orientation can produce significant variations in the dynamic reaction of the board. To control the board's orientation prior to impact a drop-tower was developed that provided the necessary orientation control. The repeatability of the test was quantified by measuring the angle of board prior to impact and relative displacement of the board after impact. Board displacement after impact includes both oscillation and rigid body rotation. Because rigid body rotation does not contribute strains, relative displacement was the field quantity used to for monitor the board's dynamic response after impact as well as the repeatability of the impact event. A series of ten tests, conducted to measure the displacement–time response after impact, showed that the impact event was reproducible.

The orientation of the board prior-to and after-impact was measured relative to a known vertical stationary reference, which was positioned in the field of view of the high-speed camera. The test boards were dropped vertically; hence an angle of impact of about zero is required. A mean angle of orientation of the board prior to impact of 0.2° to 0.34° was measured, indicating good control of the board's orientation before and after impact.

13.8.3 Board Survivability Data after Impact

Lall et al [12] investigated various design approaches for enhancing impact survivability of the test boards described previously. The configurations studied included; corner-bonded reflow encapsulants and capillary flow underfills. The data was been benchmarked against non-underfilled devices. Components mounted with 62Sn36Pb2 Ag and 95.5Sn4Ag0.5Cu solders were tested. A total of 50 devices were tested for each configuration. The results are presented in Figs. 13.34 to 13.36.

The earliest group to fail was the CSPs with no underfill. There was no significant difference in the percent of failures between the Sn/Pb and the Sn/Ag/Cu soldered CSPs. The second group to fail was the corner bonded CSPs with a factor of three improvement over the CSPs with no underfill. This finding is consistent with earlier results with Sn/Pb and corner bond underfill by other researchers [13]. Again, there was no difference between Sn/Pb and Sn/Ag/Cu. The next group to fail was the boards with CSPs mounted with capillary flow underfill. CSPs mounted with Sn/Ag/Cu solder and capillary underfill had only two CSPs failures out of 50 CSPs after enduring 100 drop tests. Failure modes include pad I/O PCB side resin cracks, failed solder joints at package interface, cracking of copper traces and underfill fillet cracking.

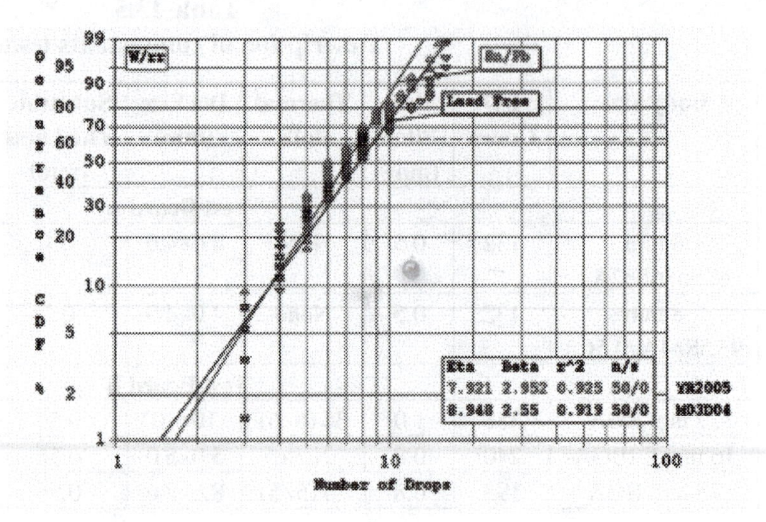

Fig. 13.34 Percent failures as a function of the number of drops with 62Sn36Pb2 Ag and 95.5Sn4Ag0.5Cu solders and no underfill.

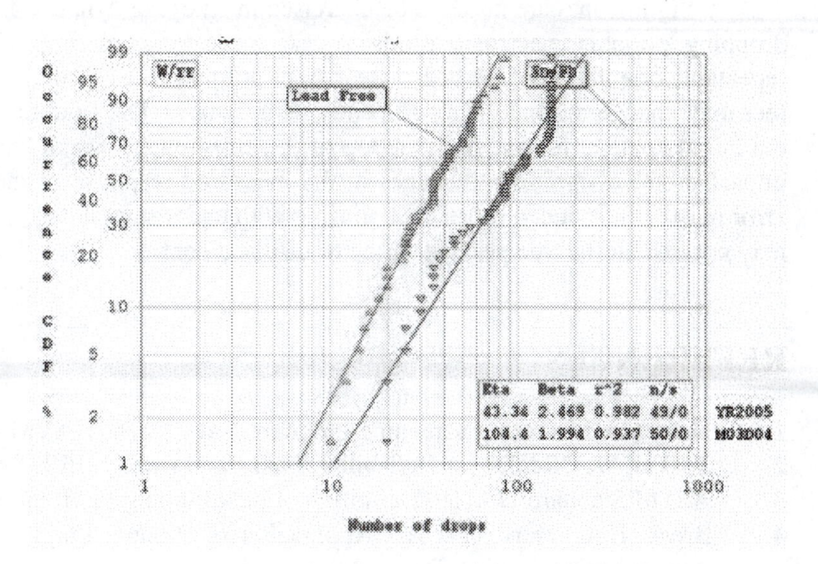

Fig. 13.35 Percent failures as a function of the number of drops with 62Sn36Pb2 Ag and 95.5Sn4Ag0.5Cu solders and capillary underfill,

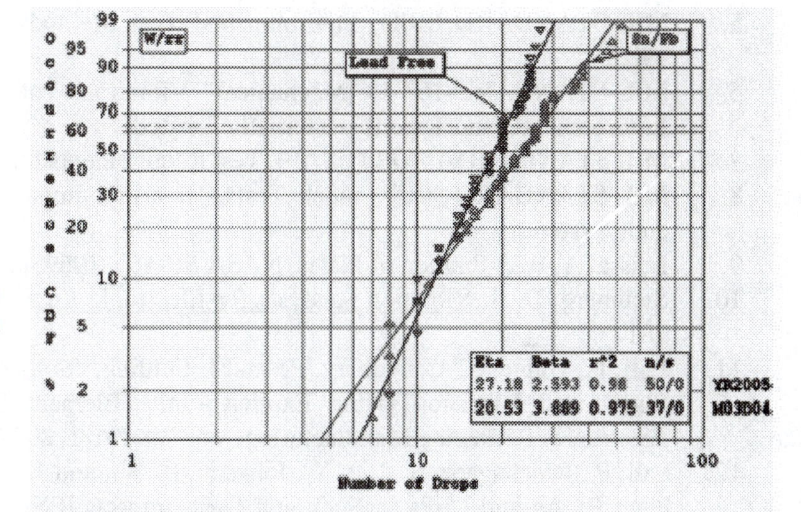

Fig. 13.36 Percent failures as a function of the number of drops with 62Sn36Pb2 Ag and 95.5Sn4Ag0.5Cu solders and corner underfill,

13.8.4 Conclusions

The results of an extensive series of drop-impact tests are summarized below:

1. CSP packages assembled with SnPb and SnAgCu solders and without underfill do not exhibit a significant difference in impact reliability.
2. Area array chip carriers with capillary-flow underfill show significantly improved impact-survivability.
3. The CSP packages with corner-bonding performed better than the CSPs without underfill but worse than CSP packages with capillary-flow underfill.
4. The corner bonded CSP packages have the advantage of eliminating the additional step of thermal cure, which adds time to the assembly process.
5. Failure analysis of the package architectures indicates that several failure modes exist including, package-to-solder interconnect failures, board-to-solder interconnect failures, board dielectric cracks under the pad, copper trace-cracks, and underfill cracking. The data indicates that the use of underfill and corner bonding, may shift the failures from solder interconnects to copper traces and underfill fillets.

Significant care needs to be exercised in studying dynamic failure mechanisms that occur when dropping portable electronic products. A well designed drop test tower is essential to produce repeatable orientation of the board assemblies at impact. Electronic assemblies dropped from five or six feet may change orientation significantly as they fall. The final orientation at impact may differ from the initial angle at release from the top of the drop-tower. Small changes in the angle of orientation at impact produce significant changes in the measured response of the product. In addition, significant error may be introduced because of aliasing with under sampled experimental data. Under sampled data may not exhibit the true peaks in strain or displacement.

REFERENCES

1. Navy Manufacturing Screening Program, Navy Dept., NAVMAT P-9492, May 1979.
2. Tuskin, W. "Recipe for Reliability: Shake and Bake," IEEE Spectrum, Dec. 1986, pp. 37-42.
3. ASTM Standard D-4169, Definition of the Shipping Environment.
4. Bresk, F. C. and Irving K. “Application of Product Fragility Information in Product Design,” Lansmont Corporation, Pacific Grove, CA.
5. MIL-STANDARD 810D, Environmental Test Methods and Engineering Guidelines, July 19, 1983.
6. MIL-STANDARD-167-1, Mechanical Vibrations of Shipboard Equipment (Type I - Environmental and Type II – Internally Excited).
7. MIL STANDARD-1540B (USAF), Test Requirements for Space Vehicles, October 10, 1982.
8. MIL-STANDARD-901D, shock Tests, HI (High Impact): Ship Board Machinery, Equipment ands Systems.
9. Leissa, A. W. Vibration of Plates, NASA SP-160, 1969, pp 41-160.
10. Steinberg, D. S. Vibration Analysis for Electronic Equipment, John Wiley, New York, 1973 p. 281.
11. Lall, P., Gupte, S., Choudhary, P., and J. Suhling, “Solder Joint Reliability in Electronics Under Shock and Vibration Using Explicit Finite Element Submodeling,” IEEE Transaction on Electronics Packaging Manufacturing, Vol. 30, No. 1, 2007, pp. 74-83.
12. Lall, P., Panchagade, S., Liu, Y., Johnson, R. W., and J. Suhling, Models for Reliability of Fine-Pitch BGAs and CSPs in Sock and Drop Impact, IEEE Transaction on Electronics Packaging Manufacturing, Vol. 29, No. 3, 2006, pp. 464-474.
13. G. Tian, Y. Liu, P. Lall, W. Johnson, and J. Suhling, “Drop Reliability of Corner Bonded CSP in Portable Products,” Proceedings ASME InterPACK Conference, Maui, HI, July 6–11, 2003.
14. Caletka, D. V., R. N. Caldwell and J. T. Vogelman, “Damage Boundary Curves: A Finite Element (FEM) Approach, Transactions of the ASME, Journal of Electronic Packaging, Vol. 112, 1990.
15. J. M. Pitarresi, D. V. Caletka, R. N. Caldwell and D. E. Smith, “The Smeared Property Technique for the FE Vibration Analysis of Printed Circuit Cards,” Transactions of the ASME, Journal of Electronic Packaging, Vol. 113, 1991.
16. Engle, P. A. and C. T. Lim, “Finding the Stress with Finite Elements, Mechanical Engineering, Vol. 108, No. 10, 1986, pp. 46-50.
17. Engle, P. A. and C. T. Lim, “Stress Analysis in Electronic Packaging,” Finite Element Analysis and Design, Vol. 4, 1988, pp. 9-18.
18. J.-P. Clech, “Solder reliability Solutions: A PC-based Design-for-Reliability Tool,” Proceedings Surface Mount Institute Conference, Vol. I, San Jose, CA, September 8–12, 1996, pp. 136–151.
19. J.-P. Clech, “Flip-chip/CSP Assembly Reliability and Solder Volume Effects,” in Proceedings Surface Mount Institute Conference, San Jose, CA, August 25–27, 1998, pp. 315–324.

20. S. Goyal and E. Buratynski, "Methods for Realistic Drop Testing," International journal Microcircuits Electronic Packaging, Vol. 23, No. 1, 2000, pp. 45–51.
21. JEDEC Solid State Technology Association, Board-Level Drop Test Method of Components for Handheld Electronic Products, Std. JESD22-B111, 2003.
22. P. Lall, D. Panchagade, Y. Liu, W. Johnson, and J. Suhling, "Models for reliability prediction of fine-pitch BGA's and CSP's in shock and drop-impact," Proceedings 54th Electronic Components Technology Conference, 2004, pp. 1296–1303.
23. C. T. Lim and Y. J. Low, "Investigating the Drop Impact of Portable Electronic Products," in Proceedings 52nd Electronic Components Technology Conference, 2002, pp. 1270–1274.
24. J. Pitaressi, B. Roggeman, and S. Chaparala, "Mechanical shock testing and modeling of PC motherboards," Proceedings 54th Electronic Components Technology Conference, 2004, pp. 1047–1054.
25. T. Y. Tee, H. S. Ng, C. T. Lim, E. Pek, and Z. Zhong, "Board Level Drop Test and Simulation of TFBGA Packages for Telecommunication Applications," Proceedings 53rd Electronic Components Technology Conference, 2003, pp. 121–129.
26. E. H. Wong, C. T. Lim, J. E. Field, V. B. C. Tan, V. P. M. Shim, K. T. Lim, and S. K. W. Seah, "Tackling the Drop Impact Reliability of Electronic Packaging," in Proceedings ASME International. Electronic Packaging. Technology Conference, Maui, HI, Jul. 6–11, 2003, pp. 1–9.
27. J. Wu, G. Song, C.-P. Yeh, and K. Wyatt, "Drop/impact simulation and test validation of telecommunication products," in Proceedings Intersociety Conference on Thermal Phenomena, 1998, pp. 330–336.
28. D. Xie, M. Arra, S. Yi, and D. Rooney, "Solder joint behavior of area array packages in board level drop for handheld devices," Proceedings 53rd Electronic Components Technology Conference, 2003, pp. 130–135.
29. L. Zhu, "Modeling Technique for Reliability Assessment of Portable Electronic Products Subjected to Drop Impact Loads," Proceedings 53rd Electronic Components Technology Conference, 2003, pp. 100–104.

EXERCISES

13.1 Write an engineering brief describing a manufacturing screening test. Indicate the advantages and disadvantages of the test. Outline the test procedures.

13.2 You are to implement a temperature cycling test as a part of a manufacturing screen. The product being screened has nearly 1,000 devices and/or parts. Specify the number of cycles and indicate the number of failures anticipated on each cycle. If it takes 2 hours to cycle the temperature and another 1.5 hours to identify, locate and replace the defective part, estimate the average time from the start of the test until the product is deemed to be defect free.

13.3 Will a manufacturing screening test eliminate all of the latent defects?

13.4 You are shipping 1,000 units of a product each week by truck to a distribution point. One of the circuit boards in the product has a natural frequency of 20 Hz. Determine the acceleration in Gs that the circuit board may be subjected to in transit.

13.5 Describe an application of an electronic system with which you are familiar that is exposed to severe shock or vibrations on a daily basis. Examine that product and indicate any special packaging features that have been used to mitigate the effects of this environment.

13.6 A transformer is mounted to an enclosure using a bracket that can be modeled as a cantilever beam. The static deflection of the beam due to the weight of the transformer is 0.50 mm. Determine the natural frequency f_n for this assembly.

13.7 Beginning with Eq. 13.7, verify Eq. 13.9.

13.8 Beginning with Eq. 13.7, verify Eq. 13.10.

13.9 Beginning with Eq. 13.7, verify Eq. 13.11.

13.10 Describe why the critically dampened and over-dampened conditions are not a major concern in protecting electronic equipment from vibratory environments.

13.11 Beginning with Eq. 13.14, verify Eq. 13.15.

13.12 Prepare a graph showing ω_{nd}/ω_n for $0 < \xi < 1$ if $\omega_{nd} \leq \omega_n$.

13.13 Beginning with Eq. 13.18, verify Eqs. 13.19 a and b.

13.14 Prepare a graph of the transmission coefficient ***A*** as a function of the dampening ratio ξ if $r = \omega/\omega_n = 1$.

13.15 Prepare a graph showing ***A*/*F*** as a function of r if ξ is equal to 0.01, 0.02, 0.05, 0.10, 0.2 and 0.5. Comment on the significance of these curves.

13.16 Why do we need two transmission coefficients to describe resonance effects on the response of the systems.

13.17 Beginning with Eq. (13.34), verify Eq. (13.38), Eq. (13.39) and Eq. (13.40).

13.18 Using data found on the Internet for an isolation system, specify the isolators and their spring rate if the system consists of four isolators that supports a 40 kg assembly. Determine the natural frequency of this isolated system. Find the transmissibility coefficient ***A*** if the forced disturbance is at a frequency of 170 Hz.

13.19 Prepare an engineering brief describing the results of the analysis of the isolator defined in Exercise 13.18.

13.20 The 40 kg electronic assembly of Exercise 13.18 is housed in a case 100 × 500 × 400 mm in size, and is permanently mounted on a shelf that is located in a trailer. Prepare a sketch showing the arrangement of the isolators and the mounting of the isolators to the shelf and the case.

13.21 Customers are returning a product that malfunctions because of broken wires. An inspection of these units reveals that the failures are due to nicks that occur on about 1 wire in 200. The wires that fail are always solid conductors. Prepare an engineering brief describing two or more approaches for resolving this problem.

13.22 Prepare a sketch showing four different boundary constraint combinations that are commonly used to model circuit boards in vibration. Why do you believe that these are support conditions are so popular. Identify these support conditions in Table 13.3.

13.23 Would you model an edge connector as a simple support or as a clamped condition? Explain why? What type of support. does a wedge lock retainer provide?

13.24 Determine the flexural rigidity D of a circuit board 2.0 mm thick that is fabricated from epoxy-glass with E = 24 GPa.

13.25 Prepare a graph showing the flexural rigidity D as a function of thickness h which ranges from 0.2 to 4.0 mm. Let E = 5, 10, 20, and 50 GPa.

13.26 If the circuit board of Exercise 13.24 is clamped along its 100 mm edge, simply supported along the opposite edge and free along its two 300 mm edges, determine its natural frequency.

13.27 Examine Table 13.3 and identify the case which gives the highest ω_n if a/b = 3. Also, identify the case which gives the lowest natural frequency.

13.28 Determine the natural frequency for the circuit board illustrated in Case No. 16. Let a/b = 4. a = 300 mm, E = 20 GPa, h = 3.0 mm and $\gamma = 0.065$ lb/in^3.

13.29 Write an equation comparable to Eq. (13.44) that describes the deflection for a circuit board with two opposing edges free, one edge clamped and the other simply supported.

13.30 Write the equation describing the deflection of the plate corresponding to Case No. 10 in Table 13.3. Reduce this equation to show the deflection as a function of position along each center line.

13.31 For the boundary conditions given in Exercise 13.26 find the acceleration $\ddot{w}_{max}$.

13.32 For the circuit board defined in Exercises 13.24 and 13.26, determine the maximum displacement, if $\boldsymbol{A} = \sqrt{f_n}$. Repeat the determination using another method for establishing $\boldsymbol{A}$, by letting ξ = 0.04.

13.33 Use Eq. (13.51) in Eq. (13.50) to eliminate $\boldsymbol{A}$. Then describe the dependency of $w_{\omega n}$ on ω_n at the resonance condition.

13.34 Examine Fig. 13.17 and explain the why the accelerometers are mounted adjacent to the supports. Describe the signal you would expect from these accelerometers.

13.35 A vendor describes tests of a new conformal coating that has increased the dampening ratio ξ from 0.04 to 0.07. Prepare an engineering brief for management describing the advantages and disadvantages of this coating.

13.36 A heat frame consisting of a sheet of aluminum 1.0 mm thick is bonded to the circuit board of Exercise 13.24. Determine the change in the flexural rigidity of the board. If the boundary conditions are the same as those described in Exercise 13.26, find the new value of f_n. If the edges of the heat frame are clamped so that the free edges are treated as fixed, find the value of f_n.

13.37 Prepare an engineering brief describing the effects of heat frames on the vibration of PCB's.

13.38 Verify Eq. (13.52).

13.39 Examine the effect of the aspect ratio (a/b) of a circuit board having an area A. Consider the case where the board is simply supported on all four edges. Prepare a graph showing the normalized natural frequency as a function of (a/b).

13.40 The customer is the U. S. Navy with a ship board application where the maximum driving frequency is 50 Hz. Prepare an engineering brief which describes a design plan for the circuit boards to withstand a 2 hour vibration dwell at 50 Hz and 2G's. Introduce the concept of fragility level in your plan.

13.41 Determine the increase in the flexural rigidity of the circuit board shown in Fig. 13.18, if h = 0.060 in. and the ribs are aluminum strips 0.125 in. wide and 0.5 in. high. The width of the circuit board is b = 6 in. and its elastic modulus E is 4×10^6 psi. Determine the enhancement in ω_n due to the ribs.

13.42 Prepare an engineering brief which explains the dampening enhancement provided by viscous coatings on vibrating plates. The brief should refer to a library or Internet search where the effect of coatings on vibrating metal panels has been characterized.

13.43 Write the equation for the displacement w(x) of the circuit boards described in Case Nos. 6, 12, 15 and 18 of Table 13.3.

13.44 Begin with Eq. (13.55) and verify Eq. (13.57).

13.45 Derive an equation for the maximum stress in a circuit board (similar to Eq. (13.61) for Case No. 6 in Table 13.3.

13.46 Consider a circuit board supported as in Case No. 6 and subject to a 5 G input acceleration. The board thickness is set at 2.0 mm and its aspect ratio is set at a/b = 2. Determine the size of the board if the maximum stresses are to be limited to 20 MPa. Where on the board will this maximum stress occur?

13.47 Determine the natural frequency of a module consisting of a ceramic card (alumina) $1.2 \times 100 \times 120$ mm in size bonded to a heat frame that is made from copper sheet 1.5 mm thick. The heat frame is tightly clamped to a cold rail with wedge lock retainers. Constraint conditions are closely described by Case No. 5, Table 13.3.

13.48 A strain gage mounted on the glass reinforced portion of a circuit board indicates a peak strain of 0.0032 when the system is subjected to a dwell at resonance frequency. The resonance frequency is 90 Hz. Estimate the time of dwell prior to circuit malfunctions due to fatigue failure of the copper traces on this circuit board.

13.49 Equation (13.71) gives the frequency f_n of an axial leaded component soldered to two rigid posts. Derive a similar relation for an axial leaded component with bent leads soldered to a circuit board.

13.50 Determine the force P for a 5W resistor supported as shown in Fig. 13.17b with L/2 = 3.5 mm for an input acceleration ranging from 1 to 10 G's.
13.51 Verify Eqs.(13.73).
13.52 Verify Eqs.(13.74).
13.53 Prepare a graph showing the stresses developed in a soldered joint as a function of M_1. Consider wire diameters of 20, 25 and 30 mils in this solution.
13.54 A solder joint is exposed to shear stresses of 800 psi and bending stresses of 1,200 psi. Determine the maximum principal stress and the maximum shear stress. If the solder joint has an endurance strength of S_e = 2,000 psi determine the safety factor based on infinite-life.
13.55 Determine the normal and shearing stresses in a BGA if it is subjected to a normal force P = 120 N and a shearing force V = 80 N. The BGA has 360 mounting pads on 1.00 mm pitch with a diameter of 0.60 mm.
13.56 Begin with Eq. (13.90) and verify Eq. (13.93) and Eq. (13.94).
13.57 Prepare instructions for a junior engineer to design a shock isolation system to accommodate a half-sine acceleration pulse with a peak acceleration of 10 Gs and a time interval of 12 ms.
13.58 Prepare instructions for a junior engineer to design a shock isolation system to accommodate a rectangular acceleration pulse with a peak acceleration of 12 Gs and a time interval of 10 ms.
13.59 Prepare instructions for a junior engineer to design a shock isolation system to accommodate a triangular acceleration pulse with a peak acceleration of 15 Gs and a time interval of 8 ms.
13.60 Prepare instructions for a junior engineer to design a shock isolation system to accommodate a versed-sine acceleration pulse with a peak acceleration of 8 Gs and a time interval of 20 ms.
13.61 Prepare an engineering specification for a drop test of a shipping container for an electronic system. Consider different package weights and sizes in writing the test procedure included in the specification.
13.62 Prepare an engineering brief describing the advantages of using finite element methods early in the design process to insure that a product designed for a military application will succeed in passing vibration and or shock qualification tests.
13.63 Describe four modeling approaches used in finite element analysis of BGAs mounted on PCBs.
13.64 What are the advantages of employing finite element analyzes over conducting tests with actual hardware during a development cycle?
13.65 Describe some of the difficulties in constructing finite element models of a PCB assembly.
13.66 Why is impact more important with portable electronic equipment versus stationary equipment?
13.67 Prepare test plan for verifying the impact resistance of a new digital camera your company is planning to introduce.
13.68 Prepare test plan for verifying the impact resistance of a new cell phone your company is planning to introduce.
13.69 Prepare test plan for verifying the impact resistance of a small portable hard drive your company is planning to introduce.
13.70 Describe the problems that may occur when conducting impact tests with recording instruments that have limited sampling rate capabilities.

PART 4

RELIABILITY

CHAPTER 14

THEORY OF RELIABILITY

14.1 INTRODUCTION

Digital systems usually contain a large number hundreds or thousands of components such as IC's, transistors, diodes, resistors etc. Often hundreds of these components are connected together in a series circuit. The successful operation of this circuit depends upon the reliability of each component and with so many components connected in series the reliability of each component must be extremely high. Even with high reliability components, system failures occur and it is necessary to provide for fault location in designing digital logic circuits and the operating software. With fault location, the digital system is periodically monitored and failures are identified and located within the system almost as soon as they occur. The location of the malfunctioning part is extremely important as identifying the faulty circuit board or a small set of suspect circuit boards greatly reduces the time necessary to replace the failed board and to return the system to an operating condition.

Analog systems usually have a much smaller number of components and adequate availability can usually be achieved without the need for fault location procedures. However, very high individual component reliability is still required to provide relatively trouble free operation in even simple analog circuits.

14.2 RELIABILITY THEORY

Semiconductor devices, passive components, switches and lights fail with time in operation. Life tests conducted by many manufactures with thousands of components indicate that the functional relationship between the number of failed components N_f is non-linear with respect to time. A typical graph showing the relation between N_f and time t is presented in Fig. 14.1.

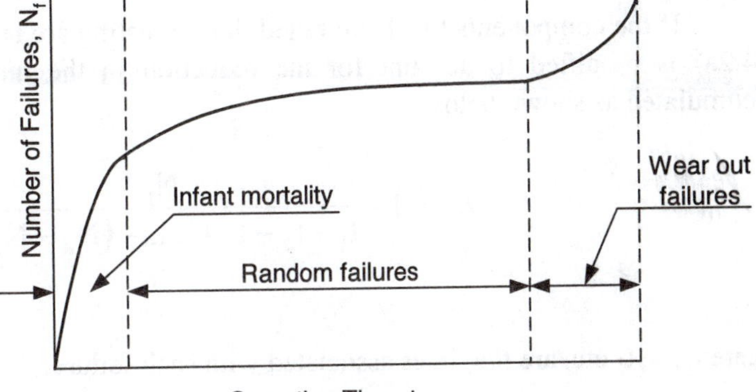

Fig. 14.1 The number of component failures N_f as a function of time during a life test.

Initially the rate of component failure is very high due to latent defects introduced in the manufacturing process. This phase of the life test is termed the infant mortality region. After these defective components have failed, failures continue at a much lower rate due to random causes. After a considerable period of time, usually several years for semiconductor devices, the failure rate increases markedly. These late failures are due to aging, wear and fatigue. Electronic systems are design with components that will not reach the wear-out phase during their warranty period.

14.2.1 Failure Rate

The failure rate $\lambda(t)$ is the measure of the number of failures that will occur in a specified number of hours of operation for a given component. As we are concerned with extremely low failure rates, 10^6 hrs of continuous operation is the usual time considered in specifying an average failure rate. The instantaneous value of the failure rate varies during the life of a given component and may increase or decrease with time. The failure rate is also a strong function of operating temperature and the voltage applied to semiconductors.

Let us consider a life test of a specified component, which is conducted to determine its failure rate λ with respect to time t. If a large number N_0 of components are placed in operation at time $t = 0$, then after some time t a certain number N_f will have failed and the remainder N_s will have survived. Thus:

$$N_f(t) + N_s(t) = N_0 \tag{14.1}$$

The average failure rate λ_{ave} is determined from:

$$\lambda_{ave}(t) = \frac{N_f}{N_0 t_{ave}} \tag{14.2a}$$

Where, t_{ave} is the average time between failures. An incremental failure rate determined over some time period Δt is given by:

$$\lambda(\Delta t) = \frac{1}{N_0}\frac{\Delta N_f(t)}{\Delta t} \tag{14.2b}$$

An instantaneous failure rate $\lambda(t)$ is given by:

$$\lambda(t) = \frac{d}{dt}\left(\frac{N_f(t)}{N_0}\right) \tag{14.2c}$$

If the components that have failed during testing are not replaced during the test time t, then Eq (14.2a) is modified to account for the reduction in the number of component hours which have accumulated as shown below:

$$\lambda_{ave}(t) = \frac{N_f}{t_1 + t_2 + t_3 + \ldots\ldots + (N_0 - N_f)t} = \frac{N_f}{\sum_{k=1}^{N} t_k} \tag{14.3}$$

where t_1, t_2, t_3 etc. are the times associated with each failure.

The results of testing a large number of components and applying either Eq (14.2) or (14.3) gives the well known bathtub mortality curve presented in Fig. 14.2. Examination of this curve shows three distinct regions of interest. During the initial period, $0 \le t \le t_1$, the failure rate is relatively high. Those components with inherent manufacturing flaws, which have not yet been eliminated by process adjustments in production, fail early in service. For $t_1 \leq t \leq t_2$, the failure rate is relatively low, constant or declining slightly with time. The failures in this phase of the life of the device are due to random defects or events. This is the best period of operation from a reliability viewpoint. For $t > t_2$ the failure rate increases rapidly with time as the components fail due to the effects of aging and fatigue.

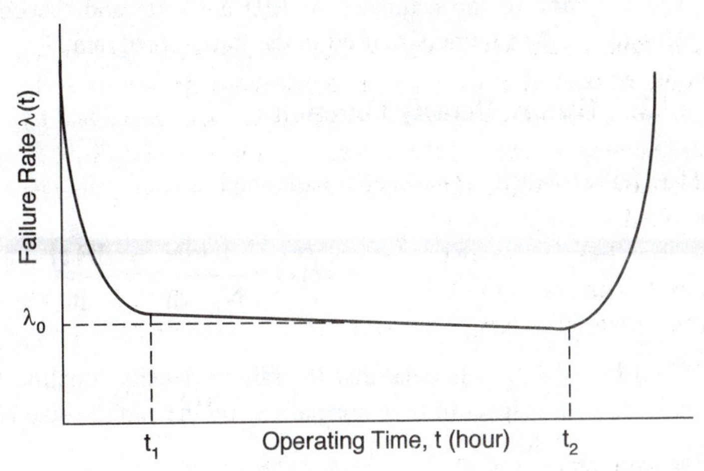

Fig. 14.2 Bathtub shaped mortality curve showing the change in failure rate with respect to time for electronic components.

For high reliability systems it is essential to eliminate the inherently defective components associated with the initial period of operation. This elimination is accomplished by improving the manufacturing processes or by subjecting the components to a burn-in period of simulated operation. Burn-in is the final phase of the manufacturing process prior to the usage of the component. Assuming adequate burn-in procedures, $t_1 = 0$, and the useful life of the component will be t_2. During this period of operation, $0 \leq t \leq t_2$, the failures will occur at a constant rate λ_0 due to random events. The satisfactory operation of the system during this period depends on λ_0, the number of components in both series and in parallel circuits and the requirements for system availability. After t_2 hours of operation the failure rate $\lambda(t)$ increases rapidly, system reliability degrades and satisfactory performance cannot be anticipated.

14.2.2 Reliability

Reliability is the statistical probability that a component, subsystem or system will perform satisfactorily for a specified time. The reliability is usually specified with some prescribed confidence level subject to certain operating and environmental conditions. In this treatment we will use probability of success for the reliability function R(t) and probability of failure for the unreliability function F(t) to express component, sub system or system malfunction.

From the test data used to establish failure rate described in the previous section, it is clear that the reliability R(t) at some time t during the test is given by:

$$R(t) = \frac{N_s(t)}{N_0} = \frac{N_0 - N_f(t)}{N_0} \tag{14.4}$$

The unreliability function (probability of failure) at some time during the test is given by:

$$F(t) = \frac{N_f(t)}{N_0} \tag{14.5}$$

From Eq. (14.4) and Eq. (14.5) it is clear that:

$$R(t) + F(t) = 1 \tag{14.6}$$

The accuracy of the estimates of R(t) and F(t) and the confidence limits placed on these estimates depends on the sample size used in the testing program.

14.2.3 Failure Density Function

The failure density function f(t) is defined as:

$$f(t) = \frac{1}{N_0}\frac{dN_f}{dt} = \frac{dF(t)}{dt} \tag{14.7}$$

From Eq. (14.7), it is clear that the failure density function f(t) is the rate of change of the number of failures with respect to time normalized by the sample size N_0. Integrating Eq. (14.7) and recalling and Eq. (14.5) yields:

$$\int_0^t f(t)dt = F(t) = \frac{N_f(t)}{N_0} \tag{14.8}$$

With increasing time under test $N_f \Rightarrow N_0$ and Eq. (14.8) can be written as:

$$\int_0^\infty f(t)\, dt = 1 \tag{14.9}$$

A graph showing f(t) as a function of testing time is presented in Fig. 14.3. Note that the area under the curve from t = 0 to $t = t_1$ is equal to $F(t_1)$ and the area under the curve for $t \geq t_1$ is equal to $R(t_1)$. The reliability function (probability of success) $R(t_1)$ can be written in terms of the failure density function as:

$$R(t_1) = \int_{t_1}^\infty f(t)\, dt \tag{14.10}$$

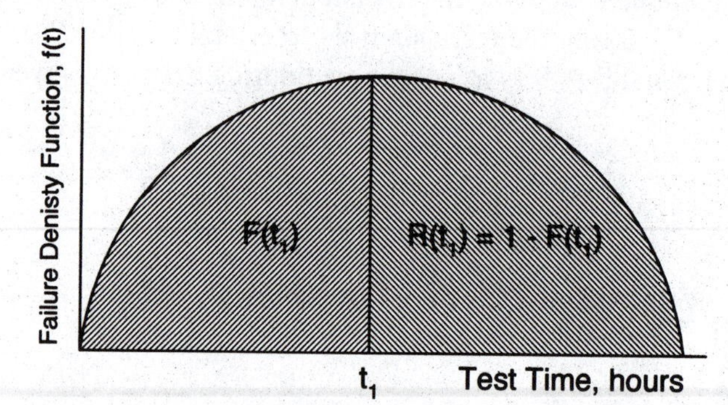

Fig. 14.3 The failure density function associated with component failure.

From Eq. (14.4), Eq. (14.5) and Eq. (14.7), it is evident that the failure rate λ(t) can be expressed as:

$$\lambda(t) = \frac{1}{R(t)}\frac{dF(t)}{dt} = \frac{1}{R(t)}\frac{d[1-R(t)]}{dt} = -\frac{1}{R(t)}\frac{dR(t)}{dt} = \frac{f(t)}{R(t)} = \frac{f(t)}{1-F(t)} \tag{14.11}$$

From Eq. (14.11) and Eq. (14.7), it is possible to show that:

$$R(t) = e^{-\int_0^t \lambda(t)dt} \tag{14.12}$$

14.2.4 Hazard Rate

The hazard rate HR is used to describe the probability of failure over a specified time period of a component that cannot be repaired. It is defined by:

$$HR(t_1 + \Delta t/2) = \frac{1}{N_s(t_1)}\frac{\Delta N_f}{\Delta t} \tag{14.13}$$

where $N_s(t_1)$ is the number of parts that have not failed in test at the time t_1; Δt is the duration of the specified time period and ΔN_f is the number of failures occurring during that period.

From Eq. (14.13) it is evident that the hazard rate HR is the number of failures per unit time normalized by the number of components still in operation. If we let $\Delta t \Rightarrow 0$ and take the limit of the right side of Eq. (14.13), the instantaneous value of the hazard rate HR(t) becomes:

$$HR(t) = \frac{1}{N_s(t)}\frac{dN_f(t)}{dt} \tag{14.14}$$

Substituting Eq. (14.7) and Eq. (14.11) into Eq. (14.14) and rearranging terms yields;

$$HR(t) = \frac{f(t)}{R(t)} = \lambda(t) \tag{14.15}$$

Integrating Eq. (14.15) yields the cumulative hazard function $\boldsymbol{H}(t)$ as:

$$\boldsymbol{H}(t) = \int_{t_1}^{t_2} \lambda(t)\, dt \tag{14.16}$$

If the failure rate is a constant (λ_0) over the time interval from t_1 to t_2, then Eq. (14.16) can be written as:

$$\boldsymbol{H} = \lambda_0\,(t_2 - t_1) \tag{14.17}$$

where $\boldsymbol{H}$ is the number of failures that are expected to occur over the time period $\Delta t = t_2 - t_1$.

14.2.5 Conditional Probability for Reliability

The probability of an event B occurring, given that event A has already occurred is determined by dividing the probability that both event A and event B will occur by the probability that event A has already occurred. This concept is illustrated in the diagram shown below:

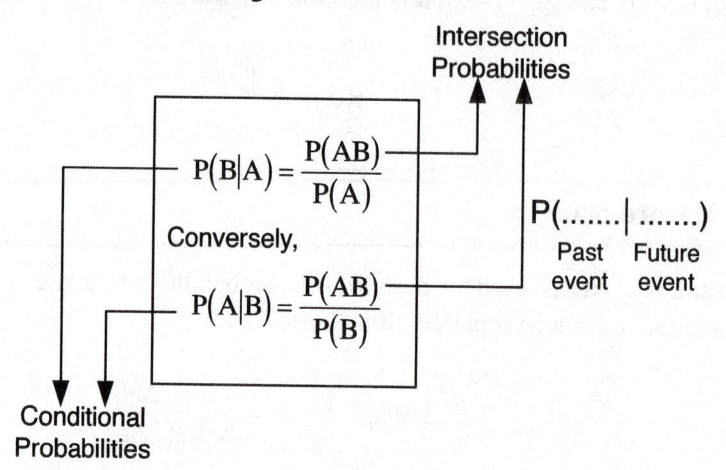

The conditional reliability is used to estimate the probability for successful operation of a system for a time period Δt after it has been operating without failure for some time t_1. The conditional reliability is defined by the expression:

$$R(t_1+\Delta t, t_1) = \frac{R(t_1+\Delta t)}{R(t_1)} \qquad (14.18)$$

The conditional reliability can be employed to predict if a system will operate successfully for some period of time Δt if it has operated successfully for t_1 hours. If the system exhibits a constant rate of failure, it can be shown that the conditional probability is independent of t_1. However, if the failure rate is increasing with respect to time, the conditional probability for success will degrade as t_1 increases.

14.2.6 MTBF, FIT and MTTF

Another measure of component failure rate is the mean time to failures (MTTF), which is given by:

$$MTTF = \int_0^\infty t\, f(t)dt \qquad (14.19a)$$

From Eq. (14.10) and Eq. (14.19a) it can be shown that the expression for MTTF can be written as:

$$MTTF = \int_0^\infty R(t)dt \qquad (14.19b)$$

If the failure density function exhibits an exponential distribution **then** the failure rate is a constant λ_0, and we can write[1]:

[1] For other distributions such as the Weibull, the failure rate may not be constant.

$$f(t) = \lambda_0 e^{-(\lambda_0 t)} \tag{14.20}$$

From Eq. (14.11) we can show that:

$$R(t) = e^{-(\lambda_0 t)} \tag{14.21}$$

Then from Eq. (14.19) and Eq. (14.21) it is evident that:

$$MTTF = \int_0^{\infty} R(t)dt = \int_0^{\infty} e^{-\lambda_0 t}dt = \frac{1}{\lambda_0} \tag{14.22}$$

The value of the MTTF is often calculated by dividing the total operating time of the units tested by the total number of failures encountered. A related measure often used by semiconductor manufactures is the "failure in time" or FIT. One FIT corresponds to:

- Failure of 1 component from a population of 1,000,000,000 operating for 1 hour.
- Failure of 1ppm in 1,000 hours of operation.
- Failure of 1 component from a population of 100,000 operating for 10,000 hours.

A similar measure, MTTF is the mean time for failure and is used to indicate the expected time for the first failure. It is used for components that cannot be repaired. MTBF, the mean time between failures, is used to indicate the time between repairs for those devices that can be repaired.

Although the mean life function denoted as either MTTF or MTBF is useful it does not provide a complete reliability metric. Instead, a reliability value with an associated time, along with an associated confidence requirement should be specified. Graphs showing reliability as a function of time based on statistical distribution functions are more suitable metrics for describing a product's reliability because they reflects the importance of time and provide a means for establishing confidence limits in characterizing the reliability of a system.

This MTTF is valid only when the data is exponentially distributed, which assumes that the failure rate is constant. One common problem that occurs when using MTTF is the difference between the mean and the median values from a set of data. The mean is the average, or the most probable value to be expected in a group of data. The median is the value that divides the data into two halves. One half of the data will be greater than the median and the other half will be less than the median. If the data in question is from a symmetrical distribution, such as the normal or Gaussian distribution, the values for the mean and median are equal. However, if the data is from an asymmetrical distribution, such as the exponential or Weibull distributions, large differences between the mean and median can occur.

14.3 ACCELERATED TESTING

14.3.1 Burn-in Testing and Infant Mortality

The bathtub curve, presented in Figure 14.2, describes the relative failure rate of an entire population of semiconductors over time. A small number of devices will fail soon after the test begins (infant mortality failures), while most operate successfully until they wear-out. A few devices will fail during the relatively long period called normal life due to random causes. Infant mortality failures are due to material defects, design and assembly errors, processing problems, etc. Wear-out failures are due to corrosion of die metallization, electro-migration, gate oxide breakdown, etc.

The bathtub curve is used to illustrate the three periods of semiconductor failure and it is not scaled with values on its axes to quantitatively describe of the expected failure rate for a specific device. Obtaining sufficient information to model a population of semiconductors with a properly scaled bathtub curve is a difficult an time consuming task. There are several reasons for the difficulty and the cost of characterizing the failure rate of a population over time.

1. The actual time periods for these three characteristic failure distributions often vary greatly.
2. Infant mortality is not the number of failures over a specific time period. Instead, infant mortality is the time over which the failure rate of a product is decreasing, and may last for years.
3. Failures due to wear-out may occur early in the life of a device or much later. It is a period when the failure rate is increasing.

The rate of failure of a device is important over its entire life cycle. A significant number of failures due to infant mortality are undesirable because they occur soon after the product is released, causing early customer dissatisfaction and significant warranty costs. Failures due to ransom causes during normal product life occur at a relatively constant and low rate. However, these failures may incur warranty costs and require service support. Accordingly, the bottom of the bathtub mortality curve should be as low as possible. Finally, the expected life of the product should be less that the time when wear-out failures begin to occur.

To avoid a significant number of product failures due to infant mortality, manufacturers employ the best design practices and perform stress testing early in development to evaluate design weaknesses and discover assembly and materials problems. For many semiconductors testing to determine the rate of failure associated with infant mortality is difficult because the time period involved is prohibitively long. For example the results of testing a large number of semiconductors presented in Fig. 14.4, show that the infant mortality period is six years with nearly 3% of the components failing during this period. Clearly, it is imperative to eliminate these components from the population prior to assembly of the circuit board. However, a burn in test at normal operating conditions would require too much time.

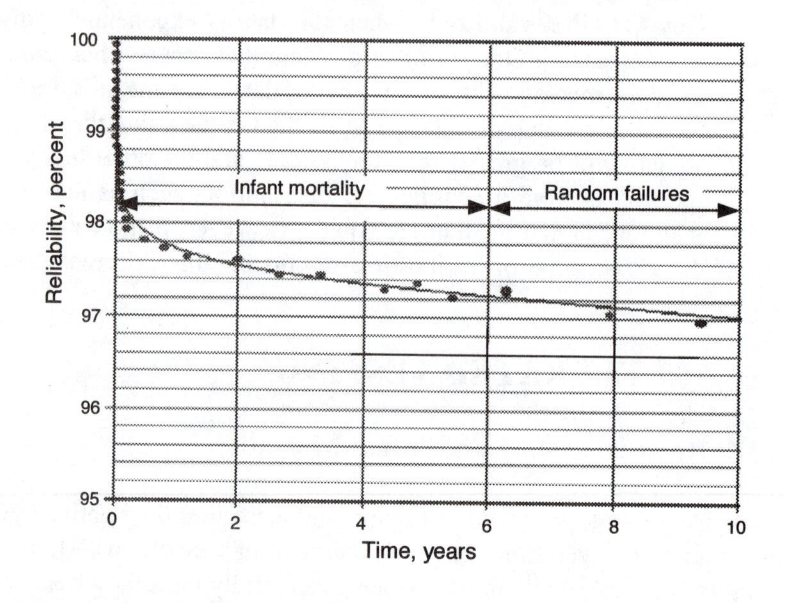

Fig. 14.4 Percent reliability as a function of time showing an extended infant mortality period. After Dennis J. Wilkins.

Semiconductor manufacturers employ two stresses to accelerate the burn-in process—temperature and voltage. Increased test temperature, relative to normal the operating temperature provides a reduction of 10 to 30 in the required for testing. Increasing the applied voltages, relative to normal operating voltages, provides even higher acceleration factors on many types of integrated circuits. Combined acceleration factors in the range of one thousand to one are common for many burn-in processes. Using acceleration methods reduce burn-in times to about ten hours, which is equivalent to operating times of about five years. The burn-in process significantly reduces the number of components with infant mortality defects that may fail in service.

14.3.2 Accelerated Testing to Determine Failure Rate.

One of the important parameters in ascertaining a product's reliability requires the determination of a component's failure rate. The usual method of determining a component's failure rate is by using accelerated high temperature life tests performed on a sample of components randomly selected from its population. The failure rate obtained from these life tests is then used as one of the metrics in predicting the failure rate in an actual application. Although there are many other techniques used by semiconductor manufacturers to characterize their component's reliability, the data from life testing samples is the most common method used for estimating the failure rate of a semiconductor devices in field service.

The correlation of accelerated test performance-to-field life under accelerated and field environmental-stress or operational-stress magnitudes respectively, is often determined by an acceleration factor. The acceleration factor may only be determined on a failure mechanisms specific basis. This functional relation of the failure rate with **steady state** temperature is often described with a thermal acceleration factor defined as:

$$A(T) = \frac{\lambda(T)}{\lambda(T_0)} \tag{14.23}$$

where T_0 is a reference temperature often taken as 55 °C.

Failure rates reported by semiconductor manufactures are usually based on a reference temperature of 55 °C and a confidence limit of 60%. The effect of steady state temperature is usually modeled with an Arrhenius law that characterizes each failure mechanism with activation energy and a temperature according to:

$$A(T) = \frac{t_1}{t_2} = e^{\left[\frac{E_a}{k}\left(\frac{1}{T_1} - \frac{1}{T_2}\right)\right]} \tag{14.24}$$

where A(T) is the acceleration factor, t_1 and t_2 correspond to time of failure at absolute temperatures T_1 and T_2 , respectively. E_a is the activation energy and $k = 8.63 \times 10^{-5}$ eV/°K is Boltzmann's constant. **Equation (14.24) is applicable to steady state temperature dependent mechanisms with an Arrhenius dependence on temperature.**

When T_1 is greater than T_2, the ratio of t_1/t_2 and the acceleration factor are decreasing. This fact implies that the time to failure at the higher temperature T_1 is less than the time to failure at a lower temperature T_2. Using this approach the time to failure t_f is often written as:

$$t_f = A_0 e^{\left(\frac{E_a}{kT}\right)} \tag{14.25}$$

Taking the logarithm of both sides of Eq. (14.25) yields:

$$\log_e(t_f) = \log_e(A_0) + \left(\frac{E_a}{kT}\right) \tag{14.26}$$

Equation (14.26) shows a linear relation for the time to failure as a function of the inverse of the absolute junction temperature when the results are plotted on a log-log graph as shown in Fig. 14.5.

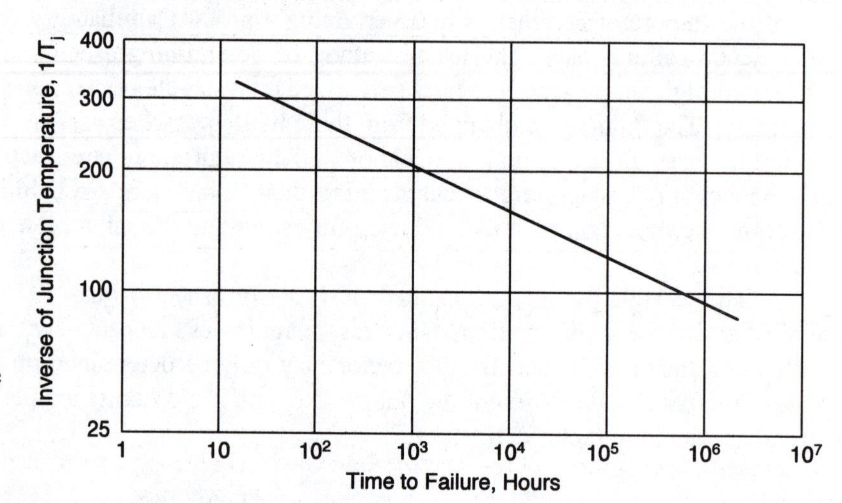

Fig. 14.5 Graph showing time to failure as a function of the inverse junction temperature.

It is clear from Eq. (14.25) and the results from Fig. 14.5 that the time to failure of a component can be maximized by minimizing the junction temperature.

The thermal activation energy E_a corresponding to a particular **steady state dependent** failure mechanism is determined by performing life tests at a minimum of two different temperatures. The life tests provide the time to failure t_{f1} and t_{f2} for the two operating conditions. This data is substituted into Eq. (14.26) to obtain:

$$\log_e(t_{f1}) = \log_e(A_0) + \left(\frac{E_a}{kT_1}\right)$$

$$\log_e(t_{f2}) = \log_e(A_0) + \left(\frac{E_a}{kT_2}\right) \tag{a}$$

Subtract the second of these equations from the first and solve for the activation energy to obtain:

$$E_a = \left(k \frac{\log_e t_{f1} - \log_e t_{f2}}{\left(\frac{1}{T_1} - \frac{1}{T_2}\right)} \right) \tag{14.27}$$

The activation energy E_s depends upon the failure mechanism and their cause. If a failure is not recorded during the life test of a sample, the activation energy is taken as 1.0 eV. If the failure mechanism is not known, an activation energy of 0.7 eV is assumed. The activation energies associated with several different failure mechanisms are presented in Table 14.1.

Table 14.1
Activation energies associated with different failure mechanisms.

Failure Mechanism	Activation Energy, eV	Failure Mechanism	Activation Energy, eV
Oxide Defects	0.3 – 0.5	Corrosion	0.45
Silicon Defects	0.3 – 0.5	Contamination	1.0
Assembly Defects	0.5 – 0.7	Electro Migration Al Circuit Trace	0.6
Mask Defects	0.7	Electro Migration Contact of Via	0.9
Charge Injection	1.3		
Hot Electrons	-0.06 eV		

14.4 RELIABILITY MODELS

14.4.1 Component Reliability

To illustrate component reliability consider a very simple circuit containing N_0 parallel components all exhibiting the same failure rate λ. In this case the component reliability R(t) is identical to the probability of survival of the individual components given by:

$$R(t) = \frac{N_s(t)}{N_0} \tag{14.4}$$

Note, that the probability of failure F(t) is given by:

$$F(t) = \frac{N_f(t)}{N_0} \tag{14.5}$$

From Eq. (14.4) and Eq. (14.5) it is clear that:

$$R(t) + F(t) = 1 \tag{14.6}$$

and

$$R(t) = 1 - \frac{N_f(t)}{N_0} \tag{14.28}$$

Differentiating Eq (14.28) with respect to time gives:

$$\frac{dR(t)}{dt} = -\frac{1}{N_0}\frac{dN_f(t)}{dt} \tag{14.29}$$

Equation (14.29) shows that that the rate of change of the probability of component survival is related to the rate of component failure. Rearranging Eq (14.29) and dividing both sides by $N_s(t)$ gives:

$$\frac{N_0}{N_s(t)}\frac{R(t)}{dt} = -\frac{1}{N_s(t)}\frac{dN_f(t)}{dt} \tag{14.30}$$

From Eq (14.2) the instantaneous value component failure rate $\lambda(t)$ can be written as:

$$\lambda(t) = \frac{1}{N_s(t)}\frac{dN_f(t)}{dt} = -\frac{N_0}{N_s(t)}\frac{R(t)}{dt} = -\frac{1}{R(t)}\frac{R(t)}{dt} \tag{14.31}$$

Note that in writing Eq. (14.31), we set $N_s = N_0$, which implies that the components were replaced upon failure in the sample tested to determine failure rate.

From by Eqs (14.11), (14.30) and (14.31) it can be shown that:

$$-\lambda(t)dt = \frac{dR(t)}{R(t)} \tag{14.32}$$

Integrating Eq (14.32) yields:

$$\log_e(R(t)) = -\int \lambda(t)dt \tag{14.33}$$

or

$$R(t) = e^{-\int \lambda(t)dt} \tag{14.34}$$

It is clear from Eq. (14.15) that the hazard rate and the failure rate are identical except for the manner in which the time period is specified. With the hazard rate the time period is specified as Δt beginning after a prescribed operational time t_1. However, the failure rate is more general because the time can be any value from zero to infinity.

For a constant failure rate with $\lambda(t) = \lambda_0$, consistent with Fig. 14.2, Eq. (14.34) reduces to:

$$R(t) = e^{-\lambda_0 t} \tag{14.35}$$

Failure rates for single components are usually quite low, of the order of 10^{-8} or 10^{-9} /hour. The rate depends on the scale of integration of the components considered and the manufacturing processes used in producing the component. This fact indicates that component reliability is usually very high. For example, Eq (14.35) predicts that a component with $\lambda_0 = 10^{-9}$ failures/hour can operate continuously for ten years with $R(87{,}600) = 0.9999124$.

14.4.2 System Reliability

Series Connected Components

The reliability of a system depends on the reliability of each component, the number of components and whether they are arranged in series or parallel. Consider first a series arrangement of four components as shown in Fig. 14.6, and note that the system fails in a series arrangement if anyone of the components fails. It is clear, that the reliability of the system R^s is the same as the probability of all four components surviving for the specified time period. Thus,

$$R^s = R_1 R_2 R_3 R_4 \tag{14.36}$$

where the superscript s refers to a system with four components in a series arrangement.

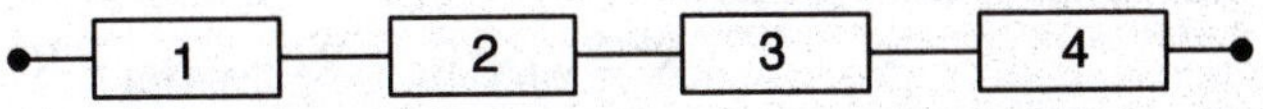

Fig. 14.6 Four components connected in a series arrangement.

Substitution of Eq. (14.35) into Eq. (14.36) gives:

$$R^s(t) = e^{-\lambda_1 t} e^{-\lambda_2 t} e^{-\lambda_3 t} e^{-\lambda_4 t} \tag{14.37}$$

When λ(t) is a constant with respect to time for each of the four components, Eq. (14.37) reduces to:

$$R^s(t) = e^{-(\lambda_1+\lambda_2+\lambda_3+\lambda_4)t} \tag{14.38}$$

If the number of components in the series arrangement is increased to N, then it is clear that:

$$R^s(t) = e^{-\left[\sum_{i=1}^{N} \lambda_i\right] t} \tag{14.39}$$

The effect of a large series circuit is to degrade reliability. For example, a series circuit with N components all exhibiting the reliability R is degraded in accordance with the number of components as indicated in Table 14.2. A system with 10^4 components in series requires component reliabilities in excess of 0.99999 to insure a system reliability of 90 percent.

Table 14.2
System reliability R^s as a function of component reliability R and the number of components connected in series. $R^s = (R)^N$

	R					
N	0.900000	0.990000	0.999000	0.999900	0.999990	0.999999
10^0	0.9000	0.9900	0.9990	0.9999	0.99999	0.999999
10^1	0.3487	0.9044	0.9900	0.9990	0.9999	0.99999
10^2	2.6×10^{-5}	0.3660	0.9048	0.9900	0.9990	0.9999
10^3	*	4.3×10^{-5}	0.3679	0.9048	0.9900	0.9990
10^4	*	*	4.5×10^{-5}	0.3679	0.9048	0.9900
10^5	*	*	*	4.5×10^{-5}	0.3679	0.9048
10^6	*	*	*	*	4.5×10^{-5}	0.3679

* Negligible

Parallel Connected Components

Consider next a parallel arrangement of three components defining the system illustrated in Fig. 14.7. Failure of a parallel system requires failure of all of the components; thus,

$$F^s = F_{f1}\, F_{f2}\, F_{f3} \tag{14.40}$$

Substituting Eq (14.6) into Eq (14.40) gives:

$$1 - R^s = (1 - R_1)(1 - R_2)(1 - R_3)$$

this reduces to:

$$R^s = R_1 + R_2 + R_3 - R_1R_2 - R_2R_3 - R_1R_3 + R_1R_2R_3 \qquad (14.41)$$

Evaluation of Eq (14.41) shows that the parallel arrangement of three components enhances system reliability. The extent of the improvement in system reliability R^s as a function of component or sub-system reliability R is shown in Fig. 14.8 for N = 1, 2 and 3.

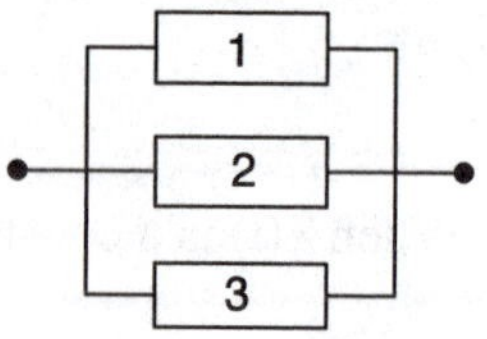

Fig. 14.7 Three components connected in a parallel arrangement.

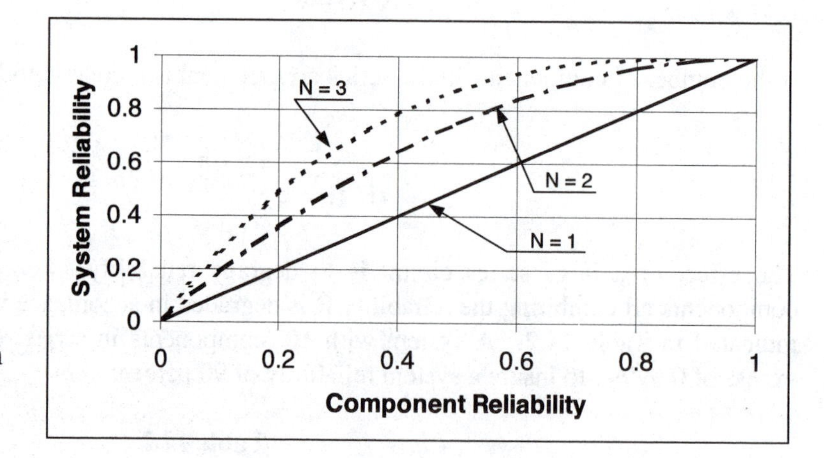

Fig. 14.8 Improvement in reliability by parallel connection of components.

The enhanced reliability of a parallel circuit is a distinct advantage; however, the corresponding disadvantage is the added cost of the components and added power, weight and size of the system. In the design of extremely reliable systems, parallel circuits, which provide redundancy by a factor of two or three, are often used. However, to minimize the added costs of the redundancy parallel circuits are only used with sub-systems that exhibit lower reliabilities. This practice leads to systems which incorporate combinations of both parallel and series circuits as shown in Fig. 14.9.

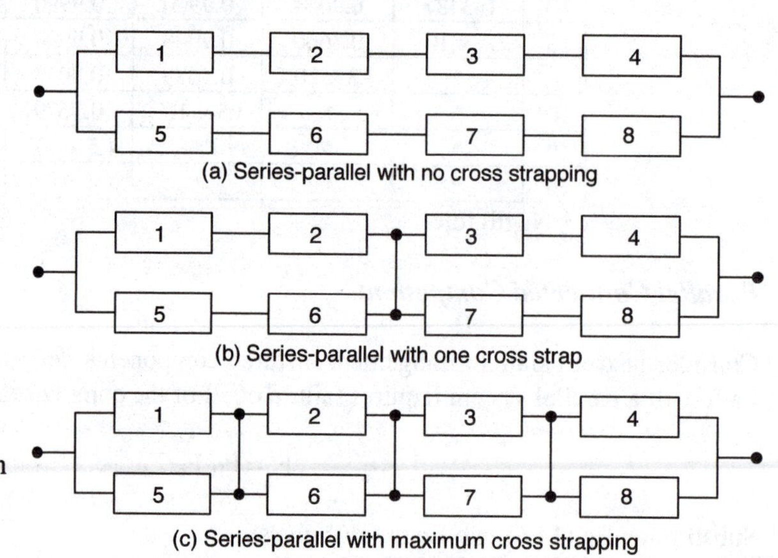

Fig. 14.9 Series parallel circuits with different degrees of cross strapping.

Series-Parallel Connected Components

Examine the series-parallel arrangement shown in Fig. 14.9a and assume the reliability of each component or sub-system is the same value R. Then the probability of survival of the series line of N components may be written from Eq (14.36) as:

$$R_1 = R^N \tag{14.42}$$

From Eq (14.6) and (14.42), it is clear that the probability of survival R^s of a series parallel system with N series components and M parallel lines is given by:

$$1 - R^s = (1 - R^N)^M \tag{14.43}$$

In this example, (Fig. 14.9a), N = 4 and M = 2 and Eq (14.43) reduces to:

$$R^s = R^4\,(2 - R^4) \tag{14.44}$$

Taking R = 0.9, Eq (14.44) gives $R^s = 0.8817$. It is evident that the two parallel lines in the system largely compensate for the four components in series and the overall system reliability is nearly the same as the component reliability.

Next, consider a series parallel arrangement with a single cross strap as illustrated in Fig. 14.9b. In this case Eq (14.43) applies to both the left and the right sides of the symmetric circuit; hence, we can write:

$$R_L = R_R = 1 - (1 - R^N)^M \tag{14.45}$$

Note, that the left and the right side of the circuit are series connected so that the system reliability is:

$$R^s = R_L\,R_R = [1 - (1 - R^N)^M]^Q \tag{14.46}$$

For the diagram shown in Fig. 14.9b, N = M = Q = 2 and Eq (14.46) reduces to:

$$R^s = R^4\,(4 - 4R^2 + R^4) \tag{14.47}$$

Again taking $P_s = 0.9$, Eq (14.47) gives $R^s = 0.9291$. This example indicates that the addition of a single cross strap has increased the reliability of the system from 0.8817 to 0.9291.

Finally, consider the case where the maximum number of cross straps is employed in the system configuration shown in Fig. 14.9c. With N = 1, M = 2 and Q = 4, Eq. (14.46) reduces to:

$$P_s^{\,s} = P_s^{\,4}\,(2 - P_s)^4 \tag{14.48}$$

Continue the study of the effects of cross strapping by taking R = 0.9, and note that Eq (14.48) gives $R^s = 0.9606$. This example shows that the maximum reliability is achieved when the cross strapping is used to the maximum extent possible.

These simple concepts in probability theory show the effects of series and parallel circuits on system reliability. It is evident that the series connections degrade reliability and that the parallel connections enhance reliability. Parallel or redundant circuits add to costs of the product, but in those cases where system availability is extremely critical, such as on-line banking transactions and airline reservations, the cost and inconvenience of a computer malfunction is so significant that redundancy in sub-systems prone to failure is certainly warranted.

Complex System Connections

In some designs, the connection arrangement is complex and it cannot be divided into combinations of series and parallel connections. An example of such a complex system is illustrated in Fig. 14.10 where the strap connecting the upper and lower lines contains a component that may fail. Analysis of complex circuits is not difficult, but it is tedious. In this example, there are $2^4 = 16$ possible component states (operational or failed) to consider in ascertaining if the system will remain operational in the event of failure of one or more components. These 16 different component states are presented in Table 14.3 under status headings.

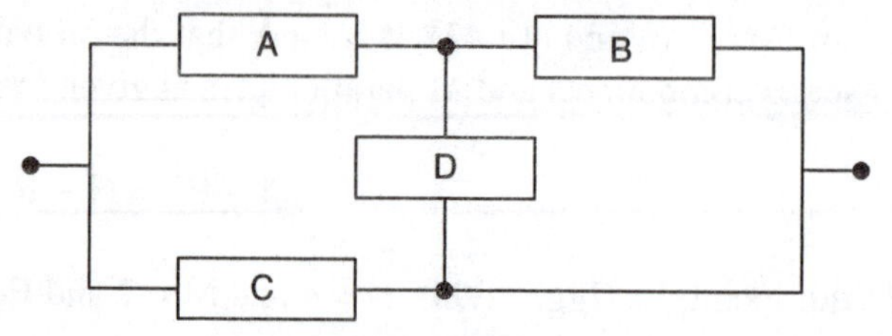

Fig. 14.10 A complex connection arrangement.

The system reliability is determined by considering the probability of each of the four components for the 11 different states that result in successful operation of the system. The expression for the probability of successful operation of the complex system shown in Fig. 14.10 is written by considering the operational states in rows 1-7, 9, 10, 13, 14, of Table 14.3. Accordingly, we may write the expression for R^s as the sum of 11 terms given by:

$$R^s = R_A R_B R_C R_D + R_A R_B R_C F_D + R_A R_B F_C R_D + R_A R_B F_C F_D + R_A F_B R_C R_D + R_A F_B R_C F_D$$

$$+ R_A F_B F_C R_D + F_A R_B R_C R_D + F_A R_B R_C F_D + F_A F_B R_C R_D + F_A F_B R_C F_D \qquad (14.49)$$

where R_A, R_B, R_C and R_D are the probabilities of successful operation of components A, B, C and D. F_A, F_B, F_C and F_D are the probabilities of failure of components A, B, C and D.

Table 14.3
Component status for the system shown in Fig. 14.10

	Component A Status	Component B Status	Component C Status	Component D Status	System Status
1	Operating	Operating	Operating	Operating	Operating
2	Operating	Operating	Operating	Failure	Operating
3	Operating	Operating	Failure	Operating	Operating
4	Operating	Operating	Failure	Failure	Operating
5	Operating	Failure	Operating	Operating	Operating
6	Operating	Failure	Operating	Failure	Operating
7	Operating	Failure	Failure	Operating	Operating
8	Operating	Failure	Failure	Failure	Failure
9	Failure	Operating	Operating	Operating	Operating
10	Failure	Operating	Operating	Failure	Operating
11	Failure	Operating	Failure	Operating	Failure
12	Failure	Operating	Failure	Failure	Failure
13	Failure	Failure	Operating	Operating	Operating
14	Failure	Failure	Operating	Failure	Operating
15	Failure	Failure	Failure	Operating	Failure
16	Failure	Failure	Failure	Failure	Failure

Recall from Eq. (14.6) that F = 1 – R and substitute this relation for F into Eq. (14.49) and simplify to obtain:

$$R^s = R_AR_BR_CR_D + R_AR_BR_C(1\text{-}R_D) + R_AR_B(1\text{-}R_C)R_D + R_AR_B(1\text{-}R_C)(1\text{-}R_D) + R_A(1\text{-}R_B)R_CR_D$$

$$+ R_A(1\text{-}R_B)R_C(1\text{-}R_D) + R_A(1\text{-}R_B)(1\text{-}R_C)R_D + (1\text{-}R_A)R_BR_CR_D + (1\text{-}R_A)R_BR_C(1\text{-}R_D)$$

$$+ (1\text{-}R_A)(1\text{-}R_B)R_CR_D + (1\text{-}R_A)(1\text{-}R_B)R_C(1\text{-}R_D) \quad (14.50a)$$

$$R^s = R_C + R_AR_B + R_AR_D - R_AR_BR_C - R_AR_BR_D - R_AR_CR_D + R_AR_BR_CR_D \quad (14.50b)$$

If we let R = 0.9 and substitute this value into Eq. (14.50) we obtain $R^s = 0.9891$. This calculation shows that high reliability can be achieved with complex connections.

Standby Systems

Standby systems employ parallel architecture with two or more independent circuits; however, only one circuit is on line and operational. The remaining circuits are dormant. The operation of the active circuit is monitored continuously and in the event of a malfunction another circuit is activated immediately. The time to detect a circuit malfunction and activate another circuit is short enough to be considered negligible.

The probability of success of a standby circuit depends on the degree of redundancy. Let's consider a system with two parallel circuits similar to that shown in Fig. 14.7. The probability F_1 that the first of the two circuits will fail after operating successfully for time t_1 is given by Eq. (14.8) as:

$$F_1(t_1) = \int_0^{t_1} f(t)dt \quad (14.51)$$

The probability F_2 that the second and final circuit will fail at time t_2 is also given by Eq. (14.8) as:

$$F_2(t_2 - t_1) = \int_0^{(t_2-t_1)} f(t)dt \quad (14.52)$$

The probability of a system failure with both circuits failing sequentially after time t_2 is given by the product shown below:

$$F^s(t_2) = F_1(t_1)F_2(t_2 - t_1) = \int_0^{t_1} f(t)dt \int_0^{(t_2-t_1)} f(t)dt \quad (14.53)$$

The reliability $R^s(t_2)$ of the same system is given by Eq. (14.6) as:

$$R^s(t_2) = 1 - F^s(t_2) = \int_{t_1}^{\infty} f(t)dt \int_{(t_2-t_1)}^{\infty} f(t)dt \quad (14.54)$$

Let's consider a simple example to demonstrate the effectiveness of standby systems with only two degrees of redundancy. Let's assume that the failure density function f(t) = C a constant. Then using Eq. (14.53) e can write:

$$F^s(t_2) = \int_0^{t_1} Cdt \int_0^{(t_2-t_1)} Cdt \quad (14.55)$$

Integrating Eq. (14.55) yields:

$$F^s(t_2) = C^2\left(t_1 t_2 - t_1^2\right) \tag{a}$$

Next assume that $t_2 = 2t_1$ indicating that the time of operation of both of the circuits before failure is identical. Substituting this relation into Eq. (a) gives:

$$F^s(t_2) = C^2 t_1^2 \tag{b}$$

Consider the circuits with $C = 10^{-5}$/hour, which implies that the first of these circuits will fail with a probability of 1 after 10^5 hours (in excess of 114 years). However, there is a 5% probability that this circuit will fail after 5,000 hours (less than 7 months) as shown in Fig. 14.11.

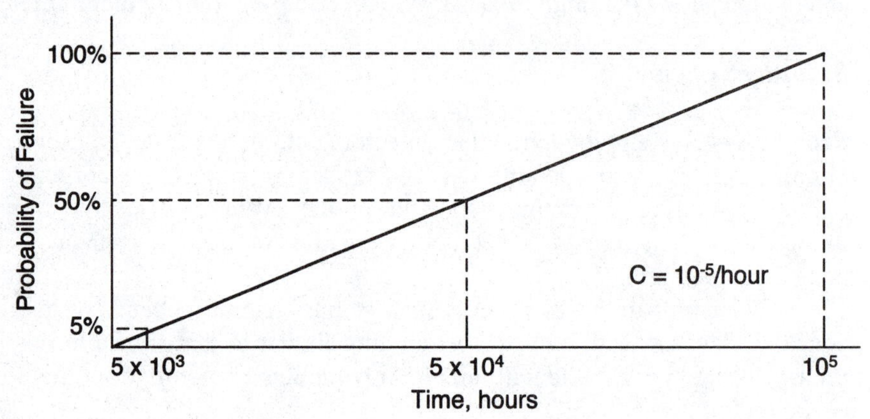

Fig. 14.11 Probability of failure of a circuit with $C = 10^{-5}$ as a function of time.

Let's use Eq. (b) to determine the probability of failure of the system with standby control and two degrees of redundancy if the system is to operate for 10,000 hours (about 14 months). We set t_2 = 10,000 and note that t_1 = 5,000 hours. Then from Eq. (b) we obtain:

$$F^s(t_2) = C^2 t_1^2 = \left(10^{-5}\right)^2 (5{,}000)^2 = 25 \times 10^6 \times 10^{-10} = 0.0025 = 0.25\% \tag{c}$$

From these results it is evident that adding a second circuit and operating in the standby mode decreased the probability of failure from 5% to 0.25% while doubling the operating time from 5,000 hours to 10,000 hours

14.5 STATISTICAL METHODS

Experimental measurements of any quantity will exhibit some variation even if the measurements are repeated a number of times with precise instruments. This variability, which is fundamental to all measuring systems, is due to two different causes. First, the quantity being measured may exhibit significant variation. For example, in a test to determine the failure rate of semiconductors operating at a specified temperature, large differences in the number of hours to failure are noted when a large number of components are tested. This variation is inherent in the life of semiconductors and it is observed in all life tests. Second, the measuring system, which often includes a transducer, signal conditioning equipment, A/D converter, recording instrument, and operator may introduce error in the measurement. This error may be systematic or random, depending upon its source. An instrument operated out of calibration produces a systematic error, whereas, reading errors due to interpolation on a chart are random. The accumulation of random errors in a measuring system produces a variation that must be examined in relation to the magnitude of the quantity being measured.

The data obtained from repeated measurements represent an array of readings not an exact result. Maximum information can be extracted from such an array of readings by employing statistical methods. The first step in the statistical treatment of data is to establish the statistical distribution. A graphical representation of the distribution is usually the most useful form for initial evaluation. Next, the statistical distribution is characterized with a measure of its central value, such as the mean, median or mode. Finally, the spread or dispersion of the distribution is determined in terms of the variance or the standard deviation.

With elementary statistical methods, the experimentalist can reduce a large amount of data to a very compact and useful form by defining the type of distribution, establishing the single value that best represents the central value of the distribution (mean), and determining the variation from the mean value (standard deviation). Summarizing data in this manner is the most meaningful form of presentation for application to design problems or for communication to others who need the results of the experiments.

The treatment of statistical methods presented in this chapter is relatively brief; therefore, only the most commonly employed techniques for representing and interpreting data are presented. A formal course in statistics, which covers these techniques in much greater detail as well as many other useful techniques, should be included in the program of study of all engineering students.

14.6 CHARACTERIZING STATISTICAL DISTRIBUTIONS

For purposes of this discussion, consider that an experiment has been conducted N times to determine the ultimate tensile strength of a copper alloy used for lead fames for J-leaded chip carriers. The data obtained represent a sample of size N from an infinite population of all possible measurements that could have been made. The simplest way to present these data is to list the strength measurements in order of increasing magnitude, as shown in Table 14.4.

Table 14.4
The ultimate tensile strength of a copper alloy, listed in order of increasing magnitude

Sample number	Strength ksi (MPa)	Sample number	Strength ksi (MPa)
1	170.5 (1175)	21	176.2 (1215)
2	171.9 (1185)	22	176.2 (1215)
3	172.6 (1190)	23	176.4 (1217)
4	173.0 (1193)	24	176.6 (1218)
5	173.4 (1196)	25	176.7 (1219)
6	173.7 (1198)	26	176.9 (1220)
7	174.2 (1201)	27	176.9 (1220)
8	174.4 (1203)	28	177.2 (1222)
9	174.5 (1203)	29	177.3 (1223)
10	174.8 (1206)	30	177.4 (1223)
11	174.9 (1206)	31	177.7 (1226)
12	175.0 (1207)	32	177.8 (1226)
13	175.4 (1210)	33	178.0 (1228)
14	175.5 (1210)	34	178.1 (1228)
15	175.6 (1211)	35	178.3 (1230)
16	175.6 (1211)	36	178.4 (1230)
17	175.8 (1212)	37	179.0 (1236)
18	175.9 (1213)	38	179.7 (1239)
19	176.0 (1214)	39	180.1 (1242)
20	176.1 (1215)	40	181.6 (1252)

These data can be arranged into seven groups to give a frequency distribution as shown in Table 14.5. The advantage of representing data in a frequency distribution is that the central tendency is more clearly illustrated.

Table 14.5
Frequency distribution of ultimate tensile strength

Group intervals ksi (MPa)	Observations in the group	Relative frequency	Cumulative frequency
169.0-170.9 (1166-1178)	1	0.025	0.025
171.0-172.9 (1179-1192)	2	0.050	0.075
173.0-174.9 (1193-1206)	8	0.200	0.275
175.0-176.9 (1207-1220)	16	0.400	0.675
177.0-178.9 (1221-1234)	9	0.225	0.900
179.0-180.9 (1235-1248)	3	0.075	0.975
181.0-182.9 (1249-1261)	1	0.025	1.000
Total	40		

14.6.1 Graphical Representations of the Distribution

The shape of the distribution function representing the ultimate tensile strength of this copper alloy is indicated by the data groupings of Table 14.5. A graphical presentation of this group data, known as a **histogram**, is shown in Fig. 14.12. The histogram method of presentation shows the central tendency and variability of the distribution much more clearly than the tabular method of presentation of Table 14.5. Superimposed on the histogram is a curve showing the relative frequency of occurrence of a group of measurements. Note that the points for the relative frequency are plotted at the midpoint of the group interval.

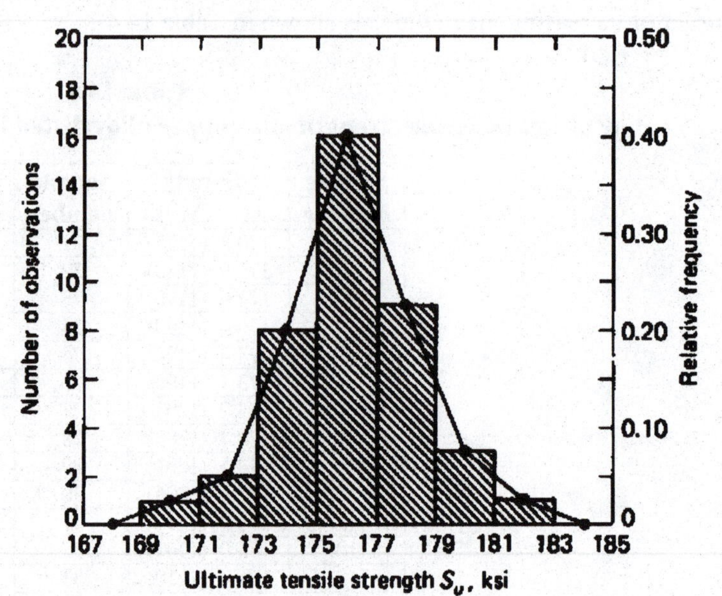

Figure 14.12 Histogram with superimposed relative-frequency diagram.

A cumulative frequency-diagram, shown in Fig. 14.13, is another way of representing the ultimate-strength data from the experiments. The cumulative frequency is the number of readings having a value less than a specified value of the quantity being measured (ultimate strength) divided by the total number of measurements. As indicated in Table 14.5, the cumulative frequency is the running sum of the relative frequencies. When the graph of cumulative frequency versus the quantity being measured is prepared, the end value for the group intervals is used to position the point along the abscissa.

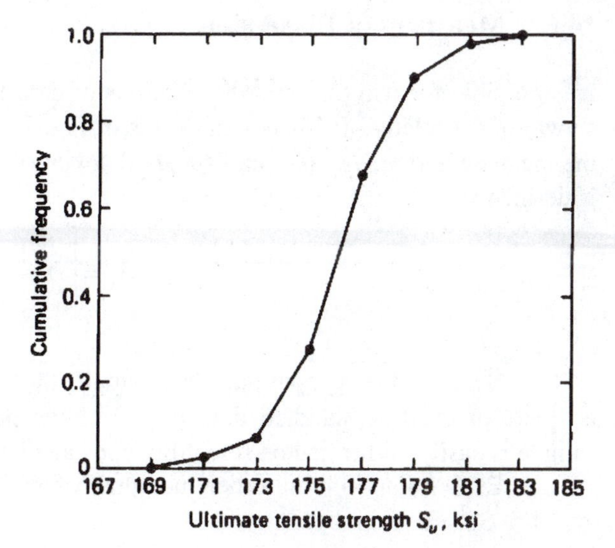

Figure 14.13 Cumulative frequency diagram.

14.6.2 Measures of Central Tendency

While histograms or frequency distributions are used to provide a visual representation of a distribution, numerical measures are used to define the characteristics of the distribution. One basic characteristic is the central tendency of the data. The most commonly employed measure of central tendency of a distribution of data is the sample mean $\overline{x}$, which is defined as:

$$\overline{x} = \sum_{i=1}^{N} \frac{x_i}{N} \tag{14.56}$$

where x_i is the i th value of the quantity being measured and N is the total number of measurements.

Because of time and costs involved in conducting tests, the number of measurements is usually limited; therefore, the sample mean $\overline{x}$ is only an estimate of the true arithmetic mean μ of the population. It is evident that $\overline{x}$ approaches μ as the number of measurements increases. The mean value of the ultimate-strength data presented in Table 14.4 is $\overline{x} = 176.1$ ksi (1,215 MPa).

The median and mode are also measures of central tendency. The median is the central value in a group of ordered data. For example, in an ordered set of 41 readings, the 21st reading represents the median value with 20 readings lower than the median and 20 readings higher than the median. In instances when an even number of readings are taken, the median is obtained by averaging the two middle values. For example, in an ordered set of 40 readings, the median is the average of the 20th and 21st readings. Thus, for the ultimate tensile strength data presented in Table 14.4, the median is ½ (176.1 + 176.2) = 176.15 ksi (1,215 MPa).

The mode is the most frequent value of the data; therefore, it is located at the peak of the relative-frequency curve. In Fig. 14.12, the peak of the relative probability curve occurs at an ultimate tensile strength S_u = 176.0 ksi (1,214 MPa); therefore, this value is the mode of the data set presented in Table 14.4.

It is clear that a typical set of data may give different values for the three measures of central tendency. There are two reasons for this difference. First, the population from which the sample was drawn may not be Gaussian where the three measures are expected to coincide. Second, even if the population is Gaussian, the number of measurements N is usually small and deviations due to a small sample size are to be expected.

14.6.3 Measures of Dispersion

It is possible for two different distributions of data to have the same mean but different dispersions, as shown in the relative-frequency diagrams of Fig. 14.14. Different measures of dispersion are the range, the mean deviation, variance and standard deviation. The standard deviation S_x is the most popular and is defined as:

$$S_x = \left[\sum_{i=1}^{N} \frac{(x_i - \bar{x})^2}{N-1} \right]^{1/2} \tag{14.57}$$

Because the sample size N is small, the standard deviation S_x of the sample represents an estimate of the true standard deviation σ of the population. Computation of S_x and $\bar{x}$ from a data sample is easily performed on scientific type calculators.

Expressions for the other measures of dispersion, namely, range $\mathcal{R}$, mean deviation d_x, and variance S_x^2 are given by:

$$\mathcal{R} = x_L - x_S \tag{14.58}$$

$$d_x = \sum_{i=1}^{N} \frac{|x_i - \bar{x}|}{N} \tag{14.59}$$

$$S_x^2 = \sum_{i=1}^{N} \frac{(x_i - \bar{x})^2}{N-1} \tag{14.60}$$

where x_L is the largest value of the quantity in the distribution and x_S is the smallest value.

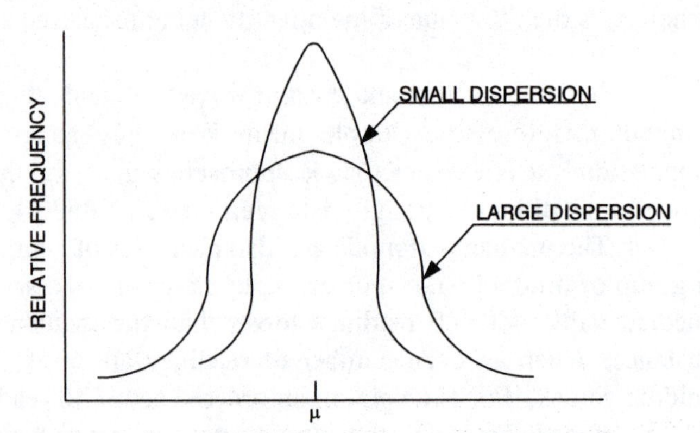

Figure 14.14 Relative frequency diagrams with large and small dispersions.

Equation (14.59) indicates that the deviation of each reading from the mean is determined and summed. The average of the N deviations is the mean deviation. The absolute value of the difference $(x_i - \bar{x})$ must be used in the summing process to avoid cancellation of positive and negative deviations. The variance of the population σ^2 is estimated by S_x^2 where the denominator (N – 1) in Eqs. (14.57) and (14.60) serves to reduce error introduced by approximating the true mean μ with the estimate of the mean $\bar{x}$. As the sample size N is increased the estimates of $\bar{x}$, S_x, and S_x^2 improve. Variance is an important measure of dispersion because it is used in defining the normal distribution function.

Finally, a measure known as the coefficient of variation C_v is used to express the standard deviation S_x as a percentage of the mean $\bar{x}$. Thus:

$$C_v = \frac{S_x}{\bar{x}}(100) \tag{14.61}$$

The coefficient of variation represents a normalized parameter that indicates the variability of the data in relation to its mean.

14.7 STATISTICAL DISTRIBUTION FUNCTIONS

As the sample size is increased, it is possible in tabulating the data to increase the number of group intervals and to decrease their width. The corresponding relative-frequency diagram, similar to the one illustrated in Fig. 14.12, will approach a smooth curve (a theoretical distribution curve) known as a **distribution function**.

A number of different distribution functions are used in statistical analyses. The best-known and most widely used distribution in experimental mechanics is the Gaussian or normal distribution. This distribution is extremely important because it describes random errors in measurements and variations observed in many experimental determinations. Other useful distributions include binomial, exponential, hypergeometric, chi-square χ^2, F, Gumbel, Poisson, Student's t, and Weibull distributions. The reader is referred to references [1-7] for a complete description of these distributions. Emphasis here will be on Gaussian and Weibull distribution functions because of their wide range of application in mechanical design of electronic systems.

14.7.1 Gaussian Distribution

The Gaussian or normal distribution function, as represented by a normalized relative-frequency diagram, is shown in Fig. 14.15. The Gaussian distribution is completely defined by two parameters; the mean μ and the standard deviation σ. The equation for the relative frequency f(z) in terms of these two parameters is given by:

$$f(z) = \frac{1}{\sqrt{2\pi}} e^{-(z^2/2)} \tag{14.62}$$

where

$$z = \frac{\bar{x} - \mu}{\sigma} \tag{14.63}$$

Experimental data (with finite sample sizes) can be analyzed to obtain $\bar{x}$ as an estimate of μ and S_x as an estimate of σ. This procedure permits the experimentalist to use data drawn from small samples to represent an entire population.

The method for predicting population properties from a Gaussian (normal) distribution function utilizes the normalized relative-frequency diagram shown in Fig. 14.15. The area A under the entire curve is given by Eq. (14.62) as:

$$A = \frac{1}{\sqrt{2\pi}} \int_{-\infty}^{\infty} e^{-(z^2/2)} dz = 1 \tag{14.64}$$

Equation (14.64) implies that the population has a value z between $-\infty$ and $+\infty$ and that the probability of making a single observation from the population with a value $-\infty \leq z \leq +\infty$ is 100%. While the previous statement may appear trivial and obvious, it serves to illustrate the concept of using the area under the normalized relative-frequency curve to determine the probability P of observing a measurement within a specific interval. Figure 14.16 shows graphically, with the shaded area under the curve, the probability that a measurement will occur within the interval between z_1 and z_2. Thus, from Eq. (14.62) it is evident that:

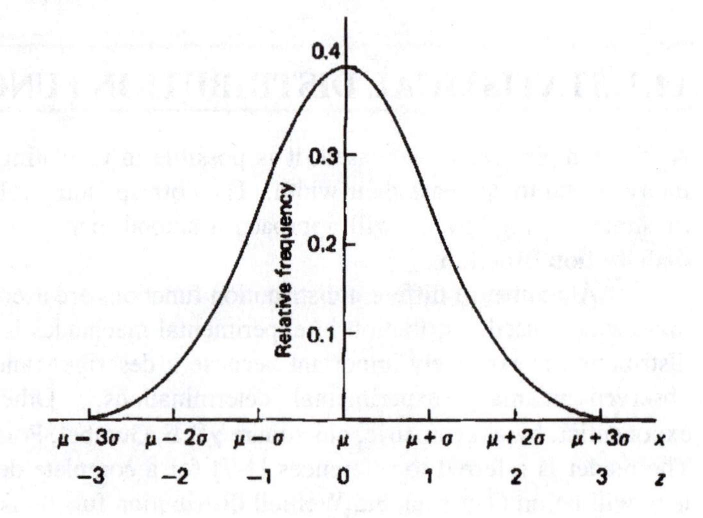

Figure 14.15 The normal or Gaussian distribution function.

$$P(z_1,z_2) = \int_{z_1}^{z_2} f(z)dz = \frac{1}{\sqrt{2\pi}} \int_{z_1}^{z_2} e^{-(z^2/2)}dz \tag{14.65}$$

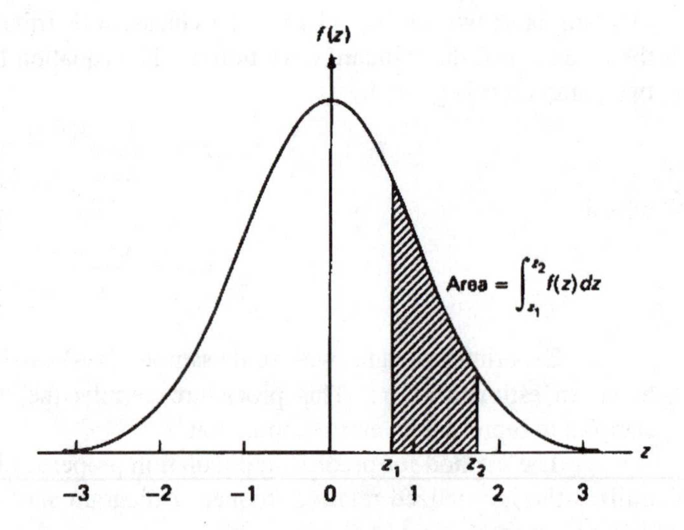

Figure 14.16 Probability of a measurement of x between limits of z_1 and z_2. The total area under the curve f(z) is 1.

Evaluation of Eq. (14.65) is most easily accomplished by using tables that list the areas under the normalized relative-frequency curve as a function of z. Table 14.6 lists one-side areas between limits of $z_1 = 0$ and z_2 for the normal distribution function.

Table 14.6
Areas under the normal distribution curve from $z_1 = 0$ to z_2 (one side)

z_2	0.00	0.01	0.02	0.03	0.04	0.05	0.06	0.07	0.08	0.09
0.0	0.0000	0.0040	0.0080	0.0120	0.0160	0.0199	0.0239	0.0279	0.0319	0.0359
0.1	0.0398	0.0438	0.0478	0.0517	0.0557	0.0596	0.0636	0.0675	0.0714	0.0753
0.2	0.0793	0.0832	0.0871	0.0910	0.0948	0.0987	0.1026	0.1064	0.1103	0.1141
0.3	0.1179	0.1217	0.1255	0.1293	0.1331	0.1368	0.1406	0.1443	0.1480	0.1517
0.4	0.1554	0.1591	0.1628	0.1664	0.1700	0.1736	0.1772	0.1808	0.1844	0.1879
0.5	0.1915	0.1950	0.1985	0.2019	0.2054	0.2088	0.2123	0.2157	0.2190	0.2224
0.6	0.2257	0.2291	0.2324	0.2357	0.2389	0.2422	0.2454	0.2486	0.2517	0.2549
0.7	0.2580	0.2611	0.2642	0.2673	0.2704	0.2734	0.2764	0.2794	0.2823	0.2852
0.8	0.2881	0.2910	0.2939	0.2967	0.2995	0.3023	0.3051	0.3078	0.3106	0.3233
0.9	0.3159	0.3186	0.3212	0.3238	0.3264	0.3289	0.3315	0.3340	0.3365	0.3389
1.0	0.3413	0.3438	0.3461	0.3485	0.3508	0.3531	0.3554	0.3577	0.3599	0.3621
1.1	0.3643	0.3665	0.3686	0.3708	0.3729	0.3749	0.3770	0.3790	0.3810	0.3830
1.2	0.3849	0.3869	0.3888	0.3907	0.3925	0.3944	0.3962	0.3980	0.3997	0.4015
1.3	0.4032	0.4049	0.4066	0.4082	0.4099	0.4115	0.4131	0.4147	0.4162	0.4177
1.4	0.4192	0.4207	0.4222	0.4236	0.4251	0.4265	0.4279	0.4292	0.4306	0.4319
1.5	0.4332	0.4345	0.4357	0.4370	0.4382	0.4394	0.4406	0.4418	0.4429	0.4441
1.6	0.4452	0.4463	0.4474	0.4484	0.4495	0.4505	0.4515	0.4525	0.4535	0.4545
1.7	0.4554	0.4564	0.4573	0.4582	0.4591	0.4599	0.4608	0.4616	0.4625	0.4633
1.8	0.4641	0.4649	0.4656	0.4664	0.4671	0.4678	0.4686	0.4693	0.4699	0.4706
1.9	0.4713	0.4719	0.4726	0.4732	0.4738	0.4744	0.4750	0.4758	0.4761	0.4767
2.0	0.4772	0.4778	0.4783	0.4788	0.4793	0.4799	0.4803	0.4808	0.4812	0.4817
2.1	0.4821	0.4826	0.4830	0.4834	0.4838	0.4842	0.4846	0.4850	0.4854	0.4857
2.2	0.4861	0.4864	0.4868	0.4871	0.4875	0.4878	0.4881	0.4884	0.4887	0.4890
2.3	0.4893	0.4896	0.4898	0.4901	0.4904	0.4906	0.4909	0.4911	0.4913	0.4916
2.4	0.4918	0.4920	0.4922	0.4925	0.4927	0.4929	0.4931	0.4932	0.4934	0.4936
2.5	0.4938	0.4940	0.4941	0.4943	0.4945	0.4946	0.4948	0.4949	0.4951	0.4952
2.6	0.4953	0.4955	0.4956	0.4957	0.4959	0.4960	0.4961	0.4962	0.4963	0.4964
2.7	0.4965	0.4966	0.4967	0.4968	0.4969	0.4970	0.4971	0.4972	0.4973	0.4974
2.8	0.4974	0.4975	0.4976	0.4977	0.4977	0.4978	1.4979	0.4979	0.4980	0.4981
2.9	0.4981	0.4982	0.4982	0.4983	0.4984	0.4984	0.4985	0.4985	0.4986	0.4986
3.0	0.49865	0.4987	0.4987	0.4988	0.4988	0.4988	0.4989	0.4989	0.4989	0.4990
z_2	0.00	0.01	0.02	0.03	0.04	0.05	0.06	0.07	0.08	0.09

Because the distribution function is symmetric about z = 0, this one-sided table is sufficient for all evaluations of the probability. For example, A(−1,0) = A(0,+1) leads to the following determinations:

$$A(-1,+1) = P(-1,+1) = 0.3413 + 0.3413 = 0.6826$$
$$A(-2,+2) = P(-2,+2) = 0.4772 + 0.4772 = 0.9544$$
$$A(-3,+3) = P(-3,+3) = 0.49865 + 0.49865 = 0.9973$$
$$A(-1,+2) = P(-1,+2) = 0.3413 + 0.4772 = 0.8185$$

Because the normal distribution function has been well characterized, predictions can be made regarding the probability of a specific strength value or measurement error. For example, one may anticipate that 68.3% of the data will fall between limits of $\bar{x} \pm 1.0\ S_x$, 95.4% between limits of $\bar{x} \pm 2.0$

S_x, and 99.7% between limits of $\bar{x} \pm 3.0\ S_x$. Also, 81.9% of the data should fall between limits of $\bar{x}$ − 1.0 S_x and $\bar{x}$ + 2.0 S_x.

In many problems, the probability of a single sample exceeding a specified value z_2 must be determined. It is possible to determine this probability by using Table 14.6 together with the fact that the area under the entire curve is unity (A = 1); however, Table 14.7, which lists one-sided areas between limits of $z_1 = z$ and $z_2 \Rightarrow \infty$, yields the results more directly.

Table 14.7
Areas under the normal distribution curve from z_1 to $z_2 \Rightarrow \infty$ (one side)

z_1	0.00	0.01	0.02	0.03	0.04	0.05	0.06	0.07	0.08	0.09
0.0	0.5000	0.4960	0.4920	0.4880	0.4840	0.4801	0.4761	0.4721	0.4681	0.4641
0.1	0.4602	0.4562	0.4522	0.4483	0.4443	0.4404	0.4364	0.4325	0.4286	0.4247
0.2	0.4207	0.4168	0.4129	0.4090	0.4052	0.4013	0.3974	0.3936	0.3897	0.3859
0.3	0.3821	0.3783	0.3745	0.3707	0.3669	0.3632	0.3594	0.3557	0.3520	0.3483
0.4	0.3446	0.3409	0.3372	0.3336	0.3300	0.3264	0.3228	0.3192	0.3156	0.3121
0.5	0.3085	0.3050	0.3015	0.2981	0.2946	0.2912	0.2877	0.2843	0.2810	0.2776
0.6	0.2743	0.2709	0.2676	0.2643	0.2611	0.2578	0.2546	0.2514	0.2483	0.2451
0.7	0.2430	0.2389	0.2358	0.2327	0.2296	0.2266	0.2236	0.2206	0.2177	0.2148
0.8	0.2119	0.2090	0.2061	0.2033	0.2005	0.1977	0.1949	0.1922	0.1894	0.1867
0.9	0.1841	0.1814	0.1788	0.1762	0.1736	0.1711	0.1685	0.1660	0.1635	0.1611
1.0	0.1587	0.1562	0.1539	0.1515	0.1492	0.1469	0.1446	0.1423	0.1401	0.1379
1.1	0.1357	0.1335	0.1314	0.1292	0.1271	0.1251	0.1230	0.1210	0.1190	0.1170
1.2	0.1151	0.1131	0.1112	0.1093	0.1075	0.1056	0.1038	0.1020	0.1003	0.0985
1.3	0.0968	0.0951	0.0934	0.0918	0.0901	0.0885	0.0869	0.0853	0.0838	0.0823
1.4	0.0808	0.0793	0.0778	0.0764	0.0749	0.0735	0.0721	0.0708	0.0694	0.0681
1.5	0.0668	0.0655	0.0643	0.0630	0.0618	0.0606	0.0594	0.0582	0.0571	0.0559
1.6	0.0548	0.0537	0.0526	0.0516	0.0505	0.0495	0.0485	0.0475	0.0465	0.0455
1.7	0.0446	0.0436	0.0427	0.0418	0.0409	0.0401	0.0392	0.0384	0.0375	0.0367
1.8	0.0359	0.0351	0.0344	0.0336	0.0329	0.0322	0.0314	0.0307	0.0301	0.0294
1.9	0.0287	0.0281	0.0274	0.0268	0.0262	0.0256	0.0250	0.0244	0.0239	0.0233
2.0	0.0228	0.0222	0.0217	0.0212	0.0207	0.0202	0.0197	0.0192	0.0188	0.0183
2.1	0.0179	0.0174	0.0170	0.0166	0.0162	0.0158	0.0154	0.0150	0.0146	0.0143
2.2	0.0139	0.0136	0.0132	0.0129	0.0125	0.0122	0.0119	0.0116	0.0113	0.0110
2.3	0.0107	0.0104	0.0102	0.00990	0.00964	0.00939	0.00914	0.00889	0.00866	0.00840
2.4	0.00820	0.00798	0.00776	0.00755	0.00734	0.00714	0.00695	0.00676	0.00657	0.00639
2.5	0.00621	0.00604	0.00587	0.00570	0.00554	0.00539	0.00523	0.00508	0.00494	0.00480
2.6	0.00466	0.00453	0.00440	0.00427	0.00415	0.00402	0.00391	0.00379	0.00368	0.00357
2.7	0.00347	0.00336	0.00326	0.00317	0.00307	0.00298	0.00288	0.00280	0.00272	0.00264
2.8	0.00256	0.00248	0.00240	0.00233	0.00226	0.00219	0.00212	0.00205	0.00199	0.00193
2.9	0.00187	0.00181	0.00175	0.00169	0.00164	0.00159	0.00154	0.00149	0.00144	0.00139

The use of Tables 14.6 and 14.7 can be illustrated by considering the ultimate-tensile-strength data presented in Table 14.4. By using Eq. (14.56) and Eq. (14.57), it is easy to establish estimates for the mean $\bar{x}$ and standard deviation S_x as $\bar{x}$ = 176.1 ksi (1,215 MPa) and S_x = 2.25 ksi (15.5 MPa). The values of $\bar{x}$ and S_x characterize the population from which the data of Table 14.4 were drawn. It is possible to establish the probability that the ultimate tensile strength of a single specimen drawn randomly from the population will be between specified limits (by using Table 14.6), or that the ultimate tensile strength of a single sample will not be above or below a specified value (by using Table

14.7). For example, one determines the probability P that a single sample will exhibit an ultimate tensile strength between 175 and 178 ksi by computing z_1 and z_2 and using Table 14.6. Thus:

$$z_1 = \frac{175-176.1}{2.25} = -0.489 \qquad z_2 = \frac{178-176.1}{2.25} = 0.844$$

$$P(-0.489, 0.844) = A(-0.489, 0) + A(0, 0.844) = 0.1875 + 0.3006 = 0.4981$$

This simple calculation shows that the probability of obtaining an ultimate tensile strength between 175 and 178 ksi from a single specimen is 49.8%. The probability of the ultimate tensile strength of a single specimen being less than 173 ksi is determined by computing z_1 and using Table 14.7. Thus:

$$z_1 = \frac{173-176.1}{2.25} = -137$$

and

$$P(-\infty, -1.37) = A(-\infty, -1.37) = A(1.37, \infty) = 0.0853$$

Clearly, the probability of drawing a single sample with an ultimate tensile strength less than 173 ksi is 8.5%.

14.7.2 Weibull Distribution

In investigations of the failure of semiconductors with time of operation, strength of brittle materials, crack-initiation toughness, and fatigue life, researchers often find that the Weibull distribution provides a more suitable approach to the statistical analysis of the available data. The Weibull failure density function f(t) is defined as:

$$f(t) = \frac{\beta}{\alpha}\left(\frac{t-t_0}{\alpha}\right)^{\beta-1} e^{-\left(\frac{t-t_0}{\alpha}\right)^{\beta}} \tag{14.66}$$

where t is the time of operation and α, β and t_0 are the three parameters which define this important function The symbol t_0 is the called the **location** parameter because it shifts the distribution relative to the abscissa (the time scale). The constant β is known as the **shape** or **slope** or **modulus** parameter. The symbol α represents the **scale** parameter.

In some applications of the Weibull failure density function, it is not necessary to shift the distribution relative to the abscissa and t_0 is set equal to zero giving the two parameter representation shown below:

$$f(t) = \frac{\beta}{\alpha}\left(\frac{t}{\alpha}\right)^{\beta-1} e^{-\left[\frac{t}{\alpha}\right]^{\beta}} \tag{14.67}$$

Three curves representing the Weibull failure density function are presented in Fig. 14.17 for the case where β = 0.50, 1.0 and 3.0. These curves show that shape of the Weibull distribution is sensitive to the selection of all three parameters. This fact is important because the Weibull distribution can be fitted to a wide range of experimental data by adjusting the value of one or more of these three parameters. The parameter t_0 shifts the curves to the right along the abscissa and is employed when no failures occur for some extended period during a test run. The scale factor α affects both the ordinate and the abscissa and is selected to fit the data for the choice of units used for the variable t. The shape factor β has a marked effect on the appearance of the failure density function as is evident in Fig. 14.17.

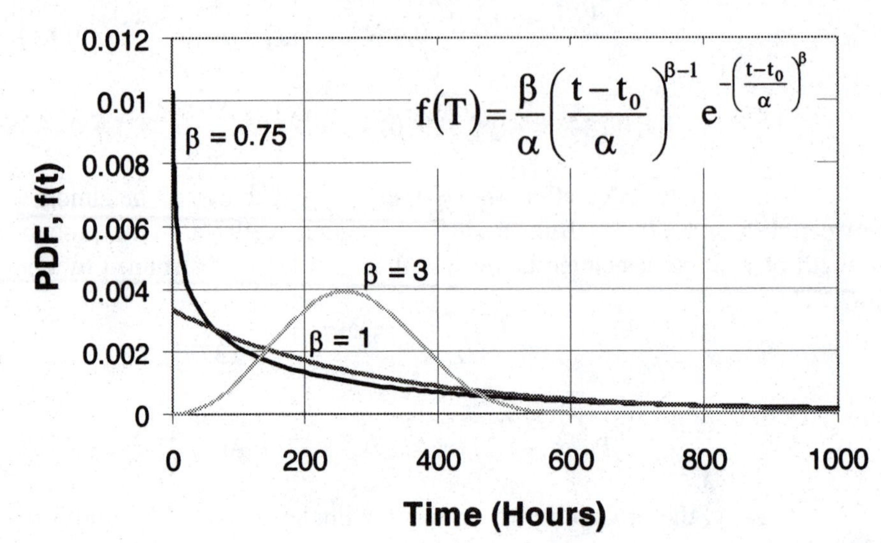

Figure 14.17 The Weibull failure density function for different shape parameters β.

The reliability R(t) of a component in terms of the Weibull parameters is given by:

$$R(t) = e^{-\left[\frac{t-t_0}{\alpha}\right]^{\beta}} \tag{14.68}$$

The probability of failure F(t) (ofte4n called the cumulative density function) for a component in terms of the Weibull parameters is given by:

$$F(t) = 1 - e^{-\left[\frac{t-t_0}{\alpha}\right]^{\beta}} \tag{14.69}$$

A graph showing the probability of failure as a function of time for different values of β is presented in Fig. 14.18.

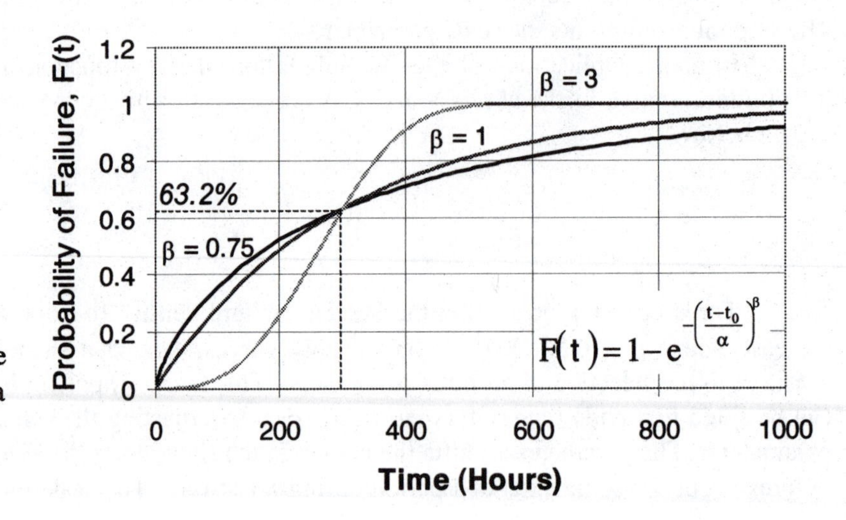

Fig. 14.18 Probability of failure as a function of time with β as a parameter.

The failure rate $\lambda(t)$ of a component in terms of the Weibull parameters is given by:

$$\lambda(t) = \frac{f(t)}{R(t)} = \frac{\beta}{\alpha}\left(\frac{t-t_0}{\alpha}\right)^{\beta-1} \qquad (14.70)$$

The mean value of t for a specific Weibull distribution is the MTTF, which is given by:

$$MTTF = t_0 + \alpha\Gamma\left(\frac{1}{\beta}+1\right) \qquad (14.71)$$

where the gamma function $\Gamma(n)$ is given by:

$$\Gamma(n) = \int_0^{\infty} e^{-x}x^{n-1}dx \qquad (14.72)$$

and $n = (1+\beta)/\beta$. Note for $\beta = 1$; $\Gamma(n) = \Gamma(2) = 1$ and Eq. (14.71) reduces to $MTTF = t_0 + \alpha$.

The median value of t for a specific Weibull distribution is given by:

$$t_{median} = t_0 + \alpha(\log_e 2)^{1/\beta} \qquad (14.73)$$

Finally, value for the mode of t for a specific Weibull distribution is given by:

$$t_{mode} = t_0 + \alpha\left(\frac{\beta-1}{\beta}\right)^{1/\beta} \qquad (14.74)$$

To utilize the Weibull failure density function requires knowledge of the Weibull parameters. In experimental investigations, it is necessary to conduct experiments and obtain a relatively large data set to accurately determine α, β and t_0. Consider as an illustration, a life test of 10 components conducted under accelerated test condition. The data presented in Table 14.8 shows the time of failure of each of the ten components[2].

Table 14.8
Time of failure of components in an accelerated life test

Failure Number, k	Median Rank %	Time t (hours)
1	6.73	14
2	16.6	36
3	26.0	78
4	35.6	97
5	45.2	142
6	54.8	200
7	64.4	325
8	74.0	440
9	83.6	550
10	93.2	670

[2] The data presented in this table is fictitious.

The median rank for each failure listed in the Table 14.7 can be estimated from:

$$MR = \frac{k-0.3}{N+0.4}(100) \qquad (14.75)$$

where k is the sequential number of the component that has failed and N is the sample size.

Next obtain a sheet of two cycle Weibull probability paper and plot the failure times as a function of their median rank as shown in Fig. 14.19. After constructing a best fit line through the points, measure its slope. Construct another line with this same slope through the slope indicator at the top of the graph paper. The intersection of this line with the slope indicator axis gives the value of $\beta = 0.85$.

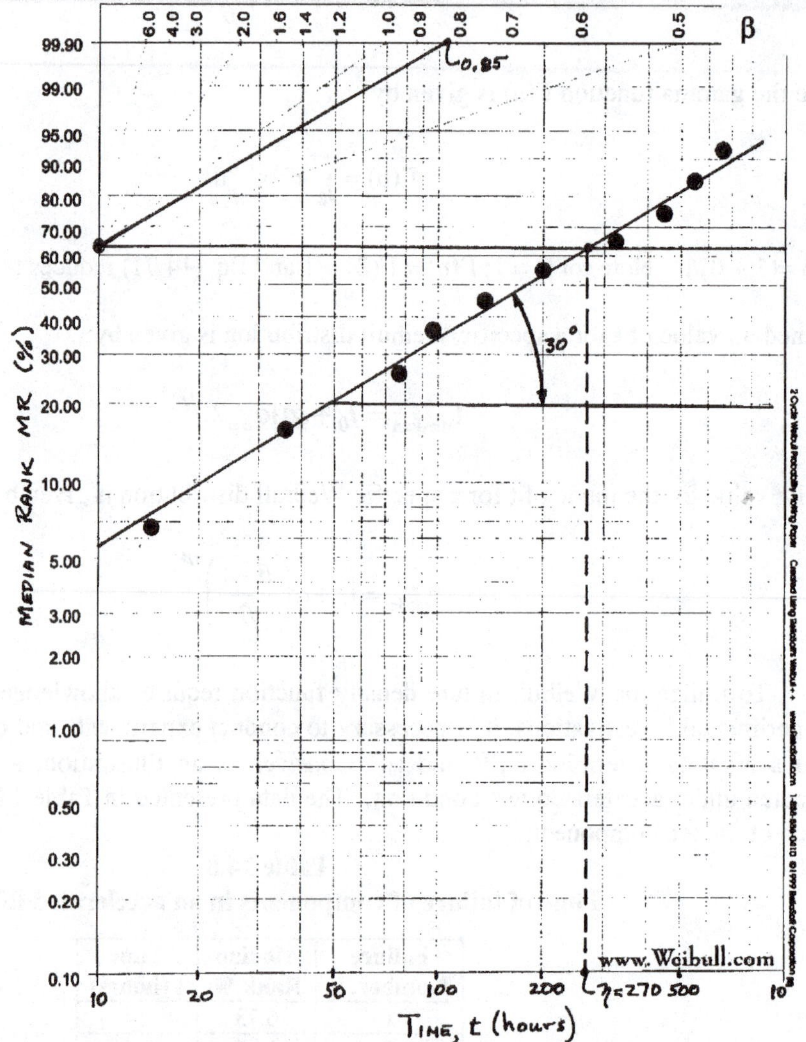

Fig. 14.19 Determining the Weibull parameters β and α using Weibull graph paper.

At the ordinate point where F(t) = 63.2%, construct a horizontal line that intersects the fitted line through the data points. Construct a vertical line through this intersection point and extend it until it intersects the abscissa. The value of this intersection point gives η that is related to the scale parameter α by:

$$\alpha = \eta^{\beta} \qquad (14.76)$$

Substituting values for η and β from Fig. 14.19 into Eq. (14.76) yields:

$$\alpha = (270)^{0.85} = 116.6$$

In this example, a two parameter Weibull failure density function was adequate because the data when plotted yielded a straight line on the Weibull graph paper. However, in some cases a three parameter failure density function is required to obtain a reasonable fit of a straight line with the data points. When the two parameter function is not sufficient, the line through the data points of MR versus t exhibit curvature. If the curvature is concave down, subtract a positive t_0 value from all the times to failure and replot the data. Repeat this process changing t_0 with each attempt until a straight line can be constructed through the modified data points.

If the curvature is concave upward, subtract a negative t_0 value from all the times to failure and replot the data. Repeat this process changing t_0 with each attempt until a straight line can be constructed through the modified data points. When adjusting t_0 to obtain a better fit of the Weibull distribution, the scale for the abscissa becomes $(t - t_0)$.

An example of two-cycle Weibull graph paper is presented in Fig.14.20.

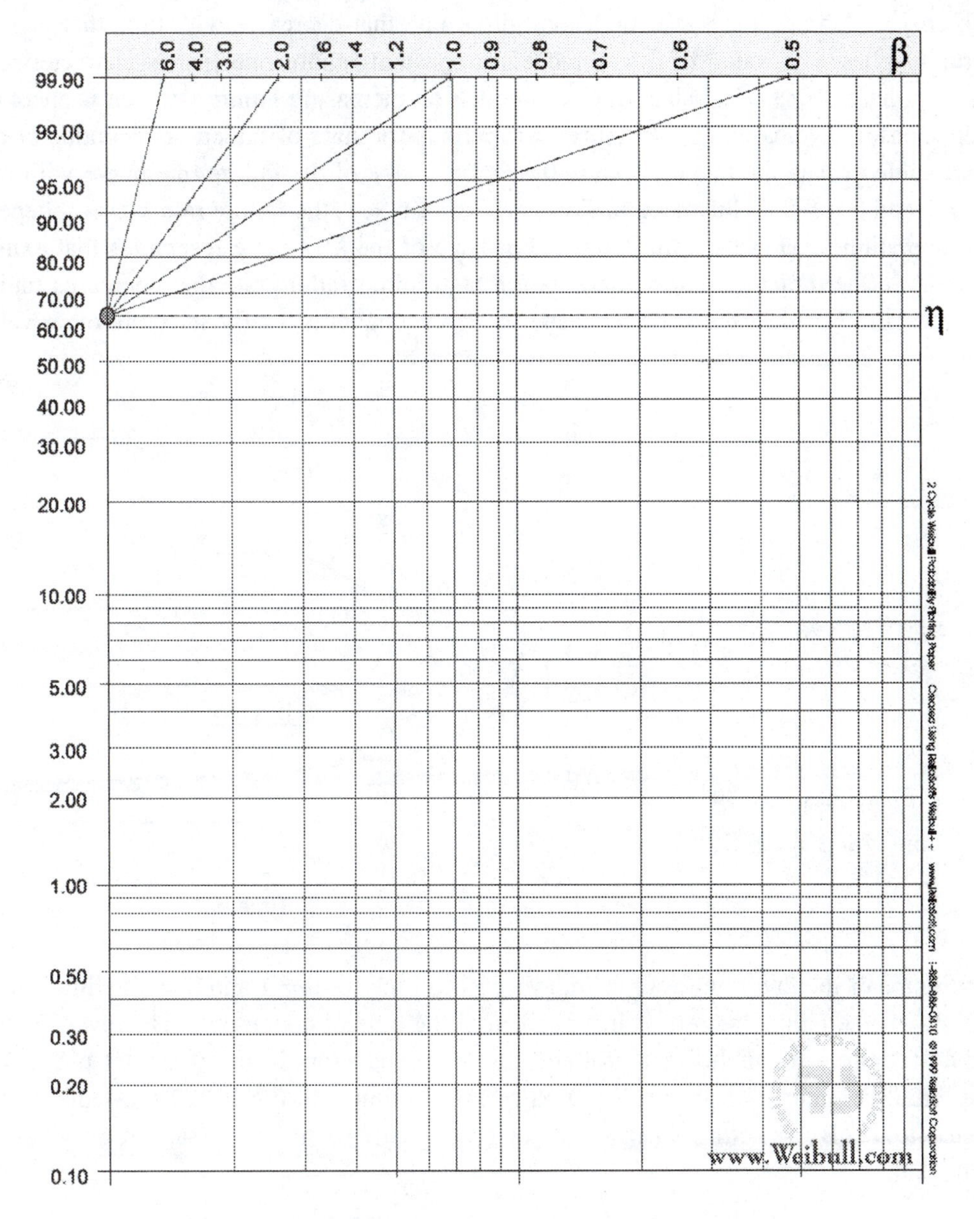

Fig. 14.20 An example of two-cycle Weibull graph paper.

14.7.3 Studying Infant Mortality Failures with the Weibull Failure Density Function

When modeling the bathtub curve the shape of the failure rate function λ(t) is considered. In life tests of semiconductors failures occur soon after the test is initiated and t_0 is usually set equal to zero. In this case a two parameter Weibull failure density function is sufficient. For the two parameter Weibull distribution, the relation for λ(t) is given by:

$$\lambda(t) = \frac{\beta}{\alpha}\left(\frac{t}{\alpha}\right)^{\beta-1} \tag{14.77}$$

The scale parameter α establishes the time when a specified percentage of the population will fail. The shape parameter β is the parameter that permits its application to each of the three phases of the bathtub curve. A value for β < 1 models a failure rate that decreases with time that occurs in the infant mortality phase. A value of β = 1 models a constant failure rate that is characteristic of the random failure phase. Finally, a value for β > 1 models an increasing failure rate that is encountered in the wear-out phase. Typical infant mortality distributions for state-of-the-art semiconductor chips are fitted with a value of β in the range of 0.2 to 0.6. The shape of the failure rate curve with respect to time for β = 1 and β = 0.5 is illustrated in Fig. 14.21. For β = 1, the failure rate λ(t) is independent of the time of operation t. However, for β = 0.5 the shape of the λ(t) curve resembles that exhibited by semiconductors in the infant mortality phase with a high initial failure rate that decreases rapidly with respect to time. In Fig. 14.21 the scale parameter α was set equal to 1. By selecting other values of α the scale of the abscissa can be adjusted to fit experimental data.

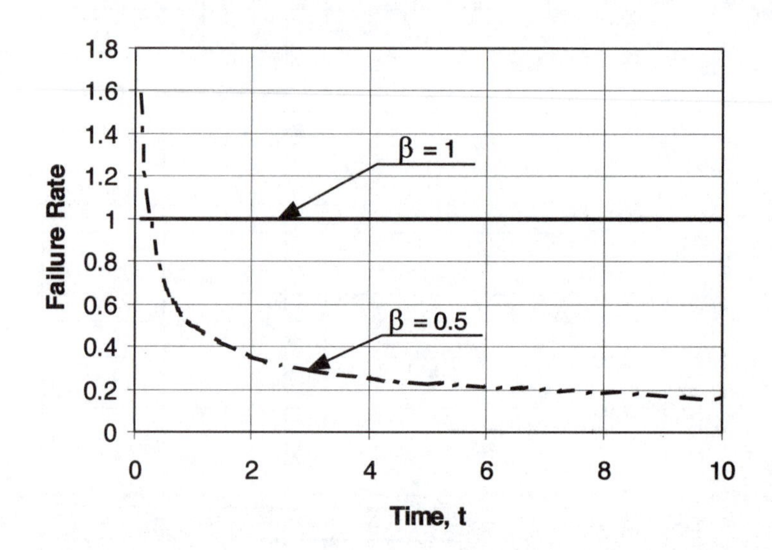

Fig. 14.21 Failure rate as a function of time for β = 1 and β = 0.5.

The shape of the failure rate curve with respect to time for β = 1 and β = 3 is illustrated in Fig. 14.22. For β = 1, the failure rate λ(t) is independent of the time of operation t. However, for β = 3 the shape of the λ(t) curve resembles that exhibited by semiconductors in the wear out phase with zero failure rate initially, which increase rapidly with respect to time. In Fig. 14.22 the scale parameter α was set equal to 1. By selecting other values of α the scale of the abscissa can be adjusted to fit experimental data.

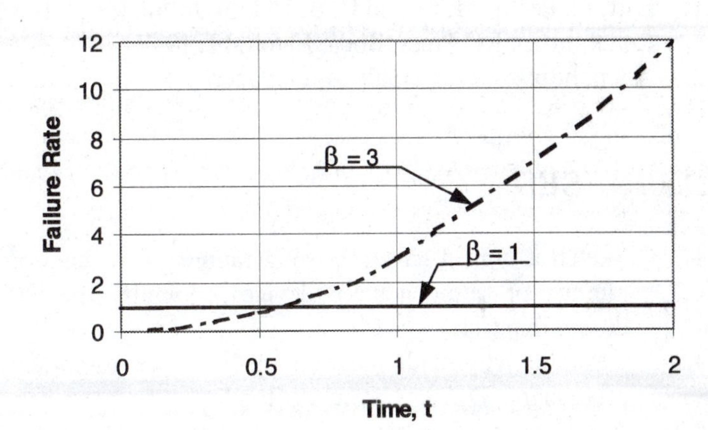

Fig. 14.22 Failure rate as a function of time for $\beta = 1$ and $\beta = 3$.

There are several techniques for characterizing the failure data using the Weibull density function including graphs of F(t) as a function of time, R(t) with respect to time and the failure rate $\lambda(t)$ versus time. However it should be recognized that each of the three phases of semiconductor life must be modeled with a different Weibull failure density function.

14.7.4 Weibull Conditional Reliability

Let's determine the reliability for a new mission of duration t, having already accumulated T hours of operation up to the start of this new mission. Will the reliability of the part meet requirements? The Weibull conditional reliability permits us to determine the probability of successful completion of the mission of duration t from the following relation.

$$R(t|T) = \frac{R(T+t)}{R(T)} = \frac{e^{-\left(\frac{T+t-t_0}{\alpha}\right)^{\beta}}}{e^{-\left(\frac{T-t_0}{\alpha}\right)^{\beta}}} = e^{-\left[\left(\frac{T+t-t_0}{\alpha}\right)^{\beta} - \left(\frac{T-t_0}{\alpha}\right)^{\beta}\right]} \tag{14.78}$$

REFERENCES

1. Bethea, R. M. and R. R. Rhinehart: Applied Engineering Statistics, Dekker, New York, 1991.
2. McClave, J. T. and T. Sincich, Statistics, 10th edition, Prentice Hall, Upper Saddle River, NJ, 2006.
3. Agresti, A. and C. Franklin, Statistics: The Art and Science of Learning from Data, Prentice Hall, Upper Saddle River, NJ, 2006.
4. Hogg, R., Craig, A. and J. McKean, Introduction to Mathematical Statistics, 6th Edition, Prentice Hall, Upper Saddle River, NJ, 2005.
5. Milton, J. S. and J. C. Arnold, Introduction to Probability and Statistics, McGraw-Hill, New York, NY, 2006.
6. Mann, P.S., Introductory Statistics, 5th Edition, John Wiley, New York, NY, 2005.
7. Snedecor, G. W. and W. G. Cochran: Statistical Methods, 8th Edition, Iowa State University Press, Ames, IA, 1989.
8. Navidi, W. C., Statistics for Engineers and Scientists, McGraw-Hill, New York, NY 2006.
9. Weibull, W.: Fatigue Testing and Analysis of Results, Pergamon Press, New York, 1961.
10. Pecht, M., Handbook of Electronic Package Design, Marcel Dekker, 1991.

11. Lall, P., Pecht, M. G. and E. B. Hakim, Influence of Temperature on Microelectronics and System Reliability, CRC Press, Boca, Raton, FL 1997.
12. Anon, http://www.weibull.com/GPaper/

EXERCISES

14.1 Sketch a typical curve showing failures as a function time during a long term test of a large number of semiconductor devices. Identify the three different types of failures that occur during this test.

14.2 Describe why fault location methods are important in system availability in large digital processors. Define system availability.

14.3 A sample of 500 components is tested under continuous operation for a year. The three components which fail at 32, 60, and 290 days are replaced during the test. Determine $\lambda(t)$ and comment on the results.

14.4 Determine $\lambda(t)$ if the components which failed in Exercise 14.3 were not replaced during the test.

14.5 Sketch the bathtub curve that is used to characterize the failure rate as a function of time for semiconductor devices. Identify the three different phases in the life of a semiconductor device on this curve.

14.6 Explain the reasons for using components that have been subjected to a burn-in operation. Why not perform the burn-in after the system is assembled?

14.7 Determine the reliability R(t) for the components tested in Exercise 14.3.

14.8 Prepare a data presentation showing R(t) for a component with $\lambda_0 = 10^{-3}, 10^{-5}, 10^{-7}, 10^{-9}, 10^{-12}$ and t = 1, 2, 5, 10,20 and 50 yrs. It may be helpful to use a spread sheet in performing these computations and displaying the results.

14.9 Prepare an engineering brief which interprets the results of Exercise 14.8.

14.10 Verify Eq. (14.11).

14.11 Write an engineering brief describing the difference between hazard rate and MTTF.

14.12 Write an engineering brief describing the difference among the terms MTTF, MTBF and FIT.

14.13 Why is accelerated testing used to determine failure rates? Describe the effect of voltage and temperature on the time to determine the failure rate in life tests.

14.14 A system consists of four series connected sub systems. The failure rate of the systems are 10^{-4}, 10^{-5}, 10^{-6} and 10^{-9}/hr. Prepare a graph showing the system reliability as a function of time for a period of 20 yrs.

14.15 Using one redundant sub-system with the system described in Exercise 14.14, improve the system reliability. Indicate where the subsystem would be placed, give reasons for the choice of the sub-system and prepare a revised reliability time graph which illustrates the improvement.

14.16 Derive the system reliability for the series-parallel arrangement with the single cross strap that is shown in Fig. Ex14.16. All of the devices have the same probability of failure F(t) = 0.001.

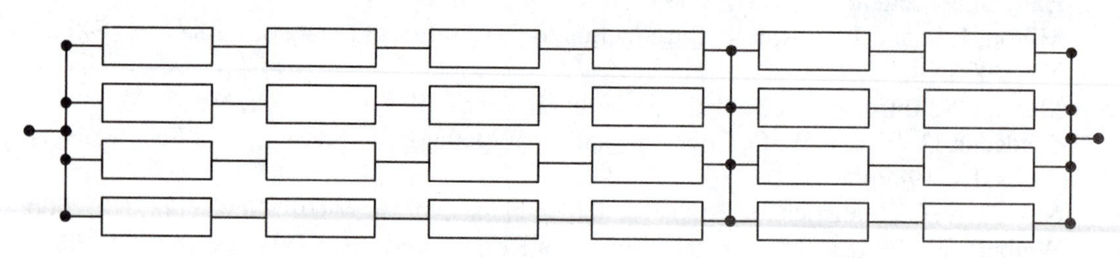

Fig. Ex14.16

14.17 Progressively add cross straps to the block arrangement described in Exercise 14.16 and determine the enhanced system reliability. Discuss the merits of adding addition cross straps. Also, describe any disadvantages associated with cross strapping.

14.18 For the complex system connections shown in Fig. 14.10, determine the reliability of the system R^s if $R_A = R_B = R_C = R_D = 0.95$.

14.19 Consider a standby systems with only two degrees of redundancy. Assume that the failure density function $f(t) = C = 10^{-4}$ /hr (a constant). Determine the probability of failure of the first circuit in 120 days. Then determine the probability of failure of the system in 60 months.

14.20 Ten measurements of the fracture strength (ksi) of an aluminum alloy are:

25.0	25.2	24.9	25.5	24.6
24.8	25.2	25.0	24.8	25.0

Determine the mean, median, and mode which represent the central tendency of this data.

14.21 Verify the mean value $\bar{x}$ of the data listed in Table 14.4.

14.22 Determine the range, mean deviation and variance S^2_x for the data given in Exercise 14.20.

14.23 Determine the range, mean deviation and variance S^2_x for the data listed in Table 14.4.

14.24 Find the coefficient of variation for the data given in Exercise 14.20.

14.25 Find the coefficient of variation for the data listed in Table 14.4.

14.26 Consider a Gaussian population with a mean $\mu = 100$ and a standard deviation $\sigma = 10$. Determine the probability of selecting a single sample with a value in the interval between:

(a) 75 – 80 (b) 98 – 102 (c) 92 – 97

(d) 115 – 123 (e) greater than 125

Also determine the percent of data which will probably be within limits of:

(a) $\bar{x} \pm 2.5\ S_x$ (b) $\bar{x} - 1.0\ S_x$ and $\bar{x} + 1.5\ S_x$

(c) $\bar{x} - 1.5\ S_x$ and $\bar{x} + 1.0\ S_x$ (d) $\bar{x} \pm 0.55\ S_x$

14.27 A manufacturing process yields aluminum rods with a mean yield strength of 35,000 psi and a standard deviation of 1,000 psi. A customer places a very large order for rods with a minimum yield strength of 32,000 psi. Prepare a letter for submission to the customer that describes the yield strength to be expected and outline your firm's procedures for assuring that this quality level will be achieved and maintained.

14.28 Determine the Weibull failure density function corresponding to the parameters $t_0 = 5$, $\alpha = 5$ and $\beta = 10$. Plot f(t) for $0 < t < 50$.

14.29 For the Weibull failure density function of Exercise 14.28, prepare a graph of R(t) for $0 < t < 50$.

14.30 For the Weibull failure density function of Exercise 14.28, prepare a graph of F(t) for $0 < t < 50$.

14.31 For the Weibull failure density function of Exercise 14.28, prepare a graph of $\lambda(t)$ for $0 < t < 50$.

14.32 For the data presented in the table below determine the Weibull parameters.

Failure Number, k	Median Rank %	Time t (hours)
1	6.73	17
2	16.6	32
3	26.0	70
4	35.6	91
5	45.2	132
6	54.8	205
7	64.4	310
8	74.0	425
9	83.6	532
10	93.2	688

14.33 Derive Eq. (14.77) and construct a graph of $\lambda(t)$ versus time for $\beta = 0.6, 1.0, 1.4$ and 4.0. Let $\alpha = 2$ for all values of β in this exercise.

14.34 An electronic module inside a military vehicle has 23 mm BGAs in one of its modules. The vehicle has been in operation for 2 years, during which it has sustained 1,400 temperature cycles from –40 to 125°C. Reliability data on the BGAs is shown in Fig. Ex14.34. What is the probability of BGA failure over the next 6 months assuming 4 cycles per day? Should the vehicle be sent into battle without replacing the module?

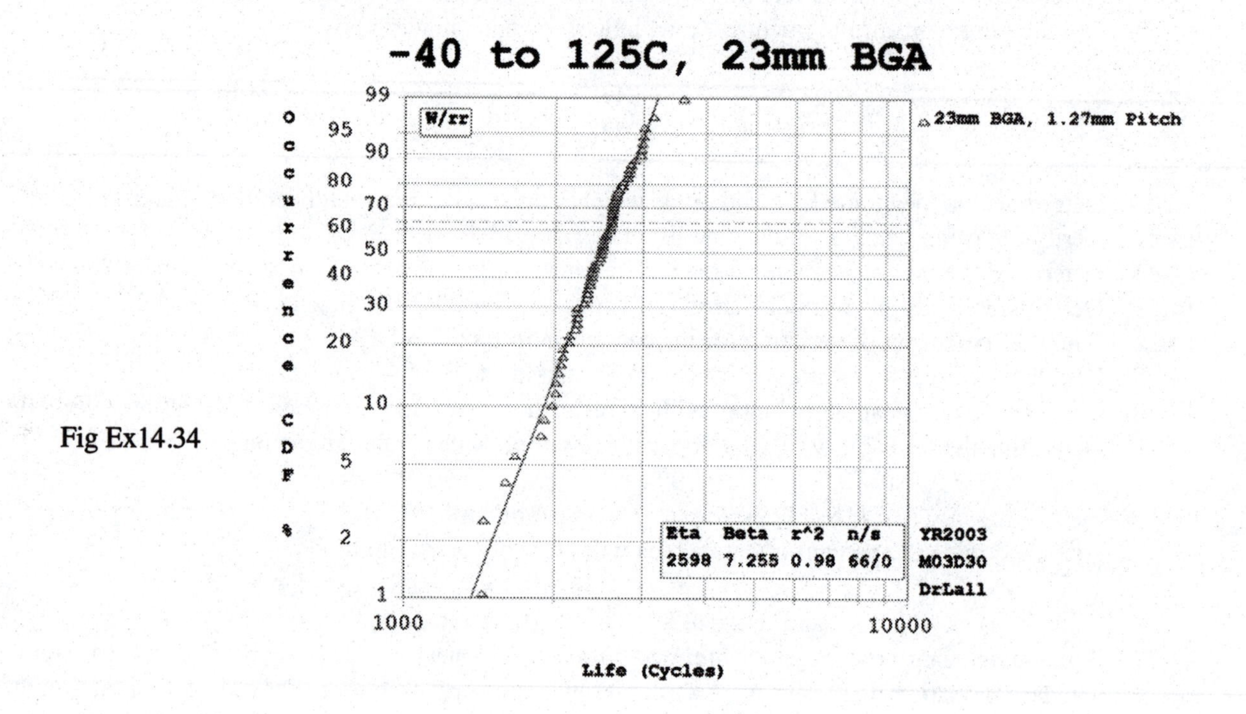

Fig Ex14.34

CHAPTER 15

DESIGN TO IMPROVE RELIABILITY

15.1 INTRODUCTION

Semiconductor devices and other electronic components have very low failure rates with λ of the order of 1×10^{-9}/hr or failures in ten-to-the-nine hours (FITS). A large percentage of electronic products are non-redundant systems, where failure of one device will cause failure of the system. Non-redundant electronic systems are equivalent to large number of components connected in series, and the system reliability is usually seriously degraded with reduction in component reliability as shown in Table 15.2. To improve reliability, tests are conducted at different stages during the development process for a new semiconductor device and the manufacturing processes used in its production are modified to eliminate latent defects associated with infant mortality. Accelerated life tests are also conducted later as the electronic systems are assembled. These tests are intended to eliminate failure mechanisms that will cause the system to malfunction or identify devices which do not meet the perform specifications. The relative costs of part replacement at various points of the product life-cycle are presented in Table 15.1.

Table 15.1
Relative costs to locate and replace a single semiconductor device.
(System with 5 printed circuit boards)

Production Stage	Relative Cost
Functional Circuit Test	0.10
Operational Test (Burn-in)	0.20
Visual PCB Inspection	2.00
PCB Functional Test	10.00
System Test	40.00
On-Site Test and Repair	300.00

Examination of the results in Table 15.1 shows that it becomes expensive to locate a malfunctioning component after the chip is packaged and even more expensive after the chip carriers are soldered into a printed circuit board. It is clear that the design of an electronic system must address reliability early in the development process to avoid substantial repair and cost penalties that are imposed when a component fails in system test or in the field after a product has been placed in service.

There are several factors to consider when developing and electronic systems with suitable[1] reliability. These factors include:

---

[1] All products cannot be designed with the same level of reliability. For example, the reliability required for an electronic system installed in a multi-million dollar fighter aircraft will be higher than the reliability required in an electronic system used in a $15,000.00 automobile. Of course, the cost of the two electronic systems will differ markedly with relatively high reliability achieved in automotive electronics at relatively low cost..

1. Product environmental profile including, shipping and storage, operating conditions, temperature, humidity, and airborne containments.
2. Product usage profile including, frequency of use, power-cycle, and long-term dormant storage.
3. Dominant failure mechanisms driven by environmental and usage stresses and the effect of competing failure modes on overall system reliability.
4. Packaging and assembly materials and processes, their interaction with extended storage, environmental stresses, damage initiation, progression and eventual failure.
5. Production process windows required to achieve high quality output from manufacturing and assembly operations at high production volumes, often referred to as six-sigma processes.
6. Expected design-life based on results of accelerated tests and the acceleration factors related to field stresses.

15.2 FAILURE MECHANISMS

Devices, packaging architectures, printed wiring boards, subsystems and complete electronic systems may fail due one or more different failure mechanisms. Dominant failure mechanisms will depend on the package element, materials, environmental and operating conditions. These environmental conditions include temperature, humidity, shock, vibration, electrostatic voltage and temperature cycling. The operating conditions involve supply voltage, current flow and power dissipation.

Failure mechanisms can be broadly classified into four categories including, on-chip, device-packaging, printed circuit board and the complete system assembly. Common failure mechanisms at the chip-level include electrical overstress, electrostatic discharge, oxide failure and electro-migration of the chip's metallization. Dominant failure mechanisms in packaging architectures include corrosion in non-hermetically sealed packages, and thermo-mechanical fatigue of second-level interconnects due to temperature cycling and spatial temperature gradients. Failures on the printed circuit board may occur at in-board interconnects such as blind vias, buried vias, on-board traces and delamination. System-level failures are dominated by inter-system interconnects including connectors, cables, and support structure or fasteners. In addition, mechanical stresses imposed to due to shock or vibration in shipping, storage, service or in qualification testing may cause failure of zero-level, first-level, second-level interconnects and the inter-system connectors. Examples include fretting corrosion of contact systems, pad cratering, and solder joint failure.

Several of these modes of failure will be discussed in the following subsections.

15.2.1 Failure Mechanisms in the Chip Metallization

Corrosion

Corrosion failures occur when the metal circuit lines (aluminum or copper) or bonding pads are subjected to a chemical reaction with the environment. Corrosion can occur during manufacturing, storage or when the product is in service. The type of chip carrier and the humidity level control the transport of moisture into the package, and markedly affect the rate of the corrosive reactions.

There are several different corrosion processes—dry, wet and galvanic. Dry corrosion, which occurs in the absence of moisture, does not normally cause problems because it forms a very thin oxide film that is self passivating. Wet corrosion, as the name implies, occurs in the presence of moisture and an ionic contaminant such as chlorine. The combination of moisture and the contaminant produces an electrolytic solution that promotes corrosion when an electrical potential is applied to the chip. The corrosion products that result provide a conductive path between adjacent circuit lines resulting in charge leakage and dendritic whisker growth. Corrosion of the aluminum wiring traces and the bonding pads can produce open circuits. If whisker growth is excessive, shorts can develop. Higher temperatures usually accelerate the wet corrosion processes.

Galvanic corrosion occurs when two dissimilar metals, in contact with each other in the presence of moisture, develop a potential difference and form a galvanic couple. This potential difference provides the force for the anode to oxidize. The metal with the more negative electrical potential will spontaneously undergo oxidation. The metal with the more positive electrical potential serves as the cathode in the couple and is passive. The oxidation of the anode is called galvanic corrosion. The amount of corrosion is affected by the environment, the difference in their electric potentials and the ratio of the cathode to the anodic areas.

The role of anode or cathode for a given couple depends on their electric potentials. The metal with the more negative potential becomes the anode and begins to oxidize. The electric potentials for several metals used in chip manufacturing and packaging are presented in Table 15.2. It is clear from the results of Table 15.2 that aluminum often used for circuit lines and bonding pads will become an anode in the presence of copper, silver or gold.

Table 15.2
Electrical potential for several metals at 25 °C

Metal	Electrode Potential Volts
Aluminum	– 1.670
Chromium	– 0.744
Tin-lead Solder	– 0.325
Nickel	– 0.250
Tin	– 0.136
Lead	– 0.126
Hydrogen Reference	0.000
Copper	+ 0.340
Silver	+ 0.799
Gold	+ 1.420

The galvanic corrosion rate depends on four different factors. First it is a function of the potential difference between the two metals forming the galvanic couple. A large difference in the electrical potential between the two metals results in a more rapid corrosion rate at the anode. Second the conductivity of the electrolyte (the degree of contamination of the moisture) affects the corrosion rate. Electrolytes with high conductivity promote more rapid corrosion over larger areas of the anode. The third factor affecting the corrosion rate is the ratio of the areas of the cathode and anode. When the area of the cathode is large relative to the area of the anode, the corrosion rate is increased. The increase in corrosion rate in this case is due to the increase in the current density for given current flow at the anode with its smaller area. The fourth factor is the distance from the contact point of the junction of the galvanic couple. The greatest amount of corrosion will occur at the contact point between the two dissimilar metals. The amount of corrosion decreases with distance from this contact point and rarely extends more than a few millimeters.

Electromigration

Electromigration refers to the migration (movement) of metal in the circuit lines of a semiconductor while it is operational. The movement of the metal is due to momentum transfer between the electrons conducting the current and the metal ions. In single crystal structures, the momentum transfer between the conduction electrons and the metal ions is negligible because of the uniform lattice structure of the metal ions. However, the lattice symmetry does not exist at grain boundaries and under certain conditions the momentum transferred is sufficient to cause atoms to separate from these grain boundaries. These atoms move in the direction of the current, although their direction is also influenced by the orientation of the grain boundary because the atoms often slide along the grain boundaries. The momentum transfer is a function of the current density and the temperature, with increasing electromigration as both of these parameters increase.

The damage produced by electromigration is due to the formation of voids or hillocks resulting from the migration of metal in conductor lines. The voids formed are initially small compared to the width and thickness of the circuit lines. However, the voids grow in number and size with time, and evenually coalese produce an open circuit failure. Hillocks, as the name imples, are the formation of small protuberances that form along the conductor. When the hillocks extend laterally they can contact an adjacent conductor and produce a short circuit failure. Hillocks that grow vertically can extend through the passivation layer on the surface of a chip that protect the circuit lines from the environment. In some cases, the damage to the protective layer leads to corrosion failures. Open circuit failures due to voids usually occurs prior to short circuit failures cause by hillocks. When voids form, they reduce the cross sectional area of the conductor increasing the current density until this point in the conductor fails.

The role of temperature and current density are both important in the rate of electromigration. To assess the influence of these two quantities on the mean time to failure MTTF an equation developed by J. R. Black in 1969 is often employed.

$$MTTF = \frac{A}{J^n} e^{\left(\frac{E_a}{kT}\right)} \tag{15.1}$$

where A is a constant dependent on the cross sectional area of the conductor
J is the current density
E_a is the activation energy taken as 0.7 eV for aluminum
k is Boltzmann's constant
T is the temperature in °K
n is a scale factor usually taken as 2

Note that the temperature of the conductor appears in the exponent; hence it markedly affects the MTTF. For a conductor to serve reliability at elevated temperatures, the maximum current density of the conductor must be lowered.

With increasing reduction in the feature size on chips the probability of failure due to electromigration increases because both the power density and the current density increase. However, in more advanced manufacuring processes, copper has replaced the aluminum conductors. Although it is more difficult to employ in production, copper is preferred because of its better conductivity and its reduced susceptibility to electromigration.

In most electronic products, chips rarely fail due to electromigration induced defects. Modern semiconductor design practices account for the effects of electromigration in the conductor design. Automated design software is employed to check and correct electromigration problems from the layout of the transistors to the routing of the circuit lines. When operated within the specified temperature and voltage range, semiconductor devices usually fail due to other environmental effects.

15.2.2 Failure Mechanisms in the Chip's Oxide Layers

Another mechanism for failure is the time dependent dielectric breakdown of the oxide films used to insulate adjacent conductors, features and gates on the surface of a chip. Failure is due to wearing-out the insulating properties of the silicon dioxide film and the formation of a conducting path through the oxide to the substrate. With a conducting path between the gate and the substrate, control of the current flow between the drain and source by the gate's electric field is not possible. The time to oxide failure is dependent on the number of defects in the gate oxide that is produced during fabrication. To reduce the number of defects chip manufacturers attempt to produce a nearly defect free oxide in their process to maximize the MTTF. Even with high quality oxide films, failure of the oxide remains a concern for chip designers, because oxide failures occur at all gate voltages. The goal of the chip manufacturers is to trade-off gate oxide thickness with operating voltage to achieve both speed and life. The MTTF of a specified gate oxide thickness is determined by the charge that flows through the gate oxide by the tunneling current. This tunneling current density J is estimated with the Fowler-Nordheim equation that is given by:

$$J = AE^2 e^{\left(\frac{B}{E}\right)} \tag{15.2}$$

where A and B are constants to account for the effective mass and barrier height of the oxide film.
E is the electric field and J is the current density.

This equation shows that operating a semiconductor device at voltages greater than the manufacturer's specification results in a significant increase in the oxide current density. After the electrons have penetrated the oxide potential barrier, the electric field, which depends on the applied voltage and the thickness of the oxide film, accelerates the electrons through the film. After passing through the oxide, the electrons transfer their energy at the oxide silicon interface. This interface is already subjected to thermally induced strains due to the difference in the thermal coefficient of expansion between oxide film SiO_2 and the silicon substrate. The strained chemical bonds in the oxide film are often broken by the accelerated electrons. When the bonds fracture, the resulting sites trap electrons (charge). This charge affects the channel carrier mobility in transistors and reduces their gain. It also increases the local electric field, which in turn increases the tunneling current. This positive feedback leads to a still larger trapped charge causing the tunneling current to increase until it is sufficient to burn a hole through the gate oxide.

Predicting the MTTF for oxide film failure is a difficult task. The process details used in manufacturing the chips are complex, and consequently simple models often are misleading if oxide purity is not factored into the model. Also errors occur in predicting MTTF if the thickness of the gate oxide film differs from the thickness used in collecting the empirical data employed in developing the constants in the relationship for predicting the MTTF. Some controversy exists regarding the effect of the electric field on the MTTF. Some investigators believe the MTTF is related to the electric field (either E or 1/E) and others believe the gate voltage controls the MTTF. It appears that simple predictive equations are valid only over limited ranges of gate oxide thickness. The MTTF in moderately thick oxides are related to 1/E (for high electric fields), E (for low electric fields) and to the applied voltage for very thin oxide films (thinner than 5 nm). All of the constants in the predictive equations are based upon empirical data gathered from experiments with different electric fields and different oxide film thickness. Tests to failure are performed with very high electric fields and at elevated temperatures to reduce testing time to a few weeks instead of tens of years. These equations all include the influence of temperature that reduces the time to failure. The reader is referred to reference [1] for an extensive discussion of equations controlling the breakdown of oxide films.

15.2.3 Failure Mechanisms with Packaging Architectures

In some cases, failures occur when the chip is packaged in a chip carrier or when the chip carrier is mounted to the printed circuit board. Most of these failures are due to thermal stresses generated during the temperature cycling that occurs during the production processes. These failure mechanisms include:

1. Parameter shifts in the chip induced by high thermal stresses occurring during the molding process in plastic encapsulated chips.
2. So called popcorn failures that occur during solder reflow operations when the chip carrier is mounted to the printed circuit board.
3. Chip interface delamination that occurs due to thermal cycling.
4. Solder bump and/or solder joint failures due to cyclic thermal stresses.

Thermal stresses are generated during temperature changes ΔT required in the molding, bonding and soldering processes. A typical plastic encapsulated chip carrier with a metallic lead frame, silicon chip and polymeric mold compound must accommodate three different temperature coefficients of expansion. Silica filled molding compound typically exhibits a temperature coefficient of expansion $\alpha_{mc} = 20 \times 10^{-6}/°C$, the silicon chip has a $\alpha_{Si} = 2.3 \times 10^{-6}/°C$, and the copper alloy lead frame has a $\alpha_{Cu} = 17 \times 10^{-6}/°C$. The thermal stresses σ_T generated by uniform temperature changes (ΔT) in a totally constrained uniaxial arrangement of two materials may be approximated by:

$$\sigma_T = E\Delta T(\alpha_{Cu} - \alpha_{Si}) \tag{15.3}$$

where E is the modulus of elasticity.

Thermally induced shear stresses are also generated at the bond lines where the chip is attached to the lead frame. These shear stresses produce delamination that begins at the edges and corners of the chip as illustrated in Fig. 15.1.

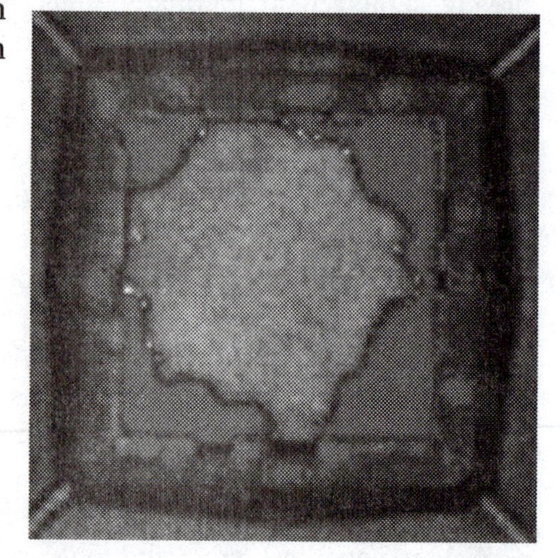

Fig. 15.1 Delamination of the bond line between the chip and the chip carrier caused by thermally induced shear stresses.

Moisture absorption over time leads to two additional failure mechanisms with plastic chip carriers. It is well known that water molecules will slowly diffuse into polymers with time. The rate of the diffusion depends on the humidity, temperature and the diffusion characteristics of the molding compound. The problem develops during the solder reflow operation if the temperature of the chip carrier exceeds the boiling point of water causing the trapped water molecules in the plastic molding compound to vaporize. Pressures are developed within the chip carrier that cause delamination of the bond lines between the chip and molding compound and between the lead frame and the molding compound. These delamination failures are illustrated in Fig. 15.2. In some instances, the pressure is sufficient to cause cracks to develop in the polymeric material that encases the chip. Bonding wire failures may occur due to the cracking, and the ingress of moisture to the chip face is accelerated. This moisture in turn leads to corrosion of the bonding pads. An example of the cracks that extend from the chip through the plastic molded chip carrier is shown in Fig. 15.3.

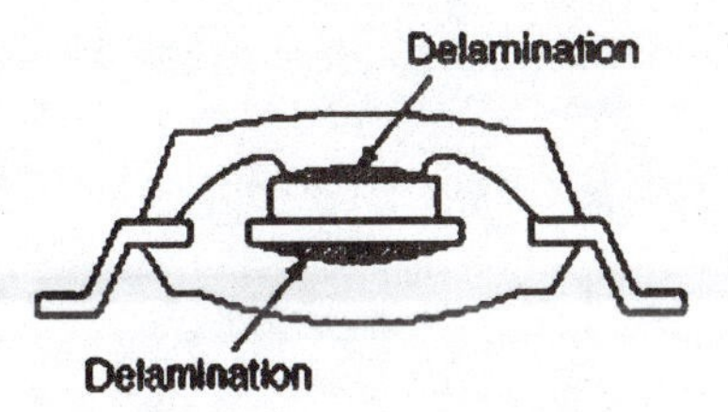

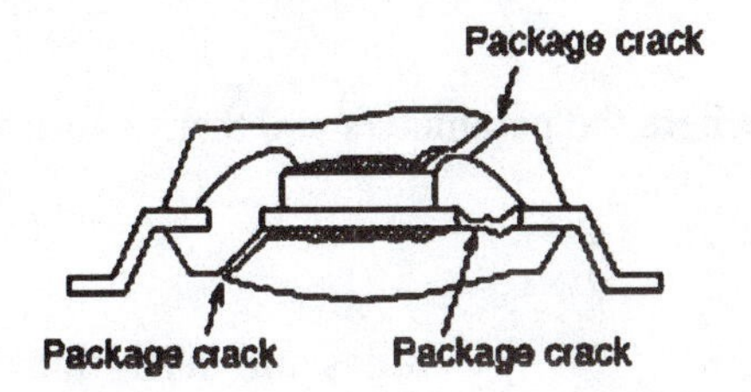

Fig. 15.2 Failure mechanisms in plastic encapsulated chip carriers due to moisture absorption and subsequent heating in a solder reflow operation.

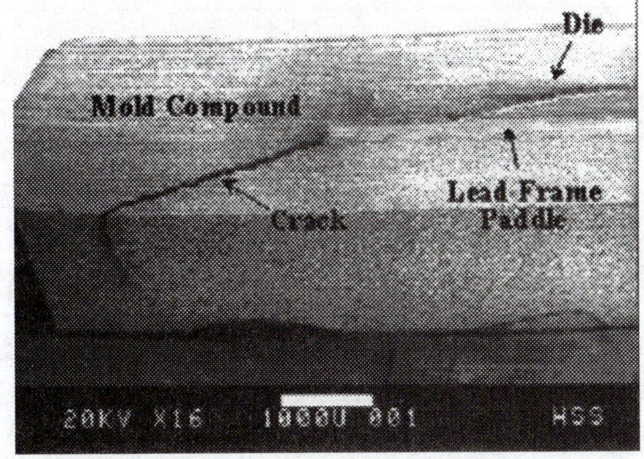

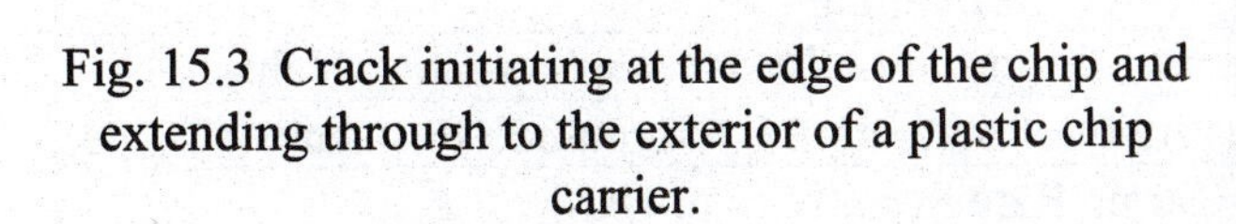

Fig. 15.3 Crack initiating at the edge of the chip and extending through to the exterior of a plastic chip carrier.

Thermal stresses produced in services by temperature changes that occur as a result of power on/off cycles or due to temperature changes that result from operation or storage in a changing environment produce failure of the solder bumps within the chip carrier or solder joints between the chip carrier and the printed circuit board. The thermal fatigue failure of electronic packages is associated with combined plastic-deformation and creep of solder joints. The Anand viscoplasticity model, a standard material in the finite element code ANSYS, has been used by several engineers to model the constitutive behavior of solder. This constitutive law has been used by Darveaux [2, 3, 4, 5]] to develop damage relationships. Anand's model is divided into a flow equation and three evolution equations that describe the strain hardening or softening of the materials. Constants used in applying these relations are presented in Table 15.3.

The Anand flow equation is written as:

$$\frac{d\varepsilon_p}{dt} = A\left(\sinh(\xi\sigma / s_0\right)^{\frac{1}{m}} \exp\left(\frac{-Q}{kT}\right) \tag{15.4}$$

And the three evolution equations are given by:

$$\frac{ds_0}{dt} = \left\{ h_0 \left(|B|\right)^a \frac{B}{|B|} \right\} \frac{d\varepsilon_p}{dt} \tag{15.5}$$

$$B = 1 - \frac{s_0}{s^*} \tag{15.6}$$

$$s^* = s\wedge \left[\frac{\frac{d\varepsilon_p}{dt}}{A} \exp\left(\frac{Q}{kT} \right) \right]^n \tag{15.7}$$

where the parameters and their values are defined Table 15.3.

Table 15.3
Values and definitions for Anand's parameters for 62Sn36Pb2Ag Solder.

Parameter	Value	Definition
S_0 (MPa)	12.41	Initial Value of Deformation Resistance
Q/k (1/K)	9400	Activation Energy/ Boltzmann's Constant
A (1/sec)	4.0E6	Pre-Exponential Factor
ξ (dimensionless)	1.5	Multiplier of Stress
m (dimensionless)	0.303	Strain Rate Sensitivity of Stress
h_0 (MPa)	1378.95	Hardening Constant
s^ (MPa)	13.79	Coefficient of Deformation Resistance Saturation Value
n (dimensionless)	0.07	Strain Rate Sensitivity of Saturation (Deformation Resistance) Value
a (dimensionless)	1.3	Strain Rate Sensitivity of Hardening

A typical crack in a solder bump used in flip chip mounting is shown in Fig. 15.4. These solder bump cracks are produced by cyclic shear stresses that cause fatigue damage. The solder bumps that fail initially are located at the corners of the chip at the point farthest from its neutral point. With time and additional thermal cycles, the solder balls located closer to the neutral point will also fail.

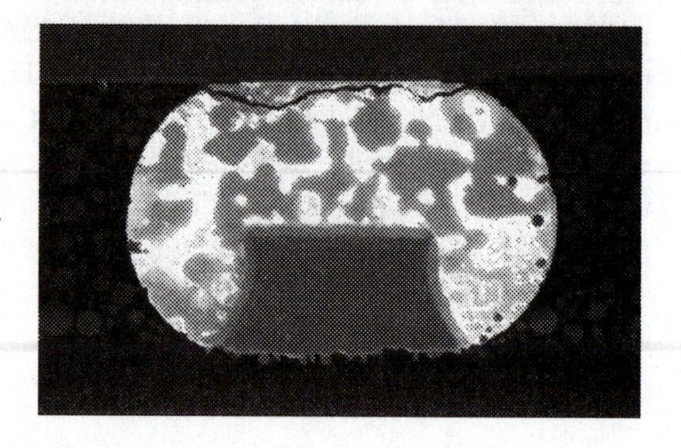

Fig. 15.4 Failure by a fatigue crack in a solder bump on a flip chip mounted die.

The progression of a crack in a solder bump on a flip-chip mounted die is presented in Fig. 15.5. the number of thermal cycles associated with the growth of the crack is shown in each photomicrograph.

Fig. 15.5: Solder Joint Progressive Crack 17mm BGA,

Solder joint failure under cyclic thermo-mechanical stresses is a dominant failure mechanism in first-level interconnects. These failures can occur with leaded chip carriers, leadless chip carriers and with ball grid arrays. Again these failures are due to thermal cycles that occur in operation and or storage. With the differences in the temperature coefficient of expansion of the chip carriers and the PCBs, solder joint failures will eventually occur with a sufficient number of thermal cycles. A solder joint failure of a gull wing chip carrier is depicted in Fig. 15.6. The crack initiated at the reentrant corner on the left side of the solder joint and propagated completely about the lower lead-solder interface. This is an unusual failure as the leads on a gull wing chip carrier normally are sufficiently compliant to prevent high stress due to differential expansion to develop. In this case, the solder encasing the inclined portion of the gull wing leads reduced their compliance and increased the thermal stresses. Also a stress concentration at the reentrant corner of the solder joint elevated the stresses that initiated the crack.

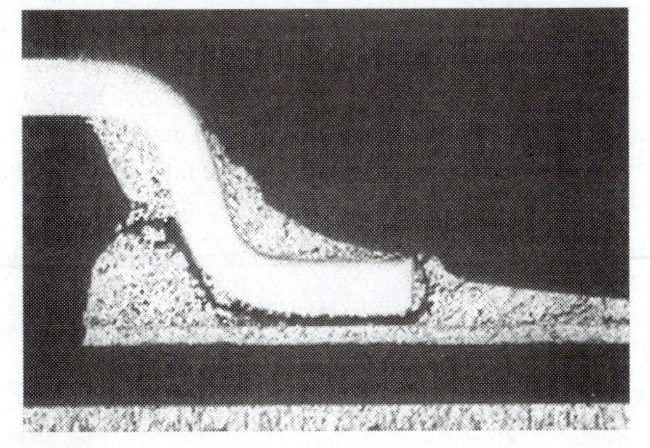

Fig. 15.6 Fatigue crack in a solder joint of a gull wing chip carrier.

Solder joint failure in leadless chip carriers and ceramic resistors and capacitors is also a serious concern. Without leads, the chip carriers exhibit very little compliance. Moreover, the temperature coefficient of expansion of the PCB is much higher than that of the ceramic chip carriers. As a result high thermal stresses are generated for relatively small temperature changes. The failures due to cyclic thermal stresses begin at the solder joints located at the corners of the package and proceed inward as the cyclic count increases. A typical fatigue failure of a solder joint on a 28 lead (leadless) chip carrier is presented in Fig. 15.7. The crack initiated in the fillet and extended through the joint and through the bond line between the chip carrier and the pad on the PCB.

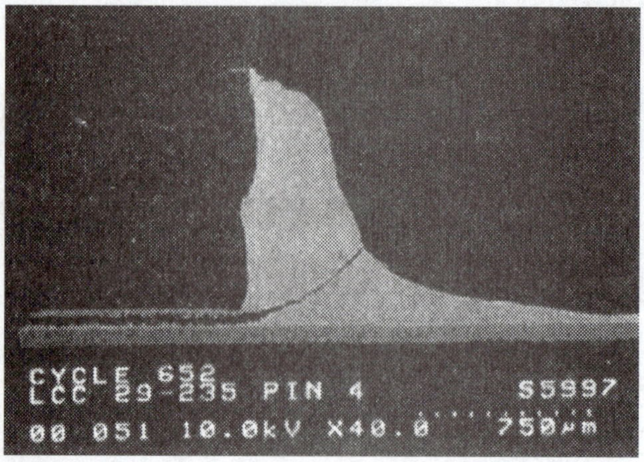

Fig. 15.7 Photomicrograph of a fatigue crack in a solder joint on a leadless chip carrier.

Another example of a crack propagating through a solder joint on leadless chip resistor is presented in Fig. 15.8

Fig. 15.8 Photomicrograph image of crack propagation of a solder joint on a 2512 chip resistor.

Some solder joint failures are produced by imperfections such as notches at reentrant corners or relatively large voids at critical locations in the solder. A relatively large void is evident in the solder joint presented in Fig. 15.9. In this instance the void did not initiate a fatigue crack, but if it had been located closer to the fillet it would have interacted with the fillet geometry increasing the stress concentration factor and reducing the fatigue life of the solder joint.

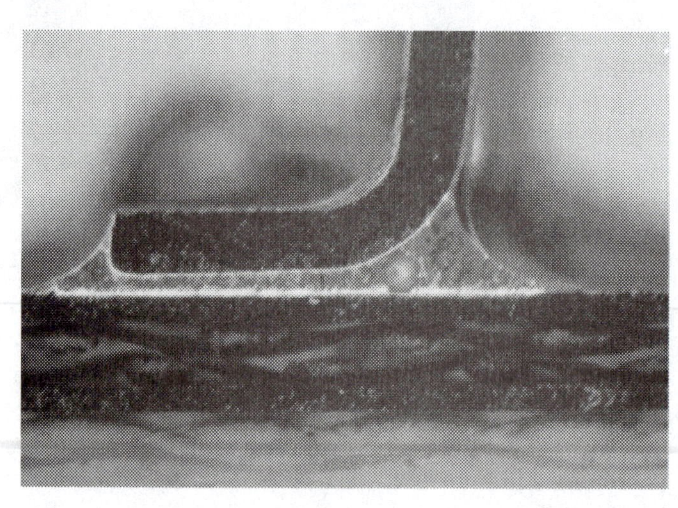

Fig. 15.9 Large voids in solder joints serve to concentrate stresses and may initiate fatigue cracks.

15.2.4 Failure Mechanisms with the Printed Circuit Board

Dominant common mechanisms for failure with printed circuit board include, blistering, pad lift off, plated-through-hole rupture, and on-board trace failures. Blistering is due to the diffusion of water molecules into the polymeric layers of the PCB while it is in storage. When the PCB is exposed to high temperature during the solder reflow operation, the water molecules vaporize. The pressure generated in the layer near the surface of the board is, in some cases, sufficient to locally delaminate a small inter laminar region. The top layer of the PCB bulges over this delamination producing a so called blister.

Pad lift-off as the name implies occurs when the solder pads adhered to the PCB fail. The copper cladding on the circuit board is bonded to the glass reinforced plastic laminate during a lamination process. Under normal processing conditions the pad to board bond produced in this manner is adequate. However, if the temperature of the pad during a repair operation is excessive, the epoxy adhesive fails and the pad lifts off the surface of the board.

Rupture of plated through hole or vias due to thermally induced stresses is another mechanism for failure of PCBs. A typical plated through hole, presented in Fig. 15.10, shows the barrel-like structure of the copper plated to the side of a hole through a glass reinforced plastic (GRP) circuit board. Note the roughness of the hole. When the assembly is subjected to an increase in temperature both the GRP and the copper plating expand. However, the GRP laminate is constrained by the presence of the glass cloth and as a consequence its out-of-plane coefficient to expansion α_z is much larger that the coefficient of expansion of the copper. Due to the differences in the expansion of the GRP and the copper, a large tensile thermal strain is imposed on the copper lining the hole. The stresses are acerbated by the surface roughness of the cladding because this roughness produces stress concentrations that reduce the resistance of the hole to cyclic fatigue. With a sufficient number of thermal cycles, fatigue cracks develop in the copper barrel of the plated through hole and failure results.

Fig. 15.10 Cross section showing a plated through hole in a glass reinforced plastic (GRP) laminate.

15.3 PREFERRED PART SELECTION

High quality and high reliability have always been the dominate factors in the design of electronic systems for military and space applications. Costs of the electronic components have always been secondary for most of these system developments. Several decades ago, performance requirements for many systems exceeded the manufacturing capability of the semiconductor industry. In the early years of the industry, production processes were inadequate and quality control practices were primitive and the product produced exhibited unacceptably high defects counts. To enhance the quality of incoming semiconductor devices, government agencies such as DOD and NASA began screening them in the 1960s according to procedures defined in MIL-STD-883, Quality and Reliability Assurance Procedures for Monolithic Microcircuits[2], a document still in use today. In the 1970s, additional MIL-STDs were adopted that specified environmental-stress screening for space-rated, (Class-S) components. This screening effectively eliminated defective devices and the problems related to defective semiconductor devices dropped dramatically.

For products designed for commercial applications such as automotive and consumer electronics, reliability requirements often were equivalent to those for military and space applications; however, for commercial applications cost was and is still a major factor. Manufacturers of semiconductors met the commercial demand for improved reliability by automating their processes to improved quality and replacing ceramic chip carriers with plastic encapsulated chip carriers to markedly reduce cost. The cost of packaging chips has dropped dramatically over the past three decades[3].

For many decades semiconductor manufactures have improved the quality and the reliability of commercial electronics. Failure rates for highly integrated electronics have also improved as is shown in Fig. 15.11. Today unscreened commercial-grade semiconductor devices exhibit failure rates essentially the same as those found for screened components, as indicated in Fig. 15.12. However, these measurements of production quality and reliability are valid only for automated production and assembly processes. Lower quality and reliability usually results when parts are handled and assembled by hand.

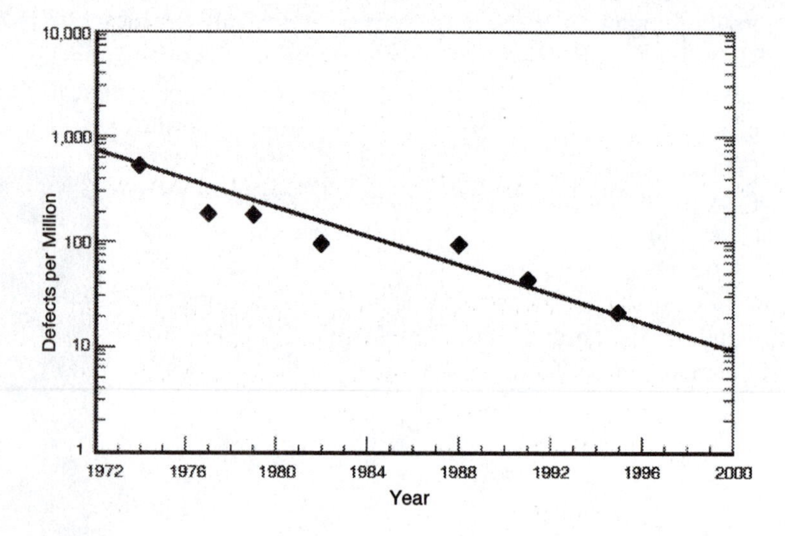

Fig. 15.11 Reduction in defects achieved with improved and automated production processes over the past three decades.

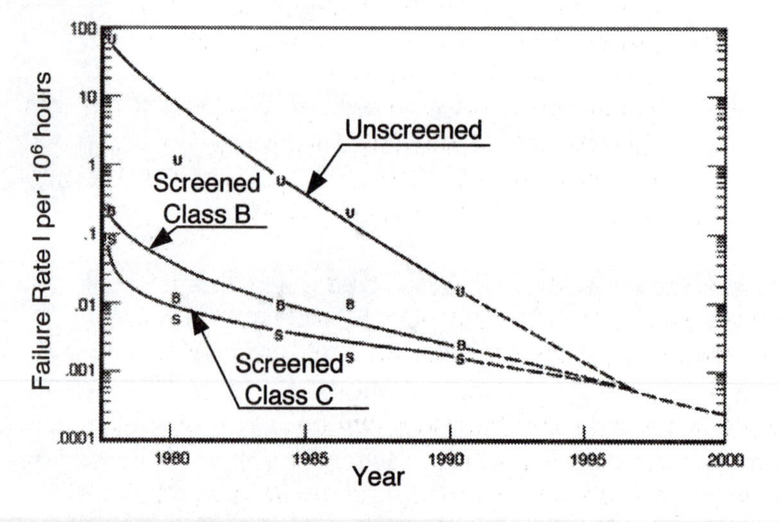

Fig 15.12 Comparison of failure rates for screened and unscreened semiconductor devices over the past two decades.

---

[2] MIL-STD-883 is still used today.

[3] In 1995, a digital signal processor cost $947 encapsulated in military grade ceramic, $400 in military-grade plastic, $182 in commercial-grade ceramic, and $73 in commercial-grade plastic.

The market for semiconductor devices was dominated by the government in the 1970s, but improved reliability of higher powered components lead to increased demand from the commercial sector of the market. Today government purchase of electronic components represents less than 2% of the market. If fact many prominent companies[4] have dropped military grade products offering only commercial products that are often employed in military applications. However, the use of commercial semiconductors in military and space applications has caused concerns for system designers that included:

- Providing adequate resistance to the effects of ionizing radiation in space.
- Surviving the stress imposed in qualification testing and shock and vibration in service.
- Demonstrating the reliability demanded of ceramic parts intended for military use.
- Surviving long-term storage in harsh environments.

Commercial semiconductor devices packaged in plastic chip carriers have been shown to be extremely reliable in demanding applications. The automobile industry has used plastic encapsulated components in harsh environments for about two decades. In 1995 a major supplier of automotive electronics reported an under-the-hood reliability for microprocessors of approximately 40 failures per million devices in a 5-year, 50,000 mile warranty period. It is expected that the failure rate today is less than 10 failures per million devices over a longer warranty period.

Radiation tolerance remains a concern for semiconductor devices housed in plastic chip carriers because they are susceptibility to gamma rays. Moreover, CMOS devices are manufactured with extremely small feature sizes (25 to 30 nm gate lengths). Radiation susceptibility increases as feature size decreases so this concern remains for space applications where radiation may be encountered. Shielding the system or the component is the recommended approach to avoid the detrimental effects of a gamma ray striking a nanometer scale feature on a CMOS chip.

15.3.1 Selecting Commercial Components with High Reliability

The objective of component selection is to ensure that those used in an electronic system will enable it to meet its requirements in terms of functionality, quality, reliability, schedule and cost. Important elements of a component selection program are:

1. Standardization of types of components
2. Definition of quality and testing levels to insure conformance to specifications.
3. Qualification of components and manufacturer's facilities and processes.
4. Testing, screening, lot acceptance and periodic inspections.
5. Well written procurement specifications.
6. Process control and inspection by the manufacturer.
7. Documentation and data definition during product development and component manufacture.

The semiconductor business is extremely competitive. To met the quality and reliability requirements of their customers they have adopted international standards for quality assurance such as ISO-9000 that covers the entire process from the product planning stage to development, design, trial production, evaluation, mass production, shipping and servicing. A comprehensive flow diagram showing the quality assurance system of NEC Electronics is presented in Fig. 15.13.

---

[4] Other manufacturers, such as Texas Instruments and Analog Devices, remain in the billion-dollar military market.

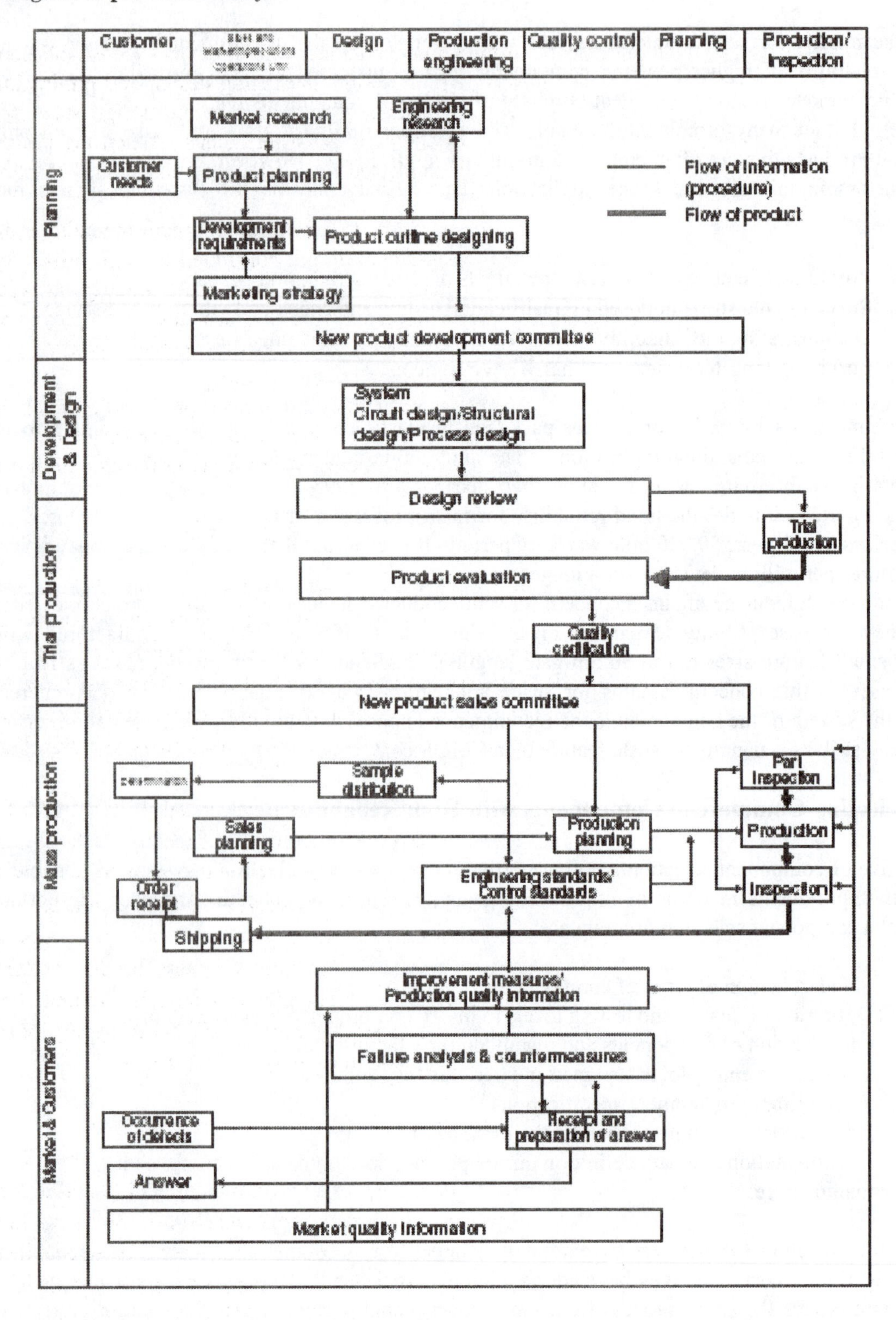

Fig. 15.13 Flow diagram showing NEC electronics quality assurance program for developing and manufacturing a new semiconductor device.

In the planning stage, the requirements for the semiconductor device or devices are established. In the development and design stage, system design, circuit design, packaging design and production process design are conducted in parallel. Testing is particularly important in the design stage for complex and advanced devices with small feature sizes and high transistor counts. When the design stage is completed, a comprehensive design review is conducted by a reliability and quality control department.

After the design review, trial production runs are made to determine optimum manufacturing conditions. After the specified level of quality has been achieved and confirmed through reliability evaluation and electrical testing, the operation moves to mass production. In the mass-production stage, the processes are computer controlled and quality information collection and feedback are employed to maintain uniform quality. Production maintenance schedules and statistical process control are implemented to maintain a high level of equipment reliability and product quality.

Verification that problems related to reliability are not introduced in mass-produced products is accomplished with long term reliability testing using selected products and by analysis of defects reported by customer. The results are used to modify the mass production processes or the chip design.

Reliability Design Concepts

The reliability of a semiconductor device is dependent on each feature on the chip, the chip and its package. To achieve high reliability, the reliability of each of these three elements must be insured. Generally, reliability is divided into three stages—early failure, random failure, and wear out failure—as expressed by the well-known bathtub curve. Early (infant) failures are usually eliminated prior to packaging by wafer level screen testing, and random failures are eliminated through production process control. Failures due to wear-out are eliminated by chip design with features sized so as to assure that wear-out failures do not occur during its specified life.

The quality approval program for the semiconductor devices involves frequent inspections to determine if the functions, performance, electrical characteristics, reliability, and quality of the initial devices meet the specifications written during the planning stage. Mass production of a device is not initiated until the tests for quality approval are completed. Quality certification is often conducted in-house by the manufacturer and by the customer. The procedure for quality approval differs depending on the maturity of each part of the device. If elements such as the production process, production line and chip carrier are completely new, the most stringent reliability and quality evaluation is required. On the other hand, if the reliability and quality of the process and chip carrier have been already been established with devices already in the marketplace, the approval procedures are less stringent.

Predicting Failure Rates of Off-the-Shelf Components

The description in the previous subsection pertained to the design of an electronic system requiring the development of new semiconductor devices. However, in many applications the electronic systems can be designed with off-the-shelf components. In this situation, the procedure is to select components that are commercially available and to perform failure analysis for each component and for the system as a whole. The methods for predicting both component and system MTTF were described in Chapter 15. Other methods for predicting reliability are described in the Military Handbook for Reliability Prediction of Electronic Equipment MIL-HDBK-217-F2. This handbook is published by the Department of Defense, based on work done by the Reliability Analysis Center and Rome Laboratory at Griffiss AFB, New York. This handbook contains failure rate models for various components used in electronic products including: integrated circuits, transistors, diodes, resistors, capacitors, relays, switches, connectors, etc.

MIL-HDBK-217 provides a common basis for predicting reliability when the DOD is in an acquisition process for contracting for electronic systems intended for military applications. The handbook provides a common basis for comparing and evaluating reliability prediction of related and competing designs. The data in MIL-HDBK-217 has been acquired from the field data of a large number of military electronic components procured according to a MIL standard. Moreover, because the DOD requires contractors to use MIL-HDBK-217 procedures when designing military electronic equipment, large allowances have been made for the coefficients used to determine the system reliability. The predicted failure rate using the methods described in MIL-HDBK-217F is usually higher than the failure rate calculated by the methods described in Chapter 15 by a factor of 10 to 100.

If the methods described in Chapter 15 for predicting the reliability or failure rate of a system are employed, data on the failure rate of each component must be established. Manufactures of off-the-shelf components have listings that provide the required data. Examples of listings showing failure rate data[5] for integrated circuits and for transistors and diodes are given in Table 15.4 and 15.5, respectively. The failure rates are based on an operating temperature of 55 °C and a confidence level of 60%. If the operating temperature is higher than this value the activation energy is used to calculate an acceleration factor and the value of the failure rate is adjusted accordingly.

Table 15.4
Failure rates for select integrated circuits

Part Number	Activation Energy, E_a	Failure Rate FIT
UPB551C	0.7	20
UPC1093T	0.7	15
UPC141G2	0.7	35
UPC1854CT	0.7	35
UPD100500D	0.7	75
UPD16261GS	0.7	5
UPD16431GC	0.7	40

Table 15.5
Failure rates for select transistors and diode circuits

Part Number	Activation Energy, E_a	Failure Rate FIT
03P2M	0.7	0.4
1S2192	0.7	0.2
1S2836	0.7	0.1
1SS221	0.7	0.1
1SZ45A	0.7	0.3
2N4392	0.7	0.6

[5] The data listed was taken from the NEC Electronic Corporation publication titled Failure Rates, Document No. C11178EJCV0IF00, 12th edition, June 2005.

15.4 DERATING AND STRESS MANAGEMENT

Derating components is a well established procedure used to develop long-life, high-reliability electronic systems. Experience has shown that derating increases both life and system reliability; however, the enhanced life and reliability is achieved at a cost. To minimize this cost it is necessary optimize the derating factors. Excessive derating leads to over-design, over-cost and over-sizing of components and additional volume and weight. The objective of derating is to obtain reliable, high performance systems with the minimum increase in sizing the semiconductor and passive devices.

The term derating refers to the intentional reduction of electrical, thermal and mechanical stresses on components to levels below their specified rating. Derating provides a safety margin between the applied voltage, current, operating temperature and temperature differentials that limit the system performance. Derating also improves life and reliability by protecting components from unexpected changes in voltage or temperature encounter in service and in small variations in the capabilities (reliability) of the components.

15.4.1 Principles of Derating

The parameter (voltage, current, power or temperature) strength of a specific component defines the upper limit where it can perform satisfactorily in a prescribed application. The parametric strength of a component, which varies from manufacturer to manufacturer, from type to type, and from lot to lot, can be represented by a statistical distribution as shown in Fig. 15.14. In a similar manner, the stress applied to this component can be represented by a statistical distribution that is also shown in Fig. 15.14. In this figure both the applied stress and parametric strength are represented by probability density functions. A component operates in a reliable way if its parametric strength is greater than the applied parametric stress. The object of a designer is to ensure that the applied stress applied is less than the component's strength for each parameter (voltage, current, temperature, etc.). Because of the variance in the statistical distribution functions there is a possibility that the two distributions will overlap as indicated in Fig. 15.14. This overlap is shown as a shaded area. Large shaded areas imply a high failure probability. To decrease the shaded area the designer decreases the applied stress, which moves the stress distribution to the left or increases the component's strength parameter by selecting an over-sized component, which moves the strength distribution to the right.

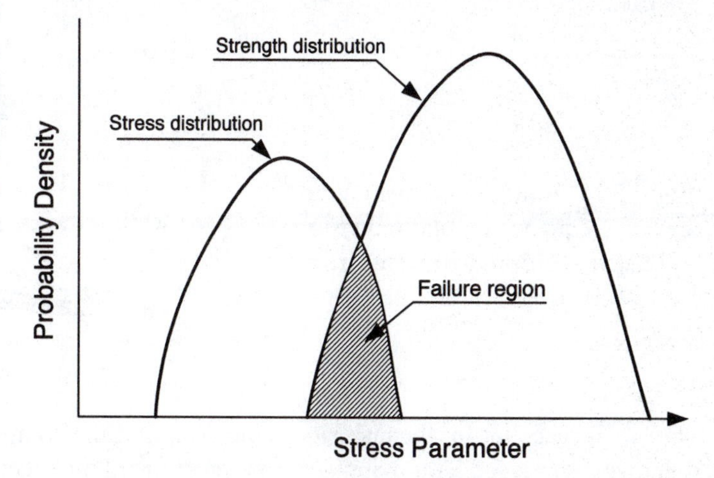

Fig. 15.14 Statistical distribution functions representing the component's parametric strength and applied stress.

The objective in design is to maximize the strength-to-stress ratio of each component in the system. Derating reduces the applied stress and moves the stress distribution to the left. Selecting over-size components increases their strength the strength distribution to the right. Derating or over-sizing components reduces the probability of failure, improves the end-of-life performance and provides additional safety margins.

15.4.2 Derating Parameters

Derating requirements are usually specified for each component family (i.e. integrated circuits, capacitors, resistors, diodes, etc.) The parameters that are derated include:

- The junction or case temperature at maximum operating conditions
- Power
- Voltage
- Current

Some larger companies provide derating ratios for different applications and different life expectancies. For example, Toshiba lists the following derating factors for discrete semiconductors and power devices.

- Junction temperature T_j = 80% of T_j maximum or less for 10 years of use at 3 hours/day.
- Junction temperature T_j = 50% of T_j maximum or less for 10 years of use at 24hours/day.
- Voltage 80% of maximum manufacturer's rating or less.
- Average current 80% of maximum manufacturer's rating or less.
- Average current 50% of maximum manufacturer's rating or less for rectifying component.
- Peak current 80% of maximum manufacturer's rating or less.
- Power 50% of maximum manufacturer's rating or less.

Derating curves are also provided in MIL standard MIL-HDBK-217F. An example of such a derating curve for a low frequency bipolar transistor is presented in Fig. 15.15. The derating data in MIL standard MIL-HDBK-217F classifies semiconductor devices into large groups according to application and does not consider maturity of the components or the process stability both of which differ for each device.

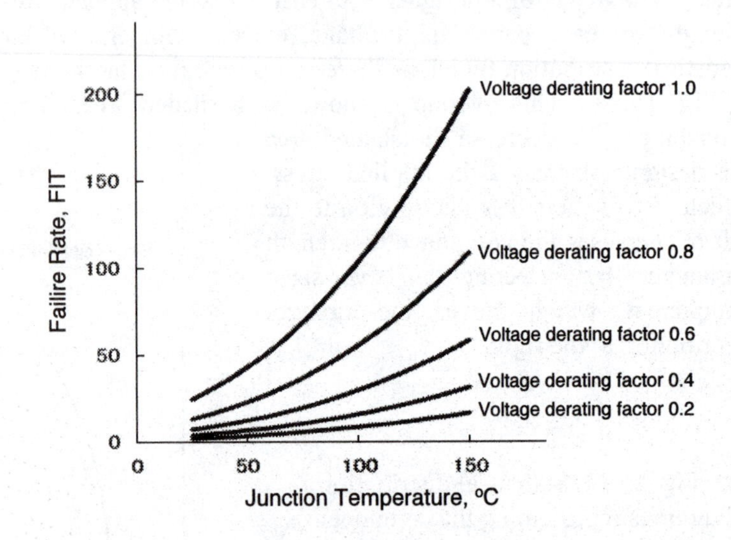

Fig. 15.15 Voltage derating curves for low frequency bipolar transistors. Data taken from MIL standard MIL-HDBK-217F.

To account for the maturity of the design of the components and the stability of their production processes, semiconductor manufacturers recommend different derating curve data based on independent studies. An example of the relative failure rate as a function of junction temperature for a silicon transistor is presented in Fig. 15.16. The data used in constructing this figure is based on in-house testing and field experiences recorded by the manufacturer (Toshiba).

Inspection of Fig. 15.16 shows that a reduction of 40 °C in the junction temperature of this specific silicon transistor results in decrease in the relative failure rate of a factor of 10. Clearly derating the operating temperature of a device markedly enhances its reliability.

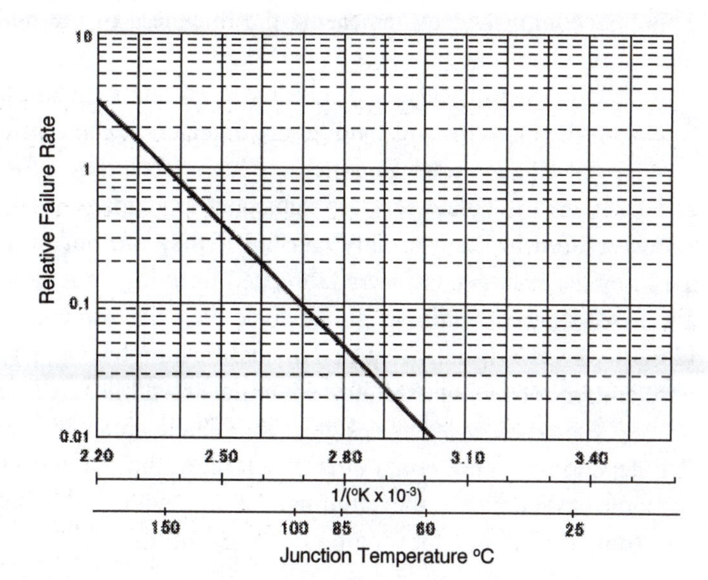

Fig. 15.16 Relative failure rate as a function of junction temperature for a silicon transistor. (Toshiba data).

Temperature Derating

Derating voltage, current and power is usually accomplished by component selection. However, derating temperature and temperature differentials requires analysis and design of the cooling system. Usually the design of the cooling system precedes the thermal analysis to provide information pertaining to the thermal resistances associated with the conduction and convection heat flow. Then a thermal analysis is performed using methods described in Chapters 9 and 10. If the predicted temperatures exceed those required for the specified amount of derating, it is necessary to redesign the cooling system to lower the thermal resistances in the heat flow path. In many cases, experiments are conducted to verify either the case or the junction temperatures that can be achieved with the newly designed thermal cooling system.

Temperature differentials are also important as the thermal stresses imposed on the solder joints are a function of the temperature history of the chip carrier and the PCB from a cold start to the highest operating temperature that is reached some time after full power is applied. As the temperature cycles each time a system is powered on or off, fatigue failures of the solder joints or solder bumps becomes a serious concern. Derating in this case is achieved by reducing the maximum temperature of the chip carrier and the PCB to limit the temperature differential ΔT.

Mechanical Stresses

Mechanical stresses are imposed on a system by vibration and shock. These stresses occur in almost every system during shipment. In some systems for military and space applications, very high levels of vibratory and shock forces are imposed during qualification testing according to contract specifications. In other systems mounted on vehicles, the vibratory and shock forces transmitted to the electronic system are determined by the support system used to mount the electronic enclosure to the frame of the vehicle. Several different approaches are employed to limit the exposure of the components to excessively high mechanical stresses imposed under these conditions.

Vibration and shock forces that are encountered in transportation are mitigated by properly designed shipping containers that crush when dropped. The electronic products housed in their enclosures and shipping containers are also tested to insure that none of their internal components under go resonance conditions during shipment. If a circuit board exhibits a resonance condition at some frequency associated with shipping, the circuit board is redesigned to increase its natural frequency.

This is accomplished by increasing the thickness of the board, adding stiffening ribs or by providing more rigid clamps at the edges of the board.

Qualification testing of electronic products for military and space applications imposes severe forces on the circuit boards and their connectors. The semiconductor devices and most of the resistors and capacitors can under go extremely high acceleration forces without incurring damage, but the pins and pads on the connectors are subjected to damage by bending or fretting. Under shock loading, fasteners used in the assembly can fail if they are not properly designed. These failures are easy to avoid by using larger fasteners fabricated from high strength steel. Again design of the circuit board to avoid resonance conditions is an objective. If resonance conditions cannot be avoided during the vibratory sweeps usually required in qualification testing, support structures are introduced that limit the acceleration forces imposed on the internal assemblies by increasing the amount of damping.

Electronic systems designed for vehicle applications are subjected to vibratory and shock forces that depend upon the terrain over which the vehicle traverses and the suspension system of the vehicle. In some applications, the environment is severe, and the design approach is to shock mount the electronic system at a location on the frame of the vehicle where the acceleration forces are a minimum. Again the failures are not encountered with well packaged semiconductors, which can undergo accelerations of many thousand Gs; however, connectors, fasteners and heavy components like transformers are prone to damage at much lower G levels.

Radiation Environments

In many military and space applications failures due to radiation is a serious concern. Radiation sensitive semiconductor devices employed in the system must be identified. When possible, semiconductor devices that are more tolerant of radiation are selected[6]. The usual approach to avoid radiation damage is to shield the complete system with a radiation tight enclosure. If the number of components that are sensitive to radiation is limited, the shielding is placed within the enclosure is designed to protect only the sensitive components. In almost all radiation resistant designs, the frame of the vehicle is incorporated into the radiation shield. Finally, standards exist for electrical derating of radiation sensitive components [6].

15.5 SCREENING COMPONENTS AND ASSEMBLIES

Screening is a process employed to eliminate defective components and assemblies from production. The concept is based on the fact that high failure rates observed with electronic components are due to defects introduced in the manufacturing process (the infant mortality phase). Screening tests are design to eliminate manufacturing defects than may be introduced at each step in the production process. Usually the defects can be eliminated with screening tests of relatively short duration. It is assumed that the detrimental effects of the tests on the surviving members of the population are negligible. Elimination of the components with manufacturing defects results in improved reliability and life of systems fabricated with screened components.

Screens are designed to test newly design components at each step in the production process until it has been demonstrated that these processes are mature and infant mortality failures are not a significant concern. For example, in the design of a new chip housed in a newly design chip carrier, the

[6] Selection of radiation tolerant semiconductor devices is become more difficult as the feature size is reduced. A single gamma ray striking a P/N junction can disable a circuit on a chip.

chip manufacturer would perform screening tests at the wafer level and adjust production processes to enhance yield and to eliminate the causes of different types of failures. The manufacturer of the chip carrier would test a large population of these carriers and adjust the design and production processes until defects associated with infant mortality were largely eliminated. The process used to mount the chip in the chip carrier and to attach the leads to the bonding pads would be adapted to produce essentially defect free packaging by using a battery of screening tests. Modules, subassemblies and the final product would be subjected to screening until the manufacturer is assured that the reliability of the product is adequate to introduce the product (electronic system) to the market without incurring significant warranty costs.

After production processes associated with a specific chip, component or module have become mature, screening tests are no longer required to insure reliable operation. Results from the initial screening tests enabled production processes to be modified and the cause for the defects associated with infant mortality failures corrected. However, new designs and new processes require extensive screening until test results indicate that the mechanisms for failure have been identified and all of the causes of the failures in production have been corrected.

Temperature cycling is often used as screening test to eliminate manufacturing defects associated with infant mortality failures. Temperatures imposed on modules and complete systems during a typical screening test are shown in Table 15.6. The tests are conducted in programmed ovens with temperature-time profiles similar to the one shown in Fig. 15.17.

Table 15.6
Temperatures imposed on modules and complete systems during a typical screening test

Temperature Cycling	Severity Level
Module Screen	
Temperature Range	Max: –55 to 125 °C (ΔT = 180 °C) Nom: –40 to 95 °C (ΔT = 135 °C) Min: –40 to 75 °C (ΔT = 115 °C)
Temperature Ramps	Max: 20 °C/minute Nom: 15 °C/minute Min: 5 °C/minute
Number of Cycles	Max: 40 Nom: 30 Min: 20
Power Condition	Power On in Development Phase Power Off in Production Phase
System Screen	
Temperature Range	Max: –55 to 125 °C (ΔT = 180 °C) Nom: –40 to 95 °C (ΔT = 135 °C) Min: –40 to 75 °C (ΔT = 115 °C)
Temperature Ramps	Max: 20 °C/minute Nom: 15 °C/minute Min: 5 °C/minute
Number of Cycles	Max: 12 Nom: 10 Min: 8
Power Condition	Power On

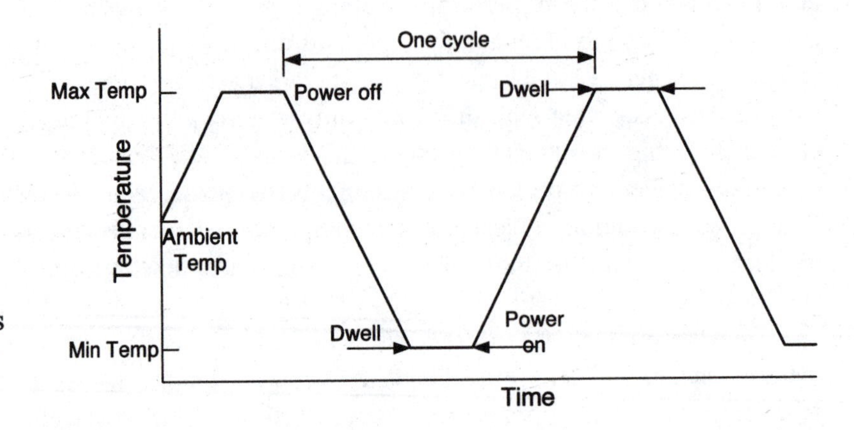

Fig. 15.17 Temperature-time profiles used in screening tests for ambient cooled electronic systems.

15.5.1 Screening Test Procedures

There are a number of different tests that are use to develop new designs and their production processes, which include:

1. Burn-in tests consist of baking, for a minimum of 48 hours, electrically active components at relatively high temperature—125 °C for military components and at 70 °C for components intended for commercial applications. The components are operated under power during these tests. The intent of the high temperature exposure is to eliminate manufacturing defects and reduce the failure rate due to infant mortality. The effectiveness of this procedure depends on the type of defect introduced during production. High temperature screening tests are not always effective, and in some cases the effects of prolonged exposure to high temperature is detrimental.
2. High temperature tests without electrical activation. These tests consist of baking hermetically sealed ceramic components at 250 °C for several hours or baking plastically encapsulated components at 150 °C. The purpose of the high temperature baking is to identify problems related to contamination, moisture ingress and inadequate metallization.
3. Electrical stressing consists of operating the semiconductor devices at either higher than specified voltage in continuous operation for several hours or by applying a number of cycles of voltage over-stress. The higher voltages are applied while the chip is operated at relatively high temperature. The electrical stressing is effective in initiating gate oxide failures in CMOS type chips.
4. Screening tests involving temperature cycling between – 55 and 125 °C for components intended for military applications are employed to identify failure mechanisms related to bonding, solder bumps and solder joints. The range of temperatures used in cycling components intended for commercial applications often are more moderate with temperatures varying from – 40 to 95 °C.
5. Thermal shock tests consist of immersing components (usually packaged chips) into a hot oil bath. This extreme exposure with very high temperature gradients is effective in identifying die attach (bonding) failures and for locating defects in the polymeric molding materials used in encapsulating the chip.

6. Mechanical shock tests that impose high acceleration forces[7] (100 or more Gs) are conducted to identify structural problems related to assembly of a system in some type of an enclosure. These problems include failure of fasteners, impact with adjacent modules, damaged connector pins, permanent deformation of reinforcing ribs, failure of clamps, etc.
7. Screening tests involving vibration sweeps between two or more specified frequencies are used to determine the resonant frequencies and the transmission coefficients associated with each resonant frequency. These tests are usually conducted with either complete systems housed in their enclosure or with relatively large subsystems. When excessive vibratory forces are measured in these tests, design of new supporting structures is usually required.
8. Humidity tests are conducted with complete systems or subsystems to identify problems that may develop due to the presence of water vapor penetrating the enclosure. These tests are often conducted with 85% relative humidity and a temperature of 85 °C.
9. Leak tests are conducted with hermetically sealed devices to establish the adequacy of their seals. One of the screening tests involves immersing the device in a hot fluorocarbon liquid. If the chip carrier leaks, a trail of bubbles is observed. A more sensitive test for leaks involves immersing the device in radioactive gas. When the device is removed from the vessel containing the gas, it is examined with a Geiger counter. Detection of radioactive gas escaping from the cavity of the chip carrier indicates leakage.
10. High G tests are conducted to screen systems intended for applications involving high acceleration forces. Shock tests described above are employed if the acceleration forces are of relatively short duration. However, for longer term effects of high acceleration forces, the completely assembled systems are placed on a centrifuge and subjected to specified acceleration forces that may be as large as several tens of thousands of Gs.

15.6 ACCELERATED TESTING

The design life of electronic systems for many commercial applications usually ranges from 5 to 10 years with warranties offered for much shorter periods of time. However, design life for military and industrial applications is for much longer periods of time. This extended life is often achieved with maintenance procedures that specify replacement of components and or modules that may exhibit relatively high failure rates. The time available to design a new electronic system in the competitive environment that exists today is extremely limited. The question the development team must address is how to insure reliability of their products in a few months. Accelerated testing is an approach that is effective in reducing the time required to insure reliable operation the product's specified life during its development phase. Several different accelerated test procedures are used in practice to reduce the testing time to a few months or less.

15.6.1 Accelerated Testing by Elevating the Junction Temperatures

We introduced the effect of temperature on the time to failure for semiconductor devices in Chapter 15. The time to failure is modeled with an Arrhenius relationship that characterizes each failure mechanism with an activation energy and a temperature according to:

[7] Semiconductor devices are capable of extremely high acceleration forces if they are properly bonded to a well supported circuit board. Electronics mounted in artillery projectiles function satisfactorily after being subjected to over 50,000 Gs that is imposed as the projectile passes through a concrete wall.

$$A(T) = \frac{t_1}{t_2} = e^{\left[\frac{E_a}{k}\left(\frac{1}{T_1} - \frac{1}{T_2}\right)\right]} \qquad T_1 > T_2 \qquad (15.8)$$

where A(T) is the acceleration factor, t_1 and t_2 correspond to time of failure at absolute temperatures T_1 and T_2 , respectively. E_a is the activation energy and $k = 8.63 \times 10^{-5}$ eV/°K is Boltzmann's constant.

When T_1 is greater that T_2, then the ratio of t_1/t_2 and the acceleration factor is less than one. This fact implies that the time to failure t_1 at the higher temperature T_1 is less than the time to failure at a lower temperature T_2. Using this approach the time to failure t_f is often written as:

$$t_f = A_0 e^{\left(\frac{E_a}{kT}\right)} \qquad (15.9)$$

It is clear from Eq. (15.9) that the time to failure of a component is minimized by maximizing the junction temperature. The activation energy E_s depends upon the failure mechanism and its cause. In most cases, the activation energy is taken as some value between 0.3 and 0.9 eV. If the failure mechanism is not known, an activation energy of 0.7 eV is assumed. The activation energies associated with several different failure mechanisms are presented in Table 15.1.

Studies have shown that the time to failure is a linear function of the log-normal cumulative probability of failure as indicated in Fig. 15.18. These results clearly indicate that the time to develop confidence in a new design can be reduced by an order of magnitude or more by increasing the temperature of the tests required in determining the probability of failure. The parallel straight line, shown in Fig. 15.18, indicates that the statistical distribution of the failures is a log-normal function for all test temperatures.

While increasing testing temperature markedly reduces test time, care must be exercised to avoid excessive temperatures that may cause uncharacteristic failures. A common practice to establish test temperatures, which reduces testing time without damaging components, is to increase the testing temperature with small samples in incremental steps. The number of failures is recorded as a function of time during the test and the reason for failure is ascertained by inspecting the failed devices. The incremental step in temperature where uncharacteristic failures are observed is deemed excessive. Test temperatures one increment below this level are consider the optimum for accelerating the test and for minimizing the time required to gain the confidence in a new design prior to marketing the product.

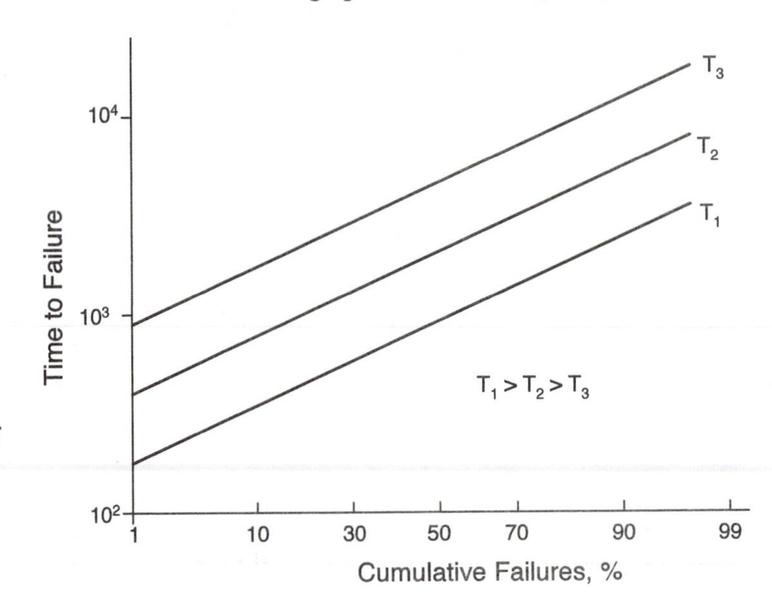

Fig. 15.18 Time to failure as a function of cumulative probability of failure with testing temperature as a parameter.

15.6.2 Accelerated Testing by Applying Electrical Stresses

Operating a semiconductor device at voltage higher than specified is also a means of reducing the time required to determine the failure rate λ. However, care must be exercised because excessively high voltages, approaching the breakdown voltage, initiate several different failure mechanisms. A relationship for the failure rate as a function of collector base voltage V_{CB} and junction temperature T_j is shown below:

$$\lambda = e^{\left(C_o - \frac{E_a}{kT_j}\right)} e^{\left(C_1 \frac{V_{CB}}{(V_{CB})_{max}}\right)} \tag{15.10}$$

where $(V_{CB})_{max}$ is the maximum base collector voltage and C_0 and C_1 are empirically determined constants.

The first term in Eq. (15.10) is the Arrhenius relation that accounts for the effects of junction temperature, while the second term accounts for the acceleration due of excessive base collector voltage. The constant C_1 is taken at 1.5 for most semiconductor devices with a limit of 60 V for the maximum base collector voltage. If C_1 and the activation energy are known for a semiconductor device, then the constant C_0 can be established with Eq. (15.10) and the data from a single life test. Kemeny [7] has shown the acceleration factor for failure rates as a function of the ratio of $V_{CB}/(V_{CB})_{max}$ with $(V_{CB})_{max}$ as a parameter. These results, presented in Fig. 15.19, indicate acceleration factors ranging from 1 to 6.

In screening tests, the ambient temperature is usually maintained at about 25 °C, so that the failures induced by elevated base collector voltages are the dominant failure mechanism.

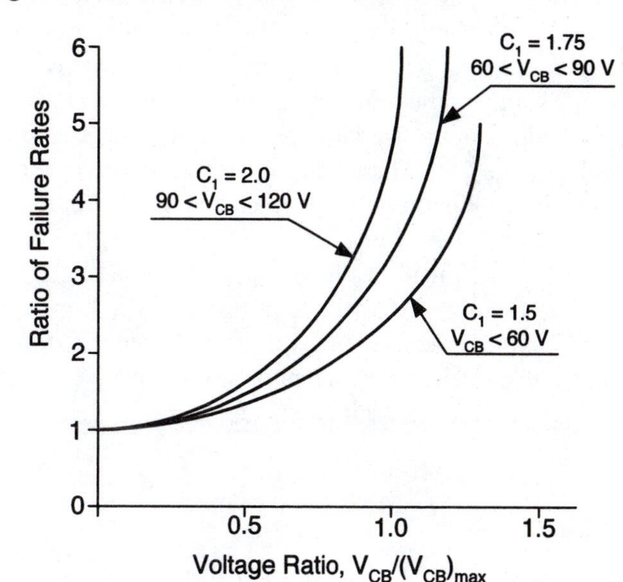

Fig. 15.19 Failure rate λ as a function of base collector voltage with transistor voltage limits as a parameter. [7]

16.6.3 Accelerated Testing Under High Humidity Conditions

Moisture that gains access to electronic components has several detrimental effects including corrosion of the metallization often by galvanic action, electrolytic conduction among electrically biased metal surfaces and the transfer of metal and mobility of surface charges that effectively extends the gate potentials over adjacent surfaces. Because any of these effects, tests are conducted in the development phase of a new product to insure that the packaging of individual components, subsystems and complete systems is adequate to prevent the ingress of water vapor into their sensitive areas.

There are two different test procedures often used to verify the performance of electronic systems when operating under humid conditions. The first is referred to as the THB 85-85 test because it involves operating the component or system at 85% relative humidity, at a temperature of 85 °C and under reverse bias. In this test, the power dissipated by the components is usually relatively low; hence, the humidity adjacent to the components remains near 85%. While the THB 85-85 test is effective in locating components that fail due to moisture ingress, the test often requires several hundred or a thousand hours to complete. If the power to the components is cycled, the time to complete the test is increased and is dependent on the cycling rate.

The second procedure employed to verify resistance against moisture is a highly accelerated stress test (HAST). In this test the ambient temperatures are increased to some value between 100 and 175 °C, while the humidity is controlled at some specified value in the range of 50 to 85%. When possible, reverse electrical bias is also employed as a means of stressing the components. The tests at these high temperatures are conducted in a pressurized vessel to avoid saturated water vapor from condensing on the surfaces of the leads. Studies of moisture induced failures using both test procedures indicated that they initiated the same failure mechanism. It appears then that the use of the HAST procedure reduces the time to verify a new design against the effects of moisture by a significant factor.

15.6.4 Accelerated Fatigue Tests

When electronic products are exposed to vibrations failures of lead wires, connector pins, fasteners, clamps, etc. often occur, particularly if the system undergoes one or more resonances during exposure to a vibratory environment. These failures are due to stresses that result from the application of vibratory forces to the components mounted within the enclosure. To verify the adequacy of the mechanical systems supporting and enclosing the circuit boards, vibration tests are conducted where the frequency sweeps simulate the vibratory environment to which the equipment will be exposed. The amplitude of the vibratory input, which is related to the stresses imposed on the internal hardware, is increased to accelerate the test. The number of cycles to failure for metallic materials is dependent on the magnitude of the alternating stresses as indicated in the S-N curves presented in Fig. 15.20.

Inspection of Fig. 15.20 shows that the life of a metallic component is reduced significantly by relatively small increases in the applied alternating stresses[8]. An approximate rule for ferrous materials is about an order of magnitude decrease in life (time require to test) for an increase in alternating stress of 15%. Similar acceleration factors are achieved with non-ferrous materials. Note in Fig. 15.20 that S_u is the single cycle tensile strength of the metallic material. Note also that the ferrous materials exhibit an endurance limit, which implies infinite life when the alternating stresses are lower than about 0.5 S_u.

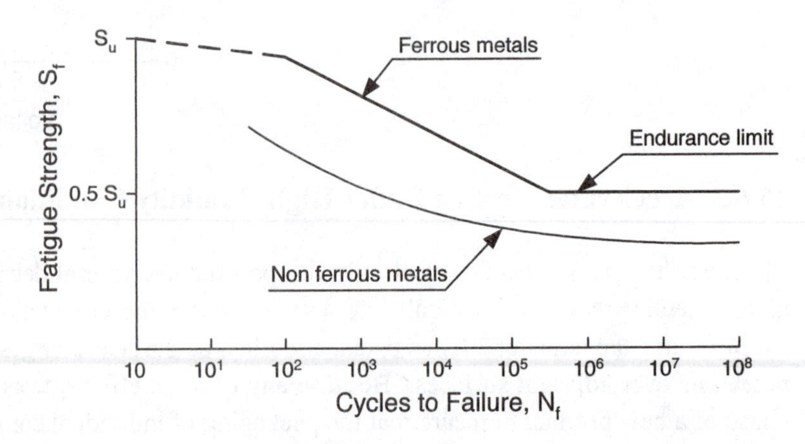

Fig. 15.20 Fatigue strength as a function of the number of cycles to failure for ferrous and non-ferrous metals.

[8] Fatigue failure occurs when the alternating component of the applied stress equals the fatigue strength $\sigma_a = S_f$.

Care must be exercised to insure that the components fail in the long-life fatigue regime where the applied stresses are lower than the yield strength of the metallic materials. If very high stresses are imposed, fatigue life of a component is very short (usually 100 cycles or less). The mechanism for fatigue failure in the low-life regime is different, as plastic flow and ductility as well as the material's yield strength control the number of cycles to failure.

15.7 RELIABILITY IMPROVEMENT

The design of a totally new product is a rare event. Most "new" products are modifications of existing products that have been redesigned to remain competitive in the market place. The redesign usually involves adding new features[9], reducing size, weight and power and cutting cost of production. Improving reliability of the product is also a design objective, although a low failure rate is rarely used in advertising the product. However, improving reliability is important because reducing field failures cuts warranty costs and enhances the company's reputation as a builder of durable products.

Let's consider a program to improve the reliability of a single[10], mid-priced model of a cell phone with annual sales exceeding 300,000 units. What steps are involved in a reliability improvement program for this model cell phone?

Step 1: Establish the failure rate during the reliability period. When a product fails during the warranty period, it is returned and dates of sale and of return are known. Using this data, it is easy to determine the failure rate as a function of time in service. In some cases, an explanation of the problem encountered by the customer is available.

Step 2: Determine the manufacturer of the components that failed. In some products with very large production quantities more than one manufacturer of a specific component is involved and failure rates differ between manufacturers.

Step 3: Determine the failure mechanisms from phones returned during the warranty period. Collect a sample that is sufficiently large to provide exposure to all of the different types of failures encountered in the field. Test and disassemble the product to isolate the cause of failure. Was the failure caused by abuse such as dropping the phone on a hard surface or spilling coffee or cola on it? If a component failed, identify it and establish the cause of failure. Examine the circuit boards and the wiring and rank the quality of the workmanship. This is a critical step in the program because it identifies the weaknesses in the design, selection of components, production processes and overall packaging quality.

Step 4: Take action to rectify the problems identified in Step 3.

Abuse:

Enclosures for cell phones can be designed to be tolerant of a significant degree of abuse. Hardening a cell phone includes providing for water and dust resistant as well as surviving the impact forces developed when it is dropped on a hard surface. Water and dust resistance are provided by employing membrane switches for the keyboard, by sealing all the openings, and by specifying waterproof input connections. Impact resistance is achieved by coating the case with a relatively thick compliant material and by increasing the shock tolerance of the internal assembly of the circuit board, battery and display.

---

[9] For example, adding digital cameras to cell phones.

[10] The author explored Nokia's website and found 13 different models of cell phones with prices ranging from $700 to free depending upon the service provider selected. The spread in features found across this product line was too great to consider in a single reliability improvement program.

Component Failure:

If the failure rate for a specific component is found to be excessive, it should be replaced by a component that exhibits a lower failure rate. This replacement can be achieved by switching to a manufacturer that provides the same component with a lower failure rate or by selecting a component with more capacity and derating it. Component failure should be a rare event in the relatively short life of a cell phone in today's rapidly changing market for this product.

Quality Issues:

Failures related to quality issues should not occur if a sound quality assurance program is in place at the manufacturing companies supplying components and at the companies performing assembly operations. In today's global economy, components are produced in many different countries and the product may be assembled by different manufacturers located anywhere in the world. Certification of each of their quality assurance programs is essential. Site visits to inspect facilities and to check on statistical process control procedures should be performed periodically. Training programs for new employees at each of these manufacturing facilities must be maintained. Samples drawn from production lots of components and completely assembled products should be inspected at frequent intervals to insure that quality is maintained at the highest possible level.

REFERENCES

1. Lall, P, M. G. Pecht and E. B. Hakim, Influence of Temperature on Microelectronics and System Reliability, CRC Press, Boca Raton, FL, 1997.
2. Darveaux, R., "Effect of Simulation Methodology on Solder Joint Crack Growth Correlation," Proceedings of 50th ECTC, pp.1048-1058, May 2000.
3. Darveaux, R., How to use Finite Element Analysis to Predict Solder Joint Fatigue Life, Proceedings of the VIII International Congress on Experimental Mechanics, Nashville, Tennessee, June 10-13, pp. 41-42, 1996.
4. Darveaux, R., Banerji, K., Mawer, A., and Dody, G., "Reliability of Plastic Ball Grid Array Assembly, "Ball Grid Array Technology, J. Lau, ed., McGraw-Hill, Inc. New York, pp. 379-442, 1995.
5. Darveaux, R., and Banerji, K., "Constitutive Relations for Tin-Based Solder Joints," IEEE Trans-CPMT-A, Vol. 15, No. 6, pp. 1013-1024, 1992.
6. R. Rausch, "Electronic Components & Systems and their Radiation Qualification for Use in the LHC Machine," European Laboratory for Particle Physics, CERN Report SL 99-004 (CO), Geneva, Switzerland, 1999.
7. Kemeny, A. P. "Experiments Concerning Life Testing of Semiconductor Devices," Microelectronic Reliability, Vol. 10, No. 3, PP. 169-193, 1971.
8. Pecht, M. G., Handbook of Electronic Package Design, Chapter 11, Marcel Dekker, New York, NY, 1991.
9. LaCombe, D. J., Reliability Control for Electronic Systems, Marcel Dekker, New York, NY, 1999.
10. Ali, S. R., Digital Switching Systems: System Reliability and Analysis, McGraw-Hill, New York, NY, 1998.
11. Billington, R., Power System Reliability Evaluation, Gordon and Breach, New York, NY, 1970.

12. Dumin, D. J., Oxide Reliability: A Summary of Silicon Oxide Wearout, Breakdown and Reliability, World Science Publishing, Singapore, 2002.
13. Hnatek, E. R., Practical Reliability of Electronic Systems and Products, Marcel Dekker, New York, NY, 2003.
14. Mourand, S. and Y. Zorian, Principles of Testing Electronic Systems, John Wiley & Sons, New York, NY, 2000.
15. Jennsen, F. Electronic Component Reliability: Fundamentals, Evaluation and Assurance, John Wiley & Sons, New York, NY, 1995.

EXERCISES

15.1 Write an engineering brief explaining why it is important to locate and eliminate a defective component as soon as possible in the production of an electronic product.
15.2 Write an engineering brief describing corrosion as a failure mechanism.
15.3 Write an engineering brief describing electromigration as a failure mechanism.
15.4 Explain why the oxide layer on semiconductor devices fails.
15.5 Describe several causes for failure of chips package in polymeric chip carriers.
15.6 Describe the common failure mechanisms found on printed circuit boards.
15.7 Reference Fig. 15.6 and explain why the solder joint on a chip carrier with gull wing leads failed. Would you expect this type of failure with a gull wing leaded chip carrier.
15.8 Explain why plated though holes and vias fail on printed circuit boards subjected to thermal cycling.
15.9 What is a preferred component and why would you use a part classified in this manner?
15.10 Describe the methods used to produce preferred parts for military applications. Also explain the methods used to produce preferred parts for commercial applications.
15.11 Write an engineering brief for your manager explaining the need to shield comments from radiation damage if the product is to be used for space applications.
15.12 Describe he procedure used n selecting commercial components with high reliability.
15.13 Explain the procedure for selecting off-the-shelf components with high reliability.
15.14 Prepare a listing showing the failure rate or FIT for five different types of components used in electronic systems.
15.15 Write an engineering brief explaining the principles of derating to your instructor.
15.16 List common derating parameters.
15.17 Prepare an instruction to a new circuit designer for derating the components to be used in a new product.
15.18 Explain the procedure used to derate temperature of several components located within an electronic enclosure.
15.19 Explain the circumstances when electronic systems are exposed to high mechanical stresses.
15.20 What are screening tests and why are they employed?
15.21 Prepare a specification for a temperature type screening test for an electronic system.
15.22 Prepare a list of five different screening tests and explain the test procedures followed in conducting these tests.
15.23 Why is it important to accelerate the tests conducted to insure the quality of components, subsystems and a complete system?
15.24 Describe a test using temperature to accelerate the failure mechanism.
15.25 Describe a test using electrical stress to accelerate the failure mechanism.
15.26 Describe a test using temperature and humidity to accelerate the failure mechanism.

15.27 Describe a test using increased alternating stresses to reduce the time required to conduct fatigue tests.

15.28 Prepare a plan to improve the quality of an IPod.

Appendix A: Acronyms

Electronic Devices

ASIC = application-specific integrated circuit
CCD = charge-control device
CMOS = complimentary MOS
CPU = central processing unit
DRAM = dynamic RAM
DSP = digital signal processing
ECL = emitter-coupled logic
EEPROM (or E' PROM) = electrically erasable and programmable
EPROM = erasable and programmable ROM
ESD = electrostatic discharge
FET = field effect transistor
FTR = functional throughput rate
GBW = gain-bandwidth
HIC = hybrid integrated circuit
1C = integrated circuit
I/O = input/output
IRED = infrared emitting diode
LCD = liquid crystal display
LD = laser diode
LED = light-emitting diode
LSI = large-scale integration
MLC - multilayer ceramic, multilayer capacitor
MOS = metal-oxide semiconductor
MOSFET = MOS FET
MS = microstrip
MSI = medium-scale integration
MSM = metal-semiconductor-metal
NMOS = n-channel MOS
NVRAM = nonvolatile RAM
PEL = pixel
PIN diode = p-intrinsic-diode
PMOS = p-channel MOS
PROM = programmable read-only memory
RAM = random-access memory
RISC = reduced-instruction set computer
ROM = read-only memory
SRAM = static RAM
SSC = spread-stacked capacitor
SSI = small-scale integration
STL = Schottky transistor logic
TTL = transistor-transistor logic
ULSI = ultra-large-scale integration
VHSIC = very high speed 1C
VLSI = very large scale integration
WSCL = water-soluble conductive layer
XMOS = high-speed MOS
ZMR = zone-melting recrystallization

Materials Processing (Semiconductors)

CMP = chemical-mechanical polishing
CVD = chemical-vapor deposition
LPE = liquid-phase epitaxy
LTE = low-temperature epitaxial [growth]
MBE = molecular beam epitaxy
RTF = rapid thermal processing
SCC = stress-corrosion cracking

Testing

AQL = acceptable quality level
ATE = automatic test equipment
CTE = coefficient of thermal expansion
DPA = destructive physical analysis
DTA = differential thermal analysis
ENR = excess noise ratio
FA = failure analysis
FACI = first article configuration inspection
FMEA = failure mode and effects analysis
FR = failure rate
HAST = highly accelerated stress testing
IRS = infrared scan
MTBF = mean time between failures
MTTF = mean time to failure
NDT = nondestructive test
NTL = low-level noise tolerance
PDA = percent defect allowable
PPM = parts per million
RGA = residual gas analysis
RI = receiving inspection

Testing (Continued)

RTD = resistance-temperature detector [probe]
SAM = scanning acoustic microscope
SEM = scanning electron microscopy
SLAM = scanning laser acoustic microscope
SRP = spreading resistance profiling [technique]
STM = scanning tunneling microscopy
TCC = temperature coefficient of capacitance
TCR = temperature coefficient of resistance
TEM = transmission electron microscopy
Tg = glass transition temperature
TGA = thermal gravimetric analysis
TLC = thin-layer chromatography
TMA = thermo-mechanical analysis
TSM = top side metallurgy
TTF = time to failure

Packaging

ATAB = area-array tape-automated bonding
BCW = bare-chip-and-wire [hybrid]
BLM = ball-limiting metallurgy
BSM = backside metallurgy
BTAB = bumped tape-automated bonding
C and W = chip and wire
CC = chip carrier
CCC = ceramic CC
CEEE = common electronics equipment enclosure
CERDIP = ceramic DIP
C4 = controlled-collapse chip connector
COB = chip-on-board
CSI = compliant solderless interface
DCA = direct chip attach
DIP = dual in-line [package]
DWF = dice-in-wafer form
FPC = fine-pitch CC
FRU = field replaceable unit
HCC = hermetic chip carrier
HDCM = high-density ceramic module
HDI = high-density interconnect
ILB = inner lead bond
LCC = leadless CC
LCCC = leadless ceramic CC
LMCH = leadless multiple-chip hybrids
LID = leadless inverted device
LMCH = leadless multiple-chip hybrid
LRU = line replacement unit or lowest repairable unit
MC = metallized ceramic
MCC = miniature CC
MCM = multichip module
MLB = multilayer board
MLC = multilayer ceramic
OLE = outer lead bond
PAA = pad area array
PCB = printed circuit board
PCC = same as PLCC
PCI = pressure contact interconnection
PCR = plastic CC, rectangular
PD1P = plastic DIP
PES = porcelain-enamel-steel
PET = porcelain-enamel technology
PGA = pin grid array [package]
PIP = pin insertion [package]
PLCC = plastic-leaded CC
PRN = priority ranking number
POS = porcelain-on-steel
PQFP = plastic quad flat package
PTF = polymer thick film
PTH = plated-through hole
PWB = printed wiring board
QFB = quad flat butt-leaded package
QUIP = quad in-line package
SCM = single-chip module
SIP = single in-line package
SLC = single-layer ceramic
SMA = surface-mounted assembly
SMD = surface-mounted device
SOB = small-outline butt leaded package
SMT = surface mount technology
SOIC = small-outline
SOJ = small-outline J-lead package
SOP = small-outline package
SOI = silicon-on-insulator
TAB = tape-automated-bonding
TC = thermo-compression bond
TCM = thermal conduction module
TS = thermosonic bond
TSOP = thin small outline package
TO = transistor outline package
US = ultrasonic bond
VSO = very small outline package
WSI = wafer-scale integration
ZIP = zero insertion force
ZIP = zigzag in-line package

Materials

AlGaAs = aluminum gallium arsenide
AlInAs = aluminum indium arsenide
AlN = aluminum nitride
$A1_2O_3$ = aluminum oxide (alumina)
AlSb = aluminum antimonide
As = arsenic
Au = gold
AuGe = gold-germanium
AuSi = gold-silicon
AuSn = gold-tin
B = boron
BN = boron nitride
Ba = barium
$BaTiO_3$ — barium titanate
Be = beryllium
BeO = beryllium oxide (beryllia)
C = carbon; graphite; diamond
CDA = clean dry air
CFC = chlorinated fluorocarbon
DI = deionized water
FR4 = epoxy resin glass laminate
Ga = gallium
GaAs = gallium arsenide
GaAsP = gallium arsenide phosphide
GalnAs = gallium indium arsenide
GaP = gallium phosphide
GaSb = gallium antimonide
Ge = germanium
H^+ = atomic hydrogen
H_2 = hydrogen molecule
H_2O = water
H_2O_2 = hydrogen peroxide
In = indium
MnO_2 = manganese dioxide
N, = nitrogen molecule
NO_2 = nitrogen peroxide
N_2O = nitrous oxide
0_2 = oxygen molecule
0_3 = ozone molecule
OH" = hydroxyl ion
P = phosphorus
PLZT = lead-lanthanum-zirconate-titanate
PMMA = polymethyl methacrylate
PSG = phosphosilicate glass
PVC = polyvinyl chloride
PX = paraxylylene
PZT = lead-zirconate-titanate
Sb = antimony
Si = silicon
SiC = silicon carbide
SiN = silicon nitride
SiO = silicon monoxide
SiO_2 = silicon dioxide (silica)
= quartz (crystalline silica)
= fused silica (silica glass)
Sn = tin
Ta = tantalum
Ta_2O_5 = tantalum pentoxide
TEOS = tetraethoxysilane
Ti = titanium
TiN = titanium nitride
TiO_2 = titanium dioxide (titania)
$TiSi_2$ = titanium silicide
Y = yttrium
YAG = yttrium aluminum garnet
YIG = yttrium iron garnet
Zn = zinc
ZnO = zinc oxide
ZnS = zinc sulfide

Miscellaneous

ASME = American Soc. of Mechanical Engineers
BIU = bus interface unit
CAD = computer-aided design
CAE = computer-aided engineering
CIM = computer-integrated manufacturing
CRT = cathode ray tube
DoD = Department of Defense
HDTV = high-definition television
IR = infrared
IEEE = Inst. of Electrical and Electronics Engineers
IEPS = Intm;. Electronics Packaging Society
ISHM = Intnl. Soc. Hybrid Microelectronics
JEDEC = Joint Electronics Device Engr. Council
MIL-STD = military standard
RADC = Rome Air Development Center
SMTA = Surface Mount Technology Association

Index